BOLYAI SOCIETY
MATHEMATICAL STUDIES

28

More information about this series at http://www.springer.com/series/4706

BOLYAI SOCIETY MATHEMATICAL STUDIES

Editor-in-Chief:
Gábor Fejes Tóth

Series Editor:
Gergely Ambrus

Publication Board:
Gyula O. H. Katona · László Lovász · Dezső Miklós · Péter Pál Pálfy
András Recski · András Stipsicz · Domokos Szász

1. **Combinatorics, Paul Erdős is Eighty, Vol. 1**
   D. Miklós, V. T. Sós, T. Szőnyi (Eds.)
2. **Combinatorics, Paul Erdős is Eighty, Vol. 2**
   D. Miklós, V. T. Sós, T. Szőnyi (Eds.)
3. **Extremal Problems for Finite Sets**
   P. Frankl, Z. Füredi, G. Katona, D. Miklós (Eds.)
4. **Topology with Applications**
   A. Császár (Ed.)
5. **Approximation Theory and Function Series**
   P. Vértesi, L. Leindler, Sz. Révész, J. Szabados, V. Totik (Eds.)
6. **Intuitive Geometry**
   I. Bárány, K. Böröczky (Eds.)
7. **Graph Theory and Combinatorial Biology**
   L. Lovász, A. Gyárfás, G. Katona, A. Recski (Eds.)
8. **Low Dimensional Topology**
   K. Böröczky Jr., W. Neumann, A. Stipsicz (Eds.)
9. **Random Walks**
   P. Révész, B. Tóth (Eds.)
10. **Contemporary Combinatorics**
    B. Bollobás (Ed.)
11. **Paul Erdős and His Mathematics I+II**
    G. Halász, L. Lovász, M. Simonovits, V. T. Sós (Eds.)
12. **Higher Dimensional Varieties and Rational Points**
    K. Böröczky Jr., J. Kollár, T. Szamuely (Eds.)
13. **Surgery on Contact 3-Manifolds and Stein Surfaces**
    B. Ozbagci, A. I. Stipsicz
14. **A Panorama of Hungarian Mathematics in the Twentieth Century, Vol. 1**
    J. Horváth (Ed.)
15. **More Sets, Graphs and Numbers**
    E. Győri, G. Katona, L. Lovász (Eds.)
16. **Entropy, Search, Complexity**
    I. Csiszár, G. Katona, G. Tardos (Eds.)
17. **Horizons of Combinatorics**
    E. Győri, G. Katona, L. Lovász (Eds.)
18. **Handbook of Large-Scale Random Networks**
    B. Bollobás, R. Kozma, D. Miklós (Eds.)
19. **Building Bridges**
    M. Grötschel, G. Katona (Eds.)
20. **Fete of Combinatorics and Computer Science**
    G. Katona, A. Schrijver, T. Szonyi (Eds.)
21. **An Irregular Mind**
    I. Bárány, J. Solymosi (Eds.)
22. **Cylindric-like Algebras and Algebraic Logic**
    H. Andréka, M. Ferenczi, I. Németi (Eds.)
23. **Deformations of Surface Singularities**
    A. Némethi, Á. Szilárd (Eds.)
24. **Geometry – Intuitive, Discrete, and Convex**
    I. Bárány, K. Böröczky, G. Fejes Tóth, J. Pach (Eds.)
25. **Erdős Centennial**
    L. Lovász, I. Ruzsa, V. T. Sós (Eds.)
26. **Contact and Symplectic Topology**
    F. Bourgeois, V. Colin, A. Stipsicz (Eds.)
27. **New Trends in Intuitive Geometry**
    G. Ambrus, I. Bárány, K. J. Böröczky, G. Fejes Tóth, J. Pach (Eds.)

Imre Bárány · Gyula O. H. Katona ·
Attila Sali
Editors

Building Bridges II

Mathematics of László Lovász

Editors
Imre Bárány
MTA Alfréd Rényi Institute of Mathematics
Budapest, Hungary

Gyula O. H. Katona
MTA Alfréd Rényi Institute of Mathematics
Budapest, Hungary

Attila Sali
MTA Alfréd Rényi Institute of Mathematics
Budapest, Hungary

ISSN 1217-4696
Bolyai Society Mathematical Studies
ISBN 978-3-662-59203-8   ISBN 978-3-662-59204-5   (eBook)
https://doi.org/10.1007/978-3-662-59204-5

Mathematics Subject Classification (2010): 05-02, 05-06, 68-02, 68-06

© János Bolyai Mathematical Society and Springer-Verlag GmbH Germany, part of Springer Nature 2019

This work is subject to copyright. All rights are reserved by the Publisher, whether the whole or part of the material is concerned, specifically the rights of translation, reprinting, reuse of illustrations, recitation, broadcasting, reproduction on microfilms or in any other physical way, and transmission or information storage and retrieval, electronic adaptation, computer software, or by similar or dissimilar methodology now known or hereafter developed.

The use of general descriptive names, registered names, trademarks, service marks, etc. in this publication does not imply, even in the absence of a specific statement, that such names are exempt from the relevant protective laws and regulations and therefore free for general use.

The publisher, the authors and the editors are safe to assume that the advice and information in this book are believed to be true and accurate at the date of publication. Neither the publisher nor the authors or the editors give a warranty, expressed or implied, with respect to the material contained herein or for any errors or omissions that may have been made. The publisher remains neutral with regard to jurisdictional claims in published maps and institutional affiliations.

Cover illustration: László Lovász, April 2019, photographer László Mudra, © MTA (Hungarian Academy of Sciences)

This Springer imprint is published by the registered company Springer-Verlag GmbH, DE part of Springer Nature.
The registered company address is: Heidelberger Platz 3, 14197 Berlin, Germany

László Lovász and his wife Katalin Vesztergombi, in Beijing 2010. Photo taken by Jianshe Zhang. With kind permission of Prof. Lovász

Preface

László Lovász turned seventy on 9 March 2018. To celebrate his birthday, a mathematical conference was held in Budapest, 2 to 6 July 2018, with invited speakers only and with well over three hundred participants. On this meeting, several world-class mathematicians and computer scientists gave talks about their latest results and paid tribute to Laci (as he is called by his friends). Some of them also dedicated a research or survey paper on this occasion. This volume is the collection of their articles.

László Lovász is a towering mathematician of our time. He has been a defining force in the field of combinatorics over the past half-century, ever since his highly influential early gems appeared. His influence has decisively contributed to the general appeal and expansion of the field and its closer integration with classical mathematics. He solved a long list of major open problems, each time tackling the problem with powerful new techniques whose significance went far beyond their original target. Lovász's work has changed the landscape not only in combinatorics but in a number of related fields, including discrete and convex optimization, the theory of algorithms, and complexity theory, and has contributed significant new tools to other fields, including the geometry of numbers, information theory, and functional analysis. Lovász has been one of the pioneers of the interplay between discrete and continuous mathematics, a master at establishing unexpected connections, "building bridges" between seemingly distant fields. His invariably elegant and powerful ideas have produced new subfields in many areas.

Here is a list of some of his breakthrough results and achievements: (1) the perfect graph conjecture; (2) topological methods in graph theory, Kneser's conjecture; (3) geometric graph theory, Shannon capacity, Lovász number; (4) semidefinite programming and its effect on the theory of algorithms; (5) Lovász Local Lemma; (6) the ellipsoid method in combinatorial optimization; (7) the lattice basis reduction algorithm; (8) hardness of approximation; (9) rapid mixing and algorithms; (10) graph homomorphisms and graph limits. Several items from this list have become very active and central research areas in mathematics, in computer science, and even in quantum physics. On pages 595–611 of this volume, his list of publications (up to 2019) is given. It shows the variety and depth of his achievements.

The present volume is dedicated to Laci on his 70th birthday. Several articles in it are connected to his achievements. For instance, Noga Alon's paper "Lovász, vectors, graphs and codes" is related to topic (3) above. It is mainly a survey, but it contains several new results about triangle-free pseudo-random Cayley graphs. Anders Björner's article "Continuous matroids revisited" is about infinite or continuous matroids, which is connected to topic (10). Christian Borgs, Jennifer T. Chayes, Henry Cohn, and László Miklós Lovász write about graphexes and the weak kernel metric, which is related to (10) again. This applies to Péter Csikvári's paper "Statistical matching theory" which deals with statistical properties of matchings in very large and infinite graphs. Persi Diaconis writes about his "ongoing conversations with Laci": estimating the number of perfect matchings in bipartite graphs, a favourite subject of Lovász. Sequential importance sampling offers an alternative way to approximately evaluate the permanent. Uriel Feige's paper "Tighter bounds for online bipartite matching" is about matchings again. The paper by David Conlon, Jacob Fox, Andrey Grinshpun, and Xiaoyu He "Online Ramsey numbers and the subgraph query problem" is about an online Ramsey-type game. It is also related to topic (5) above as it contains an exponential improvement of the lopsided Lovász Local Lemma. The contribution of Tibor Jordán and András Mihálykó is connected to (8) and combinatorial optimization plus rigidity. Peter Keevash writes about "Coloured and directed designs" with several illustrative applications of his recent result on decompositions of labelled complexes. The article by Jaroslav Nešetřil and Patrice Ossona de Mendez "Approximations of mappings" is related to topic (10), graph limits, from the point of view of first-order convergence and logic. Lex Schrijver writes about the partially disjoint paths problem and shows, among other things, that it is solvable in polynomial time if the directed graph is planar. Miklós Simonovits and Endre Szemerédi survey results on "Embedding graphs into larger graphs". It emphasizes the surprising connections of this problem to extremal graph theory and to other areas of mathematics. The paper by Yin Tat Lee, Aaron Sidford, and Santosh S. Vempala "Efficient convex optimization with oracles" is about a basic algorithmic problem: minimizing a convex function over a convex set and is related to topics (6) and (9). Avi Wigderson and Orit E. Raz's article "Derandomization through submodular optimization" presents a deterministic, strongly polynomial time algorithm for computing the matrix rank for a class of symbolic matrices and is connected to Lovász's flats problem.

Besides being a superb mathematician, Laci is an extremely nice person and a pleasant colleague. He is friendly and helpful and has inspired several young and not so young mathematicians. His modesty is legendary. As Victor Klee put it once "among the best mathematicians he is the most modest, and among the modest ones, he is the best".

Lovász has had a distinguished academic career. His employments and visiting positions include Szeged, Budapest, Waterloo, Bonn, Princeton, Yale, and Microsoft Research. He returned to Budapest again in 2016 and became Director of the Mathematical Institute at Eötvös University and later President of the

Hungarian Academy of Sciences (from 2014), in a difficult and turbulent time in Hungarian politics.

In 2006, Lovász was elected President of the International Mathematical Union for the period 2007–2010. His international reputation is stellar as shown by prestigious prizes and honours given to him. They include the Grünwald, Pólya, Fulkerson (twice), Knuth, Gödel, von Neumann, Kyoto, Bolyai, and Széchenyi prizes and various honorary degrees and professorships. A very brief Curriculum Vitae is presented on pages 593–594 of this volume.

Laci has always been a family man, a loving husband and father and by now grandfather. His wife, Kati Vesztergombi, has been a reference point in Laci's life since their highschool years in the Fazekas Mihály Gimnázium in Budapest. Together they shared family and friends and also the love for mathematics for the last fifty or more years. We dedicate this volume to Laci's 70th birthday and wish him many happy returns and many more beautiful results.

Budapest, Hungary
March 2019

Imre Bárány
Gyula O. H. Katona
Attila Sali

Contents

Lovász, Vectors, Graphs and Codes 1
Noga Alon

Continuous Matroids Revisited 17
Anders Björner

Identifiability for Graphexes and the Weak Kernel Metric 29
Christian Borgs, Jennifer T. Chayes, Henry Cohn
and László Miklós Lovász

Online Ramsey Numbers and the Subgraph Query Problem 159
David Conlon, Jacob Fox, Andrey Grinshpun and Xiaoyu He

Statistical Matching Theory 195
Péter Csikvári

Sequential Importance Sampling for Estimating the Number
of Perfect Matchings in Bipartite Graphs: An Ongoing
Conversation with Laci 223
Persi Diaconis

Tighter Bounds for Online Bipartite Matching 235
Uriel Feige

Minimum Cost Globally Rigid Subgraphs 257
Tibor Jordán and András Mihálykó

Coloured and Directed Designs 279
Peter Keevash

Efficient Convex Optimization with Oracles 317
Yin Tat Lee, Aaron Sidford and Santosh S. Vempala

Approximations of Mappings 337
Jaroslav Nešetřil and Patrice Ossona de Mendez

Subspace Arrangements, Graph Rigidity and Derandomization Through Submodular Optimization 377
Orit E. Raz and Avi Wigderson

Finding k Partially Disjoint Paths in a Directed Planar Graph 417
Alexander Schrijver

Embedding Graphs into Larger Graphs: Results, Methods, and Problems 445
Miklós Simonovits and Endre Szemerédi

Curriculum Vitae of László Lovász 593

Publications of László Lovász 595

The List of the Former Volumes 613

Lovász, Vectors, Graphs and Codes

Noga Alon

Dedicated to László Lovász, for his seventieth birthday

Abstract A family of high-degree triangle-free pseudo-random Cayley graphs has been constructed in (Alon, Electro J Combin 1(R12):8, 1994 [2]), motivated by a geometric question of Lovász. These graphs turned out to be useful in tackling a variety of additional extremal problems in Graph Theory and Coding Theory. Here we describe the graphs and their applications, and mention several intriguing related open problems. This is mainly a survey, but it contains several new results as well. One of these is a construction showing that the Lovász θ-function of a graph cannot be bounded by any function of its Shannon capacity.

Keywords Cayley graphs · The θ-function · Shannon capacity of graphs · Ramsey graphs · Maxcut · List decodable codes

Subject Classifications 05C35 · 05C50

1 Introduction

- What is the maximum possible (Euclidean) norm of a sum of n unit vectors so that any 3 of them contain 2 which are orthogonal?

N. Alon (✉)
Department of Mathematics, Princeton University, Princeton, NJ 08544, USA
e-mail: nalon@math.princeton.edu

Schools of Mathematics and Computer Science, Tel Aviv University, 69978 Tel Aviv, Israel

© János Bolyai Mathematical Society and Springer-Verlag GmbH Germany, part of Springer Nature 2019
I. Bárány et al. (eds.), *Building Bridges II*, Bolyai Society Mathematical Studies 28, https://doi.org/10.1007/978-3-662-59204-5_1

- What is the minimum possible size of the maxcut of a triangle-free graph with m edges?
- What is the maximum possible number of words in a binary code of length n so that there is no Hamming ball of radius $(1/4 + \varepsilon)n$ containing more than two words?

The first question is geometric, and was posed by Lovász motivated by the study of the θ-function of a graph. The second question is in Extremal Graph Theory, it was first considered by Erdős and Lovász. The third question is in Coding theory, and was first studied by Blinovskii, extending earlier results of Plotkin and Levenshtein. Somewhat surprisingly it turns out that all three questions, and several related ones, can be solved asymptotically using a single construction of a family of triangle-free Cayley graphs with extremal spectral properties. Here we describe this construction, show how it is used in the solution of these problems and more, and describe their connection to Ramsey theory and to questions about the Shannon capacity of graphs.

2 The Graphs

For a positive integer k, let $F_k = GF(2^k)$ denote the finite field with 2^k elements whose elements are represented, as usual, by binary vectors of length k. If a, b and c are three such vectors, let (a, b, c) denote their concatenation. Suppose k is not divisible by 3 and put $n = 2^{3k}$. Let W_0 be the set of all nonzero elements $\alpha \in F_k$ so that the leftmost bit in the binary representation of α^7 is 0, and let W_1 be the set of all nonzero elements $\alpha \in F_k$ for which the leftmost bit of α^7 is 1. Since 3 does not divide k, 7 does not divide $2^k - 1$ and hence $|W_0| = 2^{k-1} - 1$ and $|W_1| = 2^{k-1}$, as when α ranges over all nonzero elements of F_k so does α^7.

Let G_n be the Cayley graph of the elementary abelian 2-group Z_2^{3k} with the generating set $S = U_0 + U_1 = \{u_0 + u_1 : u_0 \in U_0, u_1 \in U_1\}$, where $U_0 = \{(w_0, w_0^3, w_0^5) : w_0 \in W_0\}$, and $U_1 = \{(w_1, w_1^3, w_1^5) : w_1 \in W_1\}$ with the powers computed in the finite field F_k.

The following theorem is proved in [2].

Theorem 2.1 *If k is not divisible by 3 and $n = 2^{3k}$ then G_n is a Cayley graph of Z_2^{3k}, it has n vertices, is regular of degree*

$$d_n = 2^{k-1}(2^{k-1} - 1) = (\frac{1}{4} + o(1))n^{2/3},$$

and satisfies the following properties

1. *G_n is triangle-free.*
2. *Every eigenvalue μ of G_n, besides the largest, satisfies*

$$-9 \cdot 2^k - 3 \cdot 2^{k/2} - 1/4 \leq \mu \leq 4 \cdot 2^k + 2 \cdot 2^{k/2} + 1/4.$$

The detailed proof can be found in [2]. Here is a sketch. The graph G_n is the Cayley graph of Z_2^{3k} with respect to the generating set $S = S_n = U_0 + U_1$, where U_i are defined as above. As the elements of $U_0 \cup U_1$ are the columns of the parity check matrix of a binary BCH-code of designed distance 7 (see, e.g., [40], Chap. 9), every set of six of them is linearly independent. Therefore the elements of S_n are distinct and G_n is regular of degree $|S_n| = |U_0||U_1|$.

The fact that G_n is triangle-free is equivalent to the fact that the sum (in Z_2^{3k}) of any set of 3 elements of S_n is not the zero-vector. Let $u_0 + u_1$, $u_0' + u_1'$ and $u_0'' + u_1''$ be three distinct elements of S_n, where $u_0, u_0', u_0'' \in U_0$ and $u_1, u_1', u_1'' \in U_1$. If the sum (modulo 2) of these six vectors is zero then, since every set of six members of $U_0 \cup U_1$ is linearly independent, every vector must appear an even number of times in the sequence $(u_0, u_0', u_0'', u_1, u_1', u_1'')$. However, since U_0 and U_1 are disjoint this implies that every vector must appear an even number of times in the sequence (u_0, u_0', u_0'') and this is clearly impossible. This proves part 1 of the theorem.

The proof of part 2 is based on the fact that the eigenvalues of G_n are given by the following character sums:

$$\sum_{s \in S_n} \chi(s),$$

where χ ranges over all characters of the group Z_2^{3k}. Indeed, such an expression holds for any Cayley graph of an abelian group (see, e.g., [39]), where the eigenvectors are the characters. The bounds in part 2 can now be deduced from the known results about the weight distribution of dual BCH codes, proved using the Carlitz-Uchiyama bound (see [40], pp. 280–281). The details can be found in [2].

An (n, d, λ)-graph is a d regular graph on n vertices in which all eigenvalues but the first are of absolute value at most λ. This notation was introduced by the author in the late 80s, motivated by the fact that if λ is much smaller than d, then the graph exhibits strong pseudo-random properties. In particular, as shown in [8], the average degree of every induced subgraph on a set of xn vertices deviates from xd by less than λ. By considering the trace of the square of the adjacency matrix of any (n, d, λ)-graph, which is nd and is also the sum of squares of its eigenvalues, it is easy to see that $\lambda \geq \sqrt{d(n-d)/(n-1)}$ which is $\Omega(\sqrt{d})$ whenever, say, $d < n/2$. Thus the smallest possible value of λ is $\Theta(\sqrt{d})$. The graph $G = G_n$ described above is an (n, d, λ) where $d = \Theta(n^{2/3})$ and $\lambda = \Theta(\sqrt{d})$, that is, λ is as small as possible up to a constant factor. Note that by the above fact about the distribution of edges in subsets of (n, d, λ)-graphs, it follows that any set of $cn^{2/3}$ vertices of G spans many edges, provided $c > 36$, implying that such a graph with somewhat larger degrees which are still $\Theta(n^{2/3})$ cannot be triangle-free. Note also that in a random graph with degrees $\Theta(n^{2/3})$, every edge is typically contained in $\Theta(n^{1/3})$ triangles, that is, the graph includes lots of triangles. The fact that the graphs G_n are triangle-free and yet have strong pseudo-random properties derived from their spectrum make them useful in tackling various extremal problems. Some of these are described in the following sections.

3 Shannon Capacity and the Lovász θ-Function

3.1 Shannon and Ramsey

The (and)-product of two undirected graphs $G = (V, E)$ and $G' = (V', E')$ is the graph whose vertex set is $V \times V'$ in which two distinct vertices (u, u') and (v, v') are adjacent iff (either $u = v$ or uv are adjacent in G) and (either $u' = v'$ or u', v' are adjacent in G'). The power G^n of G is defined with respect to this product. It is thus the graph whose vertex set is V^n in which two distinct vertices $(u_1, u_2, \ldots, u_n)$ and $(v_1, v_2, \ldots, v_n)$ are adjacent if and only if for all i between 1 and n either $u_i = v_i$ or $u_i v_i \in E$. The *Shannon capacity* $S(G)$ of G is the limit $\lim_{n \to \infty} (\alpha(G^n))^{1/n}$, where $\alpha(G^n)$ is the maximum size of an independent set of vertices in G^n. This limit exists, by super-multiplicativity, it is equal to the supremum over n of $(\alpha(G^n))^{1/n}$ and hence is always at least $\alpha(G)$. The Shannon capacity of a graph may be significantly larger than its independence number. In particular, there are graphs on n vertices with independence number smaller than $2 \log_2 n$ and Shannon capacity at least $\sqrt{n}$, see [12, 25]. It is not known, however, if the Shannon capacity is bounded by any function of the independence number, that is, whether or not the maximum possible value of the Shannon capacity of a graph whose independence number is a constant c is finite. This is equivalent to a well known question on multicolored Ramsey numbers. Let $r(c + 1 : \ell)$ denote the maximum number r so that there is a coloring of the edges of the complete graph K_r on r vertices by ℓ colors with no monochromatic copy of K_{c+1}. As shown in [25], (see also [12]), the maximum possible value of $\alpha(G^\ell)$ as G ranges over all graphs with independence number c is exactly $r(c + 1 : \ell)$. It follows that the maximum possible Shannon capacity of a graph with independence number c is exactly the limit as ℓ tends to infinity of $r(c + 1 : \ell)^{1/\ell}$. In particular, the question of deciding whether or not the maximum possible Shannon capacity of a graph with independence number 2 is finite is equivalent to an old problem of Erdős (see, e.g., [17]) asking whether or not the Ramsey number $r(3 : \ell)$ grows faster than any exponential in ℓ.

This question is wide open. Indeed, our understanding of the Shannon capacity of graphs is very limited. In view of this fact it is natural to replace in the question the Shannon capacity invariant by the best known upper bound for it, which is much better understood, and can be computed efficiently, namely by the Lovász θ-function of the graph.

3.2 The θ-Function and Nearly Orthogonal Vectors

If $G = (V, E)$ is a graph, an orthonormal labeling (also called orthogonal representation) of G is a family $(b_v)_{v \in V}$ of unit vectors in an Euclidean space so that if u and v are distinct non-adjacent vertices, then $b_u^T b_v = 0$, that is, b_u and b_v are orthogonal. The θ-function $\theta(G)$ of G is the minimum, over all orthonormal labelings b_v of G

and over all unit vectors c (called here a handle), of

$$max_{v \in V} \frac{1}{(c^T b_v)^2}.$$

It is easy to check that for every G, $\alpha(G) \leq \theta(G)$. Indeed, in any orthonormal labeling of G the vectors b_v assigned to the vertices of any independent set are pairwise orthogonal, and therefore for any unit vector c the square of the inner product of at least one of them with c is at most the reciprocal of the size of the set. It is also not difficult to check that for any two graphs G and G', $\theta(G \cdot G') \leq \theta(G) \cdot \theta(G')$. (It is a bit more difficult to show that in fact equality holds, see [38].) This is proved by considering the tensor product of orthogonal representations of G and G' and the tensor product of the two handles. Therefore for every n, $\alpha(G^n) \leq (\theta(G))^n$ implying that the Shannon capacity of G satisfies $S(G) \leq \theta(G)$.

The following lemma is proved in [38].

Lemma 3.1 *Let $G = (V, E)$ be a d-regular graph on n vertices and suppose that the most negative eigenvalue of the adjacency matrix A of G is at least $-\lambda$. Then*

$$\theta(G) \leq \frac{n\lambda}{d + \lambda}.$$

Proof The matrix $B = (A + \lambda I)/\lambda$ is positive semi-definite and hence it is the gram matrix of vectors $(b_v)_{v \in V}$. It is easy to check that these vectors form an orthogonal representation of G. Define

$$c = \frac{\sum_{v \in V} b_v}{\|\sum_{v \in V} b_v\|}.$$

Then for every vector b_v

$$(c^T b_v)^2 = \frac{(1 + d/\lambda)^2}{n + nd/\lambda} = \frac{\lambda + d}{n\lambda},$$

completing the proof. □

By Theorem 2.1 and the above lemma, for the graph G_n in the theorem, $\theta(G_n) \leq (1 + o(1))36n^{2/3}$. The complement $\overline{G_n}$ of G_n is a graph with n vertices, and independence number 2. Since the product $G_n \cdot \overline{G_n}$ contains an independent set of size n (consisting of all vertices (v, v) for $v \in V(G_n)$), it follows that

$$n \leq \alpha(G_n \cdot \overline{G_n}) \leq \theta(G_n \cdot \overline{G_n}) \leq \theta(G_n)\theta(\overline{G_n}) \leq (1 + o(1))36n^{2/3}\theta(\overline{G_n}).$$

Therefore $\theta(\overline{G_n}) \geq (1/36 + o(1))n^{1/3}$. We have thus shown that the maximum possible value of the θ-function of an n-vertex graph with independence number 2 is at least $\Omega(n^{1/3})$. This is tight, up to a multiplicative constant, answering a question of Lovász and improving earlier estimates of Konyagin [34] and of Kashin and Konyagin [32]. See [2] for more details.

The n unit vectors b_v described above have the following interesting geometric property. Among any three of them, some two are orthogonal (since the graph G_n is triangle-free). On the other hand, the square of the norm of their sum is $n + n\frac{d}{\lambda}$ where $d = (1/4 + o(1))n^{2/3}$ and $\lambda = (9 + o(1))n^{1/3}$. This square norm is thus $(\frac{1}{36} + o(1))n^{4/3}$. Therefore the norm of this sum is $\Omega(n^{2/3})$ which is also tight, up to a multiplicative constant, improving the estimates in [32, 34].

Here is a quick proof of the tightness (see [34] for another proof). Let $v_1, \ldots, v_n$ be n unit vectors in an Euclidean space so that among any three of them some two are orthogonal. Let A be the gram matrix of these vectors and let $\lambda_1 \geq \cdots \geq \lambda_n \geq 0$ be its eigenvalues (which are all nonnegative as A is positive semi-definite). Then the square of the norm of the sum of the vectors is $j^t A j$ where j is the all 1 vector. This is at most $\lambda_1 n$. The assumption implies that the trace of $(A - I)^3$ is 0, that is, $\sum_i (\lambda_i - 1)^3 = 0$. As $\lambda_i - 1 \geq -1$ for all i this implies that $\lambda_1 \leq (n-1)^{1/3}$ implying the required bound.

3.3 Lovász and Shannon

As described above, for every graph G, $\alpha(G) \leq S(G) \leq \theta(G)$ where $\alpha(G)$ is the independence number of G, $S(G)$ is its Shannon capacity, and $\theta(G)$ is the Lovász θ-function of G. As mentioned it is not known whether or not the Shannon capacity $S(G)$ is bounded by any function of the independence number $\alpha(G)$. On the other hand by the discussion in the previous subsection the Lovász θ function is not bounded by any function of the independence number, and can be as large as $\Omega(n^{1/3})$ for an n-vertex graph with independence number 2. Can it be bounded by any function of the Shannon capacity $S(G)$? The next result shows that the answer is negative.

Theorem 3.2 *There is a sequence of graphs H_n with the following properties. H_n has n vertices, its Shannon capacity is 3 and its θ-function is at least $(1 + o(1))n^{1/4}$.*

Proof Let $F = GF(2^k)$ be the finite field with $q = 2^k$ elements, and let $U = U_n$ be the set of all vectors $x = (x_0, x_1, x_2) \in F^3$ so that the sum $x_0 + x_1 + x_2$ (computed in F) is nonzero, and x is not of the form (y, y, y) for some $y \in F$. Define an equivalence relation on U by calling two vectors equivalent if one is a multiple of the other by a field element. The vertex set $V = V_n$ of the graph H_n is the equivalence classes of U with respect to this relation. Therefore $|V| = n = (q^3 - q^2 - q + 1)/(q - 1) = q^2 - 1$. Two vertices $x = (x_0, x_1, x_2)$ and $y = (y_0, y_1, y_2)$ are **not** connected iff $x_0 y_0 + x_1 y_1 + x_2 y_2 = 0$, where the sum and product are computed in F and x, y are any two representatives of the corresponding equivalence classes. Note that this is an induced subgraph of the complement of the Erdős-Rényi graph (which is the polarity graph of a projective plane) considered in [26]. For our purpose here it is convenient to define it over a field of characteristic 2, see [9] for a close variant.

Claim *The Shannon capacity of H_n is at most 3.*

Proof We use a variant of the argument in [5, 31]. By definition we can assign to each vertex v of H_n a vector x_v in F^3 so that the inner product of each vector with itself (over F) is nonzero and for any two nonadjacent vertices u, v, the inner product of x_u and x_v is zero. By taking tensor powers this supplies, for every k, an assignment with similar properties for the vertices of the power H_n^k. For each vertex we get a vector in F^{3^k} so that the inner product of any vector with itself is nonzero and the inner product of any two vectors associated to non-adjacent vertices is 0. This implies that the vectors corresponding to an independent set are linearly independent and hence the size of each such set is at most 3^k, establishing the claim. □

Claim *The θ-function of H_n is at least $\sqrt{q} > n^{1/4}$.*

Proof The complement of H_n is an induced subgraph of the polarity graph of the projective plane of order q. The eigenvalues of this polarity graph are easy to compute, as for its adjacency matrix A, $A^T A = qI + J$ where I is the identity matrix and J is the all 1 matrix. Thus the eigenvalues of $A^t A$ are $q + 1 + q^2 + q = (q+1)^2$ (with multiplicity 1) and q (with multiplicity $q^2 + q$). It follows that the smallest eigenvalue of A is $-\sqrt{q}$, and by eigenvalues interlacing, the smallest eigenvalue of the adjacency matrix of the complement of H_n is at least $-\sqrt{q}$. It is not difficult to check that this complement is regular of degree q. Thus, by Lemma 3.1,

$$\theta(\overline{H_n}) \leq \frac{n\sqrt{q}}{q + \sqrt{q}}.$$

It follows that

$$\theta(H_n) \geq \frac{q + \sqrt{q}}{\sqrt{q}} = \sqrt{q} + 1 > n^{1/4}.$$

This completes the proof of the claim, which together with the previous claim imply the assertion of the theorem. □

4 Ramsey Graphs and Maxcut

4.1 The Ramsey Number $r(3, m)$

Let $r(3, m)$ denote the maximum number of vertices of a triangle-free graph whose independence number is at most m. The problem of determining or estimating this function is a well studied Ramsey type problem. Ajtai, Komlós and Szemerédi proved in [1] that $r(3, m) \leq O(m^2/\log m)$, (see also [43] for an estimate with a better constant). Improving a result of Erdős who showed in [22] that $r(3, m) \geq \Omega((m/\log m)^2)$, Kim [33] proved that the upper bound is tight up to a constant factor, that is: $r(3,m) = \Theta(m^2/\log m)$. Proofs providing a better constant appear in [14, 28]. All these lower bound proofs are probabilistic, and do not supply any explicit construction of the corresponding graphs.

The problem of finding an explicit construction of triangle-free graphs of independence number m and many vertices has also received a considerable amount of attention. Erdős [23] gave an explicit construction of such graphs with

$$\Omega(m^{(2\log 2)/3(\log 3 - \log 2)}) = \Omega(m^{1.13})$$

vertices. This has been improved by Cleve and Dagum [16], and further improved by Chung, Cleve and Dagum in [15], where the authors present a construction with

$$\Omega(m^{\log 6/\log 4}) = \Omega(m^{1.29})$$

vertices. A better explicit construction is given in [3], where the number of vertices is $\Omega(m^{4/3})$.

The graphs G_n described in Sect. 2 provide the best known explicit construction, as shown in [2]. Indeed, the graph G_n is triangle-free, and as described in the previous section its Shannon capacity is at most $m = O(n^{2/3})$, where n is the number of its vertices. As the Shannon capacity is an upper bound for the independence number, these are explicit graphs showing that $r(3,m) \geq \Omega(m^{3/2})$. A different construction providing the same asymptotic bound has been given a few years later in [18]. See also [19, 35] for more recent variants.

4.2 Maxcut in Triangle-Free Graphs

For a graph G, let $f(G)$ denote the maximum number of edges in a bipartite subgraph of G, that is, the size of the maxcut of G. Edwards [20, 21] proved that for any graph G with m edges,

$$f(G) \geq \frac{m}{2} + \frac{-1+\sqrt{8m+1}}{8} = \frac{m}{2} + \Omega(m^{1/2}).$$

This is tight for every $m = \binom{s}{2}$ where s is an integer.

Erdős and Lovász (see [24]) showed that if G is a triangle-free graph with m edges, then

$$f(G) \geq m/2 + \Omega\left(m^{2/3}\left(\frac{\log m}{\log \log m}\right)^{1/3}\right).$$

This has been improved by a logarithmic factor by Poljak and Tuza [41], and further improved by Shearer [44], who proved that if G is a triangle-free graph with m edges then

$$f(G) \geq \frac{m}{2} + \Omega(m^{3/4}). \tag{1}$$

In [4] the exponent 3/4 is improved to 4/5. Moreover, it is shown that this is tight up to the multiplicative constant in the error term. That is, there exists a constant $C > 0$ so that for every m there exists a triangle-free graph G with m edges satisfying

$$f(G) \leq \frac{m}{2} + Cm^{4/5}.$$

This is proved using the graphs G_n described in Sect. 2 together with the following simple lemma, whose proof can be found, for example, in [4].

Lemma 4.1 *Let $G = (V, E)$ be a d-regular graph with n vertices and $m = nd/2$ edges, and let $\lambda_1 \geq \lambda_2 \geq \ldots \geq \lambda_n$ be the eigenvalues of G. Then*

$$f(G) \leq (d - \lambda_n)n/4 = \frac{m}{2} - \lambda_n n/4.$$

The graph $G = G_n$ is triangle-free, has n vertices, is $d = (\frac{1}{4} + o(1))n^{2/3}$-regular and its most negative eigenvalue is $\lambda_n = -\lambda$ where $\lambda \leq (9 + o(1))n^{1/3}$. Therefore the number of edges of G is $m = \Theta(n^{5/3})$ and

$$f(G) \leq \frac{m}{2} + O(n^{4/3}) = \frac{m}{2} + O(m^{4/5}).$$

5 List Decodable Zero-Rate Codes

A binary code $C \subset \{0, 1\}^n$ is $< L$-list decodable with normalized radius τ if any Hamming ball with radius τn contains less than L codewords.

Define

$$\tau_L = \frac{1}{2} - \frac{\binom{2k}{k}}{2^{2k+1}} \quad \text{if } L = 2k \text{ or } L = 2k + 1. \tag{2}$$

Blinovskii [13] proved that for any fixed radius $\tau < \tau_L$ the largest possible $< L$-list decodable code with normalized radius τ in $\{0, 1\}^n$ is exponentially large in n, that is of size at least 2^{bn} for some $b = b(\tau, L) > 0$. On the other hand he showed that for any fixed radius $\tau > \tau_L$ the largest $< L$-list decodable code with normalized radius τ (of any length n) is of constant size, that is, of size at most some $b' = b'(\tau, L)$. Therefore, the maximum possible rate is positive for $\tau < \tau_L$, and is zero for $\tau > \tau_L$. How large can C be when τ is just above the threshold τ_L? Let $m(L, \varepsilon)$ denote the maximum possible size of a $< L$ list decodable code with normalized radius at least $\tau_L + \varepsilon$, where the maximum is taken over all values of the length n.

Levenshtein [36] showed that the so-called Plotkin bound is sharp in the unique decoding case ($L = 2$), namely

$$m(2, \varepsilon) = \frac{1}{4\varepsilon} + O(1).$$

For larger values of L the situation is more complicated. The result of [13] is proved by iterating Ramsey's theorem, providing a very large (finite) bound for $m(L, \varepsilon)$. In a recent paper with Bukh and Polyanskiy [7] it is proved that for every even L, $m(L, \varepsilon) = \Theta(1/\varepsilon)$. This implies that for every L, $m(L, \varepsilon) \geq \Omega(1/\varepsilon)$. In addition, the value of $m(3, \varepsilon)$ is determined up to a constant factor, as stated in the following theorem.

Theorem 5.1 ([7])
$$m(3, \varepsilon) = \Theta(\frac{1}{\varepsilon^{3/2}}).$$

The lower bound is proved using the graphs described in Sect. 2. Here is the argument.

Proof of the lower bound: Let $G = G_m = (V, E)$ be the graph described in Sect. 2, where m is the number of its vertices. Recall it is a Cayley graph of an elementary abelian 2-group Z_2^r, let A be its adjacency matrix, and let $d = \lambda_1 \geq \lambda_2 \geq \cdots \lambda_m = -\lambda$ be its eigenvalues, where d is the degree of regularity and $-\lambda$ is the smallest eigenvalue. Thus $d = (1/4 + o(1))m^{2/3}$, $\lambda = (9 + o(1))m^{1/3}$ and G is triangle-free. As it is a Cayley graph of an elementary abelian 2-group, it has an orthonormal basis of eigenvectors $v_1, v_2, \ldots, v_m$ in which each coordinate of each vector is in $\{-1/\sqrt{m}, 1/\sqrt{m}\}$. Indeed, the eigenvectors are simply the (normalized) characters of the group. Define $B = (A + \lambda I)/\lambda$ where I is the m-by-m identity matrix. Then B is a positive semidefinite matrix, its diagonal is the all-1 vector, its eigenvalues are $\mu_i = (\lambda_i + \lambda)/\lambda$ and the corresponding eigenvectors are the vectors v_i. Let P be the m-by-m orthogonal matrix whose columns are the vectors v_i, and note that the first v_1 is the constant vector $1/\sqrt{m}$. Let D be the diagonal matrix whose diagonal entries are the eigenvalues μ_i and let $\sqrt{D}$ denote the diagonal matrix whose entries are $\sqrt{\mu_i}$. Then $P^T B P = D$ and thus $B = (P\sqrt{D})(\sqrt{D}P^T)$.

The rows of the matrix $P\sqrt{D}$ are vectors $x_1, x_2, \ldots, x_m$ where $x_i = (x_{i1}, x_{i2}, \ldots, x_{im})$. Note that for each j, $x_{ij} \in \{-\sqrt{\mu_j/m}, \sqrt{\mu_j/m}\}$ for all i, and that x_{i1} is positive for all i. In addition $x_i^T x_j = B_{ij}$ for all i, j implying that the ℓ_2-norm of each vector x_i is 1 and that among any three vectors x_i there is an orthogonal pair. Let y_i be the vector obtained from x_i by removing its first coordinate (the one which is $\sqrt{\mu_1/m} = \sqrt{(d+\lambda)/m\lambda}$). Then each y_i is a vector of ℓ_2-norm $\sqrt{1 - \mu_1/m}$ and among any three of them there is a pair with inner product $-\mu_1/m$. We can normalize the vectors by dividing each entry by $\sqrt{1 - \mu_1/m}$ to get m unit vectors $z_1, z_2, \ldots, z_m$, where any three of them contain a pair with inner product $-\delta$, where $\delta = \mu_1/(m - \mu_1)$. Moreover, for the vectors $z_i = (z_{ij})$, for each fixed j the absolute value of all z_{ij} is the same for all i. Denote this common value by t_j. We can now use the vectors z_i to define functions mapping $[0, 1]$ to $\{1, -1\}$ as follows. Split $[0, 1]$ into disjoint intervals I_j of length t_j^2 and define f_i to be sign(z_{ij}) on the interval I_j. It is clear that the ℓ_2-norm of each f_i is 1 and the inner product between f_i and f_j is exactly that between z_i and z_j. In particular, each three functions f_i contain a pair whose inner product is at most $-\delta$.

One can replace the functions by vectors of $1, -1$ with essentially the same property, using an obvious rational approximation to the lengths of the intervals. Let n denote the length of these vectors.

Put, say, $\varepsilon = \frac{\delta}{4.01}$. Plugging $d = (1/4 + o(1))m^{2/3}$ and $\lambda = (9 + o(1))m^{1/3}$ we get $\varepsilon = \Theta(m^{-2/3})$ and hence the number of vectors is $m = \Theta\big((1/\varepsilon)^{3/2}\big)$. This gives a binary code with $m = \Theta\big((1/\varepsilon)^{3/2}\big)$ codewords of length n so that among any three codewords there are two such that the Hamming distance between them exceeds $(1/2 + 2\varepsilon)n$. Thus no Hamming ball of radius $(1/4 + \varepsilon)n = (\tau_3 + \varepsilon)n$ can contain three vectors, completing the proof. □

6 Extensions and Open Problems

As described in the previous sections, if G is a graph with independence number $\alpha(G)$, Shannon capacity $S(G)$ and θ-function $\theta(G)$, then $\alpha(G) \leq S(G) \leq \theta(G)$. Already in his original paper introducing $S(G)$ Shannon [42] proved that if $\chi^*(\overline{G})$ is the fractional chromatic number of the complement of G, and $\chi(\overline{G})$ is the chromatic number of this complement, then $S(G) \leq \chi^*(\overline{G}) \leq \chi(\overline{G})$. Lovász showed that $\theta(G) \leq \chi^*(\overline{G})$. Therefore, for every graph G,

$$\alpha(G) \leq S(G) \leq \theta(G) \leq \chi^*(\overline{G}) \leq \chi(\overline{G}).$$

As mentioned in Sect. 3, it is not known whether or not $S(G)$ can be bounded by any function of $\alpha(G)$. On the other hand, for any other pair of invariants among the above five, the larger one is not bounded by any function of the smaller one. Indeed as shown in Sect. 3, there are graphs G on n vertices where $\theta(G) \geq \Omega(n^{1/4})$ and $S(G) \leq 3$, and graphs G on n vertices with $\theta(G) = \Theta(n^{1/3})$ and $\alpha(G) = 2$.

We next show that for any $\varepsilon > 0$ there is a $\delta > 0$ and n-vertex graphs for which $\chi^*(\overline{G}) \geq n^\delta$ and $\theta(G) \leq (2 + \varepsilon)$. Such graphs are constructed in [10], based on a theorem of Frankl and Rödl [29].

For a pair of integers $q > s > 0$ let $G(q, s)$ denote the graph on $n = \binom{2q}{q}$ vertices corresponding to all q-subsets of the $2q$-element set $Q = \{1, 2, \ldots, 2q\}$, where two vertices are adjacent iff the intersection of their corresponding subsets is of cardinality precisely s. By the main result of Frankl and Rödl in [29], for every $\gamma > 0$ there is a $\mu = \mu(\gamma) > 0$ so that if $(1 - \gamma)q > s > \gamma q$ then every family of more than $2^{2q(1-\mu)}$ subsets of cardinality q of Q contains some pair of subsets whose intersection is of cardinality s. This means that the independence number of the graph $G(q, s)$ for q and s that satisfy $(1 - \gamma)q > s > \gamma q$ satisfies

$$\alpha(G(q, s)) \leq n^c \quad (3)$$

for some $c = c(\gamma) < 1$, where $n = \binom{2q}{q}$ is the number of vertices. Therefore, the fractional chromatic number of $G(q, s)$ is at least $n^{1-c} = n^{\delta}$.

It is shown in [10] that the parameter γ can be chosen to ensure that $\theta(G(q, s)) \geq \frac{n}{2+\varepsilon}$, where n is the number of vertices of $G(q, s)$. Lovász proved in [38] that if a graph has a vertex transitive automorphism group then the product of its θ-function with that of its complement is the number of vertices. Since the graph $G(q, s)$ is clearly vertex transitive, this implies that the θ-function of its complement is at most $2 + \varepsilon$. Thus, this complement is a graph showing that θ may be fixed (in fact close to 2) while the fractional chromatic number of the complement grows as a small fixed power of the number of vertices.

The existence of graphs with a fixed fractional chromatic number and large chromatic number is well known. Here the gap can be only logarithmic in the number of vertices. The Kneser graphs provide examples of graphs with fractional chromatic number $2 + \varepsilon$ and chromatic number $\Omega(\log n)$ where n is the number of vertices. The Kneser graph $K(m, r)$ is the graph whose vertices are all subsets of cardinality r of an m-element set, where two are adjacent if they are disjoint. Lovász proved in [37] that the chromatic number of $K(m, r)$ is $m - 2r + 2$, and it is easy to see that its fractional chromatic number is m/r. Taking $r = \frac{m}{2+\varepsilon}$ we get the required example.

It will be interesting to find a construction of K_k-free graphs with extremal spectral properties for $k > 3$, extending that of the graphs G_n described in Sect. 2. It is not difficult to show (see [9]) that if $d^{k-1} > n^{k-2}\lambda$ then any (n, d, λ)-graph G contains a clique of size k. Therefore, if $\lambda = O(\sqrt{d})$ and G contains no copy of K_k, then

$$d \leq O(n^{1-\frac{1}{2k-3}}).$$

This is tight for $k = 3$, as shown by the graphs G_n. Is it tight for larger values of k as well?

What is the largest possible value of the θ-function of an n vertex graph with independence number smaller than k? In [10] it is shown that this maximum is at most $O(n^{1-2/k})$. This is tight for $k = 3$ but is not known to be tight for any larger value of k. The results in [27] imply that this maximum is at least $\Omega(n^{1-O(1/\log k)})$. It will be interesting to close the gap here. In a somewhat different direction it is proved in [10] that if the odd girth of the complement of an n vertex graph G exceeds $2s + 1$, then its θ-function is at most $O(n^{1/(2s+1)})$. As mentioned in [10], this is tight for all values of s, by a natural extension of the construction of the graphs G_n.

What is the maximum possible Euclidean norm of a sum of n unit vectors in an Euclidean space (of any dimension) so that among any k of them some two are orthogonal? This extends the question discussed in Sect. 3.2 and is closely related to the question about the maximum possible θ-function of a graph on n vertices with independence number smaller than k. Denote this maximum possible norm by $f(n, k)$. It is clear that $f(n, 2) = \sqrt{n}$ and as discussed in Sect. 3, $f(n, 3) = \Theta(n^{2/3})$. In [10] it is shown that $f(n, k) \leq O(n^{1-1/k})$ for all k. The following theorem can be proved following the approach in [27].

Theorem 6.1 *For any $k > k_0$, $n > n(k)$,*

$$f(n,k) \geq n^{1-O(1/\log k)}.$$

Here is an outline of the proof. Let $t = 4p$, p a prime, and let $\mathcal{F}$ be the set of all vectors in $\{-1, 1\}^t$ with an even number of -1 entries which is at most $n/3$. Let $\mathcal{G}$ denote the tensor product of s copies of $\mathcal{F}$, normalized to be unit vectors. Each vector in $\mathcal{G}$ has projection at least $(1/3)^s$ in the direction of the all 1 vector. Put $q = 2^{-0.85st}$ and let X be a random subset of $\mathcal{G}$ obtained by taking each member of $\mathcal{G}$, randomly and independently, to be a member of X with probability q. Let n denote the number of vectors in X. Clearly their sum is of norm at least $n/3^s$, and n is at least $2^{st/25}$ (say), with high probability. By a result of [30], any set of more than $2^{H(1/4)t}$ vectors in $\mathcal{F}$ contains an orthogonal pair. Now any set in the tensor power of s copies of $\mathcal{F}$ that contains no such pair is a subset of a product of its projections on the copies of $\mathcal{F}$, namely of a box of the form $\mathcal{F}_1 \times \mathcal{F}_2 \cdots \times \mathcal{F}_s$, with $\mathcal{F}_i \subset \mathcal{F}$ with no pair of orthogonal vectors. The number of choices for such a product is smaller than $2^{2^t s}$ and the probability that for, say, $k = 30 \cdot 2^t/t$, k members of such a product belong to X is small, by the union bound, as

$$2^{2^t s} \binom{2^{H(1/4)ts}}{k} q^k < 1.$$

This completes the proof. □

As described in Sect. 4, every triangle free graph with m edges contains a bipartite subgraph with at least $\frac{m}{2} + cm^{4/5}$ edges. The graphs G_n show that this is tight up to the absolute constant c. It is natural to extend the question for other forbidden graphs H. Let $f(G)$ denote the maximum number of edges in a bipartite subgraph of G and let $f(m, H)$ denote the minimum possible value of $f(G)$, as G ranges over all H-free graphs with m edges. It is proved in [11] that $f(m, H) = \frac{m}{2} + c(H)m^{4/5}$ for all graphs H obtained by joining a vertex to all vertices of any nontrivial forest, and this is tight up to the value of $c(H)$. Here, too, the tightness follows from the graphs G_n. It is also proved in the same paper that

$$f(m, C_{2r}) \geq \frac{m}{2} + c(r)m^{\frac{r}{r+1}} \qquad (4)$$

for every even cycle C_{2r}, and this is tight for $2r \in \{4, 6, 10\}$. For complete bipartite graphs with 2 or 3 vertices in the smaller color class it is shown that

$$f(m, K_{2,s}) \geq \frac{m}{2} + c(s)m^{5/6}$$

and

$$f(m, K_{3,s}) \geq \frac{m}{2} + b(s)m^{4/5}$$

and both results are tight up to the constants $c(s)$, $b(s)$. See also [6] for some related results. An intriguing conjecture raised in [6] is that for every fixed graph H there is an $\varepsilon = \varepsilon(H)$ so that $f(m, H) \geq \frac{m}{2} + \Omega(m^{3/4+\varepsilon})$. This, as well as the conjecture that for every even cycle the estimate (4) is tight, remain open.

Recall that the function $m(L, \varepsilon)$ defined in Sect. 5 is the maximum possible size of a binary code (of any length) in which every Hamming ball of normalized radius $\tau_L + \varepsilon$ contains less than L codewords. Here τ_L, defined in (2), is the threshold normalized radius between positive and zero rate for $< L$-list decodable codes. While it is proved in [7] that for every even L, $m(L, \varepsilon) = \Theta_L(1/\varepsilon)$ and that $m(3, \varepsilon) = \Theta(1/\varepsilon^{3/2})$, the problem of determining or estimating $m(L, \varepsilon)$ for odd values of $L > 3$ is open. The lower bound is $\Omega_L(1/\varepsilon)$ and the upper bound is an iterated exponential in $1/\varepsilon$. It seems plausible to conjecture that $m(n, \varepsilon)$ is bounded by a polynomial in ε, for any fixed L. This remains open.

Thucydides, who is widely considered to be the father of scientific history, wrote in the introduction to his book on the History of the Peloponnesian War between Sparta and Athens (431–404 BC): "With reference to the speeches in this history; some I heard myself, others I got from various quarters; it was in all cases difficult to carry them word for word in one's memory, so my habit has been to make the speakers say what was in my opinion demanded of them by the various occasions."

In analogy, let me conclude this short paper stating that many of the results described here are due to Lovász, others are inspired by his questions and proofs. Regarding the statements that are difficult to derive directly by following his work word for word, my habit has been to try to find out how Laci would have established them. I hope this has been at least somewhat successful.

Acknowledgements Research supported in part by NSF grant DMS-1855464, ISF grant 281/17, GIF grant G-1347-304.6/2016 and the Simons Foundation.

References

1. M. Ajtai, J. Komlós and E. Szemerédi, A note on Ramsey numbers, J. Combinatorial Theory Ser. A 29 (1980), 354–360.
2. N. Alon, Explicit Ramsey graphs and orthonormal labelings, The Electronic J. Combinatorics 1 (1994), R12, 8pp.
3. N. Alon, Tough Ramsey graphs without short cycles, J. Algebraic Combinatorics 4 (1995), 189–195.
4. N. Alon, Bipartite subgraphs, Combinatorica 16 (1996), 301–311.
5. N. Alon, The Shannon capacity of a union, Combinatorica 18 (1998), 301–310.
6. N. Alon, B. Bollobás, M. Krivelevich and B. Sudakov, Maximum cuts and judicious partitions in graphs without short cycles, J. Combinatorial Theory, Ser. B 88 (2003), 329–346.
7. N. Alon, B. Bukh and Y. Polyanskiy, List-decodable zero-rate codes, IEEE Transactions on Information Theory 65 (2019), 1657–1667.
8. N. Alon and F. R. K. Chung, Explicit construction of linear sized tolerant networks, Discrete Math. 72(1988), 15–19.
9. N. Alon and M. Krivelevich, Constructive bounds for a Ramsey-type problem, Graphs and Combinatorics 13 (1997), 217–225.

10. N. Alon and N. Kahale, Approximating the independence number via the θ-function, Math. Programming 80 (1998), 253–264.
11. N. Alon, M. Krivelevich and B. Sudakov, MaxCut in H-free graphs, Combinatorics, Probability and Computing 14 (2005), 629–647.
12. N. Alon and A. Orlitsky, Repeated communication and Ramsey graphs, IEEE Transactions on Information Theory 41 (1995), 1276–1289.
13. V. M. Blinovskiĭ, Bounds for codes in decoding by a list of finite length, Problemy Peredachi Informatsii 22 (1986),11–25.
14. T. Bohman and P. Keevash, Dynamic Concentration of the Triangle-Free Process, The Seventh European Conference on Combinatorics, Graph Theory and Applications, 489–495, CRM Series, 16, Ed. Norm., Pisa, 2013.
15. F. R. K. Chung, R. Cleve and P. Dagum, A note on constructive lower bounds for the Ramsey numbers $R(3, t)$, J. Combinatorial Theory Ser. B 57 (1993), 150–155.
16. R. Cleve and P. Dagum, A constructive $\Omega(t^{1.26})$ lower bound for the Ramsey number $R(3, t)$, Inter. Comp. Sci. Inst. Tech. Rep. TR-89-009, 1989.
17. F. Chung and R. L. Graham, Erdős on Graphs: His Legacy of Unsolved Problems, A. K. Peters, Ltd., Wellesley, MA, 1998.
18. B. Codenotti, P. Pudlák, and G. Resta, Some structural properties of low-rank matrices related to computational complexity, Theoret. Comput. Sci. 235 (2000), 89–107.
19. D. Conlon, A sequence of triangle-free pseudorandom graphs, Combin. Probab. Comput. 26 (2017), no. 2, 195–200.
20. C. S. Edwards, Some extremal properties of bipartite subgraphs, Canadian Journal of Mathematics 3 (1973), 475–485.
21. C. S. Edwards, An improved lower bound for the number of edges in a largest bipartite subgraph, Proc. 2nd Czechoslovak Symposium on Graph Theory, Prague, (1975), 167–181.
22. P. Erdős, Graph Theory and Probability, II, Canad. J. Math. 13 (1961), 346–352.
23. P. Erdős, On the construction of certain graphs, J. Combinatorial Theory 17 (1966), 149–153.
24. P. Erdős, Problems and results in Graph Theory and Combinatorial Analysis, in: *Graph Theory and Related Topics*, J. A. Bondy and U. S. R. Murty (Eds.), Proc. Conf. Waterloo, 1977, Academic Press, New York, 1979, 153–163.
25. P. Erdős, R. J. McEliece and H. Taylor, Ramsey bounds for graph products, Pacific Journal of Mathematics, 37 (1971),45–46.
26. P. Erdős and A. Rényi, On a problem in the theory of graphs (in Hungarian), Publ. Math. Inst. Hungar. Acad. Sci. 7 (1962), 215–235.
27. U. Feige, Randomized graph products, chromatic numbers, and the Lovász θ-function, Proc. of the 27th ACM STOC, ACM Press (1995), 635–640.
28. G. Fiz Pontiveros, S. Griffiths and R. Morris, The triangle-free process and R(3,k), arXiv:1302.6279
29. P. Frankl and V. Rödl, Forbidden intersections, Trans. AMS 300 (1987), 259–286.
30. P. Frankl and R. Wilson, Intersection theorems with geometric consequences, Combinatorica 1 (1981), 259–286.
31. W. Haemers, An upper bound for the Shannon capacity of a graph, Colloq. Math. Soc. János Bolyai 25, Algebraic Methods in Graph Theory, Szeged, Hungary (1978), 267–272.
32. B. S. Kashin and S. V. Konyagin, On systems of vectors in a Hilbert space, Trudy Mat. Inst. imeni V. A. Steklova 157 (1981), 64–67. English translation in: Proc. of the Steklov Institute of Mathematics (AMS 1983), 67–70.
33. J. H. Kim, The Ramsey number $R(3, t)$ has order of magnitude $t^2/\log t$, Random Structures Algorithms 7 (1995), 173–207.
34. S. V. Konyagin, Systems of vectors in Euclidean space and an extremal problem for polynomials, Mat. Zametki 29 (1981), 63–74. English translation in: Mathematical Notes of the Academy of the USSR 29 (1981), 33–39.
35. S. Kopparty, A constructive lower bound on R(3,k), see: http://sites.math.rutgers.edu/~sk1233/courses/graphtheory-F11/cayley.pdf

36. V. I. Levenshtein, The application of Hadamard matrices to a problem in coding, Problemy Kibernetiki, 5 (1961), 123–136. English translation in Problems of Cybernetics 5, 1964 pp. 166–184.
37. L. Lovász, Kneser's conjecture, chromatic number, and homotopy, J. Combin. Theory Ser. A 25 (1978), no. 3, 319–324.
38. L. Lovász, On the Shannon capacity of a graph, IEEE Transactions on Information Theory IT-25, (1979), 1–7.
39. L. Lovász, Combinatorial Problems and Exercises, North Holland, Amsterdam, 1979, Problem 11.8.
40. F. J. MacWilliams and N. J. A. Sloane, The Theory of Error-Correcting Codes, North Holland, Amsterdam, 1977.
41. S. Poljak and Zs. Tuza, Bipartite subgraphs of triangle-free graphs, SIAM J. Discrete Math. 7 (1994), 307–313
42. C. E. Shannon, The zero–error capacity of a noisy channel, IRE Trans. Inform. Theory 2 (1956), 8–19.
43. J. B. Shearer, A note on the independence number of a triangle-free graph, Discrete Math. 46 (1983), 83–87.
44. J. B. Shearer, A note on bipartite subgraphs of triangle-free graphs, Random Structures and Algorithms 3 (1992), 223–226.

Continuous Matroids Revisited

Anders Björner

Dedicated to László Lovász on the occasion of his 70th birthday

Abstract Here we look back at some work done in the mid-1980s in collaboration with László Lovász. Our main concern at that time was to provide conditions that make it possible to pass to the limit of a class of finite matroids. With the current flurry of interest in limits of combinatorial objects, a review of such matroid limits seems timely. The characteristic property of a continuous matroid is the existence of a rank function taking as values the full real unit interval. Known examples of such rank functions include Lebesgue measure on the unit interval and the dimension function of certain von Neumann algebras. In both these cases the lattice property of modularity plays a crucial role. A more general concept, *pseudomodularity*, makes possible the construction of e.g. continuous field extensions (algebraic matroids) and continuous partition lattices (graphic matroids).

Keywords Continuous geometry · von Neumann geometry · Matroid · Geometric lattice · Modular lattice · Pseudomodular lattice · von Neumann factor of type II_1 · Hyperfinite · Lebesgue measure · Continuous partition lattice · Continuous algebraic matroid

2010 Mathematics Subject Classification Primary 05B35 · Secondary 06C10 · 06C20 · 28A99 · 37A99 · 46L10 · 51A99

September 26, 2018. Research supported by Swedish Research Council (Vetenskapsrådet), grant 2015-05308.

A. Björner (✉)
Institut Mittag-Leffler, Auravägen 17, 182 60 Djursholm, Sweden
e-mail: bjorner@kth.se

Kungl. Tekniska Högskolan, Matematiska Inst., 100 44 Stockholm, Sweden

1 Introduction

From the world of finite sets $[n] := \{1,\ldots,n\}$ and structures there are two basic ways of "going to infinity": either by stepwise extensions $[n] \hookrightarrow [n+1]$ (leading to the natural numbers $\mathbb{N}$, formal power series, infinite permutation group S_∞, etc) or by subdivision (leading to the reals $\mathbb{R}$, the unit interval $[0, 1]$, measure, integral,...).

In the case of matroids, the first road leads to what is usually meant by an "infinite matroid", namely a matroid on an infinite ground set having a bounded integer-valued semimodular rank function. In this paper we are interested in taking the second road, leading to continuous matroids.

What then *is* a continuous matroid? While we leave open a precise answer to this question (see the discussion in Sect. 6), a quick answer is that a continuous matroid is a semimodular lattice with a real-valued rank function such that maximal chains are isometric to the unit interval.

Various investigations in incidence geometry, by Dedekind, Menger, Veblen, Young, and others, culminated in a theorem of Birkhoff from 1935 [1], which characterises projective geometries in lattice-theoretic terms. By "projective geometry" is here understood the full subspace lattice of some finite-dimensional vector space over a skew field. The main content of Birkhoff's theorem is that the characteristic lattice-theoretic feature of these geometries is modularity.

Generally speaking, two lines of development were inspired by Birkhoff's theorem and its roots. These lines of research, both extending the class $\mathcal{PG}$ of finite-dimensional projective geometries, were motivated by different concerns and headed in different directions. On the one hand, $\mathcal{PG}$ was embedded into the class of continuous geometries, in the sense of von Neumann, the motivation here coming from operator theory and quantum physics. In another direction, $\mathcal{PG}$ was embedded into the class of geometric lattices or matroids, an important class of combinatorial structures.

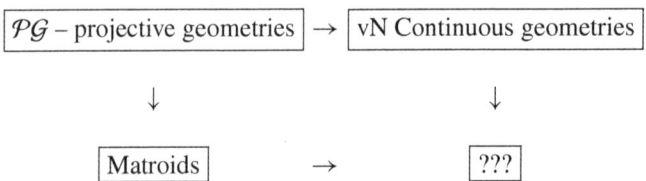

A natural question then is, can the two directions be merged in any fruitful way? Is there a common concept that allows both the semimodularity characteristic of matroids and the continuous rank functions of von Neumann geometries??

Along these lines, in the papers [2, 3] we presented conditions guaranteeing, for certain classes C of finite matroid lattices, the existence of embedding schemes that allow passing to the limit C_∞. The purpose of the present paper is to review this work from a viewpoint 30 years hence. This will be done in Sects. 4–6, following a short recapitulation of background material in Sects. 2 and 3. With the current flurry of interest in limits of combinatorial objects, much of it due to Lovász [6], a review of such matroid limits seems timely.

A basic and beautiful example of the kind we have in mind is when C is the class of finite Boolean lattices. Then C_∞ is the sigma algebra of Lebesgue measurable subsets of the unit interval, and the rank function is Lebesgue measure. Another example is the class $\mathcal{PG}$ of finite projective geometries over a finite field, for which $\mathcal{PG}_\infty$ is the corresponding hyperfinite von Neumann geometry. These measure-theoretic and dimension-theoretic interpretations suggest probabilistic and geometric connections.

From a combinatorial point of view, these examples, both due to von Neumann, can be thought of as continuous analogues of free matroids and linear matroids. We show that also partition lattices and field extensions have such continuous analogs, corresponding to the classes of graphic and algebraic matroids, respectively.

The main problem one faces when trying to construct embedding schemes and limits for matroids is the absence of a key technical property: modularity. We introduced in [3] a somewhat weaker notion, called *pseudomodularity*, that does the job for us in several cases where modularity is missing.

2 Geometry, Matroids and Lattices

Matroids, a concept generalising configurations in projective geometry, were introduced around 1935 by Birkhoff, MacLane and Whitney. As is well known, the concept has been spectacularly successful. We assume some familiarity with the concepts of lattices and matroids, and refer to Birkhoff [1] and Oxley [7] as standard references.

Recall that a lattice is said to be *modular* if it satisfies the *modular law*:

$$x \leq z \implies (x \vee y) \wedge z = x \vee (y \wedge z)$$

for all $x, y, z \in \mathcal{L}$. A lattice of finite rank is called *semimodular* if

$$x \text{ and } y \text{ cover } x \wedge y \implies x \vee y \text{ covers } x \text{ and } y,$$

for all $x, y \in \mathcal{L}$. In a semimodular lattice all maximal chains are of equal length, which induces a rank function. This satisfies the *the semimodular inequality*:

$$r(x \vee y) + r(x \wedge y) \leq r(x) + r(y)$$

A semimodular lattice is modular if and only if for all x, y:

$$r(x \vee y) + r(x \wedge y) = r(x) + r(y)$$

There are many ways to define matroids. We use the characterisation via geometric lattices [1, 7], that is via the poset of its closed subsets. These lattices, sometimes also called *matroid lattices*, are defined by being semimodular and atomic.

As was mentioned in the Introduction, the following characterization of projective geometries, also known as full linear matroids, has been of great importance.

Theorem 2.1 (Birkhoff, [1]) *A lattice $\mathcal{L}$ of finite rank $r \geq 4$ is a projective geometry of dimension $r - 1$ over a skew field $\mathbf{k}$ if and only if*

$\mathcal{L}$ is complete, complemented, irreducible, and modular.

In rank 3 there are also the non-Desarguesian planes.

In this characterization, the first three conditions mean that
- (complete) joins and meets exist for arbitrary subsets (not only for pairs),
- (complemented) for each $x \in \mathcal{L}$ there is $y \in \mathcal{L}$ such that $x \vee y = \hat{1}$ and $x \wedge y = \hat{0}$
- (irreducible) $\mathcal{L}$ is not a direct product.

It is however the fourth condition—modularity—that is the core of the result. The absense of modularity initially posed a serious technical challenge to our attempts at creating continuous limits of matroids.

It can be convenient to present geometric facts in lattice-theoretic terms. For instance, consider the following property of a line configuration:

Suppose that we have 3 lines x, y and z in 3-dimensional space, and that they are pairwise coplanar but all three are not coplanar. Then the 3 lines intersect in a point.

This property, depicted in Fig. 1, was shown by Ingleton and Main [7] to hold for all algebraic matroids. However, it is not true for linear matroids in general. As we shall see, this kind of structure is of interest for matroid limits.

Definition 2.2 A semimodular lattice $\mathcal{L}$ is *pseudomodular* if it satisfies:

Let $x, y, z, u \in \mathcal{L}$ and assume that u covers x, y, z, z covers $x \wedge z$ and z covers $y \wedge z$. Then, $r(x \wedge y) - r(x \wedge y \wedge z) \leq 1$.

This definition is illustrated in Fig. 2, which is perhaps easier to take in and remember than the verbal description. Notice the similarity with the Ingleton-Main property.

Why do we need this peculiar-looking concept? The reason is the trouble that comes from lack of modularity, and in particular from the fact that meets $x \wedge y$ are not necessarily continuous, see Theorem 4.1. Pseudomodularity has another characterization in terms of certain "pseudointersections" $x \vdash y$ that helps repair the shortcomings of the meet function, see [3] for this.

Thus, we now have these lattice properties:

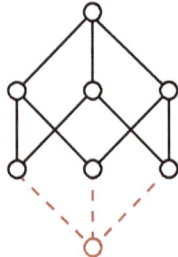

Fig. 1 Ingleton-Main property

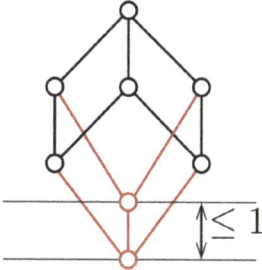

Fig. 2 Pseudomodularity

Boolean ⇒ Modular ⇒ Pseudomodular ⇒ Semimodular

Lattices in the following classes were shown to be pseudomodular in [3].

- rank ≤ 3
- Boolean
- Full linear
- Full algebraic (due to Dress and Lovász [4])
- Full graphic (partition lattices)
- Full transversal

3 Work of von Neumann

The idea of continuous geometries is due to John von Neumann. He gave an axiomatic definition, building on that of Birkhoff. To help keeping distinct but similar-sounding concepts apart, we add a prefix "JvN-". These are his axioms:

Definition 3.1 A lattice $\mathcal{L}$ is a *JvN-geometry* if it is complete, complemented, irreducible and modular, and if in addition it satisfies $x \vee (\wedge C) = \wedge (x \vee C)$ for all chains (totally ordered subsets) $C \subseteq \mathcal{L}$ and for all $x \in \mathcal{L}$.

For this class of lattices, von Neumann proved two main theorems, concerning the existence of a dimension function and concerning coordinatization.

Theorem 3.2 (von Neumann) *For a JvN-geometry $\mathcal{L}$ there exists a* dimension function $d : \mathcal{L} \to [0, 1]$ *with range either*

$$\{0, \frac{1}{n}, \frac{2}{n}, \ldots, \frac{n-1}{n}, 1\} \quad or \quad [0, 1]$$

The first case is, of course, projective geometry with normalized dimension function. The other case is something new: *continuous geometry*. Just as Birkhoff's theorem provides an algebraic representation of projective geometries, so algebraic representations for all JvN-geometries were given by von Neumann.

Theorem 3.3 (von Neumann) *For a JvN-geometry $\mathcal{L}$ there exists a regular ring $\mathcal{R}$ such that $\mathcal{L}$ is isomorphic to the lattice of left ideals in $\mathcal{R}$.*

We refer to the literature[1] for definitions and ramifications surrounding von Neumann's concepts of continuous geometries and of regular ring. Suffice it to say that his results put classical projective geometry of euclidean space and newer operator geometry in Hilbert space under one hat.

It is an intriguing fact that the background to von Neumann's interest in continuous geometry came from quantum mechanics. Around 1930 he initiated study of the algebra of bounded linear operators on Hilbert space and its subalgebras. The reason for this study was the belief that the right mathematical apparatus for quantum mechanics was in terms of operators on Hilbert space, a belief that he later came to question. Operator theory was then developed in a series of papers by von Neumann and Murray (1936–1943) and expanded by Gelfand, Naimark (mid-1940s on) and others.

So called "von Neumann algebras" are central objects in operator theory. There exists a class of such algebras, called *factors*, such that the study of von Neumann algebras reduces to the study of factors. Murray and von Neumann gave a classification of factors and showed that every factor has a *dimension function*. For factors of type II_1 the range of its dimension function is the full unit interval.

The relevance of all this for JvN-geometries is the following. The lattice of projections in a factor of type II_1 is a continuous JvN-geometry. It is concrete in the sense that it is not reached as a limit. See Sect. 6 for more about this.

4 Hyperfinite Continuous Matroids

In [3] we described a method for constructing continuous analogues, that works for several important classes of matroids. They are obtained by the following procedure. We begin by assuming that $(\mathcal{L}_n)_{n=1,2,...}$ is a sequence of matroid lattices, one for each positive integer n. Suppose also that $\text{rank}(\mathcal{L}_n) = n$.

Suppose that $n = qm$. We define a *stretch embedding* of $\mathcal{L}_m$ in $\mathcal{L}_n$ to be a lattice embedding $\phi = \phi_m^n : \mathcal{L}_m \to \mathcal{L}_n$ such that $r(\phi(x)) = qr(x)$ for each $x \in \mathcal{L}_m$. That is, a stretch embedding preserves joins, meets and normalized rank.

By an *embedding scheme* we mean a system of stretch embeddings $\phi_m^n : \mathcal{L}_m \to \mathcal{L}_n$ such that the mappings ϕ_m^n form a directed system, i.e., if $k \mid m$ and $m \mid n$ then $\phi_m^n \circ \phi_k^m = \phi_k^n$. With this one can pass to the direct limit $\mathcal{L}_{(\infty)}$.

To finish the construction of a "continuous limit" $\mathcal{L}_\infty$ we need to make a metric completion. For this, define distance in terms of the rank function by

$$d(x, y) = 2r(x \vee y) - r(x) - r(y),$$

---

[1] The extended abstracts in Proc. Natl. Acad. Sci. [10, 11] are very informative.

which is a valid metric for any semimodular lattice. Note that in case of a modular lattice this specializes to

$$d(x, y) = r(x \vee y) - r(x \wedge y),$$

which is the metric used by von Neumann.

Summarizing, we have a procedure with the following steps, beginning from some suitable class of finite matroid lattices $\mathcal{L}$.

Step 1. Construct scheme of stretch embeddings $\phi_n^m : \mathcal{L}_n \to \mathcal{L}_m$
Step 2. Direct limit: $\lim(\mathcal{L}_n, \phi_n^m) = \mathcal{L}_{(\infty)}$
Step 3: Metric completion: $\mathcal{L}_{(\infty)} \to \mathcal{L}_\infty$.
We call the lattices $\mathcal{L}_\infty$ produced by this method *hyperfinite*.

Theorem 4.1 *A hyperfinite continuous matroid $\mathcal{L}_\infty$ has an induced rank function $r : \mathcal{L}_\infty \to [0, 1]$ and satisfies the following properties:*

- $\mathcal{L}_\infty$ *is a complete lattice.*
- *It is strongly complemented: for any $x \in \mathcal{L}_\infty$ there is $y \in \mathcal{L}_\infty$ such that $x \vee y = \hat{1}$, $x \wedge y = \hat{0}$, and $r(x) + r(y) = 1$.*
- *Semimodular inequality: $r(x \vee y) + r(x \wedge y) \leq r(x) + r(y)$.*
- *Monotonicity: $x < y$ implies $r(x) < r(y)$.*
- *Metric: $d(x, y) = 2r(x \vee y) - r(x) - r(y)$ is a metric on $\mathcal{L}_\infty$.*
- *The rank function $r : \mathcal{L}_\infty \to [0, 1]$ is surjective and continuous (in the metric topology).*
- *Restricted to any maximal chain C in $\mathcal{L}_\infty$, the rank function is an isometric bijection $r : C \to [0, 1]$.*
- *The join function is uniformly continuous, since $d(x \vee y, z \vee u) \leq d(x, z) + d(y, u)$.*

The meet function is in general not continuous. It may happen that in $\mathcal{L}_\infty \setminus \{\hat{1}\}$ there are two sequences $a_1, a_2, \ldots$ and $b_1, b_2, \ldots$ such that $\lim a_j = \hat{1}$, $\lim b_j = \hat{1}$ and $a_j \wedge b_j = \hat{0}$ for all a_j.

To produce stretch embeddings is simpler than one might think, using the following result.

Theorem 4.2 *Suppose that $\mathcal{L}$ is a semimodular lattice and $a_1, \ldots, a_k$ elements of $\mathcal{L}$ such that $r(a_1) + \cdots + r(a_k) = r(a_1 \vee \cdots \vee a_k)$. Suppose furthermore that either
(i) $\mathcal{L}$ is pseudomodular, or
(ii) all the elements a_i are modular.
Then the sublattice generated by the intervals $[\hat{0}, a_i]$ is isomorphic to the direct product of these intervals.*

Consider again a sequence $\mathcal{L}_1, \mathcal{L}_2, \ldots$ of pseudomodular geometric lattices such that $\mathcal{L}_n$ has height n. Assume that when $n = qm$, there exist in $\mathcal{L}_n$ some q elements $a_1, \ldots, a_q$ of rank m such that $a_1 \vee \cdots \vee a_q = \hat{1}$ and $[\hat{0}, a_i] \cong \mathcal{L}_m$, for all i. We call these elements the *representatives* of $\mathcal{L}_m$ in $\mathcal{L}_n$. Let $\phi_i : \mathcal{L}_m \to [\hat{0}, a_i]$ ($i = 1, \ldots, q$) be any isomorphisms, and define

$$\phi(x) = \phi_1(x) \vee \cdots \vee \phi_q(x).$$

By Theorem 4.2 this is a stretch embedding.

As was argued before, to construct the "continuous limit" $\mathcal{L}_\infty$ of this sequence of geometric lattices, we have to assume that the stretch embeddings ϕ_m^n form a directed system, i.e., if $k \mid m$ and $m \mid n$ then $\phi_m^n \circ \phi_k^m = \phi_k^n$. One may assure this by choosing the representatives compatibly.

5 Examples

5.1 The Hyperfinite Continuous Matroids We Know

To make the machinery of the previous section work and produce hyperfinite continuous lattices one needs pseudomodularity and a good choice of representatives.

All the hyperfinite matroids we know, listed in Table 1, are pseudomodular, and are constructed upon the universal complete geometries of its kind. By this we mean that for each rank there is a "full" matroid into which the other matroids embed. Think e.g. of graphs: all graphs embed into the complete graph on the same number of vertices.

5.2 Continuous Free Matroid

Let $\mathcal{B}_n$ denote the lattice of subsets of an n-element set, which we take to be $[n] = \{1, 2, \ldots, n\}$. This is the unique Boolean lattice of rank n.

Let $n = qm$. The subsets $G_i = \{(i-1)m + 1, (i-1)m + 2, \ldots, im\}$, for $i = 1, 2, \ldots, q$, are good representatives of $\mathcal{B}_m$ in $\mathcal{B}_n$. Thus, via the machinery of stretch embeddings a continuous limit $\mathcal{B}_\infty$ is reached. Boolean lattices are modular, so these lattices are covered by the theory of JvN-geometries.

Table 1 Hyperfinite continuous matroids

Continuous matroid	Limit of this class of finite matroids
Lebesgue meas. σ-algebra $\mathcal{B}_\infty$ rank = Lebesgue measure	Free (Boolean) rank = normalized cardinality
Hyperfinite $\mathcal{L}_\infty(\mathbf{k})$ (vNeumann) rank = dimension (oper alg)	Linearly representable over $\mathbf{k}$ rank = normalized dimension
Algebraic $\mathcal{A}_\infty(\mathbf{k})$	Algebraically representable over $\mathbf{k}$
Partition lattice Π_∞	Graphs
Transversal	Transversal matroids

A nice fact, see von Neumann [11], is that $\mathcal{B}_\infty$ is isomorphic to the lattice of measurable subsets of the unit interval modulo subsets of measure zero. The continuous rank function of $\mathcal{B}_\infty$ is Lebesgue measure.

This can be "seen" by letting the ground set be the n-element set of half-open subintervals of the unit interval

$$(0, \frac{1}{n}], \; (\frac{1}{n}, \frac{2}{n}], \; \ldots, \; (\frac{n-1}{n}, 1].$$

Label these intervals by their right endpoint, construct the representative sets G_i as before, and identify the subsets of $[n]$ with the union of the respective intervals which is a measurable subset of the unit interval.

Another nice fact, also due to von Neumann [9], is that the automorphism group of $\mathcal{B}_\infty$ is isomorphic to the group of all measure-preserving permutations of the unit interval (mod 0). Since the automorphism group of $\mathcal{B}_n$ is the symmetric group S_n, this underscores that the correct continuous analogs of finite permutations are measure-preserving mappings of the unit interval to itself (mod 0).

5.3 Continuous Linear Matroid

Let $\mathcal{L}_n(\mathbf{k})$ be the lattice of linear subspaces of an n-dimensional vector space $\mathbf{V}_n$ over some field $\mathbf{k}$. Choose a basis $\{x_1, \ldots, x_n\}$ for $\mathbf{V}_n$, and let $\mathbf{S}_i$ be the linear subspace of $\mathbf{V}_n$ generated by $\{x_{(i-1)m+1}, \ldots, x_{im}\}$ ($i = 1, \ldots, n/m$). Then $\mathbf{S}_1, \ldots, \mathbf{S}_{n/m}$ are representatives of $\mathcal{L}_m$ in $\mathcal{L}_n$, and one checks that the induced mappings form a directed system. Since $\mathcal{L}_n(\mathbf{k})$ is pseudomodular, the embedding scheme produces a continuous analogue $\mathcal{L}_\infty(\mathbf{k})$.

The continuous matroids produced this way are the hyperfinite ones coming from lattices of projections in factors of type II$_1$, as discussed in [11].

5.4 Continuous Algebraic Matroid

Suppose that $\mathbf{k}$ is an algebraically closed field and let $\mathbf{K}_n$ be an algebraically closed field extension of $\mathbf{k}$ of transcendence degree $n \geq 1$. Let $\mathcal{A}_n(\mathbf{k})$ be the geometric lattice of algebraically closed subfields of $\mathbf{K}_n$ containing $\mathbf{k}$.

Let $\{x_1, \ldots, x_n\}$ be a transcendence basis for $\mathbf{K}_n$, and let $\mathbf{G}_i$ be the algebraically closed subfield of $\mathbf{K}_n$ generated by $\{x_{(i-1)m+1}, \ldots, x_{im}\}$ ($i = 1, \ldots, n/m$). Then $\mathbf{G}_1, \ldots, \mathbf{G}_{n/m}$ are appropriate representatives of $\mathcal{A}_m$ in $\mathcal{A}_n$, and it is easy to check that the induced mappings form a directed system. So, since $\mathcal{A}_n(\mathbf{k})$ is pseudomodular, the machinery produces a continuous analogue $\mathcal{A}_\infty(\mathbf{k})$.

One is left wondering if these "continuous field extensions" have some algebraic meaning.

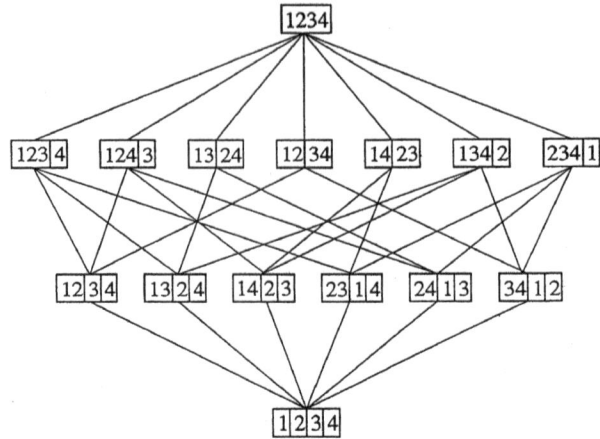

Fig. 3 The partition lattice Π_3

5.5 Continuous Partition Lattice

Let Π_n denote the lattice of partitions of an $(n + 1)$-element set, which we take to be $F_n = \{0, 1, \ldots, n\}$.[2]

The normalized rank of a partition π with k blocks is $r(\pi) = \frac{n-k}{n}$, so

$$r(\Pi_n) = \{0, \frac{1}{n}, \frac{2}{n}, \ldots, \frac{n-1}{n}, 1\}$$

To generate stretch embeddings, suppose that $n = qm$, and for $i = 1, 2, \ldots, q$ let γ_i be the partition of F_n whose only non-singleton block is $G_i = \{0, (i-1)m + 1, \ldots, im\}$. These partitions γ_i are suitable representatives of Π_m in Π_n, so a stretch embedding is induced. The system of such stretch embeddings form a scheme, so we can pass to the limit Π_∞.

There exists another approach to the problem of constructing continuous analogs of partition lattices, due to Haiman [5], see also [8]. This uses the concept of "measurable partition" from ergodic theory. Haiman defines a certain kind of measurable partitions of the unit interval, which he calls "continuous partitions", and shows that these form a lattice $\widehat{\Pi}_\infty$ with all the expected properties.

Two particularly interesting facts about the lattice $\widehat{\Pi}_\infty$ of continuous partitions are that

- the hyperfinite Π_∞ is a proper sublattice of $\widehat{\Pi}_\infty$,
- the automorphism group of $\widehat{\Pi}_\infty$ is isomorphic to the group of all measure-preserving mappings of the unit interval (mod 0), the same as for $\mathcal{B}_\infty$.

---

[2] The labeling of elements in Fig. 3 is off by one.

5.6 *Continuous Transversal Matroid*

This is the limit of matchings in full bipartite graphs. See [3] for details.

6 Epilogue

What then is a continuous matroid? Or rather, what should it be? We have seen several examples, all constructed as limits of embedding schemes – thus, all belonging to what we have called the hyperfinite case. But there are other possibilities, namely there are continuous matroids that arise via a direct construction, let us call them "concrete".

The sigma algebra of Lebesgue measurable subsets of the unit interval as usually defined is one example of a continuous matroid. The continuous-dimensional JvN-geometries coming from factors of type II_1 is another. A third example is Haiman's continuous partition lattice $\widehat{\Pi}_\infty$ [5]. In the first two cases, the presentations as hyperfinite limits and by concrete construction lead to isomorphic objects. That is not true in the third case. The hyperfinite partition lattice is a proper sublattice of Haiman's concrete lattice of measurable partitions. See [5, 8] for detailed discussions.

A well chosen definition of continuous matroids should of course keep the door open for concrete as well as hyperfinite examples. Perhaps a fruitful definition can be put together from some of the properties stated in Theorem 4.1. We choose however to refrain from any such speculation for lack of insight.

The situation reminds of a saying of the late Gian-Carlo Rota: "Starting from definitions and deducing theorems is mathematics, whereas starting from theorems and deducing definitions is philosophy."

References

1. G. Birkhoff, *Lattice theory*, American Mathematical Society Colloquium Publications, **25**, 3rd edition. American Mathematical Society, Providence, R.I. (1979).
2. A. Björner, Continuous partition lattice *Proc. Natl. Acad. Sci. U.S.A.*, **84** (1987), 6327–6329.
3. A. Björner and L. Lovász, Pseudomodular lattices and continuous matroids, *Acta Sci. Math. (Szeged)* **51** (1987), 295–308.
4. A. Dress and L. Lovász, On some combinatorial properties of algebraic matroids, *Combinatorica* **1987**, 39–48.
5. M. D. Haiman, On realization of Björner's "continuous partition lattice" by measurable partitions, *Trans. Amer. Math. Soc.* **343** (1994), 695–711.
6. L. Lovász, *Large networks and graph limits*, American Mathematical Society Colloquium Publications, **60**. American Mathematical Society, Providence, RI, 2012.
7. J. Oxley, Matroid theory. Second edition, Oxford Graduate Texts in Mathematics, **21**. Oxford University Press, 2011.

8. A.M. Vershik and Yu.V. Yakubovich, Continuous lattices of partitions, and lattices of continuous partitions, *Proceedings of the St. Petersburg Mathematical Society,* **7** *(Russian) (2000), 5–27; Amer. Math. Soc. Transl. Ser. 2, 2003*.
9. J. von Neumann, Einige Sätze über messbare Abbildungen, *Annals of Math.* **33** (1932), 574–586.
10. J. von Neumann, Continuous geometry, *Proc. Natl. Acad. Sci. USA*, **22** (1936), 92–100.
11. J. von Neumann, Examples of continuous geometries, *Proc. Natl. Acad. Sci. USA*, **22** (1936), 101–108.
12. J. von Neumann, *Continuous geometry*, Princeton Univ. Press 1960.

Identifiability for Graphexes and the Weak Kernel Metric

Christian Borgs, Jennifer T. Chayes, Henry Cohn and László Miklós Lovász

Abstract In two recent papers by Veitch and Roy and by Borgs, Chayes, Cohn, and Holden, a new class of sparse random graph processes based on the concept of graphexes over σ-finite measure spaces has been introduced. In this paper, we introduce a metric for graphexes that generalizes the cut metric for the graphons of the dense theory of graph convergence. We show that a sequence of graphexes converges in this metric if and only if the sequence of graph processes generated by the graphexes converges in distribution. In the course of the proof, we establish a regularity lemma and determine which sets of graphexes are precompact under our metric. Finally, we establish an identifiability theorem, characterizing when two graphexes are equivalent in the sense that they lead to the same process of random graphs.

Keywords Graphex · Sparse network · Weak kernel metric · Regularity · Identifiability · Expected subgraph counts · Poisson process

Subject Classifications: Primary 05C80 · Secondary 60G09 · 60G55 · 60G57

C. Borgs · J. T. Chayes (✉) · H. Cohn
Microsoft Research, One Memorial Drive, Cambridge, MA 02142, USA
e-mail: jchayes@microsoft.com

C. Borgs
e-mail: borgs@microsoft.com

H. Cohn
e-mail: cohn@microsoft.com

L. M. Lovász
Department of Mathematics, Massachusetts Institute of Technology,
77 Massachusetts Avenue, Cambridge, MA 02139, USA
e-mail: lmlovasz@mit.edu

© János Bolyai Mathematical Society and Springer-Verlag GmbH Germany, part of Springer Nature 2019
I. Bárány et al. (eds.), *Building Bridges II*, Bolyai Society Mathematical Studies 28, https://doi.org/10.1007/978-3-662-59204-5_3

1 Introduction

The theory of graph limits has been extensively developed for dense graph sequences [7–10, 21, 22], but the sparse case is not as well understood. In this paper, we study a model introduced and studied in a sequence of papers [3, 4, 11, 17, 18, 25, 26] based on the notion of graphexes. In contrast to the graphons of the dense theory, which are symmetric two-variable functions defined over a probability space, graphexes are defined over σ-finite measure spaces, and, in addition to a graphon part W, contain two other components: a function S taking values in $\mathbb{R}_+$, and a parameter $I \in \mathbb{R}_+$. Formally, the graphex is then the quadruple $\mathbb{W} = (W, S, I, \mathbf{\Omega})$, where $\mathbf{\Omega} = (\Omega, \mathcal{F}, \mu)$ is the underlying measure space.

A graphex then leads to a process $(G_T(\mathbb{W}))_{T\geq 0}$ of random graphs as follows: starting from a Poisson process over Ω with intensity $T\mu$, one attaches Poisson($TS(x_i)$) leaves to each Poisson point x_i, and in addition, joins two Poisson points x_i, x_j with probability $W(x_i, x_j)$. Finally, one adds Poisson($T^2 I$) isolated edges not connected to any of the other points. Removing isolated vertices as well as the labels of the remaining vertices gives a *graphex process* $(G_T(\mathbb{W}))_{T\geq 0}$ of unlabeled graphs *sampled from* $\mathbb{W}$.

Several notions of convergence for graphexes were introduced in [3, 25] and further studied in [18]. Among these notions, we will be particularly interested in graphex process convergence (GP-convergence), which was introduced in [25]. A sequence of graphexes is GP-convergent if the random graph processes generated by the graphexes in the sequence converge. It was pointed out in [18] that GP-convergence can be metricized using the abstract theory of probability measures over Polish spaces, but this does not give a very explicit metric on graphexes; in fact, it does not even allow us to determine when two graphexes lead to the *same* random graph process.

In this paper, we introduce a concrete notion of distance for graphexes that is equivalent to GP-convergence, which can be thought of as corresponding to the "cut distance" in the dense case. For reasons we explain in the next section, we call it the "weak kernel distance". We show that convergence in this distance is equivalent to GP-convergence.

In general, the set of all graphexes is not compact. Indeed, it is not difficult to show that for a set to be compact under GP-convergence, certain uniform boundedness assumptions are necessary on the set of graphexes, which we call "tightness". As a part of our proof that our weak kernel distance metricizes GP-convergence, we develop a (Frieze–Kannan-type) regularity lemma for graphexes and show that the sets that are precompact under the weak kernel metric are precisely those that are tight.

Finally, we prove an identifiability theorem, showing to what extent a graphex can be identified from its graphex process. Formulated differently, we give a characterization of the equivalence classes of graphexes, where two graphexes are called equivalent if they give rise to the same graphex process. Generalizing a construction that was developed by Janson for the dense case [16], we assign to each graphex

W a "canonical version" $\widehat{W}$ such that $\mathbb{W}$ is a pullback of $\widehat{\mathbb{W}}$ and show that if two graphexes are equivalent, then their canonical versions are isomorphic up to measure zero changes. This in turn will imply that two graphexes $\mathbb{W}_1$ and $\mathbb{W}_2$ are equivalent if and only if there is a third graphex $\mathbb{W}_3$ (which can be taken to be their canonical graphex) such that after restricting the two graphexes to their "support" (strictly speaking, we have to restrict them to their "degree support", a notion we will define in the next section) both $\mathbb{W}_1$ and $\mathbb{W}_2$ are pullbacks of $\mathbb{W}_3$. We note that this proves a conjecture of Janson; see Remark 5.4 in [17].

We note that in this paper we treat graphexes slightly differently from the definition in [4, 17, 18, 25, 26]. Namely, as in [3], we follow the convention from the theory of dense graph limits, and define the graphex process corresponding to a graphex as a process of graphs without loops. Indeed, we believe that a theory with loops is most naturally embedded into a more general theory of graphex processes with multi-edges and loops, which is beyond the scope of this paper.

Nonetheless, it is worth pointing out that the reader interested in the theory with loops (but not multi-edges) can derive many results for this theory from those developed here, even though some of the theorems will need to be modified to accommodate additional technical complications. For the identifiability theorem, this is done in Appendix A.2.

Finally, we note that while signed graphexes (i.e., graphexes for which W, S and I are not necessarily non-negative) do not make much sense if we want to use them to generate a random graph process, they are quite natural from an analytic point of view. Indeed, we will prove several of our results for signed graphexes. Still, the goal of this paper is to study unsigned graphexes, and our results on signed graphexes should be considered more of an aside at this point.

2 Definitions and Statements of Main Results

Definition 2.1 A graphex $\mathbb{W} = (W, S, I, \mathbf{\Omega})$ consists of a σ-finite measure space $\mathbf{\Omega} = (\Omega, \mathcal{F}, \mu)$, a symmetric measurable function $W \colon \Omega \times \Omega \to [0, 1]$, a measurable function $S \colon \Omega \to \mathbb{R}_+$, and a nonnegative real number I such that the following *local finiteness* conditions hold:

(1) $W(\cdot, x)$ is integrable for almost all $x \in \Omega$, and
(2) there exists a measurable subset $\Omega' \subseteq \Omega$ such that $\mu(\Omega \setminus \Omega') < \infty$ and $\mathbb{W}|_{\Omega'}$ is integrable.

The quadruple will be called a *signed graphex* if instead of taking values in $[0, 1]$ and $\mathbb{R}_+$, W, S and I take values in $\mathbb{R}$. The graphex $\mathbb{W} = (W, S, I, \mathbf{\Omega})$ is called *integrable* if

$$\|\mathbb{W}\|_1 := \int_{\Omega \times \Omega} |W(x, y)| \, d\mu(x) \, d\mu(y) + 2 \int_{\Omega} |S(x)| \, d\mu(x) + 2|I| < \infty,$$

and the *restriction* $\mathbb{W}|_{\Omega'}$ of $\mathbb{W}$ to $\Omega' \subseteq \Omega$ is defined as the quadruple $\mathbb{W}' = (W', S', I', \mathbf{\Omega}')$ with $\mathbf{\Omega}' = (\Omega', \mathcal{F}', \mu')$, where $\mathcal{F}' = \{A \in \mathcal{F} : A \subseteq \Omega'\}$, μ' is the restriction of μ to $\mathcal{F}'$, W' is the restriction of W to $\Omega' \times \Omega'$, S' is the restriction of S to Ω', and $I' = I$.

We often refer to $\mathbb{W}$ as a signed graphex over $\mathbf{\Omega}$, and we will refer to the function W as a *graphon*, or the *graphon part* of $\mathbb{W}$. Similarly, S will be called a *star intensity*, or the *star part* of $\mathbb{W}$, and I will be called a *dust density*, or the *dust part* of $\mathbb{W}$. (The reason for this terminology will become clear when we discuss the random graph process generated by an unsigned graphex $\mathbb{W}$; as we will see, the star part of $\mathbb{W}$ will lead to stars, and the dust part will lead to isolated edges, which we call dust following [18].) If two signed graphexes $\mathbb{W}_1, \mathbb{W}_2$ are defined on the same space $\mathbf{\Omega}$, then we say that $\mathbb{W}_1 = \mathbb{W}_2$ almost everywhere if $W_1 = W_2$ almost everywhere, $S_1 = S_2$ almost everywhere, and $I_1 = I_2$.

We define the marginal of a graphex $\mathbb{W} = (W, S, I, \mathbf{\Omega})$ over $\mathbf{\Omega} = (\Omega, \mathcal{F}, \mu)$ as the a.e. finite function $D_{\mathbb{W}} : \Omega \to \mathbb{R}_+$ defined by

$$D_{\mathbb{W}}(x) = D_W(x) + S(x) \quad \text{where} \quad D_W(x) = \int_\Omega W(x, y) \, d\mu(y).$$

We say that $\mathbb{W}$ has D-bounded marginals if $\|D_{\mathbb{W}}\|_\infty \leq D$. Finally, we define its degree support as the set

$$\operatorname{dsupp} \mathbb{W} = \{x \in \Omega : D_{\mathbb{W}}(x) > 0\}.$$

Note that $\mathbb{W}$ is integrable if and only if its marginals are integrable.

Given a graphex $\mathbb{W}$, we will define a stochastic process $(\mathcal{G}_T(\mathbb{W}))_{T \geq 0}$ indexed by $T \in \mathbb{R}_+$ and taking values in the set of graphs with labels in $\mathbb{R}_+$. To make this precise, we need to define a σ-algebra over the set of countable graphs with vertices in $\mathbb{R}_+$. To this end, we first define the adjacency measure ξ_G of a countable graph G with vertices in $\mathbb{R}_+$ as the measure ξ_G on $\mathbb{R}_+^2$ given by

$$\xi_G = \sum_{t,t' \in V(G):\{t,t'\} \in E(G)} \delta_{(t,t')}.$$

We call ξ an *adjacency measure* if there exists a countable graph G such that $\xi = \xi_G$. We then equip the set of adjacency measures with the smallest σ-algebra such that the maps $\xi \mapsto \xi(A)$ are measurable for all bounded Borel sets $A \subseteq \mathbb{R}_+^2$, and the set of countable graphs with vertices in $\mathbb{R}_+$ with the smallest σ-algebra such that the maps $G \mapsto \xi_G$ are measurable.

A graphex $\mathbb{W} = (W, S, I, \mathbf{\Omega})$ then generates a family $(\mathcal{G}_T(\mathbb{W}))_{T \geq 0}$ of random graphs as follows: we start with a Poisson point process with intensity $\lambda \times \mu$ on $\mathbb{R}_+ \times \Omega$, where λ is the Lebesgue measure on $\mathbb{R}_+$, and then connect two points (t, x) and (t', x') of the Poisson process with probability $W(x, x')$, independently for all pairs of points. For each point of the Poisson process (t, x), we take another Poisson point process on $\mathbb{R}_+$ with intensity $S(x)\lambda$, and connect (t, x) to a vertex with "birth time" t_i for each point t_i in the process. We also take a Poisson process with intensity $I(\lambda \times \lambda)$ on $\mathbb{R}_+^2$, and for each point (t_x, t_y) we take an isolated edge between vertices with birth time t_x and t_y. If we ignore the labels in the feature space Ω and delete the vertices with degree zero, this leads to an infinite graph $\mathcal{G}_\infty(\mathbb{W})$ with vertices labeled by their birth time $t \in \mathbb{R}_+$. We then define $\mathcal{G}_T(\mathbb{W})$ by first taking the induced subgraph on the set of vertices which lie in $[0, T]$ and then deleting vertices whose neighbors in $\mathcal{G}_\infty(\mathbb{W})$ all lie outside the interval $[0, T]$.

We will refer to the part of $\mathcal{G}_\infty(\mathbb{W})$ generated with the help of the dust intensity I as the *dust part* of $\mathcal{G}_\infty(\mathbb{W})$, and as the part generated with the help of the star intensity S as the *stars* in $\mathcal{G}_\infty(\mathbb{W})$. While it may not be *a priori* clear whether these parts can be inferred from just observing the infinite graph $\mathcal{G}_\infty(\mathbb{W})$, this is actually the case, a fact which was first noted in Remark 5.4 in [17]: almost surely, the dust part consists of all edges in $\mathcal{G}_\infty(\mathbb{W})$ that are isolated, the star part consists of all edges with one vertex of degree one and a second vertex of infinite degree, and the remaining edges are generated by the graphon part of $\mathbb{W}$ and have two endpoints with infinite degree.

Definition 2.2 Let $\mathbb{W}$ be a graphex, let $(\mathcal{G}_T(\mathbb{W}))_{T \geq 0}$ be the random family of graphs defined above, and let $\xi[\mathbb{W}]$ be the random adjacency measure $\xi_{\mathcal{G}_\infty(\mathbb{W})}$. We call the stochastic process $(\mathcal{G}_T(\mathbb{W}))_{T \geq 0}$ the *graphex process generated by* $\mathbb{W}$, and the adjacency measure $\xi[\mathbb{W}]$ the *adjacency measure generated by* $\mathbb{W}$. We say two graphexes are *equivalent*, if the graphex processes generated by these graphons are equal in law.

Remark 2.3 (1) Following [4], we defined a graphex process as a stochastic process taking values in a space of graphs with labels in $\mathbb{R}_+$. Alternatively, one might want to define a graphex process as a process taking values in the space of unlabeled graphs without isolated vertices. In our current context, this would correspond to ignoring the time labels of the graphs in $\mathcal{G}_T(\mathbb{W})$, leading to a graph which we denote by $G_T(\mathbb{W})$. When it is important to distinguish them, we will refer to the process $(G_T(\mathbb{W}))_{T \geq 0}$ as the unlabeled graphex process corresponding to $\mathbb{W}$, and to the process $(\mathcal{G}_T(\mathbb{W}))_{T \geq 0}$ as the labeled graphex process corresponding to $\mathbb{W}$. Note that it is easy to recover $\mathcal{G}_T(\mathbb{W})$ from $G_T(\mathbb{W})$: just assign i.i.d. labels chosen uniformly at random in $[0, T]$ to all vertices. A related observation is the fact that $G_T(\mathbb{W})$ can be generated by first choosing $(x_i)_{i \geq 1}$ according to a Poisson process with intensity $T\mu$ in Ω, then connecting i and j with probability $W(x_i, x_j)$, then adding a star whose number of leaves are chosen as a Poisson random variable with mean $TS(x_i)$ to each point of the process $(x_i)_{i \geq 1}$, and finally adding independent edges with rate IT^2. Forgetting the labels then gives us $G_T(\mathbb{W})$. Relabeling each vertex in the resulting graph independently by a uniform $t \in [0, T]$, we obtain $\mathcal{G}_T(\mathbb{W})$.

(2) It is sometimes convenient to assign a feature value to the endpoints of the isolated edges generated from the dust part I in the graphex, as well as to the leaves of the stars generated using the function S. For our purpose, we will say that these vertices have the feature label ∞, and we will extend the marginal $D_\mathbb{W}$ to $\Omega \cup \{\infty\}$ by setting

$$D_{\mathbb{W}}(\infty) = \int_\Omega S(x)\,d\mu(x) + 2I.$$

Note that with this notation, $\|\mathbb{W}\|_1 = \int_\Omega D_{\mathbb{W}}(x)\,d\mu(x) + D_{\mathbb{W}}(\infty)$.

(3) In view of (2), one might want to equip the extended feature space $\widetilde{\Omega} = \Omega \cup \{\infty\}$ with a σ-finite measure by keeping the original measure on Ω, and assigning some finite measure $Q = \widetilde{\mu}(\infty)$ to the feature value ∞, giving a new σ-finite measure space $\widetilde{\Omega}_Q$. On $\widetilde{\Omega}_Q$, one can then define a graphex of the form $\widetilde{\mathbb{W}}_Q = (\widetilde{W}_Q, 0, 0, \widetilde{\Omega})$ by setting $\widetilde{W}_Q$ equal to W on $\Omega \times \Omega$ and to $2I/Q^2$ on $\{\infty\} \times \{\infty\}$, and by setting $\widetilde{W}(x, \infty) = S(x)/Q$ and $\widetilde{W}(\infty, y) = S(y)/Q$ if only one of the two features x, y lies in Ω. With this construction, $D_{\widetilde{W}_Q}(\infty) = D_{\mathbb{W}}(\infty)/Q$, $D_{\widetilde{W}_Q}(x) = D_{\mathbb{W}}(x)$ if $x \in \Omega$, and $\|\mathbb{W}\|_1 = \|\widetilde{\mathbb{W}}_Q\|_1 = \int d\mu(x) D_{\widetilde{W}_1} + Q D_{\widetilde{W}}(\infty)$, with the notation in (2) corresponding to the case $Q = 1$. It is clear that the graphon process generated from $\widetilde{W}$ cannot have exactly the same distribution as the one generated from $\mathbb{W}$ unless I and S are zero (to see this, note that in $G_\infty(\widetilde{W})$, all vertices have infinite degrees, while $G_\infty(\mathbb{W})$ has vertices of degree one). But one might wonder whether the process generated from the "pure graphon" $\widetilde{W}$ approximates the one generated from $\mathbb{W}$. As we will see in Remark 7.7, this is indeed the case, in the sense that for any fixed T, the distribution of $G_T(\widetilde{W}_Q)$ converges to $G_T(\mathbb{W})$.

It is relatively easy to see that the local finiteness conditions (1) and (2) from Definition 2.1 imply that the adjacency measure $\xi[\mathbb{W}]$ is a.s. locally finite (i.e., $\xi[\mathbb{W}](A) < \infty$ for all bounded Borel sets $A \subset \mathbb{R}_+^2$), or equivalently, that for all $T < \infty$, the graphs $\mathcal{G}_T(\mathbb{W})$ are a.s. finite. It turns out that these conditions are also necessary for the local finiteness of $\xi[\mathbb{W}]$. This is the main statement of the following proposition, which we will prove in Appendix A.1. For graphexes over $\mathbb{R}_+$ equipped with the Lebesgue measure, a similar condition was established in [26], building on the work of [19] (the condition considered by [19, 26] is the same as our condition (E) below, specialized to the case $D = 1$, even though it is clear that both [19, 26] knew that for graphexes over $\mathbb{R}_+$, conditions (D) and (E) are equivalent). To state the proposition, we use the notation $\{D_\mathbb{W} > D\}$ for the set $\{x \in \Omega : D_\mathbb{W}(x) > D\}$, while $\{D_W > D\}$, $\{D_\mathbb{W} \leq D\}$, and $\{D_W \leq D\}$ are defined analogously.

Proposition 2.4 *Let $\mathbb{W} = (W, S, I, \Omega)$ be a 4-tuple consisting of a σ-finite measure space $\Omega = (\Omega, \mathcal{F}, \mu)$, a symmetric measurable function $W \colon \Omega \times \Omega \to [0, 1]$, a measurable function $S \colon \Omega \to \mathbb{R}_+$, and a nonnegative real number I. Then the local finite conditions (1) and (2) from Definition 2.1 are equivalent to the local finiteness of the adjacency measure generated by $\mathbb{W}$. If we assume condition (1), then following are equivalent:*

(A) *The graphex $\mathbb{W}$ obeys the local finiteness condition (2).*
(B) *For all $D > 0$, $\mu(\{D_\mathbb{W} > D\}) < \infty$ and $W|_{\{D_W \leq D\}}$ is integrable.*
(C) *There exists a $D > 0$ such that $\mu(\{D_\mathbb{W} > D\}) < \infty$ and $W|_{\{D_W \leq D\}}$ is integrable.*
(D) *For all $D > 0$, $\mu(\{D_W > D\}) < \infty$, and both $W|_{\{D_W \leq D\}}$ and $\min\{S, 1\}$ are integrable.*

(E) There exists a $D > 0$ such that $\mu(\{D_W > D\}) < \infty$, and both $W|_{\{D_W \le D\}}$ and $\min\{S, 1\}$ are integrable.

Note that this proposition implies in particular that a graphex with bounded marginals is integrable, since for graphexes with $\|D_W\|_\infty \le D$ the graphex W and the graphex $W|_{\{D_W\} \le D}$ are the same.

Having defined the graphex process associated with a graphex W, there are several natural questions one might want to answer. In particular, one might want to characterize when two graphexes lead to the same process, i.e., when $\xi[W]$ and $\xi[W']$ have the same distribution. More generally, one might want to define a metric on the set of graphexes such that the distributions of $\xi[W]$ and $\xi[W']$ are close if W and W' are close. Addressing these questions is one of the main goals of this paper.

Before discussing this further, it will be useful to embed the theory of graphex processes into the general theory of locally finite point processes. To this end, we first introduce the set $\mathcal{N} = \mathcal{N}(\mathbb{R}_+^2)$ of locally finite counting measures on $\mathbb{R}_+^2$ (i.e., the set of measures ξ such that $\xi(A)$ is a finite, non-negative integer for all bounded Borel sets $A \subset \mathbb{R}_+^2$), and equip it with the Borel σ-algebra inherited from the vague topology (defined as the coarsest topology for which the maps $\xi \mapsto \int f \, d\xi$ are continuous for all continuous functions $f : \mathbb{R}_+^2 \to \mathbb{R}_+$ with compact support). As shown in, e.g., [13], Appendix A2.6, the vague topology on $\mathcal{N}$ can be metricized in such a way that $\mathcal{N}$ becomes a complete, separable metric space, making $\mathcal{N}$ into a Polish space, and the Borel σ-algebra inherited from this topology is the smallest σ-algebra such that for all bounded Borel sets $A \subset \mathbb{R}_+^2$ the maps $\mu \mapsto \mu(A)$ are measurable.

As usual, a locally finite point process on $\mathbb{R}_+^2$ is then defined as a random measure on $\mathcal{N}(\mathbb{R}_+^2)$ equipped with this Borel algebra, and convergence in distribution is defined as weak convergence in the set of probability measures on $\mathcal{N}$, so that convergence in distribution of a sequence of locally finite point process ξ_n on $\mathbb{R}_+^2$ to a locally finite point process ξ is defined by the condition that $\mathbb{E}[F(\xi_n)] \to \mathbb{E}[F(\xi)]$ for all continuous, bounded functions F, with continuity defined with respect to the vague topology on $\mathcal{N}$. As observed in [18], the fact that $\mathcal{N}$ is Polish makes the set of probability distributions on $\mathcal{N}$ a Polish space as well (see, e.g., [1], Appendix III for a proof), showing that convergence in distribution for locally finite point processes on $\mathbb{R}_+^2$ can be metricized.

Next we consider the set $\hat{\mathfrak{G}}$ of simple graphs G with vertices in $\mathbb{R}_+$ such that (a) no vertex in G is isolated, and (b) for all $T < \infty$, the induced subgraph of G on $V(G) \cap [0, T]$ is finite. We also consider the subset $\hat{\mathfrak{G}}_0$ of finite graphs in $\hat{\mathfrak{G}}$. The map $G \mapsto \xi(G)$ then gives a one-to-one map between graphs in $\hat{\mathfrak{G}}$ and adjacency measures. In particular, $\hat{\mathfrak{G}}$ and its subset $\hat{\mathfrak{G}}_0$ inherit the vague topology and corresponding Borel σ-algebra from $\mathcal{N}$. In this language, the graphex process $(\mathcal{G}_T(W))_{T \ge 0}$ then becomes a CADLAG stochastic process with values in $\hat{\mathfrak{G}}_0$ indexed by a time $T \in \mathbb{R}_+$.

Note that $\hat{\mathfrak{G}}$ should be distinguished from the set of unlabeled countable graphs without isolated vertices, $\mathfrak{G}$. While we will not equip $\mathfrak{G}$ with any topology, the set of finite unlabeled graphs without isolated vertices, denoted by $\mathfrak{G}_0$, will be given the discrete topology. In this language, the unlabeled graphex process $(G_T(\mathbb{W}))_{T \geq 0}$ introduced in Remark 2.3 is then a CADLAG process with values in $\mathfrak{G}_0$.

In [18], various notions of convergence for graphons and graphexes (proposed originally in [3, 25]) were studied. Here we are most interested in what [25] introduces as GP-convergence, where GP stands for graphex process. This notion is closely related to the notion of sampling convergence for graphs introduced in [4]; see Lemma 5.4 in that paper, as well as the discussion at the end of this section. Janson showed that the following are equivalent if $\mathbb{W}, \mathbb{W}_1, \mathbb{W}_2, \ldots$ are graphexes:

(1) $\xi(G(\mathbb{W}_n)) \to \xi(G(\mathbb{W}))$ in distribution.
(2) For every $T < \infty, \xi(\mathcal{G}_T(\mathbb{W}_n)) \to \xi(\mathcal{G}_T(\mathbb{W}_n))$ in distribution.
(3) For every $T < \infty, G_T(\mathbb{W}_n) \to G_T(W)$ in distribution.

Following [25] we call this notion of convergence GP-convergence, and say that $\mathbb{W}_n$ is GP-convergent to $\mathbb{W}$ if one of these equivalent conditions holds.

As already alluded to above, Janson also observed that by the abstract theory of probability measures over Polish spaces, this notion of convergence can be metricized, turning the set of graphexes into a complete, separable metric space. But this abstract theory does not give a very explicit metric on the space of locally finite graphexes; in fact, it does not even address the question of when two graphexes are equivalent in the sense that the resulting point processes are equal in law.

To discuss the second question, we define measure-preserving transformations, pullbacks, and couplings. Given two σ-finite spaces $\mathbf{\Omega} = (\Omega, \mathcal{F}, \mu)$ and $\mathbf{\Omega}' = (\Omega', \mathcal{F}', \mu')$, we say that a map $\varphi \colon \Omega' \to \Omega$ is a *measure-preserving transformation* if φ is measurable and $\mu'(\varphi^{-1}(A)) = \mu(A)$ for all $A \in \mathcal{F}$. If $\mathbb{W} = (W, S, I, \mathbf{\Omega})$ is a signed graphex over $\mathbf{\Omega}$, we define its *pullback* under φ to be the graphex $\mathbb{W}^\varphi = (W^\varphi, S^\varphi, I, \mathbf{\Omega}')$ where $W^\varphi(x', y') = W(\varphi(x'), \varphi(y'))$ and $S^\varphi(x') = S(\varphi(x'))$. It is clear that for unsigned graphexes $\mathbb{W}$ and $\mathbb{W}^\varphi$ give rise to the same process of random graphs. Note that we can define the pullback even when φ is measurable but not measure-preserving, but in this case the two graphexes do not necessarily give rise to the same random process. Nevertheless, we will sometimes use pullbacks in this situation. If we do, we will write $\mathbb{W}^{\varphi, \mu'}$ to emphasize the dependence on the measure on Ω'. Given two σ-finite spaces $\mathbf{\Omega}_1 = (\Omega_1, \mathcal{F}_1, \mu_1)$ and $\mathbf{\Omega}_2 = (\Omega_2, \mathcal{F}_2, \mu_2)$, we say that μ is a coupling of μ_1 and μ_2 if μ is a measure on $\mathcal{F}_1 \times \mathcal{F}_2$ such that $\mu(\Omega_1 \times S_2) = \mu_2(S_2)$ and $\mu(S_1 \times \Omega_2) = \mu_1(S_1)$ for all $S_1 \in \mathcal{F}_1$ and all $S_2 \in \mathcal{F}_2$. Note that the existence of such a coupling implies that $\mu_1(\Omega_1) = \mu(\Omega_1 \times \Omega_2) = \mu_2(\Omega_2)$. It turns out that this condition is both necessary and sufficient for the existence of a coupling; see [3] for a proof.

Based on the known results for dense graphs, one might conjecture that two graphexes are equivalent if and only if there exists a third graphex such that both are pullbacks of this third graphex. It turns out that this is not quite correct, but that it is correct once we remove the part of the underlying space on which $D_\mathbb{W} = 0$. This is the statement of the following theorem, which is one of the main results of this paper, and will be proved in Sect. 8.

Theorem 2.5 Let $\mathbb{W}_1 = (W_1, S_1, I_1, \mathbf{\Omega}_1)$ and $\mathbb{W}_2 = (W_2, S_2, I_2, \mathbf{\Omega}_2)$ be graphexes, where $\mathbf{\Omega}_i = (\Omega_i, \mathcal{F}_i, \mu_i)$ are σ-finite spaces. Then $G_t(\mathbb{W}_1)$ and $G_t(\mathbb{W}_2)$ have the same distribution for all $t \in \mathbb{R}_+$ if and only if there exists a third graphex $\mathbb{W} = (W, S, I, \mathbf{\Omega})$ over a σ-finite measure space $\mathbf{\Omega} = (\Omega, \mathcal{F}, \mu)$ and measure preserving maps $\varphi_i \colon \operatorname{dsupp} \mathbb{W}_i \to \Omega$ such that $\mathbb{W}_i|_{\operatorname{dsupp} \mathbb{W}_i} = \mathbb{W}^{\varphi_i}$ almost everywhere.

Remark 2.6 If the two graphexes are defined over Borel spaces, we can prove an analogous statement where the measure-preserving maps are turned around. Specifically, for the case where $\mathbf{\Omega}_1$ and $\mathbf{\Omega}_2$ are σ-finite Borel spaces, we can prove that $G_t(\mathbb{W}_1)$ and $G_t(\mathbb{W}_2)$ have the same distribution for all $t \in \mathbb{R}_+$ if and only if there exists a σ-finite Borel space $\mathbf{\Omega} = (\Omega, \mathcal{F}, \mu)$ and measure preserving maps $\pi_i \colon \Omega \to \operatorname{dsupp} \mathbb{W}_i$ such that $(\mathbb{W}_1|_{\operatorname{dsupp} \mathbb{W}_1})^{\pi_1} = (\mathbb{W}_2|_{\operatorname{dsupp} \mathbb{W}_2})^{\pi_2}$ almost everywhere. In this case, the space $\mathbf{\Omega}$ can be chosen to be a coupling of $\mathbf{\Omega}_1$ and $\mathbf{\Omega}_2$, with π_i being the coordinate projections from $\Omega_1 \times \Omega_2$ to Ω_i. See Theorem 8.2 in Sect. 8. For graphexes without a dust and star part, this was independently established in [3] (using a different proof); see also [17], which establishes a similar result (with yet another proof), this time giving a coupling of the two graphexes (again without dust and star part) after trivially extending them rather than restricting them to the support of their marginals.

To address the first question, concerning the relationship between graphexes and the point processes generated by them, we would like to define an analogue of the cut distance for graphons between graphexes, so that two graphexes are close if and only if their graphex processes are close. To this end, we first define some norms of a function U over $\Omega_1 \times \Omega_2$ for two σ-finite spaces $\mathbf{\Omega}_1 = (\Omega_1, \mathcal{F}_1, \mu_1)$ and $\mathbf{\Omega}_2 = (\Omega_2, \mathcal{F}_2, \mu_2)$. We denote by $\|U\|_p$ the L^p norm of U as a function over $\Omega_1 \times \Omega_2$ (so we forget the product structure). Given two measurable functions $f \colon \Omega_1 \to \mathbb{R}$ and $g \colon \Omega_2 \to \mathbb{R}$, let

$$f \circ U(y) = \int_{\Omega_1} f(x) U(x, y)\, dx,$$

$$U \circ g(x) = \int_{\Omega_2} U(x, y) g(y)\, dy,$$

and

$$f \circ U \circ g = \int_{\Omega_1 \times \Omega_2} f(x) U(x, y) g(y)\, dx\, dy.$$

We will also use the notation U_x for the function $y \mapsto U(x, y)$.

Definition 2.7 Given a function U defined on $\Omega_1 \times \Omega_2$ for two σ-finite measure spaces $\mathbf{\Omega}_i = (\Omega_i, \mathcal{F}_i, \mu_i)$ for $i = 1, 2$, we define

$$\|U\|_{2\to 2} = \sup_{f,g : \|f\|_2 = \|g\|_2 = 1} f \circ U \circ g = \sup_{g : \|g\|_2 = 1} \|U \circ g\|_2.$$

Note that the norm $\|U\|_{2\to 2}$ is simply the operator norm when we consider U the kernel of an operator $\widehat{U}$ from $L^2(\mathbf{\Omega}_2)$ to $L^2(\mathbf{\Omega}_1)$. We will therefore call it the *kernel norm* of U. Our next norm is a modification of the standard cut norm; in the dense graph setting, it was first systematically used in [20], where it was defined as a norm for functions defined over a probability space.

Definition 2.8 Given a measurable function U defined on $\Omega \times \Omega$ for a σ-finite measure space Ω, we define the *jumble norm*

$$\|U\|_{\boxtimes} = \sup_{S,T \subseteq \Omega} \left| \frac{1}{\sqrt{\mu(S)\mu(T)}} \int_{S \times T} U(x,y) \, d\mu(x) \, d\mu(y) \right|.$$

Here the supremum is over subsets with finite and nonzero measure.

It is easy to show that these are norms; in particular, they satisfy the triangle inequality, and are equal to 0 if and only if U is zero almost everywhere. If we want to stress the dependence of these norms on the measure μ and the function U, we write $\|U\|_{*,\mu}$ instead of $\|U\|_*$, where $*$ is replaced by the appropriate norm.

We will see later that for graphexes with uniformly bounded marginals and uniformly bounded $\|\cdot\|_1$ norms, the $\|\cdot\|_{2 \to 2}$ norm and the $\|\cdot\|_{\boxtimes}$ norm are equivalent (Lemma 3.22), implying in particular that they are equivalent in the theory of dense graph limits (where Ω has bounded measure). In the dense setting, the above two norms are also equivalent to the standard cut norm, defined as

$$\|U\|_{\square} = \sup_{S,T \subseteq \Omega} \left| \int_{S \times T} U(x,y) \, d\mu(x) \, d\mu(y) \right| = \sup_{f,g \colon \Omega \to [0,1]} |f \circ U \circ g|.$$

Indeed, $\|U\|_{\square} \leq \|U\|_{\boxtimes} \mu(\Omega)$ and $\|U\|_{\boxtimes} \leq \sqrt{\|U\|_{\square} \|U\|_{\infty}}$, where the second bound follows from the fact that

$$\left| \int_{S \times T} U(x,y) \, d\mu(x) \, d\mu(y) \right| \leq \inf\{\lambda(S)\lambda(T)\|U\|_{\infty}, \|U\|_{\square}\}.$$

Therefore, in the theory of dense graph limits all three norms are equivalent. However, although the cut norm is the simplest to state, we believe that the kernel norm $\|\cdot\|_{2 \to 2}$ norm is the correct extension to graphexes.

We now define some distances between graphexes. First, we define the $\delta_{2 \to 2}$ distance, which will define a notion of convergence that is equivalent to GP-convergence for graphexes with uniformly bounded marginals. The definition of $\delta_{2 \to 2}$ will make sense for signed graphexes, provided both the graphon parts and the absolute marginals are in $L^1 \cap L^2$. We will therefore define the $\delta_{2 \to 2}$ metric in this more general[1] setting.

---

[1] To see that this setting is indeed more general than the assumption of bounded marginals for (unsigned) graphexes we recall that by Proposition 2.4, a graphex with bounded marginals is integrable. Using this, and the fact that by definition, the graphon part of a graphex is bounded, the claim is easy to verify.

Definition 2.9 A signed graphex $\mathbb{W} = (W, S, I, \mathbf{\Omega})$ over $\mathbf{\Omega} = (\Omega, \mathcal{F}, \mu)$ is said to be in $L^1 \cap L^2$ if both W and $D_{|\mathbb{W}|}$ are in $L^1 \cap L^2$. Here $|\mathbb{W}|$ is the graphex $|\mathbb{W}| = (|W|, |S|, |I|, \mathbf{\Omega})$.

Suppose $\mathbb{W}_1 = (W_1, S_1, I_1, \mathbf{\Omega})$ and $\mathbb{W}_2 = (W_2, S_2, I_2, \mathbf{\Omega})$ are defined on the same underlying space $\mathbf{\Omega}$. We then define their $d_{2\to 2}$-distance as

$$d_{2\to 2}(\mathbb{W}_1, \mathbb{W}_2) = \max\left(\|W_1 - W_2\|_{2\to 2}, \sqrt{\|D_{\mathbb{W}_1} - D_{\mathbb{W}_2}\|_2}, \sqrt[3]{|\rho(\mathbb{W}_1) - \rho(\mathbb{W}_2)|}\right), \quad (1)$$

where $\rho(\mathbb{W}_i)$ is the "edge density" of the signed graphex $\mathbb{W}_i$,

$$\rho(\mathbb{W}_i) = \int W_i + 2\int S_i + 2I. \quad (2)$$

The reason we take the roots will become clearer later when we define the general distance $\delta_\diamond$. Since $\sqrt{c_1 + c_2} \leq \sqrt{c_1} + \sqrt{c_2}$ and $\sqrt[3]{c_1 + c_2} \leq \sqrt[3]{c_1} + \sqrt[3]{c_2}$, this is indeed a metric.

Next, suppose two signed graphexes in $L^1 \cap L^2$, $\mathbb{W}_1 = (W_1, S_1, I_1, \mathbf{\Omega}_1)$ and $\mathbb{W}_2 = (W_2, S_2, I_2, \mathbf{\Omega}_2)$, are defined over two σ-finite spaces $\mathbf{\Omega}_1 = (\Omega_1, \mathcal{F}_1, \mu_1)$ and $\mathbf{\Omega}_2 = (\Omega_2, \mathcal{F}_2, \mu_2)$ with $\mu_1(\Omega_1) = \mu_2(\Omega_2)$. Let $\pi_1: \Omega_1 \times \Omega_2 \to \Omega_1$ and $\pi_2: \Omega_1 \times \Omega_2 \to \Omega_2$ be the projections. Then we define $\widetilde{\delta}_{2\to 2}(\mathbb{W}_1, \mathbb{W}_2)$ as the infimum

$$\widetilde{\delta}_{2\to 2}(\mathbb{W}_1, \mathbb{W}_2) = \inf_\mu d_{2\to 2}(\mathbb{W}_1^{\pi_1,\mu}, \mathbb{W}_2^{\pi_2,\mu}), \quad (3)$$

where the infimum is over all couplings μ of μ_1 and μ_2.

To define the $\delta_{2\to 2}$-distance we need one more notion, that of a trivial extension of $\mathbb{W} = (W, S, I, \mathbf{\Omega})$, where $\mathbf{\Omega} = (\Omega, \mathcal{F}, \mu)$ is a σ-finite measure space. It is defined as a quadruple $\mathbb{W}' = (W', S', I', \mathbf{\Omega}')$ where $\mathbf{\Omega}' = (\Omega', \mathcal{F}', \mu')$ is a σ-finite measure space such that $\Omega \in \mathcal{F}'$, $\mathcal{F} = \{A \in \mathcal{F}' : A \subseteq \Omega\}$, and μ is the restriction of μ' to $\mathcal{F}$, while W' is the extension of W that is 0 on the complement of $\Omega \times \Omega$, S' is the extension of S that is 0 on the complement of Ω, and $I' = I$. It is easy to see that taking a trivial extension of a graphex has no effect on $\mathcal{G}_T$ or $\mathcal{G}_\infty$ (since Poisson points sampled in the complement of Ω will be isolated for all T).

Definition 2.10 Let $\mathbb{W}_1$ and $\mathbb{W}_2$ be signed graphexes in $L^1 \cap L^2$. Then we define

$$\delta_{2\to 2}(\mathbb{W}_1, \mathbb{W}_2) = \widetilde{\delta}_{2\to 2}(\mathbb{W}'_1, \mathbb{W}'_2), \quad (4)$$

where $\mathbb{W}'_1$ and $\mathbb{W}'_2$ are trivial extensions of $\mathbb{W}_1$ and $\mathbb{W}_2$ to measure spaces of infinite total mass. We refer to $\delta_{2\to 2}(\mathbb{W}_1, \mathbb{W}_2)$ as the *kernel distance* of $\mathbb{W}_1$ and $\mathbb{W}_2$ and call $\delta_{2\to 2}$ the *kernel metric*.

The existence of these extensions is trivial, since we can always append an interval equipped with the Lebesgue measure. Nevertheless, it is not clear that $\delta_{2\to 2}(\mathbb{W}_1, \mathbb{W}_2)$ is well defined, since the right side of (4) could depend on the particular choice of the extensions $\mathbb{W}'_1$ and $\mathbb{W}'_2$. In a similar way, while it is clear that $\delta_{2\to 2}$ is symmetric and that $\delta_{2\to 2}(\mathbb{W}, \mathbb{W}) = 0$, it is not clear that it is a metric (even after factoring out the null space), since it is not clear that it satisfies the triangle inequality. The following theorem addresses both questions, and will be proved in Sect. 3.

Theorem 2.11 *Let $\mathbb{W}_1$ and $\mathbb{W}_2$ be signed graphexes in $L^1 \cap L^2$. Then the right side of (4) does not depend on the choice of the trivial extensions $\mathbb{W}'_1$ and $\mathbb{W}'_2$. Furthermore, given three signed graphexes $\mathbb{W}_1, \mathbb{W}_2, \mathbb{W}_3$ in $L^1 \cap L^2$,*

$$\delta_{2\to 2}(\mathbb{W}_1, \mathbb{W}_3) \leq \delta_{2\to 2}(\mathbb{W}_1, \mathbb{W}_2) + \delta_{2\to 2}(\mathbb{W}_2, \mathbb{W}_3).$$

Therefore, $\delta_{2\to 2}$ is a well-defined pseudometric.

Remark 2.12 In [3], when defining the cut distance between two graphons, it was only necessary to extend the smaller space to the larger one, and it was not necessary to extend further. It is natural to ask whether a trivial extension to a space of infinite metric is necessary, or, equivalently, whether for two graphexes $\mathbb{W}_1, \mathbb{W}_2$ defined on spaces with the same (finite) measure, $\widetilde{\delta}_{2\to 2}(\mathbb{W}_1, \mathbb{W}_2) = \delta_{2\to 2}(\mathbb{W}_1, \mathbb{W}_2)$. In contrast to the cut distance discussed in [3], for the kernel metric it is sometimes necessary to take trivial extensions of both spaces, not just an extension of the smaller space to one of the same measure as the larger one. See Example 3.11 in Sect. 3.

Our next theorem states that on sets with uniformly bounded marginals, the topology induced by the kernel metric $\delta_{2\to 2}$ is equivalent to the topology of GP-convergence. We will prove it in Sects. 6 and 7.

Theorem 2.13 *For any $D > 0$, $\delta_{2\to 2}$-convergence is equivalent to GP-convergence on the space of graphexes with D-bounded marginals.*

In general, $\delta_{2\to 2}$-convergence implies GP-convergence, but the reverse is not true. This is because if we do not assume bounded marginals, it is possible to have a very small measure set with very large degree. This will have a non-negligible effect on $\delta_{2\to 2}$ distance; however, for a fixed T, the chances of obtaining a vertex in the small set is small, and thus has a small effect on sampling. To give a more concrete example, let W_n be equal to 1 on $[0, 1/n] \times [1, 1+n]$ and $[1, 1+n] \times [0, 1/n]$, and zero everywhere else. Let $\mathbb{W}_n = (W_n, 0, 0, \mathbb{R}_+)$. Then for any fixed T, the probability of seeing a single edge in $G_T(\mathbb{W}_n)$ converges to 0, and therefore $\mathbb{W}_n$ is GP-convergent to 0. However, it is easy to see that $\delta_{2\to 2}(\mathbb{W}_n, 0)$ does not converge to 0. To address this issue, we will define a new distance such that two graphexes whose graphex processes can be obtained from each other by removing a small set of vertices are close in the new metric. Our construction is loosely motivated by the construction of the usual metric of weak convergence. For that reason, we will refer to the new metric as the weak kernel metric.

Before defining this distance, we introduce the notation $\mu - r \leq \mu' \leq \mu$ whenever μ, μ' are two measures over the same measurable space $(\Omega, \mathcal{F})$ such that

$$\mu(B) - r \leq \mu'(B) \leq \mu(B)$$

for all measurable sets B. Note that this property is equivalent to the existence of a function $h: \Omega \to [0, 1]$ such that $\mu'(B) = \int_B h \, d\mu$ and $\|1 - h\|_{1,\mu} \leq r$. An example of such a function, which we will often use, is the indicator function of a set $\Omega' \subseteq \Omega$ such that $\mu(\Omega \setminus \Omega') \leq r$.

We will define the weak kernel metric for arbitrary graphexes (removing the condition that they are in $L^1 \cap L^2$), and in fact will again allow for signed graphexes. We will assume that the graphon parts of these signed graphexes are bounded in the L^∞ norm, a condition which is true for unsigned graphexes, since for these, the graphon part takes values in $[0, 1]$.

Definition 2.14 Let $\mathbb{W}_1 = (W_1, S_1, I_1, \boldsymbol{\Omega}_1)$ and $\mathbb{W}_2 = (W_2, S_2, I_2, \boldsymbol{\Omega}_2)$ be signed graphexes, where $\boldsymbol{\Omega}_i = (\Omega_i, \mathcal{F}_i, \mu_i)$ and $\|W_i\|_\infty < \infty$ for $i = 1, 2$. We define $\delta_\diamond(\mathbb{W}_1, \mathbb{W}_2)$ as the infimum of the set of real numbers c such that there exist two measures $\widetilde{\mu}_1$ and $\widetilde{\mu}_2$ over $(\Omega_1, \mathcal{F}_1)$ and $(\Omega_2, \mathcal{F}_2)$ that satisfy the following: the signed graphexes $\widetilde{\mathbb{W}}_1$ and $\widetilde{\mathbb{W}}_2$ obtained from $\mathbb{W}_1$ and $\mathbb{W}_2$ by replacing μ_1 and μ_2 by $\widetilde{\mu}_1$ and $\widetilde{\mu}_2$, respectively, are in $L^1 \cap L^2$, and

(1) for $i = 1, 2$, we have $\mu_i - c^2 \leq \widetilde{\mu}_i \leq \mu_i$, and
(2) $\delta_{2 \to 2}(\widetilde{\mathbb{W}}_1, \widetilde{\mathbb{W}}_2) \leq c$.

We refer to $\delta_\diamond(\mathbb{W}_1, \mathbb{W}_2)$ as the *weak kernel distance* between $\mathbb{W}_1$ and $\mathbb{W}_2$ and call $\delta_\diamond$ the *weak kernel metric*.

Note that for unsigned graphexes, the weak kernel distance is well defined and finite. Indeed, given $0 < D < \infty$, choose $\widetilde{\mu}_i$ as the restriction of μ_i to $\{D_{W_i} \leq D\}$. Proposition 2.4 then implies that $\{D_{W_i} > D\}$ has finite measure, and $W_i|_{\{D_{W_i} \leq D\}}$ is integrable and hence in $L^1 \cap L^2$. The fact that $\delta_\diamond(\mathbb{W}_1, \mathbb{W}_2)$ is well defined for signed graphexes with bounded graphon part follows from Proposition 2.4 and further arguments, and is deferred to Sect. 3; see in particular Lemma 3.12 in that section.

We will show that $\delta_\diamond$ is a pseudometric. It is clear that it is symmetric, and that $\delta_\diamond(\mathbb{W}, \mathbb{W}) = 0$. It is not obvious that it satisfies the triangle inequality. We will prove this fact in Sect. 3.

Theorem 2.15 *Given three signed graphexes $\mathbb{W}_1, \mathbb{W}_2, \mathbb{W}_3$ with bounded graphon part,*

$$\delta_\diamond(\mathbb{W}_1, \mathbb{W}_3) \leq \delta_\diamond(\mathbb{W}_1, \mathbb{W}_2) + \delta_\diamond(\mathbb{W}_2, \mathbb{W}_3).$$

Therefore, $\delta_\diamond$ is a pseudometric.

Remark 2.16 Given a signed graphex $\mathbb{W} = (W, S, I, \boldsymbol{\Omega})$ with $\boldsymbol{\Omega} = (\Omega, \mathcal{F}, \mu)$ and a measure-preserving map $\varphi: \Omega' \to \Omega$, let $W' = W^\varphi$ almost everywhere. We can take a coupling $\widetilde{\mu}$ on $\Omega' \times \Omega$ defined by $\widetilde{\mu}(A \times B) = \mu'(A \cap \varphi^{-1}(B))$. It is easy to see that then the pullbacks of the two signed graphexes to $\Omega' \times \Omega$ will be equal almost everywhere, which implies that $\delta_{2 \to 2}(\mathbb{W}, \mathbb{W}') = \widetilde{\delta}_{2 \to 2}(\mathbb{W}', \mathbb{W}) = \delta_\diamond(\mathbb{W}', \mathbb{W}) = \delta_\diamond(\mathbb{W}', \mathbb{W}) = 0$.

With this new metric, we now have a definition of distance for any pair of graphexes. Note that in general, the metrics $\delta_{2 \to 2}$ and $\delta_\diamond$ are not be the same, even if both are finite. However, we will show that for graphexes with uniformly bounded marginals, the two metrics provide the same topology. This is the content of the next proposition, which will be proved in Sect. 3.

Proposition 2.17 *Fix $D < \infty$. Then $\delta_\diamond$ and $\delta_{2\to 2}$ give an equivalent topology on the space of graphexes with D-bounded marginals.*

We will also show that convergence in the weak kernel metric $\delta_\diamond$ is indeed equivalent to GP-convergence. This is the statement of the next theorem, and is one of the two main results of this paper. It will be proved using three main ingredients: a compactness statement stemming from a suitable analogue of the Frieze–Kannan regularity lemma, a counting lemma showing that subgraph counts in the graphs $G_T(\mathbb{W})$ are close if the corresponding graphexes are close in the metric $\delta_{2\to 2}$ (and the graphexes have uniformly bounded marginals), and a sampling lemma showing that as $T \to \infty$, the suitably rescaled graphex process $G_T(\mathbb{W})$ converges to $\mathbb{W}$ in probability. These techniques are developed in Sects. 5–7, and are combined to prove the theorem at the end of Sect. 7, where we will also prove Theorem 2.13.

Theorem 2.18 *Given a sequence of graphexes $\mathbb{W}_n$ and a graphex $\mathbb{W}$, $\mathbb{W}_n$ is GP-convergent to $\mathbb{W}$ if and only if $\delta_\diamond(\mathbb{W}_n, \mathbb{W}) \to 0$.*

Remark 2.19 The reader might wonder whether instead of building our metric for GP-convergence around the kernel norm $\|\cdot\|_{2\to 2}$, one could equivalently build it around the cut norm, $\|\cdot\|_\square$. Concretely, one might want to define $d_\square$ by replacing the kernel norm in (1) by the cut norm and the L^2 norm by the L^1 norm, then proceed as in (3) and (4) to obtain a cut distance $\delta_\square$ between graphexes with bounded marginals, and finally proceed as in Definition 2.14 to obtain a "weak cut metric" for arbitrary graphexes.

The following example shows that this approach does not work, in that it will not metricize GP-convergence. Define W_n to be the graphex that is constant and equal to n^{-2} over $[0, n]^2$ and 0 everywhere else, and set $\mathbb{W}_n = (W_n, 0, 0, \mathbb{R}_+)$, where $\mathbb{R}_+$ is equipped with the Lebesgue measure. The marginal $D_{\mathbb{W}_n}$ of $\mathbb{W}_n$ is then equal to $1/n$ times the indicator function of the interval $[0, n]$, and its L^1 norm is equal to 1. It is then not hard to check that $\mathbb{W}_n$ converges to the pure dust graphex $(0, 0, 1, \mathbb{R}_+)$ in the metric $\delta_\diamond$. Indeed, $\|W_n\|_{2\to 2} \to 0$ and $\|D_{\mathbb{W}_n}\|_2 = n^{-1/2} \to 0$, while $\|\mathbb{W}_n\|_1 = 1 \to 1 = \|\mathbb{W}\|_1$, which immediately implies convergence in the metric $\delta_\diamond$ and hence GP-convergence (based on the proof of equivalence in this paper, though for this specific case it is simple to check GP-convergence directly). By contrast, $\|W_n\|_\square = \|W_n\|_1 = 1$ stays bounded away from zero, showing in particular that $\mathbb{W}_n$ does not converge to $\mathbb{W}$ in the cut metric $\delta_\square$. Since changing the Lebesgue measure to a measure μ_n such that $\lambda - \varepsilon_n \leq \mu_n \leq \lambda$ with $\varepsilon_n \to 0$ will asymptotically not change the cut norm of W_n, the graphexes $\mathbb{W}_n$ do not converge to $\mathbb{W}$ in the weak cut metric either. Note that this can't be cured by choosing a different norm for the marginal difference $D_{\mathbb{W}_1} - D_{\mathbb{W}_2}$, e.g., by keeping the L^2 norm for that part, since the above counter example works independently of the norm used for that part.

In studying the general topology of graphexes, we define a notion of tightness for sets of graphexes. Tight sets play an important role, in particular, they are the precompact sets in our topology: any sequence that is tight has a convergent subsequence, and any convergent sequence must be tight.

Definition 2.20 A set $\mathcal{S}$ of graphexes is *tight* if for every $\varepsilon > 0$, there exist C and D such that for every $\mathbb{W} \in \mathcal{S}$, $\mathbb{W} = (W, S, I, \mathbf{\Omega})$ with $\mathbf{\Omega} = (\Omega, \mathcal{F}, \mu)$, there exists $\Omega_\varepsilon \subseteq \Omega$ such that $\mu(\Omega_\varepsilon) \leq \varepsilon$ and the graphex $\mathbb{W}' = \mathbb{W}|_{\Omega \setminus \Omega_\varepsilon}$ is (C, D)-bounded. Here a graphex $\mathbb{W}'$ is called (C, D)-*bounded* if its marginals are D-bounded and $\|\mathbb{W}'\|_1 \leq C$.

Note that Proposition 2.4 implies that every finite set of graphexes is tight. In Sect. 4, we will prove that a set $\mathcal{S}$ of graphexes is tight if and only if for all fixed T, the corresponding set $\{G_T(\mathbb{W})\}_{\mathbb{W} \in \mathcal{S}}$ of unlabeled graphex processes at time T is tight (which will also be equivalent to the existence of some $T > 0$ such that $\{G_T(\mathbb{W})\}_{\mathbb{W} \in \mathcal{S}}$ is tight; see Theorem 4.1 below). Here, as usual, a collection $\mathcal{S}$ of distributions on finite graphs is called tight if for every $\varepsilon > 0$, there exists a finite set T of graphs such that for each of the random graphs in $\mathcal{S}$, the probability that the random graph is not isomorphic to a graph in T is at most ε. This is equivalent to the set of random measures being tight under the discrete topology on the set of isomorphism classes of finite graphs, or the set of distributions of the number of edges being tight.

Our main theorem concerning tightness is the following theorem. It will be proved in Sect. 5, where we will establish a version of the weak (or Frieze–Kannan) regularity lemma for graphexes. Note that while our regularity lemma will hold for signed graphexes, the following is only stated for unsigned graphexes. The reason is that our proof relies heavily on the notion of tightness, which we only develop for unsigned graphexes; see also Remark 4.8 in Sect. 4.

Theorem 2.21 *The space of all graphexes is complete under the topology induced by the weak kernel metric $\delta_\diamond$. A subset is relatively compact if and only if it is tight. In particular, for any C and D, the set of graphexes with $\|\mathbb{W}\|_1 \leq C$ is compact under $\delta_\diamond$, and the set of (C, D)-bounded graphexes is compact under both $\delta_\diamond$ and $\delta_{2 \to 2}$.*

Remark 2.22 As mentioned above, we only develop the theory of tightness for unsigned graphexes. In particular, we don't characterize the set of precompact signed graphexes. That notwithstanding, some of our compactness results do hold for signed graphexes. Here we only mention that the analogue of the statement for the set of graphexes with $\|\mathbb{W}\|_1 \leq C$ holds for signed graphexes as well, provided we restrict the L^∞ norm of the graphon part (which by definition is bounded by 1 for unsigned graphexes). To be explicit, any sequence of signed graphexes $\mathbb{W}_n = (W_n, S_n, I_n, \mathbf{\Omega}_n)$ with $\|W_n\|_\infty \leq B$ and $\|\mathbb{W}_n\|_1 \leq C$ has a subsequence converging to a signed graphex $\mathbb{W} = (W, S, I, \mathbf{\Omega})$ with $\|W\|_\infty \leq B$ and $\|\mathbb{W}\|_1 \leq C$. See Remark 5.10 in Sect. 5 below.

The advantage of (C, D)-bounded (unsigned) graphexes is that although there is no *a priori* bound on the size of $G_T(\mathbb{W})$ at any given time T, for any finite graph

F, the expected number of copies of F in $G_T(\mathbb{W})$ is finite. Furthermore, it turns out that under the assumption of (C, D)-boundedness, if two graphexes have the same subgraph densities, then they are equivalent, i.e., have $\delta_{2\to 2}$ distance 0. In this way, we can heuristically think of these subgraph densities as being analogous to moments of random variables: it is well known that moments determine the distribution of random variables, provided the moments do not grow too quickly.

To make these statements precise, we will define homomorphism densities for a graphex $\mathbb{W}$. To this end, we first consider a finite, labeled graph F and a graphon W, and define

$$t(F, W) = \int_{\Omega^{V(F)}} \prod_{(i,j)\in E(F)} W(x_i, x_j) \prod_{i\in V(F)} d\mu(x_i).$$

Given a connected multigraph $F = (V, E)$ on $k \geq 2$ vertices with no loops, and a graphex $\mathbb{W} = (W, S, I, \mathbf{\Omega})$, we define $t(F, \mathbb{W})$ as follows. First, if F consists of a single edge, we define

$$t(F, \mathbb{W}) = \int_{\Omega^2} W(x, y)\, d\mu(x)\, d\mu(y) + 2\int_\Omega S(x)\, d\mu(x) + 2I = \rho(\mathbb{W}).$$

Otherwise, let $V_{\geq 2}$ be the set of vertices of F with degree at least 2, and for each such vertex v, let $d_1(v)$ be the number of neighbors of v that have degree 1. Then

$$t(F, \mathbb{W}) = \int_{\Omega^{V_{\geq 2}}} \prod_{\{v,w\}\in E(F(V_{\geq 2}))} W(z_v, z_w) \prod_{v\in V_{\geq 2}} D_\mathbb{W}(z_v)^{d_1(v)} d\mu(z_v).$$

Finally, for any multigraph F with no isolated vertices and no loops, let $F_1, F_2, \ldots, F_k$ be the components of F. Then we define the *homomorphism density of F in $\mathbb{W}$* as

$$t(F, \mathbb{W}) = \prod_{i=1}^{k} t(F_i, \mathbb{W}).$$

As we will see in Proposition 3.24, these homomorphism densities are defined in such a way that for a simple graph F and a graphex $\mathbb{W}$, they are equal to the expected number of injective homomorphisms from F into $G_T(\mathbb{W})$ times $T^{-|V(F)|}$.

Having defined the subgraph densities $t(F, \mathbb{W})$, we can summarize the main relationship between convergence in the metric $\delta_{2\to 2}$, convergence of subgraph counts, and GP-convergence in the following theorem. Its proof will also be given at the end of Sect. 7.

Theorem 2.23 *Assume that $\mathbb{W}$ and $\mathbb{W}_n$ for $n \geq 1$ are graphexes whose marginals are D-bounded for some finite D. Then the following are equivalent.*

(1) $\delta_{2\to 2}(\mathbb{W}_n, \mathbb{W}) \to 0$.
(2) *For every graph F with no isolated vertices, $t(F, \mathbb{W}_n) \to t(F, \mathbb{W})$.*
(3) *For every connected graph F, $t(F, \mathbb{W}_n) \to t(F, \mathbb{W})$.*

(4) $G_T(\mathbb{W}_n) \to G_T(\mathbb{W})$ in distribution for every T.
(5) $G_T(\mathbb{W}_n) \to G_T(\mathbb{W})$ in distribution for some T.

Remark 2.24 The above theorem implies in particular that in order to check whether a sequence $\mathbb{W}_n$ of graphexes with uniformly bounded marginals is GP-convergent, it is enough to check convergence of $G_T(\mathbb{W}_n)$ for a single $T > 0$. In a similar way, several other properties of sequences or sets of graphexes can be equivalently stated for all $T > 0$ or some $T > 0$ (see, in particular, the already mentioned Theorem 4.1 about tightness and Theorem 9.1 about uniform integrability). But for general sequences of graphexes, we do not know whether GP-convergence is equivalent to the convergence of $G_T(\mathbb{W}_n)$ for just one $T > 0$.

It is instructive to compare our notions of convergence to the notions of graph convergence introduced in [3, 4]. Before defining these notions, we first introduce the notion of a dilated empirical graphon corresponding to a finite graph G. It involves a "dilation parameter" $\rho \in \mathbb{R}_+$ and is defined as the graphex $\mathbb{W}(G, \rho)$ consisting of a zero dust part, a zero star part, a measure space consisting of the vertex set $V(G)$ where each vertex has measure ρ, and a graphon $W(G, \rho)$ which is simply the adjacency matrix of G. The usual way to embed graphs into the space of graphons in the dense case corresponds to $\rho = 1/|V(G)|$.

By contrast, in [3], ρ was chosen to be $1/\sqrt{2|E(G)|}$; the resulting dilated empirical graphon was called the stretched empirical graphon, and a sequence was said to *converge in the stretched cut metric* if the graphons $W(G, 1/\sqrt{2|E(G)|})$ converge in the cut metric $\delta_\square$. It was then shown that this leads to completeness (every Cauchy sequence has a limit), that convergence implies a certain condition called uniform tail regularity, and that any uniformly tail regular sequence has a convergent subsequence.

The notion of convergence in [4] is slightly different. It does not start from a metric, and instead tries to emulate the notion of subgraph convergence from dense graphs. Roughly speaking, it asks that certain random subgraphs of the graphs in the sequence converge in distribution to some well-defined distribution over finite graphs. More precisely, given a parameter $p \in [0, 1]$, define $\text{Smpl}(G, p)$ as the unlabelled graph obtained by first taking each vertex i.i.d. with probability p, then removing all isolated vertices in the resulting subgraph, and finally discarding all the labels. A sequence G_n is then said to be *sampling convergent* if for all $t > 0$, the samples

$$\text{Smpl}(G_n, \min\{1, t/\sqrt{2|E(G_n)|}\})$$

converge in distribution. It was then shown that any sequence of finite graphs has a convergent subsequence, and that the limiting distribution can be expressed as $G_t(\mathbb{W})$ for some integrable graphex $\mathbb{W}$ with $\|\mathbb{W}\|_1 \leq 1$. It was also shown that this inequality holds with equality if and only if the sequence has a property called *uniform sampling regularity*.

It is instructive to relate the results and notions from [4] to those developed in this paper. To this end, we first note that—as already observed in [4]—a sequence of graphs is sampling convergent if and only if the stretched canonical graphexes

$\mathbb{W}(G_n, 1/\sqrt{2|E(G_n)|})$ are GP-convergent. Since by definition, the stretched canonical graphex has L^1 norm 1, this sequence is tight. By our compactness theorem, Theorem 2.21, it therefore has a convergent subsequence.

To relate some of the other notions and results from [3, 4] to those of this paper, we introduce a couple of definitions. The first notion is that of uniform integrability. Recall that a set S of random variables with values in $\mathbb{R}$ is called uniformly integrable if for every $\varepsilon > 0$, there exists $K \in \mathbb{R}$ such that for every $X \in S$,

$$\mathbb{E}[|X|1_{|X|>K}] < \varepsilon.$$

Note that this implies that $\mathbb{E}[|X|] \leq \varepsilon + K$, so the set of random variables consists of integrable variables with uniformly bounded integrals. This motivates the following definition.

Definition 2.25 A set of graphexes S is called *uniformly integrable* if the graphexes in S have uniformly bounded $\|\cdot\|_1$-norms, and for every $\varepsilon > 0$, there exists a D such that for all $\mathbb{W} \in S$, $\|D_\mathbb{W} 1_{D_\mathbb{W} > D}\|_1 < \varepsilon$.

As we will see in Theorem 9.1 below, uniform integrability of a set S of graphexes is equivalent to uniform integrability of the random variables $\{E(G_T(\mathbb{W})) : \mathbb{W} \in S\}$ for all $T > 0$ (which is also equivalent to uniform integrability of this set of random variables for some $T > 0$).

The notion of uniform sampling regularity from [4] is then simply uniform integrability of the stretched empirical graphexes, and the following theorem is a more or less straightforward generalization of Corollary 3.10 in [4], which states that the limiting graphex of a sampling convergent sequence of graphs has norm 1 if and only if it is uniformly sampling regular. We will prove the theorem in Sect. 9.

Theorem 2.26 *Suppose $\mathbb{W}_n$ is a sequence of integrable graphexes with uniformly bounded $\|\cdot\|_1$-norms that converges to a graphex $\mathbb{W}$ in the weak kernel metric. Then*

$$\|\mathbb{W}\|_1 \leq \liminf_{n\to\infty} \|\mathbb{W}_n\|_1.$$

In particular, $\mathbb{W}$ is integrable. We furthermore have that

$$\lim_{n\to\infty} \|\mathbb{W}_n\|_1 = \|\mathbb{W}\|_1$$

if and only if the sequence $\mathbb{W}_n$ is uniformly integrable.

Our next set of theorems relates the notion of sampling convergence from [4] to the notion of convergence in the cut metric from [3]. We start by recalling the definition of uniform tail regularity from [3] (see also Lemma 9.3 in Sect. 9 for other, equivalent definitions).

Definition 2.27 ([3]) Given a set of signed, integrable *graphons* S, we say that they are *uniformly tail regular* if for any $\varepsilon > 0$, there exists M such that for each $W \in S$

Identifiability for Graphexes and the Weak Kernel Metric

with the usual notation, there exists $\Omega_0 \subseteq \Omega$ such that $\mu(\Omega_0) \leq M$ and $\|W\|_1 - \|W|_{\Omega_0}\|_1 \leq \varepsilon$.

Note that uniform tail regularity is more restrictive than uniform integrability (for the set of graphexes obtained by setting the dust and star part to zero). For sequences of graphs, the corresponding result was shown in [4], but it holds in our more general setting as well, with essentially the same proof; see Lemma 9.2 in Sect. 9 below. More interestingly, any sequence of graphons that is convergent in cut metric has uniformly regular tails, and any sequence of graphons with uniformly regular tails has a subsequence that converges in cut metric. Motivated by this (which is one of the central results of [3]), here we prove the following.

Theorem 2.28 *Given a sequence of integrable graphexes of the form*

$$\mathbb{W}_n = (W_n, 0, 0, \mathbf{\Omega}_n),$$

the following are equivalent.

(1) *The sequence W_n converges to a graphon W in cut metric.*
(2) *The sequence $\mathbb{W}_n$ is uniformly tail regular, and in the weak kernel metric, the sequence $\mathbb{W}_n$ converges to a graphex of the form $\mathbb{W} = (W, 0, 0, \mathbf{\Omega})$.*
(3) *The sequence $\mathbb{W}_n$ is uniformly tail regular, and in the weak kernel metric, the sequence $\mathbb{W}_n$ converges to some graphex $\mathbb{W}$.*

This theorem, as well as our next theorem, will also be proved in Sect. 9.

Theorem 2.29 *Given a sequence of uniformly integrable graphexes*

$$\mathbb{W}_n = (W_n, S_n, I_n, \mathbf{\Omega}_n),$$

which converge to a graphex $\mathbb{W} = (W, S, I, \mathbf{\Omega})$ in the weak kernel metric, the following are equivalent.

(1) *The graphex $\mathbb{W}$ is of the form $\mathbb{W} = (W, 0, 0, \mathbf{\Omega})$.*
(2) *$\|S_n\|_1 \to 0$ and $I_n \to 0$, and the sequence of graphons W_n has uniformly regular tails.*
(3) *$\|S_n\|_1 \to 0$ and $I_n \to 0$, and the sequence of graphons W_n converges in the cut metric.*

Recall that a sequence of graphs is sampling convergent if and only if the stretched canonical graphexes are GP-convergent. This fact, Theorems 2.26, 2.29 together imply that given a sequence of graphs that is sampling convergent to a graphex of norm one, the sequence converges to a pure graphon if and only if the sequence is uniformly tail regular. We have therefore given a characterization of when the notion of sampling convergence from [4] reduces to the notion of convergence in the stretched cut metric from [3].

Remark 2.30 In the above theorem, the assumption of uniform integrability is necessary. To see this, let W_n be the graphon defined by being 1 on the set $[0, 1/n] \times [1, n+1]$ and its transpose, and 0 otherwise, and let $\mathbb{W}_n = (W_n, 0, 0, \mathbb{R}_+)$. Then the sequence $\mathbb{W}_n$ converges to the zero graphex in the weak kernel metric, which is a pure graphon. However, the sequence is clearly not uniform tail regular (or even uniformly integrable). We can also let $S_n(x) = 1/n$ on the set $[0, n]$, and 0 everywhere else. In that case the sequence still converges to 0 in the weak kernel metric, but $\|S_n\|_1$ does not converge to 0.

We close this section by discussing possible extensions of our theory. First, as already discussed in the introduction, it would be natural to extend the theory of graphexes to a theory that naturally generates multigraphs. Note that *a priori*, this falls plainly in the framework of exchangeable random measures on $\mathbb{R}_+^2$ as developed by Kallenberg; in fact, it falls into the framework of exchangeable random counting measures. Generalizing the approach of [3, 26], this will give a natural notion of "multigraphexes" characterizing all exchangeable multigraphs with vertices labelled by $\mathbb{R}_+$. But extending the current work to multigraphexes is beyond the scope of this paper, in particular given that it would require to generalize at least some of the results from [4, 17, 18, 25, 26] to this setting in a first step. See [6] for some very preliminary steps in this direction.

The next extension one might want to consider is the extension of our analytical results (i.e., those of our results which do not refer to the graphex process generated by a graphex) to signed graphexes. In contrast to the theory of cut metric convergence for graphons over σ-finite measure spaces developed in [3], which works as well for signed, unbounded graphons as for graphons with values in [0, 1], here we focused most of the theory of graphex convergence on unsigned graphexes (with graphon parts taking values in [0, 1]). While several of our technical proofs and results hold for signed graphexes (in particular, all of Sect. 3, as well as parts of Sects. 5 and 6 are formulated in this language), the core analytic concepts and results such as tightness, precompactness, etc., have only been formulated for unsigned graphexes. Indeed, we believe that the generalization of Theorem 2.21 to signed graphexes requires modifications to either our topology or our notions of tightness; see Remark 4.8 below. In a similar way, while the identification theorem of [3] works for signed graphons, our identification theorem requires non-negative graphexes, even though one might conjecture that when stated as a characterization of the equivalence classes under the weak kernel metric, it should hold for signed graphexes as well, at least when suitably formulated.

Finally, one might want to consider graphexes where the graphon part W is unbounded, whether non-negative or signed. For non-negative graphexes, one could, for example, follow the approach in [6] and use such a graphex to generate multigraphs by adding $\text{Pois}(W(x_i, y_i))$ many edges to a pair of Poisson points with features x_i and x_j, or one could try to generalize the approach of [5] to the setting of graphexes, by taking a decreasing "dilution probability" p_t, and then connect two Poisson points with features x_i and x_j with probability $\min\{1, p_t W(x_i, x_j)\}$ (see [2] for a related approach). But it is far from obvious what the analogue of the weak kernel metric should be, and how to generalize our other results to this setting. As the other open questions discussed here, we leave these questions as open research problems.

3 Preliminaries

In this section, we study the metrics $\delta_{2\to 2}$ and $\delta_\diamond$. In particular, we will prove Theorems 2.11 and 2.15, as well as Proposition 2.17 relating the two for graphexes with bounded marginals (in fact, we will prove its generalization to signed graphexes, stated as Proposition 3.15 below).

In addition, we study the metric $\delta_\boxtimes$ obtained from $\delta_{2\to 2}$ by replacing $d_{2\to 2}$ as defined in (1) by

$$d_\boxtimes(\mathbb{W}_1, \mathbb{W}_2) = \max\left\{\|W_1 - W_2\|_\boxtimes, \|D_{\mathbb{W}_1} - D_{\mathbb{W}_2}\|_\boxtimes, |\rho(\mathbb{W}_1) - \rho(\mathbb{W}_2)|\right\}. \quad (5)$$

Here we use the norm $\|\cdot\|_\boxtimes$ both for functions from $\Omega^2 \to \mathbb{R}$ (see Definition 2.8) and functions from $\Omega \to \mathbb{R}$, where it is defined as

$$\|F\|_\boxtimes = \sup_{S \subseteq \Omega} \left|\frac{1}{\sqrt{\mu(S)}} \int_S F(x) d\mu(x)\right|.$$

We will in particular show that the analogue of Theorem 2.11 holds for this metric.

Theorem 3.1 *Let $\mathbb{W}_1$ and $\mathbb{W}_2$ be signed graphexes in $L^1 \cap L^2$. Define $\delta_\boxtimes$ by replacing the right side of (4) with $\widetilde{\delta}_\boxtimes$, which in turn is obtained from $\widehat{\delta}_{2\to 2}$ by replacing $d_{2\to 2}$ with $d_\boxtimes$. Then the value of $\delta_\boxtimes(\mathbb{W}_1, \mathbb{W}_2)$ does not depend on the choice of the trivial extensions $\mathbb{W}'_1$ and $\mathbb{W}'_2$, and $\delta_\boxtimes$ obeys the triangle inequality, making it a well-defined pseudometric.*

This *jumble metric* will be particularly useful when establishing the regularity lemma for graphexes, which takes a nicer form when stated in terms of the distance $d_\boxtimes$ instead of the distance $d_{2\to 2}$, both because of the absence of the various roots, and because the proof of the regularity lemma leads more naturally to bounds in term of $\|\cdot\|_\boxtimes$ rather than $\|\cdot\|_{2\to 2}$. To obtain our compactness results for the metric $\delta_\diamond$ (which is derived from $d_{2\to 2}$), we will then need to compare the two. We will do this in Proposition 3.19 and Remark 3.23 below.

We will also establish a simple lemma relating the kernel norm to 4-cycle counts (Lemma 3.22). Finally, we will prove that the homomorphism densities $t(F, \mathbb{W})$ indeed describe the expected number of injective homomorphisms from F into $G_T(\mathbb{W})$ (Proposition 3.24).

Except for the last result, all results in this section are as easily derived for signed graphexes as for unsigned graphexes. We therefore formulate everything in this section in the language of signed graphexes. To do so, we need very little extra notation, except for the following.

First, we define the *absolute marginal* of a signed graphex $\mathbb{W}$ as $D_{|\mathbb{W}|}$, and say that $\mathbb{W}$ has D-bounded absolute marginals if $\|D_{|\mathbb{W}|}\|_\infty \leq D$. We say that $\mathbb{W}$ is

(C, D)-bounded if in addition $\|\mathbb{W}\|_1 \leq C$. Furthermore, we introduce the notion of (B, C, D)-boundedness of a graphex $\mathbb{W} = (W, S, I, \mathbf{\Omega})$ by requiring that

$$\|W\|_\infty \leq B, \qquad \|\mathbb{W}\|_1 \leq C, \quad \text{and} \quad \|D_{|\mathbb{W}|}\|_\infty \leq D, \tag{6}$$

and finally, we say that $\mathbb{W}$ has a bounded graphon part if $\|W\|_\infty < \infty$.

We use the following standard facts about measure-preserving transformations, which we prove for completeness.

Lemma 3.2 *Suppose $\varphi \colon \Omega \to \Omega'$ is a measure preserving map between two σ-finite measure spaces $(\Omega, \mathcal{F}, \mu)$ and $(\Omega', \mathcal{F}', \mu')$.*

(1) For any σ-algebra $\mathcal{G} \subseteq \mathcal{F}$, $L^1(\Omega, \mathcal{G}, \mu)$ is a closed subspace in $L^1(\Omega, \mathcal{F}, \mu)$.
(2) If $f \in L^1(\Omega, \varphi^{-1}(\mathcal{F}'), \mu)$, then there exists a function $f' \in L^1(\Omega', \mathcal{F}', \mu')$ such that $f = f'^\varphi = f' \circ \varphi$ almost everywhere.
(3) The map $\varphi^* \colon L^1(\Omega', \mathcal{F}', \mu') \to L^1(\Omega, \varphi^{-1}(\mathcal{F}'), \mu)$ with $f' \mapsto f'^\varphi$ is an isometric isomorphism, implying that φ^* and its inverse are continuous and hence Borel measurable.

Proof (1) $L^1(\Omega, \mathcal{G}, \mu)$ is clearly a subspace of $L^1(\Omega, \mathcal{F}, \mu)$. Since they are both Banach spaces, they are both complete. Since $L^1(\Omega, \mathcal{G}, \mu) \subseteq L^1(\Omega, \mathcal{F}, \mu)$ is an isometric embedding, the only way we can have a complete subset of a complete space is if the subset itself is closed.

(2) The map sending $A \subseteq \Omega$ to $\int_A f$ defines a finite, signed measure ν on Ω, absolutely continuous with respect to μ. This measure pushes forward to a signed measure ν' on Ω'. If $B \subseteq \Omega'$ has $\mu'(B) = 0$, then $\mu(\varphi^{-1}(B)) = 0$, so $\nu'(B) = \nu(\varphi^{-1}(B)) = 0$. Therefore ν' is absolutely continuous with respect to μ', so it has a Radon-Nikodym derivative f'. It is straightforward to check that, since f is $\varphi^{-1}(\mathcal{F}')$ measurable, we have $f = f'^\varphi$ almost everywhere. We remark that this proof is basically the same as the standard proof of the existence of conditional expectations on probability spaces, except here our space is σ-finite.

(3) This follows from the previous parts. $\square$

We will also need the following lemma.

Lemma 3.3 *Suppose that $\mathbf{\Omega} = (\Omega, \mathcal{F}, \mu)$ with $\mu(\Omega) < \infty$, and suppose that $g \in L^2(\Omega)$. Let*

$$S = \sup_{f \in L^\infty(\Omega), 0 \leq f \leq 1} \left| \frac{\int_\Omega f g \, d\mu}{\sqrt{\int_\Omega f \, d\mu}} \right|. \tag{7}$$

Then $S \leq \|g\|_2 < \infty$, and there exists $X \subseteq \Omega$ such that

$$S = \frac{|\int_X g \, d\mu|}{\sqrt{\mu(X)}}.$$

Note that this expression is the same as taking $f = 1_X$ in (7).

Proof First, note that if $0 \leq f \leq 1$, then

$$\left|\frac{\int_\Omega fg\, d\mu}{\sqrt{\int_\Omega f\, d\mu}}\right| \leq \sqrt{\frac{\int_\Omega f^2 d\mu \int_\Omega g^2 d\mu}{\int_\Omega f\, d\mu}} = \|g\|_2 \sqrt{\frac{\int_\Omega f^2 d\mu}{\int_\Omega f\, d\mu}} \leq \|g\|_2 < \infty.$$

Next, we will show that there exists a $\delta > 0$ such that in (7), it suffices to consider f with $\|f\|_1 \geq \delta$. Let $\widetilde{\Omega} = \Omega \times [0, 1]$ and $\widetilde{\mu} = \mu \times \lambda$, where λ is the Lebesgue measure, and let $\widetilde{g}(x, t) = g(x)$. Then the expression in (7) is the same as taking the supremum of

$$\frac{|\int_X \widetilde{g}\, d\widetilde{\mu}|}{\sqrt{\widetilde{\mu}(X)}}$$

over $X \subseteq \widetilde{\Omega}$. Indeed, given $X \subseteq \widetilde{\Omega}$, we can plug in the function $f(x) = \lambda(\{t : (x, t) \in X\})$ into (7), which is defined almost everywhere, and given f as in (7), we can take $X = \{(x, t) : t \leq f(x)\}$. It is straightforward to check that (7) and the above expression give the same value. Note that we have

$$\frac{|\int_X \widetilde{g}\, d\widetilde{\mu}|}{\sqrt{\widetilde{\mu}(X)}} \leq \|\widetilde{g}|_X\|_2.$$

Since $\|\widetilde{g}\|_2 < \infty$, for any $\varepsilon > 0$, there exists a K such that

$$\int_X \widetilde{g}^2\, d\widetilde{\mu} \leq \int \widetilde{g}^2 1_{\widetilde{g} > \sqrt{K}}\, d\widetilde{\mu} + K\widetilde{\mu}(X) \leq \frac{\varepsilon^2}{2} + K\widetilde{\mu}(X).$$

Thus, given $\varepsilon > 0$, we can find a $\delta > 0$ such that if $\widetilde{\mu}(X) < \delta$, then $\|\widetilde{g}|_X\|_2 \leq \varepsilon$. Taking $\varepsilon = S/2$ (note that unless $g = 0$, $S > 0$), we obtain a δ such that if $\int_\Omega f\, d\mu < \delta$, then the expression in (7) is less than $S/2$. This means that it suffices to take the supremum over f with $\|f\|_1 \geq \delta$.

Recall that we assumed that Ω has finite measure. The set of $f \in L^\infty(\Omega)$ with $0 \leq f \leq 1$ and $\|f\|_1 \geq \delta$ is weak-$*$ closed and therefore weak-$*$ compact. The expression in (7) is weak-$*$ continuous; therefore, there exists an f which maximizes the expression.

Clearly such an f is supported either on the set $\{g > 0\}$ or $\{g < 0\}$. Assume without loss of generality that it is supported on $\{g > 0\}$. Suppose that f is not equal to 0 or 1 almost everywhere. Then we can find $0 \leq h \leq \widetilde{h} \leq 1$, supported on $\{g > 0\}$, such that $f = (\widetilde{h} + h)/2$, and it is not the case that $\widetilde{h} = h$ almost everywhere. This implies that

$$\int_\Omega (\widetilde{h} - h)\, d\mu > 0$$

and

$$\int_\Omega (\widetilde{h} - h)g\, d\mu > 0.$$

Let $h_t = t\tilde{h} + (1-t)h$. Note that for $t \in [0, 1]$, $0 \le h_t \le 1$ almost everywhere. Let

$$p(t) = \frac{\int_\Omega h_t g}{\sqrt{\int_\Omega h_t}}.$$

We have

$$\frac{d}{dt} p(t) =$$

$$\frac{\int_\Omega (\tilde{h} - h) g \, d\mu \int_\Omega (t\tilde{h} + (1-t)h) \, d\mu - \frac{1}{2} \int_\Omega (\tilde{h} - h) \, d\mu \int_\Omega (t\tilde{h} + (1-t)h) g \, d\mu}{\left(\int_\Omega (t\tilde{h} + (1-t)h) \, d\mu\right)^{3/2}}.$$

Notice that the denominator above is always positive, and the numerator above is of the form $At + B$, where

$$A = \int_\Omega (\tilde{h} - h) g \, d\mu \int_\Omega (\tilde{h} - h) \, d\mu - \frac{1}{2} \int_\Omega (\tilde{h} - h) \, d\mu \int_\Omega (\tilde{h} - h) g \, d\mu$$
$$= \frac{1}{2} \int_\Omega (\tilde{h} - h) \, d\mu \int_\Omega (\tilde{h} - h) g \, d\mu > 0.$$

This means that there are three possibilities for $\frac{d}{dt} p(t)$: it can be positive for every $t \in (0, 1)$, it can be negative for every $t \in (0, 1)$, or it can be negative and then positive. Either of these cases implies that the maximum of p on $[0, 1]$ is attained at one or both of the endpoints, and therefore either $p(0)$ or $p(1)$ is strictly greater than $p(1/2)$. This contradicts the assumption that f was maximal, completing the proof of the lemma. □

Using the previous two lemmas, we prove the following proposition. We will use it for functions which arise as the difference of two graphons with bounded marginals.

Proposition 3.4 *Let $\Omega = (\Omega, \mathcal{F}, \mu)$ and $\Omega' = (\Omega', \mathcal{F}', \mu')$ be σ-finite measure spaces, and let $\varphi: \Omega' \to \Omega$ be measurable. If $U: \Omega \times \Omega \to \mathbb{R}$ and $F: \Omega \to \mathbb{R}$ are square integrable, then*

$$\|U\|_{2\to 2} = \|U^\varphi\|_{2\to 2}, \qquad \|F\|_\boxtimes = \|F^\varphi\|_\boxtimes, \qquad \text{and} \qquad \|U\|_\boxtimes = \|U^\varphi\|_\boxtimes.$$

If instead of square integrability, we assume that U is integrable, then

$$\|U\|_\square = \|U^\varphi\|_\square.$$

Proof For any $f, g \in L^2(\Omega)$, it is easy to see that $f \circ U \circ g = f^\varphi \circ U^\varphi \circ g^\varphi$, and we furthermore have that $f^\varphi, g^\varphi \in L^2(\Omega')$ with $\|f^\varphi\|_2 = \|f\|_2$ and $\|g^\varphi\|_2 = \|g\|_2$. This implies that

$$\|U\|_{2\to 2} \le \|U^\varphi\|_{2\to 2}.$$

To prove the opposite inequality, let $\widehat{f}, \widehat{g} \in L^2(\Omega')$, and assume first that $\widehat{f}, \widehat{g} \in L^1(\Omega')$ as well. Let $f' = \mathbb{E}[\widehat{f}|\varphi^{-1}(\mathcal{F})]$ and $g' = \mathbb{E}[\widehat{g}|\varphi^{-1}(\mathcal{F})]$. That is, f' is a $\varphi^{-1}(\mathcal{F})$-measurable function such that for any $\varphi^{-1}(\mathcal{F})$-measurable set $S' \subseteq \Omega'$, $\int_{S'} f' = \int_{S'} \widehat{f}$, and same for g'. These functions exist by the Radon-Nikodym theorem (since all measures are σ-finite). Then $\|f'\|_2 \leq \|\widehat{f}\|_2$, $\|g'\|_2 \leq \|\widehat{g}\|_2$. We claim that $f' \circ U^\varphi \circ g' = \widehat{f} \circ U^\varphi \circ \widehat{g}$. Indeed, for any $x' \in \Omega'$,

$$(U^\varphi)_{x'}(y') = U^\varphi(x', y') = U(\varphi(x'), \varphi(y')) = U_{\varphi(x')}(\varphi(y')) = (U_{\varphi(x')})^\varphi(y'),$$

showing that $(U^\varphi)_{x'}$ is the pullback of an $\mathcal{F}$-measurable function, and is thus $\varphi^{-1}(\mathcal{F})$-measurable. Therefore, for every $x' \in \Omega'$,

$$\int_{\Omega'} U^\varphi(x', y') \widehat{g}(y') \, d\mu'(y') = \int_{\Omega'} U^\varphi(x', y') g'(y') \, d\mu'(y'),$$

which shows that $\widehat{f} \circ U^\varphi \circ \widehat{g} = \widehat{f} \circ U^\varphi \circ g'$. We can analogously show that $\widehat{f} \circ U^\varphi \circ g' = f' \circ U^\varphi \circ g'$. Then, since f' and g' are $\varphi^{-1}(\mathcal{F})$-measurable, there exist by Lemma 3.2 $f, g \in L^1(\Omega)$ with $f' = f^\varphi$ and $g' = g^\varphi$. This implies that $\|f\|_2 \leq \|\widehat{f}\|$, $\|g\|_2 \leq \|\widehat{g}\|_2$, and $f \circ U \circ g = \widehat{f} \circ U^\varphi \circ \widehat{g}$, which shows that

$$\widehat{f} \circ U^\varphi \circ \widehat{g} \leq \|U\|_{2 \to 2}$$

whenever $\widehat{f}, \widehat{g} \in L^2(\Omega') \cap L^1(\Omega')$ and $\|\widehat{f}\|_2, \|\widehat{g}\|_2 \leq 1$. Since Ω' is σ-finite, any function in $L^2(\Omega')$ can be written as a limit of functions in $L^2(\Omega') \cap L^1(\Omega')$. A dominated convergence argument then shows that the above bound holds whenever $\widehat{f}, \widehat{g} \in L^2(\Omega')$ and $\|\widehat{f}\|_2, \|\widehat{g}\|_2 \leq 1$, proving that

$$\|U^\varphi\|_{2 \to 2} \leq \|U\|_{2 \to 2}.$$

To prove the statement for the cut norm, we use the representation $\|U\|_\square = \sup_{f,g:\Omega \to [0,1]} |f \circ U \circ g|$. Using this representation, the proof for the cut norm proceeds along the same lines as the proof for the $\|\cdot\|_{2 \to 2}$ norm.

Next, let us prove that $\|F\|_\boxtimes = \|F^\varphi\|_\boxtimes$. First, for any measurable $X \subseteq \Omega$ with $\mu(X) < \infty$, since μ is the pushforward of μ',

$$\frac{\int_{\varphi^{-1}(X)} F^\varphi \, d\mu'}{\sqrt{\mu'(\varphi^{-1}(X))}} = \frac{\int_X F \, d\mu}{\sqrt{\mu(X)}}.$$

This shows that $\|F\|_\boxtimes \leq \|F^\varphi\|_\boxtimes$. Suppose now that $X' \subseteq \Omega'$, and let $f' = \mathbb{E}[1_{X'}|\varphi^{-1}(\mathcal{F})]$. Then there exists $f \in L^1(\Omega)$ so that $f' = f^\varphi$. Since F^φ is $\varphi^{-1}(\mathcal{F})$-measurable,

$$\frac{\int_{X'} F^\varphi \, d\mu'}{\sqrt{\mu'(X)}} = \frac{\int_{\Omega'} f' F^\varphi \, d\mu'}{\sqrt{\int_\Omega f' \, d\mu'}} = \frac{\int_\Omega f F \, d\mu}{\sqrt{\int_\Omega f \, d\mu}}.$$

Fix $\varepsilon > 0$. Since Ω is σ-finite, there exists $\Omega_0 \subseteq \Omega$ with $\mu(\Omega_0) < \infty$ and

$$\frac{\int_{\Omega_0} fF\,d\mu}{\sqrt{\int_{\Omega_0} f\,d\mu}} \geq (1-\varepsilon)\frac{\int_\Omega fF\,d\mu}{\sqrt{\int_\Omega f\,d\mu}}.$$

By the previous lemma, there exists a measurable set $X \subseteq \Omega_0$ so that

$$\frac{\int_X F\,d\mu}{\sqrt{\mu(X)}} \geq \frac{\int_{\Omega_0} fF\,d\mu}{\sqrt{\int_{\Omega_0} f\,d\mu}} \geq (1-\varepsilon)\frac{\int_\Omega fF\,d\mu}{\sqrt{\int_\Omega f\,d\mu}}.$$

Since this holds for any $\varepsilon > 0$, this proves that $\|F\|_{\boxtimes} \geq \|F^\varphi\|_{\boxtimes}$.

Finally, we show that $\|U\|_{\boxtimes} = \|U^\varphi\|_{\boxtimes}$. As before, for any measurable $X, Y \subseteq \Omega$ with $\mu(X), \mu(Y) < \infty$, since μ is the pushforward of μ',

$$\frac{\int_{\varphi^{-1}(X) \times \varphi^{-1}(Y)} U^\varphi\,(d\mu')^2}{\sqrt{\mu'(\varphi^{-1}(X))}\sqrt{\mu'(\varphi^{-1}(Y))}} = \frac{\int_{X \times Y} U\,(d\mu)^2}{\sqrt{\mu(X)}\sqrt{\mu(Y)}}.$$

This shows that $\|U\|_{\boxtimes} \leq \|U^\varphi\|_{\boxtimes}$. For the other direction, let $X', Y' \subseteq \Omega'$ with finite measure. By the previous argument, there exist functions $f, g: \Omega \to [0, 1]$ such that

$$\frac{\int_{\Omega^2} f(x)U(x,y)g(y)\,d\mu(x)\,d\mu(y)}{\sqrt{\int_\Omega f\,d\mu \int_\Omega g\,d\mu}} = \frac{\int_{X' \times Y'} U^\varphi\,(d\mu')^2}{\sqrt{\mu'(X')\mu'(Y')}}.$$

Fix $\varepsilon > 0$. By the previous argument, there exists $X \subseteq \Omega$ such that

$$\frac{\int_{X \times \Omega} U(x,y)g(y)\,d\mu(x)\,d\mu(y)}{\sqrt{\mu(X) \int_\Omega g\,d\mu}} \geq (1-\varepsilon)\frac{\int_{\Omega^2} f(x)U(x,y)g(y)\,d\mu(x)\,d\mu(y)}{\sqrt{\int_\Omega f\,d\mu \int_\Omega g\,d\mu}}.$$

Then, applying it again, there exists $Y \subseteq \Omega$ such that

$$\frac{\int_{X \times Y} U(x,y)\,d\mu(x)\,d\mu(y)}{\sqrt{\mu(X)\mu(Y)}} \geq (1-\varepsilon)\frac{\int_{X \times \Omega} U(x,y)g(y)\,d\mu(x)\,d\mu(y)}{\sqrt{\mu(X) \int_\Omega g\,d\mu}}.$$

Combining these, we obtain

$$\frac{\int_{X \times Y} U(x,y)\,d\mu(x)\,d\mu(y)}{\sqrt{\mu(X)\mu(Y)}} \geq (1-\varepsilon)^2 \frac{\int_{X' \times Y'} U^\varphi\,(d\mu')^2}{\sqrt{\mu'(X')\mu'(Y')}}.$$

Since this holds for any $\varepsilon > 0$, we obtain that $\|U\|_{\boxtimes} \geq \|U^\varphi\|_{\boxtimes}$. $\square$

Next we establish a sequence of lemmas leading to the proof of Theorem 2.11, which states that $\delta_{2\to 2}$ is a pseudometric. To state the first lemma, we define the vectors

$$\Delta_{2\to 2}(\mathbb{W}_1, \mathbb{W}_2) = \left(\|W_1 - W_2\|_{2\to 2}, \|D_{\mathbb{W}_1} - D_{\mathbb{W}_2}\|_2, |\rho(\mathbb{W}_1) - \rho(\mathbb{W}_2)|\right) \text{ and}$$

$$\Delta_{\boxtimes}(\mathbb{W}_1, \mathbb{W}_2) = \left(\|W_1 - W_2\|_{\boxtimes}, \|D_{\mathbb{W}_1} - D_{\mathbb{W}_2}\|_{\boxtimes}, |\rho(\mathbb{W}_1) - \rho(\mathbb{W}_2)|\right),$$

where again we require the signed graphexes $\mathbb{W}_1$ and $\mathbb{W}_2$ to be in $L^1 \cap L^2$. Note that each coordinate satisfies the triangle inequality. We will also use the following property.

Lemma 3.5 *Let $\boldsymbol{\Omega} = (\Omega, \mathcal{F}, \mu)$ and $\boldsymbol{\Omega}' = (\Omega', \mathcal{F}', \mu')$ be σ-finite measure spaces, let $\varphi \colon \Omega \to \Omega'$ be a measure-preserving map, and let $\mathbb{W}_1, \mathbb{W}_2$ be signed graphexes in $L^1 \cap L^2$. Then*

$$\Delta_{2\to 2}(\mathbb{W}_1, \mathbb{W}_2) = \Delta_{2\to 2}(\mathbb{W}_1^\varphi, \mathbb{W}_2^\varphi) \quad \text{and} \quad \Delta_{\boxtimes}(\mathbb{W}_1, \mathbb{W}_2) = \Delta_{\boxtimes}(\mathbb{W}_1^\varphi, \mathbb{W}_2^\varphi).$$

Proof Clearly $\rho(\mathbb{W}_1) = \rho(\mathbb{W}_1^\varphi)$ and $\rho(\mathbb{W}_2) = \rho(\mathbb{W}_2^\varphi)$, which means that their differences are equal too. We also have for every $x \in \Omega$, $D_{\mathbb{W}_i}(\varphi(x)) = D_{\mathbb{W}_i^\varphi}(x)$, which implies that $\|D_{\mathbb{W}_1} - D_{\mathbb{W}_2}\|_2 = \|D_{\mathbb{W}_1^\varphi} - D_{\mathbb{W}_2^\varphi}\|_2$. Therefore it remains to show that

$$\|W_1 - W_2\|_{2\to 2} = \|W_1^\varphi - W_2^\varphi\|_{2\to 2}$$

as well as

$$\|D_{\mathbb{W}_1} - D_{\mathbb{W}_2}\|_{\boxtimes} = \|D_{\mathbb{W}_1^\varphi} - D_{\mathbb{W}_2^\varphi}\|_{\boxtimes} \quad \text{and} \quad \|W_1 - W_2\|_{\boxtimes} = \|W_1^\varphi - W_2^\varphi\|_{\boxtimes}.$$

Note that $D_{\mathbb{W}_1^\varphi} - D_{\mathbb{W}_2^\varphi} = (D_{\mathbb{W}_1} - D_{\mathbb{W}_2})^\varphi$ and $W_1^\varphi - W_2^\varphi = (W_1 - W_2)^\varphi$, so by Proposition 3.4, we are done. □

To prove the triangle inequality, we would like to take a coupling of Ω_1 and Ω_2 and a coupling of Ω_2 and Ω_3, and use them to obtain a coupling of Ω_1 and Ω_3. Unfortunately this cannot be done for general signed graphexes. We can, however do it if the signed graphexes involved are step graphexes. To define these, we first define a *subspace partition* of a measure space $\boldsymbol{\Omega} = (\Omega, \mathcal{F}, \mu)$ as a partition of a measurable subset $\Omega' \subseteq \Omega$ into countably many measurable subsets. Such a subspace partition is called finite if it is a partition into finitely many sets of finite measure. A signed graphex $\mathbb{U}$ is then called a *step graphex* over the subspace partition $\mathcal{P} = (P_1, \ldots, P_m)$ if $\mathcal{P}$ is a finite subspace partition, dsupp $\mathbb{U} \subseteq P_1 \cup P_2 \cup \cdots \cup P_m$, and for all $x, x' \in P_i$, $S(x) = S(x')$ and $W_x = W_{x'}$, where, as before, W_x is the function $y \mapsto W(x, y)$.

Remark 3.6 Given a signed step graphex $\mathbb{W} = (W, S, I, \boldsymbol{\Omega})$ over a finite subspace partition $\mathcal{P} = (P_1, P_2, \ldots, P_m)$, we can define another signed graphex $\mathbb{W}' = (W', S', I', \boldsymbol{\Omega}')$ with $\boldsymbol{\Omega}' = (\Omega', \mathcal{F}', \mu')$, where $\Omega' = [m]$, $\mathcal{F}'$ consists of all subsets, and the measure is defined by $\mu'(\{i\}) = \mu(P_i)$. Setting $I' = I$, $S'(i) = S(x)$ for any $x \in P_i$ (they are all equal), and $W'(i, j) = W(x, y)$ for $x \in P_i$, $y \in P_j$ (again the choice of x and y does not matter), we obtain that $\mathbb{W} = (\mathbb{W}')^\varphi$ where $\varphi \colon \mathbf{\Omega} \to \mathbf{\Omega}'$ is the map with $\varphi(x) = i$ for $x \in P_i$. In particular, by Remark 2.16, the distance between $\mathbb{W}$ and $\mathbb{W}'$ is zero (for any of the distance notions). Suppose now that $\mathbb{W}'' = (W'', S'', I'', \mathbf{\Omega}'')$ with $\mathbf{\Omega}'' = (\Omega'', \mathcal{F}'', \mu'')$ is another signed step graphex over a finite subspace partition $\mathcal{Q} = \{Q_1, Q_2, \ldots, Q_m\}$, with $I'' = I$, $\mu''(Q_i) = \mu(P_i)$, and $S''(x'') = S(x)$, $W''(x'', y'') = W(x, y)$ for $x \in P_i$, $x'' \in Q_i$, $y \in P_j$, $y'' \in Q_j$. Then $\mathbf{\Omega}''$ can also be mapped to $\mathbf{\Omega}'$ so that $\mathbb{W}''$ is the pullback of $\mathbb{W}'$ (by mapping Q_i to i). This implies that the distance of both $\mathbb{W}$ and $\mathbb{W}''$ from $\mathbb{W}'$ is 0, which (by the still to be proven triangle inequality) implies that their distance from each other is 0 (again for any of the notions of distance).

Returning to the proof of the triangle inequality, we will in fact consider signed graphexes that are countable step graphexes, i.e., the number of "steps" is countable, and each step has finite measure. First, however, we need the following technical lemma:

Lemma 3.7 *Let $\mathbb{W}_1 = (W_1, S_1, I_1, \mathbf{\Omega})$ and $\mathbb{W}_2 = (W_2, S_2, I_2, \mathbf{\Omega})$ be signed graphexes in $L^1 \cap L^2$. Assume that both are countable step graphexes on $\mathbf{\Omega} = (\Omega, \mathcal{F}, \mu)$ with common refinement $\mathcal{P} = \{P_1, P_2, \ldots, P_m, \ldots\}$, suppose μ' is another measure on Ω with $\mu(P_i) = \mu'(P_i)$, and let $\mathbf{\Omega}' = (\Omega, \mathcal{F}, \mu')$ and $\mathbb{W}'_i = (W_i, S_i, I_i, \mathbf{\Omega}')$. Then*

$$\Delta_{2\to 2}(\mathbb{W}_1, \mathbb{W}_2) = \Delta_{2\to 2}(\mathbb{W}'_1, \mathbb{W}'_2) \quad \text{and} \quad \Delta_\boxtimes(\mathbb{W}_1, \mathbb{W}_2) = \Delta_\boxtimes(\mathbb{W}'_1, \mathbb{W}'_2).$$

Proof Let $\mathbf{\Omega}_\mathcal{P} = (\Omega_\mathcal{P}, \mathcal{F}_\mathcal{P}, \mu_\mathcal{P})$ where $\Omega_\mathcal{P} = \{x_1, x_2, \ldots, x_m, \ldots\}$, $\mathcal{F}_\mathcal{P}$ is the set of all subsets of $\Omega_\mathcal{P}$, and $\mu_\mathcal{P}(x_i) = \mu(P_i)$. Let $\varphi, \varphi' \colon \Omega \to \Omega_\mathcal{P}$ with $\varphi(x) = \varphi'(x) = x_i$ for $x \in P_i$. Then $\varphi \colon \mathbf{\Omega} \to \mathbf{\Omega}_\mathcal{P}$ and $\varphi' \colon \mathbf{\Omega}' \to \mathbf{\Omega}_\mathcal{P}$ are both measure preserving (these are the same function on Ω but as maps between measure spaces are different). Define $\mathbb{W}_{i,\mathcal{P}} = (W_{i,\mathcal{P}}, S_{i,\mathcal{P}}, I_i)$ with $S_{i,\mathcal{P}}(x_j) = S_i(x)$ for any $x \in P_j$ (they are all equal), and $W_{i,\mathcal{P}}(x_j, x_k) = W(x, y)$ for $x \in P_j$, $y \in P_k$ (again, they are all equal). Then $\mathbb{W}^\varphi_{i,\mathcal{P}} = \mathbb{W}_i$ and $\mathbb{W}^{\varphi'}_{i,\mathcal{P}} = \mathbb{W}'_i$. Therefore, by Lemma 3.5, we have

$$\Delta_{2\to 2}(\mathbb{W}_1, \mathbb{W}_2) = \Delta_{2\to 2}(\mathbb{W}_{1,\mathcal{P}}, \mathbb{W}_{2,\mathcal{P}}) = \Delta_{2\to 2}(\mathbb{W}'_1, \mathbb{W}'_2),$$

and similarly for $\Delta_\boxtimes$. $\square$

To state the next lemma, we use the symbol $\pi_{ij,k}$ to denote the coordinate projection from a product space $\Omega_i \times \Omega_j$ to Ω_k, where $k = i$ or $k = j$.

Lemma 3.8 *Let $\mathbb{W}_i = (W_i, S_i, I_i, \mathbf{\Omega}_i)$, for $i = 1, 2, 3$, be countable step graphexes in $L^1 \cap L^2$. Let μ_{12} be a coupling measure on $\Omega_1 \times \Omega_2$, and μ_{23} be a coupling measure on $\Omega_2 \times \Omega_3$. Then there exists a coupling measure μ_{13} on Ω_1 and Ω_3 such that*

$$\Delta_{2\to 2}(\mathbb{W}_1^{\pi_{13,1},\mu_{13}}, \mathbb{W}_3^{\pi_{13,3},\mu_{13}})$$
$$\leq \Delta_{2\to 2}(\mathbb{W}_1^{\pi_{12,1},\mu_{12}}, \mathbb{W}_3^{\pi_{12,2},\mu_{12}}) + \Delta_{2\to 2}(\mathbb{W}_1^{\pi_{23,2},\mu_{23}}, \mathbb{W}_3^{\pi_{23,3},\mu_{23}})$$

and

$$\Delta_{\boxtimes}(\mathbb{W}_1^{\pi_{13,1},\mu_{13}}, \mathbb{W}_3^{\pi_{13,3},\mu_{13}})$$
$$\leq \Delta_{\boxtimes}(\mathbb{W}_1^{\pi_{12,1},\mu_{12}}, \mathbb{W}_3^{\pi_{12,2},\mu_{12}}) + \Delta_{\boxtimes}(\mathbb{W}_1^{\pi_{23,2},\mu_{23}}, \mathbb{W}_3^{\pi_{23,3},\mu_{23}}),$$

where the inequalities hold coordinate-wise.

Proof Let the steps of $\mathbb{W}_1$ be $A_1, A_2, \ldots$, the steps of $\mathbb{W}_2$ be $B_1, B_2, \ldots$, and the steps of $\mathbb{W}_3$ be $C_1, C_2, \ldots$. Without loss of generality, we may assume that each $\mu_1(A_p) > 0$, each $\mu_2(B_q) > 0$, and each $\mu_3(C_r) > 0$. First, take the measure μ_{123} on $\Omega_1 \times \Omega_2 \times \Omega_3$ where

$$\mu_{123}(E) = \sum_{p,q,r} \frac{\mu_{12}(A_p \times B_q)\mu_{23}(B_q \times C_r)}{\mu_1(A_p)\mu_2(B_q)^2\mu_3(C_r)} (\mu_1 \times \mu_2 \times \mu_3)(E \cap A_p \times B_q \times C_r).$$

Then

$$\mu_{123}(A_{p_0} \times B_{q_0} \times \Omega_3) = \sum_{p,q,r} \frac{\mu_{12}(A_p \times B_q)\mu_{23}(B_q \times C_r)}{\mu_1(A_p)\mu_2(B_q)^2\mu_3(C_r)}$$
$$\cdot (\mu_1 \times \mu_2 \times \mu_3)(A_{p_0} \times B_{q_0} \times \Omega_3 \cap A_p \times B_q \times C_r)$$
$$= \mu_{12}(A_{p_0} \times B_{q_0}) \sum_r \frac{\mu_{23}(B_{q_0} \times C_r)}{\mu_1(A_{p_0})\mu_2(B_{q_0})^2\mu_3(C_r)}\mu_1(A_{p_0})\mu_2(B_{q_0})\mu_3(C_r)$$
$$= \mu_{12}(A_{p_0} \times B_{q_0}) \sum_r \frac{\mu_{23}(B_{q_0} \times C_r)}{\mu_2(B_{q_0})} = \mu_{12}(A_{p_0} \times B_{q_0}).$$

In other words, if $\pi_{123,12}$ is the projection from $\Omega_1 \times \Omega_2 \times \Omega_3$ to $\Omega_1 \times \Omega_2$ and $\mu'_{12} = \mu_{123}^{\pi_{123,12}}$, then $\mu'_{12}(A_p \times B_q) = \mu_{12}(A_p \times B_q)$. Analogously, if $\mu'_{23} = \mu_{123}^{\pi_{123,23}}$, then $\mu'_{23}(A_p \times B_q) = \mu_{23}(A_p \times B_q)$. Furthermore, for any $F \subseteq \Omega_1$,

$$\mu_{123}(F \times \Omega_2 \times \Omega_3)$$
$$= \sum_{p,q,r} \frac{\mu_{12}(A_p \times B_q)\mu_{23}(B_q \times C_r)}{\mu_1(A_p)\mu_2(B_q)^2\mu_3(C_r)}(\mu_1 \times \mu_2 \times \mu_3)(F \times \Omega_2 \times \Omega_3 \cap A_p \times B_q \times C_r)$$
$$= \sum_{p,q,r} \frac{\mu_{12}(A_p \times B_q)\mu_{23}(B_q \times C_r)}{\mu_1(A_p)\mu_2(B_q)^2\mu_3(C_r)}\mu_1(F \cap A_p)\mu_2(B_q)\mu_3(C_r)$$
$$= \sum_{p,q,r} \frac{\mu_{12}(A_p \times B_q)\mu_{23}(B_q \times C_r)}{\mu_1(A_p)\mu_2(B_q)}\mu_1(F \cap A_p) = \sum_{p,q} \frac{\mu_{12}(A_p \times B_q)}{\mu_1(A_p)}\mu_1(F \cap A_p)$$
$$= \sum_p \frac{\mu_{12}(A_p \times \Omega_2)}{\mu_1(A_p)}\mu_1(F \cap A_p) = \sum_p \mu_1(F \cap A_p) = \mu_1(F).$$

Analogously, for any $G \subseteq \Omega_3$, $\mu_{123}(\Omega_1 \times \Omega_2 \times G) = \mu_3(G)$. Therefore, $\mu_{13} = \mu_{123}^{\pi_{123,13}}$ is a coupling measure on $\Omega_1 \times \Omega_3$ of μ_1 and μ_3. By Lemma 3.5 and the triangle inequality for the coordinates of $\Delta_{2\to 2}$, we then have

$$\Delta_{2\to 2}(\mathbb{W}_1^{\pi_{13,1},\mu_{13}}, \mathbb{W}_3^{\pi_{13,3},\mu_{13}})$$
$$= \Delta_{2\to 2}(\mathbb{W}_1^{\pi_{123,1},\mu_{123}}, \mathbb{W}_3^{\pi_{123,3},\mu_{123}})$$
$$\leq \Delta_{2\to 2}(\mathbb{W}_1^{\pi_{123,1},\mu_{123}}, \mathbb{W}_2^{\pi_{123,2},\mu_{123}}) + \Delta_{2\to 2}(\mathbb{W}_2^{\pi_{123,2},\mu_{123}}, \mathbb{W}_3^{\pi_{123,3},\mu_{123}})$$
$$= \Delta_{2\to 2}(\mathbb{W}_1^{\pi_{12,1},\mu'_{12}}, \mathbb{W}_2^{\pi_{12,2},\mu'_{12}}) + \Delta_{2\to 2}(\mathbb{W}_2^{\pi_{23,2},\mu'_{23}}, \mathbb{W}_3^{\pi_{23,3},\mu'_{23}})$$
$$= \Delta_{2\to 2}(\mathbb{W}_1^{\pi_{12,1},\mu_{12}}, \mathbb{W}_2^{\pi_{12,2},\mu_{12}}) + \Delta_{2\to 2}(\mathbb{W}_2^{\pi_{23,2},\mu_{23}}, \mathbb{W}_3^{\pi_{23,3},\mu_{23}}).$$

The proof for $\Delta_\boxtimes$ is the same. □

We are now ready to prove that $\widetilde{\delta}_{2\to 2}$ and the distance $\widetilde{\delta}_\boxtimes$ (obtained by replacing $d_{2\to 2}$ in (3) with $d_\boxtimes$) obey the triangle inequality.

Lemma 3.9 *Suppose that $\mathbb{W}_i = (W_i, S_i, I_i, \boldsymbol{\Omega}_i)$ with $\boldsymbol{\Omega}_i = (\Omega_i, \mathcal{F}_i, \mu_i)$, for $i = 1, 2, 3$, are signed graphexes in $L^1 \cap L^2$, and assume that $\mu_1(\Omega_1) = \mu_2(\Omega_2) = \mu_3(\Omega_3)$. Then*

$$\widetilde{\delta}_{2\to 2}(\mathbb{W}_1, \mathbb{W}_3) \leq \widetilde{\delta}_{2\to 2}(\mathbb{W}_1, \mathbb{W}_2) + \widetilde{\delta}_{2\to 2}(\mathbb{W}_2, \mathbb{W}_3)$$

and

$$\widetilde{\delta}_\boxtimes(\mathbb{W}_1, \mathbb{W}_3) \leq \widetilde{\delta}_\boxtimes(\mathbb{W}_1, \mathbb{W}_2) + \widetilde{\delta}_\boxtimes(\mathbb{W}_2, \mathbb{W}_3).$$

Proof We first claim that it is enough to prove that for any coupling measure μ_{12} on $\Omega_1 \times \Omega_2$, any coupling measure μ_{23} on $\Omega_2 \times \Omega_3$ and any $\varepsilon > 0$, there exists a coupling measure μ_{13} on $\Omega_1 \times \Omega_3$, such that

$$\Delta_{2\to 2}(\mathbb{W}_1^{\pi_{13,1},\mu_{13}}, \mathbb{W}_3^{\pi_{13,3},\mu_{13}})$$
$$\leq \Delta_{2\to 2}(\mathbb{W}_1^{\pi_{12,1},\mu_{12}}, \mathbb{W}_2^{\pi_{12,2},\mu_{12}}) + \Delta_{2\to 2}(\mathbb{W}_2^{\pi_{23,2},\mu_{23}}, \mathbb{W}_3^{\pi_{23,3},\mu_{23}}) + (\varepsilon, \varepsilon, \varepsilon) \quad (8)$$

and

$$\Delta_\boxtimes(\mathbb{W}_1^{\pi_{13,1},\mu_{13}}, \mathbb{W}_3^{\pi_{13,3},\mu_{13}})$$
$$\leq \Delta_\boxtimes(\mathbb{W}_1^{\pi_{12,1},\mu_{12}}, \mathbb{W}_2^{\pi_{12,2},\mu_{12}}) + \Delta_\boxtimes(\mathbb{W}_2^{\pi_{23,2},\mu_{23}}, \mathbb{W}_3^{\pi_{23,3},\mu_{23}}) + (\varepsilon, \varepsilon, \varepsilon). \quad (9)$$

Indeed, given that $\varepsilon > 0$ is arbitrary, (9) clearly implies the triangle inequality for $\widetilde{\delta}_\boxtimes$. To see that (8) implies the triangle inequality for $\widetilde{\delta}_{2\to 2}$, observe that $(x + y)^{1/k} \leq x^{1/k} + y^{1/k}$ whenever $k \geq 1$.

Next, we claim that for any $\varepsilon > 0$, any $\mathbb{W} = (W, S, I, \boldsymbol{\Omega})$ in $L^1 \cap L^2$ can be approximated by a signed step graphex $\mathbb{W}'$ such that

$$\Delta_\boxtimes(\mathbb{W}, \mathbb{W}') \leq \Delta_{2\to 2}(\mathbb{W}, \mathbb{W}') \leq (\varepsilon, \varepsilon, \varepsilon).$$

Indeed, let $\boldsymbol{\Omega} = (\Omega, \mathcal{F}, \mu)$, let $\Omega_1 \subseteq \Omega_2 \subseteq \cdots \subseteq \Omega$ be such that $\mu(\Omega_n) < \infty$ and $\Omega = \bigcup_n \Omega_n$, and let $\mathbb{W}_n = (W_n, S_n, I, \boldsymbol{\Omega})$, where $W_n = W 1_{\Omega_n \times \Omega_n}$ and $S_n = S 1_{\Omega_n}$. Using the dominated convergence theorem and the assumption that $\mathbb{W}$ is in $L^1 \cap L^2$, we then have that

$$|\rho(\mathbb{W}) - \rho(\mathbb{W}_n)| \le \|W - W_n\|_1 = \||W|(1 - 1_{\Omega_n \times \Omega_n})\|_1 \to 0,$$

$$\|W - W_n\|_{2 \to 2} \le \|W - W_n\|_2 = \|W(1 - 1_{\Omega_n \times \Omega_n})\|_2 \to 0,$$

and

$$\|S - S_n\|_2 = \|S(1 - 1_{\Omega_n})\|_2 \to 0$$

as $n \to \infty$. Next, defining χ_n by $\chi_n(x, y, z) = (1 - 1_{\Omega_n \times \Omega_n}(x, y))(1 - 1_{\Omega_n \times \Omega_n}(y, z))$, we bound

$$\|D_W - D_{W_n}\|_2 \le \sqrt{\int |W(x, y)||W(y, z)|\chi_n(x, y, z)\, d\mu(x)\, d\mu(y)\, d\mu(z)}.$$

Since χ_n goes to zero pointwise and $D_{|W|}$ is in L^2, the right side again goes to zero by the dominated convergence theorem. Therefore,

$$\|D_\mathbb{W} - D_{\mathbb{W}_n}\|_2 \le \|D_W - D_{W_n}\|_2 + \|S - S_n\|_2 \to 0.$$

This shows that for n large enough $\Delta_{2 \to 2}(\mathbb{W}, \mathbb{W}_n) \le \varepsilon/4$.

Fixing n such that this holds, we now define $W^{(k)} = W_n 1_{|W_n| \le k}$ and $S^{(k)} = S_n 1_{|S_n| \le k}$. Another application of the dominated convergence theorem then shows that for k large enough, $\Delta_{2 \to 2}(\mathbb{W}^{(k)}, \mathbb{W}_n) \le \varepsilon/4$, giving us a graphex $\mathbb{W}'' = (W'', S'', I, \boldsymbol{\Omega})$ such that the degree support of $\mathbb{W}''$ has finite measure, both W'' and $D_{\mathbb{W}''}$ are bounded, and $\Delta_{2 \to 2}(\mathbb{W}'', \mathbb{W}) \le \varepsilon/2$. But such a graphex can be approximated to arbitrary precision by a step graphex with finitely many steps, proving the claim for $\Delta_{2 \to 2}$. Since on two variable functions, $\|\cdot\|_\boxtimes$ is bounded by $\|\cdot\|_{2 \to 2}$, and on functions of one variable it is bounded by $\|\cdot\|_2$, the claim for $\Delta_\boxtimes$ follows as well.

Fix $\varepsilon > 0$, and let $\mathbb{W}'_1, \mathbb{W}'_2, \mathbb{W}'_3$ be approximations of $\mathbb{W}_1, \mathbb{W}_2, \mathbb{W}_3$ by signed step graphexes such that for $k = 1, 2, 3$,

$$\Delta_{2 \to 2}(\mathbb{W}_k, \mathbb{W}'_k) \le (\varepsilon/6, \varepsilon/6, \varepsilon/6).$$

If μ_{ij} is a coupling measure on $\Omega_i \times \Omega_j$ and $\pi_{ij,k}$ is the projection onto Ω_k, $k = i$ or j, then $\pi_{ij,k}$ is measure preserving. Therefore, by Lemma 3.5,

$$\Delta_{2 \to 2}\left((\mathbb{W}_k)^{\pi_{ij,k}, \mu_{ij}}, (\mathbb{W}'_k)^{\pi_{ij,k}, \mu_{ij}}\right) \le (\varepsilon/6, \varepsilon/6, \varepsilon/6).$$

Combined with Lemma 3.8, we conclude that there exists a coupling measure μ_{13} on $\Omega_1 \times \Omega_3$ such that

$$\Delta_{2\to 2}(\mathbb{W}_1^{\pi_{13,1},\mu_{13}}, \mathbb{W}_3^{\pi_{13,3},\mu_{13}}) \leq \Delta_{2\to 2}\left((\mathbb{W}_1')^{\pi_{13,1},\mu_{13}}, (\mathbb{W}_3')^{\pi_{13,3},\mu_{13}}\right) + (\varepsilon/3, \varepsilon/3, \varepsilon/3)$$

$$\leq \Delta_{2\to 2}\left((\mathbb{W}_1')^{\pi_{12,1},\mu_{12}}, (\mathbb{W}_2')^{\pi_{12,2},\mu_{12}}\right) + \Delta_{2\to 2}\left((\mathbb{W}_2')^{\pi_{23,2},\mu_{23}}, (\mathbb{W}_3')^{\pi_{23,3},\mu_{23}}\right)$$

$$+ (\varepsilon/3, \varepsilon/3, \varepsilon/3)$$

$$\leq \Delta_{2\to 2}(\mathbb{W}_1^{\pi_{12,1},\mu_{12}}, \mathbb{W}_2^{\pi_{12,2},\mu_{12}}) + \Delta_{2\to 2}(\mathbb{W}_2^{\pi_{23,2},\mu_{23}}, \mathbb{W}_3^{\pi_{23,3},\mu_{23}}) + (\varepsilon, \varepsilon, \varepsilon)$$

proving (8) and hence the first statement of the lemma. The proof of (9) and the second statement follows in the same way. □

The proof of Theorems 2.11 and 3.1 will be an easy corollary of Lemma 3.9 and the following extension lemma.

Lemma 3.10 *Let* $\mathbb{W} = (W, S, I, \boldsymbol{\Omega})$ *be a signed graphex in* $L^1 \cap L^2$, *with possibly unbounded graphon parts, and let* $\boldsymbol{\Omega} = (\Omega, \mathcal{F}, \mu)$.

(1) *If* $\mathbb{W}'$ *and* $\mathbb{W}''$ *are trivial extensions of* $\mathbb{W}$ *by* σ-*finite spaces of infinite measure, then*

$$\widetilde{\delta}_{\boxtimes}(\mathbb{W}', \mathbb{W}'') = \widetilde{\delta}_{2\to 2}(\mathbb{W}', \mathbb{W}'') = 0.$$

(2) *If* $\mu(\Omega) = \infty$ *and* $\widetilde{\mathbb{W}} = (\widetilde{W}, \widetilde{S}, I, \widetilde{\boldsymbol{\Omega}})$ *is obtained from* $\mathbb{W}$ *by appending an arbitrary* σ-*finite space of infinite measure, then*

$$\widetilde{\delta}_{2\to 2}(\mathbb{W}, \widetilde{\mathbb{W}}) = \widetilde{\delta}_{\boxtimes}(\mathbb{W}, \widetilde{\mathbb{W}}) = 0.$$

Proof To prove the first statement, let $\boldsymbol{\Omega}' = (\Omega', \mathcal{F}', \mu')$ and $\boldsymbol{\Omega}'' = (\Omega'', \mathcal{F}'', \mu'')$ be the spaces $\boldsymbol{\Omega}$ has been extended by. Let $\widehat{\mu}$ be the measure on $\Omega \times \Omega$ which couples μ to itself along the diagonal, choose an arbitrary coupling $\widetilde{\mu}$ of μ' and μ'', and let $\widehat{\mu}'$ be the measure on $(\Omega \cup \Omega') \times (\Omega \cup \Omega'')$ defined by

$$\widehat{\mu}'(A) = \widehat{\mu}(A \cap (\Omega \times \Omega)) + \widetilde{\mu}(A \cap (\Omega' \times \Omega'')).$$

Using the fact that $\widehat{\mu}'(\Omega \times \Omega'') = \widehat{\mu}'(\Omega' \times \Omega) = 0$, it is easy to see that

$$\widetilde{\delta}_{2\to 2}(\mathbb{W}', \mathbb{W}'') \leq d_{2\to 2}((\mathbb{W}')^{\pi_1, \widehat{\mu}'}, (\mathbb{W}'')^{\pi_2, \widehat{\mu}'}) = d_{2\to 2}(\mathbb{W}^{\pi_1, \widehat{\mu}}, \mathbb{W}^{\pi_2, \widehat{\mu}}) = 0.$$

This proves the first statement for the metric $\widetilde{\delta}_{2\to 2}$. The proof for the metric $\widetilde{\delta}_{\boxtimes}$ is identical.

To prove the second statement, let $\widetilde{\boldsymbol{\Omega}} = (\widetilde{\Omega}, \widetilde{\mathcal{F}}, \widetilde{\mu})$. Since $\boldsymbol{\Omega}$ is σ-finite, we can find a sequence of measurable subsets $\Omega_n \subseteq \Omega$ such that $\Omega = \bigcup \Omega_n$ and each Ω_n has finite measure. Replacing Ω_n by $\Omega_1 \cup \cdots \cup \Omega_n$, we may further assume that Ω_n is an increasing sequence of sets. Let W_n be equal to W on $\Omega_n \times \Omega_n$ and 0 everywhere else, and let $S_n = S$ on Ω_n and 0 outside of Ω_n. Let $\mathbb{W}_n$ be the corresponding graphex on $\boldsymbol{\Omega}$ (with the same value I), and $\widetilde{\mathbb{W}}_n$ be its trivial extension to $\widetilde{\boldsymbol{\Omega}}$. By monotone convergence, $W_n \to W$ in both L^1 and L^2, and $S_n \to S$ in L^1 and L^2, implying that $\tilde\delta_{2\to 2}(\mathbb{W}_n, \mathbb{W}) \le d_{2\to 2}(W_n, W) \to 0$. For the same reason, $\tilde\delta_{2\to 2}(\widetilde{\mathbb{W}}_n, \widetilde{\mathbb{W}}) \to 0$. But since $\widetilde{\mathbb{W}}_n$ and $\mathbb{W}_n$ can both be obtained from the restriction of W_n to Ω_n by appending a space of infinite total measure, we have $\tilde\delta_{2\to 2}(\mathbb{W}_n, \widetilde{\mathbb{W}}_n) = \tilde\delta_{2\to 2}(\mathbb{W}_n, \mathbb{W}_n) = 0$ by the first statement of the lemma. Using the triangle inequality for $\tilde\delta_{2\to 2}$, this proves the second statement for the distance $\tilde\delta_{2\to 2}$. The proof for the metric $\tilde\delta_\boxtimes$ follows from the fact that the jumble norm is bounded by the kernel norm, which in turn implies that $d_\boxtimes(W_n, W) \to 0$ whenever $d_{2\to 2}(W_n, W) \to 0$. □

We are now ready to prove Theorems 2.11 and 3.1.

Proof (*Theorems 2.11 and 3.1*) The first statement of Lemma 3.10 implies that if $\mathbb{W}'_1$ and $\mathbb{W}'_2$ are trivial extensions of $\mathbb{W}_1$ and $\mathbb{W}_2$ obtained by *appending* two σ-finite spaces of infinite measure, then $\tilde\delta_{2\to 2}(\mathbb{W}'_1, \mathbb{W}'_2)$ and $\tilde\delta_\boxtimes(\mathbb{W}'_1, \mathbb{W}'_2)$ do not depend on the choice of these extensions, and the second (combined with the triangle inequality) allows us to conclude that this remains true for extensions *to* spaces of infinite measure, which completes the proof of the first statements of the two theorems.

Since clearly $\tilde\delta_{2\to 2}$ is symmetric and $\tilde\delta_{2\to 2}(\mathbb{W}, \mathbb{W}) = 0$ for all integrable graphexes, all that remains to be proved is the triangle inequality for $\tilde\delta_{2\to 2}$, which follows from the (already established) triangle inequality for $\tilde\delta_{2\to 2}$. The same holds for $\tilde\delta_\boxtimes$. □

The following example shows that the extension to infinite spaces in the definition of $\delta_{2\to 2}$ is really needed.

Example 3.11 Let $\mathbb{W}_1 = (W_1, 0, 0, \Omega_1)$ where Ω_1 consists of just two atoms a and b, with weight p and $1 - p$, where $0 < p < 1/2$, and $W_1(a, a) = W_1(b, b) = 0$, $W_1(a, b) = W_1(b, a) = 1$. Furthermore, let $\mathbb{W}_2 = (W_2, 0, 0, \Omega_2)$ where Ω_2 consists of just one atom c with weight 1, and W_2 is the constant $a = \sqrt{p(1-p)}$. Then we have just one choice of coupling. For this coupling, $W_1^{\pi_1} - W_2^{\pi_2}$ will have two atoms, and it will be equal to $-\sqrt{p(1-p)}$ on the diagonal, and $1 - \sqrt{p(1-p)}$ off the diagonal. It is then not difficult to see that $\|W_1 - W_2\|_{2\to 2}$ is equal to the largest eigenvalue (in absolute value) of the matrix

$$\begin{pmatrix} -pa & (1-p)(1-a) \\ p(1-a) & -(1-p)a \end{pmatrix}.$$

The trace of this matrix is $-a$, and the determinant is

$$p(1-p)a^2 - p(1-p)(1-a)^2 = p(1-p)(2a-1) < 0.$$

Here we used that $a = \sqrt{p(1-p)} < 1/2$. We then have that the two eigenvalues of the matrix have opposite signs, and their sum is $-a$, which implies that the negative one must be less than $-a$; i.e., it must have larger absolute value than a.

On the other hand, clearly $\|W_1\|_{2\to 2} = a$, and $\|W_2\|_{2\to 2}$ is equal to the largest eigenvalue (in absolute value) of the matrix

$$\begin{pmatrix} 0 & 1-p \\ p & 0 \end{pmatrix},$$

which can easily be seen to be equal to $\sqrt{p(1-p)} = a$.

Therefore, if we extend W_1 and W_2 by spaces of total measure at least 1, and couple each Ω_i to the extension, then

$$\|\widetilde{W}_1 - \widetilde{W}_2\|_{2\to 2} = \max\{\|\widetilde{W}_1\|_{2\to 2}, \|\widetilde{W}_2\|_{2\to 2}\} = a.$$

Therefore, by extending, we can obtain a better coupling.

Finally, note that in (1), if we multiply the measure of the underlying space by c, then the first term is multiplied by c, the second by $c^{3/4}$, and the third by $c^{2/3}$. We can therefore take c large enough so that the maximum in the term is dominated by $\|W_1 - W_2\|_{2\to 2}$. Therefore, we obtain that also for minimizing $d_{2\to 2}$, we obtain a better coupling if we extend by trivial extensions than if we do not.

Next, we would like to prove Theorem 2.15. Before doing so, we note that the distance $\delta_\diamond$ is well defined and finite for signed graphexes as well, provided the graphon part is bounded in the L^∞ norm. As the reader may easily verify, this immediately follows from the following lemma, which is an easy corollary to Proposition 2.4.

Lemma 3.12 *Let $\mathbb{W}$ be a signed graphex with bounded graphon part, and let $0 < D < \infty$. Then the set $\{D_{|\mathbb{W}|} > D\}$ has finite measure, and $\mathbb{W}_{|\{D_{|\mathbb{W}|} \leq D\}}$ is integrable and hence in $L^1 \cap L^2$.*

Proof Let $\mathbb{W} = (W, S, I, \mathbf{\Omega})$ with $\mathbf{\Omega} = (\Omega, \mathcal{F}, \mu)$, and assume that $\|W\|_\infty \leq K$. If $K \leq 1$, the lemma follows from Proposition 2.4 applied to $\mathbb{W}' = |\mathbb{W}|$. Otherwise, we define $\mathbb{W}' = (W', S', I', \mathbf{\Omega})$ where $W' = |W|/K$, $S' = |S|$, and $I' = |I|$. Applying Proposition 2.4 to this graphex, and noting that $\frac{1}{K} D_{|\mathbb{W}|} \leq D_{\mathbb{W}'} \leq D_{|\mathbb{W}|}$, we see that $\mu(\{D_{|\mathbb{W}|} > D\}) \leq \mu(\{D_{|\mathbb{W}'|} > D/K\}) < \infty$ and $\|\mathbb{W}_{|D_{|\mathbb{W}|} \leq D}\|_1 \leq \|\mathbb{W}_{|D_{\mathbb{W}'} \leq D}\|_1 \leq K\|\mathbb{W}'_{|D_{\mathbb{W}'} \leq D}\|_1 < \infty$, as claimed. □

To prove Theorem 2.15, we need to establish the triangle inequality. The reason it is not obvious is because when we decrease the measure on the underlying set, the $\delta_{2\to 2}$ distance can increase. In the following lemma, we show that although it can increase under restrictions to subsets or decreasing of the underlying measure, it cannot increase too much.

Lemma 3.13 *Let $\mathbb{W}_1 = (W_1, S_1, I_1, \mathbf{\Omega})$ and $\mathbb{W}_2 = (W_2, S_2, I_2, \mathbf{\Omega})$ be signed graphexes in $L^1 \cap L^2$, let $\mathbf{\Omega} = (\Omega, \mathcal{F}, \mu)$, and let*

(1) $\|W_1 - W_2\|_{2\to 2,\mu} = a$,
(2) $\|D_{\mathbb{W}_1} - D_{\mathbb{W}_2}\|_{2,\mu} = b$, *and*
(3) $|\rho(\mathbb{W}_1) - \rho(\mathbb{W}_2)| = c$.

If $\mathbb{W}'_1 = (W_1, S_1, I_1, \Omega')$ and $\mathbb{W}'_2 = (W_2, S_2, I_2, \Omega')$ where $\Omega' = (\Omega, \mathcal{F}, \mu')$ for some measure μ' such that $\mu - r \le \mu' \le \mu$ for some $r < \infty$, then

(1) $\|W_1 - W_2\|_{2 \to 2, \mu'} \le a$,
(2) $\|D_{\mathbb{W}'_1} - D_{\mathbb{W}'_2}\|_{2, \mu'} \le b + a\sqrt{r}$, and
(3) $\left| |\rho(\mathbb{W}'_1) - \rho(\mathbb{W}'_2)| - c \right| \le 2b\sqrt{r} + ar$.

We recall that $\mu - r \le \mu' \le \mu$ if and only if there exists a measurable function $h \colon \Omega \to [0, 1]$ such that $\mu'(B) = \int_B h \, d\mu$ for all measurable sets B, and $\|1 - h\|_{1,\mu} \le r < \infty$. An interesting special case is the case where h is the characteristic function of $\Omega \setminus R$ for a set R of measure r, in which case $\mathbb{W}'_i$ is the restriction of $\mathbb{W}$ to $\Omega \setminus R$, after neglecting points outside the degree support.

Proof To show property (1), let f, g be such that $\|f\|_{2,\mu'} = \|g\|_{2,\mu'} = 1$. In other words,
$$1 = \int_\Omega f^2 \, d\mu' = \int_\Omega hf^2 \, d\mu,$$
and
$$1 = \int_\Omega g^2 \, d\mu' = \int_\Omega hg^2 \, d\mu.$$

Let $U = W_1 - W_2$. We then have
$$\left| \int_{\Omega \times \Omega} f(x) U(x, y) g(y) \, d(\mu' \times \mu') \right| = \left| \int_{\Omega \times \Omega} f(x) U(x, y) g(y) h(x) h(y) \, d(\mu \times \mu) \right|$$
$$\le \|U\|_{2 \to 2, \mu} \|fh\|_{2,\mu} \|gh\|_{2,\mu}.$$

We also have
$$\|fh\|_{2,\mu}^2 = \int_\Omega f(x)^2 h(x)^2 \, d\mu(x) \le \int_\Omega f(x)^2 h(x) \, d\mu(x) = 1.$$

Similarly, $\|gh\|_{2,\mu} \le 1$. Therefore,
$$\left| \int_{\Omega \times \Omega} f(x) U(x, y) g(y) \, d(\mu' \times \mu') \right| \le a.$$

Since this holds for any f, g such that $\|f\|_{2,\mu'} = \|g\|_{2,\mu'} = 1$, we have $\|U\|_{2 \to 2, \mu'} \le \|U\|_{2 \to 2, \mu} = a$.

For (2), let
$$D_i(x) = D_{\mathbb{W}_i}(x) - D_{\mathbb{W}'_i}(x) = \int_\Omega W_i(x, y)(1 - h(y)) \, d\mu(y).$$

Then

$$\|D_{\mathbb{W}_1'} - D_{\mathbb{W}_2'}\|_{2,\mu'} \leq \|D_{\mathbb{W}_1'} - D_{\mathbb{W}_2'}\|_{2,\mu}$$
$$= \sup_{g:\|g\|_{2,\mu}=1} \int_\Omega (D_{\mathbb{W}_1'} - D_{\mathbb{W}_2'})(x)g(x)\,d\mu(x)$$
$$= \sup_{g:\|g\|_{2,\mu}=1} \int_\Omega \left((D_{\mathbb{W}_1} - D_{\mathbb{W}_2})(x)g(x) - (D_1 - D_2)(x)g(x)\right) d\mu(x)$$
$$= \sup_{g:\|g\|_{2,\mu}=1} \left(\int_\Omega (D_{\mathbb{W}_1} - D_{\mathbb{W}_2})(x)g(x)\,d\mu(x) \right.$$
$$\left. - \int_{\Omega\times\Omega} g(x)(W_1 - W_2)(x,y)(1 - h(y))\,d(\mu\times\mu) \right)$$
$$\leq b + a\|g\|_{2,\mu}\|1 - h\|_{2,\mu} \leq b + a\sqrt{r}.$$

To prove (3), we use that

$$\rho(\mathbb{W}_i') = \int_{\Omega\times\Omega} h(x)W_i(x,y)h(y)\,d\mu(x)\,d\mu(y) + 2\int_\Omega S_i(x)h(x)\,d\mu(x) + I_i$$
$$= \rho(\mathbb{W}_i) - \int_{\Omega\times\Omega} (1 - h(x))W_i(x,y)\,d\mu(x)\,d\mu(y)$$
$$- \int_{\Omega\times\Omega} W_i(x,y)(1 - h(y))\,d\mu(x)\,d\mu(y)$$
$$+ \int_{\Omega\times\Omega} (1 - h(x))W_i(x,y)(1 - h(y))\,d\mu(x)\,d\mu(y)$$
$$- 2\int_\Omega (1 - h(x))S(x)\,d\mu(x)$$
$$= \rho(\mathbb{W}_i) - 2\int_\Omega (1 - h(x))D_{\mathbb{W}_i}(x)\,d\mu(x)$$
$$+ \int_{\Omega\times\Omega} (1 - h(x))W_i(x,y)(1 - h(y))\,d\mu(x)\,d\mu(y).$$

Therefore,

$$\left| |\rho(\mathbb{W}_1') - \rho(\mathbb{W}_2')| - |\rho(\mathbb{W}_1) - \rho(\mathbb{W}_2)| \right|$$
$$\leq \left| 2\int_\Omega (1 - h(x))(D_{\mathbb{W}_1} - D_{\mathbb{W}_2})(x)\,d\mu(x) \right.$$
$$\left. - \int_{\Omega\times\Omega} (1 - h(x))(W_1 - W_2)(x,y)(1 - h(y))\,d\mu(x)\,d\mu(y) \right|$$
$$\leq 2b\sqrt{r} + ar.$$

Here we used the fact that $\|1 - h\|_2 \leq \sqrt{\|1 - h\|_1} \leq \sqrt{r}$. This implies the claim. □

We also use the following equivalent representation of the weak kernel distance $\delta_\diamond$.

Lemma 3.14 *For $i = 1, 2$, let $\mathbb{W}_i$ be graphexes over $\Omega_i = (\Omega_i, \mathcal{F}_i, \mu_i)$, and let $\mathbb{W}'_i$ be trivial extensions of $\mathbb{W}_i$ to σ-finite measure spaces $\Omega'_i = (\Omega'_i, \mathcal{F}'_i, \mu'_i)$ with $\mu'_i(\Omega'_i) = \infty$. Then $\delta_\diamond(\mathbb{W}_1, \mathbb{W}_2) = \tilde{\delta}_\diamond(\mathbb{W}'_1, \mathbb{W}'_2)$, where $\tilde{\delta}_\diamond(\mathbb{W}'_1, \mathbb{W}'_2)$ is defined as the infimum over all c such that there exists a measure μ' over $\Omega'_1 \times \Omega'_2$ obeying the conditions*

(1) $\mu'_i - c^2 \leq (\mu')^{\pi_i} \leq \mu'_i$ for $i = 1, 2$, and
(2) $d_{2 \to 2}((\mathbb{W}'_1)^{\pi_1, \mu'}, (\mathbb{W}'_2)^{\pi_2, \mu'}) \leq c$.

Proof Let μ' and c be such that they obey the conditions in the statement of the lemma. For $i = 1, 2$, let $\tilde{\mu}_i$ be the restriction of $(\mu')^{\pi_i}$ to Ω_i, and let $\tilde{\mathbb{W}}_i$ be obtained from $\mathbb{W}_i$ by replacing μ_i by $\tilde{\mu}_i$. Then $\mu_i - c^2 \leq \tilde{\mu}_i \leq \mu_i$. Furthermore, $(\mu')^{\pi_i}$ extends $\tilde{\mu}_i$ to Ω'_i, and $(\mu')^{\pi_i}(\Omega'_i \setminus \Omega_i) \geq \mu'_i(\Omega'_i \setminus \Omega_i) - c^2 = \infty$, showing that this defines an extension by a space of infinite measure. Finally, μ' is a coupling of $(\mu')^{\pi_1}$ and $(\mu')^{\pi_2}$. Together, these facts imply that $\delta_{2 \to 2}(\tilde{\mathbb{W}}_1, \tilde{\mathbb{W}}_2) \leq c$, proving that $\delta_\diamond(\mathbb{W}_1, \mathbb{W}_2) \leq c$. This shows that $\delta_\diamond(\mathbb{W}_1, \mathbb{W}_2) \leq \tilde{\delta}_\diamond(\mathbb{W}'_1, \mathbb{W}'_2)$.

To prove the reverse inequality, assume that c is such that there are measures $\tilde{\mu}_1$ and $\tilde{\mu}_2$ over Ω_1 and Ω_2 such that $\delta_{2 \to 2}(\tilde{\mathbb{W}}_1, \tilde{\mathbb{W}}_2) \leq c$ and $\mu_i - c^2 \leq \tilde{\mu}_i \leq \mu_i$ for $i = 1, 2$, where $\tilde{\mathbb{W}}_i$ is again obtained from $\mathbb{W}_i$ by replacing μ_i by $\tilde{\mu}_i$. If we transform Ω'_i into a space $\tilde{\Omega}'_i$ by setting $\tilde{\mu}'_i$ to $\tilde{\mu}_i$ on Ω_i, and to μ'_i on $\Omega_i \setminus \Omega'_i$, and define $\tilde{\mathbb{W}}'_i$ as the trivial extension of $\tilde{\mathbb{W}}_i$ to $\tilde{\Omega}'_i$, then Theorem 2.11 implies that $\tilde{\delta}_{2 \to 2}(\tilde{\mathbb{W}}'_1, \tilde{\mathbb{W}}'_2) = \delta_{2 \to 2}(\tilde{\mathbb{W}}_1, \tilde{\mathbb{W}}_2) \leq c$. This means that for all $\varepsilon > 0$, there is a coupling μ' of $\tilde{\mu}'_1$ and $\tilde{\mu}'_2$ such that $d_{2 \to 2}((\mathbb{W}'_1)^{\pi_1, \mu'}, (\mathbb{W}'_2)^{\pi_2, \mu'}) \leq c + \varepsilon$. Observing that the bound $\mu_i - c^2 \leq \tilde{\mu}_i \leq \mu_i$ and our construction of $\tilde{\mu}'_i$ imply that $\mu'_i - c^2 \leq \tilde{\mu}'_i = (\mu')^{\pi_i} \leq \mu'_i$, and that $(\tilde{\mathbb{W}}'_i)^{\pi_i, \tilde{\mu}'} = (\mathbb{W}'_i)^{\pi_i, \tilde{\mu}'}$, this shows that $\tilde{\delta}_\diamond(\mathbb{W}'_1, \mathbb{W}'_2) \leq c + \varepsilon$. Since ε was arbitrary, this shows that $\delta_\diamond(\mathbb{W}_1, \mathbb{W}_2) \geq \tilde{\delta}_\diamond(\mathbb{W}'_1, \mathbb{W}'_2)$. □

We are now ready to prove Theorem 2.15.

Proof (*Theorem* 2.15) It is clear that $\delta_\diamond$ is symmetric, and that $\delta_\diamond(\mathbb{W}, \mathbb{W}) = 0$. So we have to prove the triangle inequality. By Lemma 3.14, taking trivial extensions of each graphex to a space of infinite measure, it suffices to prove the triangle inequality for $\tilde{\delta}_\diamond$.

Let $\mathbb{W}_1, \mathbb{W}_2, \mathbb{W}_3$ be three graphexes with the usual notation, defined over measure spaces which all have infinite measure. Let μ_{12} be a measure on $\Omega_1 \times \Omega_2$ that shows that $\tilde{\delta}_\diamond(\mathbb{W}_1, \mathbb{W}_2) \leq c_1$, let μ_{23} be a measure on $\Omega_2 \times \Omega_3$ that shows that $\tilde{\delta}_\diamond(\mathbb{W}_1, \mathbb{W}_2) \leq c_2$, let μ'_1 and μ'_2 be the marginals of μ_{12}, and let μ''_2 and μ''_3 be the marginals of μ_{23}. We would like to use Lemma 3.9 to create a coupling of μ'_1 and μ''_3, but unfortunately, the conditions of the lemma require that $\mu'_2 = \mu''_2$, which we cannot guarantee. To deal with this problem, we will slightly decrease μ_{12} and μ_{23} so that after this perturbation, the second marginal of the first is equal to the first marginal of the second.

Let $\pi_{ij,i}$ be the projection map from $\Omega_i \times \Omega_j$ to Ω_i for $i, j \in [3]$, and let $\mathbb{W}_{ij,i} = \mathbb{W}_i^{\pi_{ij,i}, \mu_{ij}}$. Let $\mu'_2 = \mu_{12}^{\pi_{12,2}}$ and $\mu''_2 = \mu_{23}^{\pi_{23,2}}$. Let $h' = \frac{d\mu'_2}{d\mu_2}$ and $h'' = \frac{d\mu''_2}{d\mu_2}$. Then we can assume that $0 \leq h', h'' \leq 1$, $\|1 - h'\|_{1,\mu_2} \leq c_1^2$, and $\|1 - h''\|_{1,\mu_2} \leq c_2^2$. Let $\tilde{h}(x) = \min(h'(x), h''(x))$, and let $\tilde{\mu}_2$ be the measure defined by

$$\tilde{\mu}_2(A) = \int_A \tilde{h} \, d\mu_2.$$

Then $\|h' - \tilde{h}\|_{1,\mu_2} \leq \|1 - h''\|_{1,\mu_2} \leq c_2^2$. For $x \in \Omega_1 \times \Omega_2$, let $h_{12}(x) = \frac{\tilde{h}(\pi_{12,2}(x))}{h'(\pi_{12,2}(x))} \leq 1$, and let $\tilde{\mu}_{12}$ be the measure defined by

$$\tilde{\mu}_{12}(A) = \int_A h_{12}(x) \, d\mu_{12}.$$

Note that $\tilde{\mu}_{12}^{\pi_{12,2}} = \tilde{\mu}_2$. Furthermore, since $\tilde{h}(x) \leq h'(x)$,

$$\int_{\Omega_1 \times \Omega_2} (1 - h_{12}(x)) \, d\mu_{12}(x) = \int_{\Omega_2} \left(1 - \frac{\tilde{h}(x)}{h'(x)}\right) d\mu_2'(x)$$

$$= \int_{\Omega_2} \left(1 - \frac{\tilde{h}(x)}{h'(x)}\right) h'(x) \, d\mu_2(x) = \|h' - \tilde{h}\|_{1,\mu_2} \leq c_2^2.$$

This means that for any set $A \subseteq \Omega_1 \times \Omega_2$,

$$\mu_{12}(A) - c_2^2 \leq \tilde{\mu}_{12}(A) \leq \mu_{12}(A).$$

This implies that for any $A \subseteq \Omega_1$,

$$\mu_1(A) - c_1^2 - c_2^2 \leq \mu_{12}^{\pi_{12,1}}(A) - c_2^2 \leq \tilde{\mu}_{12}^{\pi_{12,1}}(A) \leq \mu_{12}^{\pi_{12,1}}(A) \leq \mu_1(A).$$

We similarly construct $\tilde{\mu}_{23}$ and $\Omega_2 \times \Omega_3$ so that $\tilde{\mu}_{23}^{\pi_{23,2}} = \tilde{\mu}_2$ and for any set $A \subseteq \Omega_2 \times \Omega_3$,

$$\mu_{23}(A) - c_1^2 \leq \tilde{\mu}_{23}(A) \leq \mu_{23}(A),$$

which implies that for any $A \subseteq \Omega_3$,

$$\mu_3(A) - c_1^2 - c_2^2 \leq \mu_{23}^{\pi_{23,3}}(A) - c_1^2 \leq \tilde{\mu}_{23}^{\pi_{23,3}}(A) \leq \mu_{23}^{\pi_{23,3}}(A) \leq \mu_3(A).$$

Let $\tilde{\mu}_1 = \tilde{\mu}_{12}^{\pi_{12,1}}$ and $\tilde{\mu}_3 = \tilde{\mu}_{23}^{\pi_{23,3}}$, and note that $\tilde{\mu}_{12}$ is a coupling of $\tilde{\mu}_1$ and $\tilde{\mu}_2$, $\tilde{\mu}_{23}$ is a coupling of $\tilde{\mu}_2$ and $\tilde{\mu}_3$, and $\mu_1 - c_1^2 - c_2^2 \leq \tilde{\mu}_1 \leq \mu_1$ and $\mu_3 - c_1^2 - c_2^2 \leq \tilde{\mu}_3 \leq \mu_3$.

Let $\tilde{\mathbb{W}}_i$ be equal to $\mathbb{W}_i$ but with the measure μ_i replaced by $\tilde{\mu}_i$. Fix $\varepsilon > 0$. By Lemma 3.9, there exists a measure $\tilde{\mu}_{13}$ on $\Omega_1 \times \Omega_3$ such that $\tilde{\mu}_{13}^{\pi_{13,1}} = \tilde{\mu}_1$ and $\tilde{\mu}_{13}^{\pi_{13,3}} = \tilde{\mu}_3$, and we have

$$\Delta_{2\to 2}(\tilde{\mathbb{W}}_1^{\pi_{13,1}}, \tilde{\mathbb{W}}_3^{\pi_{13,3}}) \leq \Delta_{2\to 2}(\tilde{\mathbb{W}}_1^{\pi_{12,1}}, \tilde{\mathbb{W}}_2^{\pi_{12,2}}) + \Delta_{2\to 2}(\tilde{\mathbb{W}}_2^{\pi_{23,2}}, \tilde{\mathbb{W}}_3^{\pi_{23,3}}) + (\varepsilon, \varepsilon, \varepsilon).$$

Note that by the above inequalities,

Identifiability for Graphexes and the Weak Kernel Metric

and

$$\mu_1 - (c_1+c_2)^2 \le \mu_1 - c_1^2 - c_2^2 \le \widetilde{\mu}_1 = \widetilde{\mu}_{13}^{\pi_{13,1}} \le \mu_1$$

$$\mu_3 - (c_1+c_2)^2 \le \mu_3 - c_1^2 - c_2^2 \le \widetilde{\mu}_3 = \widetilde{\mu}_{13}^{\pi_{13,3}} \le \mu_3.$$

By Lemma 3.13,

$$\|\widetilde{\mathbb{W}}_1^{\pi_{12,1}} - \widetilde{\mathbb{W}}_2^{\pi_{12,2}}\|_{2\to 2,\widetilde{\mu}_{12}} \le \|W_{12,1} - W_{12,2}\|_{2\to 2,\mu_{12}} \le c_1,$$

$$\|D_{\widetilde{\mathbb{W}}_1^{\pi_{12,1}}} - D_{\widetilde{\mathbb{W}}_2^{\pi_{12,2}}}\|_{2,\widetilde{\mu}_{12}} \le \|D_{W_{12,1}} - D_{W_{12,2}}\|_{2,\mu_{12}} + \|W_{12,1} - W_{12,2}\|_{2\to 2,\mu_{12}} c_2$$
$$\le c_1^2 + c_1 c_2,$$

and finally

$$|\rho(\widetilde{\mathbb{W}}_1^{\pi_{12,1}}) - \rho(\widetilde{\mathbb{W}}_2^{\pi_{12,2}})| \le |\rho(\mathbb{W}_{12,1}) - \rho(\mathbb{W}_{12,2})| + 2\|D_{W_{12,1}} - D_{W_{12,2}}\|_{2,\mu_{12}} c_2$$
$$+ \|W_{12,1} - W_{12,2}\|_{2\to 2,\mu_{12}} c_2^2$$
$$\le c_1^3 + 2c_1^2 c_2 + c_1 c_2^2.$$

To summarize, this means that

$$\Delta_{2\to 2}(\widetilde{\mathbb{W}}_1^{\pi_{12,1}}, \widetilde{\mathbb{W}}_2^{\pi_{12,2}}) \le (c_1, c_1^2 + c_1 c_2, c_1^3 + 2c_1^2 c_2 + c_1 c_2^2).$$

Similarly,

$$\Delta_{2\to 2}(\widetilde{\mathbb{W}}_2^{\pi_{23,2}}, \widetilde{\mathbb{W}}_3^{\pi_{23,3}}) \le (c_2, c_2^2 + c_1 c_2, c_2^3 + 2c_1 c_2^2 + c_1^2 c_2).$$

Therefore,

$$\|\widetilde{\mathbb{W}}_1^{\pi_{13,1}} - \widetilde{\mathbb{W}}_3^{\pi_{13,3}}\|_{2\to 2,\widetilde{\mu}_{13}} \le c_1 + c_2 + \varepsilon,$$

$$\|D_{\widetilde{\mathbb{W}}_1} - D_{\widetilde{\mathbb{W}}_3}\|_{2,\widetilde{\mu}_{13}} \le c_1^2 + c_1 c_2 + c_2^2 + c_1 c_2 + \varepsilon = (c_1+c_2)^2 + \varepsilon,$$

and finally

$$|\rho(\widetilde{\mathbb{W}}_1) - \rho(\widetilde{\mathbb{W}}_3)| \le c_1^3 + 2c_1^2 c_2 + c_1 c_2^2 + c_2^3 + 2c_1 c_2^2 + c_1^2 c_2 + \varepsilon = (c_1+c_2)^3 + \varepsilon.$$

Since this can be done for any $\varepsilon > 0$, this completes the proof that $\widetilde{\delta}_\diamond$ is a metric. With the help of Lemma 3.14 the triangle inequality for $\widetilde{\delta}_\diamond$ implies that for $\delta_\diamond$. □

Next we prove Proposition 2.17, as well the following version for signed graphexes.

Proposition 3.15 *Fix $B, C, D < \infty$. Then $\delta_\diamond$ and $\delta_{2\to 2}$ define the same topology on the space of (B, C, D)-bounded signed graphexes.*

To prove these propositions, we need a lemma complementing the bounds from Lemma 3.13. Recall that in Lemma 3.13, we showed that the distance between two graphexes defined on the same measure space cannot increase too much when we decrease of the underlying measure. Our next lemma shows that if the graphexes involved are signed graphexes that are (B, C, D)-bounded, we can also go in the other direction.

Lemma 3.16 *Let $\mathbb{W}_i = (W_i, S_i, I_i, \Omega)$, for $i = 1, 2$, be (B, C, D)-bounded signed graphexes on the same measure space Ω, and let μ', r, $\mathbb{W}_1'$, and $\mathbb{W}_2'$ be as in Lemma 3.13. Then*

(1) $\|W_1 - W_2\|_{2 \to 2, \mu'} \leq \|W_1 - W_2\|_{2 \to 2, \mu} \leq \|W_1 - W_2\|_{2 \to 2, \mu'} + 4\sqrt{BDr}$,

(2) $\left| \|D_{W_1} - D_{W_2}\|_{2,\mu}^2 - \|D_{W_1'} - D_{W_2'}\|_{2,\mu'}^2 \right| \leq (4D^2 + 8BC)r$, and

(3) $\left| |\rho(\mathbb{W}_1') - \rho(\mathbb{W}_2')| - |\rho(\mathbb{W}_1) - \rho(\mathbb{W}_2)| \right| \leq 4Dr$.

Proof Let $U = W_1 - W_2$. Then for any f, g with $\|f\|_{2,\mu} = \|g\|_{2,\mu} = 1$,

$$\int_{\Omega \times \Omega} f(x) U(x, y) g(y) \, d\mu(x) \, d\mu(y)$$

$$= \int_{\Omega \times \Omega} f(x) h(x) U(x, y) h(y) g(y) \, d\mu(x) \, d\mu(y)$$

$$+ \int_{\Omega \times \Omega} f(x) ((1 - h(x)) U(x, y) h(y)) g(y) \, d\mu(x) \, d\mu(y)$$

$$+ \int_{\Omega \times \Omega} f(x) U(x, y) (1 - h(y)) g(y) \, d\mu(x) \, d\mu(y).$$

We have

$$\int_{\Omega \times \Omega} f(x) h(x) U(x, y) h(y) g(y) \, d\mu(x) \, d\mu(y)$$

$$= \int_{\Omega \times \Omega} f(x) U(x, y) g(y) \, d\mu'(x) \, d\mu'(y) \leq \|f\|_{2,\mu'} \|U\|_{2 \to 2, \mu'} \|g\|_{2,\mu'}$$

$$\leq \|f\|_{2,\mu} \|U\|_{2 \to 2, \mu'} \|g\|_{2,\mu} \leq \|U\|_{2 \to 2, \mu'}.$$

Furthermore,

$$\int_{\Omega \times \Omega} f(x) U(x, y) (1 - h(y)) g(y) \, d\mu(x) \, d\mu(y)$$

$$\leq \|f\|_{2,\mu} \int_{\Omega} \|U_y\|_{2,\mu} (1 - h(y)) |g(y)| \, d\mu(y)$$

$$\leq \|f\|_{2,\mu} 2\sqrt{BD} \int_{\Omega} (1 - h(y)) |g(y)| \, d\mu(y)$$

$$\leq 2\sqrt{BD} \|f\|_{2,\mu} \|1 - h\|_{2,\mu} \|g\|_{2,\mu} \leq 2\sqrt{BDr}.$$

Here we used the fact that $\|U\|_\infty \leq 2B$ and $\|D_{|U|,\mu}\|_\infty \leq 2D$, which implies that $\|U_y\|_{2,\mu} \leq 2\sqrt{BD}$. Analogously, we have

$$\int_{\Omega \times \Omega} f(x)((1-h(x))U(x,y)h(y))g(y)\,d\mu(x)\,d\mu(y)$$
$$\leq 2\sqrt{BD}\|f\|_{2,\mu}\|1-h\|_{2,\mu}\|hg\|_{2,\mu} \leq 2\sqrt{BD}r.$$

Adding this all up, we have

$$\int_{\Omega \times \Omega} f(x)U(x,y)g(y)\,d\mu(x)\,d\mu(y) \leq \|U\|_{2\to 2,\mu'} + 4\sqrt{BD}r.$$

This proves the upper bound in the first claim. The lower bound follows from Lemma 3.13.

To prove the second claim, we observe that

$$0 \leq \int_\Omega (D_{\mathbb{W}_1}(x) - D_{\mathbb{W}_2}(x))^2\,d\mu(x) - \int_\Omega (D_{\mathbb{W}_1}(x) - D_{\mathbb{W}_2}(x))^2\,d\mu'(x)$$
$$= \int_\Omega (D_{\mathbb{W}_1}(x) - D_{\mathbb{W}_2}(x))^2(1-h(x))\,d\mu(x) \leq \int_\Omega 4D^2(1-h(x))\,d\mu(x) \leq 4D^2 r.$$

Furthermore, since $\|U\|_\infty \leq 2B$, we have that for any $x \in \Omega$,

$$\left|\left(D_{\mathbb{W}_1}(x) - D_{\mathbb{W}'_1}(x)\right) - \left(D_{\mathbb{W}_2}(x) - D_{\mathbb{W}'_2}(x)\right)\right|$$
$$= \left|\int_\Omega (1-h(y))U(x,y)\,d\mu(y)\right| \leq 2Br.$$

Therefore,

$$\left|\int_\Omega \left(D_{\mathbb{W}_1}(x) - D_{\mathbb{W}_2}(x)\right)^2\,d\mu'(x) - \int_\Omega \left(D_{\mathbb{W}'_1}(x) - D_{\mathbb{W}'_2}(x)\right)^2\,d\mu'(x)\right|$$
$$= \left|\int_\Omega \left(\left(D_{\mathbb{W}_1}(x) - D_{\mathbb{W}_2}(x)\right) - \left(D_{\mathbb{W}'_1}(x) - D_{\mathbb{W}'_2}(x)\right)\right)\right.$$
$$\left.\left(\left(D_{\mathbb{W}_1}(x) - D_{\mathbb{W}_2}(x)\right) + \left(D_{\mathbb{W}'_1}(x) - D_{\mathbb{W}'_2}(x)\right)\right)d\mu'(x)\right|$$
$$\leq 2Br \int_\Omega \left(|D_{\mathbb{W}_1}(x)| + |D_{\mathbb{W}_2}(x)| + |D_{\mathbb{W}'_1}(x)| + |D_{\mathbb{W}'_2}(x)|\right)d\mu'(x) \leq 8BCr.$$

Combining these two inequalities proves the second claim.

To prove the third claim, we use the following bound, where the first inequality was already established when proving the last claim of Lemma 3.13:

$$\Big||\rho(\mathbb{W}_1') - \rho(\mathbb{W}_2')| - |\rho(\mathbb{W}_1) - \rho(\mathbb{W}_2)|\Big|$$

$$\leq \Big|2\int_\Omega (1-h(x))(D_{\mathbb{W}_1} - D_{\mathbb{W}_2})(x)\,d\mu(x)$$

$$- \int_{\Omega\times\Omega}(1-h(x))(W_1 - W_2)(x,y)(1-h(y))\,d\mu(x)\,d\mu(y)\Big|$$

$$= \Big|2\int_\Omega (1-h(x))(S_1 - S_2)(x)\,d\mu(x)$$

$$+ \int_{\Omega\times\Omega}(1-h(x))(W_1 - W_2)(x,y)(2-1+h(y))\,d\mu(x)\,d\mu(y)\Big|$$

$$\leq 2\int_\Omega (1-h(x))(D_{|\mathbb{W}_1|} + D_{|\mathbb{W}_2|})(x) \leq 4Dr.$$

$\square$

Our next lemma is an easy corollary to Lemma 3.16, and in turn immediately implies Propositions 2.17 and 3.15.

Lemma 3.17 *Suppose* $\mathbb{W}_i = (W_i, S_i, I_i, \mathbf{\Omega}_i)$, *for* $i = 1, 2$, *are signed graphexes with* $\|\mathbb{W}_i\|_\infty \leq B$ *and* $\|D_{|\mathbb{W}|}\|_\infty \leq D$, *and let* $\delta_\diamond(\mathbb{W}_1, \mathbb{W}_2) \leq \varepsilon$. *Then* $|\rho(W_1) - \rho(W_2)| \leq \varepsilon^3 + 4\varepsilon^2 D$. *If, in addition,* $\|\mathbb{W}_i\|_1 \leq C$, *then* $\delta_{2\to 2}(\mathbb{W}_1, \mathbb{W}_2) \leq f(\varepsilon)$, *where*

$$f(\varepsilon) = \max\left\{\varepsilon + 4\varepsilon\sqrt{BD},\ \left(\varepsilon^2 + 2\varepsilon\sqrt{D^2 + 2BC}\right)^{1/2},\ \left(\varepsilon^3 + 4\varepsilon^2 D\right)^{1/3}\right\}$$

Proof We first note that by Lemma 3.12, $\|\mathbb{W}_i\|_1 < \infty$, so even without the assumption that $\|\mathbb{W}_i\|_1 \leq C$, we always have that $\|\mathbb{W}_i\|_1 \leq C$ for some $C < \infty$.

Next, let $\mathbf{\Omega}_i = (\Omega_i, \mathcal{F}_i, \mu_i)$, for $i = 1, 2$, and let $c > \varepsilon$. By the definition of $\delta_\diamond$, there exist measures $\widetilde{\mu}_i$ such that $\widetilde{\mu}_i \leq \mu_i$ and $\delta_i = \mu_i(\Omega_i) - \widetilde{\mu}_i(\Omega_i) \leq c^2$ and such that $\delta_{2\to 2}(\widetilde{\mathbb{W}}_1, \widetilde{\mathbb{W}}_2) < c$ for the signed graphexes $\widetilde{\mathbb{W}}_i$ obtained from $\mathbb{W}_i$ by replacing μ_i by $\widetilde{\mu}_i$.

Consider an arbitrary σ-finite space $\mathbf{\Omega}_i''$ of infinite measure, and two intervals J_i of length $c - \delta_i$. Define $\mathbf{\Omega}_i'$ by appending $\mathbf{\Omega}_i''$ and the interval J_i equipped with the Lebesgue measure to $\mathbf{\Omega}_i$, and define $\widetilde{\mathbf{\Omega}}_i'$ by appending the same spaces, except that we equip J_i with the zero measure. By Lemma 3.10 and the definition of $\delta_{2\to 2}$, we then have that $\delta_{2\to 2}(\widetilde{\mathbb{W}}_1', \widetilde{\mathbb{W}}_2') = \delta_{2\to 2}(\widetilde{\mathbb{W}}_1, \widetilde{\mathbb{W}}_2) < c$, where $\widetilde{\mathbb{W}}_i'$ are the trivial extensions to $\widetilde{\mathbf{\Omega}}_i'$. Furthermore, by our construction, $\mathbf{\Omega}_i' = (\Omega_i, \mathcal{F}_i', \mu_i')$ and $\widetilde{\mathbf{\Omega}}_i' = (\widetilde{\Omega}_i, \widetilde{\mathcal{F}}_i', \widetilde{\mu}_i')$ are such that $(\Omega_i', \mathcal{F}_i') = (\widetilde{\Omega}_i, \widetilde{\mathcal{F}}_i')$, $\widetilde{\mu}_i' \leq \mu_i'$, and $\mu_i'(\Omega_i') - \widetilde{\mu}_i'(\Omega_i') = c^2$.

Given a coupling $\widetilde{\mu}'$ of $\widetilde{\mu}_1'$ and $\widetilde{\mu}_2'$, let $\widetilde{\mathbb{U}}_i'$ be the pullback of $\widetilde{\mathbb{W}}_i'$ under the coordinate projections onto $\widetilde{\Omega}_i'$. By the definition of the distance $\widetilde{\delta}_{2\to 2}$, we can find a coupling $\widetilde{\mu}'$ such that $d_{2\to 2}(\widetilde{\mathbb{U}}_1', \widetilde{\mathbb{U}}_2') \leq c^2$. Choose a coupling μ' of μ_1' and μ_2' by coupling

$\mu'_i - \widetilde{\mu}'_i$ arbitrarily. Then $\mu' - c \leq \widetilde{\mu}' \leq \mu'$. Defining $\mathbb{U}'_i$ to be the pullbacks of $\mathbb{W}'_i$ under the coordinate projections onto Ω'_i, we may then apply Lemma 3.16 with $r = c^2$ to conclude that $|\rho(\mathbb{W}_1) - \rho(\mathbb{W}_2)| = |\rho(\mathbb{U}'_1) - \rho(\mathbb{U}'_2)| \leq c^3 + 4c^2 D$ and

$$\delta_{2\to 2}(\mathbb{W}_1, \mathbb{W}_2) \leq d_{2\to 2}(\mathbb{U}'_1, \mathbb{U}'_2)$$
$$\leq \max\left\{ c + 4c\sqrt{BD}, \left(c^2 + 2c\sqrt{D^2 + 2BC}\right)^{1/2}, \left(c^3 + 4c^2 D\right)^{1/3}\right\}$$

Since $c > \varepsilon$ was arbitrary, this concludes the proof. □

Proof (*Propositions 2.17 and 3.15*) By the definition of $\delta_\diamond$, we clearly have that $\delta_\diamond \leq \delta_{2\to 2}$. For signed graphexes that are (B, C, D)-bounded, a bound in the opposite direction follows immediately from Lemma 3.17, proving Proposition 3.15. To prove Proposition 2.17 we note that if $\delta_\diamond(\mathbb{W}_n, \mathbb{W}) \to 0$ and both $\mathbb{W}_n$ and $\mathbb{W}$ are (unsigned) graphexes with D-bounded marginals, then $\|\mathbb{W}\|_1 \leq C$ for some $C < \infty$ by Proposition 2.4. Since $\|\mathbb{W}_n\|_1 = \rho(\mathbb{W}_n)$ converges to $\rho(\mathbb{W}) = \|\mathbb{W}\|_1 \leq C$ by the first statement of the lemma, we must have that $\|\mathbb{W}_n\|_1 \leq \widetilde{C}$ for some $\widetilde{C} < \infty$, at which point the proof proceeds as the proof for the signed case. □

Our next lemma relates the kernel norm $\|\cdot\|_{2\to 2}$ of a two variable function U to the 4-cycle counts of U.

Lemma 3.18 *Let $U : \Omega \times \Omega \to \mathbb{R}$ be a measurable function. Then*

$$\|U\|_{2\to 2}^4 \leq t(C_4, U) \leq \|U\|_{2\to 2}^2 \|U\|_2^2.$$

Proof For any f, g with $\|f\|_2 = \|g\|_2 = 1$, we have (using Cauchy's inequality)

$$\|f \circ U \circ g\|_2^4 \leq \|U \circ g\|_2^4 = (g \circ U \circ U \circ g)^2$$
$$= \left(\int_{\Omega^2} g(x) U \circ U(x, y) g(y) \, d\mu(x) \, d\mu(y)\right)^2$$
$$\leq \left(\int_{\Omega^2} g(x)^2 g(y)^2 \, d\mu(x) \, d\mu(y)\right) \left(\int_{\Omega^2} (U \circ U(x, y))^2 \, d\mu(x) \, d\mu(y)\right)$$
$$= t(C_4, U).$$

This proves the first inequality. For the second, we have

$$t(C_4, U) = \int_{\Omega^4} U(x, y) U(y, z) U(z, w) U(w, x) \, d\mu(x) \, d\mu(y) \, d\mu(z) \, d\mu(w)$$
$$= \int_\Omega \int_\Omega (U \circ U_z)(x)^2 \, d\mu(x) \, d\mu(z) = \int_\Omega d\mu(z) \|U \circ U_z\|_2^2$$
$$\leq \int_\Omega \|U\|_{2\to 2}^2 \|U_z\|_2^2 = \|U\|_{2\to 2}^2 \|U\|_2^2.$$

Our next goal is to relate the jumble and the kernel distances. The next proposition shows that for (B, C, D)-bounded graphexes, they are equivalent. As we will see, similarly to the proof of Propositions 2.17 and 3.15, the proof also gives that for (unsigned) graphexes, the metrics $\delta_{2\to 2}$ and $\delta_\boxtimes$ are equivalent on the space of graphexes with D-bounded marginals; see Remark 3.23 below.

Proposition 3.19 *Given $B, C, D < \infty$, there exists a constant $c < \infty$ such that if $\mathbb{W}_1$ and $\mathbb{W}_2$ are (B, C, D)-bounded signed graphexes, then the following hold.*

(1) *If $d_{2\to 2}(\mathbb{W}_1, \mathbb{W}_2) \leq \tilde{\varepsilon}$, then $d_\boxtimes(\mathbb{W}_1, \mathbb{W}_2) \leq \max\{\tilde{\varepsilon}, \tilde{\varepsilon}^3\}$.*
(2) *If $d_\boxtimes(\mathbb{W}_1, \mathbb{W}_2) \leq \varepsilon$, then $d_{2\to 2}(\mathbb{W}_1, \mathbb{W}_2) \leq \max\{\sqrt[3]{\varepsilon}, c\sqrt[4]{\varepsilon}\}$.*

If the graphexes are such that $\rho(\mathbb{W}_1) = \rho(\mathbb{W}_2)$, then these bounds can be replaced by $d_\boxtimes(\mathbb{W}_1, \mathbb{W}_2) \leq \tilde{\varepsilon}$ and $d_{2\to 2}(\mathbb{W}_1, \mathbb{W}_2) \leq c\sqrt[4]{\varepsilon}$.

To prove the proposition, we establish three preliminary lemmas.

Lemma 3.20 *Given a bounded, nonnegative, measurable function f on some measure space $(\Omega, \mathcal{F}, \mu)$, we have*

$$\frac{\|f\|_2^2}{\sqrt{\|f\|_1 \|f\|_\infty}} \leq \sup_{S \subseteq \Omega} \frac{1}{\sqrt{\mu(S)}} \int_S f \, d\mu \leq \|f\|_2.$$

In particular, the second and third term define equivalent norms, for any C, D, on the space of nonnegative functions with $\|f\|_1 \leq C$, $\|f\|_\infty \leq D$.

Proof The second inequality follows from Cauchy's inequality. For the first one, let $\|f\|_\infty = K$. First, note that

$$\int_0^K dc \int_{\{f \geq c\}} f \, d\mu = \int_\Omega d\mu(x) f(x) \int_0^{f(x)} dc = \int_\Omega f^2 \, d\mu.$$

We then have

$$\|f\|_2^4 = \left(\int_\Omega f^2 \, d\mu\right)^2 = \left(\int_0^K \int_{\{f \geq c\}} f \, d\mu \, dc\right)^2$$

$$= \left(\int_0^K dc \sqrt{\mu(\{f \geq c\})} \frac{\int_{\{f \geq c\}} f \, d\mu}{\sqrt{\mu(\{f \geq c\})}}\right)^2$$

$$\leq \int_0^K \frac{\left(\int_{\{f \geq c\}} f \, d\mu\right)^2}{\mu(\{f \geq c\})} dc \int_0^K \mu(\{f \geq c\}) \, dc.$$

Using the fact that

$$\int_0^K \mu(\{f \geq c\}) \, dc = \int_\Omega f \, d\mu,$$

Identifiability for Graphexes and the Weak Kernel Metric

we have that

$$\int_0^K \frac{\left(\int_{\{f \geq c\}} f \, d\mu\right)^2}{\mu(\{f \geq c\})} \, dc \geq \frac{\|f\|_2^4}{\|f\|_1},$$

which means that there exists some c such that

$$\frac{\left(\int_{\{f \geq c\}} f \, d\mu\right)^2}{\mu(\{f \geq c\})} \geq \frac{\|f\|_2^4}{\|f\|_1 K} = \frac{\|f\|_2^4}{\|f\|_1 \|f\|_\infty}.$$

Taking S to be $\{f \geq c\}$, the lemma is proved. □

The following lemma is an easy corollary of Lemma 3.20.

Lemma 3.21 *Given a bounded, measurable, not necessarily nonnegative function f on some measure space (Ω, μ), we have*

$$\frac{\|f\|_2^2}{\sqrt{2\|f\|_1 \|f\|_\infty}} \leq \sup_{S \subseteq \Omega} \frac{1}{\sqrt{\mu(S)}} \left|\int_S f \, d\mu\right| \leq \|f\|_2.$$

Proof The second inequality again follows from Cauchy's inequality, so we just have to prove the first one. Note that the left term is not affected by replacing f with $|f|$. Let S be any subset of Ω, let S^+ consist of the points in S where f is nonnegative, and let S^- be the rest. Then

$$\frac{1}{\sqrt{\mu(S)}} \int_S |f| \, d\mu = \frac{|\int_{S^+} f \, d\mu|}{\sqrt{\mu(S^+)}} \frac{\sqrt{\mu(S^+)}}{\sqrt{\mu(S)}} + \frac{|\int_{S^-} f \, d\mu|}{\sqrt{\mu(S^-)}} \frac{\sqrt{\mu(S^-)}}{\sqrt{\mu(S)}}$$

$$\leq \max\left(\frac{|\int_{S^+} f \, d\mu|}{\sqrt{\mu(S^+)}}, \frac{|\int_{S^-} f \, d\mu|}{\sqrt{\mu(S^-)}}\right) \left(\frac{\sqrt{\mu(S^+)}}{\sqrt{\mu(S)}} + \frac{\sqrt{\mu(S^-)}}{\sqrt{\mu(S)}}\right)$$

$$\leq \max\left(\frac{|\int_{S^+} f \, d\mu|}{\sqrt{\mu(S^+)}}, \frac{|\int_{S^-} f \, d\mu|}{\sqrt{\mu(S^-)}}\right) \sqrt{2} \leq \sqrt{2} \sup_{S \subseteq \Omega} \frac{1}{\sqrt{\mu(S)}} \left|\int_S f \, d\mu\right|.$$

Therefore, we have

$$\sup_{S \subseteq \Omega} \frac{1}{\sqrt{\mu(S)}} \left|\int_S f \, d\mu\right| \geq \sup_{S \subseteq \Omega} \frac{1}{\sqrt{2}} \frac{1}{\sqrt{\mu(S)}} \int_S |f| \, d\mu \geq \frac{\|f\|_2^2}{\sqrt{2\|f\|_1 \|f\|_\infty}}.$$

This completes the proof. □

Lemma 3.22 *For any measurable $U: \Omega \times \Omega \to \mathbb{R}$,*

$$\|U\|_{2 \to 2} \geq \|U\|_\boxtimes \geq \frac{\|U\|_{2 \to 2}^4}{8\|U\|_\infty^{3/4} \|D_{|U|}\|_\infty^{3/4} \|D_{|U|}\|_2^{3/2}} \geq \frac{\|U\|_{2 \to 2}^4}{8\|U\|_\infty^{3/4} \|D_{|U|}\|_\infty^{3/2} \|U\|_1^{3/4}}.$$

Proof Fix f and g with $\|f\|_2 = \|g\|_2 = 1$ and recall that we use U_x to denote the function $y \mapsto U(x, y)$. First, we have

$$\|U \circ g\|_\infty \leq \sup_x \|U_x\|_2 \leq \sqrt{\|U\|_\infty \|D_{|U|}\|_\infty}.$$

We also have

$$\|U \circ g\|_1 \leq 2 \sup_{S \subseteq \Omega} \left| \int_S dx \int_\Omega U(x,y)g(y)\,dy \right| = 2 \sup_{S \subseteq \Omega} \left| \int_\Omega dy\, g(y) \int_S U(x,y)\,dx \right|$$

$$\leq 2 \int_\Omega |g(y)| D_{|U|}(y) \leq 2\|D_{|U|}\|_2.$$

Combined with the first bound from Lemma 3.21 and the fact that $|f \circ U \circ g| \leq \|U \circ g\|_2$, this shows that

$$\sup_{S \subseteq \Omega} \frac{1}{\sqrt{\mu(S)}} \left| \int_S U \circ g(x)\,d\mu(x) \right| \geq \frac{\|U \circ g\|_2^2}{\sqrt{2\|U \circ g\|_\infty \|U \circ g\|_1}}$$

$$\geq \frac{(f \circ U \circ g)^2}{2\|U\|_\infty^{1/4} \|D_{|U|}\|_\infty^{1/4} \|D_{|U|}\|_2^{1/2}}.$$

Analogously, defining g_S as the function $x \mapsto \frac{1}{\sqrt{\mu(S)}} \mathbf{1}_{x \in S}$, and observing that $\frac{1}{\sqrt{\mu(S)}} \left| \int_S (U \circ g)\,d\mu \right| = |g \circ U \circ g_S| \leq \|U \circ g_S\|_2$, we have

$$\sup_{T \subseteq \Omega} \left| \frac{1}{\sqrt{\mu(T)\mu(S)}} \int_{S \times T} d\mu(x)d\mu(y) U(x,y) \right|$$

$$= \sup_{T \subseteq \Omega} \frac{1}{\sqrt{\mu(T)}} \left| \int_T (U \circ g_S)(y)\,d\mu(y) \right|$$

$$\geq \frac{\|U \circ g_S\|_2^2}{2\|U\|_\infty^{1/4} \|D_{|U|}\|_\infty^{1/4} \|D_{|U|}\|_2^{1/2}}$$

$$\geq \frac{(f \circ U \circ g)^4}{8\|U\|_\infty^{3/4} \|D_{|U|}\|_\infty^{3/4} \|D_{|U|}\|_2^{3/2}}.$$

Therefore,

$$\|U\|_\boxtimes \geq \frac{\|U\|_{2 \to 2}^4}{8\|U\|_\infty^{3/4} \|D_{|U|}\|_\infty^{3/4} \|D_{|U|}\|_2^{3/2}} \geq \frac{\|U\|_{2 \to 2}^4}{8\|U\|_\infty^{3/4} \|D_{|U|}\|_\infty^{3/2} \|U\|_1^{3/4}},$$

where in the last step we used that $\|D_{|U|}\|_2^2 \leq \|D_{|U|}\|_1 \|D_{|U|}\|_\infty = \|U\|_1 \|D_{|U|}\|_\infty$. □

Proof (*Proposition* 3.19) The proposition follows immediately from Lemmas 3.21 and 3.22 and the definition of the distances $d_{2\to 2}$ and $d_\boxtimes$, with c being the constant $c = \max\{\sqrt[8]{8CD}, \sqrt[4]{64B^{3/4}D^{3/2}C^{3/4}}\}$. □

Remark 3.23 The above proof can easily be modified to see that for any $0 < D < \infty$ the metrics $\delta_{2\to 2}$ and $\delta_\boxtimes$ are equivalent on the space of (unsigned) graphexes with D-bounded marginals. Indeed, if $\mathbb{W}$ has bounded marginals, it is integrable, and if either $\delta_{2\to 2}(\mathbb{W}_n, \mathbb{W}) \to 0$ or $\delta_\boxtimes(\mathbb{W}_n, \mathbb{W}) \to 0$, then $\|\mathbb{W}_n\|_1 = \rho(\mathbb{W}_n) \to \rho(\mathbb{W}) = \|\mathbb{W}\|_1$. This shows that we can assume that the sequences are (C, D)-bounded for some C, which means they are (B, C, D)-bounded for $B = 1$.

We close this section with the (straightforward) proof that the homomorphism densities $t(F, \mathbb{W})$ indeed describe the expected number of injective homomorphisms from F into $G_T(\mathbb{W})$.

Proposition 3.24 *For any simple graph F with no isolated vertices and graphex $\mathbb{W}$,*

$$\mathbb{E}\big[inj(F, G_T(\mathbb{W}))\big] = T^{|V(F)|} t(F, \mathbb{W}).$$

If the marginals of $\mathbb{W}$ are bounded, then the right side is finite, with

$$t(F, \mathbb{W}) \leq \prod_i \|\mathbb{W}\|_1 \|D_\mathbb{W}\|_\infty^{v(F_i)-2},$$

where the product runs over the components of F and v_i is the number of vertices in F_i.

Proof Recall that we extended the feature space Ω to include an additional point ∞, and that we labeled the vertices corresponding to the leaves of a star generated by S, as well as the two endpoints of the isolated edges coming from I, by ∞. Let $k = |V(F)|$. First, suppose that Ω has finite measure. Then the probability that there are n points sampled from Ω is $e^{-\mu(\Omega)T}\frac{(T\mu(\Omega))^n}{n!}$. Let V_2 be the set of vertices of F of degree at least 2, let V_1 be the set of vertices of degree 1 whose neighbor is in V_2, and let V_0 be the set of vertices that belong to an isolated edge. Each vertex in V_2 must be mapped to a vertex with feature label in Ω. Each vertex in V_1 must be mapped either to a vertex with feature label in Ω or a vertex coming from the leaves of a the star attached to such a vertex (in which case its feature label is ∞). For an isolated edge, there are three possibilities: either it is mapped to two vertices with feature label in Ω, one endpoint is mapped to such a vertex and the other to a leaf of a star whose center is the first vertex, or it is mapped to an isolated edge generated by I (in which case both feature labels are ∞). Let us fix for each vertex in V_1 and V_0 whether its feature label is ∞ or lies in Ω, noting that this uniquely determines all the choices we just discussed. Let $V_0' \subseteq V_0$ and $V_1' \subseteq V_1$ be the sets of vertices mapped to a vertex coming from Ω, and let $V' = V_0' \cup V_1' \cup V_2$. Let U be the set of remaining vertices.

Let J be the set of isolated edges, and J'' the set of isolated edges where we have fixed that they are mapped to an edge generated by I; i.e., they have both endpoints in U. For a vertex $i \in V'$, let $d_U(i)$ be its degree to U. Conditioned on V', we have

$$\mathbb{E}[\mathrm{inj}(F, G_T)|V'] = \sum_{n=0}^{\infty} e^{-\mu(\Omega)T} \frac{(T\mu(\Omega))^n}{n!} \frac{(n)_{|V'|}}{\mu(\Omega)^{|V'|}}$$

$$\int_{\Omega^{V'}} \prod_{\{i,j\} \in E(F|_{V'})} W(x_i, x_j) \prod_{i \in V'} (TS(x_i))^{d_U(i)} (2T^2 I)^{|J''|}$$

$$= \sum_{n=|V'|}^{\infty} T^{|V'|} e^{-\mu(\Omega)T} \frac{(T\mu(\Omega))^{n-|V'|}}{(n-|V'|)!}$$

$$\int_{\Omega^{V'}} T^{|U|} \prod_{\{i,j\} \in E(F|_{V'})} W(x_i, x_j) \prod_{i \in V'} (S(x_i))^{d_U(i)} (2I)^{|J''|}$$

$$= T^{|V|} \left(\sum_{n=|V'|}^{\infty} e^{-\mu(\Omega)T} \frac{(T\mu(\Omega))^{n-|V'|}}{(n-|V'|)!} \right)$$

$$\left(\int_{\Omega^{V'}} \prod_{\{i,j\} \in E(F|_{V'})} W(x_i, x_j) \prod_{i \in V'} (S(x_i))^{d_U(i)} (2I)^{|J''|} \right)$$

$$= T^{|V|} \int_{\Omega^{V'}} \prod_{\{i,j\} \in E(F|_{V'})} W(x_i, x_j) \prod_{i \in V'} (S(x_i))^{d_U(i)} (2I)^{|J''|}.$$

Therefore,

$$\mathbb{E}[\mathrm{inj}(F, G_T)] = T^{|V|} \sum_{\substack{V_0' \subseteq V_0 \\ V_1' \subseteq V_1}} \int_{\Omega^{V'}} \prod_{\{i,j\} \in E(F|_{V'})} W(x_i, x_j) \prod_{i \in V'} (S(x_i))^{d_U(i)} (2I)^{|J''|}.$$

Now, it is not difficult to check that this is multiplicative over connected components of F. Indeed, each term with fixed V_0', V_1' is multiplicative, and the choice of which vertices to put in V_0', V_1' from each of the components is independent. Therefore, we may assume that F is connected.

If F consists of a single edge $\{i, j\}$, then the above expression gives

$$\mathbb{E}[\mathrm{inj}(F, G_T)] = T^2 \left(\int_{\Omega^2} W(x, y)\, dx\, dy + 2 \int_{\Omega} S(x)\, dx + 2I \right) = T^2 t(F, \mathbb{W}),$$

as required. Otherwise, F has no isolated edges, so V_0 is empty (and so is J''). We then have

Identifiability for Graphexes and the Weak Kernel Metric

$$\mathbb{E}[\mathrm{inj}(F, G_T)] = T^{|V|} \sum_{V_1' \subseteq V_1} \int_{\Omega^{V'}} \prod_{\{i,j\} \in E(F|_{V'})} W(x_i, x_j) \prod_{i \in V_2} (S(x_i))^{d_U(i)}$$

$$= T^{|V|} \int_{\Omega^{V_2}} \prod_{\{i,j\} \in E(F_{V_2})} W(x_i, x_j) \cdot$$

$$\prod_{i \in V_2} \left(\sum_{T_i \subseteq N_{V_1}(i)} \int_{\Omega^{T_i}} \prod_{j \in T_i} W(x_i, x_j) S(x_i)^{|N_{V_1}(i)| - |T_i|} \right)$$

$$= T^{|V|} \int_{\Omega^{V_2}} \prod_{\{i,j\} \in E(F_{V_2})} W(x_i, x_j) \cdot$$

$$\prod_{i \in V_2} \left(\int_\Omega W(x_i, x_j) dx_j + S(x_i) \right)^{|N_{V_1}(i)|}$$

$$= T^{|V|} t(F, \mathbb{W}).$$

This completes the proof of the first statement if Ω has finite measure. The general σ-finite case follows by monotone convergence, with both sides being possibly infinite in the limit.

To prove the second statement we consider the components of F separately. Furthermore, given a component F' of F, we use the fact that $\|W\|_\infty \leq 1$ to delete edges from F' until F' becomes a tree. At this point, we can remove the leaves of the tree at the cost of a factor D for each leaf, getting a new tree with less edges. We continue until we are left with a single edge, at which point we bound the remaining integral by $\|\mathbb{W}\|_1$. □

4 Tightness

The goal of this section is to establish various equivalent notions of tightness, and to then use tightness to relate convergence in the kernel and the weak kernel metric. In particular, we will relate the convergence of a sequence of graphexes in the weak kernel metric $\delta_\diamond$ to convergence of a "regularized" sequence in the kernel metric $\delta_{2\to 2}$, where the regularized sequence is obtained from the original one by discarding the part of the space which has large marginals; see Proposition 4.6 below. In contrast to the last section, in this section we restrict ourselves to unsigned graphexes since we believe that the obvious generalization of the notion of tightness to signed graphexes will not be the right notion of tightness for the metric $\delta_\diamond$; see Remark 4.8 at the end of this section.

We start by establishing the equivalence of various formulations of tightness.

Theorem 4.1 *Given a set of graphexes $\mathcal{S}$, the following are equivalent:*

(1) $\mathcal{S}$ is tight. In other words, for every $\varepsilon > 0$, there exist C and D such that for every graphex $\mathbb{W} \in \mathcal{S}$, $\mathbb{W} = (W, S, I, \mathbf{\Omega})$ with $\mathbf{\Omega} = (\Omega, \mathcal{F}, \mu)$, there exists $\Omega_\varepsilon \subseteq \Omega$ such that $\mu(\Omega_\varepsilon) \leq \varepsilon$ and $\mathbb{W}' = \mathbb{W}|_{\Omega \setminus \Omega_\varepsilon}$ has $\|\mathbb{W}'\|_1 \leq C$, $\|D_{\mathbb{W}'}\|_\infty \leq D$.
(2) For every $\varepsilon > 0$, there exists a C such that for every graphex $\mathbb{W} \in \mathcal{S}$, $\mathbb{W} = (W, S, I, \mathbf{\Omega})$, there exists $\Omega_\varepsilon \subseteq \Omega$ such that $\mu(\Omega_\varepsilon) \leq \varepsilon$ and $\mathbb{W}' = \mathbb{W}|_{\Omega \setminus \Omega_\varepsilon}$ has $\|\mathbb{W}'\|_1 \leq C$.
(3) For every ε, there is a D and C such that for any $\mathbb{W} \in \mathcal{S}$, taking $\Omega_{\leq D}$ to be the set of points with $D_{\mathbb{W}}(x) \leq D$, $\mu(\Omega \setminus \Omega_{\leq D}) \leq \varepsilon$, and $\|\mathbb{W}|_{\Omega_{\leq D}}\|_1 \leq C$.
(4) For every $T > 0$, the set of random unlabeled finite graphs $G_T(\mathbb{W})$ with $\mathbb{W} \in \mathcal{S}$ is tight.
(5) There exists $T > 0$ such that the set of random unlabeled finite graphs $G_T(\mathbb{W})$ with $\mathbb{W} \in \mathcal{S}$ is tight.

Corollary 4.2 *Let $\mathcal{S}$ be a set of graphexes.*

(1) *If there exists a $C < \infty$ such that $\|\mathbb{W}\|_1 \leq C$ for all $\mathbb{W} \in \mathcal{S}$, then $\mathcal{S}$ is tight.*
(2) *If $\mathcal{S}$ is tight and has uniformly bounded marginals, then there exist $C, D < \infty$ such that $\mathcal{S}$ is (C, D)-bounded.*

Proof (1) Taking $\Omega_\varepsilon = \emptyset$ for any ε, the set $\mathcal{S}$ clearly satisfies condition (2) from the theorem.

(2) Choose ε arbitrarily, say $\varepsilon = 1$, and let C', D' be such that (3) from Theorem 4.1 holds. Furthermore, let D be such that the marginals of all graphexes in $\mathcal{S}$ are bounded by D. Then

$$\|\mathbb{W}\|_1 \leq \|\mathbb{W}|_{\Omega_{\leq D'}}\|_1 + 2\int W(x, y) 1_{D_{\mathbb{W}}(x) > D'} \, d\mu(x) d\mu(y) + 2\int_{D_{\mathbb{W}} > D'} S(x) \, d\mu(x)$$

$$\leq C' + 2\int_{D_{\mathbb{W}} > D'} D_{\mathbb{W}(x)} \, d\mu(x) \leq C' + 2D\varepsilon = C' + 2D =: C,$$

proving the claim. $\square$

In order to prove Theorem 4.1, we will use the following lemma.

Lemma 4.3 *The probability that $G_T(\mathbb{W})$ has more than $KT^2\|\mathbb{W}\|_1$ edges is at most*

$$\frac{T^2\|\mathbb{W}\|_1/2 + T^3\|D_{\mathbb{W}}\|_2^2}{(K - 1/2)^2 T^4\|\mathbb{W}\|_1^2},$$

and the probability that it has less than $T^2\|\mathbb{W}\|_1/4$ edges is at most

$$\frac{16(T^2\|\mathbb{W}\|_1/2 + T^3\|D_{\mathbb{W}}\|_2^2)}{T^4\|\mathbb{W}\|_1^2}$$

Proof Let X_T be the number of edges of $G_T(\mathbb{W})$. By Proposition 3.24, X_T has expectation $T^2\|\mathbb{W}\|_1/2$. To calculate the variance, note that we have

$$X_T^2 = \frac{\text{inj}(F_1, G_T)}{2} + \frac{\text{inj}(F_2, G_T)}{4} + \text{inj}(F_3, G_T),$$

where F_1 consists of a single edge, F_2 consists of a pair of disjoint edges, and F_3 consists of two edges joined at one vertex. Therefore, we can again use Proposition 3.24 to conclude that

$$\begin{aligned}\text{Var}(X_T) &= \mathbb{E}[X_T^2] - \mathbb{E}[X_T]^2 \\ &= \frac{T^2\|\mathbb{W}\|_1}{2} + \frac{T^4\|\mathbb{W}_1\|^2}{4} + T^3\|D_\mathbb{W}\|_2^2 - \left(\frac{T^2\|\mathbb{W}\|_1}{2}\right)^2 \\ &= \frac{T^2\|\mathbb{W}\|_1}{2} + T^3\|D_\mathbb{W}\|_2^2.\end{aligned}$$

The bounds on the probabilities of having too many or too few edges follow from Chebyshev's inequality. □

Proof *(Theorem* 4.1*)* (1) ⇒ (2) is obvious.

(2) ⇒ (1): Suppose $\mathcal{S}$ satisfies (2), and let $\varepsilon > 0$. Take C from property (2) for $\varepsilon/2$, and take $D = 2C/\varepsilon$. For each $\mathbb{W} \in \mathcal{S}$ with underlying space Ω, there is a set $\Omega' \subseteq \Omega$ with $\mu(\Omega \setminus \Omega') \leq \varepsilon/2$ so that the restriction $\mathbb{W}' = \mathbb{W}|_\Omega$ has $\|\mathbb{W}'\|_1 \leq C$. Suppose $\mu(x \in \Omega' : D_{\mathbb{W}'}(x) > D) > \varepsilon/2$. Then we would have $\|W'\|_1 > D\varepsilon/2 = C$, a contradiction. Therefore, removing the set of points with $D_\mathbb{W}(x) > D$, we have removed points with total measure at most ε, and the restricted graphex is (C, D)-bounded.

(3) ⇒ (1) is obvious.

(1) ⇒ (4): Fix $T > 0$ and $\varepsilon > 0$. Take ε' such that $e^{-T\varepsilon'} > 1 - \varepsilon/2$, and take C, D for $\mathcal{S}$ from the definition of tightness. Given $\mathbb{W} \in \mathcal{S}$, there exists $\Omega_{\varepsilon'} \subseteq \Omega$ such that $\mu(\Omega_{\varepsilon'}) \leq \varepsilon'$ and $\mathbb{W}' = \mathbb{W}|_{\Omega \setminus \Omega_{\varepsilon'}}$ has $\|\mathbb{W}'\|_1 \leq C, \|D_{\mathbb{W}'}\|_\infty \leq D$. The probability that $G_T(\mathbb{W})$ samples a point in $\Omega \setminus \Omega_{\varepsilon'}$ during the Poisson process is at most $1 - e^{-T\varepsilon'} < \varepsilon/2$. Conditioned on this not happening, the sample is equivalent to a sample from $G_T(\mathbb{W}')$. For this, we have that $\|D_{\mathbb{W}'}\|_2^2 \leq CD$; therefore we can take $K = K(C, D)$ large enough so that the probability that there are more than KT^2C edges in $G_T(\mathbb{W}')$ is at most $\varepsilon/2$ (independently of $\mathbb{W}'$). Therefore, the probability that there are more than KT^2C edges in $G_T(\mathbb{W})$ is at most ε.

(4) ⇒ (5) is obvious.

(5) ⇒ (3): Let $\varepsilon > 0$. First, we show that there exists a D so that for every $\mathbb{W} \in \mathcal{S}$, the measure of the set $\{D_\mathbb{W} > D\}$ is at most ε. Suppose not. We will show that this implies that for each M we can find a $\mathbb{W} \in \mathcal{S}$ such with probability at least $\frac{1}{2}(1 - e^{-\varepsilon T/2})$, the number of edges in $G_T(\mathbb{W})$ is at least M. This contradicts the assumption that the set of random graphs $G_T(\mathbb{W})$ is tight.

Assume thus that for every D, there exists a $\mathbb{W} = \mathbb{W}(D) \in \mathcal{S}$ such that the set $\{D_\mathbb{W} > D\}$ has measure larger than ε. Take $G_T(\mathbb{W})$ and randomly color the vertices red and blue. With probability at least $1 - e^{-\varepsilon T/2}$, there exists at least one blue point whose feature label falls into the set $\{D_\mathbb{W} > D\}$. Conditioned on this, taking a blue point with feature label $x \in \{D_\mathbb{W} > D\}$, the number of red neighbors it has is a Poisson random variable with mean $TD_\mathbb{W}(x)/2$. Given $M < \infty$, choose $D = D(M, T)$ in such a way that a Poisson random variable with mean at least $TD/2$ has probability at least $1/2$ of being greater than M. As a consequence, given T and an arbitrary large M and we can find a D and $\mathbb{W} = \mathbb{W}(D) \in \mathcal{S}$ such that with probability at least $\frac{1}{2}(1 - e^{-\varepsilon T/2})$, the number of edges in $G_T(\mathbb{W})$ is at least M, contradicting tightness.

We claim that $\|\mathbb{W}|_{\Omega_{\leq D}}\|_1$ can't be arbitrarily large. Set $\mathbb{W}' = \mathbb{W}|_{\Omega_{\leq D}}$ and assume that $\|\mathbb{W}'\|_1 = C$. Then the probability that $G_T(\mathbb{W}')$ has less than $T^2C/4$ edges is at most

$$\frac{8 + 16TD}{T^2C}.$$

If C is large enough, this is less than $1/2$. But then for large enough C with probability at least $1/2$, the number of edges is at least $T^2C/4$, contradicting the assumption of tightness. This means that C can't be arbitrarily large.

This completes the proof of the theorem. □

Remark 4.4 It will sometimes be useful to transform a graphex $\mathbb{W}$ over an arbitrary σ-finite space $\mathbf{\Omega} = (\Omega, \mathcal{F}, \mu)$ into a graphex over an atomless space by mapping $\mathbf{\Omega}$ to the product space $\Omega \times [0, 1]$ equipped with the measure $\mu \times \lambda$, with λ denoting the Lebesgue measure, and mapping $\mathbb{W}$ to $\Phi(\mathbb{W}) = \mathbb{W}^\varphi$, with $\varphi \colon \Omega \times [0, 1] \to \Omega$ denoting the coordinate projection onto Ω. It is easy to see that $\delta_\diamond(\mathbb{W}, \Phi(\mathbb{W})) = 0$, which together with the triangle inequality implies that the map Φ does not change distances between graphons. It is also easy to check that if $\mathcal{S}$ is a tight set of graphexes, then the set of graphexes obtained by mapping each graphex $\mathbb{W} \in \mathcal{S}$ to the corresponding atomless graphex $\Phi(\mathbb{W})$ is tight as well.

Let us analyze when graphexes converge under $\delta_\diamond$. To this end, we first prove a few lemmas.

Lemma 4.5 *Given $C, D, M \in (0, \infty)$, there exists a function $f \colon [0, \infty)^2 \to [0, \infty)$ such that $f(x) \to 0$ as $x \to 0$ and such that the following holds*

(1) *Let $\mathbb{W}$ be a graphex over $(\Omega, \mathcal{F}, \mu)$, and let $\widetilde{\mu}$ be a second measure over $(\Omega, \mathcal{F})$ such that $\mu - r \leq \widetilde{\mu} \leq \mu$. If $\widetilde{\mathbb{W}}$ is obtained from $\mathbb{W}$ by replacing μ with $\widetilde{\mu}$ then*

$$\mu(D_\mathbb{W} > D + r) - r \leq \widetilde{\mu}(D_{\widetilde{\mathbb{W}}} > D) \leq \mu(D_\mathbb{W} > D)$$

for all $D > 0$.

(2) *Let $\mathbb{W}_1$ and $\mathbb{W}_2$ be graphexes with bounded marginals, defined over the same space $(\Omega, \mathcal{F}, \mu)$. Suppose that $d_{2 \to 2}(\mathbb{W}_1, \mathbb{W}_2) < \varepsilon$. Then*

$$\mu(\{|D_{\mathbb{W}_1} - D_{\mathbb{W}_2}| \geq \varepsilon\}) < \varepsilon^2.$$

(3) *Let $\widetilde{\mathbb{W}}_1$ and $\widetilde{\mathbb{W}}_2$ be graphexes with bounded marginals, defined over $(\Omega_1, \mathcal{F}_1, \widetilde{\mu}_1)$ and $(\Omega_2, \mathcal{F}_2, \widetilde{\mu}_2)$. If $\delta_{2 \to 2}(\widetilde{\mathbb{W}}_1, \widetilde{\mathbb{W}}_2) < \varepsilon$ and $D > \varepsilon$, then*

$$\tilde{\mu}_1(\{D_{\widetilde{\mathbb{W}}_1} > D + \varepsilon\}) - \varepsilon^2 \leq \tilde{\mu}_2(\{D_{\widetilde{\mathbb{W}}_2} > D\}) \leq \tilde{\mu}_1(\{D_{\widetilde{\mathbb{W}}_1} > D - \varepsilon\}) + \varepsilon^2.$$

(4) *For $i = 1, 2$, let $\mathbb{W}_i$ be graphexes defined over $(\Omega_i, \mathcal{F}_i, \mu_i)$, and let $\mathbb{W}_{i, \leq D}$ be the restriction of $\mathbb{W}_i$ to the subset $\{D_{\mathbb{W}_i} \leq D\}$ of Ω_i. Assume that $\varepsilon + \varepsilon^2 < D$, $\mu_1(\{D_{\mathbb{W}_1} > D\}) + \mu_2(\{D_{\mathbb{W}_2} > D\}) \leq M$, $\|\mathbb{W}_{2, \leq D}\|_1 \leq C$, $\delta_\diamond(\mathbb{W}_1, \mathbb{W}_2) \leq \varepsilon$, and $\mu_2(\{D - \varepsilon - \varepsilon^2 < D_{\mathbb{W}_2} \leq D + \varepsilon + \varepsilon^2\}) \leq \delta$. Then*

$$\delta_{2 \to 2}(\mathbb{W}_{1, \leq D}, \mathbb{W}_{2, \leq D}) \leq f(\varepsilon, \delta).$$

Proof (1) The assumption $\mu - r \leq \tilde{\mu} \leq \mu$ clearly implies that for all $x \in \Omega$,

$$D_{\mathbb{W}}(x) - r \leq D_{\widetilde{\mathbb{W}}}(x) \leq D_{\mathbb{W}}(x).$$

As a consequence,

$$\mu(D_{\mathbb{W}} > D + r) - r \leq \tilde{\mu}(D_{\mathbb{W}} > D + r) \leq \tilde{\mu}(D_{\widetilde{\mathbb{W}}} > D)$$

and

$$\tilde{\mu}(D_{\widetilde{\mathbb{W}}} > D) \leq \tilde{\mu}(D_{\mathbb{W}} > D) \leq \mu(D_{\mathbb{W}} > D).$$

(2) By the definition of $d_{2 \to 2}$, $\|D_{\mathbb{W}_1} - D_{\mathbb{W}_2}\|_2^2 < \varepsilon^4$, which clearly implies that

$$\mu(\{|D_{\mathbb{W}_1} - D_{\mathbb{W}_2}| \geq \varepsilon\}) < \varepsilon^2.$$

(3) For $i = 1, 2$, let $(\widetilde{\Omega}'_i, \widetilde{\mathcal{F}}'_i, \tilde{\mu}'_i)$ be a measure space obtained from $(\Omega_i, \mathcal{F}_i, \tilde{\mu}_i)$ by appending some space of infinite total measure, and let $\widetilde{\mathbb{W}}'_i$ be the trivial extension of $\widetilde{\mathbb{W}}_i$ onto $(\widetilde{\Omega}'_i, \widetilde{\mathcal{F}}'_i, \tilde{\mu}'_i)$. Furthermore, let μ' be a coupling of $\tilde{\mu}'_1$ and $\tilde{\mu}'_2$ such that $d_{2 \to 2}(\mathbb{W}'_1, \mathbb{W}'_2) \leq \varepsilon$, where $\mathbb{W}'_i = (\widetilde{\mathbb{W}}'_i)^{\pi_i, \mu'}$ for $i = 1, 2$. Then by (2),

$$\tilde{\mu}_1(\{D_{\widetilde{\mathbb{W}}_1} > D + \varepsilon\}) = \tilde{\mu}'_1(\{D_{\widetilde{\mathbb{W}}'_1} > D + \varepsilon\})$$
$$= \mu'(\{D_{\mathbb{W}'_1} > D + \varepsilon\})$$
$$\leq \mu'(\{D_{\mathbb{W}'_2} > D\}) + \mu'(\{|D_{\mathbb{W}'_1} - D_{\mathbb{W}'_2}| \geq \varepsilon\})$$
$$\leq \mu'(\{D_{\mathbb{W}'_2} > D\}) + \varepsilon^2$$
$$= \tilde{\mu}_2(\{D_{\mathbb{W}_2} > D\}) + \varepsilon^2,$$

proving the first bound in (3). The second is proved analogously.

(4) For $i = 1, 2$, let $\mathbb{W}'_i$ be the trivial extension of $\mathbb{W}_i$ to a space $(\Omega'_i, \mathcal{F}'_i, \mu'_i)$ obtained from $(\Omega_i, \mathcal{F}_i, \mu_i)$ by appending some σ-finite space of infinite total mass. Recalling Lemma 3.14, we can use the assumption $\delta_\diamond(\mathbb{W}_1, \mathbb{W}_2) < \varepsilon$ to infer the existence of a measure μ' over $\Omega'_1 \times \Omega'_2$ such that $d_{2 \to 2}((\mathbb{W}'_1)^{\pi_1, \mu'}, (\mathbb{W}'_2)^{\pi_2, \mu'}) < \varepsilon$ and $\mu'_i - \varepsilon^2 \leq (\mu')^{\pi_i} \leq \mu'_i$, $i = 1, 2$. For $i = 1, 2$, define $\tilde{\mu}_i = (\mu')^{\pi_i}$, $\mathbb{U}'_i = (\mathbb{W}'_i)^{\pi_i, \mu'}$, and $\Omega'_{i, \leq D} = \{x \in \Omega'_i : D_{\mathbb{W}'_i}(x) \leq D\}$. Then $\mathbb{W}'_{1, \leq D} := (\mathbb{W}'_1)|_{\Omega'_{1, \leq D}}$ and $\mathbb{W}'_{2, \leq D} := (\mathbb{W}'_2)|_{\Omega'_{2, \leq D}}$ are extensions of $\mathbb{W}_{1, \leq D}$ and $\mathbb{W}_{2, \leq D}$ by spaces of infinite

measure. Let $\mu'_{1,D}$ and $\mu'_{2,D}$ be the marginals of the measure $\mu'|_{\Omega'_{1,\leq D} \times \Omega'_{2,\leq D}}$. We then have that $\mu'_{1,D} \leq \mu'_1$. Observing that $D_{\mathbb{W}'_i}(\pi_i(x)) - \varepsilon^2 \leq D_{\mathbb{U}'_i}(x) \leq D_{\mathbb{W}'_i}(\pi_i(x))$, we furthermore have that

$$\begin{aligned}
0 \leq (\mu'_1 - \mu'_{1,D})(\Omega'_{1,\leq D}) &\leq \varepsilon^2 + (\widetilde{\mu}'_1 - \mu'_{1,D})(\Omega'_{1,\leq D}) \\
&= \varepsilon^2 + \mu'(\pi_1^{-1}(\{D_{\mathbb{W}_1} \leq D\}) \times \pi_2^{-1}(\{D_{\mathbb{W}_2} > D\})) \\
&\leq \varepsilon^2 + \mu'(\{D_{\mathbb{U}'_1} \leq D\} \cap \{D_{\mathbb{U}'_2} > D - \varepsilon^2\}) \\
&\leq 2\varepsilon^2 + \mu'(\{D - \varepsilon^2 < D_{\mathbb{U}'_2} \leq D + \varepsilon\}) \\
&\leq 2\varepsilon^2 + \mu_2(\{D - \varepsilon^2 < D_{\mathbb{W}_2} \leq D + \varepsilon + \varepsilon^2\}) \leq 2\varepsilon^2 + \delta =: \widetilde{\delta}.
\end{aligned}$$

Here we used the fact that $d_{2 \to 2}(\mathbb{U}'_1, \mathbb{U}'_2) < \varepsilon$, which meant that we could apply (2). Similarly, $\mu'_{2,D} \leq \mu'_2$ and

$$\begin{aligned}
0 \leq (\mu'_2 - \mu'_{2,D})(\widetilde{\Omega}'_{2,\leq D}) &\\
&\leq \varepsilon^2 + \mu'(\{D_{\mathbb{U}'_1} > D - \varepsilon^2\} \cap \{D_{\mathbb{U}'_2} \leq D\}) \\
&\leq 2\varepsilon^2 + \mu'(\{D - \varepsilon - \varepsilon^2 < D_{\mathbb{U}'_2} \leq D\}) \\
&\leq 2\varepsilon^2 + \mu_2(\{D - \varepsilon - \varepsilon^2 < D_{\mathbb{W}_2} \leq D + \varepsilon^2\}) \leq \widetilde{\delta}.
\end{aligned}$$

Next we claim that we may assume without loss of generality that

$$(\mu'_1 - \mu'_{1,D})(\Omega_{1,\leq D}) = (\mu'_2 - \mu'_{2,D})(\Omega'_{2,\leq D}) \leq \widetilde{\delta}.$$

Indeed, we can trivially extend either $\mathbb{W}'_1$ or $\mathbb{W}'_2$ by appending a space of total measure $\delta' \leq \widetilde{\delta}$ (e.g., the interval $[0, \delta')$), setting μ' to zero on the additional set. This corresponds to trivially extending both $\mathbb{U}'_1$ and $\mathbb{U}'_2$ by either $[0, \delta'] \times \Omega'_2$ or $\Omega'_1 \times [0, \delta']$. Since $\mu' = 0$ on the extension, this does not change $d_{2 \to 2}(\mathbb{U}'_1, \mathbb{U}'_2)$.

Note also that

$$\begin{aligned}
\mu'(\Omega'_1 \times \Omega'_2 \setminus \Omega'_{1,\leq D} \times \Omega'_{2,\leq D}) &\leq \mu'(\{D_{\mathbb{W}_1} > D\} \times \Omega'_2) + \mu'(\Omega'_1 \times \{D_{\mathbb{W}_2} > D\}) \\
&\leq \mu_1(\{D_{\mathbb{W}_1} > D\}) + \mu_2(\{D_{\mathbb{W}_2} > D\}) \leq M.
\end{aligned}$$

If $\mathbb{U}''_1$ and $\mathbb{U}''_2$ are the restrictions of $\mathbb{U}'_1$ and $\mathbb{U}'_2$ to $\Omega'_{1,\leq D} \times \Omega'_{2,\leq D}$, and μ'' is the restriction of μ', then by Lemma 3.13,

$$\|\mathbb{U}''_1 - \mathbb{U}''_2\|_{2 \to 2, \mu''} \leq \varepsilon,$$

$$\|D_{\mathbb{U}''_1} - D_{\mathbb{U}''_2}\|_{2, \mu''} \leq \varepsilon^2 + \sqrt{M}\varepsilon,$$

and

$$|\|\mathbb{U}''_1\|_1 - \|\mathbb{U}''_2\|_1| \leq \varepsilon^3 + 2\sqrt{M}\varepsilon^2 + M\varepsilon.$$

Next, we increase μ'' to a measure μ on $\Omega'_{1,\leq D} \times \Omega'_{2,\leq D}$ by coupling $\mu'_1 - \mu'_{1,D}$ and $\mu'_2 - \mu'_{2,D}$ arbitrarily. Then μ has marginals $\mu'_1|_{\Omega'_{1,\leq D}}$ and $\mu'_2|_{\Omega'_{2,\leq D}}$ and $\mu - \tilde{\delta} \leq \mu'' \leq \mu$. If we apply Lemma 3.16, we then obtain a coupling of $\mathbb{W}'_{1,\leq D}$ and $\mathbb{W}'_{2,\leq D}$ such that the pullbacks $\mathbb{U}'''_1$ and $\mathbb{U}'''_2$ obey the bounds

$$\|\mathbb{U}'''_1 - \mathbb{U}'''_2\|_{2\to 2,\mu} \leq \varepsilon + 2\sqrt{2D\tilde{\delta}},$$

and

$$|\|\mathbb{U}'''_1\|_1 - \|\mathbb{U}'''_2\|_1| \leq \varepsilon^3 + 2\sqrt{M}\varepsilon^2 + M\varepsilon + 2D\tilde{\delta}.$$

Setting

$$\tilde{C} = C + \varepsilon^3 + 2\sqrt{M}\varepsilon^2 + M\varepsilon + 2D\tilde{\delta},$$

we then have $\max\{\|\mathbb{U}'''_1\|_1, \|\mathbb{U}'''_2\|_1\} \leq \tilde{C}$ and hence

$$\|D_{\mathbb{U}'''_1} - D_{\mathbb{U}'''_2}\|^2_{2,\mu} \leq \left(\varepsilon^2 + \sqrt{M}\varepsilon\right)^2 + 4\tilde{C}\tilde{\delta} + D^2\tilde{\delta}.$$

This completes the proof of (4). □

Suppose $(\mathbb{W}_n)_{n=1}^\infty$ and $\mathbb{W}$ are graphexes over the σ-finite measure spaces $\boldsymbol{\Omega}_n = (\Omega_n, \mathcal{F}_n, \mu_n)$ and $\boldsymbol{\Omega} = (\Omega, \mathcal{F}, \mu)$. Define, for any $D > 0$,

$$\Omega_{n,\leq D} = \{x \in \Omega_n : D_{\mathbb{W}_n}(x) \leq D\} \quad \text{and} \quad \Omega_{n,>D} = \{x \in \Omega_n : D_{\mathbb{W}_n}(x) > D\}.$$

Recall that we have $\mu_n(\Omega_{n,>D}) < \infty$. Let $\mathbb{W}_{n,\leq D}$ consist of $\mathbb{W}_n$ restricted to $\Omega_{n,\leq D}$. Define $\Omega_{\leq D}$, $\Omega_{>D}$, and $\mathbb{W}_{\leq D}$ similarly. We then have the following.

Proposition 4.6 *Given a sequence of graphexes $\mathbb{W}_n$ and a graphex $\mathbb{W}$ over the σ-finite measure spaces $\boldsymbol{\Omega}_n = (\Omega_n, \mathcal{F}_n, \mu_n)$ and $\boldsymbol{\Omega} = (\Omega, \mathcal{F}, \mu)$, respectively, define $\mathbb{W}_{n,\leq D}$ and $\mathbb{W}_{\leq D}$ as above. Then the following are equivalent.*

(1) *For all $D > 0$ such that $\mu(\{D_\mathbb{W} = D\}) = 0$, we have $\delta_{2\to 2}(\mathbb{W}_{n,\leq D}, \mathbb{W}_{\leq D}) \to 0$ and $\mu_n(\Omega_{n,>D}) \to \mu(\Omega_{>D})$.*
(2) *The sequence is tight, and for all $D > 0$ such that $\mu(\{D_\mathbb{W} = D\}) = 0$, we have $\delta_{2\to 2}(\mathbb{W}_{n,\leq D}, \mathbb{W}_{\leq D}) \to 0$.*
(3) *For every $\varepsilon > 0$ and $n \in \mathbb{N}$ there exist subsets $\Omega_{n,\varepsilon} \subseteq \Omega_n$ and a subset $\Omega_\varepsilon \subseteq \Omega$ with $\mu_n(\Omega_n \setminus \Omega_{n,\varepsilon}) \leq \varepsilon$ and $\mu(\Omega \setminus \Omega_\varepsilon) \leq \varepsilon$, such that $\delta_{2\to 2}(\mathbb{W}'_n, \mathbb{W}') \to 0$, where $\mathbb{W}'_n = (\mathbb{W}_n)|_{\Omega_{n,\varepsilon}}$ and $\mathbb{W}' = \mathbb{W}|_{\Omega_\varepsilon}$.*
(4) $\delta_\diamond(\mathbb{W}_n, \mathbb{W}) \to 0$.

Proof (1) $\Rightarrow$ (2): We have to prove tightness. For any ε, there exists a D such that $\mu(\Omega_{>D}) \leq \varepsilon/2$. Assuming without loss of generality that D is chosen in such a way that $\mu(\{D_\mathbb{W} = D\}) = 0$, we further have $\mu_n(\Omega_{n,>D}) \to \mu(\Omega_{>D})$. This means that for all but a finite set of n, $\mu_n(\Omega_{n,>D}) \leq \varepsilon$. By increasing D, we can guarantee this for all n. Since $\delta_{2\to 2}$ convergence implies in particular that $\|\mathbb{W}_{n,\leq D}\|_1 \to \|\mathbb{W}_{\leq D}\|_1$, we have that $\|\mathbb{W}_{n,\leq D}\|_1$ is bounded. This proves property 3 from Theorem 4.1.

(2) clearly implies (3), because tightness implies that for any ε, there exists a D such that the measure of points with degree greater than D is at most ε, and we can increase D to make sure that $\mu(\{D_\mathbb{W} = D\}) = 0$.

To show that (3) implies (4), we note that the conditions in (3) imply that $\limsup_{n\to\infty} \delta_\diamond(\mathbb{W}_n, \mathbb{W}) \leq \sqrt{\varepsilon}$. Since ε is arbitrary, this gives (4).

It remains to show that (4) implies (1). The assumption $\delta_\diamond(\mathbb{W}_n, \mathbb{W}) \to 0$ implies that for all $\varepsilon > 0$ there exists an n_0 such that for $n \geq n_0$, $\delta_\diamond(\mathbb{W}_n, \mathbb{W}) < \varepsilon$. Recalling Definition 2.14 and combining statements (1) and (3) of the previous lemma, this implies that for $D > \varepsilon + \varepsilon^2 > 0$ and $n \geq n_0$,

$$\mu(D_\mathbb{W} > D + \varepsilon + \varepsilon^2) - 2\varepsilon^2 \leq \mu_n(D_{\mathbb{W}_n} > D) \leq \mu(D_\mathbb{W} > D - \varepsilon^2 - \varepsilon) + 2\varepsilon^2.$$

By our assumption that $\mu(\{D_\mathbb{W} = D\}) = 0$, we have that D is a continuity point of the function $x \mapsto \mu(\{D_\mathbb{W} > x\})$, showing that the upper and lower bound converge to $\mu(\{D_\mathbb{W} > D\})$ as $\varepsilon \to 0$. This shows that $\mu_n(D_{\mathbb{W}_n} > D) \to \mu(\{D_\mathbb{W} > D\})$ as $n \to \infty$.

Next we define

$$M = \sup_n \mu_n(\Omega_{n,>D}) + \mu(\Omega_{>D}) \quad \text{and} \quad \delta(\varepsilon) = \mu(\{D - \varepsilon - \varepsilon^2 < D_\mathbb{W} \leq D + \varepsilon + \varepsilon^2\}).$$

Note that M is finite by the fact that $\mu_n(\Omega_{n,>D}) \to \mu(\Omega_{>D})$, and that $\delta(\varepsilon) \to 0$ as $\varepsilon \to 0$ by the fact that $\mu(\{D_\mathbb{W} = D\}) = 0$. We now apply the previous lemma (with $\mathbb{W}_1$ replaced by $\mathbb{W}_n$ and $\mathbb{W}_2$ replaced by $\mathbb{W}$) to conclude the proof. □

The following proposition is an easy corollary of Proposition 4.6.

Proposition 4.7 *Given two graphexes $\mathbb{W}_1, \mathbb{W}_2$, let $\mathbb{W}_{i,\leq D}$ be the graphex $\mathbb{W}_i$ restricted to $\Omega_{i,\leq D} = \{x \in \Omega_i : D_{\mathbb{W}_i}(x) \leq D\}$. Then the following are equivalent.*

(1) *For any $D > 0$,*

$$\mu_1(\Omega_1 \setminus \Omega_{1,\leq D}) = \mu_2(\Omega_2 \setminus \Omega_{2,\leq D}) \text{ and } \delta_{2\to 2}(\mathbb{W}_{1,\leq D}, \mathbb{W}_{2,\leq D}) = 0.$$

(2) *For any $D > 0$, $\delta_{2\to 2}(\mathbb{W}_{1,\leq D}, \mathbb{W}_{2,\leq D}) = 0$.*
(3) *For any $\varepsilon > 0$, there exist subsets $\Omega_{1,\varepsilon} \subseteq \Omega_1$, $\Omega_{2,\varepsilon} \subseteq \Omega_2$ with $\mu_i(\Omega_i \setminus \Omega_{i,\varepsilon}) \leq \varepsilon$, such that if $\mathbb{W}'_i$ is the restriction to $\Omega_i \setminus \Omega_{i,\varepsilon}$, then $\delta_{2\to 2}(\mathbb{W}'_1, \mathbb{W}'_2) \leq \varepsilon$.*
(4) $\delta_\diamond(\mathbb{W}_1, \mathbb{W}_2) = 0$.

It is easy to see that this is an equivalence relation.

Proof Observing that the functions $f_i : \mathbb{R}_+ \to \mathbb{R}+$ defined by $D \mapsto \mu_i(\Omega_i \setminus \Omega_{i,\leq D})$ for $i = 1, 2$ are equal if and only if $f_1(D) = f_2(D)$ for all continuity points of f_1, this follows by applying Proposition 4.6 with $\mathbb{W} = \mathbb{W}_1$ and each $\mathbb{W}_n = \mathbb{W}_2$. □

Remark 4.8 It is not *a priori* clear how to generalize the notion of tightness to signed graphexes, even if we restrict ourselves to the case where the graphon parts are uniformly bounded, for example by taking graphons that take values in $[-1, 1]$. Indeed, recalling Lemma 3.12 and the role it played in showing that $\delta_\diamond$ is finite for signed graphexes with bounded graphon part, one might want to modify Definition 2.20 for such graphexes by replacing the notion of D-bounded marginals by that of D-bounded absolute marginals, since this would, in particular, guarantee that a finite set of signed graphexes with graphon parts in $[-1, 1]$ is tight. It would also make the generalization of several of our results straightforward, since this definition just reduces the notion of tightness of a set of graphexes $\mathcal{S}$ to the set of graphexes $\mathcal{S}' = \{|\mathbb{W}| : \mathbb{W} \in \mathcal{S}\}$.

The following example shows that this straightforward generalization of Definition 2.20 to signed graphexes does not give a characterization of precompact sets with respect to the metric $\delta_\diamond$, as it did for unsigned graphexes; see Theorem 2.21.

Let W_n be equal to $n^{-3/4}$ on $[0, 1) \times [1, n+1) \cup [1, n+1) \times [0, 1)$, equal to $-n^{-3/4}$ on $[0, 1) \times [n+1, 2n+2) \cup [n+1, 2n+1) \times [0, 1)$, and zero everywhere else on $\mathbb{R}_+^2$. Define $\mathbb{W}_n$ to be the graphex with graphon part W_n and zero star and dust part. Then $D_{\mathbb{W}_n}$ is equal to $n^{-3/4}$ on $[1, n+1)$, equal to $-n^{-3/4}$ on $[n+1, 2n+1)$, and 0 everywhere else. Finally, $\rho(\mathbb{W}_n) = 0$ for all n. Since $\|W_n\|_2 = 2n^{-1/2}$ and $\|D_{\mathbb{W}_n}\|_2 = \sqrt{2}n^{-1/2}$, $\mathbb{W}_n$ tends to the zero graphex in the metric $\delta_{2\to 2}$ and hence also in $\delta_\diamond$. But $\|W_n\|_1 = \|W\|_1 = 4n^{1/4} \to \infty$, a fact which can't be changed by removing just a part of measure ε from the underlying space, $\mathbb{R}_+$. This shows that with the obvious generalization of Definition 2.20 to signed graphexes not all sequences of signed graphexes that are convergent in $\delta_{2\to 2}$ or $\delta_\diamond$ are tight.

We therefore believe that a complete theory of signed graphexes, even in the simplified case where all graphons take values in $[-1, 1]$, requires either a modification of the metric, or modification of the notion of tightness. We leave this problem as an open research problem.

5 Regularity Lemma and Compactness

In this section, we will prove a regularity lemma (Theorem 5.3 below), and use it to prove Theorem 2.21, which in turn is an important ingredient in our proof that GP-convergence and $\delta_\diamond$-convergence are equivalent. To state the regularity lemma, we recall that a finite subspace partition of a measure space $\Omega = (\Omega, \mathcal{F}, \mu)$ is a partition of a measurable subset of Ω into finitely many measurable subsets of finite measure. Throughout this section, we will use the notation $\mathscr{P} = (\Omega_\mathscr{P}, \mathcal{P})$ for a finite subspace partition, with $\Omega_\mathscr{P}$ denoting the subset of Ω, and $\mathcal{P} = (P_1, \ldots, P_m)$ denoting the partition of $\Omega_\mathscr{P}$. We will also need the notion of refinement.

Definition 5.1 Given two subspace partitions $\mathscr{P} = (\Omega_\mathscr{P}, \mathcal{P})$ and $\mathscr{Q} = (\Omega_\mathscr{Q}, \mathcal{Q})$, we say that $\mathscr{P}$ refines $\mathscr{Q}$ if $\Omega_\mathscr{Q} \subseteq \Omega_\mathscr{P}$ and $\mathcal{P}$ is a refinement of $\mathcal{Q} \cup \{\Omega_\mathscr{P} \setminus \Omega_\mathscr{Q}\}$.

Given an integrable, signed graphex $\mathbb{W}$, a subspace partition $\mathscr{P}$ naturally generates a step function $\mathbb{W}_\mathscr{P}$ by "averaging". The precise definition is as follows.

Definition 5.2 Given a signed graphex $\mathbb{W} = (W, S, I, \mathbf{\Omega})$ and a finite subspace partition $\mathscr{P} = (\Omega_{\mathscr{P}}, \mathcal{P})$, take $\mathbb{W}_{\mathscr{P}}$ to be the signed graphex $\mathbb{W}_{\mathscr{P}} = (W_{\mathscr{P}}, S_{\mathscr{P}}, I_{\mathscr{P}}, \mathbf{\Omega})$ defined by

$$I_{\mathscr{P}} = \frac{1}{2} \int_{(\Omega \setminus \Omega_{\mathscr{P}}) \times (\Omega \setminus \Omega_{\mathscr{P}})} W(x, y) \, d\mu(x) \, d\mu(y) + \int_{\Omega \setminus \Omega_{\mathscr{P}}} S(x) \, d\mu(x) + I,$$

$$S_{\mathscr{P}}(x) = \frac{1}{\mu(P_i)} \int_{P_i} \left(S(x) + \int_{\Omega \setminus \Omega_{\mathscr{P}}} W(x, y) \, d\mu(y) \right) d\mu(x) \quad \text{if } x \in P_i$$

for some $i \in \{1, \ldots, k\}$, and 0 everywhere else, and

$$W_{\mathscr{P}}(x, y) = \frac{1}{\mu(P_i)\mu(P_j)} \int_{P_i \times P_j} W(x', y') \, d\mu(x') \, d\mu(y') \quad \text{if } (x, y) \in P_i \times P_j$$

for some $i, j \in \{1, \ldots, k\}$ and $W_{\mathscr{P}}(x, y) = 0$ everywhere else.

Note that with this definition, for $x \in P_i$,

$$D_{\mathbb{W}_{\mathscr{P}}}(x) = \frac{1}{\mu(P_i)} \int_{P_i} D_{\mathbb{W}}(x) \, d\mu(x).$$

We also have $\rho(\mathbb{W}_{\mathscr{P}}) = \rho(\mathbb{W})$, as well as $\|W_{\mathcal{P}}\|_1 \leq \|W\|_1$, $\|D_{\mathbb{W}_{\mathscr{P}}}\|_\infty \leq \|D_{\mathbb{W}}\|_\infty$, and $\|W_{\mathscr{P}}\|_\infty \leq \|W\|_\infty$.

Theorem 5.3 *For any $B, C, D < \infty$, and $\varepsilon > 0$, there exists an $M(\varepsilon)$ and $N(\varepsilon)$ such that for any signed graphex $\mathbb{W}$ that is (B, C, D)-bounded there exists a partition $(\Omega_{\mathscr{P}}, \mathcal{P})$ with $\mathcal{P} = \{P_1, \ldots, P_m\}$, $m \leq M(\varepsilon)$, and $\mu(\Omega_{\mathscr{P}}) \leq N(\varepsilon)$ such that*

$$d_{\boxtimes}(\mathbb{W}, \mathbb{W}_{\mathscr{P}}) \leq \varepsilon.$$

We can take

$$M(\varepsilon) = 2^{(2BC+CD)/\varepsilon^2} \quad \text{and} \quad N(\varepsilon) = (4C^3 D + 8BC^2 D)/\varepsilon^4.$$

Given any finite subspace partition $\mathscr{Q} = (\Omega_{\mathscr{Q}}, \mathcal{Q})$, we can require the subspace partition $\mathscr{P} = (\Omega_{\mathscr{P}}, \mathcal{P})$ to be a refinement of $\mathscr{Q}$. In this case, the bound on the number of parts is $|\mathcal{Q}|M(\varepsilon)$ and the bound on $\mu(\Omega_{\mathscr{P}})$ is $\mu(\Omega_{\mathscr{Q}}) + N(\varepsilon)$.

Proof Motivated by the original proof of the weak regularity lemma (see in particular the proof of Theorem 12 in [14], which to our knowledge is the first place where a weak regularity lemma for functions $W : [0, 1]^2 \to \mathbb{R}$ was established), we construct a sequence of partitions $\mathscr{P}_0, \mathscr{P}_1, \ldots, \mathscr{P}_\ell$ such that eventually, we must have that $d_{\boxtimes}(\mathbb{W}, \mathbb{W}_{\mathscr{P}_\ell}) \leq \varepsilon$. We start with the trivial partition $\mathscr{P}_0 = (\emptyset, \emptyset)$ (so that $\mathbb{W}_{\mathscr{P}_0}$ is the graphex with zero graphon and star part, and dust part $\rho(\mathbb{W})$), and then construct a sequence of refinements $\mathscr{P}_0, \mathscr{P}_1, \ldots, \mathscr{P}_\ell$.

Identifiability for Graphexes and the Weak Kernel Metric 87

In a preliminary step, we claim that for any partition $\mathscr{P}$,

$$\langle W - W_{\mathscr{P}}, W_{\mathscr{P}} \rangle = \int_{\Omega \times \Omega} (W(x, y) - W_{\mathscr{P}}(x, y)) W_{\mathscr{P}}(x, y) \, dx \, dy = 0$$

and

$$\langle D_W - D_{W_{\mathscr{P}}}, D_{W_{\mathscr{P}}} \rangle = \int_{\Omega} (D_W(x) - D_{W_{\mathscr{P}}}(x)) D_{W_{\mathscr{P}}}(x) \, dx = 0$$

This follows from the fact that for any pair of finite parts P_i, P_j, $W_{\mathscr{P}}$ and $D_{W_{\mathscr{P}}}$ are constant, and the integral of $W - W_{\mathscr{P}}$ and $D_W - D_{W_{\mathscr{P}}}$ is zero. Since $W_{\mathscr{P}}$ is zero between pairs of parts where at least one is non-finite, and $D_{W_{\mathscr{P}}}$ is zero on non-finite parts, this implies the claim. Therefore, we have

$$\|W\|_2^2 = \|W_{\mathscr{P}}\|_2^2 + \|W - W_{\mathscr{P}}\|_2^2$$

and

$$\|D_W\|_2^2 = \|D_{W_{\mathscr{P}}}\|_2^2 + \|D_W - D_{W_{\mathscr{P}}}\|_2^2.$$

If we have a finite subspace partition $\mathscr{P} = (\Omega_{\mathscr{P}}, \mathcal{P})$ and a refinement $\mathscr{P}'$, it is then easy to check that $(\mathbb{W}_{\mathscr{P}'})_{\mathscr{P}} = \mathbb{W}_{\mathscr{P}}$; therefore, the same properties hold for $\mathbb{W}_{\mathscr{P}}$ and $\mathbb{W}_{\mathscr{P}'}$.

Suppose now that we have constructed a sequence of refinements $\mathscr{P}_0, \mathscr{P}_1, \ldots, \mathscr{P}_i$ such that $d_{\boxtimes}(\mathbb{W}, \mathbb{W}_{\mathscr{P}_j}) > \varepsilon$ for all $j \leq i$. Then we in particular have that $d_{\boxtimes}(\mathbb{W}, \mathbb{W}_{\mathscr{P}_i}) > \varepsilon$, which implies that $\|D_W - D_{W_{\mathscr{P}_i}}\|_{\boxtimes} > \varepsilon$ or $\|W - W_{\mathscr{P}_i}\|_{\boxtimes} > \varepsilon$. If the former holds, then there exists a set $S \subseteq \Omega$ (of finite measure) such that

$$\frac{1}{\sqrt{\mu(S)}} \left| \int_S D_W(x) \, d\mu(x) - \int_S D_{W_{\mathscr{P}_i}}(x) \, d\mu(x) \right| = A > \varepsilon. \tag{10}$$

Let $\mathscr{P}_{i+1} = (\Omega_{\mathscr{P}_{i+1}}, \mathcal{P}_{i+1})$ with $\Omega_{\mathscr{P}_{i+1}} = \Omega_{\mathscr{P}_i} \cup S$ and let $\mathcal{P}_{i+1}$ be the partition that refines each part of $\mathcal{P}_i$ by the intersection with S; in particular, this divides each part into at most 2 parts, and $\mu(\Omega_{\mathscr{P}_{i+1}}) \leq \mu(\Omega_{\mathscr{P}_i}) + \mu(S)$. We also have

$$\frac{1}{\sqrt{\mu(S)}} \left| \int_S D_{W_{\mathscr{P}_{i+1}}}(x) \, d\mu(x) - \int_S D_{W_{\mathscr{P}_i}}(x) \, d\mu(x) \right| = A.$$

Therefore,

$$\|D_{W_{\mathscr{P}_{i+1}}} - D_{W_{\mathscr{P}_i}}\|_2^2 \geq \left\langle D_{W_{\mathscr{P}_{i+1}}} - D_{W_{\mathscr{P}_i}}, \frac{\chi_S}{\sqrt{\mu(S)}} \right\rangle^2 = A^2.$$

Overall, this implies that we have

$$\|D_{W_{\mathscr{P}_i}}\|_2^2 + \varepsilon^2 \leq \|D_{W_{\mathscr{P}_{i+1}}}\|_2^2 \leq \|D_W\|_2^2.$$

We also have
$$\|W_{\mathscr{P}_i}\|_2^2 \le \|W_{\mathscr{P}_{i+1}}\|_2^2 \le \|W\|_2^2.$$

If $\|W - W_{\mathscr{P}_i}\|_\square > \varepsilon$, then there exist sets $S, T \subseteq \Omega$ (of finite measure) such that

$$\frac{1}{\sqrt{\mu(S)\mu(T)}}\left|\int_{S\times T} W(x,y)\,d\mu(x)\,d\mu(y) - \int_{S\times T} W_{\mathscr{P}_i}(x,y)\,d\mu(x)\,d\mu(y)\right| > \varepsilon. \tag{11}$$

Let $\mathscr{P}_{i+1} = (\Omega_{\mathscr{P}_{i+1}}, \mathcal{P}_{i+1})$ with $\Omega_{\mathscr{P}_{i+1}} = \Omega_{\mathscr{P}} \cup S \cup T$ and let $\mathcal{P}_{i+1}$ be the partition that refines each part of $\mathcal{P}_i$ by the intersection with S and T, in particular, this refines each part into at most 4 parts, and $\mu(\Omega_{\mathscr{P}_{i+1}}) \le \mu(\Omega_{\mathscr{P}_i}) + \mu(S) + \mu(T)$. Proceeding as before, we have

$$\|W_{\mathscr{P}_i}\|_2^2 + \varepsilon^2 \le \|W_{\mathscr{P}_{i+1}}\|_2^2 \le \|W\|_2^2,$$

and furthermore,
$$\|D_{\mathbb{W}_{\mathscr{P}_i}}\|_2 \le \|D_{\mathbb{W}_{\mathscr{P}_{i+1}}}\|_2 \le \|D_{\mathbb{W}}\|_2.$$

The first step can occur at most $\|D_\mathbb{W}\|_2/\varepsilon^2$ times, and the second at most $\|W\|_2^2/\varepsilon^2$ times. Since in the first step, the number of partition classes at most doubles, and in the second it goes up by at most a factor of four, this proves that there exists a partition $\mathscr{P}$ with at most

$$2^{(2\|W\|_2^2 + \|D_\mathbb{W}\|_2^2)/\varepsilon^2} \le 2^{(2BC + CD)/\varepsilon^2} = M(\varepsilon)$$

classes such that $d_\square(\mathbb{W}, \mathbb{W}_{\mathscr{P}}) \le \varepsilon$.

To prove that $\mu(\Omega_{\mathscr{P}}) \le N(\varepsilon)$, we claim that in each step, (10) implies that

$$\mu(S) \le \frac{4C^2}{\varepsilon^2},$$

and (11) implies that
$$\mu(S), \mu(T) \le \frac{4CD}{\varepsilon^2}.$$

Indeed, for any $S \subseteq \Omega$,
$$\left|\int_S D_\mathbb{W}(x) - D_{\mathbb{W}_{\mathcal{P}_i}}(x)\right| \le 2C.$$

This implies that if
$$\frac{1}{\sqrt{\mu(S)}}\left|\int_S D_\mathbb{W}(x)\,d\mu(x) - \int_S D_{\mathbb{W}_{\mathscr{P}_i}}(x)\,d\mu(x)\right| \ge \varepsilon$$

then
$$\mu(S) \leq \frac{4C^2}{\varepsilon^2}.$$

On the other hand,
$$\left|\int_{S\times T} W - W_{\mathscr{P}}\right| \leq 2C \quad \text{and} \quad \left|\int_{S\times T} W - W_{\mathscr{P}}\right| \leq 2D\mu(T).$$

Therefore,
$$\left|\int_{S\times T} W - W_{\mathscr{P}}\right| \leq 2\sqrt{CD\mu(T)}.$$

This implies that if
$$\frac{1}{\sqrt{\mu(S)\mu(T)}} \left|\int_{S\times T} W - W_{\mathscr{P}_i}\right| \geq \varepsilon$$

then
$$\mu(S) \leq \frac{4CD}{\varepsilon^2}.$$

The bound for $\mu(T)$ follows similarly.

Since first step can occur at most $\|D_{\mathbb{W}}\|_2^2/\varepsilon^2$ times, and the second at most $\|W\|_2^2/\varepsilon^2$ times, this shows that

$$\mu(\Omega_{\mathscr{P}}) \leq \frac{4C^2}{\varepsilon^2} \frac{\|D_{\mathbb{W}}\|_2^2}{\varepsilon^2} + 2\frac{4CD}{\varepsilon^2} \frac{\|W\|_2^2}{\varepsilon^2} \leq \frac{4}{\varepsilon^4}\left(C^3 D + 2BC^2 D\right) = N(\varepsilon).$$

The second statement follows by choosing $\mathscr{P}_0 = \mathscr{Q}$. □

Remark 5.4 With the help of Proposition 3.19, Theorem 5.3 can immediately be transformed into a similar statement for the kernel distance $d_{2\to 2}(\mathbb{W}, \mathbb{W}_{\mathscr{P}})$, provided $N(\varepsilon)$ and $M(\varepsilon)$ are replaced by bounds of the form $M(\varepsilon) = 2^{c/\varepsilon^8}$ and $N(\varepsilon) = d/\varepsilon^{16}$ where c and d are constants depending on B, C and D.

We would like to prove a version of this "regularity lemma" for $d_{2\to 2}$ where the parts have equal size. We first show some preliminary lemmas.

Lemma 5.5 *Let $\mathbb{W}_1$, $\mathbb{W}_2$ be two graphexes on the same space Ω, and let $\mathscr{P}$ be a finite subspace partition of Ω. Then*

$$\|W_{1,\mathscr{P}} - W_{2,\mathscr{P}}\|_{2\to 2} \leq \|W_1 - W_2\|_{2\to 2}$$

and

$$\|D_{\mathbb{W}_{1,\mathscr{P}}} - D_{\mathbb{W}_{2,\mathscr{P}}}\|_2 \leq \|D_{\mathbb{W}_1} - D_{\mathbb{W}_2}\|_2.$$

Proof Note that

$$\|W_{1,\mathscr{P}} - W_{2,\mathscr{P}}\|_{2\to 2} = \sup_{\substack{f,g\in L^2(\Omega) \\ \|f\|_2=\|g\|_2=1}} f\circ(W_{1,\mathscr{P}} - W_{2,\mathscr{P}})\circ g$$

$$= \sup_{\substack{f,g\in L^2(\Omega_\mathscr{P}) \\ \|f\|_2=\|g\|_2=1}} f\circ(W_{1,\mathscr{P}} - W_{2,\mathscr{P}})\circ g.$$

If we let $f_\mathscr{P}$ and $g_\mathscr{P}$ consist of the average values of f and g on each part of $\mathcal{P}$ and zero outside $\Omega_\mathscr{P}$, then

$$f\circ(W_{1,\mathscr{P}} - W_{2,\mathscr{P}})\circ g = f_\mathscr{P}\circ(W_{1,\mathscr{P}} - W_{2,\mathscr{P}})\circ g_\mathscr{P} = f_\mathscr{P}\circ(W_1 - W_2)\circ g_\mathscr{P},$$

which implies the first claim. The second claim follows similarly. □

Corollary 5.6 *Suppose that* $\mathbb{W}$ *is a graphex, and that* $\mathbb{U}$ *is a step graphex over the same space as* $\mathbb{W}$. *If* $\mathscr{P} = (\Omega_\mathscr{P}, \mathcal{P})$ *is a finite subspace partition such that* $\mathbb{U}$ *is constant on each part of* $\mathcal{P}$ *and zero outside* $\Omega_\mathscr{P}$, *then*

$$\|W_\mathscr{P} - W\|_{2\to 2} \le 2\|U - W\|_{2\to 2}$$

and

$$\|D_{\mathbb{W}_\mathscr{P}} - D_\mathbb{W}\|_2 \le 2\|D_\mathbb{U} - D_\mathbb{W}\|_2.$$

Proof Note that $\mathbb{U}_\mathscr{P} = \mathbb{U}$. We have

$$\|W_\mathscr{P} - W\|_{2\to 2} \le \|W_\mathscr{P} - U\|_{2\to 2} + \|U - W\|_{2\to 2}$$
$$= \|W_\mathscr{P} - U_\mathscr{P}\|_{2\to 2} + \|U - W\|_{2\to 2} \le 2\|U - W\|_{2\to 2}.$$

Similarly,

$$\|D_{\mathbb{W}_\mathscr{P}} - D_\mathbb{W}\|_2 \le \|D_{\mathbb{W}_\mathscr{P}} - D_\mathbb{U}\|_2 + \|D_\mathbb{U} - D_\mathbb{W}\|_2$$
$$= \|D_{\mathbb{W}_\mathscr{P}} - D_{\mathbb{U}_\mathscr{P}}\|_2 + \|D_\mathbb{U} - D_\mathbb{W}\|_2 \le 2\|D_\mathbb{U} - D_\mathbb{W}\|_2.$$

□

Theorem 5.7 *Given* B, C, D, *and* $\varepsilon > 0$, *there exists* $\rho_0 = \rho_0(\varepsilon, B, C, D) > 0$ *and* $N_0 = N_0(\varepsilon, B, C, D)$ *such that for any* $\rho < \rho_0$, *any* $m \ge N_0/\rho$, *and any* (B, C, D)-*bounded signed graphex* $\mathbb{W}$ *on an atomless space with infinite measure, there exists a subspace partition* $\mathscr{P} = (\Omega_\mathscr{P}, \mathcal{P})$ *with exactly* m *parts of size* ρ *such that* $d_{2\to 2}(\mathbb{W}, \mathbb{W}_\mathscr{P}) \le \varepsilon$. *If* $\mathscr{P}_0 = (\Omega_{\mathscr{P}_0}, \mathcal{P}_0)$ *is an arbitrary finite subspace partition, we can require* $\mathscr{P}$ *to refine* $\mathscr{P}_0$, *as long as each part of* $\mathcal{P}_0$ *is divisible by* ρ (*and increasing the bound on* $N_0(\varepsilon)$ *and decreasing* ρ_0 *appropriately depending on* $|\mathcal{P}_0|$ *and* $\mu(\Omega_{\mathscr{P}_0})$).

Proof Apply Theorem 5.3 and Proposition 3.19 to obtain a subspace partition $\mathscr{P}'$ with at most $M(\varepsilon/3)$ parts and size at most $N(\varepsilon/3)$ such that $d_{2\to 2}(\mathbb{W},\mathbb{W}_{\mathscr{P}'}) \leq \varepsilon/3$. We first construct a refinement $\mathscr{Q} = (\Omega_{\mathscr{Q}}, \mathcal{Q})$ as follows. Add a part from $\Omega \setminus \Omega_{\mathscr{P}'}$ so that the total measure of $\Omega_{\mathscr{Q}}$ is equal to $m\rho$ (this will require $m \geq N(\varepsilon/3)/\rho$), and divide each part of $\mathcal{P}' \cup \{\Omega_{\mathscr{Q}} \setminus \Omega_{\mathscr{P}'}\}$ into parts of size ρ, with perhaps one part remaining of smaller size. We then define $\mathscr{P}$ by combining the parts of $\mathcal{Q}$ that have size smaller than ρ, including the part added from $\Omega \setminus \Omega_{\mathscr{P}'}$, into a single set Ω', and then dividing Ω' into parts of size ρ (and keeping the remaining parts of $\mathscr{Q}$). Then $\mathbb{W}_{\mathcal{Q}}$ and $\mathbb{W}_{\mathcal{P}}$ differ only on Ω', which has size at most $(M(\varepsilon/3)+1)\rho$, implying that

$$\|W_{\mathscr{Q}} - W_{\mathscr{P}}\|_{2\to 2}^2 \leq \|W_{\mathscr{Q}} - W_{\mathscr{P}}\|_2^2$$
$$\leq 2\int_{\Omega \times \Omega'} (W_{\mathscr{Q}} - W_{\mathscr{P}})^2 \leq 2\int_{\Omega \times \Omega'} B(|W_{\mathscr{Q}}| + |W_{\mathscr{P}}|)$$
$$\leq 4B \int_{\Omega'} D_{|W|} \leq 4(M(\varepsilon/3)+1)\rho B D.$$

We also have

$$\|D_{W_{\mathscr{Q}}} - D_{W_{\mathscr{P}}}\|_2 \leq 2\sqrt{\int_{\Omega'} D_{|W|}(x)^2 \, d\mu(x)} \leq 2\sqrt{(M(\varepsilon/3)+1)\rho} D.$$

Furthermore, we know that $\mathbb{W}_{\mathscr{P}'}$ is constant on each part of $\mathscr{Q}$. Therefore, applying Corollary 5.6, we have

$$\|W - W_{\mathscr{Q}}\|_{2\to 2} \leq 2\|W - \mathbb{W}_{\mathscr{P}'}\|_{2\to 2} \leq 2\varepsilon/3$$

and

$$\|D_W - D_{W_{\mathscr{Q}}}\|_2 \leq 2\|D_W - D_{W_{\mathscr{P}'}}\|_2 \leq 2\varepsilon^2/9.$$

Therefore, if ρ is small enough, then

$$\|W - W_{\mathscr{P}}\|_{2\to 2} \leq \varepsilon$$

and

$$\|D_W - D_{W_{\mathscr{P}}}\|_2 \leq \varepsilon^2.$$

We can add parts of zero measure and the above argument still works. If we want $\mathscr{P}$ to be a refinement of a starting partition $\mathscr{P}_0$, we apply Theorem 5.3 and Proposition 3.19 and make sure to combine the leftover parts so that they are within the same part of $\mathcal{P}_0$, this can be done since each part of $\mathcal{P}_0$ is divisible by ρ. $\square$

We close this section by proving Theorem 2.21. To this end, we will first establish two lemmas.

Lemma 5.8 *Let $\mathcal{S}$ be a set of signed graphexes over atomless spaces of infinite measure such that the following condition holds:*

(a) *For every $\varepsilon > 0$, there exists B, C, D such that for any $\mathbb{W} = (W, S, I, (\Omega, \mathcal{F}, \mu)) \in \mathcal{S}$, taking $\Omega_{\leq D}$ to be the set of points with $D_{|\mathbb{W}|}(x) \leq D$, we have that $\mu(\Omega \setminus \Omega_{\leq D}) \leq \varepsilon$, $\|W|_{\Omega_{\leq D}}\|_\infty \leq B$, and $\|\mathbb{W}|_{\Omega_{\leq D}}\|_1 \leq C$.*

Then there exist strictly increasing sequences of integers $a_k \geq 2k$ and $b_k \geq k$ and sequences of positive real numbers B_k, C_k, and D_k such that the following holds. For any graphex $\mathbb{W} \in \mathcal{S}$, $\mathbb{W} = (W, S, I, \Omega)$ with $\Omega = (\Omega, \mathcal{F}, \mu)$, there exists a sequence of subsets $P_{k,0} \subseteq \Omega$ and subspace partitions $\mathscr{P}_k = (\Omega_{\mathscr{P}_k}, \mathcal{P}_k)$, $\mathcal{P}_k = \{P_{k,1}, \ldots, P_{k,m_k}\}$ with $m_k = 2^{a_k + b_k}$, such that for all k, $\Omega_{\mathscr{P}_k}$ is disjoint from $P_{k,0}$ and

(1) $P_{k+1,0} \subseteq P_{k,0}$ and $\mu(P_{k,0}) = 2^{-2k}$,
(2) $\mathbb{W}_k = \mathbb{W}|_{\Omega \setminus P_{k,0}}$ is (B_k, C_k, D_k)-bounded,
(3) $\mathscr{P}_{k+1}$ refines $\mathscr{P}_k$,
(4) $P_{k,i}$ for $i \geq 1$ has measure 2^{-a_k}, and
(5) $d_{2 \to 2}(\mathbb{W}_k, (\mathbb{W}_k)_{\mathscr{P}_k}) \leq 2^{-k}$.

Note that by property (3) from Theorem 4.1, every tight set of unsigned graphexes obeys the condition (a) (with $B = 1$), showing that the conclusions of the theorem hold for any tight set $\mathcal{S}$ of graphexes over atomless spaces of infinite measure.

Proof Define
$$D_k = \inf\{D : \text{for all } \mathbb{W} \in \mathcal{S}, \mu(\Omega \setminus \Omega_{\leq D}) \leq 2^{-2k}\},$$

and let $B_k = \sup_{\mathbb{W} \in \mathcal{S}} \|W|_{\Omega_{\leq D_k}}\|_\infty$ and $C_k = \sup_{\mathbb{W} \in \mathcal{S}} \|\mathbb{W}|_{\Omega_{\leq D_k}}\|_1$. By the condition (a) these are finite, and by construction they are monotone non-decreasing functions of k. Given a graphex, we then first set each $P'_{k,0}$ to be the set of points with degree greater than D_k. In this way we have each $P'_{k+1,0} \subseteq P'_{k,0}$; however, they may be strictly smaller than the required size. We therefore extend them one by one, starting with $P_{0,0}$, and make sure that we still have each $P_{k+1,0} \subseteq P_{k,0}$. Taking $\mathbb{W}_k$ to be the restrictions, properties (1) and (2) are satisfied.

Next, apply Theorem 5.7 with $\varepsilon = 1$ to obtain N_0 and ρ_0. Increasing N_0 (if needed) so that it is of the form 2^{b_0} for some nonnegative integer b_0, we then choose a_0 such that $2^{-a_0} < \rho_0$. For any graphex, take $\mathscr{P}_0$ according to the theorem with $\rho = 2^{-a_0}$. Keep iterating the theorem, in each step applying Theorem 5.7 with $B = B_k$, $C = C_k$, $D = D_k$, $\varepsilon = 2^{-k}$, and $\mathcal{P}_0 = (\Omega_{\mathscr{P}_{k-1}} \cup (P_{k-1,0} \setminus P_{k,0}), \mathcal{P}_k \cup \{P_{k-1,0} \setminus P_{k,0}\})$, ensuring in each step that $a_k \geq \max\{2k, a_{k-1} + 1\}$, $2^{-a_k} < \rho_k$, and $b_k > b_{k-1}$. □

Lemma 5.9 *Every tight sequence of graphexes has a subsequence that is $\delta_\diamond$-convergent. More generally, every sequence of signed graphexes obeying the condition (a) from Lemma 5.8 has a $\delta_\diamond$-convergent subsequence. If condition (a) holds with one or more of the constants B, C, D not depending on ε, then the subsequential limit inherits the corresponding bound.*

During the proof, we will use the following.

Claim *Suppose that $\mathbb{W} = (W, S, I, \mathbf{\Omega})$ is a signed graphex, $\mathscr{P} = (\Omega_{\mathscr{P}}, \mathcal{P})$ is a finite subspace partition, and $\mathcal{P}_0 \subseteq \mathcal{P}$. Let $\Omega_0 = \bigcup_{P \in \mathcal{P}_0} P$, $\mathcal{P}' = \mathcal{P} \setminus \mathcal{P}_0$, and $\mathscr{P}' = (\Omega'_{\mathscr{P}'}, \mathcal{P}')$, where $\Omega'_{\mathscr{P}'} = \Omega_{\mathscr{P}} \setminus \Omega_0$. Then $(\mathbb{W}_{\mathscr{P}})|_{\Omega'} = (\mathbb{W}|_{\Omega'})_{\mathscr{P}'}$.*

Proof Note that $\Omega \setminus \Omega_{\mathscr{P}} = \Omega' \setminus \Omega'_{\mathscr{P}'}$ (since Ω_0 is disjoint from both). First, we have

$$I_{\mathscr{P}} = \frac{1}{2} \int_{(\Omega \setminus \Omega_{\mathscr{P}}) \times (\Omega \setminus \Omega_{\mathscr{P}})} W(x, y) \, d\mu(x) \, d\mu(y) + \int_{(\Omega \setminus \Omega_{\mathscr{P}})} S(x) \, d\mu(x) + I$$

$$= \frac{1}{2} \int_{(\Omega' \setminus \Omega'_{\mathscr{P}'}) \times (\Omega' \setminus \Omega'_{\mathscr{P}'})} W(x, y) \, d\mu(x) \, d\mu(y) + \int_{(\Omega' \setminus \Omega'_{\mathscr{P}'})} S(x) \, d\mu(x) + I = I_{\mathscr{P}'}.$$

We also have for $x \in \Omega'$, if $x \in P_i$, then

$$S_{\mathscr{P}}(x) = \frac{1}{\mu(P_i)} \int_{P_i} \left(S(x) + \int_{\Omega \setminus \Omega_{\mathscr{P}}} W(x, y) \, d\mu(y) \right) d\mu(x)$$

$$= \frac{1}{\mu(P_i)} \int_{P_i} \left(S(x) + \int_{\Omega' \setminus \Omega'_{\mathscr{P}}} W(x, y) \, d\mu(y) \right) d\mu(x) = S_{\mathscr{P}'}(x),$$

and $S_{\mathscr{P}}(x) = 0 = S_{\mathscr{P}'}(x)$ if $x \in \Omega' \setminus \Omega'_{\mathscr{P}'}$. Finally, if $x, y \in \Omega'$ and $x \in P_i$, $y \in P_j$ with $P_i, P_j \in \mathcal{P}'$, then

$$W_{\mathscr{P}}(x, y) = \frac{1}{\mu(P_i)\mu(P_j)} \int_{P_i \times P_j} W(x', y') \, d\mu(x') \, d\mu(y') = (W|_{\Omega'})_{SP'},$$

and 0 otherwise. $\square$

Proof (*Lemma 5.9*) Let $\mathbb{W}_1, \mathbb{W}_2, \ldots, \mathbb{W}_n, \ldots$ be a sequence of signed graphexes obeying the condition (a), and let $\widetilde{\mathbb{W}}_i$ be obtained from $\mathbb{W}_i$ by appending an arbitrary σ-finite space of infinite measure. Then $\widetilde{\mathbb{W}}_1, \widetilde{\mathbb{W}}_2, \ldots, \widetilde{\mathbb{W}}_n, \ldots$ obeys the condition (a) as well, and arguing as in Remark 4.4, we can assume without loss of generality that $\widetilde{\mathbb{W}}_1, \widetilde{\mathbb{W}}_2, \ldots, \widetilde{\mathbb{W}}_n, \ldots$ are all defined over atomless spaces. We want to show that they have a subsequence that converges to a graphex.

We can take for each n and k sets $\widetilde{P}_{n,k,0}$ and subspace partitions $\widetilde{\mathscr{P}}_{n,k}$ as in Lemma 5.8, defining in particular $\widetilde{\mathbb{W}}_{n,k}$ as the restriction of $\widetilde{\mathbb{W}}_n$ to $\widetilde{\Omega}_n \setminus \widetilde{P}_{n,k,0}$. For $k \geq k_0$, we will also define $\widetilde{\mathbb{W}}_{n,k,k_0}$ as the restriction of $(\widetilde{\mathbb{W}}_{n,k})_{\widetilde{\mathscr{P}}_{n,k}}$ to $\widetilde{\Omega}_n \setminus \widetilde{P}_{n,k_0,0}$. By the above claim, $\widetilde{\mathbb{W}}_{n,k,k_0} = (\widetilde{\mathbb{W}}_{n,k_0})_{\widetilde{\mathscr{P}}_{n,k,k_0}}$, where $\widetilde{\mathscr{P}}_{n,k,k_0}$ consists of the classes in $\widetilde{\mathscr{P}}_{n,k}$ which are subsets of $\widetilde{\Omega}_n \setminus \widetilde{P}_{n,k_0,0}$. This implies in particular that $\widetilde{\mathbb{W}}_{n,k,k_0}$ is $(B_{k_0}, C_{k_0}, D_{k_0})$-bounded.

Furthermore, in view of Remark 3.6, we can replace each $(\widetilde{\mathbb{W}}_{n,k})_{\widetilde{\mathscr{P}}_{n,k}}$ by an equivalent step function $\mathbb{W}_{n,k}$ over $\mathbb{R}_+$, where the first part is $P_{n,k,0} := [0, 2^{-2k})$ (which is disjoint from dsupp $\mathbb{W}_{n,k}$), the remaining parts $P_{n,k,i}$ for $i \geq 1$ are of the form $[\ell/2^{a_k}, (\ell+1)/2^{a_k})$ for some nonnegative integer ℓ, and we extend $\mathbb{W}_{n,k}$ to zero above $N_k + 2^{-2k}$, where $N_k = 2^{b_k}$.

Let $\mathscr{P}'_k = \big([2^{-2k}, N_k + 2^{-2k}), \mathcal{P}'_k\big)$, where $\mathcal{P}'_k$ partitions $[2^{-2k}, N_k + 2^{-2k})$ into intervals of length 2^{-a_k}. Note that after this change, the bound (5) from Lemma 5.8 becomes the bound $\delta_{2\to 2}(\widetilde{\mathbb{W}}_{n,k}, \mathbb{W}_{n,k}) \leq 2^{-k}$. In this way, we have mapped the "steps" of each step graphex to $\mathcal{P}'_k$. For each k, the graphex $\mathbb{W}_{n,k}$ then just depends on a finite number of parameters, each bounded. We can then use a diagonalization argument to take a subsequence so that for every k, $\mathbb{W}_{n,k}$ converges to some $\mathbb{W}^k$ as $n \to \infty$, in the sense that $\mathbb{W}^k$ is a step graphex with the same parts and each value of the function converges, which also implies that $\|\mathbb{W}_{n,k}\|_1 \to \|\mathbb{W}^k\|_1$ and $\mathbb{W}_{n,k} \to \mathbb{W}^k$ in the metric $\delta_{2\to 2}$ (since there are a finite number of steps).

Given $k \geq k_0$, let $\mathbb{W}^{k,k_0}$ be equal to $\mathbb{W}^k$ restricted to $[2^{-2k_0}, \infty)$, let $\mathcal{P}_{k,k_0}$ consist of those intervals in $\mathcal{P}'_k$ which are above 2^{-2k_0}, and finally, let $\mathscr{P}_{k,k_0} = (\Omega_{k,k_0}, \mathcal{P}_{k,k_0})$ where $\Omega_{k,k_0} = [2^{-2k_0}, N_k + 2^{-2k})$. We claim that $\mathbb{W}^{k,k_0} = \mathbb{W}^{k+1,k_0}_{\mathscr{P}_{k,k_0}}$. This follows from the fact that for any n, if $\mathbb{W}_{n,k,k_0}$ is $\mathbb{W}_{n,k}$ restricted to $[2^{-2k_0}, \infty)$, then $\big(\mathbb{W}_{n,k+1,k_0}\big)_{\mathscr{P}_{k,k_0}} = \mathbb{W}_{n,k,k_0}$, by the above claim.

Next, recalling that $\widetilde{\mathbb{W}}_{n,k,k_0}$ is $(B_{k_0}, C_{k_0}, D_{k_0})$-bounded for each n and k, we have that $\mathbb{W}^{k,k_0}$ is $(B_{k_0}, C_{k_0}, D_{k_0})$-bounded as well, implying in particular that $\|\mathbb{W}^{k,k_0}\|_1 \leq C_{k_0}$. Also note that if $P_i, P_j \in \mathcal{P}_{k,k_0}$, then

$$\int_{P_i \times P_j} |W^{k,k_0}(x,y)| \, d\mu(x) \, d\mu(y) \leq \int_{P_i \times P_j} |W^{k+1,k_0}(x,y)| \, d\mu(x) \, d\mu(y).$$

Since W^{k,k_0} is supported on the union of $P_i \times P_j$ for all choices of P_i and P_j, this implies that $\|W^{k,k_0}\|_1$ cannot decrease as k increases. Together, these observations imply that the limit $\lim_{k' \to \infty} \|W^{k',k_0}\|_1$ exists and is at most C_{k_0}. Given $\varepsilon > 0$, we can therefore find $k(\varepsilon, k_0) < \infty$ such that $\lim_{k' \to \infty} \|W^{k',k_0}\|_1 - \|W^{k,k_0}\|_1 < \varepsilon$ for all $k \geq k(\varepsilon, k_0)$. Furthermore, because Ω^2_{k,k_0} is the support of W^{k,k_0},

$$\int_{\Omega^2_{k,k_0}} |W^{k,k_0}(x,y)| \, d\mu(x) \, d\mu(y) \leq \int_{\Omega^2_{k,k_0}} |W^{k+1,k_0}(x,y)| \, d\mu(x) \, d\mu(y).$$

As a consequence, for all $k' \geq k(\varepsilon, k_0)$,

$$\int_{\mathbb{R}^2_+ \setminus \Omega^2_{k,k_0}} |W^{k',k_0}(x,y)| \, d\mu(x) \, d\mu(y) < \varepsilon.$$

Therefore, the random variables W^{k,k_0} are uniformly integrable. By the martingale convergence theorem (applied to each $[2^{-2k_0}, 2^{-2k'} + N_{k'}] \times [2^{-2k_0}, 2^{-2k'} + N_{k'}]$ with $k' \geq k_0$), the graphon part W^{k,k_0} of $\mathbb{W}^{k,k_0}$ is pointwise convergent almost everywhere to a function $\widetilde{W}^{k_0}$ defined on $[2^{-k_0}, \infty)^2$, and it also converges to $\widetilde{W}^{k_0}$ in L^1. Since $\|W^{k,k_0}\|_\infty \leq B_{k_0}$, this implies convergence in L^2, and hence in the kernel metric $\|\cdot\|_{2\to 2}$. Furthermore, the graphon marginals converge in L^1, because

$$\|D_{W^{k,k_0}} - D_{\widetilde{W}^{k_0}}\|_1 \leq \|W^{k,k_0} - \widetilde{W}^{k_0}\|_1 \xrightarrow[k \to \infty]{} 0.$$

We also have

$$\int_{P_i} |D_{\mathbb{W}^{k,k_0}}(x)|\, d\mu(x) \le \int_{P_i} |D_{\mathbb{W}^{k+1,k_0}}(x)|\, d\mu(x).$$

This implies that $\|D_{\mathbb{W}^{k,k_0}}\|_1$ cannot decrease as k increases. Since Ω_{k,k_0} is the support of $D_{\mathbb{W}^{k,k_0}}$,

$$\int_{\Omega_{k,k_0}} |D_{\mathbb{W}^{k,k_0}}(x)|\, d\mu(x) \le \int_{\Omega_{k,k_0}} |D_{\mathbb{W}^{k+1,k_0}}(x)|\, d\mu(x).$$

As before, this implies that the functions $D_{\mathbb{W}^{k,k_0}}$ are uniformly integrable and uniformly bounded (by D_{k_0}). We can again use the martingale convergence theorem (applied to each $[2^{-2k_0}, 2^{-2k'} + N_{k'}]$) to show that $D_{\mathbb{W}^{k,k_0}}$ converges pointwise and in L^1, and therefore in L^2, to a function $D_{\widetilde{\mathbb{W}}^{k_0}}$ taking values in $[-D_{k_0}, D_{k_0}]$. Define

$$\widetilde{S}^{k_0}(x) = D_{\widetilde{\mathbb{W}}^{k_0}}(x) - D_{\widetilde{W}^{k_0}}(x).$$

Since we also have

$$S^{k,k_0}(x) = D_{\mathbb{W}^{k,k_0}}(x) - D_{W^{k,k_0}}(x),$$

and since both terms in the difference converge in L^1, S^{k,k_0} converges in L^1 to $\widetilde{S}^{k_0}$. By Theorem 3.12 in [24] this implies that some subsequence converges pointwise almost everywhere, showing in particular that in the unsigned case, $\widetilde{S}^{k_0} \ge 0$ almost everywhere.

We also have that $\rho := \rho(\mathbb{W}_{k,k_0})$ is constant in k. Define

$$\widetilde{I}^{k_0} = \rho - 2\int \widetilde{S}^{k_0} - \int \widetilde{W}^{k_0}.$$

Since for each k,

$$I^{k,k_0} = \rho - 2\int S^{k,k_0} - \int W^{k,k_0},$$

and since S^{k,k_0} and W^{k,k_0} converge in L^1 to $\widetilde{S}^{k_0}$ and $\widetilde{W}^{k_0}$, respectively, this implies that I^{k,k_0} converges to $\widetilde{I}^{k_0}$. Because W^{k,k_0}, S^{k,k_0}, and $D_{\mathbb{W}}^{k,k_0}$ converge in L^1 (and hence pointwise almost everywhere on some subsequence), the limit inherits $(B_{k_0}, C_{k_0}, D_{k_0})$-boundedness from $\mathbb{W}^{k,k_0}$.

Since for each k, $\mathbb{W}^{k,k_0+1}$, when restricted to $[2^{-2k_0}, \infty)$, is equal to $\mathbb{W}^{k,k_0}$, and since $\widetilde{W}^{k_0+1}$ and $\widetilde{W}^{k_0}$ are pointwise limits along some subsequence, $\widetilde{W}^{k_0}$ is also the restriction of $\widetilde{W}^{k_0+1}$ to $[2^{-2k_0}, \infty)$. Therefore, we can define a signed graphex $\mathbb{W}$ on $\mathbb{R}_+$ as the "union" of the signed graphexes $\widetilde{W}^{k_0}$. Note that the limit $\mathbb{W}$ is locally finite by the fact that $\widetilde{W}^{k_0}$ is $(B_{k_0}, C_{k_0}, D_{k_0})$-bounded. This also implies that $\mathbb{W}$ inherits any of these bounds from $\mathbb{W}^{k_0}$ that do not depend on k_0.

We claim that on the subsequence where $\mathbb{W}_{n,k}$ converges to $\mathbb{W}^k$ in $\delta_{2\to 2}$ for each k, $\mathbb{W}_n \to \mathbb{W}$ in the weak kernel metric $\delta_\diamond$. To see this, we first note that $\widetilde{W}^{k_0}$ is obtained from $\mathbb{W}$ by removing a set of measure 2^{-2k_0}, and $\mathbb{W}^{k,k_0}$ is obtained from $\mathbb{W}^k$ by removing a set of the same measure, showing that for any k_0,

$$\delta_\diamond(\mathbb{W}, \mathbb{W}^k) \leq \max\{2^{-k_0}, \delta_{2\to 2}(\widetilde{\mathbb{W}}^{k_0}, \mathbb{W}^{k,k_0})\}.$$

Given $\varepsilon > 0$, choose k_0 such that $2^{-k_0} \leq \varepsilon/2$. Since $\delta_{2\to 2}(\mathbb{W}^{k,k_0}, \widetilde{\mathbb{W}}^{k_0}) \to 0$ for each k_0 as $k \to \infty$, this shows that for k large enough, $\delta_\diamond(\mathbb{W}, \mathbb{W}^k) \leq \varepsilon/2$. In a similar way,

$$\delta_\diamond(\mathbb{W}_n, \mathbb{W}^k) = \delta_\diamond(\widetilde{\mathbb{W}}_n, \mathbb{W}^k) \leq \max\{2^{-k}, \delta_{2\to 2}(\widetilde{\mathbb{W}}_{n,k}, \mathbb{W}^k)\}.$$

Since $\delta_{2\to 2}(\widetilde{\mathbb{W}}_{n,k}, \mathbb{W}_{n,k}) \leq 2^{-k}$ and $\delta_{2\to 2}(\mathbb{W}_{n,k}, \mathbb{W}^k) \to 0$ for each k as $n \to \infty$, we can first choose k and then n large enough to guarantee that the right side is smaller than $\varepsilon/2$. Combined with the triangle inequality for $\delta_\diamond$, this shows that for all $\varepsilon > 0$ we can find an n_0 such that for $n \geq n_0$, we have $\delta_\diamond(\mathbb{W}_n, \mathbb{W}) \leq \varepsilon$, as claimed. □

Remark 5.10 It is not hard to see that a sequence of signed graphexes $\mathbb{W}_n = (W_n, S_n, I_n, \Omega_n)$ with $\|W_n\|_\infty \leq B$ and $\|W_n\|_1 \leq C$ obeys the condition (a) from Lemma 5.8. If $B \leq 1$, this follows by applying Corollary 4.2 (1) to the sequence $|W_n|$, and for $B > 1$ it follows by applying Corollary 4.2 (1) to the sequence $(|W_n|/B, |S_n|/B, |I_n|/B, \Omega_n)$. Choosing a convergent subsequence, it is clear from the last proof that the limiting graphex $\mathbb{W} = (W, S, I, \Omega)$ must obey the bound $\|W\|_\infty \leq B$. The statement from Remark 2.22 therefore is a direct consequence of Lemma 5.9.

Proof (Theorem 2.21) By Corollary 4.2 (1), any set of graphexes whose L^1 norms are bounded by C is tight, so by Lemma 5.9 any such sequence of graphexes has a subsequence with a limit $\mathbb{W}$ in the metric $\delta_\diamond$. Lemma 5.9 also implies that the limit inherits the bound on the L^1 norm, and therefore the set of graphexes whose L^1 norms are bounded by C is compact. The same proof gives the statement of the theorem for (C, D)-bounded graphexes.

Suppose now that $\mathbb{W}_1, \mathbb{W}_2, \ldots, \mathbb{W}_n, \ldots$ is a Cauchy sequence in $\delta_\diamond$. We first claim that it must be tight. Indeed, for any $\varepsilon > 0$, there exists n such that for any $m > n$, $\delta_\diamond(\mathbb{W}_n, \mathbb{W}_m) < \varepsilon$. Fix such an n and an $m > n$. By Lemma 3.14, we can then decrease the measures μ_n, μ_m by at most ε^2 such that the $\delta_{2\to 2}$ distance of the resulting graphexes $\widehat{\mathbb{W}}_n, \widehat{\mathbb{W}}_m$ is less than ε. Let $\mathbb{W}'_n, \mathbb{W}'_m$ be trivial extensions of the modified graphexes by spaces of infinite measure, and let $\widetilde{\mathbb{W}}'_n, \widetilde{\mathbb{W}}'_m$ be pullbacks according to a coupling so that $d_{2\to 2}(\widetilde{\mathbb{W}}'_n, \widetilde{\mathbb{W}}'_m) < \varepsilon$, which exists by the definition of $\delta_{2\to 2}$. Furthermore, since every finite set of graphexes is tight, we can find (C, D) such that we can remove a set of measure ε^2 from $\widehat{\mathbb{W}}_n$ to make it (C, D)-bounded (independent of m). If we remove the pullback of this set from the underlying space of $\widetilde{\mathbb{W}}'_n$ and $\widetilde{\mathbb{W}}'_m$ and replace them by the restrictions, then the two graphexes will still have $\delta_{2\to 2}$ distance at most 2ε by Lemma 3.13. In particular, this means that $\|\widetilde{\mathbb{W}}'_m\|_1 \leq C + 8\varepsilon^3$. We also have

$$\|D_{\widetilde{\mathbb{W}}'_m}\|_2 \leq \|D_{\widetilde{\mathbb{W}}'_n}\|_2 + \|D_{\widetilde{\mathbb{W}}'_m} - D_{\widetilde{\mathbb{W}}'_n}\|_2 \leq \sqrt{CD} + 4\varepsilon^2 \leq 2\sqrt{CD}.$$

Therefore, the measure of the points x for which $D_{\widetilde{\mathbb{W}}'_m}(x) > \sqrt{CD/\varepsilon}$ is at most 4ε. Taking $C' = C + \varepsilon$ and $D' = \sqrt{CD/\varepsilon}$, we have obtained that for any $m > n$ we can remove a set of measure at most 6ε so that the remainder is (C', D')-bounded. Since any finite set is tight, and the union of two tight sets is tight, this means that the entire sequence is tight. Therefore, it must have a convergent subsequence that converges to a graphex $\mathbb{W}$. But then because the original sequence was a Cauchy sequence, the entire sequence must converge to $\mathbb{W}$. This proves that the space of graphexes is complete.

The above lemma implies that every tight set is relatively compact, and the fact that any Cauchy sequence must be tight implies that every relatively compact set is tight. $\square$

6 Subgraph Counts

In this section we will prove that convergence in the weak kernel metric implies GP-convergence. The main technical tool for this proof will be the following counting lemma, which says that given any $C, D < \infty$, two (C, D)-bounded graphexes that are close in kernel metric $\delta_{2\to 2}$ must have close subgraph counts.

While this lemma and its corollary are formulated only for unsigned graphexes, we note that both have natural generalizations to signed graphexes. See Remark 6.9 at the end of Sect. 6.1 below.

Lemma 6.1 *Let F be a simple, connected graph with m edges and $n \geq 3$ vertices, and let $C, D < \infty$. Suppose that $\mathbb{W}_1$ and $\mathbb{W}_2$ are graphexes on the same underlying space Ω, with $\|\mathbb{W}_i\|_1 \leq C$ and $\|D_{\mathbb{W}_i}\|_\infty \leq D$ for $i = 1, 2$, and let $\varepsilon = \max\{\|\mathbb{W}_1 - \mathbb{W}_2\|_{2\to 2}, \|D_{\mathbb{W}_1 - \mathbb{W}_2}\|_2\}$. Then*

$$|t(F, \mathbb{W}_1) - t(F, \mathbb{W}_2)| \leq m\varepsilon \widetilde{C} D^{n-3},$$

where $\widetilde{C} = \max\{C, \sqrt{CD}\}$.

Corollary 6.2 *Suppose $\mathbb{W}_n, \mathbb{W}$ have uniformly bounded marginals. If*

$$\delta_{2\to 2}(\mathbb{W}_n, \mathbb{W}) \to 0,$$

then for any finite graph F with no isolated vertices, $t(F, \mathbb{W}_n)$ converges to $t(F, \mathbb{W})$.

Proof By the definition of $\delta_{2\to 2}$, if F is an edge, then $t(F, \mathbb{W}_n) \to t(F, \mathbb{W})$. This implies in particular that $\|\mathbb{W}\|_1$ and $\|\mathbb{W}_n\|_1$ are uniformly bounded. By Lemma 6.1, it follows that $t(F, \mathbb{W}_n) \to t(F, \mathbb{W})$ for any connected graph F. Since homomorphism densities factor over connected components of the graph F, this means that if F is a finite graph without isolated vertices, then $t(F, \mathbb{W}_n) \to t(F, \mathbb{W})$. □

In addition to the above counting lemma, we will need to show that convergence of subgraph counts implies GP-convergence. Thinking of the subgraph counts as the moments of a graphex, this result is similar to standard moment theorems for random variables that show that under suitable growth conditions, the distribution of a random variable is determined by its moments.

Theorem 6.3 *Assume that the marginals of $\mathbb{W}_n$ and $\mathbb{W}$ are bounded by some finite constant D. Then the following are equivalent:*

(1) $G_T(\mathbb{W}_n) \to G_T(\mathbb{W})$ in distribution for every T.
(2) $G_T(\mathbb{W}_n) \to G_T(\mathbb{W})$ in distribution for some T.
(3) For every graph F with no isolated vertices, $t(F, \mathbb{W}_n) \to t(F, \mathbb{W})$.
(4) For every connected graph F, $t(F, \mathbb{W}_n) \to t(F, \mathbb{W})$.

We will prove the counting lemma in Sect. 6.1 below, and Theorem 6.3 in Sect. 6.2. In the final subsection, Sect. 6.3, we use these results to first show that under the assumption of uniformly bounded marginals, $\delta_{2\to 2}$-convergence implies GP-convergence (Theorem 6.16 below). With the help of the results about tightness established in Sect. 4, this in turn allows us to show that without any assumption on the marginals, $\delta_\diamond$-convergence implies GP-convergence (Theorem 6.17 below).

6.1 Proof of the Counting Lemma

In order to prove the counting lemma, it will be convenient to consider several variants of the homomorphism densities. For these variants, it will be natural to consider signed graphexes, since we will need to consider differences of graphexes for the proof of the counting lemma anyway. Note that our proof of the counting lemma can easily be generalized to signed graphons; see Remark 6.9 below.

Definition 6.4 Suppose we have a connected multigraph F, and signed graphexes $\mathbb{W}_e$ assigned to each edge $e \in E(F)$ (refer to this vector of graphexes as $\mathbb{W}_F$), each with the same feature space Ω. Let $V_{\geq 2}$ be the set of vertices with degree at least 2. If $V_{\geq 2}$ is nonempty (i.e., F does not consist of a single edge), then we define

$$t(F, \mathbb{W}_F) = \int_{\Omega^{V_{\geq 2}}} dz_{V_{\geq 2}} \prod_{\{v,w\} \in E(F(V_{\geq 2}))} W_{\{v,w\}}(z_v, z_w)$$
$$\cdot \prod_{v \in V_{\geq 2}} \prod_{\substack{w \in V \setminus V_{\geq 2}: \\ \{v,w\} \in E(F)}} D_{\mathbb{W}_{\{v,w\}}}(z_v).$$

If F consists of just a single edge f, then

$$t(F, \mathbb{W}_F) = \rho(\mathbb{W}_f),$$

with $\rho(\mathbb{W}_f)$ as in (2). Note that for signed graphexes $t(F, \mathbb{W}_F)$ is in general only well defined if the integrals are absolutely convergent, a condition which can, e.g., be guaranteed by requiring that $t(F, \mathbb{W}_F^{\text{abs}}) < \infty$, where $\mathbb{W}_F^{\text{abs}}$ is obtained from $\mathbb{W}_F$ by replacing the graphexes $\mathbb{W}_f$ by $|\mathbb{W}_f|$.

Note that $t(F, \mathbb{W}_F)$ is a multilinear function of the signed graphexes $\mathbb{W}_e$ in $\mathbb{W}_F$. If each $\mathbb{W}_e$ is equal to some fixed $\mathbb{W}$, then this is just the previous definition.

We also define a conditional density where we fix the image of a single vertex.

Definition 6.5 Suppose we have a connected multigraph F with a labeled vertex v_0, more than one edge, and signed graphexes $\mathbb{W}_e$ assigned to each edge as before. Let $x \in \Omega$, and define $z_{v_0} = x$. Take, furthermore, $V'_{\geq 2} = V_{\geq 2} \setminus \{v_0\}$ and $\widetilde{V}_{\geq 2} = V_{\geq 2} \cup \{v_0\}$. Then we define $t_x(F, \mathbb{W}_F)$ to be

$$\int_{\Omega^{V'_{\geq 2}}} dz_{V'_{\geq 2}} \prod_{\{v,w\} \in E(F(\widetilde{V}_{\geq 2}))} W_{\{v,w\}}(z_v, z_w) \prod_{v \in V_{\geq 2}} \prod_{\substack{w \in V \setminus (\widetilde{V}_{\geq 2}): \\ \{v,w\} \in E(F)}} D_{\mathbb{W}_{\{v,w\}}}(z_v).$$

If F consists of just a single edge f adjacent to v_0, then we set $t_x(F, \mathbb{W}) = D_{\mathbb{W}}(x)$. Again, for signed graphexes, this is in general only well defined if the integrals are absolutely convergent.

Note that if $v_0 \in V_{\geq 2}$, then $t_x(F, \mathbb{W}_F)$ is obtained from $t(F, \mathbb{W}_F)$ by simply fixing the feature corresponding to v_0 to be x, implying in particular that $\int t_x(F, \mathbb{W}_F) \, dx = t(F, \mathbb{W}_F)$ (assuming the integrals defining these are absolutely convergent). If $v_0 \notin V_{\geq 2}$, i.e., if v_0 has degree 1, and if F has more than one edge, then the situation is slightly more complicated, since the "feature" of the image of v_0 could be either an element of Ω, or the special value ∞, interpreted earlier as the feature of the leaves of the star part of a graphon process. With this reinterpretation, $t_x(F, \mathbb{W}_F)$ is still obtained from $t(F, \mathbb{W}_F)$ by fixing the feature corresponding to v_0 to be x, but the integral $\int t_x(F, \mathbb{W}_F) \, dx$ now misses the contribution of $x = \infty$, and hence is in general only bounded above by $t(F, \mathbb{W}_F)$. It is, however, equal to $t(F, \mathbb{W}'_F)$, where $\mathbb{W}'_F$ is obtained from $\mathbb{W}_F$ by setting the star part of the graphex corresponding to the edge containing v_0 to 0.

Lemma 6.6 *Suppose F is a connected multigraph with no loops and a labeled vertex v_0, T is a spanning tree, and $\mathbb{W}_e$ is a signed graphex corresponding to each edge e, each with the same feature space Ω. Let $f \in T$ be an edge adjacent to v_0, and $x \in \Omega$. Then*

$$|t_x(F, \mathbb{W}_F)| \leq D_{|\mathbb{W}_f|}(x) \prod_{e \in T \setminus f} \|D_{|\mathbb{W}_e|}\|_\infty \prod_{e \in E(F) \setminus T} \|W_e\|_\infty.$$

Proof Replacing all signed graphexes $\mathbb{W}_e$ by the non-negative versions $|\mathbb{W}_e|$, and noting that $|t_x(F, \mathbb{W}_F)| \leq t_x(F, \mathbb{W}_F^{\text{abs}})$, we may without loss of generality assume that W_e and S_e are non-negative.

Next, assume that F is a tree, i.e., $F = T$. We then prove the claim by induction on the number of edges. If F consists of a single edge, then by definition $t_x(F, \mathbb{W}) = D_\mathbb{W}(x)$, which is exactly the bound in the lemma. Otherwise, we can find an edge $\{v, w\}$ not equal to f, such that v has degree at least two, and w has degree 1 and is different from v_0. The edge $\{v, w\}$ then contributes $D_{\mathbb{W}_{\{v,w\}}}(z_v)$ to the second product in the integral representing $t_x(F, \mathbb{W}_F)$. For each z_v, this is at most $\|D_{\mathbb{W}_{\{v,w\}}}\|_\infty$. Taking a factor $\|D_{\mathbb{W}_{\{v,w\}}}\|_\infty$ out of the integral, and defining F' to be the restriction of F to $V(F) \setminus \{w\}$, we therefore have that

$$t_x(F, \mathbb{W}_F) \leq \|D_{\mathbb{W}_{\{v,w\}}}\|_\infty t_x(F', \mathbb{W}_{F'}).$$

Note that this bound is actually weaker than necessary if the vertex v becomes a vertex of degree one in F', in which case we could have obtained a contribution of $D_{\mathbb{W}_{\{v,v'\}}}(z_{v'})$ for its neighbor v' instead of the contribution $D_{\mathbb{W}_{\{v,v'\}}}(z_{v'})$ implicit in the above bound.

Suppose now that F has edges outside T. Let $\{v, w\}$ be such an edge. Note that both v and w must be in $V_{\geq 2}$. Therefore, this edge contributes $W_{\{v,w\}}(z_v, z_w)$ to the product, which is at most $\|W_{\{v,w\}}\|_\infty$. We can therefore conclude the lemma by induction on the number of edges of F outside T. □

Lemma 6.7 *Suppose F is a connected multigraph with no loops, and we have a signed graphex $\mathbb{W}_e$ corresponding to each edge $e \in F$. Let T be any spanning tree in F, and $f \in T$. Then*

$$|t(F, \mathbb{W}_F)| \leq \|\mathbb{W}_f\|_1 \prod_{e \in T \setminus f} \|D_{|\mathbb{W}_e|}\|_\infty \prod_{e \in E(F) \setminus T} \|W_e\|_\infty.$$

Proof If F consists of a single edge f, then $|t(F, \mathbb{W}_F)| = |\rho(\mathbb{W}_f)| \leq \|\mathbb{W}_f\|_1$ by definition, and if F has more than one edge, then

$$|t(F, \mathbb{W}_F)| \leq \int_\Omega |t_x(F, \mathbb{W}_F)| \, d\mu(x)$$
$$\leq \int_\Omega D_{|\mathbb{W}_f|}(x) \prod_{e \in T \setminus f} \|D_{|\mathbb{W}_e|}\|_\infty \prod_{e \in E(F) \setminus T} \|W_e\|_\infty \, d\mu(x)$$
$$\leq \|\mathbb{W}_f\|_1 \prod_{e \in T \setminus f} \|D_{|\mathbb{W}_e|}\|_\infty \prod_{e \in E(F) \setminus T} \|W_e\|_\infty.$$

□

Proof (Lemma 6.1) Let $f = \{u, v\}$ be an edge in F, and let $\mathbb{W}_{F,f}$ be a vector of graphexes where we assign one of $\mathbb{W}_1$ or $\mathbb{W}_2$ to each edge $e \neq f$, and $(W_1 - W_2, S_1 - S_2, 0, \Omega)$ to f (since F is a connected graph with at least 2 edges, the dust parts of $\mathbb{W}_1$ and $\mathbb{W}_2$ don't contribute to $t(F, \mathbb{W}_1) - t(F, \mathbb{W}_2)$ and can be set to 0). We would like to bound $|t(F, \mathbb{W}_{F,f})|$.

First, assume both endpoints of f have degree at least 2. Let the components of F restricted to $V(F) \setminus \{u, v\}$ be $C_1, C_2, \ldots, C_k$, with corresponding vertex sets $V_1, \ldots, V_k$. There can be three types of components: those with at least one edge to u but none to v, those with at least one edge to v but none to u, and those with at least one edge to both. Let $\mathcal{C}_u$ be the set of components connected to u, $\mathcal{C}_v$ the set of those connected to v, and $\mathcal{C}_{uv}$ the set connected to both. For each $i \in \mathcal{C}_u$, let F_i be the labeled graph where we add u back to C_i as the labeled vertex, and for each $i \in \mathcal{C}_y$, let F_i be the labeled graph where we add v back to C_i as the labeled vertex. Furthermore, for each $i \in \mathcal{C}_u \cup \mathcal{C}_v$, choose an additional vertex $v_i \in V_i$ such that v_i is incident to an edge in F_i. Let V_{uv} consist of vertices that belong to a component in $\mathcal{C}_{uv}$.

Given a set of vertices U, let U' be the set of vertices in U that have degree at least 2 in F, and let

$$W_{F,U,u}(z_u, z_{U'}) = \prod_{\substack{w \in U' \\ \{u,w\} \in E(F)}} W_{\{u,w\}}(z_u, z_w)$$

and

$$W_{F,U,v}(z_v, z_{U'}) = \prod_{\substack{w \in U' \\ \{v,w\} \in E(F)}} W_{\{v,w\}}(z_v, z_w).$$

Let us also use the notation

$$\mathbb{W}_{F,U}(z_{U'}) = \prod_{\{w,w'\} \in E(F(U'))} W_{\{w,w'\}}(z_w, z_{w'}) \prod_{w \in U'} \prod_{\substack{w' \in U \setminus U' \\ \{w,w'\} \in E(F)}} D_{\mathbb{W}_{\{w,w'\}}}(z_w).$$

Observe that if a vertex in V_{uv} is adjacent to u or v, it must be in V'_{uv}. We therefore express $t(F, \mathbb{W}_{F,f})$ as

$$\int_{\Omega^{V'_{uv}}} \mathbb{W}_{F,V_{uv}}(z_{V'_{uv}}) \int_{\Omega^2} \bigg(W_{F,V_{uv},v}(z_v, z_{V'_{uv}}) W_{F,V_{uv},u}(z_u, z_{V'_{uv}}) W_f(z_u, z_v) \\ \prod_{i \in \mathcal{C}_u} t_{z_u}(F_i, \mathbb{W}_{F,f}) \prod_{i \in \mathcal{C}_v} t_{z_v}(F_i, \mathbb{W}_{F,f}) \bigg),$$

and bound the inner integral by

$$\left\| W_{F,V_{uv},u}(\cdot, z_{V'_{uv}}) \prod_{i \in \mathcal{C}_u} t.(F_i, \mathbb{W}_{F,f}) \right\|_2$$

$$\left\| W_{F,V_{uv},v}(\cdot, z_{V'_{uv}}) \prod_{i \in \mathcal{C}_v} t.(F_i, \mathbb{W}_{F,f}) \right\|_2 \|W_f\|_{2\to 2}$$

$$\leq \left\| W_{F,V_{uv},u}(\cdot, z_{V'_{uv}}) \prod_{i \in \mathcal{C}_u} D_{\mathbb{W}_{uv_i}} \right\|_2$$

$$\left\| W_{F,V_{uv},v}(\cdot, z_{V'_{uv}}) \prod_{i \in \mathcal{C}_v} D_{\mathbb{W}_{vv_i}} \right\|_2 \|W_f\|_{2\to 2} \prod_{i \in \mathcal{C}_v \cup \mathcal{C}_u} D^{|V_i|-1},$$

where in the last step we used Lemma 6.6 and the fact that the number of edges in a spanning tree for F_i is $|V(F_i)| - 1 = |V_i|$. Inserting this bound into the outer integral, an application of the Cauchy–Schwartz inequality then gives the bound

$$t(F, \mathbb{W}_{F,f}) \leq \|W_f\|_{2\to 2} \sqrt{\int_{\Omega^{V'_{uv}}} W_{F,V_{uv}}(z_{V'_{uv}}) \left\| W_{F,V_{uv},u}(\cdot, z_{V'_{uv}}) \prod_{i \in \mathcal{C}_u} D_{\mathbb{W}_{uv_i}} \right\|_2^2}$$

$$\cdot \sqrt{\int_{\Omega^{V'_{uv}}} W_{F,V_{uv}}(z_{V'_{uv}}) \left\| W_{F,V_{uv},v}(\cdot, z_{V'_{uv}}) \prod_{i \in \mathcal{C}_v} D_{\mathbb{W}_{vv_i}} \right\|_2^2 \prod_{i \in \mathcal{C}_v \cup \mathcal{C}_u} D^{|V_i|-1}}$$

We claim that the expressions under the square roots can be written as $t(F'_u, \mathbb{W}_{F,f})$ and $t(F'_v, \mathbb{W}_{F,f})$ for some suitable multigraphs F'_u and F'_v. Indeed, starting from $F(V_{uv} \cup \{u\})$, we first duplicate every edge in this graph that joins u to some vertex in V_{uv}, keeping the edges between vertices in V_{uv} as simple edges. The graph F'_u is obtained from this graph by adding two more edges for each component $i \in \mathcal{C}_u$: the edge uv_i, and a second edge uv'_i, with v'_i being a new vertex we should think of as a twin of v_i (in F'_u, they both have degree one and are connected to u). This gives a connected multigraph on $V'_u = V(F'_u)$ with $|V_{uv}| + 1 + 2k_u$ many vertices and $|E(F(V_{uv}))| + 2d_u + 2k_u$ many edges where d_u is the number of vertices $v' \in V_{uv}$ such that uv' is an edge in F, and $k_u = |\mathcal{C}_u|$. Define F'_v (as well as d_v and k_v) analogously. We then can reexpress the above bound as

$$t(F, \mathbb{W}_{F,f}) \leq \|W_f\|_{2\to 2} \sqrt{t(F'_u, \mathbb{W}_{F,f}) t(F'_v, \mathbb{W}_{F,f})} \prod_{i \in \mathcal{C}_v \cup \mathcal{C}_u} D^{|V_i|-1}$$

$$\leq \|W_f\|_{2\to 2} C D^{|V_{uv}|+k_u+k_v-1} \prod_{i \in \mathcal{C}_v \cup \mathcal{C}_u} D^{|V_i|-1}$$

$$= \|W_f\|_{2\to 2} C D^{n-3}.$$

If f has one endpoint with degree 1, and the other endpoint with degree at least 2, then let v be the endpoint with degree at least 2, and let $u' \neq u$ be a neighbor of v. Let $\mathbb{W}_{F-f}$ be the graphex assignment restricted to the edges in $F - f$. Then

$$|t(F, \mathbb{W}_{F,f})| = \left| \int_\Omega t_{x_v}(F - f, \mathbb{W}_{F-f}) D_{\mathbb{W}_1 - \mathbb{W}_2}(x_v) \, d\mu(x_v) \right|$$
$$\leq \|D_{\mathbb{W}_1 - \mathbb{W}_2}\|_2 \|t.(F - f, \mathbb{W}_{F-f})\|_2$$
$$\leq \|D_{\mathbb{W}_1 - \mathbb{W}_2}\|_2 \|D_{\mathbb{W}_{vu'}}\|_2 D^{n-3} \leq \sqrt{CD} D^{n-3} \|D_{\mathbb{W}_1 - \mathbb{W}_2}\|_2.$$

Now, let $e_1, e_2, \ldots, e_m$ be the edges of F. Let $\mathbb{W}_{F,i}$ be the vector of graphexes where we assign $\mathbb{W}_1 - \mathbb{W}_2$ to e_i, $\mathbb{W}_1$ to e_j with $j < i$, and $\mathbb{W}_2$ to e_j with $j > i$. Then,

$$\left| t(F, \mathbb{W}_1) - t(F, \mathbb{W}_2) \right| \leq \sum_{i=1}^m \left| t(F, \mathbb{W}_{F,i}) \right| \leq m\varepsilon \tilde{C} D^{n-3}.$$

Remark 6.8 It is instructive to note that the bound in Lemma 6.1 can be tightened to give the constant

$$\tilde{C} = \max_i \max\{\|W_i\|_2, \|D_{\mathbb{W}_i}\|_2\}$$

instead of the constant $\tilde{C} = \max\{C, \sqrt{CD}\}$. To see this, we first note that near the end of the proof, we bounded $\|D_{\mathbb{W}_{vu'}}\|_2$ by $\sqrt{CD}$, even though it is possible that the first term is finite while the second is infinite. In a similar way, bounding the integral representing $t(F'_u, \mathbb{W}_{F,f})$ and $t(F'_v, \mathbb{W}_{F,f})$ with the help of Lemma 6.7 is suboptimal. Indeed, the multigraph F'_v always contains at least one double edge, or contain at least one edge uv_i and its twin uv'_i, with both v_i and v'_i having degree one. In the first case, Lemma 6.7 can be improved to extract a factor $\|W_f\|_2$ instead of a factor $\|W_f\|_1$, in the second it can be improved to extract a factor $\|D_{\mathbb{W}_f}\|_2$. Inserted into the proof of Lemma 6.1, this gives the claimed improvement.

Remark 6.9 As the reader can easily verify, the above proof immediately generalizes to signed graphexes, showing that Lemma 6.1 holds for (B, C, D) bounded graphexes, provided we include a factor of $B^{m-(n-1)}$ on the right side. As a consequence, Corollary 6.2 holds for sequences of (B, C, D)-bounded graphexes that converge in the kernel distance $\delta_{2 \to 2}$.

6.2 GP-Convergence and Subgraph Counts

In this subsection, we prove Theorem 6.3. We start by establishing the following theorem.

Theorem 6.10 *Let $\mathbb{G}$ and $\mathbb{G}_n$, for $n \geq 1$, be random finite graphs with no isolated vertices, and let X and X_n, for $n \geq 1$, be the random variables that correspond to the* number of vertices in $\mathbb{G}$ and $\mathbb{G}_n$, respectively. If, for every $t > 0$, $\mathbb{E}[e^{tX}]$ and $\mathbb{E}[e^{tX_n}]$ are finite and uniformly bounded, then the following are equivalent:

(1) For any graph G, the probability that $\mathbb{G}_n$ is isomorphic to G converges to the probability that $\mathbb{G}$ is isomorphic to G.
(2) For every graph F, $\mathbb{E}[inj(F, \mathbb{G}_n)] \to \mathbb{E}[inj(F, \mathbb{G})]$.
(3) For every graph F with no isolated vertices, $\mathbb{E}[inj(F, \mathbb{G}_n)] \to \mathbb{E}[inj(F, \mathbb{G})]$.

To prove the theorem, we will first establish a couple of lemmas. As a preparation, note that if $\mathbb{E}[e^{tX}] \leq C$, then for any graph F on k vertices,

$$\mathbb{E}[inj(F, \mathbb{G})] \leq \sum_{n=0}^{\infty} P(X = n)n^k \leq \sum_{n=0}^{\infty} P(X = n)\frac{k!}{t^k}e^{tn} = \frac{k!}{t^k}\mathbb{E}[e^{tX}] \leq C'.$$

Here we use the fact that $e^{tn} \geq \frac{(tn)^k}{k!}$. The same bound holds for $\mathbb{G}_n$. In other words, for any graph F the values $\mathbb{E}[inj(F, \mathbb{G}_n)]$ and $\mathbb{E}[inj(F, \mathbb{G})]$ are bounded uniformly in n.

Our first lemma roughly says that if a random graph model does not have isolated vertices, and the number of vertices is not too large with high probability, then the expected number of counts of finite graphs without isolated vertices determines the expected number of counts of all graphs.

Lemma 6.11 *Suppose we have two random finite graphs with no isolated vertices, $\mathbb{G}$ and $\mathbb{G}'$. Suppose that for every finite graph F with no isolated vertices, $\mathbb{E}[inj(F, \mathbb{G})] = \mathbb{E}[inj(F, \mathbb{G}')]$. Let X be the random variable that gives the number of vertices in $\mathbb{G}$, and suppose that for every t, $\mathbb{E}[e^{tX}]$ is finite. Then $\mathbb{E}[inj(F, \mathbb{G})]$ and $\mathbb{E}[inj(F, \mathbb{G}')]$ are equal for every finite graph F.*

Proof We prove this by induction on the number of isolated vertices in F. If F has zero isolated vertices, the claim is true by the assumptions of the lemma. Otherwise, let F consist of F' plus an isolated vertex w_0. For every k, and graph G, let $inj^*(F', k, G)$ be equal to the number of ways we can take an injective image of F' in G, take a vertex v_0 not in the image of F', and take a k-term sequence of distinct neighbors of v_0 (which may or may not be in the image of F'). If G has no isolated vertices, then

$$inj(F, G) = \sum_{k=1}^{\infty} \frac{(-1)^{k-1}}{k!} inj^*(F', k, G).$$

This follows from the fact that for each injective copy of F, if v_0 is the image of w_0, then this contributes $(d(v_0))_k$ to $inj^*(F', k, G)$. Since G is finite and has no isolated vertices, we must have $0 < d(v_0) < \infty$; therefore,

$$\sum_{k=1}^{\infty} \frac{(-1)^{k-1}}{k!}(d(v_0))_k = \sum_{k=1}^{d(v_0)}(-1)^{k-1}\binom{d(v_0)}{k} = 1.$$

Next, we claim that

$$\mathbb{E}[\text{inj}(F, \mathbb{G})] = \sum_{k=1}^{\infty} \frac{(-1)^{k-1}}{k!} \mathbb{E}[\text{inj}^*(F', k, \mathbb{G})].$$

To show this, it suffices to show that

$$\sum_{k=1}^{\infty} \frac{\mathbb{E}[\text{inj}^*(F', k, \mathbb{G})]}{k!} < \infty.$$

Fix k. Then $\text{inj}^*(F', k, G)$ is the sum of terms of the form $\text{inj}(F'', G)$. The terms are obtained as follows. Let $L = V(F') \cup z_0$, where z_0 is disjoint from $V(F')$. Let $\ell = (\ell_1, \ell_2, \ldots, \ell_k)$ be a sequence of k elements of L. Suppose further that any vertex in $V(F')$ appears at most once in ℓ. Then we define $F''(\ell)$ by taking a copy of F', a disjoint vertex w_0, for each $\ell_i \in V(F')$, we add an edge from w_0 to ℓ_i, for each other ℓ_i we add an edge going to a new vertex. For example, if $\ell = (z_0, z_0, z_0, \ldots, z_0)$, then $F''(\ell)$ is the disjoint union of F' and a star with k edges. It is then not difficult to see that

$$\text{inj}^*(F', k, G) = \sum_{\ell} \text{inj}(F''(\ell), G).$$

Let $a = |L|$. For each k, the number of such sequences is at most a^k. Furthermore, for each such $F''(\ell)$, the number of vertices is at most $a + k$. Thus,

$$\text{inj}^*(F', k, G) \le a^k (|V(G)|)_{(a+k)}.$$

Therefore, recalling that $X = |\mathbb{G}|$, we have that

$$\sum_{k=1}^{\infty} \frac{\mathbb{E}[\text{inj}^*(F', k, \mathbb{G})]}{k!} \le \sum_{k=1}^{\infty} \frac{a^k \mathbb{E}[(X)_{(a+k)}]}{k!}$$

$$= \sum_{k=1}^{\infty} \mathbb{E}\left[a^k X_{(a)} \binom{X-a}{k}\right] = \mathbb{E}\left[X_{(a)}(a+1)^{X-a}\right] < \infty.$$

Finally, we claim that $\mathbb{E}[\text{inj}^*(F', k, \mathbb{G})] = \mathbb{E}[\text{inj}^*(F', k, \mathbb{G}')]$ for every k. This follows from the fact that the graphs $F''(\ell)$ above each have fewer isolated vertices than F, so $\mathbb{E}[\text{inj}(F''(\ell), \mathbb{G})] = \mathbb{E}[\text{inj}(F''(\ell), \mathbb{G}')]$ for each $F''(\ell)$, and $\text{inj}^*(F', k, \mathbb{G})$ and $\text{inj}^*(F', k, \mathbb{G}')$ are each a finite sum of such terms. Therefore, for $\mathbb{G}'$,

$$\sum_{k=1}^{\infty} \frac{\mathbb{E}[\text{inj}^*(F', k, \mathbb{G}')]}{k!} < \infty.$$

We therefore have

$$\mathbb{E}[\text{inj}(F,\mathbb{G})] = \sum_{k=1}^{\infty} \frac{(-1)^{k-1}}{k!} \mathbb{E}[\text{inj}^*(F',k,\mathbb{G})]$$

$$= \sum_{k=1}^{\infty} \frac{(-1)^{k-1}}{k!} \mathbb{E}[\text{inj}^*(F',k,\mathbb{G}')] = \mathbb{E}[\text{inj}(F,\mathbb{G}')].$$

□

Lemma 6.12 *Suppose we have two random finite graphs, $\mathbb{G}$ and $\mathbb{G}'$. Suppose that for every finite graph F, $\mathbb{E}[\text{inj}(F,\mathbb{G})] = \mathbb{E}[\text{inj}(F,\mathbb{G}')]$. Let X be the random variable that gives the number of vertices in $\mathbb{G}$, and suppose that for some $\varepsilon > 0$, $\mathbb{E}[(2+\varepsilon)^X]$ is finite. Then $\mathbb{G}$ and $\mathbb{G}'$ give rise to the same distribution on graphs (up to isomorphism).*

We would like to emphasize that in these two lemmas, it suffices to assume the finiteness condition for $\mathbb{G}$, not $\mathbb{G}'$ (for which it follows).

Proof For graphs F and G, let $X(F,G)$ be equal to the random variable which is equal to $\text{inj}(F,G)$ if G has the same number of vertices as F, and gives 0 otherwise. Fix a graph F with k vertices, and let F_i be the graph obtained by adding i isolated vertices to F. We can then write

$$X(F,G) = \sum_{i=0}^{\infty} \frac{(-1)^i}{i!} \text{inj}(F_i, G).$$

Indeed, if G has the same number of vertices as F, then each term with $i > 0$ is zero, and the $i = 0$ term gives $\text{inj}(F_0, G) = \text{inj}(F, G)$. If G has fewer vertices, then the entire expression is zero. If G has $k + \ell$ vertices where $\ell > 0$, then each injective copy of G contributes (since $\binom{\ell}{i} = 0$ if $i > \ell$)

$$\sum_{i=0}^{\ell} (-1)^i \binom{\ell}{i} = 0;$$

therefore the entire expression is 0. Next, we claim that

$$\sum_{i=0}^{\infty} \frac{\mathbb{E}[\text{inj}(F_i, \mathbb{G})]}{i!} < \infty.$$

To see this, note that for every i,

$$\mathbb{E}[\text{inj}(F_i, \mathbb{G})] \le \sum_n \mathbb{P}(X = n)(n)_{k+i}.$$

By the condition on X, there exists an $\varepsilon > 0$ and c such that $\mathbb{P}(X = n) \le c(2+\varepsilon)^{-n}$. Therefore,

$$\sum_{i=0}^{\infty} \frac{\mathbb{E}[\operatorname{inj}(F_i, \mathbb{G})]}{i!} \leq \sum_i \sum_n c(2+\varepsilon)^{-n}(n)_k \binom{n-k}{i}$$
$$= \sum_n c(2+\varepsilon)^{-n} 2^{n-k}(n)_k < \infty.$$

We therefore have that

$$\mathbb{E}[X(F, \mathbb{G})] = \sum_{i=0}^{\infty} \frac{(-1)^i}{i!} \mathbb{E}[\operatorname{inj}(F_i, \mathbb{G})] = \sum_{i=0}^{\infty} \frac{(-1)^i}{i!} \mathbb{E}[\operatorname{inj}(F_i, \mathbb{G}')] = \mathbb{E}[X(F, \mathbb{G}')].$$

Thus, by an inclusion-exclusion formula, we can express the probability that $\mathbb{G}$ and $\mathbb{G}'$ is isomorphic to a graph F for any F, and the two probabilities must also be equal. This completes the proof. □

Proof (Theorem 6.10) We first show that (1) implies (2). Recall that for any graph F, $\mathbb{E}[\operatorname{inj}(F, \mathbb{G}_n)]$ and $\mathbb{E}[\operatorname{inj}(F, \mathbb{G})]$ are uniformly bounded. Fix a graph F on k vertices. We claim that $\mathbb{E}[\operatorname{inj}(F, \mathbb{G}_n)^2]$ and $\mathbb{E}[\operatorname{inj}(F, \mathbb{G})^2]$ are uniformly bounded. This follows from the fact that for a graph G, $\operatorname{inj}(F, G)^2$ is a linear combination of the form $\sum_{F'} c_{F'} \operatorname{inj}(F', G)$, where $c_{F'}$ is independent of G and zero for all but a finite number of graphs F'. We also know that $\operatorname{inj}(F, \mathbb{G}_n)$ converges to $\operatorname{inj}(F, \mathbb{G})$ in distribution. Since their second moments are uniformly bounded, their expectations must converge as well.

It is clear that (2) implies (3). Let us show that (3) implies (1). Assume $\mathbb{G}_n$ satisfies (3). We first claim that the sequence $\mathbb{G}_n$ is tight. That is, we claim that for every ε, there exists a finite set of graphs such that for each n, with probability at least $1 - \varepsilon$, $\mathbb{G}_n$ is in this set. Indeed, the expected number of edges $\mathbb{E}[\operatorname{inj}(K_2, \mathbb{G}_n)]$ is uniformly bounded, which means that for any ε, there exists an M such that the probability of having more than M edges is at most ε. But the number of graphs with M edges and no isolated vertices is finite; therefore $\mathbb{G}_n$ is tight.

This means that there is a subsequence that converges to a random graph $\mathbb{H}$ in the sense of (1), which also has no isolated vertices. First, we claim that for any F with no isolated vertices, $\mathbb{E}[\operatorname{inj}(F, \mathbb{H})]$ is finite, and $\mathbb{E}[\operatorname{inj}(F, \mathbb{G}_n)]$ converges to it. This again follows from the fact that $\mathbb{E}[\operatorname{inj}(F, \mathbb{G}_n)^2]$ is uniformly bounded. But then this implies that $\mathbb{E}[\operatorname{inj}(F, \mathbb{H})] = \mathbb{E}[\operatorname{inj}(F, \mathbb{G})]$ for every graph F with no isolated vertices. Therefore, the two distributions are equal by Lemmas 6.11 and 6.12. □

Lemma 6.13 *Given $t > 0$, let $t' = 2t + \log 4$. Then the following holds for any positive integer n and any graph G on n vertices with no isolated vertices. Suppose we randomly color the vertices red and blue, and let X be the number of vertices that are colored red, and have at least one blue neighbor. Then*

$$\mathbb{E}[e^{t'X}] \geq e^{tn}.$$

Proof Without loss of generality, we may assume G is the disjoint union of stars, since otherwise we can delete an edge and G will still have no isolated vertices.

Suppose that G is the disjoint union of stars with edge count $s_1, s_2, \ldots, s_k$, where each $s_i \geq 1$ and $\sum_i s_i = n - k$. Note that $k \leq n/2$. Then

$$\mathbb{E}\left[e^{t'X}\right] = \prod_{i=1}^{k} \left(\frac{1}{2}\left(\frac{e^{t'}+1}{2}\right)^{s_k} + \frac{1}{2}\left(\frac{e^{t'}+2^{s_k}-1}{2^{s_k}}\right)\right)$$

$$\geq \left(\frac{e^{t'}+1}{2}\right)^{n-k} 2^{-k} \geq (e^{t'}+1)^{n/2} 2^{-n} \geq e^{tn},$$

given that our choice of t' implies that $(e^{t'}+1)/4 \geq e^{2t}$. □

Lemma 6.14 *For any $T, t, C, D \in \mathbb{R}_+$, there exists a finite B such that the following holds. Suppose we have a graphex $\mathbb{W}$, with $\|D_{\mathbb{W}}\|_\infty \leq D$ and $\|\mathbb{W}\|_1 \leq C$. Let X be the number of vertices (that are not isolated) of $G_T(\mathbb{W})$. Then $\mathbb{E}[e^{tX}] \leq B$.*

Proof Note that since X is nonnegative, we only need to worry about $t > 0$. First, let X' be obtained by randomly coloring the vertices of G_T red and blue, and taking the red vertices with at least one blue neighbor. By the above lemma, for $t' = 2t + \log 4 > 0$, we have

$$\mathbb{E}[e^{t'X'}] \geq \mathbb{E}[e^{tX}].$$

We claim that

$$\mathbb{E}[e^{t'X'}] \leq e^{\frac{T^2 I}{2}(e^{t'}-1)} \exp\left(\frac{T}{2}\int_\Omega \left(e^{\frac{T}{2}D_{\mathbb{W}}(x)(e^{t'}-1)} - 1\right) d\mu(x)\right). \quad (12)$$

Let us show how this implies uniform boundedness. We know that $D_{\mathbb{W}}$ is bounded by D. Since the function $z \to e^z - 1$ is 0 at $z = 0$ and convex, there exists a constant K depending only on t', T, and D such that

$$e^{\frac{T}{2}D_{\mathbb{W}}(x)(e^{t'}-1)} - 1 \leq \frac{D_{\mathbb{W}}(x)}{D} e^{\frac{T}{2}D(e^{t'}-1)} = K D_{\mathbb{W}}(x).$$

Therefore,

$$\frac{T^2 I}{2}(e^{t'}-1) + \frac{T}{2}\int_\Omega \left(e^{\frac{T}{2}D_{\mathbb{W}}(x)(e^{t'}-1)} - 1\right) d\mu(x)$$

$$\leq \frac{T^2 I}{2}(e^{t'}-1) + \frac{T}{2}\int_\Omega K D_{\mathbb{W}}(x) d\mu(x)$$

$$\leq \frac{T^2 C}{4}(e^{t'}-1) + \frac{KTC}{2}.$$

In order to prove (12), we first show that it is enough to consider the case where $\mu(\Omega) < \infty$. To see this, we write a general σ-finite measure space as the union of finite spaces, $\Omega = \bigcup_n \Omega_n$ where Ω_n is an increasing sequence with $\mu(\Omega_n) < \infty$ for all n. For a fixed T, let $G_{T,n}$ consist of the induced graph on those vertices which were born in Ω_n, come from a star of a vertex born in Ω_n, or come from a dust edge, and define X'_n to be the number of red vertices in $G_{T,n}$ with at least one blue neighbor in $G_{T,n}$. It is easy to see that $0 \leq X'_1 \leq X'_2 \leq \cdots \leq X'_n \leq \cdots$. By monotone convergence, the expectation of $e^{t'X'_n}$ converges to $\mathbb{E}[e^{t'X'}]$. The bound we obtain also converges, giving the required bound on $\mathbb{E}[e^{t'X'}]$.

Assume thus that $\mu(\Omega) < \infty$. In this case, almost surely, a finite number of blue points will be created. Suppose that these are $x_1, x_2, \ldots, x_k$. Conditioned on this, red vertices that have blue neighbors can be created as follows. They can be created by the Poisson process on Ω, and then be connected to at least one of the x_i. If a point is created at x, the probability that it is connected to at least one of the x_i is $1 - \prod(1 - W(x_i, x))$. It can also be created as a leaf of a star created at one of the x_i. Finally, it can be created by a dust edge being colored red and blue. Since the number of red vertices coming from each of these cases is independent, the number of red vertices with at least one blue neighbor is a Poisson distribution with expectation

$$f(x_1, \ldots, x_k) := \frac{T}{2}\int_\Omega \left(1 - \prod_{i=1}^k (1 - W(x_i, x))\right) d\mu(x) + \frac{T}{2}\sum_{i=1}^k S(x_i) + \frac{T^2 I}{2}$$

$$\leq \frac{T}{2}\int_\Omega \left(\sum_{i=1}^k W(x_i, x)\right) d\mu(x) + \frac{T}{2}\sum_{i=1}^k S(x_i) + \frac{T^2 I}{2}$$

$$= \frac{T}{2}\sum_{i=1}^k D_\mathbb{W}(x_i) + \frac{T^2 I}{2}.$$

In particular, this means that (for $t' > 0$) we have

$$\mathbb{E}[e^{t'X'} | x_1, x_2, \ldots, x_k] = e^{f(x_1,\ldots,x_k)(e^{t'}-1)} \leq e^{\frac{T}{2}\left(TI + \sum_{i=1}^k D_\mathbb{W}(x_i)\right)(e^{t'}-1)}.$$

Therefore

$$\mathbb{E}[e^{t'X'}] = \sum_{k=0}^\infty e^{-\frac{T}{2}\mu(\Omega)} \frac{(\mu(\Omega)T/2)^k}{k!} \frac{1}{\mu(\Omega)^k} \int_{\Omega^k} e^{f(x_1,x_2,\ldots,x_k)(e^{t'}-1)} d\mu(x_1)\ldots d\mu(x_k)$$

$$\leq \sum_{k=0}^\infty e^{-\frac{T}{2}\mu(\Omega)} \frac{(T/2)^k}{k!} \int_{\Omega^k} e^{\frac{T}{2}\left(TI + \sum_{i=1}^k D_\mathbb{W}(x_i)\right)(e^{t'}-1)} d\mu(x_1)\ldots d\mu(x_k)$$

$$= e^{\frac{T^2 I}{2}(e^{t'}-1)} \sum_{k=0}^\infty e^{-\frac{T}{2}\mu(\Omega)} \frac{(T/2)^k}{k!} \left(\int_\Omega e^{\frac{T}{2}D_\mathbb{W}(x)(e^{t'}-1)} d\mu(x)\right)^k.$$

Here we think of Ω^0 as consisting of a single point on which f is 0. We then have

$$\mathbb{E}[e^{t'X'}] \le e^{\frac{T^2 I}{2}(e^{t'}-1)} e^{\frac{T}{2}\int_\Omega e^{\frac{T}{2}D_\mathbb{W}(x)(e^{t'}-1)}\,d\mu(x) - \frac{T}{2}\mu(\Omega)}$$

$$= e^{\frac{T^2 I}{2}(e^{t'}-1)} e^{\frac{T}{2}\int_\Omega \left(e^{\frac{T}{2}D_\mathbb{W}(x)(e^{t'}-1)}-1\right)d\mu(x)}.$$

So we know that (12) is true for any Ω with finite measure. This completes the proof of the lemma. □

After these preparations, the proof of Theorem 6.3 is straightforward.

Proof (*Theorem* 6.3) We first note that it is enough to prove the lemma for the case that $\mathbb{W}_n$ and $\mathbb{W}$ are (C, D)-bounded for some finite $C, D < \infty$. Indeed, both (3) and (4) clearly imply a bound on the $\|\cdot\|_1$-norm, but also (2) (and therefore (1)) does, since (2) implies that the random graphs $G_T(\mathbb{W}_n)$ are tight, which implies that the set of graphexes is tight, which by Corollary 4.2 (2) implies uniform boundedness of the $\|\cdot\|_1$-norms.

Assume thus that $\mathbb{W}_n$ and $\mathbb{W}$ are (C, D)-bounded for some finite $C, D < \infty$. The equivalence of (3) and (4) follows from the fact that t is multiplicative over components of F. (1) $\Rightarrow$ (2) is obvious. Using Lemma 6.14, we can apply the equivalence in Theorem 6.10 and Proposition 3.24 to show that (2) $\Rightarrow$ (3) and (3) $\Rightarrow$ (1). □

A slight modification of the above proof gives the following theorem.

Theorem 6.15 *Given two graphexes* $\mathbb{W}, \mathbb{W}'$ *with bounded marginals, the following are equivalent:*

(1) $G_T(\mathbb{W})$ and $G_T(\mathbb{W}')$ have the same distribution for every T.
(2) $G_T(\mathbb{W})$ and $G_T(\mathbb{W}')$ have the same distribution for some T.
(3) For every graph F with no isolated vertices, $t(F, \mathbb{W}) = t(F, \mathbb{W}')$.
(4) For every connected graph F, $t(F, \mathbb{W}) = t(F, \mathbb{W}')$.

Proof As before, the equivalence of (4) and (3) follows from the product property of t. The implication (1) $\Rightarrow$ (2) is obvious. To prove (2) $\Rightarrow$ (3), we use the fact that $t(F, \mathbb{W}) = T^{-|V(F)|}\mathbb{E}[\text{inj}(F, G_T(\mathbb{W}))]$ and the same holds for $\mathbb{W}'$. Since $\text{inj}(F, G_T(\mathbb{W}))$ and $\text{inj}(F, G_T(\mathbb{W}'))$ have the same distribution, their expectations must be equal. With the help of Proposition 3.24, this implies (3). (3) $\Rightarrow$ (1) follows from Proposition 3.24, the observation that graphexes with bounded marginals are integrable, and Lemmas 6.14, 6.11, and 6.12. □

6.3 Metric Convergence Implies GP-Convergence

We close this section by proving that under the assumption of uniformly bounded marginals, $\delta_{2\to 2}$-convergence implies GP-convergence. We then use this result to show that without any assumptions on the marginals, $\delta_\diamond$-convergence implies GP-convergence.

Identifiability for Graphexes and the Weak Kernel Metric 111

Theorem 6.16 *Suppose $\mathbb{W}_n$ and $\mathbb{W}$ have uniformly bounded marginals, and*

$$\delta_{2\to 2}(\mathbb{W}_n, \mathbb{W}) \to 0.$$

Then $\mathbb{W}_n$ is GP-convergent to $\mathbb{W}$.

Proof Note that $\delta_{2\to 2}$ convergence implies that $\|\mathbb{W}_n\|_1 \to \|\mathbb{W}\|$; therefore the sequence is (C, D)-bounded for some C, D. If F is a graph without isolated vertices, then $t(F, \mathbb{W}_n) \to t(F, \mathbb{W})$ by Corollary 6.2. Therefore, by Theorem 6.3, for any T, $G_T(\mathbb{W}_n)$ converges to $G_T(\mathbb{W})$ in distribution. □

Theorem 6.17 *Suppose that graphexes $\mathbb{W}$ and $(\mathbb{W}_n)_{n=1}^\infty$ have the property that $\delta_\diamond(\mathbb{W}_n, \mathbb{W}) \to 0$. Then $\mathbb{W}_n$ is GP-convergent to $\mathbb{W}$.*

Proof By Proposition 4.6, the sequence is tight, and for all D such that $\mu(\{D_\mathbb{W} = D\}) = 0$, we have that $\mu_n(\Omega_{n,>D}) \to \mu(\Omega_{>D})$ and $\delta_{2\to 2}(\mathbb{W}_{n,\le D}, \mathbb{W}_{\le D}) \to 0$.

Fix T and $\varepsilon > 0$, and take δ small enough so that for all sets Ω_δ of measure at most δ the probability that any of the vertices in G_T has a feature in Ω_δ is at most $\varepsilon/3$. Take D large enough so that for all n, $\mu_n(\Omega_{n,>D})$ and $\mu(\Omega_{>D})$ are at most δ. Then the total variation distance between $G_T(\mathbb{W}_{n,\le D})$ and $G_T(\mathbb{W}_n)$ is at most $\varepsilon/3$, and the same is true for $G_T(\mathbb{W}_{\le D})$ and $G_T(\mathbb{W})$. We also know that $\delta_{2\to 2}(\mathbb{W}_{n,\le D}, \mathbb{W}_{\le D}) \to 0$, which in particular implies that the sequence is uniformly (C, D)-bounded for some C. Therefore, it is GP-convergent. In particular, for n large enough, the total variation distance between $G_T(\mathbb{W}_{n,\le D})$ and $G_T(\mathbb{W}_{\le D})$ is at most $\varepsilon/3$. This implies that for n large enough, the total variation distance between $G_T(\mathbb{W}_n)$ and $G_T(\mathbb{W})$ is at most ε, which shows that the sequence is GP-convergent. □

7 Sampling

In this section, we prove that GP-convergence implies convergence in the weak kernel metric, completing the proof of the equivalence of convergence in the metric $\delta_\diamond$ and GP-convergence (Theorem 2.18). The main technical tool to establish this will be a "sampling lemma", showing that as $T \to \infty$, the graphs $G_T(\mathbb{W})$ sampled from a graphex $\mathbb{W}$ converge to the generating graphex according to $\delta_\diamond$.

To make this precise, we need a way to compare graphs to graphexes. As in [3, 4], we do this by transforming the graph into a suitable "empirical graphon" and corresponding "empirical graphex". Differing slightly from both [3, 4], where the empirical graphon was a graphon over $\mathbb{R}_+$, here we define it to be a graphon over the vertex set of the graph. Explicitly, given a finite graph G and $\rho > 0$, we define the graphon $W(G, \rho)$ as follows. Let $\mathbf{\Omega} = (\Omega, \mathcal{F}, \mu)$, where Ω is the set of vertices, $\mathcal{F}$ is the σ-algebra consisting of all subsets, and μ is the measure where each vertex has weight ρ. Set $W(x, y)$ to be 1 if there is an edge between the corresponding vertices, and 0 otherwise. This gives us the graphon $W(G, \rho)$. We then set $\mathbb{W}(G, \rho) = (W, 0, 0, \mathbf{\Omega})$. Similarly, if H is a weighted graph with countably many vertices, we define Ω to be the set of vertices, $\mathcal{F}$ to be the σ-algebra consisting of all subsets of Ω, and μ to be the σ-finite measure which gives weight ρ to each vertex; $W(H, \rho)$ and $\mathbb{W}(H, \rho)$ are then the graphon and graphex obtained by taking W according to edge weights.

With these definitions, we are ready to state the sampling lemma.

Theorem 7.1 *For every graphex $\mathbb{W}$ and $\varepsilon > 0$,*

$$\lim_{T \to \infty} \mathbb{P}[\delta_\diamond(\mathbb{W}(G_T(\mathbb{W}), 1/T), \mathbb{W}) > \varepsilon] = 0.$$

For a set of graphexes that is tight, the convergence is uniform.

Remark 7.2 The above theorem only claims convergence in probability. However, once we establish equivalence of GP-convergence and convergence in the weak kernel norm, the results of [18] imply convergence with probability one (since there convergence with probability one is proved for GP-convergence). Nevertheless, to *establish* the equivalence, all we need is convergence in probability, so this is all we will prove here.

7.1 Closeness of Graphexes Implies Closeness of Samples

In order to prove the sampling lemma, we will first prove that two graphexes with bounded marginals that are close in the kernel metric lead to samples that are close. This is formalized in the following theorem.

Theorem 7.3 *Suppose $\mathbb{W}_1, \mathbb{W}_2$ are two (C, D)-bounded graphexes on the same space $\mathbf{\Omega}$, and suppose that $d_{2 \to 2}(\mathbb{W}_1, \mathbb{W}_2) \leq c$ for some $0 < c < 1$. Then there exists a T_0 (depending only on c, C, and D) such that for any $T > T_0$, there exists a coupling of the random graphs $G_T(\mathbb{W}_1)$ and $G_T(\mathbb{W}_2)$ so that*

$$\mathbb{P}\left[\delta_{2 \to 2}(\mathbb{W}(G_T(\mathbb{W}_1), 1/T), \mathbb{W}(G_T(\mathbb{W}_2), 1/T)) > \min\left((31cC)^{1/4}, 2c^{3/4}, \sqrt[3]{3c}\right)\right] < c.$$

For *graphons*, or graphexes with only a graphon part, we can think of obtaining G_T as having two phases: first we sample the set of vertices, and then we sample the edges according to the edge probability. If we do not do the second phase, we obtain a weighted graph. We will work with this intermediate graph in this section. To make this precise, given a graphon $(W, \mathbf{\Omega})$, define $H_T(W)$ as the random weighted graph where we take a Poisson process on $\mathbf{\Omega} \times [0, T]$, set these to be the vertices of $H_T(W)$, and for each pair of vertices (x_i, t_i) and (x_j, t_j), put a weighted edge with weight $W(x_i, x_j)$ (with 0 weights on the diagonals).

In order to prove the theorem, we need to find a coupling of the random processes that provide $G_T(\mathbb{W}_1)$ and $G_T(\mathbb{W}_2)$. Since $\mathbb{W}_1$ and $\mathbb{W}_2$ have the same underlying space, it is natural to couple the Poisson processes that generate the vertices into a single Poisson process. Conditioned on this, we generate the two random graphs independently (this is not optimal but it is satisfactory for our purposes). Let $\mathbb{W}_i' = \mathbb{W}_i(G_T(\mathbb{W}_i), 1/T) = (W_i', S_i', I_i', \Omega_i')$. The underlying space of $\mathbb{W}_i'$ consists of the vertex set of $G_T(\mathbb{W}_i)$, everything with weight $1/T$. We couple the two underlying spaces by matching vertices that correspond to the same point in Ω, and couple the other vertices arbitrarily (adding points with degree 0 if necessary). We will show that in this way, all three components of our distance will be close.

Let us first show that $\|W_1' - W_2'\|_{2\to 2}$ is small, with high probability. Note that $G_T(\mathbb{W})$ consists of the edges in $G_T(W)$, and the edges generated by the stars and the independent edges. In the following lemma, we show that the extra edges generated have a small effect on this distance.

Lemma 7.4 *Let $\mathbb{W} = (W, S, I, \Omega)$ be a (C, D)-bounded graphex, and $T > 1/D$. Let $G_T(\mathbb{W})$ be the usual sample at time T, and let $\widetilde{G}_T(\mathbb{W})$ consist of only those edges which come from I or S. Then*

$$\mathbb{P}\left[\|W(\widetilde{G}_T(\mathbb{W}), 1/T)\|_{2\to 2} > \left(\frac{2CD}{\sqrt{T}}\right)^{1/4}\right] < \frac{1}{\sqrt{T}}.$$

Proof Suppose we have sampled stars with $s_1, s_2, \ldots, s_\ell$ leaves, and we have sampled m isolated edges. Let $U = W(\widetilde{G}_T(\mathbb{W}), 1/T)$. Then

$$t(C_4, U) = \frac{1}{T^4}\left(2\sum_i s_i^2 + 2m\right).$$

Therefore,

$$\mathbb{E}[t(C_4, U)] = \frac{2T\int_\Omega (T^2 S(x)^2 + TS(x))\,d\mu(x)}{T^4} + \frac{2T^2 I}{T^4} \leq \frac{C(D+1/T)}{T} \leq \frac{2CD}{T},$$

and hence

$$\mathbb{P}\left[t(C_4, U) > \frac{2CD}{\sqrt{T}}\right] < \frac{1}{\sqrt{T}}.$$

Using the fact that $\|U\|_{2\to 2} \leq t(C_4, U)^{1/4}$ (Lemma 3.18), the lemma follows. □

This lemma implies that for the $2 \to 2$ component of the distance, we can compare $\widetilde{G}_T(\mathbb{W}_1)$ and $\widetilde{G}_T(\mathbb{W}_2)$ instead of $G_T(\mathbb{W}_1)$ and $G_T(\mathbb{W}_2)$. The following lemma will imply that it in fact suffices to compare $H_T(\mathbb{W}_1)$ and $H_T(\mathbb{W}_2)$, because $\widetilde{G}_T$ is close to H_T, as long as $H_T(\mathbb{W}_i)$ satisfies certain boundedness conditions (which, by the boundedness of the $\mathbb{W}_i$, will be satisfied with high probability).

Lemma 7.5 *Suppose H is a weighted graph on $\mathbb{N}$ with weights $H_{i,j} \in [0, 1]$, and $H_{i,i} = 0$. Suppose that G is generated by taking an edge between i and j with probability $H_{i,j}$, independently for every pair of vertices. Suppose that $\sum_{i,j} H_{i,j} \leq E$ and $\sum_{i,j,k} H_{i,j} H_{j,k} \leq F$ where the sum goes over pairwise distinct vertices. Let $0 < \rho$. Then*

$$\mathbb{P}[\|\mathbb{W}(G, \rho) - \mathbb{W}(H, \rho)\|_{2\to 2} > \rho^{7/8}(E + 2F)^{1/4}] < \sqrt{\rho}.$$

Proof Let $X_{i,j} = G_{i,j} - H_{i,j}$. Notice that $\mathbb{E} X_{i,j} = 0$ and $X_{i,j}$ over different pairs are independent. Also, each $X_{i,i} = 0$. Therefore,

$$\mathbb{E}[t(C_4, \mathbb{W}(G, \rho) - \mathbb{W}(H, \rho))] = \rho^4 \mathbb{E}\left[\sum_{i,j} X_{i,j}^4 + 2\sum_{i,j,k} X_{i,j}^2 X_{j,k}^2 + \sum_{i,j,k,l} X_{i,j} X_{j,k} X_{k,l} X_{l,i}\right],$$

where in each of the sums, all indices are pairwise distinct. Here

$$\sum_{i,j} \mathbb{E}[X_{i,j}^4] = \sum_{i,j} \left(H_{i,j}(1 - H_{i,j})^4 + (1 - H_{i,j}) H_{i,j}^4\right) \leq \sum_{i,j} H_{i,j} \leq E.$$

Also,

$$\sum_{i,j,k} \mathbb{E}\left[X_{i,j}^2 X_{j,k}^2\right] = \sum_{i,j,k} \left(H_{i,j} - 2H_{i,j}^2 + H_{i,j}^2\right)\left(H_{j,k} - 2H_{j,k}^2 + H_{j,k}^2\right)$$
$$= \sum_{i,j,k} H_{i,j}(1 - H_{i,j}) H_{j,k}(1 - H_{j,k}) \leq \sum_{i,j,k} H_{i,j} H_{j,k}$$
$$\leq F.$$

Finally, for any pairwise distinct i, j, k, ℓ,

$$\mathbb{E}[X_{i,j} X_{j,k} X_{k,\ell} X_{\ell,i}] = 0.$$

Therefore,
$$0 \leq \mathbb{E}[t(C_4, \mathbb{W}(G, \rho) - \mathbb{W}(H, \rho))] \leq \rho^4(E + 2F). \tag{13}$$

This implies that

$$\mathbb{P}\left[t(C_4, \mathbb{W}(G, \rho) - \mathbb{W}(H, \rho)) > \rho^{7/2}(E + 2F)\right] < \sqrt{\rho}.$$

Using the fact that $\|U\|_{2\to 2} \leq (t(C_4, U))^{1/4}$ (Lemma 3.18), the lemma follows. □

Proof (*Theorem* 7.3) We are now ready to show that with high probability, $\|W_1' - W_2'\|_{2\to 2}$ is small. Recall that we have a coupling of $H_T(W_1)$ and $H_T(W_2)$ such that $H_T(W_1) - H_T(W_2) = H_T(W_1 - W_2)$ with probability one. Let $U = W_1 - W_2$, so that $H_T(W_1) - H_T(W_2) = H_T(U)$.

Let us first show that $\|W(H_T(U), 1/T)\|_{2\to 2}$ is small. First, suppose that $\mu = \mu(\Omega)$ is finite. Let

$$X = \int_{\Omega^2} U(x_1, x_2)^4 \, d\mu(x_1) \, d\mu(x_2) \leq 2C,$$

$$Y = \int_{\Omega^3} U(x_1, x_2)^2 U(x_2, x_3)^2 \, d\mu(x_1) \, d\mu(x_2) \, d\mu(x_3) \leq 4CD,$$

and

$$Z = \int_{\Omega^4} U(x_1, x_2) U(x_2, x_3) U(x_3, x_4) U(x_4, x_1) = t(C_4, U)$$
$$\leq \|W_1 - W_2\|_{2\to 2}^2 \|W_1 - W_2\|_2^2$$
$$\leq \|W_1 - W_2\|_{2\to 2}^2 \|W_1 - W_2\|_1$$
$$\leq 2c^2 C,$$

where we used Lemma 3.18, the fact that both graphexes are (C, D)-bounded, and the fact that $d_{2\to 2}(\mathbb{W}_1, \mathbb{W}_2) \leq c$. If $T > \max\{16\frac{D}{c^2}, 2/c\}$, then

$$\mathbb{E}[t(C_4, H_T(U))] = \sum_{n=0}^{\infty} e^{-T\mu} \frac{(T\mu)^n}{n!} \left(\frac{n(n-1)}{T^4 \mu^2} X + 2\frac{n(n-1)(n-2)}{T^4 \mu^3} Y \right.$$
$$\left. + \frac{n(n-1)(n-2)(n-3)}{T^4 \mu^4} Z \right)$$
$$\leq \frac{2C}{T^2} + \frac{8CD}{T} + 2c^2 C \leq 3c^2 C.$$

In general, we can take a sequence of finite measure subsets $\Omega_1 \subseteq \cdots \subseteq \Omega_n \subseteq \cdots$ with $\bigcup_n \Omega_n = \Omega$ to show that the above bound on the expectation holds for general Ω. Therefore,

$$\mathbb{P}[\|H_T(U)\|_{2\to 2} > (30cC)^{1/4}] \leq \mathbb{P}[t(C_4, H_T(U)) > 30cC] < \frac{c}{10}.$$

Next, let P_2 be the star with two leaves. If T is large enough, then

$$\mathbb{E}[t(P_2, H_T(W_1))] = \left(t(P_2, W_1)T^3 + T^2 \|W_1\|_1 \right) \leq CDT^3 + CT^2 \leq 2CDT^3.$$

Therefore,

$$\mathbb{P}\left[t(P_2, H_T(W_1)) > \frac{20CDT^3}{c}\right] \leq \frac{c}{10}.$$

Also, since
$$\mathbb{E}[\|H_T(W_1)\|_1] \leq CT^2,$$

we also have
$$\mathbb{P}\left[\|H_T(W_1)\|_1 > \frac{10CT^2}{c}\right] \leq \frac{c}{10}.$$

Conditioned on neither of these happening, we can apply Lemma 7.5 with
$$E + 2F \leq \frac{10CT^2}{c} + 2\frac{20CDT^3}{c} \leq \frac{50CDT^3}{c}.$$

This means that
$$\mathbb{P}\left[\|\mathbb{W}(G_T(W_1), 1/T) - \mathbb{W}(H_T(W_1), 1/T)\|_{2\to 2} > \left(\frac{50CD}{c\sqrt{T}}\right)^{1/4}\right] \leq \frac{1}{\sqrt{T}}.$$

Clearly the analogous statements hold for $H_T(W_2)$. Let $\widetilde{W}_i = \mathbb{W}(\widetilde{G}_T(\mathbb{W}_i), 1/T)$ (i.e., the part consisting of edges generated by the stars and independent edges). Also, let $\widehat{W}_i = \mathbb{W}(G_T(W_i), 1/T) - \mathbb{W}(H_T(W_i), 1/T)$. Assuming none of the bad events happen, if T is large enough, then

$$\begin{aligned}
\|W_1' - W_2'\|_{2\to 2} &\leq \|\widetilde{W}_1\|_{2\to 2} + \|\widetilde{W}_2\|_{2\to 2} \\
&\quad + \|\widehat{W}_1\|_{2\to 2} + \|\widehat{W}_2\|_{2\to 2} + \|\mathbb{W}(H_T(U), 1/T)\|_{2\to 2} \\
&\leq 2\left(\frac{2CD}{\sqrt{T}}\right)^{1/4} + 2\left(\frac{(50CD)^{1/4}}{c\sqrt{T}}\right)^{1/4} + (30cC)^{1/4} \leq (31cC)^{1/4}.
\end{aligned}$$

The probability of one of the bad events happening is at most
$$\frac{2}{\sqrt{T}} + 4\frac{c}{10} + \frac{2}{\sqrt{T}} + \frac{c}{10} \leq \frac{6c}{10}.$$

Let us now bound the probability that $\|D_{\mathbb{W}_1} - D_{\mathbb{W}_2}\|_2$ is large. For $x \in \Omega$, let
$$D_{\mathbb{W}_1 \mathbb{W}_2}(x) = \int_\Omega W_1(x, y) W_2(x, y) \, d\mu(y).$$

With our coupling,

$$\mathbb{E}\left[\sum_{v \in V_T} d_{G_T(\mathbb{W}_1)}(v)^2\right] = \int_\Omega T\left((TD_{\mathbb{W}_1}(x))^2 + TD_{\mathbb{W}_1}(x)\right) d\mu(x),$$

$$\mathbb{E}\left[\sum_{v \in V_T} d_{G_T(\mathbb{W}_1)}(v) d_{G_T(\mathbb{W}_2)}(v)\right] =$$
$$\int_\Omega T\left((TD_{\mathbb{W}_1}(x))(TD_{\mathbb{W}_2}(x)) + TD_{\mathbb{W}_1\mathbb{W}_2}(x)\right) d\mu(x),$$

$$\mathbb{E}\left[\sum_{v \in V_T} d_{G_T(\mathbb{W}_2)}(v)^2\right] = \int_\Omega T\left((TD_{\mathbb{W}_2}(x))^2 + TD_{\mathbb{W}_2}(x)\right) d\mu(x).$$

Therefore,

$$\mathbb{E}\left[\sum_{v \in V_T} \left(d_{G_T(\mathbb{W}_1)}(x) - d_{G_T(\mathbb{W}_2)}(x)\right)^2\right]$$
$$= T^3 \int_\Omega \left(D_{\mathbb{W}_1}(x) - D_{\mathbb{W}_2}(x)\right)^2 d\mu(x)$$
$$+ T^2 \int_\Omega \left(D_{\mathbb{W}_1}(x) + D_{\mathbb{W}_2}(x) - 2D_{\mathbb{W}_1\mathbb{W}_2}(x)\right) d\mu(x)$$
$$\leq T^3 \|D_{\mathbb{W}_1} - D_{\mathbb{W}_2}\|_2^2 + T^2 \|\mathbb{W}_1\|_1 + T^2 \|\mathbb{W}_2\|_1,$$

This means that if T is large enough,

$$\mathbb{E}\left[\|D_{\mathbb{W}_1'} - D_{\mathbb{W}_2'}\|_2^2\right] \leq c^4 + 2C/T \leq 2c^4.$$

Therefore,

$$\mathbb{P}[\|D_{\mathbb{W}_1'} - D_{\mathbb{W}_2'}\|_2 > 4c^{3/2}] \leq \frac{c}{8}.$$

Finally, recall that by Lemma 4.3, the number of edges of $G_T(\mathbb{W}_i)$ has expectation $T^2 \|\mathbb{W}_i\|_1/2$ and variance $T^2 \|\mathbb{W}_i\|_1/2 + T^3 \|D_{\mathbb{W}_i}\|_2^2$. Therefore, the probability that $G_T(\mathbb{W}_i)$ has more than $T^2(\|\mathbb{W}_i\|_1 + c^3)/2$ or less than $T^2(\|\mathbb{W}_i\|_1 - c^3)/2$ edges is less than
$$\frac{T^2 \|\mathbb{W}_i\|_1/2 + T^3 \|D_{\mathbb{W}_i}\|_2^2}{c^6 T^4/4} \leq \frac{2C + 4TCD}{c^6 T^2} \leq \frac{c}{8}.$$

Here we used the fact that $\|D_{\mathbb{W}_i}\|_2^2 \leq \|D_{\mathbb{W}_i}\|_1 \|D_{\mathbb{W}_i}\|_\infty = \|\mathbb{W}_i\|_1 \|D_{\mathbb{W}_i}\|_\infty$, and we are assuming that T is large. Assuming neither of these events happens, $\|\mathbb{W}_i'\|_1$ is between $\|\mathbb{W}_i\|_1 - c^3$ and $\|\mathbb{W}_i\|_1 + c^3$. Since $|\|\mathbb{W}_1\|_1 - \|\mathbb{W}_2\|_1| \leq c^3$, we have that $|\|\mathbb{W}_1'\|_1 - \|\mathbb{W}_2'\|_1| \leq 3c^3$.

To summarize, we have that with high probability,

$$d_{2\to 2}(\mathbb{W}_1', \mathbb{W}_2') \leq \min(31cC)^{1/4}, 2c^{3/4}, \sqrt[3]{3}c).$$

The probability that this does not happen is at most

$$\frac{6c}{10} + \frac{c}{8} + 2\frac{c}{8} \leq c.$$

This completes the proof. □

7.2 Samples Converge to Graphex

In this subsection, we prove the sampling lemma, Theorem 7.1. To this end, we will first establish two lemmas. The first one states that each (C, D)-bounded graphex can be approximated by a step graphon, i.e., a graphex where the star and dust part is zero, and the graphon part is a step graphon.

Lemma 7.6 *For every ε, C, and D, there exist M, N, and ρ such that the following holds. For every (C, D)-bounded graphex $\mathbb{W}$, there exists a (C, D)-bounded graphex $\mathbb{W}_\varepsilon = (W_\varepsilon, 0, 0, \boldsymbol{\Omega}_\varepsilon)$, where $\boldsymbol{\Omega}_\varepsilon = (\Omega_\varepsilon, \mathcal{F}_\varepsilon, \mu_\varepsilon)$ and $\mu_\varepsilon(\Omega_\varepsilon) \leq N$, and furthermore the graphon W_ε is a step function with at most M steps, with each part having size equal to ρ, and $\delta_{2 \to 2}(\mathbb{W}, \mathbb{W}_\varepsilon) \leq \varepsilon$.*

Proof By Remark 4.4, we may assume that $\mathbb{W}$ is a graphex over an atomless measure space $\boldsymbol{\Omega} = (\Omega, \mathcal{F}, \mu)$. By Theorem 5.7, there exists $M(\varepsilon)$ and ρ such that there is a partition $\mathcal{P} = \{P_1, P_2, \ldots, P_m\}$ of $\Omega_{\mathscr{P}} \subseteq \Omega$ with $m \leq M(\varepsilon)$ such that $\delta_{2 \to 2}(\mathbb{W}_{\mathscr{P}}, \mathbb{W}) \leq \varepsilon/2$ and each part has size ρ. Let $\Omega_\varepsilon = \Omega_{\mathscr{P}} \cup Q$ where Q is any set disjoint from $\Omega_{\mathscr{P}}$, and obtain μ_ε by extending μ to Q (with measure to be determined later). Let $\mathbb{W}_\varepsilon = (W_\varepsilon, 0, 0, \boldsymbol{\Omega}_\varepsilon)$ with

$$W_\varepsilon(x, y) = \begin{cases} W_{\mathscr{P}}(x, y) & \text{if } x \in P_i, y \in P_j, \\ \frac{S_{\mathscr{P}}(x)}{\mu(Q)} & \text{if } x \in P_i, y \in Q, \\ \frac{S_{\mathscr{P}}(y)}{\mu(Q)} & \text{if } x \in Q, y \in P_i, \text{ and} \\ \frac{2I_{\mathscr{P}}}{\mu(Q)^2} & \text{if } x, y \in Q. \end{cases}$$

Extend $W_{\mathscr{P}}$ by 0 to Q. Since $\mathbb{W}_{\mathscr{P}}$ is (C, D)-bounded, there exists K depending only on ε, C, D such that if $\mu(Q) \geq K$, then $W_\varepsilon - W_{\mathscr{P}}$ is at most $\varepsilon^2/(4C)$ everywhere, which implies that

$$\|W_\varepsilon - W_{\mathscr{P}}\|_{2 \to 2} \leq \|W_\varepsilon - W_{\mathscr{P}}\|_2$$
$$\leq \sqrt{\|W_\varepsilon - W_{\mathscr{P}}\|_1 \|W_\varepsilon - W_{\mathscr{P}}\|_\infty} \leq \sqrt{C\varepsilon^2/(4C)} = \varepsilon/2.$$

For $x \in \Omega_{\mathcal{P}}$,

$$D_{\mathbb{W}_\varepsilon}(x) = D_{\mathbb{W}_{\mathcal{P}}}(x) + \mu(Q) \frac{S_{\mathcal{P}}(x)}{\mu(Q)} = D_{\mathbb{W}_{\mathcal{P}}}(x).$$

We also have that for $x \in Q$,

$$D_{\mathbb{W}_\varepsilon}(x) = \mu(Q)\frac{2I_\mathscr{P}}{\mu(Q)^2} + \sum_i \int_{P_i} \frac{S_\mathscr{P}(y)}{\mu(Q)} d\mu(y) = \frac{D_{\mathbb{W}_\mathscr{P}}(\infty)}{\mu(Q)}.$$

Therefore, there exists a K' depending only on ε, C, and D such that if $\mu(Q) \geq K'$, then

$$\|D_{\mathbb{W}_\varepsilon} - D_{\mathbb{W}_\mathscr{P}}\|_2^2 = \int_{\Omega_\mathscr{P} \cup Q} \left(D_{\mathbb{W}_\varepsilon}(x) - D_{\mathbb{W}_\mathscr{P}}(x)\right)^2 d\mu(x)$$

$$= \int_Q \left(\frac{D_{\mathbb{W}_\mathscr{P}}(\infty)}{\mu(Q)}\right)^2 = \frac{D_{\mathbb{W}_\mathscr{P}}(\infty)^2}{\mu(Q)} \leq \varepsilon^4/16.$$

Also, by construction, $\|\mathbb{W}_\varepsilon\|_1 = \|\mathbb{W}_\mathscr{P}\|_1$. Therefore $\delta_{2\to 2}(\mathbb{W}_\varepsilon, \mathbb{W}_\mathscr{P}) \leq \varepsilon/2$, and hence $\delta_{2\to 2}(\mathbb{W}_\varepsilon, \mathbb{W}) \leq \varepsilon$. □

Remark 7.7 Using the ideas of the previous proof, it is not hard to see that in distribution, the process generated from the graphex $\widetilde{\mathbb{W}}_Q = (\widetilde{W}_Q, 0, 0, \widetilde{\Omega}_D)$ constructed in Remark 2.3 (3) converges to the one generated from $\mathbb{W}$. Indeed, we claim that

$$\delta_\diamond(\widetilde{\mathbb{W}}_Q, \mathbb{W}) \to 0 \quad \text{as} \quad Q \to \infty.$$

To see this, fix $\varepsilon > 0$ and choose D in such a way that the set $\Omega_{>D} = \{D_\mathbb{W} > D\}$ has measure at most ε^2. Let $\Omega_{\leq D} = \Omega \setminus \Omega_{>D}$ and $\widetilde{\Omega}_{\leq D} = \Omega_{\leq D} \cup \{\infty\} = \widetilde{\Omega} \setminus \Omega_{>D}$. Setting $\widetilde{\mathbb{W}}_{Q,\leq D} = (\widetilde{\mathbb{W}}_Q)_{|\widetilde{\Omega}\setminus\Omega_{>D}}$ and $\mathbb{W}_{\leq D} = \mathbb{W}_{|\Omega\setminus\Omega_{>D}}$ and defining $\widetilde{\mathbb{W}}_{\leq D}$ as the trivial extension of $\mathbb{W}_{\leq D}$ to $\widetilde{\Omega}_{\leq D}$, we will want to show that for Q large enough, $d_{2\to 2}(\widetilde{\mathbb{W}}_{\leq D}, \widetilde{\mathbb{W}}_{Q,\leq D}) \leq \varepsilon$, since this implies that $\delta_{2\to 2}(\mathbb{W}_{\leq D}, \widetilde{\mathbb{W}}_{Q,\leq D}) \leq \varepsilon$ and hence $\delta_\diamond(\widetilde{\mathbb{W}}_Q, \mathbb{W}) \leq \varepsilon$. But this follows by essentially the same argument as the one in the previous proof; all that is needed is that by Proposition 2.4, $\widetilde{\mathbb{W}}_{\leq D}$ is (C, D)-bounded for some $C < \infty$.

Our second lemma estimates the distance between the empirical graphex corresponding to a weighted graph H with weights in $[0, 1]$ and the one corresponding to the graph G obtained from H by choosing the edge in G randomly according to H. More precisely, given a finite weighted graph H with weights $H_{i,j} \in [0, 1]$ and $H_{i,i} = 0$, define $G(H)$ as the graph generated by taking an edge between i and j with probability $H_{i,j}$, independently for every pair of vertices. Our next lemma estimates the distance between the empirical graphon of H and the empirical graphon of $G(H)$.

Lemma 7.8 *For every N_0, ε, and δ, there exists n_0 such that the following holds. For any weighted graph H on $n \geq n_0$ vertices with weights in $[0, 1]$, and any $N \leq N_0$, the probability that $\delta_{2\to 2}(\mathbb{W}(H, N/n), \mathbb{W}(G(H), N/n)) > \varepsilon$ is at most δ.*

Proof We first extend both H and G trivially to $\mathbb{N}$, and then define U as the graphon $U = W(G(H), N/n) - W(H, N/n)$. Then $\|U\|_{2\to 2} \leq (t(C_4, U))^{1/4}$ by

Lemma 3.18. As a consequence, the probability that $\|U\|_{2\to 2} > \varepsilon$ is bounded by $\varepsilon^{-4}\mathbb{E}[t(C_4, U)]$. Using the bound (13) from the proof of Lemma 7.5 with $E = n^2$, $F = n^3$, and $\rho = N/n$, we get that the probability that $\|U\|_{2\to 2} > \varepsilon$ is bounded by

$$\varepsilon^{-4}\mathbb{E}[t(C_4, U)] \le \varepsilon^{-4}\frac{N^4}{n^4}\left(n^2 + 2n^3\right) \le 3\varepsilon^{-4}n^3\frac{N^4}{n^4} \le 3\varepsilon^{-4}\frac{N_0^4}{n_0} \le \delta/2,$$

provided $n_0 \ge 6\varepsilon^{-4}\delta^{-1}N_0^4$. Let us now bound the other two components of $\Delta_{2\to 2}$. For a fixed vertex v, by Hoeffding's inequality [15],

$$\mathbb{P}[|d_{G(H)}(v) - d_H(v)| > \varepsilon' n] \le 2e^{-2\varepsilon'^2 n}.$$

Therefore, by a union bound,

$$\mathbb{P}[\text{there exists a vertex } v \text{ such that } |d_{G(H)}(v) - d_H(v)| > \varepsilon' n] \le 2n e^{-2\varepsilon'^2 n}.$$

For any fixed ε', if n is large enough, this probability is less than $\delta/2$. If this does not happen, then for every vertex v,

$$\left|D_{W(G(H), \frac{N}{n})}(v) - D_{W(H, \frac{N}{n})}(v)\right| \le N\varepsilon'.$$

Therefore, for ε' small enough,

$$\|D_{W(G(H), N/n)} - D_{W(H, N/n)}\|_2 \le N^{3/2}\varepsilon' \le \varepsilon^2,$$

and

$$\left|\|W(G(H), N/n)\|_1 - \|W(H, N/n)\|_1\right| = \left|\|D_{W(G(H), N/n)}\|_1 - \|D_{W(H, N/n)}\|_1\right|$$
$$\le N^2\varepsilon' \le \varepsilon^3.$$

This completes the proof of the lemma. □

With these preparations, we are ready to prove the sampling lemma.

Proof (*Theorem 7.1*) Fix $\varepsilon > 0$. We know from the definition of tightness that there exist C and D so that we can remove a set Ω_ε of measure at most $\varepsilon^2/2$ to obtain a (C, D)-bounded graphex. Then the expected number of points in G_T whose feature lies inside Ω_ε is $\varepsilon^2 T/2$. Therefore, since it is a Poisson distribution, the probability that $G_T(\mathbb{W})$ has more than $\varepsilon^2 T$ points in Ω_ε is at most

$$e^{\frac{\varepsilon^2 T}{2}(1 - 2\log 2)}.$$

This converges to 0 as $T \to \infty$. If $G_T(\mathbb{W})$ does not have more than $\varepsilon^2 T$ points, then we can remove those points from $G_T(\mathbb{W})$ and the sample is equivalent to a sample from the graphex restricted to $\Omega \setminus \Omega_\varepsilon$. (It may have isolated vertices but this does not affect our distance.) Since a set of $\varepsilon^2 T$ points in $G_T(\mathbb{W})$ corresponds to a set of measure ε^2 in $\mathbb{W}(G_T(\mathbb{W}, 1/T))$, this shows that we may assume without loss of generality that the original set is (C, D)-bounded, and prove Theorem 7.1 for $\delta_{2\to 2}$ instead of $\delta_\diamond$.

Choose $\mathbb{W}_\delta$ as in Lemma 7.6 (with δ taking the role of ε) so that in particular $\delta_{2\to 2}(\mathbb{W}, \mathbb{W}_\delta) \leq \delta$. For sufficiently small δ, Theorem 7.3 then implies that there exists a T_0 such that if $T > T_0$, then the samples from $\mathbb{W}$ and $\mathbb{W}_\delta$ can be coupled so that
$$\mathbb{P}[\delta_{2\to 2}(\mathbb{W}(G_T(\mathbb{W}), 1/T), \mathbb{W}(G_T(\mathbb{W}_\delta), 1/T)) > (31C\delta)^{1/4}] < \delta.$$

This means that it suffices to prove Theorem 7.1 for step function graphons with equal size parts, uniformly over any set of graphons with a bounded number of parts with the same size. Indeed, for any $\varepsilon > 0$, let $\delta > 0$ be such that
$$2\delta + (31\delta C)^{1/4} < \varepsilon.$$

If we then take $\mathbb{W}_\delta$ as above, then $\delta_{2\to 2}(\mathbb{W}_\delta, \mathbb{W}) \leq \delta$, so by Theorem 7.3, for large enough T, we can couple $\mathbb{W}(G_T(\mathbb{W}_\delta, 1/T))$ and $\mathbb{W}(G_T(\mathbb{W}, 1/T))$ so that the probability that they have $\delta_{2\to 2}$ distance more than $(31\delta C)^{1/4}$ is at most δ. Furthermore, we can take T large enough so that the probability that $\delta_{2\to 2}(\mathbb{W}(G_T(\mathbb{W}_\delta), 1/T), \mathbb{W}_\delta) > \delta$ is at most δ (detailed below). Overall, by the triangle inequality, this implies that the probability that $\delta_{2\to 2}(\mathbb{W}(G_T(\mathbb{W}), 1/T), \mathbb{W}) \geq \varepsilon$ is at most 2δ. Since this works for arbitrarily small δ, the theorem follows.

Suppose therefore that $\mathbb{W} = (W, 0, 0, \boldsymbol{\Omega})$, where W is a step graphon with step size ρ and m steps total. Fix $\varepsilon > 0$ and $\delta > 0$. For a fixed part P_i and T, let $X_{T,i}$ be the number of points in P_i in the Poisson process. The expectation of each $X_{T,i}$ is ρT. For $\varepsilon' > 0$, we have
$$\mathbb{P}\left[X_{T,i} > (1+\varepsilon')\rho T\right] < e^{\rho T(\varepsilon' - (1+\varepsilon')\log(1+\varepsilon'))} = e^{-\rho T c(\varepsilon')}$$
for a nonnegative number $c(\varepsilon')$. We also have
$$\mathbb{P}\left[X_{T,i} < (1-\varepsilon')\rho T\right] < e^{\rho T(-\varepsilon' + (1-\varepsilon')\log(\frac{1}{1-\varepsilon'}))} = e^{-\rho T c'(\varepsilon')}$$
for a nonnegative number $c'(\varepsilon')$. Therefore, if T is large enough, then the probability that any part P_i has more than $(1+\varepsilon')\rho T$ or less than $(1-\varepsilon')\rho T$ points is less than $\delta/2$. Note that in particular this means that the total measure of nonzero points is at most $(1+\varepsilon')\rho m$. Therefore, with probability at least $1 - \delta/2$, we can add or delete points with total measure at most $\varepsilon' \rho m$ to obtain W from $W(H_T(W), \rho)$. This means that we can couple $\mathbb{W}(H_T(W), \rho)$ and $\mathbb{W}$ so that they differ on points with total measure at most $\varepsilon' \rho m$, and hence
$$\|W(H_T(W), \rho) - W\|_1 \leq 2\varepsilon'(1+\varepsilon')\rho^2 m^2.$$

Therefore, we have the same bound for $|\|W(H_T(W), \rho)\|_1 - \|W\|_1|$. Since both graphons are between 0 and 1, we also have

$$\|W(H_T(W), \rho) - W\|_{2\to 2} \leq \|W(H_T(W), \rho) - W\|_2$$
$$\leq \sqrt{\|W(H_T(W), \rho) - W\|_1} \leq \sqrt{2\varepsilon'(1+\varepsilon')\rho m}.$$

Finally, we have

$$\|D_{\mathbb{W}(H_T(W), \rho)} - D_{\mathbb{W}}\|_2^2 \leq \rho m(\varepsilon' \rho m)^2 + \varepsilon' \rho m((1+\varepsilon')\rho m)^2 = (\varepsilon'^3 + 3\varepsilon'^2 + \varepsilon')\rho^3 m^3.$$

We can therefore take ε' small enough that $\delta_{2\to 2}(\mathbb{W}(H_T(W), \rho), \mathbb{W}) < \varepsilon/2$ with probability at least $1 - \delta/2$.

Using Lemma 7.8 for $\varepsilon/2$ and $\delta/2$, we have that with probability at least $1 - \delta/2$,

$$\delta_{2\to 2}(\mathbb{W}, H_T(W, 1/T)) \leq \frac{\varepsilon}{2},$$

and with probability at least $1 - \delta/2$,

$$\delta_{2\to 2}(\mathbb{W}(H_T(W), 1/T), \mathbb{W}(G_T(\mathbb{W}), 1/T)) \leq \frac{\varepsilon}{2}.$$

Therefore, with probability at least $1 - \delta$,

$$\delta_{2\to 2}(\mathbb{W}, \mathbb{W}(G_T(\mathbb{W}), 1/T))$$
$$\leq \delta_{2\to 2}(\mathbb{W}, H_T(W, 1/T)) + \delta_{2\to 2}(\mathbb{W}(H_T(W), 1/T), \mathbb{W}(G_T(\mathbb{W}), 1/T))$$
$$\leq \frac{\varepsilon}{2} + \frac{\varepsilon}{2} = \varepsilon.$$

This completes the proof of Theorem 7.1. □

7.3 Proofs of Theorem 2.18, Proposition 2.13, and Theorem 2.23

Having completed the proof of Theorem 7.1, we are finally ready to establish that $\delta_\diamond$ convergence is equivalent to GP-convergence, together with several of the other equivalences stated in Sect. 2. To this end, we first prove the following theorem.

Theorem 7.9 *Given a pair of graphexes* $\mathbb{W}, \mathbb{W}'$, *we have* $\delta_\diamond(\mathbb{W}, \mathbb{W}') = 0$ *if and only if for every* $T > 0$, $G_T(\mathbb{W})$ *and* $G_T(\mathbb{W}')$ *have the same distribution.*

Proof If $\delta_\diamond(\mathbb{W}', \mathbb{W}) = 0$, then taking $\mathbb{W}_n = \mathbb{W}'$ for each n, Theorem 6.17 implies that $G_T(\mathbb{W})$ and $G_T(\mathbb{W}')$ must have the same distribution for every T. Suppose now that $G_T(\mathbb{W})$ and $G_T(\mathbb{W}')$ have the same distribution for every T. By Theorem

7.1, we can choose T such that with probability at least 0.99, $\delta_\diamond(G_T(\mathbb{W}), \mathbb{W}) < \varepsilon/2$ and $\delta_\diamond(G_T(\mathbb{W}'), \mathbb{W}') < \varepsilon/2$. Since $G_T(\mathbb{W})$ and $G_T(\mathbb{W}')$ have the same distribution, the two graphexes have distance at most ε. Since this holds for every ε, the lemma follows. $\square$

Proof (*Theorems* 2.18, *Proposition* 2.13, *and Theorem* 2.23) We start with the proof of Theorem 2.18. One direction follows from Theorem 6.17. Suppose now that $\mathbb{W}_n$ is GP-convergent to $\mathbb{W}$. We know by Theorem 4.1 that then the set $\mathbb{W}_n$ is tight. By Theorem 2.21, $\mathbb{W}_n$ therefore has a subsequence that converges according to $\delta_\diamond$ to a graphex $\mathbb{W}'$, which in turn implies the subsequence is GP-convergent to $\mathbb{W}'$. This implies that for any $T > 0$, $G_T(\mathbb{W})$ and $G_T(\mathbb{W}')$ have the same distribution. By Theorem 7.9, $\delta_\diamond(\mathbb{W}, \mathbb{W}') = 0$, so $\delta_\diamond(\mathbb{W}_n, \mathbb{W}) \to 0$. Next recall that by Proposition 2.17, the distances $\delta_\diamond$ and $\delta_{2\to 2}$ give equivalent topologies on sets with uniformly bounded marginals, showing that Theorem 2.18 implies Proposition 2.13. We conclude by noting that Theorem 2.23 follows from Corollary 6.2 and Proposition 2.13. $\square$

8 Identifiability

In this section, we prove Theorem 2.5. In fact, we will prove the following version, which by Theorem 7.9 is equivalent.

Theorem 8.1 *Let* $\mathbb{W}_1 = (W_1, S_1, I_1, \boldsymbol{\Omega}_1)$ *and* $\mathbb{W}_2 = (W_2, S_2, I_2, \boldsymbol{\Omega}_2)$ *be graphexes, where* $\boldsymbol{\Omega}_i = (\Omega_i, \mathcal{F}_i, \mu_i)$ *are σ-finite spaces. Suppose* $\delta_\diamond(\mathbb{W}_1, \mathbb{W}_2) = 0$. *Then there exists a third graphex* $\mathbb{W} = (W, S, I, \boldsymbol{\Omega})$ *over a σ-finite measure space* $\boldsymbol{\Omega} = (\Omega, \mathcal{F}, \mu)$ *and measure preserving maps* $\varphi_i \colon \mathrm{dsupp}\, W_i \to \Omega$ *such that* $\mathbb{W}_i|_{\varphi^{-1}(\Omega)} = \mathbb{W}^{\varphi_i}$ *(and* $W_i, S_i = 0$ *everywhere else) for* $i = 1, 2$.

To prove the theorem, we will first prove the following theorem, which may be of independent interest. We recall that a Borel measure space is a measure space that is isomorphic to a Borel subset of a complete separable metric space equipped with a Borel measure, where, as usual, two measure spaces $\boldsymbol{\Omega} = (\Omega, \mathcal{F}, \mu)$ and $\boldsymbol{\Omega}' = (\Omega', \mathcal{F}', \mu')$ are called isomorphic if there exists a bijective map $\varphi \colon \Omega \to \Omega'$ such that both φ and its inverse are measure preserving.

Theorem 8.2 *Let* $\mathbb{W}_1 = (W_1, S_1, I_1, \boldsymbol{\Omega}_1)$ *and* $\mathbb{W}_2 = (W_2, S_2, I_2, \boldsymbol{\Omega}_2)$ *be graphexes, where* $\boldsymbol{\Omega}_i = (\Omega_i, \mathcal{F}_i, \mu_i)$ *are σ-finite Borel spaces. Suppose further that* $D_{\mathbb{W}_i} > 0$ *everywhere for* $i = 1, 2$, *and* $\delta_\diamond(\mathbb{W}_1, \mathbb{W}_2) = 0$. *Then* $\mu_1(\Omega_1) = \mu_2(\Omega_2)$, $I_1 = I_2$, *and there exists a coupling of* $\boldsymbol{\Omega}_1$ *and* $\boldsymbol{\Omega}_2$, *that is, a measure* ν *on* $(\Omega_1 \times \Omega_2, \mathcal{F}_1 \times \mathcal{F}_2)$ *with marginals* μ_1 *and* μ_2, *such that if* $\pi_i \colon \Omega_1 \times \Omega_2 \to \Omega_i$ *is the projection map, then* $W_1^{\pi_1} = W_2^{\pi_2}$ *ν-almost-everywhere, and* $S_1^{\pi_1} = S_2^{\pi_2}$ *ν-almost-everywhere.*

Theorem 8.2 should be compared to Proposition 8 from [3] which states the analogous result for integrable Borel graphons that have cut distance zero (without the assumption that W_1 and W_2 are non-negative), using a different proof technique. Using still different proof techniques, Janson proved a similar result (again without assuming non-negativity), showing that after trivially extending two integrable Borel graphons with cut distance zero they can be coupled so that the projections are equal almost everywhere; see [17]. We will prove Theorem 8.2 in Sect. 8.1.

Remark 8.3 Throughout this paper, we have considered graphexes where all three parts are non-negative. While this makes sense when considering graphexes as generators of a graphex process, from an analytical point of view, it is less natural. Indeed, it is easy to define the kernel and weak kernel distance for graphexes where the three parts take values in $\mathbb{R}$. Taking, e.g., the kernel distance $d_{2\to 2}$ defined in (1), all we need to do is replace the L^1 norms in the third part by a signed "edge density" $\rho(\mathbb{W}_i) = \int W \, d\mu \times d\mu + 2\int S \, d\mu + 2I$, and then use the third root of $|\rho(\mathbb{W}_1) - \rho(\mathbb{W}_2)|$ instead of the third root of $|\|\mathbb{W}_1\|_1 - \|\mathbb{W}_2\|_1|$. In particular in view of the just discussed results from [3, 17], we conjecture that Theorem 8.2 holds for signed graphexes as well, provided the condition $D_{\mathbb{W}_i} > 0$ is replaced by the condition $D_{|\mathbb{W}_i|} > 0$, where $|\mathbb{W}_i|$ is obtained from $\mathbb{W}_i$ by replacing all three components of $\mathbb{W}_i$ by their absolute values. We leave the proof of this conjecture as an open problem.

Once we have established Theorem 8.2, we will then prove Theorem 8.1 by generalizing a construction which was developed by Janson for the dense case in [16]. To this end, we will assign to each graphex $\mathbb{W}$ a "canonical version" $\widehat{\mathbb{W}}$ such that $\mathbb{W}$ is a pullback of $\widehat{\mathbb{W}}$ and show that if two graphexes are equivalent, then their canonical versions are isomorphic up to measure zero changes. This will be carried out in Sect. 8.2.

Remark 8.4 Section 8.2 does not use nonnegativity in any essential way, and should be easily generalizable to signed graphexes. This should give a relatively straightforward proof of the analogue of Theorem 8.1 for graphons of cut distance zero, and also allow for the more general setting of signed graphexes, once the above conjectured generalization of Theorem 8.2 is established. Again we leave this as an open problem.

8.1 Infimum Is Minimum

In this subsection, we will prove Theorem 8.2. The proof will be based on a series of lemmas.

Let $\widetilde{\mathbb{W}}_i = (\widetilde{W}_i, \widetilde{S}_i, \widetilde{I}_i, \widetilde{\mathbf{\Omega}}_i)$, for $i = 1, 2$, be trivial extensions of $\mathbb{W}_i$ to spaces of infinite measure, where $\widetilde{\mathbf{\Omega}}_i = (\widetilde{\Omega}_i, \widetilde{\mathcal{F}}_i, \widetilde{\mu}_i)$, and let $\varepsilon > 0$. By Proposition 4.7 (3), there exist $\widetilde{\Omega}_i^\varepsilon \subseteq \widetilde{\Omega}_i$ such that $\widetilde{\Omega}_i \setminus \widetilde{\Omega}_i^\varepsilon$ has measure at most ε, and a measure ν_ε on $\widetilde{\Omega}_1^\varepsilon \times \widetilde{\Omega}_2^\varepsilon$ with marginals $\widetilde{\mu}_1|_{\widetilde{\Omega}_1^\varepsilon}$ and $\widetilde{\mu}_2|_{\widetilde{\Omega}_2^\varepsilon}$ such that for the restricted graphexes $\widetilde{\mathbb{W}}_{i,\varepsilon}$,

$$\|\widetilde{W}_{1,\varepsilon}^{\pi_1} - \widetilde{W}_{2,\varepsilon}^{\pi_2}\|_{2\to 2, \nu_\varepsilon} \leq \varepsilon$$

and
$$\int_{\tilde{\Omega}_1^\varepsilon \times \tilde{\Omega}_2^\varepsilon} \left(D_{\tilde{\mathbb{W}}_{1,\varepsilon}}(x) - D_{\tilde{\mathbb{W}}_{2,\varepsilon}}(y)\right)^2 dv_\varepsilon(x,y) \leq \varepsilon^4.$$

With a slight abuse of notation, we extend v_ε to $\tilde{\Omega}_1 \times \tilde{\Omega}_2$ by zero. This means that in fact
$$\|\tilde{W}_1^{\pi_1} - \tilde{W}_2^{\pi_2}\|_{2\to 2, v_\varepsilon} \leq \varepsilon$$
and
$$\|D_{\tilde{\mathbb{W}}_1^{\pi_1, v_\varepsilon}} - D_{\tilde{\mathbb{W}}_2^{\pi_2, v_\varepsilon}}\|_2 \leq \varepsilon^2.$$

We first prove the following lemma.

Lemma 8.5 *For any c,*

(1) $\lim_{\varepsilon \to 0} v_\varepsilon(\Omega_{1,>c} \times \Omega_{2,>c}) = \mu_i(\Omega_{i,>c})$ *for $i=1,2$ (regardless of the choice of v_ε), and*

(2) $\mu_1(\Omega_{1,>c}) = \mu_2(\Omega_{2,>c})$.

Proof We first prove (1). By symmetry, it suffices to prove it for $i=1$. Note that for any $x \in \tilde{\Omega}_i^\varepsilon$,
$$D_{\tilde{\mathbb{W}}_{i,\varepsilon}}(x) \leq D_{\tilde{\mathbb{W}}_i}(x) \leq D_{\tilde{\mathbb{W}}_{i,\varepsilon}}(x) + \varepsilon$$
and that $\tilde{\Omega}_{i,>c} = \Omega_{i,>c}$. Therefore,
$$\varepsilon v_\varepsilon\left(\Omega_{1,>c} \times (\tilde{\Omega}_2 \setminus \Omega_{2,>c-(\sqrt{\varepsilon}+\varepsilon)})\right)$$
$$\leq \int_{\Omega_{1,>c} \times (\tilde{\Omega}_2 \setminus \Omega_{2,>c-(\sqrt{\varepsilon}+\varepsilon)})} \left(D_{\mathbb{W}_{1,\varepsilon}}(x) - D_{\mathbb{W}_{2,\varepsilon}}(y)\right)^2 \leq \varepsilon^4,$$
which implies that
$$v_\varepsilon(\Omega_{1,>c} \times (\tilde{\Omega}_2 \setminus \Omega_{2,>c-(\sqrt{\varepsilon}+\varepsilon)})) < \varepsilon^3.$$

Now, for any ε,
$$|\mu_1(\Omega_{1,>c}) - v_\varepsilon(\Omega_{1,>c} \times \Omega_{2,>c})| \leq \varepsilon + |v_\varepsilon(\Omega_{1,>c} \times \tilde{\Omega}_2) - v_\varepsilon(\Omega_{1,>c} \times \Omega_{2,>c})|$$
$$= \varepsilon + v_\varepsilon(\Omega_{1,>c} \times (\tilde{\Omega}_2 \setminus \Omega_{2,>c}))$$
$$= \varepsilon + v_\varepsilon(\Omega_{1,>c} \times (\tilde{\Omega}_2 \setminus \Omega_{2,>c-(\sqrt{\varepsilon}+\varepsilon)})) + v_\varepsilon(\Omega_{1,>c} \times (\Omega_{2,>c-(\sqrt{\varepsilon}+\varepsilon)} \setminus \Omega_{2,>c}))$$
$$< \varepsilon + \varepsilon^3 + \mu_2(\Omega_{2,>c-(\sqrt{\varepsilon}+\varepsilon)} \setminus \Omega_{2,>c}).$$

This last expression is finite, and tends to 0 as $\varepsilon \to 0$, so
$$\lim_{\varepsilon \to 0} v_\varepsilon(\Omega_{1,>c} \times \Omega_{2,>c}) = \mu_1(\Omega_{1,>c}).$$

This proves (1). From this, (2) is obvious. □

Lemma 8.6 *For any $n \geq 1$, there exists a measure ν_n on $\Omega_{1,>1/n} \times \Omega_{2,>1/n}$ such that the following hold:*

(1) ν_n is a coupling of $\Omega_{1,>1/n}$ and $\Omega_{2,>1/n}$,
(2) $\nu_{n+1}|_{\Omega_{1,>1/n} \times \Omega_{2,>1/n}} = \nu_n$,
(3) $W_1^{\pi_1}$ and $W_2^{\pi_2}$ are equal when restricted to $(\Omega_{1,>1/n} \times \Omega_{2,>1/n})^2$, $\nu_n \times \nu_n$-almost everywhere, and
(4) $D_{W_1^{\pi_1}}$ and $D_{W_2^{\pi_2}}$ are equal when restricted to $\Omega_{1,>1/n} \times \Omega_{2,>1/n}$, ν_n-almost everywhere.

Proof It is well known that any two Borel measurable spaces with the same cardinality are isomorphic; see, e.g., Theorem 8.3.6 in [12]. As a consequence, each Borel measure space $(\Omega, \mathcal{F}, \mu)$ with $\mu(\Omega) < \infty$ is either empty or isomorphic to a finite set (with the discrete topology), the countable set $\{0\} \cup \{1/n : n \in \mathbb{N}\}$ (with the induced topology from $\mathbb{R}$), or the Cantor cube $\mathcal{C} = \{0, 1\}^\infty$ (with the product topology), equipped with the Borel σ-algebras generated by the topologies, and a measure that is a finite Borel measure with full support.

We can therefore assume without loss of generality that for each i, $\Omega_{i,>1}$ and each set $\Omega_{i,>1/(n+1)} \setminus \Omega_{i,>1/n}$ are of this form. This means we may without loss of generality assume the following properties:

(1) Each $\Omega_{i,>1/n}$, and thus each $\Omega_{i,>1/n}^2$, is compact.
(2) For any $i_1, i_2 = 1, 2$ and $n_1, n_2 \in \mathbb{N}^+$, and any finite Borel measure ν on $\Omega_{i_1,>1/n_1} \times \Omega_{i_2,>1/n_2}$, the set of all step functions on $\Omega_{i_1,>1/n_1} \times \Omega_{i_2,>1/n_2}$ corresponding to partitions of $\Omega_{i_j,>1/n_j}$ into clopen sets for $j = 1, 2$ is dense in $L^1(\Omega_{i_1,>1/n_1} \times \Omega_{i_2,>1/n_2})$.

Now, take a sequence $\varepsilon_k \to 0$, and recall that we have an almost coupling measure ν_{ε_k} on $\widetilde{\Omega}_1 \times \widetilde{\Omega}_2$ with

$$\|\widetilde{W}_1^{\pi_1} - \widetilde{W}_2^{\pi_2}\|_{2 \to 2, \nu_{\varepsilon_k}} \leq \varepsilon_k$$

and

$$\|D_{\widetilde{W}_1^{\pi_1, \nu_{\varepsilon_k}}} - D_{\widetilde{W}_2^{\pi_2, \nu_{\varepsilon_k}}}\|_2 \leq \varepsilon_k^2.$$

We know that for each n, $\Omega_{1,>1/n} \times \Omega_{2,>1/n}$ is compact, and for any $K > 0$, the set of measures on it bounded by K is compact under the topology of weak convergence of measures. Since for any c,

$$\lim_{k \to \infty} \nu_{\varepsilon_k}(\Omega_{1,>c} \times \Omega_{2,>c}) = \mu_1(\Omega_{1,>c}) = \mu_2(\Omega_{2,>c}) < \infty,$$

we can take a subsequence of ν_{ε_k} such that for each n, the measure is convergent when restricted to $\Omega_{1,>1/n} \times \Omega_{2,>1/n}$. Without loss of generality we assume that the original sequence has this property. For each n, we then define ν_n as the limit measure on $\Omega_{1,1/n} \times \Omega_{2,>1/n}$. Having defined ν_n we now prove (1)–(4).

(1) We have seen in Lemma 8.5 that

$$\nu_n(\Omega_{1,>1/n} \times \Omega_{2,>1/n}) = \mu_1(\Omega_{1,>1/n}) = \mu_2(\Omega_{2,>1/n}).$$

For any clopen $F \subseteq \Omega_{1,>1/n}$, $F \times \Omega_{2,>1/n}$ is clopen in $\Omega_{1,>1/n} \times \Omega_{2,>1/n}$. Therefore,

$$\lim_{k\to\infty} \nu_{\varepsilon_k}(F \times \Omega_{2,>1/n}) = \nu_n(F \times \Omega_{2,>1/n}).$$

On the other hand, by Lemma 8.5,

$$\nu_{\varepsilon_k}(F \times (\widetilde{\Omega}_2 \setminus \Omega_{2,>1/n})) \leq \nu_{\varepsilon_k}(\Omega_{1,>1/n} \times (\widetilde{\Omega}_2 \setminus \Omega_{2,>1/n})) \xrightarrow{k\to\infty} 0.$$

We also have that
$$|\nu_{\varepsilon_k}(F \times \widetilde{\Omega}_2) - \mu_1(F)| \leq \varepsilon_k \xrightarrow{k\to\infty} 0.$$

Therefore for any clopen set, $\mu_1(F) = \nu_n(F \times \Omega_{2,>1/n})$. This implies that μ_1 and the projection of ν_n onto $\Omega_{1,>1/n}$ are the same, which proves (1).

(2) Since $\Omega_{1,>1/n} \times \Omega_{2,>1/n}$ is clopen in $\Omega_{1,>1/(n+1)} \times \Omega_{2,>1/(n+1)}$,

$$\nu_{n+1}(\Omega_{1,>1/n} \times \Omega_{2,>1/n}) = \lim_{k\to\infty} \nu_{\varepsilon_k}(\Omega_{1,>1/n} \times \Omega_{2,>1/n}) = \nu_n(\Omega_{1,>1/n} \times \Omega_{2,>1/n}).$$

Furthermore, for any closed $F \subseteq \Omega_{1,>1/n} \times \Omega_{2,>1/n}$, F is also closed in $\Omega_{1,1/(n+1)} \times \Omega_{2,1/(n+1)}$, which implies that

$$\limsup_{k\to\infty} \nu_{\varepsilon_k}(F) \leq \nu_{n+1}(F).$$

This implies that ν_{ε_k} converges weakly to $\nu_{n+1}|_{\Omega_{1,>1/n} \times \Omega_{2,>1/n}}$, but since it also converges to ν_n, the two must be equal.

(3) Let $W_{i,n} = \widetilde{W}_i|_{(\Omega_{i,>1/n})^2}$. Since

$$\|\widetilde{W}_1^{\pi_1} - \widetilde{W}_2^{\pi_2}\|_{2\to 2, \nu_{\varepsilon_k}} \leq \varepsilon_k,$$

we have that in particular, for any n,

$$\|W_{1,n}^{\pi_1} - W_{2,n}^{\pi_2}\|_{2\to 2, \nu_{\varepsilon_k}} \leq \varepsilon_k.$$

This implies that
$$\|W_{1,n}^{\pi_1} - W_{2,n}^{\pi_2}\|_{\square, \nu_{\varepsilon_k}} \leq \varepsilon_k \mu_1(\Omega_{1,>1/n}).$$

Since $\Omega_{i,>1/n}$ each have finite measure, we can use Janson's argument in [16]. We present the argument for completeness. Fix ε. We can find step graphons $U_{1,n}$ and $U_{2,n}$ on $(\Omega_{1,>1/n})^2$ and $(\Omega_{2,>1/n})^2$, with each part in the partition being a clopen set, such that
$$\|U_{i,n} - W_{i,n}\|_1 \leq \varepsilon.$$

This means that for any coupling measure on $\Omega_{1,>1/n} \times \Omega_{2,>1/n}$,

$$\|U_{i,n}^{\pi_i} - W_{i,n}^{\pi_i}\|_1 \leq \varepsilon.$$

Now, $U_{1,n}^{\pi_1} - U_{2,n}^{\pi_2}$ is a step function on $(\Omega_{1,>1/n} \times \Omega_{2,>1/n})^2$ with a partition into clopen parts. This means that since the restrictions of ν_{ε_k} weakly converge to ν_n,

$$\|U_{1,n}^{\pi_1} - U_{2,n}^{\pi_2}\|_{\square, \nu_{\varepsilon_k}} \xrightarrow{k \to \infty} \|U_{1,n}^{\pi_1} - U_{2,n}^{\pi_2}\|_{\square, \nu_n}.$$

Take k large enough so that $\varepsilon_k \mu_1(\Omega_{1,>1/n}) \leq \varepsilon$ and

$$\left| \|U_{1,n}^{\pi_1} - U_{2,n}^{\pi_2}\|_{\square, \nu_{\varepsilon_k}} - \|U_{1,n}^{\pi_1} - U_{2,n}^{\pi_2}\|_{\square, \nu_n} \right| \leq \varepsilon.$$

We then have

$$\|W_{1,n}^{\pi_1} - W_{2,n}^{\pi_2}\|_{\square, \nu_n}$$
$$\leq \|W_{1,n}^{\pi_1} - U_{1,n}^{\pi_1}\|_{1, \nu_n} + \|U_{1,n}^{\pi_1} - U_{2,n}^{\pi_2}\|_{\square, \nu_n} + \|U_{2,n}^{\pi_2} - W_{2,n}^{\pi_2}\|_{1, \nu_n}$$
$$\leq 3\varepsilon + \|U_{1,n}^{\pi_1} - U_{2,n}^{\pi_2}\|_{\square, \nu_{\varepsilon_k}}$$
$$\leq 3\varepsilon + \|W_{1,n}^{\pi_1} - U_{1,n}^{\pi_1}\|_{1, \nu_{\varepsilon_k}} + \|W_{1,n}^{\pi_1} - W_{2,n}^{\pi_2}\|_{\square, \nu_{\varepsilon_k}} + \|U_{2,n}^{\pi_2} - W_{2,n}^{\pi_2}\|_{1, \nu_{\varepsilon_k}}$$
$$\leq 6\varepsilon.$$

Since this holds for any ε, this proves (3).

(4) Fix $\varepsilon > 0$ and assume that $\varepsilon_k \leq \varepsilon$. Since $D_{\widetilde{W}_{i,\varepsilon_k}}(x) \leq D_{W_i}(x) \leq D_{\widetilde{W}_{i,\varepsilon_k}}(x) + \varepsilon_k$ for all $x \in \Omega_i \cap \widetilde{\Omega}_i^\varepsilon$,

$$\nu_\varepsilon \left(\left\{ x \in \Omega_{1,>1/n} \times \Omega_{2,>1/n} : \left| D_{W_1^{\pi_1}}(x) - D_{W_2^{\pi_2}}(x) \right| \geq 2\varepsilon \right\} \right)$$
$$\leq \nu_\varepsilon \left(\left\{ x \in \Omega_{1,>1/n} \times \Omega_{2,>1/n} : \left| D_{\widetilde{W}_1^{\pi_1, \nu_{\varepsilon_k}}}(x) - D_{\widetilde{W}_2^{\pi_2, \nu_{\varepsilon_k}}}(x) \right| \geq \varepsilon \right\} \right)$$
$$\leq \varepsilon^{-2} \|D_{\widetilde{W}_1^{\pi_1, \nu_{\varepsilon_k}}} - D_{\widetilde{W}_2^{\pi_2, \nu_{\varepsilon_k}}}\|_2^2 \leq \varepsilon^{-2} \varepsilon_k^4.$$

Since for all $\varepsilon > 0$ the right side converges to 0 as $k \to \infty$ this shows that ν_n is supported on $\{x \in \Omega_{1,>1/n} \times \Omega_{2,>1/n} : D_{W_1^{\pi_1}}(x) = D_{W_2^{\pi_2}}(x)\}$. $\square$

Proof (*Theorem* 8.2) After these preparations, we are ready to define the measure μ on $\Omega_1 \times \Omega_2$. Note that since before the extensions $D_{W_i} > 0$ almost everywhere, we have that $\bigcup_n \Omega_{i,>1/n} = \Omega_i$. For $A \subseteq \Omega_1 \times \Omega_2$, let

$$\nu(A) = \lim_{n \to \infty} \nu_n(A \cap (\Omega_{1,>1/n} \times \Omega_{2,>1/n}))$$
$$= \sum_{n=1}^{\infty} \nu_n(A \cap ((\Omega_{1,>1/n} \times \Omega_{2,>1/n}) \setminus (\Omega_{1,>1/(n-1)} \times \Omega_{2,>1/(n-1)}))).$$

Here with a slight abuse of notation we think of $\Omega_{i,1/0}$ as the empty set. To show that this is a coupling, note that for any measurable set $X \subseteq \Omega_1$,

$$\mu_1(X) = \lim_{n\to\infty} \mu_1(X \cap \Omega_{1,>1/n})$$
$$= \lim_{n\to\infty} \nu_n((X \cap \Omega_{1,>1/n}) \times \Omega_{2,>1/n})$$
$$= \lim_{n\to\infty} \nu_n((X \times \Omega_2) \cap (\Omega_{1,>1/n} \times \Omega_{2,>1/n})) = \nu(X \times \Omega_2),$$

where in the second step we used that for each n, the restriction of ν to $\Omega_{1,>1/n} \times \Omega_{2,>1/n}$ is equal to ν_n, which is a coupling. Clearly the analogous argument works for subsets $Y \subseteq \Omega_2$.

Now, let

$$N = \{(x,y) = ((x_1,x_2),(y_1,y_2)) \in (\Omega_1 \times \Omega_2)^2 : W_1^{\pi_1}(x,y) \neq W_2^{\pi_2}(x,y)\}$$

and

$$M = \{x = (x_1,x_2) \in (\Omega_1 \times \Omega_2) : D_{\mathbb{W}_1^{\pi_1}}(x) \neq D_{\mathbb{W}_2^{\pi_2}}(x)\}.$$

Let

$$N_n = \{(x,y) \in (\Omega_{1,>1/n} \times \Omega_{2,>1/n})^2 : W_1^{\pi_1}(x,y) \neq W_2^{\pi_2}(x,y)\}$$

and

$$M_n = \{x = (x_1,x_2) \in (\Omega_{1,>1/n} \times \Omega_{2,>1/n}) : D_{\mathbb{W}_1^{\pi_1}}(x) \neq D_{\mathbb{W}_2^{\pi_2}}(x)\}.$$

Since $\bigcup_n \Omega_{i,>1/n} = \Omega_i$ and $\Omega_{i,>1/n} \subseteq \Omega_{i,1/(n+1)}$, we have that $N = \bigcup_n N_n$ and $M = \bigcup_n M_n$. By Lemma 8.6 (3,4), $(\nu \times \nu)(N_n) = (\nu_n \times \nu_n)(N_n) = 0$ and $\nu(M_n) = 0$, which implies that $(\nu \times \nu)(N) = 0$ and $\nu(M) = 0$, and hence $W_1^{\pi_1} = W_2^{\pi_2}$ and $D_{\mathbb{W}_1^{\pi_1}} = D_{\mathbb{W}_2^{\pi_2}}$ almost everywhere. Since $W_1^{\pi_1} = W_2^{\pi_2}$ almost everywhere implies that $D_{W_1^{\pi_1}} = D_{W_2^{\pi_2}}$ almost everywhere, this in turn implies that $S_1^{\pi_1} = S_2^{\pi_2}$ ν-almost everywhere.

To prove that $I_1 = I_2$, we again use Proposition 4.7, but instead of (3) we this time use (2). Fix $D > 0$. Since $\delta_{2\to 2}(\mathbb{W}_{1,\leq D}, \mathbb{W}_{2,\leq D}) = 0$, we in particular have that $\|\mathbb{W}_{1,\leq D}\|_1 = \|\mathbb{W}_{2,\leq D}\|_1$. But since $W_1^{\pi_1} = W_2^{\pi_2}$ and $D_{\mathbb{W}_1^{\pi_1}} = D_{\mathbb{W}_2^{\pi_2}}$ almost everywhere,

$$\int_{(\Omega_1,\leq D)^2} W_1 \, d\mu_1 \times d\mu_1 = \int_{(\Omega_1 \times \Omega_2)^2} W_1^{\pi_1} 1_{D_{\mathbb{W}_1^{\pi_1}} \leq D} \, d\nu \times d\nu$$
$$= \int_{(\Omega_1 \times \Omega_2)^2} W_2^{\pi_2} 1_{D_{\mathbb{W}_2^{\pi_2}} \leq D} \, d\nu \times d\nu$$
$$= \int_{(\Omega_2,\leq D)^2} W_2 \, d\mu_2 \times d\mu_1.$$

In a similar way, $\int_{\Omega_1,\leq D} S_1 = \int_{\Omega_2,\leq D} S_2$. Therefore $\|W_{1,\leq D}\|_1 = \|W_{2,\leq D}\|_1$ implies $I_1 = I_2$. □

Corollary 8.7 *Let* $\mathbb{W} = (W, S, I, \mathbf{\Omega})$ *and* $\mathbb{W}' = (W', S', I', \mathbf{\Omega}')$ *be graphexes that are equivalent, and suppose that* $D_\mathbb{W}, D_{\mathbb{W}'} > 0$ *everywhere. Then there exist a positive integer n and a chain of graphexes* $\mathbb{W}_i = (W_i, S_i, I_i, \mathbf{\Omega}_i)$ *for* $i = 0, \ldots, n$, *with* $D_{\mathbb{W}_i} > 0$ *everywhere for each* $i = 0, \ldots, n$, $\mathbb{W}_0 = \mathbb{W}$, $\mathbb{W}_n = \mathbb{W}'$, *and for each* $i \geq 1$, *either* $\mathbb{W}_{i-1} = \mathbb{W}_i^{\varphi_i}$ *almost everywhere for some measure preserving map* φ_i *from* $\mathbf{\Omega}_{i-1}$ *to* $\mathbf{\Omega}_i$, *or* $\mathbb{W}_i = \mathbb{W}_{i-1}^{\varphi_i}$ *almost everywhere for some* φ_i *from* $\mathbf{\Omega}_i$ *to* $\mathbf{\Omega}_{i-1}$. *In fact, we can take* $n = 4$.

Proof By the construction in the next section (which itself does not use Corollary 8.7), there exists $\mathbb{W}_1 = (W_1, S_1, I_1, \mathbf{\Omega}_1)$ such that $\mathbf{\Omega}_1$ is Borel and a measure-preserving map φ_1 from $\mathbf{\Omega}_0$ to $\mathbf{\Omega}_1$ with $W = W_0 = W_1^{\varphi_1}$ almost everywhere. Then $\mathbf{\Omega}_0 = \text{dsupp } W_0 = \varphi^{-1}(\text{dsupp } W_1)$, so by replacing $\mathbf{\Omega}_1$ by its restriction to dsupp W_2, we may assume that $D_{W_1} > 0$ everywhere. Similarly, there exists $(W_3, \mathbf{\Omega}_3)$ with $\mathbf{\Omega}_3$ Borel and a measure preserving map φ_4 from $\mathbf{\Omega}_4$ to $\mathbf{\Omega}_3$ with $W' = W_4 = W_3^{\varphi_4}$ almost everywhere and $D_{W_3} > 0$ everywhere. Now, we can apply Theorem 7.1 to show that $\delta_\diamond(\mathbb{W}_1, \mathbb{W}_3) = 0$. We can then apply Theorem 8.2 to $(\mathbb{W}_1, \mathbf{\Omega}_1)$ and $(\mathbb{W}_3, \mathbf{\Omega}_3)$ to find a Borel space $\mathbf{\Omega}_2$ and measure preserving maps φ_2 from $\mathbf{\Omega}_2$ to $\mathbf{\Omega}_1$ and φ_3 from $\mathbf{\Omega}_2$ to $\mathbf{\Omega}_3$ such that $W_3^{\varphi_3} = W_1^{\varphi_2}$ almost everywhere. We can then take W_2 to be, say, $W_1^{\varphi_2}$. □

8.2 Canonical Graphex

In this section, we prove Theorem 8.1. We follow the approach of Janson in [16], based on the construction of Lovász and Szegedy in [23].

Concretely, given a graphex $\mathbb{W} = (W, S, I, \mathbf{\Omega})$, where $\mathbf{\Omega} = (\Omega, \mathcal{F}, \mu)$, define a map $\psi_W: \Omega \to L^1(\Omega, \mathcal{F}, \mu)$ by $x \mapsto W(x, \cdot)$, and define a map $\psi_\mathbb{W}: \Omega \to L^1(\Omega, \mathcal{F}, \mu) \times \mathbb{R}$ by $x \mapsto (W(x, \cdot), S(x))$, where we equip $L^1(\Omega, \mathcal{F}, \mu)$ with the standard Borel σ-algebra and $L^1(\Omega, \mathcal{F}, \mu) \times \mathbb{R}$ with the standard product Borel σ-algebra. Note that in general, we only know that $\psi_W(x) \in L^1(\Omega, \mathcal{F}, \mu)$ for almost all $x \in \Omega$, but by changing W on a set of measure zero, we may assume that this holds for all $x \in \Omega$; we will assume that throughout this section. We will see in Lemma 8.9 that ψ_W, and thus $\psi_\mathbb{W}$, is measurable. Defining μ_W and $\mu_\mathbb{W}$ as the pushforward of μ under ψ_W and $\psi_\mathbb{W}$ respectively, and Ω_W and $\Omega_\mathbb{W}$ as the corresponding supports, we then construct a graphex $\widehat{\mathbb{W}}$ over $\Omega_\mathbb{W}$ (equipped with the Borel σ-algebra and the measure $\mu_\mathbb{W}$) such that $\mathbb{W}$ is almost everywhere equal to a pullback of $\widehat{\mathbb{W}}$. Furthermore, we will show that if $\mathbb{W}'$ is a.e. equal to a pullback of $\mathbb{W}$, then we can find a measure preserving bijection φ^* from $\Omega_\mathbb{W}$ to $\Omega'_\mathbb{W}$ such that $\widehat{\mathbb{W}} = (\widehat{\mathbb{W}'})^{\varphi^*}$ a.e. Combined with Corollary 8.7, this will establish Theorem 8.1.

We first state some preliminary lemmas. Recall that if $(\Omega, \mathcal{F}, \mu)$ is a measure space, then $L^1(\Omega, \mathcal{F}, \mu)$ is a Banach space where each point is an equivalence class consisting of integrable measurable functions from Ω to $\mathbb{R}$, where two functions are equivalent if they are equal almost everywhere (or equivalently their L^1 distance is 0). Note that in general, the space $L^1(\Omega, \mathcal{F}, \mu)$ is not separable, a fact which will lead to technical complications when considering measurable functions into $L^1(\Omega, \mathcal{F}, \mu)$ (e.g., the sum of two such functions is in general not measurable). As in [16], we will avoid these difficulties by carefully constructing separable subspaces of $L^1(\Omega, \mathcal{F}, \mu)$ such that the functions of interest take values in these subspaces. See Lemma 8.8 below.

Throughout this section, we will frequently consider two measure spaces $(\Omega, \mathcal{F}, \mu)$ and $(\Omega', \mathcal{F}', \mu')$, where $\Omega = \Omega'$, $\mathcal{F}' \subseteq \mathcal{F}$, and μ' is the restriction of μ to $\mathcal{F}'$. With a slight abuse of notation, we will often denote the second space by $(\Omega, \mathcal{F}', \mu)$, rather than $(\Omega, \mathcal{F}', \mu|_{\mathcal{F}'})$.

As already noted, the space $L^1(\Omega, \mathcal{F}, \mu)$ is in general not separable (if Ω is a Borel space, it is, but we want to define this construction for general Ω). We will therefore need the following lemma.

Lemma 8.8 *Suppose that $(\Omega, \mathcal{F}, \mu)$ and $(\Omega', \mathcal{F}', \mu')$ are σ-finite measure spaces, $W: \Omega \times \Omega' \to \mathbb{R}$ is measurable, and for all $x \in \Omega$, the function $W(x, \cdot)$ is integrable. Let $\psi_W: \Omega \to L^1(\Omega, \mathcal{F}, \mu)$ be the map $x \mapsto W(x, \cdot)$. Then we can find a separable closed subspace B of $L^1(\Omega', \mathcal{F}', \mu')$ such that $\psi_W(x) \in B$ for all $x \in \Omega$.*

Proof First, assume that W is bounded and both $\mu(\Omega) < \infty$ and $\mu(\Omega') < \infty$. The statement of the lemma then clearly holds for all step functions, and by a monotone class argument it holds for all bounded W.

Next, relax the condition that $\mu'(\Omega') < \infty$. Let $\Omega'_1 \subseteq \Omega'_2 \subseteq \cdots \subseteq \Omega'_n \subseteq \cdots$ be a sequence of measurable subsets of Ω' with finite measure and $\bigcup_{n=1}^{\infty} \Omega'_n = \Omega'$. Then for every n we can find a separable closed subspace $B_n \subseteq L^1(\Omega'_n, \mathcal{F}'|_{\Omega'_n}, \mu'|_{\Omega'_n})$ such that for every $x \in \Omega$, $W(x, \cdot)|_{\Omega'_n} \in B_n$. Let $\widetilde{B}_n$ consist of those $f \in L^1(\Omega', \mathcal{F}', \mu')$ that have $f|_{\Omega'_n} \in B_n$ and $f|_{\Omega'-\Omega'_n} \equiv 0$. Clearly $\widetilde{B}_n$ is isomorphic to B_n, and thus separable. Let B be the closure of the space generated by $\bigcup_n \widetilde{B}_n$; this is separable. We claim that for any $x \in \Omega$, the function $W(x, \cdot)$ is contained in B. It suffices to show that for any ε, there is an $n \in \mathbb{N}$ and a $g \in \widetilde{B}_n$ such that $\|W(x, \cdot) - g\|_1 < \varepsilon$. Since $W(x, \cdot) \in L^1(\Omega', \mathcal{F}', \mu')$, we can take n large enough that $\|W(x, \cdot) - W(x, \cdot)\chi(\Omega'_n)\|_1 < \varepsilon$. But then by the definition of $\widetilde{B}_n$, we have $W(x, \cdot)\chi(\Omega'_n) \in \widetilde{B}_n$, so taking $g = W(x, \cdot)\chi(\Omega'_n)$, we are done.

In a similar way, we can approximate an unbounded W by the function $W 1_{|W| \leq n}$ to relax the condition that W is bounded.

Finally, for general Ω and Ω', we can write Ω as the disjoint union of finite measure sets $\Omega_1, \Omega_2, \ldots, \Omega_n, \ldots$. We know that the image of each Ω_n is contained in a separable closed subspace $B_n \subseteq L^1(\Omega', \mathcal{F}')$. Therefore, taking B to be the closure of the subspace generated by $\bigcup_n B_n$, the image of the map ψ_W is contained in B. $\square$

Lemma 8.9 *Suppose $(\Omega, \mathcal{F}, \mu)$ and $(\Omega', \mathcal{F}', \mu')$ are σ-finite measure spaces, $W: \Omega \times \Omega' \to \mathbb{R}$ is $\mathcal{F} \times \mathcal{F}'$-measurable, and that for all $x \in \Omega$, the function $W(x, \cdot)$ is integrable. Then the map $\psi_W: \Omega \to L^1(\Omega', \mathcal{F}', \mu')$ with $x \mapsto W(x, \cdot)$ is measurable with respect to the standard Borel σ-algebra on $L^1(\Omega, \mathcal{F}, \mu)$.*

Proof Our goal is to show that for any $c \in [0, \infty)$ and $f \in L^1(\Omega', \mathcal{F}', \mu')$, the set

$$B(f, W, c) = \{x \in \Omega : \|W(x, \cdot) - f\|_1 \leq c\}$$

is measurable.

Assume that $x \in B(f, W, c)$, and let $F_W \subset L^1(\Omega', \mathcal{F}', \mu')$ be a countable set such that $W(x, \cdot)$ lies in the closure of F_W for all $x \in \Omega$ (the existence of such a set follows from Lemma 8.8). Given any $\varepsilon > 0$, we can then find an $\widehat{f} \in F_W$ such that $x \in B(\widehat{f}, W, \varepsilon)$, and $\|f - \widehat{f}\|_1 \leq c + \varepsilon$. If, on the other hand, $\|f - \widehat{f}\|_1 \leq c + \varepsilon$ and $x \in B(\widehat{f}, W, \varepsilon)$ then $x \in B(f, W, c + 2\varepsilon)$. Since $B(f, W, c) = \bigcap_i B(f, W, c + \varepsilon_i)$ whenever $\varepsilon_i \to 0$, this proves that it is enough to prove measurability of $B(\widehat{f}, W, \varepsilon_i)$ for all $\widehat{f} \in F_W$ and an arbitrary sequence $\varepsilon_i \in (0, \infty)$ such that $\varepsilon_i \to 0$.

Using this observation, it is easy to see that if W_1 and W_2 obey the conclusions of the lemma, then so does any linear combination. The lemma is clearly also true for all step functions. A standard monotone class argument then implies that the lemma holds for all bounded, measurable W.

If W is unbounded, we use that by assumption, $\psi_W(x) \in L^1(\Omega', \mathcal{F}', \mu')$ for all $x \in \Omega$. Using this fact, one easily shows that

$$B(f, W, c) = \bigcap_{n=1}^{\infty} B(f, W(x, \cdot)1_{|W(x,\cdot)|\leq n}, c).$$

Since $B(f, W(x, \cdot)1_{|W(x,\cdot)|\leq n}, c)$ is measurable, this proves the statement for unbounded W. □

We will also use the following technical lemma, which is Lemma G.1 in [16] (the proof also works for σ-finite measures):

Lemma 8.10 *Let $(\Omega, \mathcal{F}, \mu)$ be any σ-finite measure space, and $B \subseteq L^1(\Omega, \mathcal{F}, \mu)$ a closed separable subspace. Then there exists a measurable evaluation map*

$$\Phi : B \times \Omega \to \mathbb{R}$$

such that for any $f \in B$, for almost every $x \in \Omega$, $f(x) = \Phi(f, x)$ (note that $f(x)$ is only defined almost everywhere). In particular, if $L^1(\Omega, \mathcal{F}, \mu)$ is separable, we can take $B = L^1(\Omega, \mathcal{F}, \mu)$.

We need one more lemma.

Lemma 8.11 *Suppose that* $(\Omega, \mathcal{F}, \mu)$ *and* $(\Omega', \mathcal{F}', \mu')$ *are σ-finite measure spaces, $W\colon \Omega \times \Omega' \to \mathbb{R}$ is measurable, and for all $x \in \Omega$ and $x' \in \Omega'$, both $W(x, \cdot)$ and $W(\cdot, x')$ are integrable. Let ψ_W be the map $\Omega \to L^1(\Omega', \mathcal{F}', \mu')$ with $x \mapsto W(x, \cdot)$, and ψ'_W be the map $\Omega' \to L^1(\Omega, \mathcal{F}, \mu)$ with $y \mapsto W(\cdot, y)$. Further, let $\mathcal{G}$ be the Borel σ-algebra on $L^1(\Omega, \mathcal{F}, \mu)$, and let*

$$\mathcal{F}'_W = \psi'^{-1}_W(\mathcal{G}).$$

Then for almost every $x \in \Omega$, $\psi_W(x) \in L^1(\Omega', \mathcal{F}'_W, \mu')$.

Proof By Lemma 8.8, the image of ψ'_W is contained in a separable subspace $B \subseteq L^1(\Omega, \mathcal{F}, \mu)$. By Lemma 8.10, there exists a measurable map $\Phi\colon B \times \Omega \to \mathbb{R}$ such that for any $f \in B$ and almost any $x \in \Omega$, $f(x) = \Phi(f, x)$. Define $\widetilde{W}$ on $\Omega \times \Omega'$ by

$$\widetilde{W}(x, y) = \Phi(\psi'_W(y), x).$$

Then for every y and almost every x, $\widetilde{W}(x, y) = W(x, y)$, so $\widetilde{W} = W$ almost everywhere on $\Omega \times \Omega'$. Thus, if we define $\psi_{\widetilde{W}}$ analogously to ψ_W, then for almost every $x \in \Omega$, $\|\psi_{\widetilde{W}}(x) - \psi_W(x)\|_{L^1(\Omega', \mathcal{F}', \mu')} = 0$; that is, for almost all x, $\psi_W(x) = \psi_{\widetilde{W}}(x)$ as elements of $L^1(\Omega', \mathcal{F}', \mu')$. However, since Φ is $B \times \Omega$-measurable, and $\psi'_W\colon \Omega' \to B$ is measurable, $\widetilde{W}$ is $\mathcal{F} \times \mathcal{F}'_W$-measurable, so in particular $\psi_{\widetilde{W}}(x)$ is $\mathcal{F}'_W$-measurable for all x. Since $\psi_W(x) = \psi_{\widetilde{W}}(x) \in L^1(\Omega', \mathcal{F}', \mu')$ for almost every x, it follows that $\psi_W(x) \in L^1(\Omega', \mathcal{F}'_W, \mu')$ for almost every $x \in \Omega$. $\square$

Let $(W, S, I, \boldsymbol{\Omega})$ be a graphex over $\boldsymbol{\Omega} = (\Omega, \mathcal{F}, \mu)$ such that $W(x, \cdot)$ is integrable for all x. For each $x \in \Omega$, we have the section $W_x \in L^1(\Omega, \mathcal{F}, \mu)$ defined by $W_x(y) = W(x, y)$, giving us the map

$$\psi_W\colon \Omega \to L^1(\Omega, \mathcal{F}, \mu)$$

defined by $x \mapsto W_x$. Let

$$\psi_\mathbb{W}\colon \Omega \to L^1(\Omega, \mathcal{F}, \mu) \times \mathbb{R}$$

be defined by $x \mapsto (W_x, S(x))$. By Lemma 8.9, ψ_W, and thus $\psi_\mathbb{W}$, is measurable. Let $\mu_W = \mu^{\psi_W}$ and $\mu_\mathbb{W} = \mu^{\psi_\mathbb{W}}$, and let $\Omega_W \subseteq L^1(\Omega, \mathcal{F}, \mu)$ and $\Omega_\mathbb{W} \subseteq L^1(\Omega, \mathcal{F}, \mu) \times \mathbb{R}$ be the supports of μ_W and $\mu_\mathbb{W}$, respectively, i.e.,

$$\Omega_W = \{f \in L^1(\Omega, \mathcal{F}, \mu) : \mu_W(U) > 0 \text{ for every open } U \subseteq L^1(\Omega, \mathcal{F}, \mu) \text{ with } f \in U\},$$

and

$$\Omega_\mathbb{W} = \{(f, c) \in L^1(\Omega, \mathcal{F}, \mu) \times \mathbb{R} :$$
$$\mu_\mathbb{W}(U) > 0 \text{ for every open } U \subseteq L^1(\Omega, \mathcal{F}, \mu) \times \mathbb{R} \text{ with } (f, c) \in U\}.$$

Alternatively, we can also define ψ_W and μ_W as follows. Let $(\widetilde{\Omega}, \widetilde{\mathcal{F}}, \widetilde{\mu})$ be defined as $\widetilde{\Omega} = \Omega \cup \{\Omega_\infty\}$, where Ω_∞ is an atom with measure 1. Then we can think of $\psi_{\mathbb{W}}(x)$ as a function in $L^1(\widetilde{\Omega}, \widetilde{\mathcal{F}}, \widetilde{\mu})$, with $\psi_{\mathbb{W}}(x)(y) = W(x, y)$ if $y \in \Omega$ and $\psi_{\mathbb{W}}(x)(y) = S(x)$ if $y = \Omega_\infty$, and $\|\psi_{\mathbb{W}}(x)\|_1 = D_{\mathbb{W}}(x)$. This gives a bijection $L^1(\widetilde{\Omega}, \widetilde{\mathcal{F}}, \widetilde{\mu}) \equiv L^1(\Omega, \mathcal{F}, \mu) \times \mathbb{R}$ and $\psi_{\mathbb{W}}$ as a map from Ω to $L^1(\widetilde{\Omega}, \widetilde{\mathcal{F}}, \widetilde{\mu})$. Note that μ_W is the projection of $\mu_{\mathbb{W}}$, and thus Ω_W is the closure of the projection of $\Omega_{\mathbb{W}}$. Equipping Ω_W with the standard Borel σ-algebra $\mathcal{G}_W$, this gives us a measure space $(\Omega_W, \mathcal{G}_W, \mu_W)$, and similarly we obtain $\Omega_{\mathbb{W}} = (\Omega_{\mathbb{W}}, \mathcal{G}_{\mathbb{W}}, \mu_{\mathbb{W}})$.

Let $\mathcal{G}$ and $\widetilde{\mathcal{G}}$ be the Borel σ-algebra on $L^1(\Omega, \mathcal{F}, \mu)$ and $L^1(\widetilde{\Omega}, \widetilde{\mathcal{F}}, \widetilde{\mu})$, respectively. Via the maps ψ_W and $\psi_{\mathbb{W}}$ they induce two different σ-algebras on Ω, the σ-algebras

$$\mathcal{F}_W = \psi_W^{-1}(\mathcal{G})$$

and

$$\mathcal{F}_{\mathbb{W}} = \psi_{\mathbb{W}}^{-1}(\widetilde{\mathcal{G}}).$$

We also define $\widetilde{\mathcal{F}}_W = \mathcal{F}_W \times \mathcal{B}$ and $\widetilde{\mathcal{F}}_{\mathbb{W}} = \mathcal{F}_{\mathbb{W}} \times \mathcal{B}$. Note that $\mathcal{F}_W \subseteq \mathcal{F}_{\mathbb{W}}$, with the example of a zero graphon but a nonconstant S function showing that strict inequality is possible.

It is easy to see that if $\mathbb{W}'$ is equal to $\mathbb{W}$ almost everywhere, then $\mu_{\mathbb{W}} = \mu_{\mathbb{W}'}$ and hence $\Omega_{\mathbb{W}} = \Omega_{\mathbb{W}'}$. Indeed, if $\mathbb{W} = \mathbb{W}'$ a.e., then for almost all x, $\psi_{\mathbb{W}}(x) = \psi_{\mathbb{W}'}(x)$ when viewed as vectors in L^1. This implies that there exists a set $N \subseteq \Omega$ of measure zero such that for all $A \in \widetilde{\mathcal{G}}$, the symmetric difference of $\psi_{\mathbb{W}}^{-1}(A)$ and $\psi_{\mathbb{W}'}^{-1}(A)$ lies in N, which shows that $\mu_{\mathbb{W}} = \mu_{\mathbb{W}'}$. Furthermore, under the same change, $\mathcal{F}_W$, $\mathcal{F}_{\mathbb{W}}$ and $\mathcal{F}_{W'}$, $\mathcal{F}_{\mathbb{W}'}$ only change on a set of measure zero, implying that the Banach spaces $L^1(\Omega, \mathcal{F}_W, \mu)$ and $L^1(\Omega, \mathcal{F}_{\mathbb{W}}, \mu)$ remain unchanged.

Note that in general, μ_W is not σ-finite. Indeed, choosing W to be the graphon 0 over any space of infinite measure, and I, S to be 0, we have that $\psi_{\mathbb{W}}^{-1}(A) = \Omega$ for every A containing the origin $(0) \in L^1(\widetilde{\Omega}, \widetilde{\mathcal{F}}, \widetilde{\mu})$, so $\mu_{\mathbb{W}}(A) = \mu(\Omega) = \infty$ if $0 \in A$, and $\mu_{\mathbb{W}}(A) = 0$ otherwise. So in particular $\Omega_{\mathbb{W}} = \{(0)\}$ and $\mu_{\mathbb{W}}(\Omega_{\mathbb{W}}) = \infty$. This also means μ_W is not σ-finite.

It turns out, however, that this problem can be avoided if we require that the set where $D_{\mathbb{W}} = 0$ has measure zero; see Lemma 8.13 below. Before stating the lemma, we prove the following:

Proposition 8.12 *The space $\Omega_{\mathbb{W}}$ as defined above is a complete, separable metric space, with the metric induced by $L^1(\widetilde{\Omega}, \widetilde{\mathcal{F}}, \widetilde{\mu})$, and $\mu_{\mathbb{W}}$ has full support in $\Omega_{\mathbb{W}}$. Furthermore, $\Omega_{\mathbb{W}} \subseteq L^1(\widetilde{\Omega}, \widetilde{\mathcal{F}}_W, \widetilde{\mu}) \subseteq L^1(\widetilde{\Omega}, \widetilde{\mathcal{F}}_{\mathbb{W}}, \widetilde{\mu})$. Finally, after modifying W and S on a set of measure zero $\psi_{\mathbb{W}}$ becomes an everywhere defined, measure-preserving map from $(\Omega, \mathcal{F}, \mu)$ to $(\Omega_{\mathbb{W}}, \mathcal{G}_{\mathbb{W}}, \mu_{\mathbb{W}})$, and we furthermore have $\psi_{\mathbb{W}}^{-1}(\mathcal{G}_{\mathbb{W}}) = \mathcal{F}_{\mathbb{W}}$.*

Proof By Lemma 8.8, the image of ψ_W is contained in a closed separable subspace B of $L^1(\Omega, \mathcal{F}, \mu)$. This means that the image of $\psi_{\mathbb{W}}$ is contained in $\widetilde{B} = B \times \mathbb{R}$, which is a closed separable subspace of $L^1(\widetilde{\Omega}, \widetilde{\mathcal{F}}, \widetilde{\mu})$. We will show that in fact

$$\Omega_W = \{f \in B : \mu_W(U) > 0 \text{ for every open } U \subseteq B \text{ with } f \in U\}.$$

and

$$\Omega_W = \{(f, c) \in \widetilde{B} : \mu_W(U) > 0 \text{ for every open } U \subseteq \widetilde{B} \text{ with } (f, c) \in U\}.$$

Indeed, since $\psi_W^{-1}(L^1(\Omega, \mathcal{F}, \mu) \setminus B) = \emptyset$, we have $\mu_W(L^1(\Omega, \mathcal{F}, \mu) \setminus B) = 0$, which in turn implies that support of μ_W is contained in B. Let, for a moment, the above defined set be Ω'_W. First, if $f \in \Omega'_W$, then for any open set $U \subseteq L^1(\Omega, \mathcal{F}, \mu)$ with $f \in U$, we have $\mu_W(U) \geq \mu_W(U \cap B) > 0$. Conversely, if $f \in \Omega_W$, we know we must have $f \in B$, and for any open $U \subseteq B$ with $f \in U$, we can find an open $V \subseteq L^1(\Omega, \mathcal{F}, \mu)$ with $U = V \cap B$, so in particular $f \in V$. Then $\mu_W(V) > 0$, and since $\mu_W(L^1(\Omega, \mathcal{F}, \mu) \setminus B) = 0$, we have $\mu_W(U) = \mu_W(V) > 0$, showing that $f \in \Omega'_W$. A similar, argument shows the second claim, noting that $\Omega_W \subseteq \widetilde{B} = B \times \mathbb{R}$.

Now take the union V of all open sets $U \subseteq \widetilde{B}$ with $\mu_W(U) = 0$. Then $\Omega_W = \widetilde{B} \setminus V$. Since $\widetilde{B}$ is separable and thus second countable, we can find a countable collection $U_1, U_2, \ldots, U_n, \ldots$ with $\mu_W(U_n) = 0$ and $\bigcup_n U_n = V$. This means that $\mu_W(V) = 0$, so $\mu(\psi_W^{-1}(V)) = 0$, and thus almost every point in Ω is mapped to Ω_W. Since V is open, Ω_W is closed in $\widetilde{B}$, so it is a closed subset of a separable Banach space; thus it is a complete, separable metric space.

To see that μ_W has full support, consider an open subset $U \subseteq \Omega_W$ with $\mu_W(U) = 0$. Since Ω_W is closed in $\widetilde{B}$ and $\mu_W(\widetilde{B} \setminus \Omega_W) = 0$, we can find an open subset $\widetilde{U} \subseteq \widetilde{B}$ such that $U = \widetilde{U} \cap \Omega_W$ and $\mu_W(\widetilde{U}) = \mu_W(U) = 0$, so in particular $\widetilde{U} \subseteq V = \widetilde{B} \setminus \Omega_W$. But this implies $U = \widetilde{U} \cap \Omega_W = \emptyset$, as required.

Next, we use Lemma 8.11 to infer that $\psi_W(x) \in L^1(\Omega, \mathcal{F}_W, \mu)$ for almost all $x \in \Omega$, which in turn implies that $\psi_W(x) \in L^1(\widetilde{\Omega}, \widetilde{\mathcal{F}}_W, \widetilde{\mu})$ for almost all $x \in \Omega$. As a consequence, the open set $U = L^1(\widetilde{\Omega}, \widetilde{\mathcal{F}}, \widetilde{\mu}) \setminus L^1(\widetilde{\Omega}, \widetilde{\mathcal{F}}_W, \widetilde{\mu})$ has measure zero:

$$\mu_W(U) = \mu(\psi_W^{-1}(U)) = \mu(\{x \in \Omega : \psi_W(x) \notin L^1(\widetilde{\Omega}, \widetilde{\mathcal{F}}_W, \widetilde{\mu})\}) = 0.$$

Since μ_W has full support on Ω_W, the open set $U \cap \Omega_W \subseteq \Omega_W$ is empty, showing that $\Omega_W \subseteq L^1(\widetilde{\Omega}, \widetilde{\mathcal{F}}_W, \widetilde{\mu})$.

Let $N = \psi_W^{-1}(V)$, where as above $V = \widetilde{B} \setminus \Omega_W$. We have seen that $\mu(N) = 0$. Furthermore, for $A \in \mathcal{G}_W$, $A \subseteq \Omega_W$ and hence $\psi_W^{-1}(A) \subseteq \Omega \setminus N$. Finally, $\mu(\psi_W^{-1}(A)) = \mu_W(A)$ by the definition of μ_W. Fix some $f \in \Omega_W$. On $N \times (\Omega \setminus N)$, change $W(x, y)$ to $f(y)$, and change it on $(\Omega \setminus N) \times N$ to make it symmetric. Finally, change it to 0 on $N \times N$, and on N, change S to $f(\Omega_\infty)$. Clearly it is still the case that W and S are measurable and $W(x, \cdot)$ integrable for every x. We have changed W and S on a set of measure zero, so Ω_W and μ_W did not change, and we now have that ψ_W is an everywhere defined, measure-preserving map from $(\Omega, \mathcal{F}, \mu)$ to $(\Omega_W, \mathcal{G}_W, \mu_W)$.

To complete the proof, we need to show that $\psi_W^{-1}(\mathcal{G}_W) = \mathcal{F}_W$. To this end, we note that for each open $A \subseteq L^1(\widetilde{\Omega}, \widetilde{\mathcal{F}}_W, \widetilde{\mu})$, $\psi_W^{-1}(A) = \psi_W^{-1}(A \cap \Omega_W)$, which shows that $\psi_W^{-1}(\widetilde{\mathcal{G}}) = \psi_W^{-1}(\mathcal{G}_W)$. This proves the last claim. $\square$

Lemma 8.13 *Let $\Omega_{\mathbb{W}}$ and $\mu_{\mathbb{W}}$ be as defined above, and let $\mathcal{G}_{\mathbb{W}}$ be the Borel σ-algebra over $\Omega_{\mathbb{W}}$. If $\mathbb{W}$ is locally finite, $W(x, \cdot)$ is integrable for all $x \in \Omega$, and $\mu(\Omega \setminus \mathrm{dsupp}\,\mathbb{W}) < \infty$, then $(\Omega_{\mathbb{W}}, \mathcal{G}_{\mathbb{W}}, \mu_{\mathbb{W}})$ is σ-finite.*

Remark 8.14 Note that the condition $\mu(\Omega \setminus \mathrm{dsupp}\,\mathbb{W}) < \infty$ is necessary, since otherwise the point $\{(0, 0)\}$ becomes an atom with infinite measure.

Proof By Proposition 8.12, we can change W on a set of measure zero such that it still satisfies the conditions of the lemma, and $\psi_{\mathbb{W}}$ becomes a measure-preserving map from $(\Omega, \mathcal{F}, \mu)$ to $(\Omega_{\mathbb{W}}, \mathcal{G}_{\mathbb{W}}, \mu_{\mathbb{W}})$. Let $\Omega_0 \subset L^1(\Omega, \mathcal{F}, \mu) \times \mathbb{R}$ be the subspace consisting of just the origin, i.e., $\Omega_0 = \{(0, 0)\}$. Then

$$\mu_{\mathbb{W}}(\Omega_0) = \mu(\psi_{\mathbb{W}}^{-1}(\Omega_0)) = \mu(\{x \in \Omega : \|\psi_{\mathbb{W}}(x)\|_1 = 0\}) = \mu(\Omega \setminus \mathrm{dsupp}\,\mathbb{W}) < \infty.$$

Next define $\Omega_n = \{(f, c) \in \Omega_{\mathbb{W}} : \|f\|_1 + c \geq 1/n\}$ for $n \geq 1$. Then

$$\mu_{\mathbb{W}}(\Omega_n) = \mu(\{x \in \Omega : \|\psi_{\mathbb{W}}(x)\|_1 + S(x) \geq 1/n\}) = \mu(\{x \in \Omega : D_{\mathbb{W}}(x) \geq 1/n\}) < \infty.$$

Here the last inequality follows from Proposition 2.4. Since $\Omega_{\mathbb{W}} = \Omega_0 \cup \Omega_1 \cup \cdots$, this proves that $\mu_{\mathbb{W}}$ is σ-finite. $\square$

Next, we would like to show that we can define a graphex $\widehat{\mathbb{W}}$ on $\Omega_{\mathbb{W}}$ such that its pullback is equal almost everywhere to $\mathbb{W}$. Using Proposition 8.12, we can without loss of generality assume that $\psi_{\mathbb{W}}$ is a measure-preserving map from $(\Omega, \mathcal{F}, \mu)$ to $(\Omega_{\mathbb{W}}, \mathcal{G}_{\mathbb{W}}, \mu_{\mathbb{W}})$. By Lemma 3.2, this implies that the map $\psi_{\mathbb{W}}^* : L^1(\Omega_{\mathbb{W}}, \mathcal{G}_{\mathbb{W}}, \mu_{\mathbb{W}}) \to L^1(\Omega, \mathcal{F}_{\mathbb{W}}, \mu)$ with $f \mapsto f^{\psi_{\mathbb{W}}}$ and $\mathcal{F}_{\mathbb{W}} = \psi_{\mathbb{W}}^{-1}(\mathcal{G}_{\mathbb{W}})$ is an isometric isomorphism. Note that this in particular implies that $\psi_{\mathbb{W}}^*$ and $(\psi_{\mathbb{W}}^*)^{-1}$ are continuous, and hence measurable.

Now, since $\Omega_{\mathbb{W}}$ is a separable metric space, $L^1(\Omega_{\mathbb{W}}, \mathcal{G}_{\mathbb{W}}, \mu_{\mathbb{W}})$ is separable, so there exists an evaluation map

$$\Phi : L^1(\Omega_{\mathbb{W}}, \mathcal{G}_{\mathbb{W}}, \mu_{\mathbb{W}}) \times \Omega_{\mathbb{W}} \to \mathbb{R}$$

such that for every $\alpha \in L^1(\Omega_{\mathbb{W}}, \mathcal{G}_{\mathbb{W}}, \mu_{\mathbb{W}})$ and almost every $g \in \Omega_{\mathbb{W}}, \alpha(g) = \Phi(\alpha, g)$. Note that by definition, we also have that for every fixed α and almost every $y \in \Omega$, $\psi_{\mathbb{W}}^*(\alpha)(y) = \alpha(\psi_{\mathbb{W}}(y)) = \Phi(\alpha, \psi_{\mathbb{W}}(y))$.

By Proposition 8.12, $\Omega_{\mathbb{W}} \subseteq L^1(\widetilde{\Omega}, \widetilde{\mathcal{F}}_{\mathbb{W}}, \widetilde{\mu}) = L^1(\Omega, \mathcal{F}_{\mathbb{W}}, \mu) \times \mathbb{R}$, which means that $(\psi_{\mathbb{W}}^*)^{-1}(f|_\Omega)$ is well defined for all $f \in \Omega_{\mathbb{W}}$. We therefore may define

$$\widehat{W}_0(f, g) = \Phi\left((\psi_{\mathbb{W}}^*)^{-1}(f|_\Omega), g\right)$$

and

$$\widehat{W}(f, g) = \frac{1}{2}\left(\widehat{W}_0(f, g) + \widehat{W}_0(g, f)\right).$$

Since $(\psi_{\mathbb{W}}^*)^{-1}$ is measurable, $\widehat{W}_0$ and hence $\widehat{W}$ is measurable.

Suppose $\psi_\mathbb{W}(x) \in \Omega_\mathbb{W}$. Then, noting that for all x, $\psi_\mathbb{W}(x)|_\Omega = \psi_W(x)$, we have for almost all y,

$$\widehat{W}_0(\psi_\mathbb{W}(x), \psi_\mathbb{W}(y)) = \Phi((\psi_\mathbb{W}^*)^{-1}(\psi_\mathbb{W}(x)|_\Omega), \psi_\mathbb{W}(y))$$
$$= \psi_\mathbb{W}^*\Big((\psi_\mathbb{W}^*)^{-1}(\psi_\mathbb{W}(x)|_\Omega)\Big)(y) = \psi_W(x)(y) = W(x, y),$$

where the third and the fourth terms are only defined for almost all y. Thus $\widehat{W}_0^{\psi_\mathbb{W}}$ and hence $\widehat{W}^{\psi_\mathbb{W}}$ is equal to W almost everywhere on $\Omega \times \Omega$. We also define $\widehat{S}(f) = f(\Omega_\infty)$, which gives us that for $x \in \Omega$,

$$\widehat{S}(\psi_\mathbb{W}(x)) = \psi_\mathbb{W}(x)(\Omega_\infty) = S(x),$$

implying that $\widehat{S}^{\psi_\mathbb{W}} = S$. Finally, we take $\widehat{I} = I$, giving us a graphex $\widehat{\mathbb{W}} = (\widehat{W}, \widehat{S}, \widehat{I}, \Omega_\mathbb{W})$ such that $\widehat{\mathbb{W}}^{\psi_\mathbb{W}} = \mathbb{W}$ almost everywhere.

Note that this implies in particular that $\widehat{\mathbb{W}}$ inherits the local finiteness property from $\mathbb{W}$, so $\widehat{\mathbb{W}}$ is a bona fide graphex over the σ-finite Borel space $(\Omega_\mathbb{W}, \mathcal{G}_\mathbb{W}, \mu_\mathbb{W})$.

Note that the requirement that $\widehat{W}^{\psi_\mathbb{W}} = W$ and $\widehat{S}^{\psi_\mathbb{W}} = S$ almost everywhere uniquely determines $\widehat{\mathbb{W}}$ up to changes on a set of measure zero. Indeed, if $\widehat{W}'$ is another graphon with $\widehat{W}'^{\psi_\mathbb{W}} = \widehat{W}^{\psi_\mathbb{W}}$ ($\mu \times \mu$)-almost everywhere, then by the definition of pullbacks and the definition of $\mu_\mathbb{W}$, the equality $\widehat{W}' = \widehat{W}$ must hold ($\mu_\mathbb{W} \times \mu_\mathbb{W}$)-almost everywhere. Similarly, if $\widehat{S}'$ is another function with $\widehat{S}'^{\psi_\mathbb{W}} = \widehat{S}^{\psi_\mathbb{W}}$ μ-almost everywhere, then $\widehat{S}' = \widehat{S}$ $\mu_\mathbb{W}$-almost everywhere. Also by definition we must have $\widehat{I}' = \widehat{I}$.

On the other hand, suppose we have two graphexes $\mathbb{W}_1$ and $\mathbb{W}_2$ on the same space Ω with $W_1 = W_2$ almost everywhere, $S_1 = S_2$ almost everywhere, and $I_1 = I_2$. We have seen that $\Omega_{\mathbb{W}_1} = \Omega_{\mathbb{W}_2}$ and $\mu_{\mathbb{W}_1} = \mu_{\mathbb{W}_2}$. Since their pullbacks are equal almost everywhere, we must have $\widehat{W}_1 = \widehat{W}_2$ almost everywhere for any choices of $\widehat{W}_1$ and $\widehat{W}_2$, and $\widehat{S}_1 = \widehat{S}_2$ and $\widehat{I}_1 = \widehat{I}_2$ by definition.

Finally, if the graphex $\mathbb{W}$ only has the property that $W(x, \cdot)$ is integrable for almost every x, we can still define $\Omega_\mathbb{W}$ and $\mu_\mathbb{W}$ in the same way, and find a $\widehat{W}$ such that the pullback is defined almost everywhere on $\Omega \times \Omega$ and equal to W almost everywhere. Again, it is easy to see that we obtain the same $\Omega_\mathbb{W}$ and $\mu_\mathbb{W}$ if we first modify W on a set of measure zero to make $W(x, \cdot)$ integrable for every x, and any choice of $\widehat{W}$ will be equal almost everywhere. Therefore, this construction gives a graphex $\widehat{\mathbb{W}}$ on $(\Omega_\mathbb{W}, \mathcal{G}_\mathbb{W}, \mu_\mathbb{W})$ for any graphex $\mathbb{W}$.

Next, we show the following:

Lemma 8.15 *For $i = 1, 2$, let $\mathbb{W}_i = (W_i, S_i, I_i, \mathbf{\Omega}_i)$ be graphexes with $\mathbf{\Omega}_i = (\Omega_1, \mathcal{F}_i, \mu_i)$ and $\mu_i(\Omega_i \setminus \mathrm{dsupp}\, W_i) = 0$. Suppose that there exists a measure-preserving map $\varphi \colon \Omega_1 \to \Omega_2$ such that $\mathbb{W}_1 = \mathbb{W}_2^\varphi$ almost everywhere. Extend φ to $\widetilde{\varphi} \colon \widetilde{\Omega}_1 \to \widetilde{\Omega}_2$ by $\widetilde{\varphi}(\Omega_{1,\infty}) = \Omega_{2,\infty}$. Then the map $\widetilde{\varphi}^* \colon L^1(\widetilde{\mathbf{\Omega}}_2) \to L^1(\widetilde{\mathbf{\Omega}}_1)$ defined by $f \mapsto f \circ \widetilde{\varphi}$ restricts to a map $\Omega_{\mathbb{W}_2} \to \Omega_{\mathbb{W}_1}$, which is an isometric measure-preserving bijection between $(\Omega_{\mathbb{W}_2}, \mathcal{G}_{\mathbb{W}_2}, \mu_{\mathbb{W}_2})$ and $(\Omega_{\mathbb{W}_1}, \mathcal{G}_{\mathbb{W}_1}, \mu_{\mathbb{W}_1})$, and $\widehat{\mathbb{W}}_2 = \widehat{\mathbb{W}}_1^{\widetilde{\varphi}^*}$ almost everywhere, for any choices of $\widehat{\mathbb{W}}_1$ and $\widehat{\mathbb{W}}_2$.*

Proof By the remarks before the lemma, we may assume that $\mathbb{W}_1 = \mathbb{W}_2^{\varphi}$ everywhere, not just almost everywhere, and $W_1(x,\cdot)$ and $W_2(x',\cdot)$ are always integrable. Since φ and thus $\tilde{\varphi}$ is measure preserving, $\tilde{\varphi}^*$ is isometric and injective from $L^1(\tilde{\Omega}_2, \tilde{\mathcal{F}}_2, \tilde{\mu}_2)$ to $L^1(\tilde{\Omega}_1, \tilde{\varphi}^{-1}(\tilde{\mathcal{F}}_2), \tilde{\mu}_2)$ by Lemma 3.2. If $x \in \Omega_1$, then for almost every $y \in \Omega_1$ (note that the first two terms below are only defined for almost every y),

$$(\varphi^* \circ \psi_{W_2} \circ \varphi)(x)(y) = (\psi_{W_2} \circ \varphi)(x)(\varphi(y)) = W_2(\varphi(x), \varphi(y)) = W_1(x, y).$$

Therefore $\varphi^* \circ \psi_{W_2} \circ \varphi = \psi_{W_1}$ a.e. Furthermore,

$$(\mathrm{id} \circ S_2 \circ \varphi)(x) = S_2(\varphi(x)) = S_1(x),$$

which implies that $(\mathrm{id} \circ S_2 \circ \varphi) = S_1$. Since $\tilde{\varphi}^* = \varphi^* \times \mathrm{id}$ and $\psi_{W_i} = \psi_{W_i} \times S$, this implies that $\tilde{\varphi}^* \circ \psi_{W_2} \circ \varphi = \psi_{W_1}$ almost everywhere. Now let $A \subseteq L^1(\tilde{\Omega}_1)$ be Borel measurable. Then

$$\mu_{W_2}((\tilde{\varphi}^*)^{-1}(A)) = \mu_2(\psi_{W_2}^{-1}((\tilde{\varphi}^*)^{-1}(A)))$$
$$= \mu_1(\varphi^{-1}(\psi_{W_2}^{-1}((\tilde{\varphi}^*)^{-1}(A)))) = \mu_1(\psi_{W_1}^{-1}(A)) = \mu_{W_1}(A).$$

So $\varphi^* \times \mathrm{id} : L^1(\Omega_2) \times \mathbb{R} \to L^1(\Omega_1) \times \mathbb{R}$ is a measure preserving isometry. Since it is an isometry, in particular, it is continuous. Thus, $(\varphi^* \times \mathrm{id})^{-1}(L^1(\Omega_1) \times \mathbb{R} \setminus \Omega_{W_1})$ is an open set with measure zero, so it is disjoint from Ω_{W_2}. This implies that φ^* restricts to a measure-preserving injection $\Omega_{W_2} \to \Omega_{W_1}$. Since Ω_{W_2} is complete, $(\varphi^* \times \mathrm{id})(\Omega_{W_2})$ is a complete subset of Ω_{W_1}, which is itself complete. Therefore $\varphi^*(\Omega_{W_2})$ is a closed subset of Ω_{W_1}. However, we also have that

$$\mu_{W_1}(\Omega_{W_1} \setminus \varphi^*(\Omega_{W_2})) = \mu_{W_2}((\varphi^*)^{-1}(\Omega_{W_1} \setminus \varphi^*(\Omega_{W_2}))) = \mu_{W_2}((\varphi^*)^{-1}(\Omega_{W_1}) \setminus \Omega_{W_2}) = 0.$$

But $\Omega_{W_1} \setminus \varphi^*(\Omega_{W_2})$ is an open subset of Ω_{W_1} of measure 0, which means it must be the empty set because μ_{W_1} has full support in Ω_{W_1}. Therefore, $\varphi^* : \Omega_{W_2} \to \Omega_{W_1}$ is a measure-preserving isometry of metric measure spaces.

Now, we want to show that $\widehat{W}_1^{\tilde{\varphi}^*} = \widehat{W}_2$. We have that almost everywhere on $\Omega_1 \times \Omega_1$,

$$((\widehat{W}_1^{\tilde{\varphi}^*})^{\psi_{W_2}})^{\varphi} = \widehat{W}_1^{\tilde{\varphi}^* \circ \psi_{W_2} \circ \varphi} = \widehat{W}_1^{\psi_{W_1}} = W_1 = W_2^{\varphi}.$$

Therefore $(\widehat{W}_1^{\tilde{\varphi}^*})^{\psi_{W_2}} = W_2$ almost everywhere, but then $\widehat{W}_1^{\tilde{\varphi}^*} = \widehat{W}_2$ almost everywhere. By definition, we also have $\widehat{S}_1^{\varphi^*} = \widehat{S}_2$, and $\widehat{I}_1 = I_1 = I_2 = \widehat{I}_2$. □

Now, suppose $\mathbb{W}_1$ and $\mathbb{W}_2$ are equivalent. Then their restrictions to their respective degree supports are also equivalent. By Corollary 8.7, there exists a chain of pullbacks that link $\mathbb{W}_1$ and $\mathbb{W}_2$. We have seen that if a graphex is a pullback of another, then the construction above yields an isomorphism between the corresponding graphexes, up to almost everywhere changes. This clearly extends to chains of pullbacks; thus, we may find an isomorphism between Ω_{W_1} and Ω_{W_2} so that $\widehat{\mathbb{W}}_1$ and $\widehat{\mathbb{W}}_2$ are equal almost everywhere. We can extend the map $\psi_{\mathbb{W}_i}: \Omega_i \to \Omega_{\mathbb{W}_i}$, which is defined almost everywhere, to be defined everywhere, by mapping the rest of the points in Ω_i to an arbitrary point.

9 Uniform Integrability and Uniform Tail Regularity

9.1 Uniform Integrability

The goal of this subsection is to prove Theorem 2.26. Before doing this we establish that several alternative definitions of uniform integrability are equivalent to Definition 2.25.

Theorem 9.1 *Given a set of integrable graphexes $\mathcal{S}$, the following are equivalent.*

(1) $\mathcal{S}$ is uniformly integrable.
(2) The graphexes in $\mathcal{S}$ have uniformly bounded $\|\cdot\|_1$-norms, and for every $\varepsilon > 0$, there exists a D such that for all $\mathbb{W} \in \mathcal{S}$, $\|\mathbb{W}\|_1 - \|\mathbb{W}_{\leq D}\|_1 < \varepsilon$.
(3) For any $T > 0$, the random variables $E(G_T(\mathbb{W}))$ with $\mathbb{W} \in \mathcal{S}$ are uniformly integrable.
(4) There exists $T > 0$ such that the random variables $E(G_T(\mathbb{W}))$ with $\mathbb{W} \in \mathcal{S}$ are uniformly integrable.

Proof Throughout this proof, let $\Omega_{>D}, \Omega_{\leq D}, \mathbb{W}_{>D}, \mathbb{W}_{\leq D}$, etc, be defined as before. Let us first show (1) $\Rightarrow$ (2). We have that

$$\|\mathbb{W}_1\|_1 - \|\mathbb{W}_{\leq D}\|_1 = 2\int_{\Omega_{>D}} S(x)\,d\mu(x) + 2\int_{\Omega_{>D}\times\Omega_{\leq D}} W(x,y)\,d\mu(x)\,d\mu(y)$$
$$+ \int_{\Omega_{>D}\times\Omega_{>D}} W(x,y)\,d\mu(x)\,d\mu(y)$$
$$= 2\int_{\Omega_{>D}} D_\mathbb{W}(x)\,d\mu(x) - \int_{\Omega_{>D}\times\Omega_{>D}} W(x,y)\,d\mu(x)\,d\mu(y)$$
$$\leq 2\int_\Omega D_\mathbb{W} 1_{D_\mathbb{W}>D}\,d\mu.$$

Therefore, taking D for $\varepsilon/2$ in uniform integrability gives a good D for ε in (2).

To show that (2) $\Rightarrow$ (3), we first show that if a set of graphexes has uniformly bounded marginals, then the set of random variables is uniformly integrable. Let E_T be the random variable for a fixed $\mathbb{W} \in \mathcal{S}$ that gives the number of edges of $G_T(\mathbb{W})$. Recall that by Lemma 4.3, E_T has expectation $T^2\|\mathbb{W}\|_1/2$ and variance $T^2\|\mathbb{W}\|_1/2 + T^3\|D_\mathbb{W}\|_2^2$. Let C be a bound on $\|\mathbb{W}\|_1$ for $\mathbb{W} \in \mathcal{S}$. We then have that for any $K > T^2\|\mathbb{W}\|_1/2$,

$$\mathbb{P}[E_T > K] \leq \frac{T^2\|\mathbb{W}\|_1/2 + T^3\|D_{\mathbb{W}}\|_2^2}{(K - T^2\|\mathbb{W}\|_1/2)^2} \leq \frac{T^2 C/2 + T^3 CD}{(K - T^2 C/2)^2}.$$

If $K \geq T^2 C$, then this gives

$$\mathbb{P}[E_T > K] \leq \frac{T^2 C/2 + T^3 CD}{(K - T^2 C/2)^2} \leq \frac{T^2 C/2 + T^3 CD}{(K/2)^2} = \frac{2T^2 C + 4T^3 CD}{K^2}.$$

Therefore, for $K_0 \geq T^2 C$,

$$\mathbb{E}[E_T 1_{E_T > K_0}] = \sum_{K=K_0+1}^{\infty} \mathbb{P}[E_T \geq K]$$

$$\leq \sum_{K=K_0+1}^{\infty} \frac{2T^2 C + 4T^3 CD}{K^2} \leq \frac{2T^2 C + 4T^3 CD}{K_0}.$$

Suppose now that instead of uniformly bounded marginals, we have only (2). For $D > 0$, let $E_{T,D}$ be the number of edges that either have both endpoints labeled with a vertex in $\Omega_{\leq D}$, one endpoint is labeled with a vertex in $\Omega_{\leq D}$ and the edge is generated as a star from that vertex, or the edge is a dust edge. We then have that for all $D > 0$,

$$\mathbb{E}[E_T 1_{E_T > 2K_0}] = \mathbb{E}[E_T 1_{E_T > 2K_0, E_{T,D} > K_0}] + \mathbb{E}[E_T 1_{E_T > 2K_0, E_{T,D} \leq K_0}]$$
$$\leq \mathbb{E}[E_T - E_{T,D}] + \mathbb{E}[E_{T,D} 1_{E_{T,D} > K_0}]$$
$$\quad + \mathbb{E}[E_{T,D} 1_{E_{T,D} \leq K_0, E_T - E_{T,D} > K_0}]$$
$$\leq \mathbb{E}[E_T - E_{T,D}] + \frac{2T^2 C + 4T^3 CD}{K_0} + K_0 \mathbb{P}[E_T - E_{T,D} > K_0]$$
$$\leq 2\mathbb{E}[E_T - E_{T,D}] + \frac{2T^2 C + 4T^3 CD}{K_0},$$

provided $K_0 \geq T^2 C$. Condition (2) now implies that for any $\varepsilon > 0$, there exists a D such that $\mathbb{E}[E_{T,D} - E_T] < \varepsilon$. Given such a D, we choose K_0 in such a way that the last term in the above bound is at most ε, implying that for each $\varepsilon > 0$ we can find a K_0 such that $\mathbb{E}[E_T 1_{E_T > 2K_0}] \leq 3\varepsilon$. This proves that the set of random variables E_T are indeed uniformly integrable.

It is clear that (3) implies (4). Suppose now that (4) holds. Since the expectation of E_T is $T^2 \|\mathbb{W}\|_1/2$, $\|\mathbb{W}\|_1$ must be uniformly bounded for $\mathbb{W} \in \mathcal{S}$. Let C be an upper bound. Suppose that (1) is false. Then there exists a fixed $\varepsilon > 0$, such that for any D, there exists a graphex $\mathbb{W} \in \mathcal{S}$ such that

$$\int_{\Omega_{>D}} D_{\mathbb{W}} 1_{D_{\mathbb{W}} > D} \, d\mu \geq \varepsilon.$$

Since $\mathbb{E}[D_\mathbb{W}] \leq \|\mathbb{W}\|_1 \leq C$, we have that $\mu(\Omega_{>D}) \leq C/D$. By taking D large enough, we can assume that $C/D \leq D/2$. Let $F_{T,D}$ be the number of edges in $G_T(\mathbb{W})$ that have exactly one endpoint in $\Omega_{>D}$. Then

$$\mathbb{E}[F_{T,D}] \geq \int_{\Omega_{>D}} T^2 (D_\mathbb{W}(x) - C/D) \, d\mu(x) \geq T^2 \int_{\Omega_{>D}} (D_\mathbb{W}(x)/2) \, d\mu(x) \geq T^2 \varepsilon/2.$$

If there are no points sampled in $\Omega_{>D}$, then $F_{T,D} = 0$. Conditioned on there being at least one point sampled in $\Omega_{>D}$, the number of neighbors of a point whose feature is $x \in \Omega_{>D}$ is a Poisson random variable with mean equal to $T D_\mathbb{W}(x)/2 \geq TD/2$. Therefore,

$$\mathbb{E}[F_{T,D} | F_{T,D} > 0] \geq TD/2.$$

We also have that
$$\mathbb{E}[F_{T,D} 1_{F_{T,D} \leq TD/4} | F_{T,D} > 0] \leq TD/4.$$

Therefore,
$$\mathbb{E}[F_{T,D} 1_{F_{T,D} > TD/4} | F_{T,D} > 0] \geq \frac{1}{2} \mathbb{E}[F_{T,D} | F_{T,D} > 0].$$

We then have
$$\mathbb{E}[F_{T,D} 1_{F_{T,D} > TD/4}] = \mathbb{E}[F_{T,D} 1_{F_{T,D} > TD/4} | F_{T,D} > 0] \mathbb{P}[F_{T,D} > 0]$$
$$\geq \frac{1}{2} \mathbb{E}[F_{T,D} | F_{T,D} > 0] \mathbb{P}[F_{T,D} > 0] = \frac{1}{2} \mathbb{E}[F_{T,D}] \geq T^2 \varepsilon/4.$$

Since D can be arbitrary (above some D_0), this contradicts Condition (4). □

Theorem 2.26 is an easy corollary of Theorem 9.1.

Proof (*Theorem* 2.26) By Theorem 2.18, $\mathbb{W}_n$ is GP-convergent to $\mathbb{W}$. Fix a subsequence n_i such that $\liminf_{n \to \infty} \|\mathbb{W}_n\|_1 = \lim_{i \to \infty} \|\mathbb{W}_{n_i}\|_1$, and fix $T > 0$. Let e_i be the number of edges in $G_T(\mathbb{W}_{n_i})$, and let e be the number of edges in $G_T(\mathbb{W})$. Following the proof of Corollary 3.10 in [4], for $\lambda > 0$ define $f_\lambda \colon \mathbb{R}_+ \to \mathbb{R}_+$ by $f_\lambda(x) = x 1_{x \leq \lambda}$. Then $\mathbb{E}[f_\lambda(e_i)] \leq \mathbb{E}[e_i] = T^2 \|\mathbb{W}_{n_i}\|_1$. Since $e_i \to e$ in distribution, $\mathbb{E}[f_\lambda(e)] = \lim_{i \to \infty} \mathbb{E}[f_\lambda(e_i)] \leq T^2 \lim_{i \to \infty} \|\mathbb{W}_i\|_1 = \liminf_{n \to \infty} \|\mathbb{W}_n\|_1$. The monotone convergence theorem then gives that $T^2 \|\mathbb{W}\|_1 = \mathbb{E}[e] = \lim_{\lambda \to \infty} \mathbb{E}[f_\lambda(e)] \leq T^2 \liminf_{n \to \infty} \|\mathbb{W}_n\|_1$, proving the first part of the theorem.

To prove the second part, assume first that $\mathbb{W}_n$ is uniformly integrable, and fix $\varepsilon > 0$. By Theorem 9.1 (2), for every $\varepsilon > 0$, there exists a D such that each $\mathbb{W}_n$ has

$$\big| \|\mathbb{W}_n\|_1 - \|\mathbb{W}_{n, \leq D}\|_1 \big| \leq \varepsilon.$$

Since $\mathbb{W}$ is integrable, after possibly increasing D, we can also assume that

$$\big| \|\mathbb{W}\|_1 - \|\mathbb{W}_{\leq D}\|_1 \big| \leq \varepsilon.$$

Increasing D further, we may also assume that $\mu(\{D_\mathbb{W} = D\}) = 0$. By Proposition 4.6, $\delta_{2\to 2}(\mathbb{W}_{n,\leq D}, \mathbb{W}_{\leq D}) \to 0$, which implies in particular that $\|\mathbb{W}_{n,\leq D}\|_1 \to \|\mathbb{W}_{\leq D}\|_1$. Therefore, we can take n_0 so that if $n \geq n_0$, then

$$\left| \|\mathbb{W}_{\leq D}\|_1 - \|\mathbb{W}_{n,\leq D}\|_1 \right| \leq \varepsilon.$$

These three inequalities imply that if $n \geq n_0$, then

$$\left| \|\mathbb{W}_n\|_1 - \|\mathbb{W}\|_1 \right| \leq 3\varepsilon.$$

Since ε was arbitrary, this completes the proof of the first direction.

For the other direction, fix ε. Since $\mathbb{W}$ is integrable, there exists $D > 0$ such that

$$\|\mathbb{W}_{\leq D}\|_1 \geq \|\mathbb{W}\|_1 - \varepsilon/2.$$

By increasing D, we can assume that $\mu(\{D_\mathbb{W} = D\}) = 0$. By Proposition 4.6,

$$\|\mathbb{W}_{n,\leq D}\|_1 \to \|\mathbb{W}_{\leq D}\|_1.$$

We then have that

$$\limsup_{n\to\infty} \left(\|\mathbb{W}_n\|_1 - \|\mathbb{W}_{n,\leq D}\|_1 \right) = \limsup_{n\to\infty} \left((\|\mathbb{W}_n\|_1 - \|\mathbb{W}\|_1) + (\|\mathbb{W}\|_1 - \|\mathbb{W}_{\leq D}\|_1) \right.$$
$$\left. + (\|\mathbb{W}_{\leq D}\|_1 - \|\mathbb{W}_{n,\leq D}\|_1) \right) \leq \varepsilon/2.$$

Therefore, there exists an n_0 such that if $n > n_0$, then

$$\|\mathbb{W}_n\|_1 - \|\mathbb{W}_{n,\leq D}\|_1 < \varepsilon.$$

Since $\mathbb{W}_1, \mathbb{W}_2, \ldots, \mathbb{W}_{n_0}$ is a finite set of graphexes, each of which is integrable, we can increase D so that the above inequality holds for each n, which by Theorem 9.1 means that they are uniformly integrable. $\square$

9.2 Uniform Tail Regularity

The goal of this subsection is to prove Theorem 2.28. Before doing this, we show that uniform tail regularity implies uniform integrability.

Lemma 9.2 *Suppose that a set of graphexes consisting only of graphons is uniformly tail regular. Then the set is uniformly integrable.*

Proof Fix $\varepsilon > 0$. By the definition of tail regularity, we can find an $M < \infty$ such that for each graphon W in the set there exists a subset Ω_0 of measure at most M such that $\|W\|_1 - \|W|_{\Omega_0}\|_1 \leq \varepsilon/3$. Note that clearly $\|W\|_1 \leq M^2 + \varepsilon/3$, so in particular the set of graphons has uniformly bounded L^1 norm. Let

$$A = \{x \in \Omega_0, D_W(x) > 2M\}.$$

Note that for any $x \in A$,

$$\int_{\Omega \setminus \Omega_0} W(x,y)\,d\mu(y) \geq D_W(x) - \int_{\Omega_0} W(x,y)\,d\mu(y) \geq D_W(x) - M \geq M,$$

which implies that

$$\mu(A) \leq \frac{1}{M}\int_{A \times \Omega \setminus \Omega_0} W\,d\mu^2 \leq \frac{1}{M}\int_{\Omega \setminus \Omega_0} D_W\,d\mu \leq \frac{\varepsilon}{3M}.$$

We then have

$$\int_\Omega D_W 1_{D_W > 2M}\,d\mu \leq \int_{\Omega \setminus \Omega_0} D_W\,d\mu + \int_A D_W\,d\mu$$

$$= \int_{\Omega \setminus \Omega_0} D_W\,d\mu + \int_{A \times (\Omega \setminus \Omega_0)} W(x,y)\,d\mu(x)\,d\mu(y)$$

$$+ \int_{A \times \Omega_0} W(x,y)\,d\mu(x)\,d\mu(y)$$

$$\leq 2\int_{\Omega \setminus \Omega_0} D_W\,d\mu + \int_{A \times \Omega_0} W(x,y)\,d\mu(x)\,d\mu(y)$$

$$\leq 2\varepsilon/3 + \mu(A)\mu(\Omega_0) \leq \varepsilon.$$

$\square$

Next, we show the following lemma. As before, $\Omega_{>\delta}$ is the set $\{x \in \Omega : D_{\mathbb{W}}(x) > \delta\}$.

Lemma 9.3 *Given a set of graphons $\mathcal{S}$, the following are equivalent:*

(1) *The set of graphons is uniformly tail regular.*
(2) *The set of graphons has a uniform bound on their $\|\cdot\|_1$-norm, and for every ε, there exists a δ such that $\|W\|_1 - \|W|_{\Omega_{>\delta}}\|_1 \leq \varepsilon$.*
(3) *The set of graphons has a uniform bound on their $\|\cdot\|_1$-norm, and for every ε, there exists a δ such that $\|D_W 1_{D_W \leq \delta}\|_1 \leq \varepsilon$.*

Corollary 9.4 *Given a set of graphons $\mathcal{S}$, suppose that we replace each graphon with a pullback. Let $\mathcal{S}'$ be the new set. Then $\mathcal{S}'$ is uniformly tail regular if and only if $\mathcal{S}$ is uniformly tail regular.*

Proof Property (2) in Lemma 9.3 is unaffected by taking pullbacks. $\square$

Proof (*Lemma* 9.3) We first show that (1) implies (2). Fix $\varepsilon > 0$. Take M for $\varepsilon/2$ as in the definition of uniform tail regularity, let $\delta = \varepsilon/4M$. Fix an arbitrary graphon $W \in \mathcal{S}$, and let Ω_0 be a set of measure M such that

$$\|W\|_1 - \|W|_{\Omega_0}\|_1 \leq \varepsilon/2.$$

Note that W has L^1 norm at most $M^2 + \varepsilon/2$, which proves that the graphs have a uniform bound on their $\|\cdot\|_1$-norm. Now, we have

$$\|W\|_1 - \|W|_{\Omega_{>\delta}}\|_1 \leq \|W\|_1 - \|W|_{\Omega_0}\|_1 + \|W|_{\Omega_0}\|_1 - \|W|_{\Omega_0 \cap \Omega_{>\delta}}\|_1 \leq \varepsilon/2 + 2\delta M \leq \varepsilon.$$

This shows that (1) implies (2).

The fact that (2) implies (1) follows from the observation that

$$\mu(\Omega_{>\delta}) = \int d\mu(x) 1_{D_W(x) > \delta} \leq \int d\mu(x) \frac{D_W(x)}{\delta} = \frac{1}{\delta}\|W\|_1.$$

Finally, (2) and (3) are equivalent by the fact that

$$\int_{\Omega \times \Omega \setminus \Omega_{>\delta}} W \leq \int_{\Omega \times \Omega} W - \int_{\Omega_{>\delta} \times \Omega_{>\delta}} W \leq 2 \int_{\Omega \times \Omega \setminus \Omega_{>\delta}} W.$$

$\square$

To prove Theorem 2.28 we establish three more lemmas.

Lemma 9.5 *Suppose that a sequence of integrable graphons W_n converges to a graphon W in the cut metric. Then for any $D > 0$ such that $\mu(D_W = D) = 0$, the graphons $W_{n, \leq D}$ converge to $W_{\leq D}$ in the cut metric.*

Proof Let $\widetilde{\mu}_n$ be a coupling of trivial extensions of W_n and W, and let $\widetilde{\Omega}_n$ be the product space on which the coupling is defined. Let $\widetilde{W}_n$ and $\widetilde{W}$ be the pullbacks of the trivial extensions to $\widetilde{\Omega}_n$, and suppose that

$$\|\widetilde{W}_n - \widetilde{W}\|_\square < \varepsilon.$$

Defining $A = \{D_{\widetilde{W}_n} - D_{\widetilde{W}} > 0\}$ and $B = \{D_{\widetilde{W}} - D_{\widetilde{W}_n} > 0\}$, we then have that

$$\widetilde{\mu}_n(\{|D_{\widetilde{W}_n} - D_{\widetilde{W}}| > \sqrt{\varepsilon}\}) \leq \frac{1}{\sqrt{\varepsilon}} \|D_{\widetilde{W}_n} - D_{\widetilde{W}}\|_1$$

$$= \frac{1}{\sqrt{\varepsilon}} \left(\int_{A \times \widetilde{\Omega}_n} (\widetilde{W}_n - \widetilde{W}) + \int_{B \times \widetilde{\Omega}_n} (\widetilde{W} - \widetilde{W}_n) \right) < 2\sqrt{\varepsilon}.$$

As a consequence,

$$\widetilde{\mu}_n(\{D_{\widetilde{W}_n} > D, D_{\widetilde{W}} \leq D\})$$
$$\leq \widetilde{\mu}_n(\{|D_{\widetilde{W}_n} - D_{\widetilde{W}}| > \sqrt{\varepsilon}\}) + \widetilde{\mu}_n(\{D - \sqrt{\varepsilon} < D_{\widetilde{W}} \leq D\})$$
$$< 2\sqrt{\varepsilon} + \mu(\{D - \sqrt{\varepsilon} < D_W \leq D\}).$$

For any $\delta > 0$, we can take ε small enough so that this is at most δ. Similarly, we have

$$\widetilde{\mu}_n(\{D_{\widetilde{W}_n} \leq D, D_{\widetilde{W}} > D\})$$
$$\leq \widetilde{\mu}_n(\{|D_{\widetilde{W}_n} - D_{\widetilde{W}}| > \sqrt{\varepsilon}\}) + \widetilde{\mu}_n(\{D < D_{\widetilde{W}} \leq D + \sqrt{\varepsilon}\})$$
$$< 2\sqrt{\varepsilon} + \mu(\{D < D_W \leq D + \sqrt{\varepsilon}\}),$$

which is also at most δ if ε is small enough.

Next we trivially extend $W_{n, \leq D}$ and $W_{\leq D}$ first to the spaces W_n and W are defined on, and then to the spaces used in the coupling $\widetilde{\mu}_n$. Let $\widetilde{W}_{n, \leq D}$ and $\widetilde{W}_{\leq D}$ be the pullbacks, let $\widetilde{W}'_{n, \leq D}$ be equal to $\widetilde{W}_{n, \leq D}$ on $\{D_{\widetilde{W}} \leq D\}^2$ and 0 otherwise, and let $\widetilde{W}'_{\leq D}$ be equal to $\widetilde{W}_{\leq D}$ on $\{D_{\widetilde{W}_n} \leq D\}^2$ and 0 otherwise. Then $\widetilde{W}_{n, \leq D}$ and $\widetilde{W}'_{n, \leq D}$ differ only on $\{D_{\widetilde{W}_n} \leq D, D_{\widetilde{W}} > D\} \times \{D_{\widetilde{W}_n} \leq D\}$ and its transpose. Indeed, if $D_{\widetilde{W}_n} > D$ in either coordinate, then both graphons are zero, and if $D_{\widetilde{W}} \leq D$ in both coordinates, then by the definition they are the same. Since $\widetilde{W}_{n, \leq D}$ has maximum degree D, this implies that

$$\|\widetilde{W}_{n, \leq D} - \widetilde{W}'_{n, \leq D}\|_\square \leq \|\widetilde{W}_{n, \leq D} - \widetilde{W}'_{n, \leq D}\|_1 \leq 2\widetilde{\mu}_n(\{D_{\widetilde{W}_n} \leq D, D_{\widetilde{W}} > D\})D \leq 2\delta D.$$

Analogously,

$$\|\widetilde{W}_{\leq D} - \widetilde{W}'_{\leq D}\|_\square \leq 2\delta D.$$

Note that $\widetilde{W}'_{n, \leq D}$ and $\widetilde{W}'_{\leq D}$ are equal to $\widetilde{W}_n$ and $\widetilde{W}$, respectively, on $\{D_{\widetilde{W}_n} \leq D, D_{\widetilde{W}} \leq D\}^2$, and zero everywhere else, which implies that $\widetilde{W}'_{n, \leq D} - \widetilde{W}'_{\leq D}$ is the restriction of $\widetilde{W}_n - \widetilde{W}$ to $\{D_{\mathbb{W}_n} \leq D, D_{\mathbb{W}} \leq D\}^2$. This implies that

$$\|\widetilde{W}'_{n, \leq D} - \widetilde{W}'_{\leq D}\|_\square \leq \|\widetilde{W}_n - \widetilde{W}\|_\square < \varepsilon,$$

which in turn implies that

$$\|\widetilde{W}_{n, \leq D} - \widetilde{W}_{\leq D}\|_\square \leq \|\widetilde{W}_{n, \leq D} - \widetilde{W}'_{n, \leq D}\|_\square + \|\widetilde{W}'_{n, \leq D} - \widetilde{W}'_{\leq D}\|_\square$$
$$+ \|\widetilde{W}'_{\leq D} - \widetilde{W}_{\leq D}\|_\square$$
$$\leq 4\delta D + \varepsilon.$$

Taking ε small enough, this can be made arbitrarily small, which completes the proof. $\square$

We also have the following:

Lemma 9.6 *Suppose that a sequence of integrable graphons W_n has uniformly bounded marginals, and converges to a (necessarily integrable) graphon W in the cut metric. Then W has the same bound on its marginals, and $\delta_{2\to 2}(W_n, W) \to 0$.*

Proof Suppose that for each n, $D_{W_n} \leq D$ almost everywhere, but $D_W > D$ on a set of positive measure. Then there exists $D' \geq D$ such that $\mu(\{D_W > D'\}) > 0$ and $\mu(\{D_W = D'\}) = 0$. By Lemma 9.5, $W_{n,\leq D'}$ converges to $W_{\leq D'}$ in the cut metric, but the cut distance of $W_{n,\leq D'}$ from W_n is 0, since $D_{W_n} \leq D$ almost everywhere. Therefore, the cut distance of $W_{\leq D'}$ and W is 0, which is a contradiction.

Now, we have the following. Since $\|W_n\|_1 \to \|W\|_1$, there exists a uniform bound C on $\|W_n\|_1$ and $\|W\|_1$, which implies that $\|W_n - W\|_1 \leq 2C$. Furthermore, for any x, $|D_{W_n - W}(x)| \leq D_{W_n}(x) + D_W(x) \leq 2D$. By Lemma 3.22, and recalling that $\|U\|_{\boxtimes} \leq \sqrt{\|U\|_{\square}\|U\|_\infty}$, we have

$$\|W_n - W\|_{2\to 2} \leq \left(8\|W_n - W\|_{\boxtimes}\|W_n - W\|_\infty^{3/4}\|D_{|W_n - W|}\|_\infty^{3/2}\|W_n - W\|_1^{3/4}\right)^{1/4}$$

$$\leq \left(8\|W_n - W\|_{\square}^{1/2}\|W_n - W\|_\infty^{5/4}\|D_{|W_n - W|}\|_\infty^{3/2}\|W_n - W\|_1^{3/4}\right)^{1/4}$$

$$\leq \left(100 D^{3/2} C^{3/4} \|W_n - W\|_{\square}^{1/2}\right)^{1/4} \to 0.$$

Furthermore, note that

$$\|D_{W_n} - D_W\|_1 \leq 2 \sup_S \left\{\left|\int_S D_{W_n} - D_W\right|\right\} \leq 2\|W_n - W\|_{\square}.$$

Therefore,

$$\|D_{\mathbb{W}_n} - D_{\mathbb{W}}\|_2 = \|D_{W_n} - D_W\|_2$$
$$\leq \sqrt{\|D_{W_n} - D_W\|_1 \|D_{W_n} - D_W\|_\infty} \leq \sqrt{4D\|W_n - W\|_{\square}} \to 0.$$

Finally,
$$\|\mathbb{W}_n\|_1 = \|W_n\|_1 \to \|W\|_1 = \|\mathbb{W}\|_1.$$

$\square$

The last lemma we need to prove Theorem 2.28 is the following.

Lemma 9.7 *Suppose that $\delta_\diamond(\mathbb{W}, \mathbb{W}') = 0$ for two integrable graphexes, and suppose that $\mathbb{W} = (W, 0, 0, \Omega)$. Then $\mathbb{W}' = (W', 0, 0, \Omega')$ and $\delta_\square(W, W') = 0$.*

Proof Since $\delta_\diamond(\mathbb{W}, \mathbb{W}') = 0$, $\xi(G(\mathbb{W}))$ and $\xi(G(\mathbb{W}'))$ have the same distribution. But, as already observed in Remark 5.4 in [17], almost surely, the dust part of $\mathbb{W}'$ generates edges which are isolated, the star part generates edges with one vertex of degree one and a second vertex of infinite degree, and the graphon part generates edges with two endpoints of infinite degree. Since $\xi(G(\mathbb{W}))$ has no star or dust edges,

$\xi(G(\mathbb{W}'))$ doesn't have these either, showing that $\mathbb{W}' = (W', 0, 0, \mathbf{\Omega}')$. Finally, since the graphon process generated by W and W' have the same distribution, $\delta_\square(W, W') = 0$ by Theorem 27 in [3]. □

We are now ready to prove Theorem 2.28.

Proof (*Theorem* 2.28) First, we show that if a sequence converges in the cut metric, then it converges in $\delta_\diamond$. We show property (2) from Proposition 4.6. Since the graphons converge in cut metric, we must have in particular that $\|\mathbb{W}_n\|_1 = \|W_n\|_1 \to \|W\|_1 = \|\mathbb{W}\|_1$, which implies that the set $\{\|\mathbb{W}_n\|\}_n$ is uniformly bounded; therefore, the sequence is tight by Corollary 4.2 (1). By Lemma 9.5, for any $D > 0$ with $\mu(\{D_W = D\}) = 0$, $W_{n, \leq D}$ converges to $W_{\leq D}$ in cut metric, and by Lemma 9.6, they must also converge in $\delta_{2\to 2}$, which completes the proof that cut metric convergence implies weak kernel convergence. Since we know that any cut metric convergent sequence is uniformly tail regular, this completes that proof that (1) implies (2). It is clear that (2) is stronger than (3), so it remains to show that (3) implies (1).

To this end, we first note that uniform tail regularity implies uniformly bounded L^1 norms, which by Theorem 2.26 implies that $\mathbb{W}$ is integrable. Suppose that $\mathbb{W}$ does not consist of only a graphon part, or that $\mathbb{W}_n$ does not converge to it in the cut metric. Since the sequence $\mathbb{W}_n$ is uniformly tail regular, we may choose a subsequence that converges to an integrable graphon W', such that either $\delta_\square(W', W) \neq 0$, or $\mathbb{W}$ is not just a graphon. In either case, letting $\mathbb{W}' = (W', 0, 0, \mathbf{\Omega}')$, we have by Lemma 9.7 that $\delta_\diamond(\mathbb{W}, \mathbb{W}') \neq 0$. However, since $\delta_\square(\mathbb{W}_n, W') \to 0$, we must have that $\delta_\diamond(\mathbb{W}_n, \mathbb{W}') \to 0$, which implies that $\delta_\diamond(\mathbb{W}', \mathbb{W}) = 0$, which is a contradiction. This completes the proof that (3) implies (1), and thus we have proven the theorem. □

Proof (*Theorem* 2.29) First, assume that $\mathbb{W} = (W, 0, 0, \mathbf{\Omega})$, and let us prove (2). Assume first that the sequence has uniformly bounded marginals. In this case, $\delta_{2\to 2}(\mathbb{W}_n, \mathbb{W}) \to 0$. Take a sequence of couplings of trivial extensions of $\mathbb{W}_n$ and $\mathbb{W}$ which show that their kernel distance goes to zero, let $\mathbf{\Omega}'_n$ be the space for each n, and $\mathbb{W}'_n$ and $\mathbb{W}^n$ the pulled back graphexes, and let W'_n and W^n be their graphon parts. By Corollary 9.4, it is enough to prove uniform tail regularity for W'_n. Given $\varepsilon > 0$, let $\delta > 0$ be such that

$$\|W\|_1 - \|W|_{M_\delta}\|_1 \leq \varepsilon,$$

where M_δ is the set
$$M_\delta = \{x \in \Omega : D_W(x) \geq \delta\}.$$

Let M_δ^n be the pullback to Ω'_n. We then have that

$$\|W'_n - W^n\|_{2\to 2} \to 0.$$

Since M_δ^n has finite measure, this implies that

$$\int_{M_\delta^n \times M_\delta^n} \mathbb{W}'_n \to \int_{M_\delta \times M_\delta} \mathbb{W}.$$

Since $\|\mathbb{W}'_n\|_1 = \|\mathbb{W}_n\|_1 \to \|\mathbb{W}\|_1 = \|W\|_1$, this implies that

$$\limsup_{n\to\infty}\left(\|\mathbb{W}'_n\|_1 - \int_{M_\delta^n \times M_\delta^n} \mathbb{W}'_n\right) \le \limsup_{n\to\infty}\left(\|\mathbb{W}'_n\|_1 - \int_{M_\delta^n \times M_\delta^n} \mathbb{W}'_n\right) \le \varepsilon.$$

This can be made arbitrarily small by taking δ small enough, which proves uniform tail regularity under the assumption of uniformly bounded marginals. On the other hand,

$$2\limsup_{n\to\infty}\left(\|S_n\|_1 + I_n\right) = \limsup_{n\to\infty}\left(\|\mathbb{W}_n\|_1 - \|\mathbb{W}'_n\|_1\right)$$

$$\le \|W\|_1 - \lim_{n\to\infty}\int_{M_\delta^n \times M_\delta^n} \mathbb{W}'_n \le \varepsilon,$$

which implies that $\|S_n\|_1 \to 0$ and $I_n \to 0$, completing the proof of (2) under the assumption of uniformly bounded marginals.

If instead of uniformly bounded marginals, we have uniform integrability, then the claims follow from the fact that for each $D > 0$, $\delta_{2\to 2}(\mathbb{W}_{n,\le D}, \mathbb{W}_{\le D}) \to 0$, and we can take D large enough so that each $\int_{\Omega_{n,>D}} S_n$ is less than ε and

$$\int_{\Omega \times \Omega_{>D}} W_n < \varepsilon,$$

and I_n is unaffected by the restriction.

Conversely, assume (2). Let, for each n, $\mathbb{W}'_n = (W_n, 0, 0, \mathbf{\Omega})$ (so we are replacing S_n and I_n with 0). Since $\|S_n\|_1 \to 0$ and $I_n \to 0$,

$$\delta_\circ(\mathbb{W}_n, \mathbb{W}'_n) \to 0$$

Indeed, clearly $|\|\mathbb{W}_n\|_1 - \|\mathbb{W}'_n\|_1| \to 0$, and for any $D > 0$, we have that $\int_{\Omega_{\le D}} S_n^2 \le D\|S_n\|_1 \to 0$. Taking D large enough that $\Omega_{>D}$ has small measure, we can show that $\delta_\circ(\mathbb{W}_n, \mathbb{W}'_n)$ is arbitrarily small for large enough n. Now, the statement follows from Theorem 2.28; specifically, we have shown that (3) holds for the sequence $\mathbb{W}'_n$, which by (2) implies that the limit is a pure graphon.

The equivalence of (2) and (3) follows from Theorem 2.28 applied to the sequence $\mathbb{W}'_n$. □

Acknowledgements László Miklós Lovász thanks Microsoft Research New England for an internship in the summer of 2016, when most of the research part of this work was done. László Miklós Lovász was also supported by NSF Postdoctoral Fellowship Award DMS 1705204 for part of this work. All of us thank Svante Janson and Nina Holden for various discussions about the work presented here, and the anonymous referee for comments and suggestions.

Appendices

A.1: Local Finiteness

In this appendix, we prove Proposition 2.4.

Throughout this appendix, $\boldsymbol{\Omega} = (\Omega, \mathcal{F}, \mu)$ will be a σ-finite measure space, $S \colon \Omega \to \mathbb{R}_+$ will be measurable, $W \colon \Omega \times \Omega \to [0, 1]$ will be a symmetric, measurable function, $\eta = \sum_i \delta_{x_i}$ will be a Poisson point process on Ω with intensity μ, and

$$\eta(S) = \sum_i S(x_i) \quad \text{and} \quad \eta^2(W) = \sum_{i \neq j} W(x_i, x_j).$$

We start with the following lemma, which is the analogue of Lemma A.3.6 from [19] for general measure spaces. We use $\mathbb{E}$ to denote expectations with respect to the Poisson point process and $W \circ W$ to denote the function $(x, y) \mapsto \int W(x, z) W(z, y) \, d\mu(z)$.

Lemma A.1.1 *Let $\psi(x) = 1 - e^{-x}$. Then the following hold, with both sides of the various identities being possibly infinite:*

(1) $\mathbb{E}[\eta(S)] = \|S\|_1$ *and* $\mathbb{E}[\eta^2(W)] = \|W\|_1$,
(2) $\mathbb{E}[\psi(\eta(S))] = \psi(\|\psi(S)\|_1)$, *and*
(3) $\mathbb{E}[(\eta^2(W))^2] = \|W\|_1^2 + 4\|W \circ W\|_2 + 2\|W^2\|_1$.

Proof We first assume that $m = \mu(\Omega)$ is finite and S is bounded. Then η can be generated by first choosing N as a Poisson random variable with rate m and then choosing $x_1, \ldots, x_N$ i.i.d. according to the distribution $\frac{1}{m}\mu$. Conditioned on N, the expectations of $\eta(S)$ and $\eta^2(W)$ are $\frac{N}{m}\|S\|_1$ and $\frac{N(N-1)}{m^2}\|W\|_1$, respectively, and the expectation of $\psi(\eta(S))$ is

$$\mathbb{E}[\psi(\eta(S)) \mid N] = 1 - \mathbb{E}[e^{-\sum_{i=1}^N S(x_i)}]$$
$$= 1 - \prod_{i=1}^N \frac{1}{m} \int_\Omega d\mu(x_i) e^{-S(x_i)} = 1 - \left(\frac{1}{m} \int_\Omega d\mu(x) e^{-S(x)}\right)^N.$$

Therefore,

$$\mathbb{E}[\eta(S)] = \sum_{N=0}^{\infty} e^{-m} \frac{m^N}{N!} \frac{N}{m} \|S\|_1 = \sum_{N=1}^{\infty} e^{-m} \frac{m^{N-1}}{(N-1)!} \|S\|_1 = \|S\|_1.$$

Also,

$$\mathbb{E}[\eta^2(W)] = \sum_{N=0}^{\infty} e^{-m} \frac{m^N}{N!} \frac{N(N-1)}{m^2} \|W\|_1 = \sum_{N=2}^{\infty} e^{-m} \frac{m^{N-2}}{(N-2)!} \|W\|_1 = \|W\|_1.$$

Finally,

$$\mathbb{E}[\psi(\eta(S))] = \sum_{N=0}^{\infty} e^{-m} \frac{m^N}{N!} \left(1 - \left(\frac{1}{m} \int_\Omega d\mu(x) e^{-S(x)}\right)^N\right)$$

$$= 1 - \sum_{N=0}^{\infty} e^{-m} \frac{m^N}{N!} \left(\frac{1}{m} \int_\Omega d\mu(x) e^{-S(x)}\right)^N$$

$$= 1 - \exp\left(\int_\Omega e^{-S(x)} d\mu(x) - m\right) = 1 - \exp\left(\int_\Omega -\psi(S(x)) d\mu(x)\right)$$

$$= 1 - e^{\|\psi(S)\|_1} = \psi(\|\psi(S)\|_1).$$

To calculate the expectation of

$$(\eta^2(W))^2 = \sum_{i \neq j} \sum_{k \neq \ell} \mathbb{E}[W(x_i, x_j) W(x_k, x_\ell)]$$

we distinguish whether $\{i, j\}$ and $\{k, \ell\}$ intersect in 0, 1, or 2 elements, leading to the expression

$$\mathbb{E}[(\eta^2(W))^2 \mid N] = \frac{N(N-1)(N-2)(N-3)}{m^4} \|W\|_1^2$$
$$+ \frac{4N(N-1)(N-2)}{m^3} \|W \circ W\|_1 + \frac{2N(N-1)}{m^2} \|W^2\|_1.$$

Taking the expectation over N gives the expression in the lemma similarly. This completes the proof for spaces of finite measure and bounded functions S. The general case follows by the monotone convergence theorem. □

Using Lemma A.1.1, we now prove the following proposition, which is the analogue of the relevant parts for us of Theorem A3.5 from [19] for general σ-finite measure spaces.

Proposition A.1.2 *Let $S: \Omega \to \mathbb{R}_+$ be measurable, and let $W: \Omega \times \Omega \to [0, 1]$ be symmetric and measurable. Then the following hold:*

(1) $\eta(S) < \infty$ a.s. if and only if $\|\min\{S, 1\}\|_1 < \infty$, and

(2) $\eta^2(W) < \infty$ a.s. if and only if there exists a finite $D > 0$ such that the following three conditions hold:

(a) $D_W < \infty$ almost surely,
(b) $\mu(\{x \in \Omega : D_W(x) > D\}) < \infty$, and
(c) $\|W|_{\{x \in \Omega : D_W(x) \leq D\}}\|_1 < \infty$.

Proof Since $\frac{1}{2}\min\{1, x\} \leq \psi(x) \leq \min\{1, x\}$, the condition $\|\min\{S, 1\}\|_1 < \infty$ in (1) is equivalent to the statement that $\|\psi(S)\|_1 < \infty$, which is equivalent to the statement that $\psi(\|\psi(S)\|_1) < 1$. By Lemma A.1.1 (2), this is equivalent to saying that $\mathbb{E}[\psi(\eta(S))] < 1$, which holds if and only if $\eta(S) < \infty$ with positive probability. By Kolmogorov's zero-one law, we either have $\eta(S) < \infty$ almost surely, or $\eta(S) = \infty$ almost surely; therefore we have obtained that $\|\min\{S, 1\}\|_1 < \infty$ if and only if $\eta(S) < \infty$ almost surely.

To prove the second statement, assume first that the conditions (a)–(c) hold. Condition (a) then implies that a.s., no Poisson point falls into the set $\{D_W = \infty\}$, which means we may replace Ω by a space such that $D_W(x) < \infty$ for all $x \in \Omega$. Let $\Omega_{>D} = \{x \in \Omega : D_W(x) > D\}$ and $\Omega_{\leq D} = \Omega \setminus \Omega_{>D}$. Since $\Omega_{>D}$ has finite measure by assumption (b), we have that a.s., only finitely many Poisson points fall into this set, which in particular implies that the contribution of the points $x_i, x_j \in \Omega_{>D}$ to $\eta^2(W)$ is a.s. finite. Next let us consider the contributions to $\eta^2(W)$ from pairs of points x_i, x_j such that one lies in $\Omega_{>D}$ and the other one lies in $\Omega_{\leq D}$. Observing that the Poisson process in $\Omega_{>D}$ and $\Omega_{\leq D}$ are independent, and that a.s., there are only finitely many points in $\Omega_{>D}$, it will clearly be enough to show that for all $x \in \Omega_{>D}$, a.s. with respect to the Poisson process in $\Omega_{\leq D}$,

$$\sum_{j : x_j \in \Omega_{\leq D}} W(x, x_j) < \infty.$$

But by Lemma A.1.1 (1) applied to the function $S' : \Omega_{\leq D} \to \mathbb{R}_+$ defined by $S'(y) = W(x, y)$, the expectation of this quantity is equal to

$$\int_{\Omega_{\leq D}} S'(y)\, d\mu(y) = \int_{\Omega_{\leq D}} W(x, y)\, d\mu(y).$$

This is bounded by $D_W(x)$ and hence finite, which proves that the sum is a.s. finite. We are thus left with estimating $\eta^2(W|_{\Omega_{\leq D}})$. Again by Lemma A.1.1 (1), we have that $\mathbb{E}[\eta^2(W|_{\Omega_{\leq D}})] = \|W|_{\Omega_{\leq D}}\|_1$ which is finite by assumption (c), showing that $\eta^2(W|_{\Omega_{\leq D}})$ is a.s. finite.

Conversely, let us assume that a.s., $\eta^2(W) < \infty$. First we will prove that this implies $\mu(\{D_W = \infty\}) = 0$. Assume for a contradiction that this is not the case. Since μ is σ-finite, we can find a measurable set $N \subseteq \Omega$ such that $D_W(x) = \infty$ for all $x \in N$ and $0 < \mu(N) < \infty$. Consider the contribution to $\eta^2(W)$ by all Poisson points (x_i, x_j) such that $x_i \in N$ and $x_j \in N^c = \Omega \setminus N$. Since the Poisson processes on N and N^c are independent, the finiteness of $\eta^2(W)$ implies that for almost all $x \in N$, the sum $\sum_{j : x_j \in N^c} W(x, x_j)$ is a.s. finite. Applying statement (1) of the current proposition to $W(x, \cdot)$ (and recalling that W is bounded by 1), we conclude that for almost all $x \in N$, $\int_{N^c} W(x, y) \, d\mu(y) < \infty$, which implies that for almost all $x \in N$, $\int_N W(x, y) \, d\mu(y) = D_W(x) - \int_{N^c} W(x, y) \, d\mu(y) = \infty$. This is a contradiction since $\mu(N) < \infty$ and $W \le 1$.

We next prove (b) (for any value of D). Suppose for a contradiction that $\mu(\{x \in \Omega : D_W(x) > D\}) = \infty$. We then claim that almost surely, $\eta^2(W) = \infty$. After obtaining the Poisson process, color each point randomly red or blue, with equal probability, independently. We can then obtain the red and blue points equivalently by taking two independent Poisson processes, both with intensity $\mu/2$. We claim that almost surely, the sum of $W(x, y)$ just over red-blue pairs is already ∞. We know that almost surely, there are an infinite number of red points x_i with $D_W(x_i) > D$. Let x_n be such a sequence, and given $y \in \Omega$, let $S'(y) = \sum_{n=1}^{\infty} W(x_n, y)$. Then the sum of W over red-blue edges is equal to $\eta(S')$ for the Poisson process with intensity $\mu/2$. Therefore, it suffices to prove that $\|\min\{S', 1\}\|_{1, \mu/2} = \infty$. First, note that if either $\mu(\{y \in \Omega : S'(y) = \infty\}) > 0$ or $\mu(\{y \in \Omega : S'(y) > 1\}) = \infty$, then it clearly holds. Otherwise, we have that as $D' \to \infty$, $\mu(\{y \in \Omega : S'(y) > D'\}) \to 0$; therefore, there exists some D' (without loss of generality, we may assume $D' \ge 1$) such that $\mu(\{y \in \Omega : S'(y) > D'\}) < D/2$. Let Ω' be the complement of $\{y \in \Omega : S'(y) > D'\}$. We then have that for each x_n,

$$\int_{\Omega'} W(x_n, y) \frac{d\mu(y)}{2} = \int_{\Omega} W(x_n, y) \frac{d\mu(y)}{2} - \int_{\Omega \setminus \Omega'} W(x_n, y) \frac{d\mu(y)}{2}$$
$$\ge \frac{1}{2} D_W(x_n) - \frac{1}{2} \mu(\Omega \setminus \Omega') \ge \frac{D}{2} - \frac{D}{4}.$$

We also have that

$$\int_{\Omega'} S'(y) \frac{d\mu(y)}{2} = \int_{\Omega'} \sum_{n=1}^{\infty} W(x_n, y) \frac{d\mu(y)}{2}$$
$$= \sum_{n=1}^{\infty} \int_{\Omega'} W(x_n, y) \frac{d\mu(y)}{2} \ge \sum_{n=1}^{\infty} D/4 = \infty.$$

Therefore,

$$\int_{\Omega} \min\{S'(y), 1\} \frac{d\mu(y)}{2} \ge \int_{\Omega'} \min\{S'(y), 1\} \frac{d\mu(y)}{2}$$
$$\ge \frac{1}{D'} \int_{\Omega'} \min\{S'(y), D'\} \frac{d\mu(y)}{2} = \frac{1}{D'} \int_{\Omega'} S'(y) \frac{d\mu(y)}{2} = \infty.$$

This contradiction completes the proof.

We are left with proving (c) (we will again prove it for any value of D). Assume the opposite, and let $\Lambda_n \subseteq \Lambda$ be an increasing sequence such that $\mu(\Lambda_n) < \infty$ and $\bigcup_n \Lambda_n = \Omega_{\le D}$. Let $U_n = W|_{\Lambda_n}$. Then $\|U_n\|_1 < \infty$, $\|U_n\|_1 \uparrow \|W|_{\Omega_{\le D}}\|_1 = \infty$, and $\|D_{U_n}\|_\infty \le D$, implying in particular that $\|U_n \circ U_n\|_1 = \|D_{U_n}\|_2^2 \le D\|D_{U_n}\|_1 =$

$D\|U_n\|_1$. Given an arbitrary constant λ, we claim that

$$\mathbb{P}\Big(\eta^2(W|_{\Omega_{\leq D}}) > \lambda\Big) \geq \frac{(\|U_n\|_1 - \lambda)^2}{\|U_n\|_1^2 + (4D+2)\|U_n\|_1}, \tag{14}$$

provided n is large enough to ensure that $\|U_n\|_1 > \lambda$. Indeed, writing

$$\mathbb{E}[\eta^2(U_n)] = \mathbb{E}[\eta^2(U_n)1_{\eta^2(U_n)\leq\lambda}] + \mathbb{E}[\eta^2(U_n)1_{\eta^2(U_n)>\lambda}],$$

we can bound the first term by λ and the second by $\sqrt{\mathbb{E}[(\eta^2(U_n))^2]\mathbb{P}[\eta^2(U_n) > \lambda]}$, using Cauchy's inequality. We therefore obtain that

$$\mathbb{E}[\eta^2(U_n)] \leq \lambda + \sqrt{\mathbb{E}[(\eta^2(U_n))^2]\mathbb{P}[\eta^2(U_n) > \lambda]}.$$

Rearranging, we obtain the bound

$$\mathbb{P}\Big(\eta^2(U_n) > \lambda\Big) \geq \frac{(\mathbb{E}[\eta^2(U_n)] - \lambda)^2}{\mathbb{E}[(\eta^2(U_n))^2]} = \frac{(\|U_n\|_1 - \lambda)^2}{\|U_n\|_1^2 + 4\|U_n \circ U_n\|_1 + \|U_n^2\|_1},$$

where we used Lemma A.1.1 (1) and (3) in the last step. Observing that

$$\Pr(\eta^2(W|_{\Omega_{\leq D}}) > \lambda) \geq \Pr(\eta^2(U_n) > \lambda)$$

and bounding $4\|U_n \circ U_n\|_1 + \|U_n^2\|_1$ by $(4D+2)\|U_n\|_1$, we obtain (14). Since the right side of (14) goes to 1 as $n \to \infty$, we get that with probability one, $\eta^2(W|_{\Omega_{\leq D}}) > \lambda$ for all λ, which contradicts the assumption that $\eta^2(W|_{\Omega_{\leq D}}) < \infty$ a.s. □

Proof (*Proposition* 2.4) We first prove the equivalence of (A) – (E). Clearly $(B) \Rightarrow (C) \Rightarrow (A)$ and $(D) \Rightarrow (E)$. It is also not hard to see that $(E) \Rightarrow (A)$. Indeed, note first that for any D, the condition on S is equivalent to the condition that $\min\{S, D\}$ is integrable (which implies that $\mu(\{S > D\}) < \infty$.) Set $\Omega' = \{D_W \leq D\} \cap \{S \leq D\}$. Then (E) implies that

$$\|\mathbb{W}|_{\Omega'}\|_1 \leq 2I + \|\mathbb{W}_{\{D_W \leq D\}}\|_1 + 2\|S1_{S \leq D}\|_1$$
$$\leq 2I + \|\mathbb{W}_{\{D_W \leq D\}}\|_1 + 2\|\min\{S, D\}\|_1 < \infty$$

and $\mu(\Omega \setminus \Omega') \leq \mu(\{D_W > D\}) + \mu(\{S > D\}) < \infty$, proving (A). So it will be enough to show $(A) \Rightarrow (B)$ and $(A) \Rightarrow (D)$.

Suppose that (A) holds, and let Ω' be a set such that $\mu(\Omega \setminus \Omega') < \infty$, $\mathbb{W}' = \mathbb{W}|_{\Omega'}$, and $\|\mathbb{W}'\|_1 = C < \infty$. Let $D > 0$. First, assume that $D > D_0 = \mu(\Omega \setminus \Omega')$. Then

$$\{x \in \Omega : D_\mathbb{W}(x) > D\} \subseteq (\Omega \setminus \Omega') \cup \{x \in \Omega', D_{\mathbb{W}'}(x) > D - D_0\}.$$

Since $\|D_{\mathbb{W}'}\|_1 \leq \|\mathbb{W}'\|_1 = C$, this set has measure at most

$$D_0 + \frac{C}{D - D_0}.$$

Now, let $\mathbb{W}'' = \mathbb{W}|_{\{x : D_\mathbb{W}(x) \leq D\}}$. Then

$$\|\mathbb{W}''\|_1 \leq \|\mathbb{W}'\|_1 + 2 \int_{\{x \in \Omega \setminus \Omega' : D_\mathbb{W}(x) \leq D\}} D_\mathbb{W}(x) \leq \|\mathbb{W}'\|_1 + 2DD_0.$$

We have thus proven that (B) holds for all D larger than some D_0, and more generally for any D for which there exists an $\Omega' \subseteq \Omega$ with $\mu(\Omega \setminus \Omega') < D$ and $\|\mathbb{W}|_{\Omega'}\|_1 < \infty$.

Note that if $\mathbb{W}|_{\{x : D_\mathbb{W}(x) \leq D\}}$ is integrable for $D > D_0$, then it must remain integrable if we decrease D, since that is just a restriction to a subset. Therefore, this implies that $\mathbb{W}|_{\{x : D_\mathbb{W}(x) \leq D\}}$ is integrable for all D. Since $D_\mathbb{W} < \infty$ almost everywhere, we further have that $\mu(\{x \in \Omega : D_\mathbb{W}(x) \geq \lambda\})$ tends to 0 as λ tends to ∞ (since we at least know that it is finite for large enough λ). Fixing $D > 0$, we can therefore take D' large enough so that $\mu(\{x \in \Omega : D_\mathbb{W}(x) \geq D'\}) < D$. Taking $\Omega' := \Omega \setminus \{x \in \Omega : D_\mathbb{W}(x) \geq D'\}$, we get a set Ω' such that $\mu(\Omega \setminus \Omega') < D$ and $\|\mathbb{W}|_{\Omega'}\|_1 < \infty$ proving that (B) holds for all $D > 0$.

On the other hand if (A) holds for some Ω', then $\|W|_{\Omega'}\|_1 < \infty$ and $\|S1_{\Omega'}\|_1 < \infty$. Proceeding exactly as above we conclude that for all D, $\mu(\{D_W > D\}) < \infty$ and $\|W|_{\{D_W \leq D\}}\|_1 < \infty$, as well as $\mu(\{S > D\}) < \infty$ and $\|S1_{\{S \leq D\}}\|_1 < \infty$. Since

$$\|\min\{S, D\}\|_1 = D\mu(\{S > D\}) + \|S1_{\{S \leq D\}}\|_1 < \infty,$$

the latter condition is equivalent to $\|\min\{S, D\}\|_1 < \infty$, as required.

We are left with proving that the local finiteness conditions in Definition 2.1 are necessary and sufficient for the almost sure finiteness of $G_T(\mathbb{W})$ for all $T < \infty$. It is easy to check that the local finiteness conditions are not affected if we multiply the underlying measure by T and S by T. We therefore assume that $T = 1$. Let $\eta = \sum_i \delta_{x_i}$ be a Poisson process of intensity μ on Ω, let Y_i be Poisson random variable with mean $S(x_i)$, and let Y_{ij} be Bernoulli with mean $W(x_i, x_j)$, all of them independent of each other. We will have to show that the local finiteness conditions on $\mathbb{W}$ are equivalent to the a.s. finiteness of the sums

$$e_S = \sum_i Y_i \quad \text{and} \quad e_W = \sum_{i > j} Y_{ij}.$$

We next use the fact that a sum of independent, non-negative random variables $\sum_k Z_k$ is a.s. finite if and only if $\sum_i \mathbb{E}[\min\{Z_i, 1\}] < \infty$. In the case of e_W, $Y_{i,j}$ is bounded, and therefore we immediately have that e_W is a.s. finite if and only if $\eta^2(W)$ is a.s. finite. Proposition A.1.2 (b) then proves this case. In the case of e_S, setting $S' = \min\{S, 1\}$, applying Proposition A.1.2 to S', and noting that S' is bounded, we have that $\sum_i S'(x_i)$ is almost surely finite if and only if $\|S'\|_1 < \infty$. This is exactly the condition on S. □

A.2: Sampling with Loops

In this section, we discuss how to handle samples with loops. The sampling process is adjusted as follows. We follow the same process as for $\mathcal{G}_T(\mathbb{W})$ and $\mathcal{G}_\infty(\mathbb{W})$; however, for each vertex labeled as (t, x), with probability $W(x, x)$, we add a loop to the vertex. Deleting isolated vertices as before, and then removing the feature labels from the vertices, we obtain a family $(\widetilde{\mathcal{G}}_T(\mathbb{W}))_{T \geq 0}$ of labelled graphs with loops, as well as the infinite graph $\widetilde{\mathcal{G}}_\infty(\mathbb{W}) = \bigcup_{T \geq 0} \widetilde{\mathcal{G}}_T(\mathbb{W})$.

Note that a vertex that was previously isolated may not be isolated anymore if it receives a loop, so a vertex may have been deleted from $\mathcal{G}_T(\mathbb{W})$ but not from $\widetilde{\mathcal{G}}_T(\mathbb{W})$. We add a further condition for local finiteness:

$$\int_\Omega W(x, x)\, d\mu(x) < \infty.$$

Note that if $\mathbb{W}$ is atomless, then the values $W(x, x)$ do not have an effect on $\mathcal{G}_T$ and $\mathcal{G}_\infty$, and the diagonal constitutes a zero measure set in $\Omega \times \Omega$.

As stated, Theorem 2.5 is false for sampling with loops. Since the diagonal may be a zero measure set, almost everywhere equal pullbacks do not imply having the same looped samples. We could further add the condition that $W(x, x)$ is equal to the pullback almost everywhere, but the theorem would still be false. This is demonstrated by the following example. Let $\Omega_1 = \Omega_2 = [0, 1]$. Take W_1 to be constant $1/2$ on $[0, 1] \times [0, 1]$, and let W_2 be constant $1/2$ off the diagonal, 0 if $x < 1/2$, 1 otherwise. Let $\mathbb{W}_i = (W_i, 0, 0, \Omega_i)$. Then we claim that $\widetilde{\mathcal{G}}_T(\mathbb{W}_1)$ and $\widetilde{\mathcal{G}}_T(\mathbb{W}_2)$ have the same distribution. Indeed, both are equivalent to taking Poisson(T) vertices, adding a loop to each vertex with probability $1/2$, independently, and also taking an edge between each pair of vertices with probability $1/2$, independently over different pairs.

It turns out that in general, allowing diagonal values strictly between 0 and 1 is not necessary, because we could extend the feature space to determine whether each vertex has loops. For graphexes where the diagonal is 0 or 1, we can then conclude an analogous theorem from Theorem 2.5.

We first show the following:

Proposition A.2.1 *For any graphex* $\mathbb{W} = (W, S, I, \Omega)$, *there exists a graphex* $\widetilde{\mathbb{W}} = (\widetilde{W}, \widetilde{S}, \widetilde{I}, \widetilde{\Omega})$ *on an atomless space* $\widetilde{\Omega}$ *such that on the diagonal,* $\widetilde{W}$ *is* $\{0, 1\}$ *valued and such that* $\widetilde{\mathcal{G}}_\infty(\widetilde{\mathbb{W}})$ *and* $\widetilde{\mathcal{G}}_T(\widetilde{\mathbb{W}})$ *are equivalent to* $\widetilde{\mathcal{G}}_\infty(\mathbb{W})$ *and* $\widetilde{\mathcal{G}}_T(\mathbb{W})$, *respectively.*

Proof Let $\widetilde{\Omega} = \Omega \times [0, 1]$, and let π_1, π_2 be the projection maps. Note that we can obtain a Poisson process on $\widetilde{\Omega} \times \mathbb{R}_+$ by taking a Poisson process on $\Omega \times \mathbb{R}$, and independently labeling each point with a uniform random real number from $[0, 1]$, which becomes the second coordinate. Clearly $\widetilde{\Omega}$ is atomless, so the diagonal values only affect the generation of the loops. Define $\widetilde{I} = I$, $\widetilde{S} = S \circ \pi_1$, $\widetilde{W}(x, y) = W(\pi_1(x), \pi_1(y))$ if $x \neq y$, and

$$\widetilde{W}(x,x) = \begin{cases} 1, & \text{if } \pi_2(x) \le W(\pi_1(x), \pi_1(x)) \\ 0, & \text{otherwise.} \end{cases}$$

Then the sampling of edges between vertices is not affected by the second coordinate of a vertex. Note that the probability that there exist two vertices corresponding to the same point in $\widetilde{\Omega}$ is zero, since $\widetilde{\Omega}$ is atomless. For the loops, since we can obtain the vertices by first taking the Poisson process on $\Omega \times \mathbb{R}$ and then randomly labeling each vertex with a $[0,1]$ real number, we can see that for a point $y \in \Omega$, if it ends up as a point, there is a $W(y,y)$ probability that the point x corresponding to it has $\widetilde{W}(x,x) = 1$, and $1 - W(y,y)$ that $\widetilde{W}(x,x) = 0$, and this is independent over different points. Therefore, the distribution of loops is the same. □

Using this proposition, sampling loops according to the diagonal is equivalent to the following theory. The objects are graphexes with special subsets $\mathbb{W} = (W, S, I, \Omega, A)$ where W, S, I, and Ω are as before, and the *special set* $A \subseteq \Omega$ is a measurable subset with finite measure. We sample $\widetilde{\mathcal{G}}_\infty(\mathbb{W})$ in the same way as $\mathcal{G}_\infty(\mathbb{W})$, except that we add a loop to each vertex with a feature label in A. We then take the non-isolated vertices with time label at most T for $\widetilde{\mathcal{G}}_T(\mathbb{W})$. We can extend the definition of measure-preserving map by requiring that points in the special set be mapped to points in the special set, and points not in the special set be mapped to points not in the special set. We also define dsupp as earlier, except it contain all points in A (even if otherwise they would not be included).

Theorem A.2.2 *Let $\mathbb{W}_1$ and $\mathbb{W}_2$ be graphexes with special subsets as above. Then $\widetilde{\mathcal{G}}_T(\mathbb{W}_1)$ and $\widetilde{\mathcal{G}}_T(\mathbb{W}_2)$ have the same distribution for all $T \in \mathbb{R}_+$ if and only if there exists a third graphex with special subset $\mathbb{W}$ such that $\mathbb{W}_1$ and $\mathbb{W}_2$ are pullbacks of $\mathbb{W}$.*

Proof It is clearly enough to prove the only if direction. Suppose therefore that $\mathbb{W}_1$ and $\mathbb{W}_2$ have the same distribution. Then for any $0 < c < 1$, $c\mathbb{W}_1$ and $c\mathbb{W}_2$ have the same distributions (i.e., W, S, I are all multiplied by c, and the special set stays the same). Then let $\widehat{\mathbb{W}}_i$ be obtained by taking $\mathbb{W}_i/2$, adding a set B_i of measure 1 to Ω_i, and extending W_i to be 1 on $B_i \times B_i$, 1 between B_i and A_i, and 0 between B_i and $\Omega_i \setminus A_i$. Then we can obtain $G_T(\widehat{\mathbb{W}}_i)$ from $\widetilde{G}_T(\mathbb{W}_i)$ by the following process. We first keep each edge that is not a loop with probability $1/2$, and delete it otherwise, independently. We keep all the loops. Then we take Poisson(T) new vertices, put an edge between every pair, and put an edge between each new vertex and each vertex that had a loop (and delete loops). It is clear that in this way, the distributions $G_T(\widehat{\mathbb{W}}_1)$ and $G_T(\widehat{\mathbb{W}}_2)$ are the same for every T. Therefore, there exists a graphex $\widehat{\mathbb{W}} = (\widehat{W}, \widetilde{S}, \widetilde{I}, \widetilde{\Omega})$ such that $\widehat{\mathbb{W}}_1$ and $\widehat{\mathbb{W}}_2$ are both pullbacks of $\widehat{\mathbb{W}}$. It is clear that $\widehat{\mathbb{W}}$ must have a set of measure 1, call it B, which has $\widehat{W}(x,y) = 1$ if $x, y \in B$, and $\widehat{W}(x,y)$ is either 0 or 1 if $x \in B$, $y \notin B$, and only depends on y, and $\widehat{W}(x,y) \le 1/2$ if $x, y \notin B$, and B must pullback to exactly B_1 and B_2. If we let A be the set of points x with $\widehat{W}(x,y) = 1$ for any and all $y \in B$, then A must pullback to A_1 and A_2. If we therefore let $\mathbb{W}$ have underlying set $\widetilde{\Omega} \setminus B$, and be equal to $2\widehat{\mathbb{W}}$ restricted to this set, and special set A, then $\mathbb{W}$ pulls back to both $\mathbb{W}_1$ and $\mathbb{W}_2$. □

References

1. P. Billingsley, *Convergence of Probability Measures*, Wiley, New York, 1968.
2. C. Borgs, J. T. Chayes, H. Cohn, and N. Holden, in preparation, 2018.
3. C. Borgs, J. T. Chayes, H. Cohn, and N. Holden, *Sparse exchangeable graphs and their limits via graphon processes*, J. Mach. Learn. Res. **18** (2018), Paper No. 210, 71 pp.
4. C. Borgs, J. T. Chayes, H. Cohn, and V. Veitch, *Sampling perspectives on sparse exchangeable graphs*, arXiv:1708.03237.
5. C. Borgs, J. T. Chayes, H. Cohn, and Y. Zhao, An L^p theory of sparse graph convergence I: limits, sparse random graph models, and power law distributions, arXiv:1401.2906, to appear in Transactions of the American Mathematical Society.
6. C. Borgs, J. T. Chayes, S. Dhara, and S. Sen, *Limits of sparse configuration models and beyond: graphexes and multi-graphexes*, in preparation, 2018.
7. C. Borgs, J. T. Chayes, L. Lovász, V. Sós, and K. Vesztergombi, *Counting graph homomorphisms*, Topics in Discrete Mathematics (M. Klazar, J. Kratochvíl, M. Loebl, J. Matoušek, R. Thomas, and P. Valtr, eds.), Springer, 2006, pp. 315–371.
8. C. Borgs, J. T. Chayes, L. Lovász, V. Sós, and K. Vesztergombi, *Convergent graph sequences I: subgraph frequencies, metric properties, and testing*, Advances in Math. **219** (2008), 1801–1851.
9. C. Borgs, J. T. Chayes, L. Lovász, V. Sós, and K. Vesztergombi, *Convergent graph sequences II: multiway cuts and statistical physics*, Ann. of Math. **176** (2012), 151–219.
10. C. Borgs, J. Chayes, and L. Lovász, *Moments of two-variable functions and the uniqueness of graph limits*, Geom. Funct. Anal. **19** (2010), 1597–1619.
11. F. Caron and E. B. Fox, *Sparse graphs using exchangeable random measures*, J. R. Stat. Soc. Ser. B. Stat. Methodol. **79** (2017), 1295–1366.
12. D. L. Cohn, *Measure Theory*, second edition. Birkhäuser Advanced Texts: Basler Lehrbücher. Birkhäuser/Springer, New York, 2013.
13. D. J. Daley and D. Vere-Jones, *An Introduction to the Theory of Point Processes: Volume I: Elementary Theory and Methods*, second edition, Springer, 2003.
14. A. Frieze and R. Kannan, *Quick approximation to matrices and applications*, Combinatorica **19** (1999), 175–220.
15. W. Hoeffding, *Probability inequalities for sums of bounded random variables*, Journal of the American Statistical Association **58** (1963), 13–30.
16. S. Janson, *Graphons, cut norm and distance, couplings and rearrangements*, New York Journal of Mathematics. NYJM Monographs, vol. 4, State University of New York, University at Albany, Albany, NY, 2013.
17. S. Janson, *Graphons and cut metric on sigma-finite measure spaces*, arXiv:1608.01833, 2016.
18. S. Janson, *On convergence for graphexes*, arXiv:1702.06389, 2017.
19. O. Kallenberg, *Probabilistic Symmetries and Invariance Principles*, Springer, 2005.
20. D. Kunszenti-Kovács, L. Lovász, and B. Szegedy, *Multigraph limits, unbounded kernels, and Banach space decorated graphs*, arXiv:1406.7846, 2014.
21. L. Lovász and B. Szegedy, *Limits of dense graph sequences*, J. Combin. Theory Ser. B **96** (2006), 933–957.
22. L. Lovász and B. Szegedy, *Szemerédi's lemma for the analyst*, Geom. Funct. Anal. **17** (2007), 252–270.
23. L. Lovász and B. Szegedy, *Regularity partitions and the topology of graphons*, in *An Irregular Mind*, Bolyai Soc. Math. Stud., vol. 21, János Bolyai Math. Soc., Budapest, 2010, pp. 415–446.
24. W. Rudin, *Real and Complex Analysis*, third edition, McGraw-Hill Book Co., New York, 1987.
25. V. Veitch and D. M. Roy, *Sampling and estimation for (sparse) exchangeable graphs*, arXiv:1611.00843, 2016.
26. V. Veitch and D. M. Roy, *The class of random graphs arising from exchangeable random measures*, arXiv:1512.03099, 2015.

Online Ramsey Numbers and the Subgraph Query Problem

David Conlon, Jacob Fox, Andrey Grinshpun and Xiaoyu He

Abstract The (m, n)-online Ramsey game is a combinatorial game between two players, Builder and Painter. Starting from an infinite set of isolated vertices, Builder draws an edge on each turn and Painter immediately paints it red or blue. Builder's goal is to force Painter to create either a red K_m or a blue K_n using as few turns as possible. The online Ramsey number $\tilde{r}(m, n)$ is the minimum number of edges Builder needs to guarantee a win in the (m, n)-online Ramsey game. By analyzing the special case where Painter plays randomly, we obtain an exponential improvement $\tilde{r}(n, n) \geq 2^{(2-\sqrt{2})n + O(1)}$ for the lower bound on the diagonal online Ramsey number, as well as a corresponding improvement $\tilde{r}(m, n) \geq n^{(2-\sqrt{2})m + O(1)}$ for the off-diagonal case, where $m \geq 3$ is fixed and $n \to \infty$. Using a different randomized Painter strategy, we prove that $\tilde{r}(3, n) = \tilde{\Theta}(n^3)$, determining this function up to a polylogarithmic factor. We also improve the upper bound in the off-diagonal case for $m \geq 4$. In connection with the online Ramsey game with a random Painter, we study the problem of finding a copy of a target graph H in a sufficiently large unknown Erdős–Rényi random graph $G(N, p)$ using as few queries as possible, where each query reveals whether or not a particular pair of vertices are adjacent. We call this problem the Subgraph Query Problem. We determine the order of the number of queries needed for complete graphs up to five vertices and prove general bounds for this problem.

D. Conlon
Mathematical Institute, Oxford OX2 6GG, UK
e-mail: david.conlon@maths.ox.ac.uk

J. Fox · X. He (✉)
Stanford University, Stanford, CA 94305, USA
e-mail: alkjash@stanford.edu

J. Fox
e-mail: jacobfox@stanford.edu

A. Grinshpun
Teza Technologies, Austin, TX 78701, USA
e-mail: agrinshp@gmail.com

© János Bolyai Mathematical Society and Springer-Verlag GmbH Germany, part of Springer Nature 2019
I. Bárány et al. (eds.), *Building Bridges II*, Bolyai Society
Mathematical Studies 28, https://doi.org/10.1007/978-3-662-59204-5_4

Keywords Online Ramsey numbers · Ramsey numbers · Combinatorial games

Subject Classifications 05C57 · 05C55 · 91A60 · 91A05 · 05D10 · 05D40

1 Introduction

The *Ramsey number* $r(m, n)$ is the minimum integer N such that every red/blue-coloring of the edges of the complete graph K_N on N vertices contains either a red K_m or a blue K_n. Ramsey's theorem guarantees the existence of $r(m, n)$ and determining or estimating Ramsey numbers is a central problem in combinatorics. Classical results of Erdős–Szekeres and Erdős imply that $2^{n/2} \leq r(n, n) \leq 2^{2n}$ for $n \geq 2$. The only improvements to these bounds over the last seventy years have been to lower order terms (see [9, 26]), with the best known lower bound coming from an application of the Lovász local lemma [14].

Off-diagonal Ramsey numbers, where m is fixed and n tends to infinity, have also received considerable attention. In progress that has closely mirrored and often instigated advances on the probabilistic method, we now know that

$$r(3, n) = \Theta(n^2/\log n).$$

The lower bound here is due to Kim [21] and the upper bound to Ajtai, Komlós and Szemerédi [1]. Recently, Bohman and Keevash [8] and, independently, Fiz Pontiveros, Griffiths and Morris [18] improved the constant in Kim's lower bound via careful analysis of the triangle-free process, determining $r(3, n)$ up to a factor of $4 + o(1)$.

More generally, for $m \geq 4$ fixed and n growing, the best known lower bound is

$$r(m, n) = \Omega_m(n^{\frac{m+1}{2}}/(\log n)^{\frac{m+1}{2} - \frac{1}{m-2}}),$$

proved by Bohman and Keevash [7] using the H-free process, while the best upper bound in this setting is

$$r(m, n) = O_m(n^{m-1}/(\log n)^{m-2}),$$

again due to Ajtai, Komlós and Szemerédi [1]. Here the subscripts denote the variable(s) that the implicit constant is allowed to depend on.

There are many interesting variants of the classical Ramsey problem. One such variant is the *size Ramsey number* $\hat{r}(m, n)$, defined as the smallest N for which there exists a graph G with N edges such that every red/blue-coloring of the edges of G contains either a red K_m or a blue K_n. It was shown by Chvátal (see Theorem 1 in the foundational paper of Erdős, Faudree, Rousseau and Schelp [13]) that $\hat{r}(m, n)$ is just the number of edges in the complete graph on $r(m, n)$ vertices, that is,

$$\hat{r}(m,n) = \binom{r(m,n)}{2}.$$

We will be concerned with a much-studied game-theoretic variant of the size Ramsey number, introduced independently by Beck [4] and by Kurek and Ruciński [25]. The (m, n)-online Ramsey game is a game between two players, Builder and Painter, on an infinite set of initially isolated vertices. Each turn, Builder places an edge between two nonadjacent vertices and Painter immediately paints it either red or blue. The *online Ramsey number* $\tilde{r}(m, n)$ is then the smallest number of turns N that Builder needs to guarantee the existence of either a red K_m or a blue K_n.

It is a simple exercise to show that $\tilde{r}(m, n)$ is related to the usual Ramsey number $r(m, n)$ by

$$\frac{1}{2}r(m,n) \leq \tilde{r}(m,n) \leq \binom{r(m,n)}{2}. \qquad (1)$$

In the diagonal case, the upper bound in (1) has been improved by Conlon [10], who showed that for infinitely many n,

$$\tilde{r}(n,n) \leq 1.001^{-n} \binom{r(n,n)}{2}.$$

The main result of this paper is a new lower bound for online Ramsey numbers.

Theorem 1.1 *If, for some $m, n, N \geq 1$, there exist $p \in (0, 1)$, $c \leq \frac{1}{2}m$, and $d \leq \frac{1}{2}n$ for which*

$$p^{\binom{m}{2}-c(c-1)}(2N)^{m-c} + (1-p)^{\binom{n}{2}-d(d-1)}(2N)^{n-d} \leq \frac{1}{2},$$

then $\tilde{r}(m, n) > N$.

In particular, if $\tilde{r}(n) := \tilde{r}(n, n)$ is the diagonal online Ramsey number, Theorem 1.1 can be used to improve the classical bound $\tilde{r}(n) \geq 2^{n/2-1}$ by an exponential factor. Indeed, taking $p = \frac{1}{2}$ and $c = d \approx (1 - \frac{1}{\sqrt{2}})n$ in Theorem 1.1, we get the following immediate corollary.

Corollary 1.2 *For the diagonal online Ramsey numbers $\tilde{r}(n)$,*

$$\tilde{r}(n) \geq 2^{(2-\sqrt{2})n - O(1)}.$$

As for the off-diagonal case, when m is fixed and $n \to \infty$, Theorem 1.1 can be also used to substantially improve the best-known lower bound. In this case, we take $c \approx (1 - \frac{1}{\sqrt{2}})m$, $d = 0$, and $p = C\frac{m \log n}{n}$ for a sufficiently large $C > 0$ to obtain the following corollary.

Corollary 1.3 *For fixed $m \geq 3$ and n sufficiently large in terms of m,*

$$\tilde{r}(m,n) \geq n^{(2-\sqrt{2})m - O(1)}.$$

For general m, Corollary 1.3 gives the best known lower bounds for the off-diagonal online Ramsey number. However, it is possible to do better for $m = 3$ by using a smarter Painter strategy which deliberately avoids building red triangles.

Theorem 1.4 *For $n \to \infty$,*

$$\tilde{r}(3, n) = \Omega\left(\frac{n^3}{\log^2 n}\right).$$

Roughly speaking, Painter's strategy is to paint every edge blue initially, but to switch to painting randomly if both endpoints of a freshly built edge have high degree. Also, when presented with an edge that would complete a red triangle, Painter always paints it blue. The bound given in Theorem 1.4 is n times the bound on the usual Ramsey number that comes from applying the Lovász Local Lemma [14]. However, our argument is closer in spirit to an earlier proof of the same bound given by Erdős [12] using alterations. This method for lower bounding $r(3, n)$ was later generalized to all $r(m, n)$ by Krivelevich [22] and we suspect that Theorem 1.4 can be generalized to $\tilde{r}(m, n)$ in the same way.

In the other direction, we prove a new upper bound on the off-diagonal online Ramsey number.

Theorem 1.5 *For any fixed $m \geq 3$,*

$$\tilde{r}(m, n) = O_m\left(\frac{n^m}{(\log n)^{\lfloor m/2 \rfloor - 1}}\right).$$

In particular, note that Theorems 1.4 and 1.5 determine the asymptotic growth rate of $\tilde{r}(3, n)$ up to a polylogarithmic factor, namely,

$$\Omega\left(\frac{n^3}{\log^2 n}\right) \leq \tilde{r}(3, n) \leq O\left(n^3\right).$$

Theorem 1.5 has a similar flavor to the improvement on diagonal online Ramsey numbers made by the first author [10] and work on the so-called vertex online Ramsey numbers due to Conlon, Fox and Sudakov [11]. It is obtained by adapting the standard Erdős–Szekeres proof of Ramsey's theorem to the online setting and applying a classical result of Ajtai, Komlós and Szemerédi [1] bounding $r(m, n)$.

In order to prove Theorem 1.1, we specialize to the case where Painter plays randomly. This is sufficient because Builder, who we may assume has unlimited computational resources, will always respond in the best possible manner to Painter's moves. Therefore, if a random Painter can stop this perfect Builder from winning within a certain number of moves with positive probability, an explicit strategy exists by which Painter can delay the game up to this point. This motivates the following key definition.

Definition 1.6 For $m, n \geq 3$ and $p \in (0, 1)$, define $\tilde{r}(m, n; p)$ to be the number of turns Builder needs to win the (m, n)-online Ramsey game with probability at least $\frac{1}{2}$ against a Painter who independently paints each edge red with probability p and blue with probability $1 - p$. The *online random Ramsey number* $\tilde{r}_{\mathrm{rand}}(m, n)$ is the maximum value of $\tilde{r}(m, n; p)$ over $p \in (0, 1)$.

We note that there is a rich literature on simplifying the study of various combinatorial games by specializing to the case where one or both players play randomly (see [5, 20, 23]). For example, a variant of the online Ramsey game with random Builder instead of random Painter was studied by E. Friedgut, Y. Kohayakawa, V. Rödl, A. Ruciński and P. Tetali [19].

We make the following conjectures about the growth rate of $\tilde{r}_{\mathrm{rand}}(m, n)$.

Conjecture 1.7 *(a) The diagonal online random Ramsey numbers satisfy*

$$\tilde{r}_{\mathrm{rand}}(n, n) = 2^{(1+o(1))\frac{2}{3}n}.$$

(b) The off-diagonal online random Ramsey numbers ($m \geq 3$ fixed and $n \to \infty$) satisfy

$$\tilde{r}_{\mathrm{rand}}(m, n) = n^{(1+o(1))\frac{2}{3}m}.$$

These conjectures are motivated by a connection with another problem, which we now describe.

Let $p \in (0, 1)$ be a fixed probability and suppose Builder plays the following one-player game, which we call the *Subgraph Query Game*, on the random graph $G(\mathbb{Z}, p)$ with infinitely many vertices. The edges of the graph are initially hidden. At each step, Builder queries a single pair of vertices and is told whether the pair is an edge of the graph or not. Equivalently, the graph starts out empty and each edge is successfully built by Builder with probability p (each edge may be queried at most once). In what follows, we use the terms "query" and "build" interchangeably.

Builder's goal is to find a copy of a given graph H in the ambient random graph as quickly as possible. We call this problem of minimizing the number of steps in the Subgraph Query Game the *Subgraph Query Problem*. When $H = K_m$, this may be seen as a variant of the online random Ramsey game, but where Builder is only interested in finding a red copy of K_m.

A version of this problem was studied independently by Ferber, Krivelevich, Sudakov and Vieira [16, 17], although they were interested in querying for long paths and cycles in $G(n, p)$. For instance, they showed that if $p \geq \frac{\log n + \log \log n + \omega(1)}{n}$, then it is possible to find a Hamiltonian cycle with high probability in $G(n, p)$ after $(1 + o(1))n$ positive answers. In contrast, we are mainly interested in the setting where H is a fixed graph to be found in a much larger random graph.

Definition 1.8 If $p \in (0, 1)$, define $f(H, p)$ to be the minimum (over all Builder strategies) number of turns Builder needs to be able to build a copy of H with probability at least $1/2$ in the Subgraph Query Game, if each edge is built successfully with probability p.

It might appear equally reasonable to study the minimum number of turns in which one can build at least one copy of H *in expectation*. However, for certain H, such as a clique K_m together with many leaves off a single vertex, it is possible to describe a strategy which has a tiny probability of successfully constructing copies of H, but upon success immediately builds a large number of copies, attaining low success probability but high expectation. Such a strategy is undesirable for application to online random Ramsey numbers, so we use the first definition instead.

Conjecture 1.7 is motivated by the following conjecture regarding $f(K_m, p)$. The upper bound in this conjecture is proved in Sect. 5.2.

Conjecture 1.9 *For any $m \geq 4$,*

$$f(K_m, p) = 2^{o(m)} p^{-\frac{2}{3}m + c_m},$$

where

$$c_m = \begin{cases} \frac{m}{2m-3} & m \equiv 0 \pmod{3} \\ \frac{2}{3} & m \equiv 1 \pmod{3} \\ \frac{2m+8}{6m-3} & m \equiv 2 \pmod{3}. \end{cases}$$

The following result shows that the Subgraph Query Problem and the online random Ramsey game are closely related.

Theorem 1.10 *For any $m, n \geq 3$ and $p \in (0, 1)$,*

$$\tilde{r}(m, n; p) \leq \min\{f(K_m, p), f(K_n, 1-p)\} \leq 3\tilde{r}(m, n; p).$$

Using Theorem 1.10, we can show that Conjecture 1.9 implies both cases of Conjecture 1.7. We can also determine an approximately optimal value for the probability parameter p in the online Ramsey game with random Painter.

Theorem 1.11 *For $m \geq 3$ fixed and $n \to \infty$, there exists a $p = \Theta(m/n \log(n/m))$ for which*

$$\tilde{r}_{\text{rand}}(m, n) \leq 3\tilde{r}(m, n; p).$$

We say that a graph has a *k-matching* if it contains k vertex-disjoint edges. Our main result on the Subgraph Query Problem shows that graphs with large matchings are hard to build in few steps. We write $V(H)$ and $E(H)$ for the vertices and edges of H and let $v(H) = |V(H)|$ and $e(H) = |E(H)|$.

Theorem 1.12 *If H is a graph that contains a k-matching, then*

$$f(H, p) = \Omega_H(p^{-(e(H)-k(k-1))/(v(H)-k)}).$$

Together with the upper bound construction described in Sect. 5.2, this is enough to settle the growth rate of $f(K_m, p)$ for $m \leq 5$. In particular, it proves Conjecture 1.9 for $m = 4, 5$.

Theorem 1.13 *The asymptotic growth rates of $f(K_m, p)$ for $m = 3, 4, 5$ are*

$$f(K_3, p) = \Theta(p^{-3/2})$$
$$f(K_4, p) = \Theta(p^{-2})$$
$$f(K_5, p) = \Theta(p^{-8/3}).$$

Asymptotically, $k = (1 - 1/\sqrt{2})m$ is the optimal k to pick in Theorem 1.12. With this value, we get the following bound on $f(K_m, p)$ which corresponds to Corollaries 1.2 and 1.3 in the online Ramsey number setting.

Corollary 1.14 *For all $m \geq 3$,*

$$f(K_m, p) = \Omega_m(p^{-(2-\sqrt{2})m + O(1)}).$$

In studying the function $f(H, p)$, we were naturally led to consider the following function. When H is a graph with no isolated vertices, define $t(H, p, N)$ to be the maximum expected number of copies of H that can be built in N moves in the Subgraph Query Game with parameter p, the maximum taken over all possible Builder strategies.

However, if H has isolated vertices, the expected value is zero or infinite. Instead, if H has exactly k isolated vertices $v_1, \ldots, v_k$, we define

$$t(H, p, N) := (2N)^k t(H \setminus \{v_1, \ldots, v_k\}, p, N)$$

to capture the fact that the game with N turns involves at most $2N$ vertices and therefore might as well be played on $2N$ fixed vertices.

Studying the threshold value of N for which $t(H, p, N) \geq 1$ leads to Theorem 1.12 above. Intuitively, we expect the best strategy for building a copy of H to be the same as the one which expects to build a single copy of H in as few turns as possible.

Another natural question about the function $t(H, p, N)$ is: if N is very large, what is the maximum number of copies Builder can expect to build in the Subgraph Query Game? Here we show that for N sufficiently large the strategy of taking $O(\sqrt{2N})$ vertices and building all pairs of edges between them is asymptotically optimal for maximising $t(K_m, p, N)$, even though it is decidedly suboptimal for trying to build a single copy of K_m.

Theorem 1.15 *For all $m \geq 2$, $p \in (0, 1)$, $\varepsilon > 0$, there exists $C > 0$ such that if $N \geq Cp^{-(2m-1)}(\log(p^{-1}))^2$, then*

$$t(K_m, p, N) = (1 \pm \varepsilon) p^{\binom{m}{2}} (2N)^{\frac{m}{2}}.$$

The rest of the paper is organized as follows. In Sect. 2, we motivate and prove Theorem 1.1, our lower bound on the online Ramsey number, via the method of conditional expectations. In Sect. 3, we prove the lower bound Theorem 1.4 for $\tilde{r}(3, n)$ using a Painter strategy designed to avoid red triangles. We prove the upper bound Theorem 1.5 in Sect. 4. Then, in Sect. 5, we study the Subgraph Query Problem for its own sake, proving the upper bound in Conjecture 1.9 as well as Theorems 1.10, 1.11, 1.12, 1.13 and 1.15. We include a handful of open problems raised by our research in the closing remarks.

Unless otherwise indicated, all logarithms are base e. For clarity of presentation, we omit floor and ceiling signs when they are not crucial. We also do not attempt to optimize constant factors in the proofs.

2 General Lower Bounds

2.1 Motivation

In this section, we prove Theorem 1.1 via a weighting argument, motivated by the method of conditional expectations and a result of Alon [2] on the maximum number of copies of a given graph H in a graph with a fixed number of edges.

The first idea, the derandomization technique known as the method of conditional expectations (see Alon and Spencer [3]), can be used to give the following "deterministic" proof of the classical lower bound on diagonal Ramsey numbers. We will show that
$$\binom{r(n,n)}{n} 2^{-\binom{n}{2}+1} \geq 1.$$

Suppose that for some N,
$$\binom{N}{n} 2^{-\binom{n}{2}+1} < 1. \tag{2}$$

Paint the edges of K_N one at a time as follows. To each vertex subset U of order n, assign a weight $w(U)$ which is the probability that U becomes a monochromatic clique if the edges which remain uncolored at that time are colored uniformly at randomly. That is, writing $e(U)$ for the number of edges already colored in U,

$$w(U) = \begin{cases} 2^{-\binom{n}{2}+1} & e(U) = 0 \\ 2^{-\binom{n}{2}+e(U)} & e(U) > 0 \text{ and all colored edges in } U \text{ are the same color} \\ 0 & \text{otherwise.} \end{cases}$$

At every step, the total weight $\sum_U w(U)$ is equal to the expected number of monochromatic cliques if the remaining edges are painted uniformly at random. It is therefore possible to paint each edge so as not to increase the total weight. Since the condition $\sum_U w(U) < 1$ is initially guaranteed by (2), we can maintain this condition throughout the course of the game, ending with a coloring where there is no monochromatic clique of order n.

We now wish to apply such a weighting argument to the online Ramsey game. The key observation is that if $\tilde{r}(n, n)$ is close to $r(n, n)$, then, since the graph built by Builder has at least $r(n, n)$ vertices, it must be extremely sparse. In particular, most of the weight should be concentrated on sets U almost none of whose edges are ever built.

This is where the idea behind Alon's result [2] comes in. For any fixed graph H, that paper solves the problem of determining the maximum possible number of copies of H in a graph with a prescribed number of edges. Roughly speaking, Alon showed that the maximum number of copies of H can be controlled by the size of the maximum matching in H. We show that this heuristic also applies to the online Ramsey game, though it will be more convenient for our calculations to work with minimum vertex covers instead of maximum matchings.

To make this idea work, instead of controlling the total weight function $\sum_U w(U)$, we restrict the sum to subsets U with a large minimum vertex cover, which are comparatively few in number. Even if the total weight $\sum_U w(U)$ becomes large, the amount of weight supported on sets U with a large vertex cover is much smaller, and this is the only weight that stands a chance to make it to the finish line and complete a monochromatic clique.

2.2 The Proof

Using the weighting argument described informally above, we now prove a lower bound on the value of $\tilde{r}(m, n; p)$, where Painter plays randomly, independently coloring each edge red with probability p and blue with probability $1 - p$.

Theorem 2.1 *If, for some $m, n, N \geq 1$ and $p \in (0, 1)$, there exist $c \leq \frac{1}{2}m$ and $d \leq \frac{1}{2}n$ for which*

$$p^{\binom{m}{2}-c(c-1)}(2N)^{m-c} + (1-p)^{\binom{n}{2}-d(d-1)}(2N)^{n-d} \leq \frac{1}{2},$$

then $\tilde{r}(m, n; p) > N$.

We would like to show that regardless of Builder's strategy, the online random Ramsey game lasts for more than N steps with probability at least $1/2$.

Suppose the game ends in at most N turns and, without loss of generality, is played on $2N$ vertices. Let G_t, for $0 \leq t \leq N$, be the state of the graph after t turns. Assign to each subset $U \subset V(G)$ an evolving weight function

$$w(U, t) = \begin{cases} p^{\binom{|U|}{2}-e(G_t[U])} & G_t[U] \text{ is monochromatic red} \\ 0 & \text{otherwise.} \end{cases}$$

The value of $w(U, t)$ is the probability that U becomes a red clique if the remaining edges are built.

We say that $C \subset V(G)$ is a *vertex cover* of G if every edge is incident to some vertex $v \in C$. If $U \subset V(G)$, let $c(U, t)$ be the size of the minimum vertex cover of $G_t[U]$. Note that $c(U, t)$ is a nondecreasing function of t. For each pair (k, c) with $k \geq 2c$, we will be interested in the total weight supported on sets of order k with $c(U, t) \geq c$,

$$w_{k,c}(t) = \sum_{|U|=k, c(U,t) \geq c} w(U, t).$$

Since $w(U, N)$ is nonnegative and $w(U, N) = 1$ if and only if U is a red clique, we see that for all $c \leq m/2$, $w_{m,c}(N)$ is an upper bound for the number of red copies of K_m built after N turns. We would like to upper bound the expected value of $w_{m,c}(N)$.

Lemma 2.2 *With $w_{m,c}(t)$ as above, regardless of Builder's strategy,*

$$\mathbb{E} w_{m,c}(N) \leq p^{\binom{m}{2} - c(c-1)} (2N)^{m-c}.$$

Proof Each U with the property $c(U, N) \geq c$ first achieves this property at a time $t_c(U)$. We say that U is c-critical at this time. Write

$$w^*_{k,c}(t) = \sum_{|U|=k, t_c(U)=t} w(U, t)$$

to be the contribution of the c-critical sets U to $w_{k,c}(t)$. Crucially, if we focus on the family of U for which $t_c(U) = t$, their expected total weight will remain $w^*_{k,c}(t)$ indefinitely. Thus,

$$\mathbb{E} w_{k,c}(N) = \sum_{t \leq N} \mathbb{E} w^*_{k,c}(t).$$

Now, a set U which is c-critical at time t must be the vertex-disjoint union of the edge e_t that Builder builds at time t and a set U' of size $k - 2$ with a vertex cover of order $c - 1$. Also, because U has a vertex cover of order $c - 1$ before adding this edge e_t, the edges incident to e_t must also be incident to one of the $c - 1$ vertices in the vertex cover of U', so e_t is incident to a total of at most $2c - 2$ edges in U. It follows that after turn $t = t_c(U)$,

$$w(U, t) \leq p^{2k-2c-2} w(U', t),$$

where in particular if U' is already not monochromatic then neither is U. The exponent comes from the fact that among the total $2(k - 2)$ edges between e_t and U' at least $2(k - 2) - 2(c - 1) = 2k - 2c - 2$ are thus far unbuilt and still contribute factors of p to the weight of $w(U, t)$. Thus, since each U' completes at most one set U which is c-critical at time t,

$$w^*_{k,c}(t) \leq p^{2k-2c-2} w_{k-2,c-1}(t).$$

Further, note that there can only be c-critical sets at time t if e_t is colored red, which occurs with probability p. Otherwise, $w^*_{k,c}(t) = 0$. Taking expectations and using the fact that $\mathbb{E}w_{k,m}(t)$ is nondecreasing in t gives

$$\mathbb{E}w^*_{k,c}(t) \leq p \cdot \mathbb{E}[p^{2k-2c-2} w_{k-2,c-1}(t)]$$
$$\leq p^{2k-2c-1} \mathbb{E}w_{k-2,c-1}(N).$$

Summing over all t,

$$\mathbb{E}w_{k,c}(N) \leq N \cdot p^{2k-2c-1} \mathbb{E}w_{k-2,c-1}(N).$$

Iterating this last inequality, we conclude that

$$\mathbb{E}w_{m,c}(N) \leq N^c \cdot p^{2mc-3c^2} \mathbb{E}w_{m-2c,0}(N)$$
$$\leq N^c \cdot p^{2mc-3c^2} \cdot (2N)^{m-2c} p^{\binom{m-2c}{2}}$$
$$\leq p^{\binom{m}{2} - c(c-1)} (2N)^{m-c},$$

as desired. $\square$

The same analysis with the blue weight function

$$w'(U, t) = \begin{cases} (1-p)^{\binom{|U|}{2} - e(G_t[U])} & G_t[U] \text{ is monochromatic blue} \\ 0 & \text{otherwise} \end{cases}$$

leads to the conclusion that $\mathbb{E}w'_{n,d}(N) \leq (1-p)^{\binom{n}{2} - d(d-1)} (2N)^{n-d}$ for all $n \geq 2d$. The assumption of Theorem 2.1 then implies that the expected number of red K_m plus the expected number of blue K_n is at most $1/2$. This implies that the probability of containing either is at most $1/2$, completing the proof of Theorem 2.1. Theorem 1.1 follows as an immediate corollary.

3 Lower Bound via Alterations

In this section, we improve the lower bound for the off-diagonal online Ramsey numbers $\tilde{r}(3, n)$ using a different Painter strategy. Our proof extends an alteration argument of Erdős [12] which shows that

$$r(3, n) \geq \frac{cn^2}{\log^2 n},$$

for some constant $c > 0$. The main idea of Erdős' proof was to show that in a random graph $G(r, p)$ with $p \approx r^{-1/2}$, only a small fraction of the edges need to be removed to destroy all triangles. Moreover, with high probability, removing these edges doesn't significantly affect the graph's independence number.

Our proof involves a randomized strategy which pays particular attention to avoiding red triangles. Instead of painting entirely randomly, Painter's strategy is modified in two ways to avoid creating red triangles. First, if an edge is built incident to a vertex of degree less than $(n-1)/4$, Painter always paints it blue. Second, if painting an edge red would create a red triangle, Painter again always paints it blue. In all other cases, Painter paints edges red with probability p and blue with probability $1 - p$.

In order to show that this Painter strategy works, we first prove a structural result about Erdős–Rényi random graphs. Roughly speaking, this lemma implies that if an edge is removed from each triangle in $G(r, p)$, the remaining graph still has small independence number.

Lemma 3.1 *Suppose n is sufficiently large, $p = 20 \log n / n$, $r = 10^{-6} n^2 / (\log n)^2$ and $G \sim G(r, p)$ is an Erdős–Rényi random graph. Then, with high probability, there does not exist a set $S \subset V(G)$ of order n such that more than $\frac{n^2}{10}$ pairs of vertices in S have a common neighbor outside S.*

Proof Let E_1 be the event that the maximum degree of G is at most $2rp$. For a given vertex subset S of order n, let $E_1(S)$ be the event that every vertex outside S has at most $2rp$ neighbors in S. Thus, E_1 implies $E_1(S)$ for all S.

For a set S of size n, let $E_2(S)$ be the event that at most $\frac{n^2}{10}$ pairs of vertices in S have a common neighbor outside S and let E_2 be the event that $E_2(S)$ holds for all S. We will show $E_1 \wedge E_2$ occurs w.h.p. which in turn implies that E_2 itself occurs w.h.p.

The distribution of $\deg(v)$ for a single vertex $v \in G$ is the binomial distribution $B(r-1, p)$. Using the Chernoff bound (see, e.g., Appendix A in [3]), we find that

$$\Pr[\deg(v) > 2rp] < \left(\frac{e}{4}\right)^{rp} < \exp\left(-\frac{n}{5 \cdot 10^5 \log n}\right).$$

Taking the union over all vertices of G, it follows that

$$\Pr[\overline{E_1}] < r \exp\left(-\frac{n}{5 \cdot 10^5 \log n}\right),$$

so E_1 occurs w.h.p.

Fix a set S of n vertices. For $v \in V(G) \setminus S$, define $\deg_S(v)$ to be the number of neighbors of v in S. Since E_1 implies $E_1(S)$, we have

$$\Pr[E_1 \wedge \overline{E_2(S)}] \leq \Pr[E_1(S) \wedge \overline{E_2(S)}].$$

We will show that this last probability is so small that we may union bound over all S.

For $E_1(S)$ to occur, the possible values of $\deg_S(v)$ range through $[0, 2rp]$. We will cut off the bottom of this range and divide the rest into dyadic intervals. Let $D_0 = -1$, $D_1 = 4enp$, $D_2 = 8enp$, $D_3 = 16enp, \ldots, D_k = 2rp$ so that $D_i = 2D_{i-1}$ for each $2 \leq i \leq k-1$ and $D_k \leq 2D_{k-1}$. The number of intervals k satisfies $k \leq \log_2(r/n) \leq 2 \log n$.

Define d_i to be the number of $v \in V(G) \setminus S$ satisfying $D_{i-1} < \deg_S(v) \leq D_i$. For $E_2(S)$ to occur, it must be the case that

$$\sum_{v \notin S} \binom{\deg_S(v)}{2} \geq \frac{n^2}{10},$$

as the left hand side counts each pair in S with a common neighbor outside S at least once. In particular,

$$\sum_{i=1}^{k} d_i \binom{D_i}{2} \geq \frac{n^2}{10}. \qquad (3)$$

Notice that since $D_1 = 4enp = 80e \log n$ and $d_1 \leq r$,

$$d_1 \binom{D_1}{2} \leq r \cdot D_1^2 = \frac{64e^2}{10^4} n^2 < \frac{n^2}{20},$$

so at least half the contribution of (3) must come from $i \geq 2$. Thus,

$$\sum_{i=2}^{k} d_i \binom{D_i}{2} \geq \frac{n^2}{20}. \qquad (4)$$

We would like to bound the probability that $E_1(S)$ and (4) occur simultaneously. Let T be the family of all sequences $(d_i)_{i=1}^{k}$ which sum to $r - n$ and satisfy (4). Given the choice of $(d_i)_{i=1}^{k}$, the number of ways to assign vertices to dyadic intervals $(D_{i-1}, D_i]$ is at most $\binom{r-n}{d_1, d_2, \ldots, d_k}$.

If $i \geq 2$ and a vertex v is assigned to $(D_{i-1}, D_i]$, the probability that $\deg_S(v)$ lies in that interval is at most

$$\Pr[\deg_S(v) > D_{i-1}] \leq \binom{n}{D_{i-1}} p^{D_{i-1}} \leq \left(\frac{enp}{D_{i-1}}\right)^{D_{i-1}}.$$

If $i = 1$, then we simply use the trivial bound $\Pr[\deg_S(v) \in (D_0, D_1]] \leq 1$. Thus,

$$\Pr[E_1(S) \wedge \overline{E_2(S)}] \leq \sum_{(d_i) \in T} \binom{r-n}{d_1, d_2, \ldots, d_k} \prod_{i=1}^{k} \Pr[\deg_S(v) \in (D_{i-1}, D_i]]$$

$$\leq \sum_{(d_i) \in T} \binom{r-n}{d_1, d_2, \ldots, d_k} \prod_{i=2}^{k} \left(\left(\frac{enp}{D_{i-1}} \right)^{D_{i-1}} \right)^{d_i}$$

$$\leq \sum_{(d_i) \in T} \prod_{i=2}^{k} \left(r \cdot \left(\frac{enp}{D_{i-1}} \right)^{D_{i-1}} \right)^{d_i},$$

where we used $\binom{r-n}{d_1, d_2, \ldots, d_k} < r^{d_2 + \cdots + d_k}$. Next, the number of compositions of $r - n$ into k parts is at most r^k, so $|T| \leq r^k$ and we have

$$\Pr[E_1(S) \wedge \overline{E_2(S)}] \leq r^k \max_{(d_i) \in T} \prod_{i=2}^{k} \left(r \cdot \left(\frac{enp}{D_{i-1}} \right)^{D_{i-1}} \right)^{d_i}$$

$$\leq r^k \max_{(d_i) \in T} \exp\left(\sum_{i=2}^{k} d_i \log A_i \right), \qquad (5)$$

where

$$A_i = r \cdot \left(\frac{enp}{D_{i-1}} \right)^{D_{i-1}}.$$

It remains to maximize the exponent in (5) subject to (4). Consider the function

$$f(D) = \frac{1}{D^2} \log\left(r \cdot \left(\frac{enp}{D} \right)^D \right) = \frac{\log r}{D^2} + \frac{\log(enp)}{D} - \frac{\log D}{D}.$$

Notice that $D_1 = 4enp = 80e \log n$ so that for $D \geq D_1$,

$$r \cdot \left(\frac{enp}{D} \right)^D \leq r \cdot \left(\frac{enp}{D_1} \right)^{D_1} \leq r \cdot 2^{-80e \log n} < 1.$$

Thus, $f(D)$ takes negative values on $[D_1, D_k]$. Its derivative is

$$f'(D) = -\frac{2 \log r}{D^3} - \frac{\log enp}{D^2} + \frac{\log D}{D^2} - \frac{1}{D^2} = \frac{D(\log D - \log(e^2 np)) - 2 \log r}{D^3}.$$

Since $r \leq n^2$, we find that whenever $D \geq D_1 = 4enp = 80e \log n$,

$$f'(D) \geq \frac{D \log(4/e) - 2 \log r}{D^3} \geq \frac{80e \log(4/e) \cdot \log n - 4 \log n}{D^3} > 0,$$

and so $f(D)$ is monotonically increasing on $[D_1, D_k]$ and attains its maximum value at $D_k = 2rp$. With $2rp = 4 \cdot 10^{-5} n / \log n$ and n sufficiently large, observe that

$$\left(\frac{enp}{2rp}\right)^{2rp} = \left(\frac{10^6 e (\log n)^2}{2n}\right)^{4 \cdot 10^{-5} n / \log n} \leq \exp(-2 \cdot 10^{-5} n),$$

so that this maximum value is

$$f(2rp) \leq \frac{10^{10} (\log n)^2}{16 n^2} \cdot \log(n^2 \cdot \exp(-2 \cdot 10^{-5} n)) \leq -\frac{10^5 (\log n)^2}{16 n}.$$

In particular, because $\binom{D}{2} \geq D^2/3$ for $D \geq 3$ and $f(D)$ is always negative,

$$\sum_{i=2}^{k} d_i \log A_i = \sum_{i=2}^{k} d_i \binom{D_i}{2} \cdot \frac{\log A_i}{\binom{D_i}{2}}$$

$$\leq 3 \sum_{i=2}^{k} d_i \binom{D_i}{2} \cdot f(D_i)$$

$$\leq 3 f(D_k) \sum_{i=2}^{k} d_i \binom{D_i}{2}$$

$$\leq 3 f(2rp) \cdot \frac{n^2}{20}$$

$$\leq -n (\log n)^2$$

for any $(d_i) \in T$.

Returning to (5), it follows that

$$\Pr[E_1(S) \wedge \overline{E_2(S)}] \leq r^k \max_{(d_i) \in T} \exp\left(\sum_{i=2}^{k} d_i \log A_i\right) \leq r^k \exp(-n (\log n)^2).$$

There are at most $\binom{r}{n} \leq e^{2n \log n}$ subsets S of size n to consider and $r^k \leq r^n \leq e^{2n \log n}$ as well, so

$$\Pr[\overline{E_1 \wedge E_2}] = \Pr[\overline{E_1} \vee \bigvee_S \overline{E_2(S)}]$$

$$\leq \Pr[\overline{E_1}] + \sum_S \Pr[E_1 \wedge \overline{E_2(S)}]$$

$$\leq \Pr[\overline{E_1}] + \sum_S \Pr[E_1(S) \wedge \overline{E_2(S)}]$$

$$\leq \Pr[\overline{E_1}] + \exp(4n \log n) \cdot \exp\left(-n (\log n)^2\right).$$

Both summands on the right vanish rapidly, so E_2 holds w.h.p., as desired. □

With this lemma in hand, we are now ready to prove Theorem 1.4.

Proof of Theorem 3.2 Let $p = 20\log n/n$, $r = 10^{-6}n^2/(\log n)^2$ and $N = \frac{(n-1)r}{8}$.

We will give a randomized strategy for Painter such that, regardless of Builder's strategy, after N edges are colored there is neither a red K_3 nor a blue K_n w.h.p. Thus, there exists a strategy for Painter which makes the game last more than N steps and the desired bound $\tilde{r}(3, n) > N$ follows. Note that proving the result with positive probability suffices, but our argument shows it w.h.p. for no additional cost.

We now describe Painter's strategy. Initially, all vertices are considered inactive; a vertex is activated when its degree reaches at least $(n-1)/4$. The active vertices are labeled with the natural numbers in $[r]$ when they reach degree at least $(n-1)/4$, using an arbitrary underlying order on the vertices to break ties. Since $N = (n-1)r/8$, there will never be more than r active vertices.

When Builder builds an edge (u, v), this edge is considered inactive if either u or v is inactive immediately after (u, v) is built and active otherwise. The status of an edge remains fixed once it is built, so that inactive edges remain inactive even if both of its incident vertices are active at a later turn. Painter automatically colors inactive edges blue.

If Builder builds an active edge (u, v), Painter first checks if u and v have a common neighbor w such that (u, w) and (v, w) are both red. For brevity's sake, we call such a vertex w a *red common neighbor* of u and v. If so, Painter paints (u, v) blue so as to not build a red triangle and we call such an edge *altered*. Otherwise, Painter paints it red with probability p and blue with probability $1 - p$. Following this strategy, Painter guarantees that no red triangles are built. It suffices to show that w.h.p. no blue K_n is built either.

Here is an equivalent formulation of Painter's strategy. At the start of the game, Painter samples an Erdős–Rényi graph $G = G([r], p)$ on the labels which he keeps hidden from Builder. Inactive edges are painted blue. When an active edge between vertices labelled i and j is built, it is painted red if and only if $i \sim j$ in G and these two vertices currently have no red common neighbor.

Now, we apply Lemma 3.1 to the graph G. Letting $E_2(S)$ be the event that an n-set S has at most $n^2/10$ pairs with outside common neighbors and $E_2 = \bigwedge_S E_2(S)$, we see that $\Pr[\overline{E_2}] \to 0$ as $n \to \infty$.

For a set $S \subseteq [r]$ of labels, write $T(S)$ for the set of active vertices with labels in S. We seek to bound the probability of the event $B(T(S))$ that $T(S)$ is a blue n-clique at the end of the game. Because any blue n-clique would have all of its vertices active (as each vertex of the n-clique would have degree at least $n - 1 \geq (n-1)/4$), if none of the events $B(T(S))$ occurs, then no blue K_n is ever built. Once we show that the probability of a single $B(T(S))$ is sufficiently small, we will apply the union bound over all S to show that w.h.p. no blue K_n is built.

First, note that if any edge (u, v) in $T(S)$ is altered (and hence blue), we may assume that their red common neighbors lie outside $T(S)$. Otherwise, there must be two red edges inside $T(S)$ already and $T(S)$ can never become a blue n-clique.

With this in mind, conditioning on the event $E_2(S)$, at most $n^2/10$ altered blue edges are built in $T(S)$. Within $T(S)$ there can be at most $n^2/4$ inactive edges. Assuming $B(T(S))$ occurs, there are at least

$$\binom{n}{2} - \frac{n^2}{4} - \frac{n^2}{10} \geq \frac{n^2}{8}$$

edges between vertices of $T(S)$ that are both active and unaltered. For $B(T(S))$ to occur, each of these active and unaltered edges would have to be colored blue on its turn. On the other hand, each of these edges has a chance p of being colored red on that turn.

Thus, we find that

$$\Pr[B(T(S))|E_2(S)] \leq (1-p)^{\frac{n^2}{8}},$$

with one factor of $1 - p$ for each unaltered active edge built in $T(S)$. Thus,

$$\Pr[\bigvee_S B(T(S))] \leq \Pr[E_2 \wedge \bigvee_S B(T(S))] + \Pr[\overline{E_2}].$$

The second summand goes to zero, so it suffices to show the first does as well. We have

$$\Pr[E_2 \wedge \bigvee_S B(T(S))] \leq \sum_S \Pr[E_2 \wedge B(T(S))]$$

$$\leq \sum_S \Pr[E_2(S) \wedge B(T(S))]$$

$$\leq \sum_S \Pr[B(T(S))|E_2(S)]$$

$$\leq \binom{r}{n}(1-p)^{\frac{n^2}{8}}.$$

Using $1 - p \leq e^{-p}$, the right-hand side is at most

$$r^n e^{-pn^2/8} \leq e^{n \log r - pn^2/8} = e^{-(\frac{1}{2}+o(1))n \log n},$$

also tending to zero as $n \to \infty$. Thus, the probability that either $\overline{E_2}$ or some $B(T(S))$ occurs tends to zero. Therefore, with high probability no blue K_n is built.

4 Off-Diagonal Upper Bounds

In Sect. 2, we proved lower bounds of the form $\tilde{r}(m, n) \geq \Omega(n^{(2-\sqrt{2})m+o(m)})$ on the off-diagonal online Ramsey numbers through an analysis of the online random Ramsey number. It is easy to give an upper bound of the form $\tilde{r}(m, n) \leq O(n^{2m-2})$ simply by applying the Erdős–Szekeres bound for classical Ramsey numbers and the trivial observation that $\tilde{r}(m, n) \leq \binom{r(m,n)}{2}$.

However, the simple inductive proof of the Erdős–Szekeres bound suggests a Builder strategy that does considerably better. Namely, build many edges from one vertex until it has a large number of edges of one color, then proceed inductively in that neighborhood. This strategy is particularly well suited to the online Ramsey game because the number of edges built is only slightly more than linear in the number of vertices used, allowing us to derive a bound of the form $\tilde{r}(m, n) \leq O(n^m)$.

A slight variation on this argument allows us to bound the online Ramsey number in terms of the bounds for classical Ramsey numbers.

Lemma 4.1 *Let $m \leq n$ be positive integers with m fixed. Let $m_0 = \lfloor m/2 \rfloor + 1$ and $n_0 = \lfloor \sqrt{n} \rfloor$. Suppose L is a positive real such that for all $m_0 \leq m' \leq m$ and $n_0 \leq n' \leq n$,*

$$r(m_0, n') \leq \frac{1}{L}\binom{m_0 + n' - 2}{m_0 - 1},$$
$$r(m', n_0) \leq \frac{1}{L}\binom{m' + n_0 - 2}{m' - 1}.$$

Then

$$\tilde{r}(m, n) \leq \frac{C_m n}{L}\binom{m + n - 2}{m - 1}$$

for a constant C_m depending only on m.

Proof We describe a general Builder strategy for the online Ramsey game with parameters m and n and some savings parameter L. Let $f(m, n) = \frac{1}{L}\binom{m+n-2}{m-1}$, so we have $f(m - 1, n) + f(m, n - 1) = f(m, n)$ by Pascal's identity.

Begin by building $f(m, n) - 1$ edges out of a given initial vertex v_1. If $f(m - 1, n)$ of these edges are colored red, we proceed to the red neighborhood of v_1; otherwise, we proceed to the at least $f(m, n - 1)$ vertices in the blue neighborhood of v_1. If at some step we reach a neighborhood with $f(m - i, n - j)$ vertices, we build $f(m - i, n - j) - 1$ edges inside this neighborhood from one of the vertices, which we label v_{i+j+1}. If $f(m - i - 1, n - j)$ of these edges are colored red, we proceed to the red neighborhood of v_{i+j+1}; otherwise, we proceed to the at least $f(m - i, n - j - 1)$ vertices in the blue neighborhood of v_{i+j+1}. We stop once m reaches m_0 or n reaches n_0, ending up with either $f(m_0, n')$ vertices for some $n_0 \leq n' \leq n$ or $f(m', n_0)$ vertices for some $m_0 \leq m' \leq m$. Once we reach this stage, we build all edges in the remaining set.

Suppose now that we arrive at a set S of order $f(m_0, n')$. By construction, there are $\ell = m + n - m_0 - n'$ vertices $v_1, \ldots, v_\ell$ such that $m - m_0$ of the vertices v_i are joined in red to every v_j with $j > i$ and every $w \in S$. The remaining $n - n'$ vertices v_i are joined in blue to every v_j with $j > i$ and every $w \in S$. But since

$$r(m_0, n') \le \frac{1}{L}\binom{m_0 + n' - 2}{m_0 - 1} = f(m_0, n'),$$

the complete graph on S contains either a red K_{m_0} or a blue $K_{n'}$, either of which can be completed to a red K_m or a blue K_n by using the appropriate subset of $v_1, \ldots, v_\ell$. If we had instead arrived at a set of order $f(m', n_0)$, a similar analysis would have applied.

Note that the total number of edges built in the branching phase is at most $(m + n) f(m, n)$, while the number built by filling in the final clique is at most $\max(f(m_0, n)^2, f(m, n_0)^2)$. Using the choice of m_0 and n_0, the total number of edges built is easily seen to be at most a constant in m times the previous expression. □

From here we derive Theorem 1.5.

Proof of Theorem 4.2 We apply the bound

$$r(m, n) = O_m(n^{m-1}/\log^{m-2} n),$$

due to Ajtai, Komlós and Szemerédi [1]. In particular, suppose $m_0 = \lfloor m/2 \rfloor + 1$, $n_0 = \lfloor \sqrt{n} \rfloor$ and m', n' satisfy $m_0 \le m' \le m$ and $n_0 \le n' \le n$. Then, for some constants $C, C' > 0$ depending only on m, we have

$$r(m_0, n') \le \frac{C}{\log^{m_0 - 2} n'} (n')^{m_0 - 1} \le \frac{C'}{\log^{\lfloor m/2 \rfloor - 1} n} \binom{m_0 + n' - 2}{m_0 - 1}$$

and

$$r(m', n_0) \le \frac{C}{\log^{m' - 2} n_0} n_0^{m' - 1} \le \frac{C'}{\log^{\lfloor m/2 \rfloor - 1} n} \binom{m' + n_0 - 2}{m' - 1},$$

verifying the conditions of Lemma 4.1 with $L = \Omega_m(\log^{\lfloor m/2 \rfloor - 1} n)$. It follows by that lemma that there exists another constant $C'' > 0$ depending only on m for which

$$\tilde{r}(m, n) \le \frac{C'' n}{\log^{\lfloor m/2 \rfloor - 1} n} \binom{m + n - 2}{m - 1}.$$

Fixing $m \ge 3$ and taking $n \to \infty$, this implies

$$\tilde{r}(m, n) = O_m\left(\frac{n^m}{(\log n)^{\lfloor m/2 \rfloor - 1}}\right),$$

as desired. □

We remark that while the statement and proof of Lemma 4.1 are designed for the case where m is a constant, they can be easily modified to make them meaningful for all m and n.

5 The Subgraph Query Problem

The vertex cover argument in Sect. 2 was motivated by our study of the closely-related Subgraph Query Problem. Indeed, one can view this problem as an instance of the online Ramsey game with a random Painter where Builder single-mindedly tries to build a clique in one color, ignoring the other color entirely.

Let $p \in (0, 1)$ be the probability that Builder successfully builds any given edge in the Subgraph Query Problem. We are primarily interested in the quantity $f(H, p)$, which we defined as the minimum N for which there exists a Builder strategy which builds a copy of H with probability at least $\frac{1}{2}$ in N turns. Of secondary interest is the quantity $t(H, p, N)$, which we define as the maximum, over all Builder strategies, of the expected number of copies of H that can be built in N turns. It is easy to see that

$$t(H, p, N) < \frac{1}{2} \implies f(H, p) > N.$$

Thus, upper bounds on $t(H, p, N)$ yield lower bounds on $f(H, p)$.

5.1 Connection with Online Ramsey Numbers

We first check that the Subgraph Query Problem gets easier when edges are built with higher probability.

Lemma 5.1 *For any $m \geq 3$, $f(K_m, p)$ is a nonincreasing function of $p \in (0, 1)$.*

Proof Suppose $p < q$ and $f(K_m, p) = N$. This means that in the Subgraph Query Problem with parameter p, Builder has an N-move strategy S to win with probability at least half. Strategy S is defined by Builder's choice of edge to build at each step, given the data of which edges were successfully built in previous steps.

Builder's strategy for the Subgraph Query Problem with parameter q is as follows. For each edge that Builder successfully builds, Builder then flips a biased coin that comes up heads $\frac{p}{q}$ of the time. If the coin comes up tails, Builder pretends the edge actually failed to build, and acts according to strategy S with respect to only the edges for which the coin came up heads. Just looking at the edges which come up heads, Builder is exactly following strategy S, and so builds a K_m with probability at least $1/2$ in N steps. □

We now prove Theorem 1.10, which connects the Subgraph Query Problem to the online Ramsey game. Recall the statement:

$$\tilde{r}(m,n;p) \leq \min\{f(K_m, p), f(K_n, 1-p)\} \leq 3\tilde{r}(m,n;p).$$

Proof of Theorem 5.2 We first show the left side of the inequality. Let $N = \min\{f(K_m, p), f(K_n, 1-p)\}$ and suppose that $f(K_m, p)$ is the smaller of the two. Then there exists an N-move Builder strategy which builds a K_m with probability at least half. Now, let Builder play the online Ramsey game against a random Painter with the same probability parameter p. Builder's strategy will be to treat red edges as successfully built and blue edges as failed. In this way, Builder wins the online Ramsey game in N moves with probability at least half, by constructing a red K_m. Similarly, if $f(K_n, 1-p)$ were smaller, Builder would instead treat blue edges as successfully built and red edges as failed. This would then guarantee the construction of a blue K_n with probability at least half.

Now we show the right side of the inequality. Suppose $N = \tilde{r}(m, n; p)$, so in the online Ramsey game against random Painter with parameter p, there exists an N-move Builder strategy which builds a red K_m or blue K_n with probability at least half. In particular, this same strategy guarantees either a red K_m with probability at least $\frac{1}{4}$ or a blue K_n with probability at least $\frac{1}{4}$.

Suppose the first is true. Then Builder plays the Subgraph Query Game using this same strategy, treating red edges as successfully built and blue as failed. In N moves, he has at least a $\frac{1}{4}$ chance of successfully building a K_m. Repeating this strategy three independent times on three different vertex sets, Builder uses $3N$ moves to build a K_m with probability at least

$$1 - \left(1 - \frac{1}{4}\right)^3 = \frac{37}{64} > \frac{1}{2},$$

showing that $f(K_m, p) \leq 3\tilde{r}(m, n; p)$ in this case. Similarly, if the second case occurs, $f(K_n, 1-p) \leq 3\tilde{r}(m, n; p)$. Either way, the smaller of $f(K_m, p)$ and $f(K_n, 1-p)$ is bounded above by $3\tilde{r}(m, n; p)$.

Now we show that Conjecture 1.9 about the Subgraph Query Problem directly implies Conjecture 1.7 about online random Ramsey numbers.

Proof that Conjecture 1.9 *implies Conjecture* 1.7. Assume Conjecture 1.9, i.e., $f(K_m, p) = 2^{o(m)} p^{-\frac{2}{3}m + c_m}$ for all $m \geq 3$, $p \in (0, 1)$. By Theorem 1.10, we have

$$\tilde{r}(m, n; p) = \Theta(\min\{f(K_m, p), f(K_n, 1-p)\}). \tag{6}$$

In the diagonal case of the online Ramsey game, (6) together with Lemma 5.1 implies that $p = \frac{1}{2}$ gives the online random Ramsey number to within a constant factor. Thus,

$$\tilde{r}_{\text{rand}}(n, n) = 2^{\frac{2}{3}n + o(n)}.$$

This proves part (a).

In the off-diagonal case, a value of p nearly optimizing the right hand side of (6) satisfies $p = \Theta(\frac{m}{n} \log \frac{n}{m})$ by Theorem 1.11, which is proved in Sect. 5.3. Plugging in this value of p, we get

$$\tilde{r}_{\text{rand}}(m, n) = 2^{o(m)} \left(\Theta \left(\frac{m}{n} \log \frac{n}{m} \right) \right)^{-\frac{2}{3}m + c_m},$$

which implies case (b) of Conjecture 1.7. □

5.2 The Branch and Fill Strategy

We now prove the upper bound in Conjecture 1.9.

We will say it is possible to build a graph H in $O(T)$ turns, where $T = T(p)$ is a function of p, if for any $p \in (0, 1)$ it is possible, in the Subgraph Query Game played with probability p, to build a copy of H in $O(T)$ time with probability at least $\frac{1}{2}$. It is a simple fact about randomized algorithms that if one can achieve any constant success probability in $O(T)$ time then one can iterate the algorithm to succeed with probability $1 - \varepsilon$ in $O(T \log \varepsilon^{-1})$ time.

We describe a Builder strategy to prove the upper bound in Conjecture 1.9 and conjecture that this is essentially the optimal strategy for the Subgraph Query Problem for cliques.

Lemma 5.3 *Let $a \geq 1$, $b \geq 2$ and $n = a + b + 1$ satisfy $2a + 3 - b \geq 0$. Then*

$$f(K_n, p) = O_n(p^{-\frac{2a+b+1}{2} + \frac{\alpha}{b}}),$$

where $\alpha = \min(1, \frac{b(2a+3-b)}{2(b-1)})$.

Proof To build a clique K_n in $O(T)$ turns, where $T = p^{-\frac{2a+b+1}{2} + \frac{\alpha}{b}}$, we follow a strategy with three phases:

1. Build a clique U on a vertices. By induction, the number of turns needed will be negligible.
2. Find $p^a T$ common neighbors of U in $O_n(T)$ time with high probability. This is done by repeatedly picking a new vertex v and trying to build each of the edges between v and the vertices in U until one fails. Let W be the set of common neighbors found in this way.
3. Among the vertices of W, pick a vertex w_1 and try to build all edges incident to w_1 within W. Let $W_1 = N(w_1) \cap W$ be the neighborhood determined. Try to build all $\binom{|W_1|}{2}$ edges within W_1. Remove $\{w_1\} \cup W_1$ from W and repeat a total of $p^{-\alpha}$ times, picking $w_2, \ldots, w_{p^{-\alpha}}$, finding their neighborhoods, and filling them in. Here, $\alpha \in [0, 1]$ is a parameter which we have not yet specified.

After the process is complete, if any one of the W_i contains a b-clique W_i', then we are done, since $U \cup \{w_i\} \cup W_i'$ forms an n-clique.

It remains to determine the success probability and the number of steps taken in the above process. By the standard Chernoff bounds, the sizes of all the sets W_i concentrate around their means with high probability. Hence, with high probability,

$$|W_i| = (1 + o(1)) p^{a+1} (1-p)^{i-1} T.$$

A standard application of Janson's inequality (see Chaps. 8 and 10 of [3]) then implies

$$\Pr[W_i \text{ contains a } b\text{-clique}] = \Omega_b \left(\min(p^{\binom{b}{2}} |W_i|^b, 1) \right)$$
$$= \Omega_b \left(\min(p^{(a+1)b + \binom{b}{2}} (1-p)^{(i-1)b} T^b, 1) \right).$$

If i ranges up to $p^{-\alpha}$ and $\alpha \leq 1$, then the decay factor $(1-p)^{(i-1)b}$ is $\Theta_b(1)$ and can be safely ignored. Since the event that each W_i contains a b-clique is independent of all the others, we need only pick p, T, α for which the expression $p^{-\alpha} p^{(a+1)b + \binom{b}{2}} T^b$ is a positive constant. If this is the case, then with at least constant probability our strategy constructs an n-clique.

We also need to know that the total number of turns taken is $O_n(T)$. This is true in Phases 1 and 2 by design. With high probability, the number of turns taken in filling out each W_i is $O_a(p^a T + p^{2(a+1)} T^2)$. Since this is repeated $p^{-\alpha}$ times, it suffices to have

$$T = O_a(p^{\alpha - 2(a+1)})$$

for the number of turns to be $O(T)$. It remains to optimize the value of T subject to the constraints $T = O_a(p^{\alpha - 2(a+1)})$ and $p^{-\alpha} p^{(a+1)b + \binom{b}{2}} T^b = \Omega_b(1)$. As long as $2a + 3 - b \geq 0$, this system has solutions. Solving for α which minimizes T, we find that any

$$\alpha \leq \frac{b(2a+3-b)}{2(b-1)}$$

works, as long as the decay condition $\alpha \leq 1$ was also satisfied. □

Lemma 5.3 provides upper bounds for $f(K_m, p)$ for all $m \geq 4$, where the shape of the power of p depends on the residue class of m modulo 3.

Theorem 5.4 *If $p \in (0, 1)$, then $f(K_3, p) = O(p^{-3/2})$ and, for $m \geq 4$,*

$$f(K_m, p) = O_m(p^{-\frac{2}{3}m + c_m}),$$

where

$$c_m = \begin{cases} \frac{m}{2m-3} & m \equiv 0 \pmod{3} \\ \frac{2}{3} & m \equiv 1 \pmod{3} \\ \frac{2m+8}{6m-3} & m \equiv 2 \pmod{3}. \end{cases}$$

Proof For $m = 3$, the bound is simple. Query $\Theta(p^{-3/2})$ pairs containing a given vertex v_1 and then, among the $\Theta(p^{-1/2})$ neighbors successfully found, query all pairs. For sufficiently large implied constants, the probability that we build a triangle containing v_1 is at least $1/2$.

When $m \geq 4$, we use Lemma 5.3, taking

$$(a, b) = \begin{cases} \left(\frac{m-3}{3}, \frac{2m}{3}\right) & m \equiv 0 \pmod{3} \\ \left(\frac{m-4}{3}, \frac{2m+1}{3}\right) & m \equiv 1 \pmod{3} \\ \left(\frac{m-2}{3}, \frac{2m-1}{3}\right) & m \equiv 2 \pmod{3}. \end{cases}$$

This gives the required result. □

We conjecture that the bounds in Theorem 5.4 are best possible up to the constant factor. In the next two subsections, we prove this is the case for $m \leq 5$.

5.3 Recursive Graph Building

Recall that $f(H, p)$ is the number of queries needed in the Subgraph Query Problem to build a copy of H with probability at least $\frac{1}{2}$. When $H = K_m$, we can prove a lower bound on $f(H, p)$ by combining Theorem 1.10 with Theorem 2.1.

Proposition 5.5 *If $m \geq 3$ and $c \leq \frac{1}{2}m$, then*

$$f(K_m, p) \geq \frac{1}{4} p^{-(\binom{m}{2} - c(c-1))/(m-c)}.$$

Proof Take $N = \frac{1}{4} p^{-(\binom{m}{2} - c(c-1))/(m-c)}$, which is chosen so that

$$p^{\binom{m}{2} - c(c-1)} (2N)^{m-c} \leq \frac{1}{4}.$$

Since $(1 - p)^{\binom{n}{2}} (2N)^n \to 0$ as $n \to \infty$, there is some n sufficiently large for which

$$p^{\binom{m}{2} - c(c-1)} (2N)^{m-c} + (1 - p)^{\binom{n}{2}} (2N)^n \leq \frac{1}{2}.$$

With $d = 0$, this choice of m, n, N, p, c, d satisfies the conditions of Theorem 2.1, so $\tilde{r}(m, n; p) > N$. By Theorem 1.10, $f(K_m, p) \geq \tilde{r}(m, n; p)$, giving the required result. □

We now describe a general method for obtaining a similar lower bound on $f(H, p)$ when H is not a clique. As before, define $t(H, p, N)$ to be the maximum expected number of copies of H that can be constructed in N queries. The main result of this section bounds $t(H, p, N)$ when H contains a large matching. To this end, recall that a graph has a k-matching if it contains k disjoint edges.

Theorem 5.6 *Let H be a graph containing a k-matching. Then there exists an absolute constant $A > 1$ for which*

$$t(H, p, N) \leq (Ae(H))^{e(H)} p^{e(H)-k(k-1)} (2N)^{v(H)-k},$$

whenever $pN \geq 1$.

For any edge $e \in H$, write $H \backslash e$ for the graph formed by removing the edge e from H. If U is a subset of the vertices of H, write $H \backslash U$ for the induced subgraph of H on the complement of U. We begin by proving the following pair of recursive bounds on $t(H, p, N)$.

Lemma 5.7 *If H is a simple labeled graph, then*

$$t(H, p, N) \leq p \sum_{e \in E(H)} t(H \backslash e, p, N) \tag{7}$$

and

$$t(H, p, N) \leq (1 + o(1)) pN \min_{(u,v) \in E(H)} t(H \backslash \{u, v\}, p, N), \tag{8}$$

where the $o(1)$ term tends to 0 as $pN \to \infty$.

Proof Suppose Builder follows an optimal strategy which achieves $t(H, p, N)$ expected copies of H in N turns. For each copy H_i of H that appears during the game, distinguish the edge e_i which is built last in H_i. For each $e \in E(H)$, let $t_e(H, p, N)$ be the maximum expected number of copies of H that Builder can build, only counting those copies of H in which e is the last edge built. Then, clearly,

$$t(H, p, N) \leq \sum_{e \in E(H)} t_e(H, p, N).$$

Furthermore, $t_e(H, p, N) \leq pt(H \backslash e, p, N)$, since each copy of $H \backslash e$ can become exactly one copy of H with success rate p if e is built. Inequality (7) follows.

As for recursion (8), note simply that the number of copies of H is bounded by the number of choices for the images of the vertices u, v which are connected by an edge times the number of copies of $H \backslash \{u, v\}$. By the Chernoff bound, the number of choices of an edge is tightly concentrated around pN, so the inequality follows. □

It remains to apply these inequalities recursively.

Proof of Theorem 5.8 By (8), there is an absolute constant $A > 1$ for which

$$t(H, p, N) \leq ApN \min_{(u,v) \in E(H)} t(H \backslash \{u, v\}, p, N) \tag{9}$$

whenever $pN \geq 1$.

We proceed by induction on the number of edges in H. When H is an empty graph on m vertices, the result is trivial with $k = 0$. Let H be a labeled graph for which the induction hypothesis is true for every graph with fewer edges than H. Let $e \in E(H)$ run over all edges of H. We break into two cases:

Case 1. Every $H \setminus e$ contains a k-matching. Then, by induction and (7), it follows that

$$t(H, p, N) \leq p \sum_{e \in E(H)} t(H \setminus e, p, N)$$
$$\leq pe(H)(A(e(H) - 1))^{e(H)-1} \cdot p^{e(H)-1-k(k-1)}(2N)^{v(H)-k}$$
$$\leq (Ae(H))^{e(H)} p^{e(H)-k(k-1)}(2N)^{v(H)-k},$$

as desired.

Case 2. There exists $e \in E(H)$ for which $H \setminus e$ contains no k-matching. Then, let $e_2, \ldots, e_k$ be $k - 1$ edges which complete a k-matching of H containing e. The edges incident to e must all be incident to one of the e_i or else $H \setminus e$ would contain a k-matching. Also, e cannot form a 4-cycle with any e_i for the same reason. From these two facts one finds that e can be incident to at most $2(k - 1)$ other edges in total. Let H' be the graph obtained from H by removing the two vertices of e from H. Applying the induction hypothesis on H', which is a graph on $v(H) - 2$ vertices with at least $e(H) - (2k - 1)$ edges and a $(k - 1)$-matching, we find that

$$t(H', p, N) \leq (Ae(H'))^{e(H')} p^{e(H)-(2k-1)-(k-1)(k-2)}(2N)^{v(H)-2-(k-1)}.$$

Combining this with inequality (9), we have

$$t(H, p, N) \leq ApN \cdot t(H', p, N)$$
$$\leq (Ae(H))^{e(H)} p^{e(H)-k(k-1)}(2N)^{v(H)-k},$$

as desired.

For our purposes, we will always assume $pN \geq 1$. Otherwise, with high probability at most a constant number of edges are built successfully in the Subgraph Query Game, so $t(H, p, N)$ will be negligibly small.

Since $t(H, p, N) < 1/2$ implies $f(H, p) > N$, Theorem 5.6 immediately implies Theorem 1.12. Comparing this with Proposition 5.5, we note that while Theorem 1.12 gives a bound for all graphs H, it gives an inferior quantitative dependence on $e(H)$. While this stronger quantitative dependence in Proposition 5.5 seems to be only a minor benefit, it was needed in the proof of Theorem 1.11, which is why we retained the proof.

For large m, this bound only gives Corollary 1.14, that

$$f(K_m, p) = \Omega_m(p^{-(2-\sqrt{2})m+O(1)}),$$

which is still far from the conjectured growth rate $p^{-\frac{2}{3}m+O(1)}$. However, for $m \leq 5$, Theorem 1.12 can be used to pin down the asymptotic growth rate of $f(K_m, p)$, proving Theorem 1.13.

Proof of Theorem 5.9 The upper bounds for these cases are proved in Sect. 5.2. Apply Theorem 1.12 by taking $k = 1$ for $m = 3$ and $k = 2$ for $m = 4, 5$ to get the desired lower bounds. □

When $m \geq 6$, the matching argument of Theorem 5.6 does not seem sufficient for determining the exact growth rate of $f(K_m, p)$. Indeed, we will now exhibit an infinite family of graphs for which Theorem 5.6 is tight.

For $k \geq 1$, let H_k be the graph on $2k$ vertices a_i, b_i, $1 \leq i \leq k$, such that $a_i \sim a_j$ for all $i \neq j$, $b_i \nsim b_j$ for all $i \neq j$, and $a_i \sim b_j$ if and only if $i \leq j$. Thus H_k is a split graph consisting of a k-clique, a k-independent set, and a half graph between them. We show that Theorem 5.6 is tight for H_k up to a constant factor.

Note that the construction below requires N to grow like a tower of p^{-1}'s of height k. It is possible that the same lower bound is false in the regime $N \leq p^{-C}$ for any $C = C(k) > 0$.

Theorem 5.10 *For every $k \geq 1$, the graph H_k defined above contains a k-matching and, for any $p \in (0, 1)$,*

$$t(H_k, p, N) = \Omega_k(p^{e(H_k)-k(k-1)} N^{v(H_k)-k}),$$

provided N is sufficiently large in terms of p.

Proof In fact, H_k has k^2 edges, $2k$ vertices, and contains a unique k-matching $(a_i, b_i)_{i \leq k}$. It will suffice to show that for all $p \in (0, 1)$ and N sufficiently large in terms of p,

$$t(H_k, p, N) = \Omega_k(p^k N^k).$$

Builder's strategy will involve constructing a nested sequence of vertex sets $U_1, U_2, \ldots, U_k$. The first set U_1 is just an arbitrary set of N/k vertices. In each successive U_i, assuming $|U_i| \geq \sqrt{N}$ we can pick $N_i = N/(k|U_i|)$ vertices $a_i^{(1)}, a_i^{(2)}, \ldots, a_i^{(N_i)} \in U_i$ and try to build all edges from each $a_i^{(j)}$ to every other vertex in U_i. This step takes at most N/k turns. The set U_{i+1} is then defined to be the common neighborhood of $a_i^{(1)}, \ldots, a_i^{(N_i)}$ within U_i.

Repeating this process k times, we use at most N turns. For N sufficiently large, with high probability the edge density from $a_i^{(1)}, a_i^{(2)}, \ldots, a_i^{(N_i)}$ to the rest of $|U_i|$ is $(1 + o(1))p$. Thus, the number of copies of H_k built in this way is bounded below by

$$\prod_i (N_i \cdot p|U_i|) \geq (1 + o(1))(pN)^k/k^k,$$

since we can choose a_i out of any of the N_i vertices $a_i^{(1)}, \ldots, a_i^{(N_i)}$ and b_i out of any of its $(1 + o(1))p|U_i|$ neighbors. As long as N is large enough that $|U_k| \geq \sqrt{N}$

with high probability, there will be enough vertices in the last set U_k to perform the strategy. This argument successfully constructs $\Theta_k(p^k N^k)$ copies of H_k. Taking N to be a tower of $(2 + 2p^{-1})$'s of height k is sufficient. □

We finish the subsection with an application of the preceding results and prove Theorem 1.11. Recall that this theorem states that for m fixed and $n \to \infty$, a value of p for which $\tilde{r}_{\text{rand}}(m, n) \leq 3\tilde{r}(m, n; p)$ satisfies $p = \Theta(\frac{m}{n} \log \frac{n}{m})$.

Proof of Theorem 5.11 By Theorem 5.4,

$$f(K_m, p) = O_m(p^{-2m/3}),$$

and in fact it can be checked from the proof that the explicit dependence on m is polynomial. Moreover, using Proposition 5.5 with $c = 0$, we have that

$$f(K_m, p) \geq \frac{1}{4} p^{-\frac{m-1}{2}} \geq \frac{1}{4} p^{-m/3},$$

since $\frac{m-1}{2} \geq \frac{m}{3}$ for $m \geq 3$. Putting all this together, there exists an absolute constant $A > 0$ for which

$$\frac{1}{4} p^{-m/3} \leq f(K_m, p) \leq m^A p^{-2m/3} \tag{10}$$

for all $m \geq 3$, $p \in (0, 1)$.

By Theorem 1.10, we have

$$\tilde{r}(m, n; p) \leq \min\{f(K_m, p), f(K_n, 1 - p)\} \leq 3\tilde{r}(m, n; p).$$

Pick some $p_0 \in (0, 1)$ which maximizes the function $\min\{f(K_m, p), f(K_n, 1 - p)\}$. Such a p_0 exists because $f(K_m, p)$ is nonincreasing in p, $f(K_n, 1 - p)$ is nondecreasing, and both are integer-valued. Then, $\tilde{r}_{\text{rand}}(m, n) \leq 3 \cdot \tilde{r}(m, n; p_0)$. It remains to check that we could have chosen $p_0 = \Theta(\frac{m}{n} \log \frac{n}{m})$. By (10) and the fact that the bounds are continuous, we have

$$\frac{1}{4} p_0^{-\frac{1}{3}m} \leq n^A (1 - p_0)^{-\frac{2}{3}n}$$

and

$$\frac{1}{4} (1 - p_0)^{-\frac{1}{3}n} \leq m^A p_0^{-\frac{2}{3}m}.$$

Since $m \geq 3$ is fixed and $n \to \infty$, the first inequality implies $p_0 \to 0$. In particular, $\log(1 - p_0) = -p_0 + O(p_0^2)$. Taking the logarithm of both sides in the inequalities above, we have

$$-\frac{1}{3} m \log p_0 - \log 4 \leq A \log n + \frac{2}{3} n(p_0 + O(p_0^2))$$

and

$$\frac{1}{3}n(p_0 + O(p_0^2)) - \log 4 \le A \log m - \frac{2}{3} m \log p_0.$$

Taking $n \to \infty$ and dividing through by mp_0, these inequalities combine to show

$$\frac{\log(1/p_0)}{p_0} = \Theta\left(\frac{n}{m}\right)$$

and it follows that $p_0 = \Theta(\frac{m}{n} \log \frac{n}{m})$, as desired. □

5.4 The Value of $t(K_m, p, N)$ for Large N

In this section, we investigate the behavior of the function $t(K_m, p, N)$ as $N \to \infty$. We find that when N is very large, the essentially optimal strategy for building as many copies of K_m as possible is to fill in the edges of a clique on $\sqrt{2N}$ vertices. This is in stark contrast with the rather delicate procedure described in Sect. 5.2 to build a single copy of K_m.

5.4.1 Chernoff Bounds and Subjumbledness

We will need a standard lemma (see, for example, [24, Theorem 2.1]) saying that with high probability all moderately large induced subgraphs of a random graph $G(N, p)$ have the expected number of edges. Recall that if $U \subset V(G)$ is a vertex subset of G, we write $G[U]$ for the induced subgraph on U.

Lemma 5.12 *If $G = G(N, p)$ and $\varepsilon > 0$, then, with high probability,*

$$e(G[U]) = (1 \pm \varepsilon) p \binom{|U|}{2}$$

for all $|U| = \Omega_\varepsilon(p^{-1} \log N)$.

In the literature (see [24] and its references), this pseudorandomness property is usually called jumbledness. We also use this term, though in a slightly different way to how it is usually used.

Definition 5.13 A graph G is (p, M, ε)-*jumbled* if, for every $U \subseteq V(G)$ with $|U| \ge M$,

$$e(G[U]) = (1 \pm \varepsilon) p \binom{|U|}{2}.$$

A graph G is (p, M, ε)-*subjumbled* if it is a subgraph of some (p, M, ε)-jumbled graph.

In what follows, we will show that subjumbled graphs cannot have too many cliques. For the graph-building problem, the heuristic is that it's not possible to build more copies of H in a known jumbled graph G with pN queries than it is with N queries in $G(N, p)$.

5.4.2 Degeneracy

Define a graph to be d-degenerate if there exists an ordering of the vertices $v_1, \ldots, v_n$ such that $|N(v_i) \cap \{v_1, \ldots, v_{i-1}\}| \leq d$ for all i. The following simple lemma is well known.

Lemma 5.14 *Every graph with E edges is $\sqrt{2E}$-degenerate.*

Proof We exhibit the degenerate ordering by picking the vertices backwards from v_n to v_1. At each step, pick v_i to be the minimal degree vertex in the current graph and delete it. Note that $d(v_i) \leq i - 1$ because there are only i points left and also $d(v_i) \leq \frac{2E}{i}$ because the sum of the degrees is at most $2E$ and v_i has minimal degree. It follows that at every step $d(v_i) \leq \min(i, \frac{2E}{i}) \leq \sqrt{2E}$, as desired. □

This is not quite sufficient for our purposes, but it gives the main idea. What we really need is a better understanding of degeneracy in jumbled graphs. In what follows, given a candidate ordering $v_1, \ldots, v_n$ of the vertices, we write $N^-(v_i) = N(v_i) \cap \{v_1, \ldots, v_{i-1}\}$ and $d^-(v_i) = |N^-(v_i)|$.

Lemma 5.15 *Any (p, M, ε)-subjumbled graph on N edges is $\max((1 + \varepsilon)M\sqrt{p}, (1 + \varepsilon)\sqrt{2pN})$-degenerate.*

Proof Let H be a graph on N edges that is a subgraph of some (p, M, ε)-jumbled graph G. We again pick vertices of the graph H in order of increasing degree among the remaining vertices. Let the resulting order be $v_1, \ldots, v_n$ and write $U_i = \{v_1, \ldots, v_i\}$. The construction guarantees that v_i is of minimal degree in $G[U_i]$.

If $i \leq M$, then the subgraph $H[v_1, \ldots, v_i]$ has at most as many edges as $H[v_1, \ldots, v_M]$, which has at most $(1 + \varepsilon)pM^2/2$ edges. Thus,

$$d^-(v_i) \leq \min\left(i, \frac{(1+\varepsilon)pM^2}{i}\right)$$
$$\leq (1+\varepsilon)M\sqrt{p}.$$

Otherwise, if $i > M$, the induced subgraph $H[v_1, \ldots, v_i]$ has at most $(1 + \varepsilon)pi^2/2$ edges and it clearly cannot have more than $e(H) = N$ edges. Because v_i is of minimal degree in this induced subgraph,

$$d^-(v_i) \leq \frac{2}{i} \min\left(\frac{(1+\varepsilon)pi^2}{2}, N\right)$$
$$= \min((1+\varepsilon)pi, 2Ni^{-1})$$
$$\leq (1+\varepsilon)\sqrt{2pN}$$

and so every vertex has $d^-(v_i) \leq \max((1 + \varepsilon)M\sqrt{p}, (1 + \varepsilon)\sqrt{2pN})$, as desired. □

5.4.3 Counting Cliques

We are ready to prove the following lemma. Recall the standard notation that $t(K, H)$ is the number of labeled graph homomorphisms from K to H. Up to a lower order term, this is the same as counting labeled copies of K in H. In fact, the equality is exact in the case we care about, where K is a clique and H is a simple graph without self-loops.

Lemma 5.16 *For all $p \in (0, 1)$, $m, M \geq 2$ and $0 < \varepsilon < 1$, if H is a (p, M, ε)-subjumbled graph with N edges, then*

$$t(K_m, H) \leq (1 + O_m(\varepsilon + p^{1/2}N^{-1/2}))p^{\binom{m}{2}}(2p^{-1}N)^{\frac{m}{2}} + O_m\left(\sum_{k=2}^{m-1} p^{\frac{m+k(k-3)}{2}} \cdot M^{m-k} \cdot N^{\frac{k}{2}}\right).$$

Proof Take a degenerate ordering $v_1, \ldots, v_n$ of H such that v_i is of minimum degree in $H[v_1, \ldots, v_i]$. By Lemma 5.15,

$$d^-(v_i) \leq \max((1 + \varepsilon)M\sqrt{p}, (1 + \varepsilon)\sqrt{2pN}),$$

where the second term dominates as soon as $N \geq M^2/2$. Conditioning on whether or not v_n is in the copy of K_{m+1} we are counting, we see that

$$t(K_{m+1}, H) - t(K_{m+1}, H \setminus v_n) = (m + 1)t(K_m, H[N^-(v_n)]).$$

In particular, writing $t(m, N) = \max^*_{e(H)=N} t(K_m, H)$, where the maximum is taken over all graphs H with N edges that are subgraphs of some (p, M, ε)-jumbled graph, we find that

$$t(m+1, N) \leq \max_{d \leq U(N)} \left[t(m+1, N-d) + (m+1)t(m, e^+(d)) \right], \quad (11)$$

where $U(N) = \max((1+\varepsilon)M\sqrt{p}, (1+\varepsilon)\sqrt{2pN})$ and $e^+(d)$ is any upper bound on the number of edges in a graph on d vertices that is a subgraph of a (p, M, ε)-jumbled graph. The function e^+ we take is

$$e^+(d) = \begin{cases} \frac{d^2}{2} & d < M\sqrt{p} \\ (1+\varepsilon)\frac{pM^2}{2} & M\sqrt{p} \leq d < M \\ (1+\varepsilon)\frac{pd^2}{2} & d \geq M. \end{cases}$$

To see that $e^+(d)$ is indeed an upper bound on the number of edges in a graph on d vertices that is a subgraph of a (p, M, ε)-jumpled graph, we use the trivial bound when d is small, extend to a size M set to use jumbledness when d is somewhat close to M, and use jumbledness directly for d larger than M.

We are left to bound t using the system of inequalities (11). Write

$$t^*(m, N) = p^{\binom{m}{2}}(2p^{-1}N)^{\frac{m}{2}}$$

for the approximate optimum value of $t(m, N)$. We induct on m. The base case is $t(2, N) = 2N$. Assume, by induction, that for some $m \geq 2$,

$$t(m, N) \leq (1 + O_m(\varepsilon + p^{1/2}N^{-1/2}))t^*(m, N) + O_m\left(\sum_{k=2}^{m-1} p^{\frac{m+k(k-3)}{2}} \cdot M^{m-k} \cdot N^{\frac{k}{2}}\right).$$

We would like to show that the same inequality holds for $m + 1$. Iterating (11), there exists a sequence $(d_i)_{i \geq 1}$ of positive integers summing to N for which

$$d_i \leq U\left(N - \sum_{1 \leq j < i} d_j\right)$$

and

$$t(m + 1, N) \leq (m + 1) \sum_{i \geq 1} t(m, e^+(d_i)),$$

which implies, by the induction hypothesis, that

$$t(m+1, N) \leq (1 + O_m(\varepsilon + p^{1/2}N^{-1/2}))(m+1) \sum_{i \geq 1} t^*(m, e^+(d_i))$$

$$+ O_{m+1}\left(\sum_{k=2}^{m-1} p^{\frac{m+k(k-3)}{2}} \cdot M^{m-k} \cdot \left[\sum_{i \geq 1} e^+(d_i)^{\frac{k}{2}}\right]\right). \quad (12)$$

Since $e^+(d)$ is constant on the range $M\sqrt{p} \leq d < M$, the optimal choice of d_i will never have any points in this range. The main term of (12) can thus be separated into the sum over $d_i < M\sqrt{p}$ and the sum over $d_i \geq M$:

$$\sum_{i \geq 1} t^*(m, e^+(d_i)) \leq \sum_{d_i \geq M} t^*\left(m, (1+\varepsilon)pd_i^2/2\right) + \sum_{d_i < M\sqrt{p}} t^*\left(m, (1+\varepsilon)d_i^2/2\right). \quad (13)$$

Note that $m \geq 2$, so $t^*(m, N)$ is a convex nondecreasing function in N. Also, the function $e^+(d)$ is nondecreasing and convex in d except for the jump discontinuity at $d = M\sqrt{p}$. Therefore, in each of the ranges above, $t^*(m, \cdot)$ and $e^+(\cdot)$ are both convex nondecreasing functions.

To bound the first sum in (13), we pass to an integral. Write

$$N_i = N - \sum_{j \leq i} d_j.$$

Then
$$\sum_{d_i \geq M} t^*(m, e^+(d_i)) \leq \sum_{i \geq 1} t^*\left(m, (1+\varepsilon)pd_i^2/2\right)$$
$$= \sum_{i \geq 1} \int_{N_i}^{N_{i-1}} \frac{t^*(m, (1+\varepsilon)pd_i^2/2)}{d_i} dx.$$

Because $t^*(m, (1+\varepsilon)pd^2/2)/d$ is an increasing function of d and $d_i \leq U(N_{i-1}) = (1+\varepsilon)\sqrt{2pN_{i-1}}$, we have

$$\sum_{i \geq 1} \int_{N_i}^{N_{i-1}} \frac{t^*(m, (1+\varepsilon)pd_i^2/2)}{d_i} dx \leq \sum_{i \geq 1} \int_{N_i}^{N_{i-1}} \frac{t^*(m, (1+\varepsilon)^3 p(\sqrt{2pN_{i-1}})^2/2)}{\sqrt{2pN_{i-1}}} dx$$
$$\leq \sum_{i \geq 1} \int_{N_i+(1+\varepsilon)\sqrt{2pN}}^{N_{i-1}+(1+\varepsilon)\sqrt{2pN}} \frac{t^*(m, (1+\varepsilon)^3 p^2 x)}{(1+\varepsilon)\sqrt{2px}} dx$$
$$\leq \int_0^{N+(1+\varepsilon)\sqrt{2pN}} \frac{t^*(m, (1+\varepsilon)^3 p^2 x)}{\sqrt{2px}} dx.$$

We had to shift integrals in the second step to guarantee that every value of x in the range of integration is at least N_{i-1}.

Next, $t^*(m, N)$ is a polynomial in N, so we can absorb the $(1+\varepsilon)$ into the error term. Similarly, we can pull out an error term of $(1+(1+\varepsilon)\sqrt{2p/N})$ from the bounds of the integral to simplify. Reorganizing various error terms, we get

$$\int_0^{N+(1+\varepsilon)\sqrt{2pN}} \frac{t^*(m, (1+\varepsilon)p^2 x)}{\sqrt{2px}} dx \leq (1 + O_{m+1}(\varepsilon + p^{1/2}N^{-1/2})) \int_0^N \frac{t^*(m, p^2 x)}{\sqrt{2px}} dx.$$

Finally, explicitly evaluating the integral, we have

$$\int_0^N \frac{t^*(m, p^2 x)}{\sqrt{2px}} dx = \int_0^N p^{\binom{m}{2}} (2p^{-1} p^2 x)^{\frac{m}{2}} \frac{dx}{\sqrt{2px}}$$
$$= 2^{\frac{m-1}{2}} p^{\frac{m^2-1}{2}} \int_0^N x^{\frac{m-1}{2}} dx$$
$$= \frac{1}{m+1} 2^{\frac{m+1}{2}} p^{\frac{m^2-1}{2}} x^{\frac{m+1}{2}} \Big|_0^N$$
$$= \frac{1}{m+1} p^{\binom{m+1}{2}} (2p^{-1}N)^{\frac{m+1}{2}}$$
$$= \frac{1}{m+1} t^*(m+1, N).$$

Estimating the second sum in (13) trivially, we get

$$\sum_{i\geq 1} t^*(m, e^+(d_i)) \leq (1 + O_{m+1}(\varepsilon + p^{1/2}N^{-1/2}))t^*(m+1, N) + O_{m+1}\left(\frac{N}{M\sqrt{p}} t^*\left(m, \frac{pM^2}{2}\right)\right).$$

To check the error terms in (12) match up is similar: break up each sum into the sums over $d_i \geq M$ and $d_i < M\sqrt{p}$. The first sum is estimated by an integral and the second trivially. The result is

$$O_{m+1}\left(\sum_{k=2}^{m-1} p^{\frac{m+k(k-3)}{2}} \cdot M^{m-k} \cdot \left[\sum_{i\geq 1} e^+(d_i)^{\frac{k}{2}}\right]\right) \leq O_{m+1}\left(\sum_{k=2}^{m} p^{\frac{m+1+k(k-3)}{2}} \cdot M^{m+1-k} N^{\frac{k}{2}}\right),$$

which is the right error term for $t(m+1, N)$, completing the induction. □

In particular, and this is essential, the implicit constants in this lemma do not depend on M. As an immediate corollary, we now prove Theorem 1.15. Note that the N above is the number of edges in H, which will correspond to $(1+o(1))pN$ below if N is the number of queries made in the Subgraph Query Game.

Proof of Theorem 5.17 Applying the Chernoff bound from Lemma 5.12, we see that for any $\varepsilon > 0$ we can take some $M = Cp^{-1}\log N$ so that the random graph $G(2N, p)$ is (p, M, ε)-jumbled with high probability. Also with high probability, the number of edges built in N queries is $(1+o(1))pN$. It is easy to check that the exponentially small probabilities with which either of these are false have negligible impact on the value of $t(K_m, p, N)$. The subgraph H built by Builder must therefore satisfy the hypotheses of Lemma 5.16 with $(1+o(1))pN$ edges.

The main term dominates the error terms for N sufficiently large, giving the expected answer which is just $p^{\binom{m}{2}}(2N)^{\frac{m}{2}}$, the number of m-cliques in $G(\sqrt{2N}, p)$. This happens once the main term outgrows the largest error term, the term with $k = m - 1$. This happens at $N = \Omega(p^{-(2m-3)}M^2)$, so it suffices to have $N \geq \omega(p^{-(2m-1)}\log^2(p^{-1}))$. This proves the upper bound in Theorem 1.15. Of course, the lower bound is proved by the strategy of building all edges among $\sqrt{2N}$ vertices. □

6 Concluding Remarks

It is an interesting problem to close the gap in the bounds for the online Ramsey number $\tilde{r}(m, n)$. In particular, we know that there are positive constants c, c' for which $cn^3/(\log n)^2 \leq \tilde{r}(3, n) \leq c'n^3$ and it seems plausible that these bounds could be brought closer together. Indeed, we conjecture that the lower bound can be improved to $cn^3/\log n$ by considering the following Painter strategy motivated by the triangle-free process [6]. Painter applies the triangle-free process to obtain an auxiliary triangle-free graph G on vertex set $\{1, 2, \ldots, r\}$ with $r = c_0 n^2/\log n$.

Painter does not reveal this auxiliary graph. As before, we label vertices that reach degree $n/4$ with $1, \ldots, r$ as they arrive at degree $n/4$. When Builder adds an edge between two vertices in which both vertices have degree at least $n/4$, then these vertices have labels, say i and j, and Painter paints the edge with the color of the edge ij in G. Otherwise, they color the edge blue. This coloring clearly contains no red triangles, but it remains to show that it contains no blue K_n.

In studying the online Ramsey number, we were usually led by the idea that Builder's optimal strategy is to fill out an extremely sparse graph on the vertex set they touch. However, if Builder is restricted to play on a small vertex set, this intuition seems to go awry. If we define $\tilde{r}(m, n; N)$ in the same manner as the online Ramsey number but with the additional restriction that only N vertices are allowed, then we conjecture that the function $\tilde{r}(m, n; N)$ increases substantially as N decreases from $2\tilde{r}(m, n)$, the maximum number of vertices in a graph with $\tilde{r}(m, n)$ edges, down to its minimal meaningful value $r(m, n)$.

The order of growth of $f(K_m, p)$ is still open for $m \geq 6$. In particular, Theorems 1.12 and 5.4 show that $f(K_6, p) = \Omega(p^{-13/4})$ and $f(K_6, p) = O(p^{-10/3})$ and we conjecture that the upper bound is correct. This belief is rooted in our conviction that the upper bound for $t(H, p, N)$ given by Theorem 5.6, upon which Theorem 1.12 relies, is not tight when N is on the order of $f(H, p)$. Because of the examples in Theorem 5.10, these upper bounds can be tight when N is very large, so further progress on this problem would need to be more sensitive to the size of N. It is plausible that any advance on this question and its generalizations could also impinge on our estimates for online Ramsey numbers.

Acknowledgements The first author's research is supported by a Royal Society University Research Fellowship and by ERC Starting Grant 676632. The second author's research is supported by a Packard Fellowship and by NSF Career Award DMS-1352121. The fourth author's research is supported by a NSF Graduate Research Fellowship DGE-1656518. We are extremely grateful to Joel Spencer for pointing out a serious flaw in our previous proof of Theorem 1.4 which had been based on a generalization of the Lovász Local Lemma [15]. In the current version, we have a correct proof using a different approach. We would also like to thank the referee for some helpful remarks and Benny Sudakov for bringing the paper of Krivelevich [22] to our attention.

References

1. M. Ajtai, J. Komlós and E. Szemerédi, A note on Ramsey numbers, *J. Combin. Theory Ser. A* **29** (1980), 354–360.
2. N. Alon, On the number of subgraphs of prescribed type of graphs with a given number of edges, *Israel J. Math* **38** (1981), 116–130.
3. N. Alon and J. H. Spencer, **The Probabilistic Method**, 3rd ed., Wiley, 2008.
4. J. Beck, Achievement games and the probabilistic method, in *Combinatorics, Paul Erdős is Eighty, Vol. 1*, 51–78, Bolyai Soc. Math. Stud., János Bolyai Math. Soc., Budapest, 1993.
5. J. Beck, **Combinatorial Games: Tic-Tac-Toe Theory**, Cambridge University Press, 2008.
6. T. Bohman, The triangle-free process, *Adv. Math.* **221** (2009), 1653–1677.
7. T. Bohman and P. Keevash, The early evolution of the H-free process, *Invent. Math.* **181** (2010), 291–336.

8. T. Bohman and P. Keevash, Dynamic concentration of the triangle-free process, preprint available at arXiv:1302.5963 [math.CO].
9. D. Conlon, A new upper bound on diagonal Ramsey numbers, *Ann. of Math.* **170** (2009), 941–960.
10. D. Conlon, On-line Ramsey numbers, *SIAM J. Discrete Math.*, **23** (2009), 1954–1963.
11. D. Conlon, J. Fox and B. Sudakov, Hypergraph Ramsey numbers, *J. Amer. Math. Soc.* **23** (2010), 247–266.
12. P. Erdős, Graph theory and probability. II, *Canad. J. Math.*, **13** (1961), 346–352.
13. P. Erdős, R. J. Faudree, C. C. Rousseau and R. H. Schelp, The size Ramsey number, *Period. Math. Hungar.* **9** (1978), 145–161.
14. P. Erdős and L. Lovász, Problems and results on 3-chromatic hypergraphs and some related questions, in *Infinite and finite sets (Colloq., Keszthely, 1973), Vol. II*, 609–627, Colloq. Math. Soc. János Bolyai, Vol. 10, North-Holland, Amsterdam, 1975.
15. P. Erdős and J. Spencer, Lopsided Lovász local lemma and Latin transversals, *Discrete Appl. Math.* **30** (1991), 151–154.
16. A. Ferber, M. Krivelevich, B. Sudakov and P. Vieira, Finding Hamilton cycles in random graphs with few queries, *Random Structures Algorithms* **49** (2016), 635–668.
17. A. Ferber, M. Krivelevich, B. Sudakov and P. Vieira, Finding paths in sparse random graphs requires many queries, *Random Structures Algorithms* **50** (2017), 71–85.
18. G. Fiz Pontiveros, S. Griffiths and R. Morris, The triangle-free process and $R(3, k)$, to appear, *Mem. Amer. Math Soc.*
19. E. Friedgut, Y. Kohayakawa, V. Rödl, A. Ruciński and P. Tetali, Ramsey games against a one-armed bandit, *Combin. Probab. Comput.* **12** (2003), 515–545.
20. D. Hefetz, M. Krivelevich, M. Stojakovic and T. Szabó, **Positional Games**, Birkhäuser, 2014.
21. J. H. Kim, The Ramsey number $R(3, t)$ has order of magnitude $t^2/\log t$, *Random Structures Algorithms* **7** (1995), 173–207.
22. M. Krivelevich, Bounding Ramsey numbers through large deviation inequalities, *Random Structures Algorithms* **7** (1995), 145–155.
23. M. Krivelevich, Positional games, in *Proceedings of the International Congress of Mathematicians, Vol. 4*, 355–379, Kyung Moon Sa, Seoul, 2014.
24. M. Krivelevich and B. Sudakov, Pseudo-random graphs, in *More sets, graphs and numbers*, 199–262, Bolyai Soc. Math. Stud., 15, Springer, Berlin, 2006.
25. A. Kurek and A. Ruciński, Two variants of the size Ramsey number, *Discuss. Math. Graph Theory* **25** (2005), 141–149.
26. J. Spencer, Ramsey's theorem – a new lower bound, *J. Combin. Theory Ser. A* **18** (1975), 108–115.

Statistical Matching Theory

Péter Csikvári

Abstract In this paper, we survey some recent developments on statistical properties of matchings of very large and infinite graphs. We discuss extremal graph theoretic results like Schrijver's theorem on the number of perfect matchings of regular bipartite graphs and its variants from the point of view of graph limit theory. We also study the number of matchings of finite and infinite vertex-transitive graphs.

Keywords Matchings · Permanent · Graph limits · Lattices · Bethe approximation

1 Introduction

In this paper, we survey some recent developments on statistical properties of matchings of very large and infinite graphs.

We will focus on two topics: *extremal graph theory* and *vertex-transitive bipartite graphs*. Both topics are intimately related to graph limit theory. In the first case, when we consider extremal graph theoretical problems, it turns out that in certain extremal problems concerning matchings of d–regular bipartite graphs, the extremal graph is not a finite graph, but the infinite d–regular tree. The proper understanding of this phenomenon leads not only to new proofs of classical theorems, but also to new results such as the Lower Matching Conjecture and other new theorems. In the second case, when we study matchings of finite vertex-transitive bipartite graphs, the direction is, in some sense, exactly the opposite: we would like to understand the matchings of infinite lattices through finite graphs. These finite graphs exhibit certain properties that can be utilized to study their matchings. Then these new observations transfer to the original lattices.

P. Csikvári (✉)
Department of Computer Science, ELTE Eötvös Loránd University, Budapest, Hungary
e-mail: peter.csikvari@gmail.com

MTA–ELTE Geometric and Algebraic Combinatorics Research Group,
Pázmány P. stny. 1/C, Budapest 1117, Hungary

Extremal graph theory. To give an example for the discussed problems we offer the following theorem of A. Schrijver. This theorem asserts that if G is a d–regular bipartite graph on $v(G)$ vertices, and $pm(G)$ denotes the number of perfect matchings, then

$$pm(G)^{1/v(G)} \geq \left(\frac{(d-1)^{d-1}}{d^{d-2}} \right)^{1/2}.$$

It is a natural question whether we can improve on the constant on the right hand side. The answer is no. Then it is natural to ask whether there is a finite graph G for which

$$pm(G)^{1/v(G)} = \left(\frac{(d-1)^{d-1}}{d^{d-2}} \right)^{1/2}.$$

The answer is again no! The two negative answers together mean that

$$\inf_G pm(G)^{1/v(G)} = \left(\frac{(d-1)^{d-1}}{d^{d-2}} \right)^{1/2},$$

where the infimum is taken over all d–regular bipartite graphs, but this infimum is never achieved by a finite graph. Can we still use classical extremal graph theoretic arguments to prove Schrijver's theorem? The answer is yes, and this is what Sect. 4 is about. On the other hand, there will be a little twist in the argument, and this is where graph limit theory comes into the picture. In extremal graph theory it is a natural idea to find a graph transformation φ such that for a given graph parameter $p(\cdot)$ we have

$$p(G) \leq p(\varphi(G)),$$

and the studied class of graphs is closed under φ. An example for this strategy is Zykov's symmetrization which does not decrease the number of edges and the size of the largest clique, and so it provides a powerful tool to prove Turán's theorem, and as it turns out, many other theorems where we expect the Turán-graph to be extremal. In general, we apply this transformation as long as we can, and then we solve an optimization problem on a much restricted class of graphs. In case of Turán's theorem, this restricted class is the class of complete multipartite graphs. This is the point where we will deviate from this strategy.

We will find a graph transformation φ such that for the graph parameter $p(G) = pm(G)^{1/v(G)}$ we have $p(G) \geq p(\varphi(G))$. In general, we will be able to apply φ in many different ways, so $\varphi(G)$ refers to one of the possible applications of the graph transformation φ to G. As a next step we prepare a graph sequence G_i such that $G_0 = G$, $G_{i+1} = \varphi(G_i)$, and consequently we have

$$p(G_0) \geq p(G_1) \geq p(G_2) \geq p(G_3) \geq \ldots.$$

The point is that the sequence (G_i) will not stabilize as in the proof of Turán's theorem using Zykov's symmetrization, but it will converge in the sense of Benjamini–Schramm. We will explain this convergence in Sect. 3. This convergence enables us to extend the universe of finite graphs with some new elements which we will call *random unimodular graphs*. We will define $p(\cdot)$ for these new elements as well. It turns out that in this extended topological space there will be a minimizer of the parameter $p(G)$, namely the infinite d–regular tree $\mathbb{T}_d$. Section 2 contains a brief survey on related results and in Sect. 4 we give an almost complete proof of Schrijver's theorem along these lines.

Vertex-transitive bipartite graphs. To motivate the other main topic of this survey let us consider the following classical result of Kasteleyn [34] and independently Fisher and Temperley [50]. Let $Z_{m,n}$ be the number of perfect matchings of the $m \times n$ grid. Then

$$Z_{m,n} = \left(\prod_{j=1}^{m} \prod_{k=1}^{n} \left(4\cos^2\left(\frac{\pi j}{m+1}\right) + 4\cos^2\left(\frac{\pi k}{n+1}\right) \right) \right)^{1/4}.$$

This leads to the limit formula

$$\lim_{\substack{m,n \to \infty \\ 2 \mid mn}} \frac{1}{mn} \log Z_{m,n} = \frac{4}{\pi^2} \int_0^{\pi/2} \int_0^{\pi/2} \log(4\cos^2(x) + 4\cos^2(y)) \, dx \, dy.$$

It is intuitively clear that the grids converge to the lattice $\mathbb{Z}^2$. To make this statement precise once again we need the concept of Benjamini–Schramm convergence (see Sect. 3). Benjamini–Schramm convergence primarily grasps the local structure of a graph. It turns out that perfect matchings are especially fragile: a small change in the graph can lead to a dramatic change in the number of perfect matchings even if we restrict our attention to graphs with even number of vertices, or even if we only consider bipartite regular graphs. Fortunately, vertex-transitive bipartite graphs behave well in this respect, and so we can build a graph limit theory using them. More details can be found in Sect. 5.

This paper is organized as follows: in the next section we survey extremal graph theoretic results concerning matchings of finite (regular) graphs. In the third section we give the definition of Benjamini–Schramm convergence together with some examples. In the fourth section we give a sample proof of an extremal graph theoretic result on matchings that utilizes graph limit theory. In the fifth section we will study matchings of vertex-transitive bipartite graphs. In the final section we mention some results about matchings of dense graphs that are naturally connected to our discussion.

Basic notations: Throughout the paper, G denotes a graph, and $v(G)$ denotes the number of vertices of G. Recall that a matching of size k is set of k edges covering $2k$ vertices together. In other words, the edges $e_1, \ldots, e_k$ form a matching of size k, or shortly a k-matching if e_i and e_j have disjoint set of endpoints for any $1 \leq i, j \leq k$.

The number of matchings of size k will be denoted by $m_k(G)$. The size of the largest matching is denoted by $\nu(G)$. A matching is called perfect if it covers all vertices, that is, it has size $v(G)/2$. The number of perfect matchings will be denoted by pm(G).

2 Extremal Graph Theory

In this section we will consider lower and upper bounds for the number of (perfect) matchings of bipartite graphs. Recall that $m_k(G)$ denotes the number of matchings of size k, and pm(G) denotes the number of perfect matchings. When G is a bipartite graph with classes of size n, then the problem of counting the number of perfect matchings of G is equivalent to computing the permanent of a $0-1$ matrix of size n by n. Recall that the permanent of a matrix A is defined as follows:

$$\text{per}(A) = \sum_{\pi \in S_n} a_{1,\pi(1)} a_{2,\pi(2)} \cdots a_{n,\pi(n)}.$$

Let us suppose for a moment that all $a_{ij} \in \{0, 1\}$, and define a graph G on the vertex set $R \cup C$, where $R = \{r_1, r_2, \ldots, r_n\}$ and $C = \{c_1, c_2, \ldots, c_n\}$ correspond to the rows and columns of the matrix, respectively. If $a_{ij} = 1$, then put an edge between the vertices r_i and c_j. Now it is clear that per(A) = pm(G), the number of perfect matchings of G.

A well-known result concerning permanents of $0-1$ matrices is due to L. M. Brégman [8].

Theorem 2.1 (L. M. Brégman [8]) *Let A be a $0-1$ matrix of size $n \times n$, and let r_i denote the number of 1's in the i-th row. Then*

$$\text{per}(A) \leq \prod_{i=1}^{n} (r_i!)^{1/r_i}.$$

Since pm($K_{d,d}$) = $d!$, Theorem 2.1 immediately implies the following result about d-regular bipartite graphs.

Theorem 2.2 *Let pm(G) denote the number of perfect matchings. Then for a d-regular (bipartite) graph we have*

$$\text{pm}(G)^{1/v(G)} \leq \text{pm}(K_{d,d})^{1/v(K_{d,d})}.$$

A priori Brégman's theorem implies the above result only for bipartite graphs but using the observation

$$\text{pm}(G)^2 \leq \text{pm}(G \times K_2) \tag{1}$$

one can deduce the general case from the bipartite case. Here $G \times K_2$ is the graph with vertex set $V(G) \times \{0, 1\}$, in which (u, i) and (v, j) are adjacent if $i \neq j$ and $(u, v) \in E(G)$. This is clearly a bipartite graph. Observation 1 was rediscovered several times, see, for instance, [4]. A generalization of this observation will be proved in Theorem 4.3.

Let us mention that one can prove an analogue of Brégman's result for the number of all matchings, or even for weighted sums of matchings. Let

$$M(G, \lambda) = \sum_{k=0}^{\lfloor v(G)/2 \rfloor} m_k(G) \lambda^k.$$

It is the matching generating function of the graph G. In statistical physics it is known as the partition function of the monomer-dimer model.

Theorem 2.3 (E. Davies, M. Jenssen, W. Perkins, B. Roberts [16]) *For a d-regular graph G and $\lambda > 0$ we have*

$$M(G, \lambda)^{1/v(G)} \leq M(K_{d,d}, \lambda)^{1/v(K_{d,d})}.$$

For $\lambda = 1$ this result simplifies to the number of all matchings, and as $\lambda \to \infty$ it also implies Theorem 2.2.

Concerning lower bounds for perfect matchings of regular graphs, M. Voorhoeve ($d = 3$) and A. Schrijver (general d) proved the following result.

Theorem 2.4 (A. Schrijver [45], for $d = 3$ M. Voorhoeve [52]) *Let G be a d-regular bipartite graph on $v(G) = 2n$ vertices, and let $\mathrm{pm}(G)$ denote the number of perfect matchings of G. Then*

$$\mathrm{pm}(G) \geq \left(\frac{(d-1)^{d-1}}{d^{d-2}} \right)^n.$$

In other words, for every d-regular bipartite graph G we have

$$\frac{\ln \mathrm{pm}(G)}{v(G)} \geq \frac{1}{2} \ln \left(\frac{(d-1)^{d-1}}{d^{d-2}} \right).$$

There are various different proofs for Schrijver's theorem and its generalizations in the literature. Schrijver's original proof is elementary but tricky. L. Gurvits [27] gave another proof using stable polynomials. For an account of this proof, see also [37]. This is a beautiful proof, probably one from The Book. D. Straszak and N. Vishnoi [49] found a generalization for certain graphical models also using stable polynomials. Another proof, revealing the extremal graph, was given in [13] and is sketched in this survey. M. Lelarge [38] gave a variant of this proof for another generalization of Schrijver's theorem.

A result of Wilf [53] (see also [6, 46]) shows that the constant $\frac{1}{2}\ln\left(\frac{(d-1)^{d-1}}{d^{d-2}}\right)$ is the best possible. This can be shown by computing the expected value of $\mathrm{pm}(G)$ for d-regular random bipartite graphs. There was no explicit construction for regular bipartite graphs with small number of perfect matchings for a long time. Very recently it turned out that if a d-regular bipartite graph has a small number of short cycles, then it has asymptotically the same number of perfect matchings as a random d-regular graph, the more precise formulation is the following.

Theorem 2.5 (M. Abért, P. Csikvári, P. E. Frenkel, G. Kun [1]) *Let (G_i) be a sequence of d-regular graphs such that $g(G_i) \to \infty$, where g denotes the girth, that is, the length of the shortest cycle.*
(a) For the number of perfect matchings $\mathrm{pm}(G_i)$, we have

$$\limsup_{i\to\infty} \frac{\ln \mathrm{pm}(G_i)}{v(G_i)} \leq \frac{1}{2} \ln\left(\frac{(d-1)^{d-1}}{d^{d-2}}\right).$$

(b) If, in addition, the graphs G_i are bipartite, then

$$\lim_{i\to\infty} \frac{\ln \mathrm{pm}(G_i)}{v(G_i)} = \frac{1}{2} \ln\left(\frac{(d-1)^{d-1}}{d^{d-2}}\right).$$

We note that it is enough to assume that (G_i) converges to $\mathbb{T}_d$ in Benjamini–Schramm sense. See Sect. 3 for more details.

In [28] L. Gurvits derived a version of Schrijver's theorem for permanents using Schrijver's theorem itself.

Theorem 2.6 (L. Gurvits [28]) *Let A be an n by n non-negative matrix. Then*

$$\mathrm{per}(A) \geq \sup_{B \in \mathrm{DS}_n} \exp\left(\sum_{i,j} B_{ij} \ln \frac{A_{ij}}{B_{ij}} + \sum_{i,j}(1-B_{ij})\ln(1-B_{ij})\right),$$

where DS_n is the set of doubly stochastic matrices of size n by n.

In the same paper [28] L. Gurvits also extended Schrijver's theorem from perfect matchings to matchings of arbitrary size.

Theorem 2.7 (L. Gurvits [28]) *Let G be an arbitrary d-regular bipartite graph on $v(G) = 2n$ vertices. Let $m_k(G)$ denote the number of k-matchings. Set $p = \frac{k}{n}$. Then*

$$\frac{\ln m_k(G)}{v(G)} \geq \frac{1}{2}\left(p\ln\left(\frac{d}{p}\right) + (d-p)\ln\left(1-\frac{p}{d}\right) - 2(1-p)\ln(1-p)\right) + o_{v(G)}(1).$$

Gurvits's theorem was previously conjectured by Friedland, Krop and Markström [22] under the name Asymptotic Lower Matching Conjecture. They also had a more precise form of this conjecture known as Lower Matching Conjecture.

Statistical Matching Theory

Conjecture 2.8 (Lower Matching Conjecture [22]) *Let G be a d–regular bipartite graph on $v(G) = 2n$ vertices, and let $m_k(G)$ denote the number of matchings of size k, then*

$$m_k(G) \geq \binom{n}{k}^2 \left(\frac{d-p}{d}\right)^{n(d-p)} (dp)^{np},$$

where $p = \frac{k}{n}$.

To see the connection between the Lower Matching Conjecture and its asymptotic version, it is worth introducing two notations. The first one is the function appearing in Gurvits's theorem:

$$\mathbb{G}_d(p) = \frac{1}{2}\left(p \ln\left(\frac{d}{p}\right) + (d-p)\ln\left(1 - \frac{p}{d}\right) - 2(1-p)\ln(1-p)\right).$$

Furthermore, let $p = \frac{k}{n}$, and let $p_\mu = \binom{n}{k} p^k (1-p)^{n-k}$. This is the probability that a random variable X with distribution Binomial(n, p) takes its mean value. It turns out that

$$\binom{n}{k}^2 \left(\frac{d-p}{d}\right)^{n(d-p)} (dp)^{np} = p_\mu^2 \cdot \exp(2n\mathbb{G}_d(p)).$$

Using these notations Gurvits's theorem says that $m_k(G) \geq \exp(2n(\mathbb{G}_d(p) + o_n(1)))$, while the Lower Matching Conjecture claims that $m_k(G) \geq p_\mu^2 \cdot \exp(2n\mathbb{G}_d(p))$. It turns out that the truth is even more beautiful.

Theorem 2.9 ([13]) *Let G be a d–regular bipartite graph on $v(G) = 2n$ vertices, and let $m_k(G)$ denote the number of matchings of size k, then*

$$m_k(G) \geq p_\mu \cdot \exp(2n\mathbb{G}_d(p)),$$

where $p = \frac{k}{n}$.

Furthermore, there exists a d–regular bipartite graph G on $2n$ vertices such that

$$m_k(G) \leq \sqrt{\frac{1 - p/d}{1 - p}} \cdot p_\mu \cdot \exp(2n\mathbb{G}_d(p)).$$

Next we introduce the function $\lambda_G(p)$, called the entropy function in statistical physics, which is closely related to counting matchings. The simplest way to define it is as follows: let rG be the disjoint union of r copies of G and let the sequence k_r be chosen in such a way that

$$\lim_{r \to \infty} \frac{2k_r}{v(rG)} = p;$$

then

$$\lambda_G(p) := \lim_{r \to \infty} \frac{\ln m_{k_r}(rG)}{v(rG)}.$$

If G contains a perfect matching, then the limit indeed exists whenever $p < 1$, and for $p = 1$ one can define it as

$$\lambda_G(1) := \frac{\ln \operatorname{pm}(G)}{v(G)}.$$

The above definition for $\lambda_G(p)$ is not the original definition and is really hard to work with, but at least it is very easy to explain. Intuitively, it counts a normalized number of matchings covering p fraction of the vertices, but it has the advantage that it is meaningful even if p is irrational, thereby providing a continuous function. We also mention that it is really easy to extend $\lambda_G(p)$ to random rooted graphs $\mathcal{G}$. One can prove that for $p = \frac{k}{n}$ the following inequalities hold true:

$$p_\mu \cdot \exp(2n\lambda_G(p)) \leq m_k(G) \leq \exp(2n\lambda_G(p)).$$

This means that

$$\lambda_G(p) \approx \frac{\ln m_k(G)}{v(G)}.$$

Surprisingly, it turns out that Gurvits's theorem is equivalent to $\lambda_G(p) \geq \mathbb{G}_d(p)$ for all $0 \leq p \leq 1$ provided that G is a d–regular bipartite graph, see [13] for details. So practically Gurvits's theorem implies its more precise form.

At this moment, it may be mysterious why and how the function $\mathbb{G}_d(p)$ appears in these theorems. The mystery vanishes as soon as we realize that the function $\mathbb{G}_d(p)$ is nothing else but the entropy function of the infinite d–regular tree:

$$\lambda_{\mathbb{T}_d}(p) = \mathbb{G}_d(p).$$

We offer three more results in the spirit of Theorem 2.9. In Sect. 4 we will give a detailed sketch of the proof of Theorem 2.10. One can prove Theorems 2.11 and 2.12 with similar tools.

Theorem 2.10 ([13]) *Let G be a d–regular bipartite graph on $v(G) = 2n$ vertices, and let $m_k(G)$ denote the number of matchings of size k. For $\lambda \geq 0$ let $M(G, \lambda) = \sum_{k=0}^{n} m_k(G)\lambda^k$. Then*

$$\frac{1}{v(G)} \ln M(G, \lambda) \geq \frac{1}{2} \ln S_d(\lambda),$$

where

$$S_d(\lambda) = \frac{1}{\eta_\lambda^2}\left(\frac{d-1}{d-\eta_\lambda}\right)^{d-2} \quad \text{and} \quad \eta_\lambda = \frac{\sqrt{1+4(d-1)\lambda}-1}{2(d-1)\lambda}.$$

Alternatively, for $0 \leq p \leq 1$ we have

$$\sum_{k=0}^{n} m_k(G)\left(\frac{p}{d}\left(1-\frac{p}{d}\right)\right)^k (1-p)^{2(n-k)} \geq \left(1-\frac{p}{d}\right)^{nd}.$$

Note that this theorem directly reduces to Schrijver's theorem if $p = 1$, and strongly suggests that there might be some probabilistic proofs for some results appearing in this survey.

The following theorem shows that one can extend a few theorems from d–regular bipartite graphs to arbitrary bipartite graphs or even to the permanent of a nonnegative matrix. Before we state this result we need some notations.

The matching polytope $\mathrm{MP}(G)$ of a graph G is defined as the convex hull of incidence vectors of matchings in G. We define the fractional matching polytope as

$$\mathrm{FMP}(G) = \left\{ \underline{x} \in \mathbb{R}^{E(G)} \;\middle|\; x_e \geq 0 \; \forall e \in E(G), \sum_{e:v \in e} x_e \leq 1 \; \forall v \in V(G) \right\}.$$

It is known that $\mathrm{MP}(G) = \mathrm{FMP}(G)$ if and only if G is bipartite. Similarly, we can define by $\mathrm{MP}_k(G)$ the convex hull of incidence vectors of matchings in G of size k, and

$$\mathrm{FMP}_k(G) = \left\{ \underline{x} \in \mathrm{FMP}(G) \;\middle|\; \sum_{e \in E(G)} x_e = k \right\}.$$

Again, if G is bipartite, then $\mathrm{MP}_k(G) = \mathrm{FMP}_k(G)$. Finally, let $\nu(G)$ be the size of the largest matching in G.

Theorem 2.11 (M. Lelarge [38]) *For a vector $\underline{x} \in [0, 1]^E$ let*

$$\mathbb{F}_G(\underline{x}) = \sum_{e \in E}(-x_e \ln x_e + (1 - x_e)\ln(1 - x_e)) - \sum_{v \in V(G)} \left(1 - \sum_{e:v \in e} x_e\right) \ln\left(1 - \sum_{e:v \in e} x_e\right).$$

Then for any bipartite graph G and $\lambda \geq 0$ we have

$$\ln M(G, \lambda) \geq \max_{\underline{x} \in \mathrm{MP}(G)} \left\{ \left(\sum_{e \in E(G)} x_e\right) \ln \lambda + \mathbb{F}_G(\underline{x}) \right\}.$$

Furthermore, for all $k < \nu(G)$ we have

$$m_k(G) \geq b_{\nu(G), k}\left(\frac{k}{\nu(G)}\right) \exp\left(\max_{\underline{x} \in \mathrm{MP}_k(G)} \mathbb{F}_G(\underline{x})\right),$$

where $b_{n,k}(p) = \binom{n}{k} p^k (1 - p)^{n-k}$, that is, the probability that a binomial random variable $\mathrm{Bin}(n, p)$ takes the value k.

This theorem has a counterpart for permanents. For a non-negative matrix A of size n by n let

$$\mathrm{allper}(A) = \sum_{|I|=|J|} \mathrm{per}(A_{I,J}),$$

where $A_{I,J}$ is the submatrix of A induced by the rows I and J.

Theorem 2.12 (M. Lelarge [38]) *Let A be a non-negative matrix of size n by n. Let* SDS_n *be the set of non-negative matrices such that in each row and each column the sum of the elements is at most 1. For* $B \in SDS_n$ *let*

$$B_{i,0} = 1 - \sum_{j=1}^{n} B_{i,j} \quad and \quad B_{0,j} = 1 - \sum_{i=1}^{n} B_{i,j},$$

and

$$\mathbb{F}_A(B) = \sum_{1 \leq i,j \leq n} B_{ij} \ln \frac{A_{ij}}{B_{ij}^2} + \sum_{i=1}^{n} B_{i0} \ln \frac{1}{B_{i0}} + \sum_{j=1}^{n} B_{0j} \ln \frac{1}{B_{0j}}$$
$$+ \sum_{1 \leq i,j \leq n} (B_{ij} \ln B_{ij} + (1 - B_{ij}) \ln(1 - B_{ij})).$$

Then for any non-negative matrix A we have

$$\text{allper}(A) \geq \sup_{B \in SDS_n} \exp(\mathbb{F}_A(B))$$

Remark 2.13 Theorems 2.6 and 2.12 are special cases of a more general phenomenon. Permanents, the number of matchings or the number of homomorphisms of graphs can be expressed as partition functions of so-called graphical models. To a graphical model one can associate two objects: the partition function $Z(G)$ and the Bethe partition function $Z_B(G)$. There is no general inequality between them, but in certain cases $Z(G) \geq Z_B(G)$. This happens for attractive graphical models [44], and for certain bipartite graphical models [49]. In Theorems 2.6 and 2.12 $Z(G)$ appears on the left hand side, and $Z_B(G)$ on the right hand side. For more details on this subject see [44, 49].

3 Graph Limits and Examples

In the previous section we have seen that $\mathbb{G}_d(p) = \lambda_{\mathbb{T}_d}(p)$ gives a lower bound on $\lambda_G(p)$ if G is a d–regular bipartite graph. This establishes a claim to handle infinite graphs and connect them to the theory of finite graphs. This is exactly the goal of this section. In what follows we introduce the concept of Benjamini–Schramm convergence with some examples. Before we define this concept one more remark is in order: in this paper we will always assume that there is some Δ such that the largest degree of any graph G_i in a given sequence of graphs is at most Δ. In such a case we say that the graph sequence (G_i) is a *bounded-degree* graph sequence. This assumption simplifies our task significantly.

Statistical Matching Theory

Definition 3.1 Let L be a probability distribution on (infinite) connected rooted graphs; we will call L a *random rooted graph*. For a finite connected rooted graph α and a positive integer r, let $\mathbb{P}(L, \alpha, r)$ be the probability that the r-ball centered at the root vertex is isomorphic to α, where the root is chosen from the distribution L.

For a finite graph G, a finite connected rooted graph α and a positive integer r, let $\mathbb{P}(G, \alpha, r)$ be the probability that the r-ball centered at a uniform random vertex of G is isomorphic to α.

We say that a bounded-degree graph sequence (G_i) is *Benjamini–Schramm convergent* if for all finite rooted graphs α and $r > 0$, the probabilities $\mathbb{P}(G_i, \alpha, r)$ converge. Furthermore, we say that (G_i) *Benjamini–Schramm converges to* L, if for all positive integers r and finite rooted graphs α, $\mathbb{P}(G_i, \alpha, r) \to \mathbb{P}(L, \alpha, r)$.

The Benjamini–Schramm convergence is also called *local convergence* as it primarily grasps the local structure of the graphs (G_i).

Note that if (G_i) is a sequence of d–regular graphs such that the girth $g(G_i)$ tends to infinity, then it is Benjamini–Schramm convergent and we can even see its limit object: the rooted infinite d-regular tree $\mathbb{T}_d$, so the corresponding random rooted graph L is simply the distribution which takes a rooted infinite d-regular tree with probability 1. When L is a certain rooted infinite graph with probability 1, then we simply say that this rooted infinite graph is the limit without any further reference on the distribution.

There are other very natural graph sequences that are Benjamini–Schramm convergent, for instance if we take larger and larger boxes in the d-dimensional grid $\mathbb{Z}^d$, then it will converge to the rooted $\mathbb{Z}^d$.

The following problem is one of the main problems in the area, and will be especially crucial for us.

Problem. For which graph parameters $p(G)$ is it true that the sequence $(p(G_i))_{i=1}^{\infty}$ converges whenever the graph sequence $(G_i)_{i=1}^{\infty}$ is Benjamini–Schramm convergent?

The problem in such a generality is intractable, but there are various tools to attack it in special cases. One of the most popular tools is the so-called belief propagation. For matchings we will use another way to attack this problem using certain empirical measures called matching measures.

Concerning the general problem the reader might wish to consult with the papers [7, 17–19, 47, 48] and the book [40] and the references therein.

4 A Proof Strategy

We have seen in Sect. 2 that many results suggest that the extremal graph for matchings is not finite but the infinite d–regular tree. This raises the question: how can we attack a problem if we conjecture that the d–regular tree is the extremal graph for a given graph parameter:

Problem: Given a graph parameter $p(G)$. We would like to prove that among d-regular graphs we have
$$p(G) \geq p(\mathbb{T}_d).$$

Proof *(A possible two-step solution.)*

- Find a graph transformation φ for which $p(G) \geq p(\varphi(G))$, and for every graph G there exists a sequence of graphs (G_i) such that $G = G_0$ and $G_i = \varphi(G_{i-1})$, and $G_i \to \mathbb{T}_d$.
- Show that if (G_i) is Benjamini–Schramm-convergent, then $p(G_i)$ is convergent, and compute $p(\mathbb{T}_d)$. (Or at least, show it in the case of $G_i \to \mathbb{T}_d$.) Then

$$p(G) = p(G_0) \geq p(G_1) \geq p(G_2) \geq \cdots \geq p(\mathbb{T}_d).$$

Concerning the first step we will be more explicit: it seems that the 2-lift transformation can be used in a wide range of problems. Experience shows that the second step can be the most difficult, but the first step can also be tricky. Nevertheless, in the special case when we only consider a graph sequence converging to $\mathbb{T}_d$, there are many available tools: see for instance the paper of D. Gamarnik and D. Katz [24]. If $p(G) = \ln \tau(G)/v(G)$, where $\tau(G)$ denotes the number of spanning trees, then the second step is carried out in [41], and B. McKay proved [43] that $p(G)$ is maximized by the d-regular infinite tree among d-regular graphs. However, to prove McKay's result with our approach, the first step requires some modification. If $p(G) = \ln I(G)/v(G)$, where $I(G)$ denotes the number of independent sets, then the first step is very easy for d-regular bipartite graphs, while the second step concerning the limit theorem was established by A. Sly and N. Sun [47].

In this section we demonstrate this approach by sketching the proof of Theorem 2.10. In the following sections we study each step separately.

4.1 First Step: Graph Transformation

In this section we introduce the concept of 2-lift (Fig. 1).

Definition 4.1 A k-cover (or k-lift) H of a graph G is defined as follows. The vertex set of H is $V(H) = V(G) \times \{0, 1, \ldots, k-1\}$, and if $(u, v) \in E(G)$, then we choose a perfect matching between the vertices (u, i) and (v, j) for $0 \leq i, j \leq k-1$. If $(u, v) \notin E(G)$, then there are no edges between (u, i) and (v, j) for $0 \leq i, j \leq k-1$.

When $k = 2$ one can encode the 2-lift H by putting signs on the edges of the graph G: the $+$ sign means that we use the matching $((u, 0), (v, 0)), ((u, 1), (v, 1))$ at the edge (u, v), the $-$ sign means that we use the matching $((u, 0), (v, 1)), ((u, 1), (v, 0))$ at the edge (u, v). For instance, if we put $+$ signs to every edge, then we simply get $G \cup G$ as H, and if we put $-$ signs everywhere, then the obtained 2-cover H is simply $G \times K_2$.

The following result will be crucial for our argument.

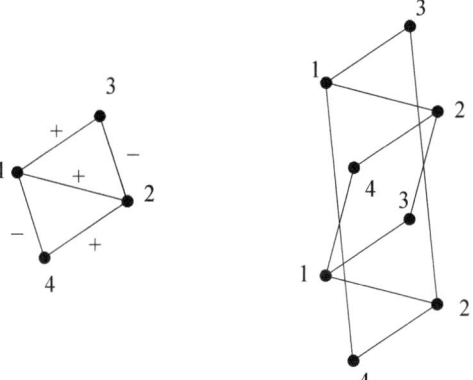

Fig. 1 A 2-lift

Lemma 4.2 (N. Linial [39]) *For any graph G, there exists a graph sequence $(G_i)_{i=0}^{\infty}$ such that $G_0 = G$, G_i is a 2-lift of G_{i-1} for $i \geq 1$, and $g(G_i) \to \infty$, where $g(H)$ is the girth of the graph H, that is, the length of the shortest cycle. In particular, if G_0 is d–regular, then $G_i \to \mathbb{T}_d$.*

Proof It is clear that if H' is a 2-lift of H, then $g(H') \geq g(H)$. Hence it is enough to show that for every H there exists an H'' obtained from H by a sequence of 2-lifts such that $g(H'') > g(H)$. We show that if the girth $g(H) = k$, then there exists a lift of H with fewer k-cycles than H. Let X be the random variable counting the number of k-cycles in a random 2-lift of H. Every k-cycle of H lifts to two k-cycles or a $2k$-cycle with probability $1/2$ each, so $\mathbb{E}X$ is exactly the number of k-cycles of H. But $H \cup H$ has two times as many k-cycles than H, so there must be a lift with strictly fewer k-cycles than H has. Choose this 2-lift and iterate this step to obtain an H'' with girth at least $k+1$. □

Note that if G is a bipartite d–regular graph, and H is a 2-lift of G, then H is again a d–regular bipartite graph.

The following theorem shows that the first step of the plan works for matchings of bipartite graphs.

Theorem 4.3 *Let G be a graph, and let H be an arbitrary 2-lift of G. Then*

$$m_k(H) \leq m_k(G \times K_2),$$

where $m_k(.)$ denotes the number of matchings of size k.

In particular, if $H = G \cup G$, then $m_k(G \cup G) \leq m_k(G \times K_2)$ for every k. It follows that $\mathrm{pm}(G)^2 \leq \mathrm{pm}(G \times K_2)$.

Furthermore, if G is a bipartite graph and H is a 2-lift of G, then

$$\frac{\ln M(G,\lambda)}{v(G)} = \frac{\ln M(G \cup G, t)}{v(G \cup G)} \geq \frac{\ln M(H,\lambda)}{v(H)},$$

where $M(G, \lambda) = \sum_k m_k(G)\lambda^k$. (Note that $M(G \cup G, \lambda) = M(G, \lambda)^2$.)

Proof Let M be any matching of a 2-lift of G. Let us consider the projection of M to G, then it will consist of cycles, paths and "double-edges" (i.e, when two edges project to the same edge). Let $\mathcal{R}$ be the set of these configurations. Then

$$m_k(H) = \sum_{R \in \mathcal{R}} |\phi_H^{-1}(R)|$$

and

$$m_k(G \times K_2) = \sum_{R \in \mathcal{R}} |\phi_{G \times K_2}^{-1}(R)|,$$

where ϕ_H and $\phi_{G \times K_2}$ are the projections from H and $G \times K_2$ to G. Note that

$$|\phi_{G \times K_2}^{-1}(R)| = 2^{k(R)},$$

where $k(R)$ is the number of cycles and paths of R. Indeed, in each cycle or path we can lift the edges in two different ways. The projection of a double-edge is naturally unique. On the other hand,

$$|\phi_H^{-1}(R)| \le 2^{k(R)},$$

since in each cycle or path if we know the inverse image of one edge, then we immediately know the inverse images of all other edges. Clearly, there is no equality in general for cycles. Hence

$$|\phi_H^{-1}(R)| \le |\phi_{G \times K_2}^{-1}(R)|$$

and consequently,

$$m_k(H) \le m_k(G \times K_2).$$

Note that if G is bipartite, then $G \times K_2 = G \cup G$, and so

$$\frac{1}{v(H)} \ln M(H, \lambda) \le \frac{1}{v(G \cup G)} \ln M(G \cup G, \lambda) = \frac{1}{v(G)} \ln M(G, \lambda).$$

This finishes the proof. □

Remark 4.4 In certain cases it is also possible to prove that for a graph parameter $p(\cdot)$ one has $p(G) \ge p(H)$ for all k-cover H of G. Such a result was given by N. Ruozzi in [44] for attractive graphical models. The advantage of using k-covers is that one can spare the graph limit step in the above approach, and replace it with a much simpler averaging argument over all k-covers of G with k converging to infinity. For homomorphisms this averaging argument was given by P. Vontobel [51]. For matchings such a result was established by C. Greenhill, S. Janson and A. Ruciński [26].

4.2 Second Step: Graph Limit Theory

In this subsection we carry out the second step of our plan. First we develop the necessary terminology.

Recall that if $G = (V, E)$ is a finite graph, then $v(G)$ denotes the number of vertices, and $m_k(G)$ denotes the number of k-matchings ($m_0(G) = 1$). Let

$$\mu(G, x) = \sum_{k=0}^{\lfloor v(G)/2 \rfloor} (-1)^k m_k(G) x^{v(G)-2k}.$$

We call $\mu(G, x)$ the matching polynomial. Clearly, the matching generating function $M(G, \lambda)$ introduced in Sect. 2 and the matching polynomial encode the same information.

The following theorem is crucial in the development of the theory of matching measure.

Theorem 4.5 (O. J. Heilmann and E. H. Lieb [32]) *The zeros of the matching polynomial $\mu(G, x)$ are real, and if the largest degree Δ is greater than 1, then all zeros lie in the interval $[-2\sqrt{\Delta - 1}, 2\sqrt{\Delta - 1}]$.*

Now we introduce a key concept of this theory, the matching measure.

Definition 4.6 (M. Abért, P. Csikvári, P. E. Frenkel, G. Kun [1]) The matching measure of a finite graph G is defined as

$$\rho_G = \frac{1}{v(G)} \sum_{z_i:\, \mu(G,z_i)=0} \delta(z_i),$$

where $\delta(s)$ is the Dirac-delta measure on s, and we take every z_i into account with its multiplicity. In other words, it is the uniform distribution on the zeros of $\mu(G, x)$.

Example: Let us consider the matching measure of the cycle on 6-vertices, C_6.

$$\mu(C_6, x) = x^6 - 6x^4 + 9x^2 - 2 =$$

$$= \left(x - \sqrt{2}\right)\left(x + \sqrt{2}\right)\left(x - \sqrt{2+\sqrt{3}}\right)\left(x + \sqrt{2+\sqrt{3}}\right)\left(x - \sqrt{2-\sqrt{3}}\right)\left(x + \sqrt{2-\sqrt{3}}\right).$$

Hence

$$\int f(z)\, d\rho_{C_6}(z) =$$

$$\frac{1}{6}\left(f\left(\sqrt{2}\right) + f\left(-\sqrt{2}\right) + f\left(\sqrt{2+\sqrt{3}}\right) + f\left(-\sqrt{2+\sqrt{3}}\right) + f\left(\sqrt{2-\sqrt{3}}\right) + f\left(-\sqrt{2-\sqrt{3}}\right)\right).$$

The following theorem enables us to consider the matching measure of a unimodular random graph which can be obtained as a Benjamini–Schramm limit of finite graphs. In particular, it provides an important tool to establish the second step in our plan.

Theorem 4.7 (M. Abért, P. Csikvári, P. E. Frenkel, G. Kun [1]) *Let (G_i) be a Benjamini–Schramm convergent bounded degree graph sequence. Let ρ_{G_i} be the matching measure of the graph G_i. Then the sequence (ρ_{G_i}) is weakly convergent, that is, there exists some measure $\rho_\mathcal{G}$ such that for every bounded continuous function f, we have*

$$\lim_{i \to \infty} \int f(z) \, d\rho_{G_i}(z) = \int f(z) \, d\rho_\mathcal{G}(z).$$

Theorem 4.7 is originated in the work of M. Abért and T. Hubai [3]. They showed a similar result for measures arising from the chromatic polynomial. This result has been generalized to a wide class of graph polynomials including the chromatic polynomial and the matching polynomial by P. Csikvári and P. E. Frenkel [14]. It turns out that for the matching polynomial, one does not need to use this general theorem as it also follows from a result of C. Godsil [25], for details see [2]. Theorem 4.5 asserts that the matching measure is supported on a bounded interval. Then to show that ρ_{G_i} is weakly convergent, it is enough to show that for every fixed k the sequence $\int z^k \, d\rho_{G_i}(z)$ is convergent. For many graph polynomials one can show that knowing only the statistics of the k-balls already determines this integral. For instance, for the matching polynomial the integral is directly related to the enumeration of the so-called tree-like walks of length k, see [25]. A better-known example is that for the spectral measure, that is, the probability measure of uniform distribution on the eigenvalues of the adjacency matrix of the graph, this integral is determined by the number of closed walks of length k.

To illustrate the power of Theorem 4.7, let us consider an application that also provides us the second step of our plan.

Theorem 4.8 (M. Abért, P. Csikvári, T. Hubai [2]) *Let (G_i) be a Benjamini–Schramm convergent graph sequence of bounded degree graphs. Then the sequences of functions*

$$\frac{\ln M(G_i, \lambda)}{v(G_i)}$$

is pointwise convergent.

Proof If G is a graph on $v(G) = 2n$ vertices and

$$M(G, \lambda) = \prod_{i=1}^{v(G)/2} (1 + \gamma_i \lambda),$$

then
$$\mu(G, x) = \prod_{i=1}^{v(G)/2} (x - \sqrt{\gamma_i})(x + \sqrt{\gamma_i}).$$

Thus
$$\frac{\ln M(G, \lambda)}{v(G)} = \frac{1}{v(G)} \sum_{i=1}^{v(G)/2} \ln(1 + \gamma_i \lambda) = \int \frac{1}{2} \ln(1 + \lambda z^2) \, d\rho_G(z).$$

Since $\frac{1}{2} \ln(1 + \lambda z^2)$ is a continuous function for every fixed positive λ, the theorem immediately follows from Theorem 4.7. □

It is worth introducing the notation
$$p_\lambda(G) = \frac{\ln M(G, \lambda)}{v(G)}.$$

We can even introduce $p_\lambda(\mathcal{G})$ if $\mathcal{G}$ is a Benjamini–Schramm-limit of sequence of finite graphs (G_i). (In fact, it is possible to define the function $p_\lambda(\mathcal{G})$ even if $\mathcal{G}$ is not the Benjamini–Schramm-limit of finite graphs.) In particular, we can speak about $p_\lambda(\mathbb{T}_d)$.

If we know the matching measure of a random unimodular graph, then it is just a matter of computation to derive various results on matchings.

In the particular case when the sequence (G_i) converges to the infinite d–regular tree $\mathbb{T}_d$, the limit measure $\rho_{\mathbb{T}_d}$ turns out to be the so-called Kesten–McKay measure. It is true in general that for any finite tree or infinite random rooted tree the matching measure coincides with the so-called spectral measure, for details see [1]. In particular, this is true for the infinite d–regular tree $\mathbb{T}_d$. Its spectral measure is computed explicitly in the papers [36, 42]. The Kesten–McKay measure is given by the density function
$$f_d(x) = \frac{d\sqrt{4(d-1) - x^2}}{2\pi(d^2 - x^2)} \chi_{[-\omega, \omega]},$$

where $\omega = 2\sqrt{d-1}$. Hence for any continuous function $h(z)$ we have
$$\int h(z) \, d\rho_{\mathbb{T}_d}(z) = \int_{-\omega}^{\omega} h(z) f_d(z) \, dz.$$

In particular,
$$p_\lambda(\mathbb{T}_d) = \int \frac{1}{2} \ln(1 + \lambda z^2) \, d\rho_{\mathbb{T}_d}(z) = \frac{1}{2} \ln S_d(\lambda),$$

where

$$S_d(\lambda) = \frac{1}{\eta_\lambda^2}\left(\frac{d-1}{d-\eta_\lambda}\right)^{d-2} \quad \text{and} \quad \eta_\lambda = \frac{\sqrt{1+4(d-1)\lambda}-1}{2(d-1)\lambda}.$$

It is worth introducing the following substitution:

$$\lambda = \frac{\frac{p}{d}\left(1-\frac{p}{d}\right)}{(1-p)^2}.$$

As p runs through the interval $[0, 1)$, λ runs through the interval $[0, \infty)$ and we have

$$\eta_\lambda = \frac{1-p}{1-\frac{p}{d}} \quad \text{and} \quad S_d(\lambda) = \frac{\left(1-\frac{p}{d}\right)^d}{(1-p)^2}.$$

One can also prove that

$$\lambda_{\mathbb{T}_d}(p) = \frac{1}{2}\left(p\ln\left(\frac{d}{p}\right) + (d-p)\ln\left(1-\frac{p}{d}\right) - 2(1-p)\ln(1-p)\right).$$

Remark 4.9 Instead of using measures one can use belief propagation to establish the convergence of certain graph parameters. For instance, M. Lelarge [38] used this method to prove Theorems 2.11 and 2.12. In general, one can choose the sequence $G = G_0, G_1, \ldots$ of covering graphs such that the sequence (G_i) converges to the universal cover tree of G. The advantage of belief propagation over matching measures is that it is not always easy to compute the matching measure of such a tree. On the other hand, when it is possible to compute the limiting measure, integration yields a wide variety of results without any difficulty.

4.3 The End of the Proof of Theorem 2.10

For every sequence of 2-covers we know from Theorem 4.3 that

$$p_\lambda(G_0) \geq p_\lambda(G_1) \geq p_\lambda(G_2) \geq p_\lambda(G_3) \geq \ldots$$

Furthermore, from Theorems 4.2 and 4.8 we know that we can choose the sequence of 2-covers such that the sequence $p_\lambda(G_i)$ converges to $p_\lambda(\mathbb{T}_d)$, hence $p_\lambda(G) \geq p_\lambda(\mathbb{T}_d)$ for any d-regular bipartite graph G. In other words,

$$\frac{1}{2n}\ln M(G, \lambda) \geq \frac{1}{2}\ln S_d(\lambda).$$

With the substitution $\lambda = \frac{\frac{p}{d}(1-\frac{p}{d})}{(1-p)^2}$ we obtain the inequality

$$M\left(G, \frac{\frac{p}{d}\left(1 - \frac{p}{d}\right)}{(1-p)^2}\right) \geq \frac{1}{(1-p)^{2n}}\left(1 - \frac{p}{d}\right)^n.$$

After multiplying by $(1-p)^{2n}$, we get that

$$\sum_{k=0}^{n} m_k(G) \left(\frac{p}{d}\left(1 - \frac{p}{d}\right)\right)^k (1-p)^{2(n-k)} \geq \left(1 - \frac{p}{d}\right)^{nd}.$$

This is true for all $p \in [0, 1)$ and so by continuity it is also true for $p = 1$, where it directly reduces to Schrijver's theorem since all but the last term vanish on the left hand side.

5 Perfect Matchings of Vertex-Transitive Bipartite Graphs and Lattices

The starting point of this section is the following theorem of R. Kenyon, A. Okounkov, S. Sheffield [35]. We will not be able to fully understand this theorem as we did not define the characteristic function of a lattice. On the other hand, we can see that this theorem provides a sufficient condition ensuring the convergence of

$$\lim_{i \to \infty} \frac{\ln \operatorname{pm}(G_i)}{v(G_i)}$$

for a given graph sequence (G_i). Moreover, it also provides a way of computing this limit explicitly (as long as we accept an integral as an explicit expression).

Theorem 5.1 (R. Kenyon, A. Okounkov, S. Sheffield [35]) *Let $\mathcal{G}$ be a $\mathbb{Z}^2$-periodic bipartite planar graph, and let G_i be the quotient graph of $\mathcal{G}$ by the action of $i\mathbb{Z}^2$. Let $P(z, w)$ be the characteristic function of $\mathcal{G}$. Assume that $P(z, w)$ has a finite number of zeros on the unit torus $\mathbb{T}^2 = \{(z, w) \in \mathbb{C}^2 : |z| = |w| = 1\}$. Then*

$$\lim_{i \to \infty} \frac{\ln \operatorname{pm}(G_i)}{v(G_i)} = \frac{1}{(2\pi i)^2} \int_{\mathbb{T}^2} \ln P(z, w) \frac{dz\, dw}{z\, w}.$$

To help one understand the setting of this theorem we provide a figure of the square-octagon lattice, and its quotient graph by the action of $(4\mathbb{Z})^2$. The fundamental domain is given by the dotted lines (Fig. 2).

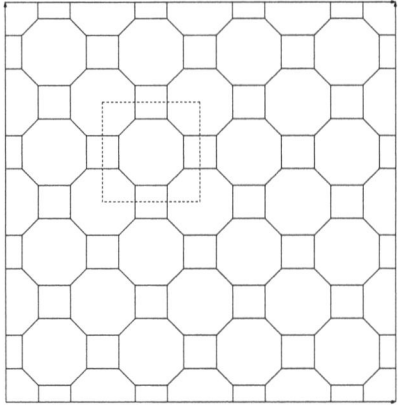

Fig. 2 The 4–8 lattice

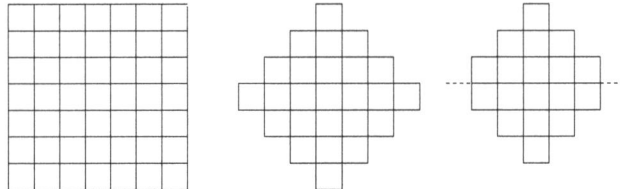

Fig. 3 Boxes, Aztec diamonds and modified Aztec diamonds

In particular cases, the characteristic function $P(z, w)$ can be computed (but not the integral!). For instance, for the square-octagon graph we have

$$P(z, w) = z + \frac{1}{z} + w + \frac{1}{w} + 5.$$

This theorem naturally raises the question why we needed such a special graph sequence, why did we simply not choose larger and larger subgraphs of the lattice? The problem is that in case of perfect matchings the boundary of the graph heavily affects the number of perfect matchings. Let us look at the following three sequences of graphs. All of them are Benjamini–Schramm convergent to $\mathbb{Z}^2$ (Fig. 3).

The first graph sequence is the sequence of boxes B_i (the graph B_8 is depicted in the figure). A classical result of Kasteleyn [34] and independently Fisher and Temperley [50] claims that

$$\lim_{i \to \infty} \frac{\ln \operatorname{pm}(B_i)}{v(B_i)} = \frac{1}{\pi} \sum_{k=0}^{\infty} \frac{(-1)^k}{(2k+1)^2} \approx 0.291.$$

The second sequence of graphs are called Aztec diamonds, A_4 is depicted in the figure. A surprising fact due to N. Elkies, G. Kuperberg, M. Larsen and J. Propp [20] is that for all i we have
$$\mathrm{pm}(A_i) = 2^{i(i+1)/2}.$$

Therefore,
$$\lim_{i \to \infty} \frac{\ln \mathrm{pm}(A_i)}{v(A_i)} = \frac{\ln \mathrm{pm}(A_i)}{v(A_i)} = \frac{\ln 2}{4} \approx 0.173.$$

In the third sequence, we slightly modify the Aztec diamonds. Now it turns out that $\mathrm{pm}(D_i) = 1$ for all i, since one has to include the dotted edges in the perfect matching and this completely determines the whole perfect matching. Hence

$$\lim_{i \to \infty} \frac{\ln \mathrm{pm}(D_i)}{v(D_i)} = \frac{\ln \mathrm{pm}(D_i)}{v(D_i)} = 0.$$

This example shows that some nice boundary conditions are required for the graphs appearing in our convergent graph sequence. Unfortunately, it is unclear what would be such a boundary condition for nonplanar graph. One way to overcome this difficulty is to consider vertex-transitive bipartite graphs.

Theorem 5.2 ([12]) *Let (G_i) be a Benjamini–Schramm convergent sequence of vertex-transitive bipartite graphs. Then the sequence*

$$\frac{\ln \mathrm{pm}(G_i)}{v(G_i)}$$

is convergent.

One might wonder whether vertex-transitivity is really necessary or it suffices to assume regularity. The next theorem shows that regularity is actually not sufficient.

Theorem 5.3 (M. Abért, P. Csikvári, P. E. Frenkel, G. Kun [1]) *Fix $d \geq 3$. Then there exists a sequence of d-regular bipartite graphs (G_i) such that (G_i) is Benjamini–Schramm convergent and*

$$\frac{\ln \mathrm{pm}(G_i)}{v(G_i)}$$

is not convergent.

Let us see what goes wrong with the proof of Theorem 4.8 if we apply it to perfect matchings. If G is a graph on $v(G) = 2n$ vertices and

$$M(G, \lambda) = \prod_{i=1}^{v(G)/2} (1 + \gamma_i \lambda),$$

then
$$\mu(G, x) = \prod_{i=1}^{v(G)/2} (x - \sqrt{\gamma_i})(x + \sqrt{\gamma_i}),$$

and therefore
$$\frac{\ln \mathrm{pm}(G)}{v(G)} = \frac{1}{v(G)} \sum_{i=1}^{v(G)/2} \ln(\gamma_i) = \int \ln|z|\, d\rho_G(z).$$

Now we see that $\ln|z|$ is not a bounded continuous function and this causes the problem.

On the other hand, we also see that the situation is not as bad as one might have previously thought. The function $\ln|z|$ is only discontinuous at 0. If we could prove that only a small measure is supported on the neighborhood of 0, then it would immediately resolve our problem. This is exactly the case when we consider vertex-transitive bipartite graphs.

Theorem 5.4 ([12]) *Let G be a d-regular vertex-transitive bipartite graph on $2n$ vertices, and*
$$M(G, t) = \prod_{i=1}^{n} (1 + \gamma_i t),$$

where $\gamma_1 \leq \gamma_2 \leq \cdots \leq \gamma_n$. Then
$$\gamma_k(G) \geq \frac{d^2}{d-1} \frac{k^2}{4n^2}.$$

Consequently,
$$\rho_G([-s, s]) \leq \frac{2\sqrt{d-1}}{d} s$$

for all $s \in \mathbb{R}$.

For vertex-transitive graphs one can also extend certain extremal graph theoretic results. For instance, the following theorem is a strengthened form of the fact that the infinite d–regular tree plays the role of the extremal graph for regular bipartite graphs.

Theorem 5.5 ([12]) *Let G be a finite d–regular vertex-transitive bipartite graph, where $d \geq 2$. Furthermore, let the gap function $g(p)$ be defined as*
$$g(p) = \lambda_G(p) - \mathbb{G}_d(p).$$

Then $g(p)$ is a monotone increasing function with $g(0) = 0$, and hence $g(p)$ is nonnegative. Furthermore, if G contains an ℓ-cycle, then

$$g(p) \geq \int_0^p f(x)^\ell \, dx,$$

where

$$f(x) = \frac{1}{4d} \min(x, (1-x)^2).$$

5.1 Computational Results

In statistical physics matchings of large and infinite graphs are studied under the name monomer-dimer model. Let B_n be a box of size $n \times n \times \cdots \times n$ in $\mathbb{Z}^d$, and let $M(B_n)$ be the number of all matchings in B_n. It has been known for a long time that the limit

$$\tilde{\lambda}(\mathbb{Z}^d) := \lim_{n \to \infty} \frac{\ln M(B_n)}{v(B_n)}$$

exists. When we count only perfect matchings, then the corresponding model is called the dimer model:

$$\lambda(\mathbb{Z}_d) := \lim_{2|n,\, n \to \infty} \frac{\ln \mathrm{pm}(B_n)}{v(B_n)}.$$

The quantities $\tilde{\lambda}(\mathbb{Z}_d)$ and $\lambda(\mathbb{Z}_d)$ are called monomer-dimer and dimer free energies.

The computation of monomer-dimer and dimer free energies has a long history. The precise value is known only in very special cases. Such an exceptional case is the Fisher-Kasteleyn-Temperley formula [34, 50] for the dimer model on $\mathbb{Z}^2$. There is no such exact result for monomer-dimer models if $d \geq 2$. The first approach for getting estimates was the use of the transfer matrix method. Hammersley [29, 30], Hammersley and Menon [31] and Baxter [5] obtained the first (non-rigorous) estimates for the free energy. Then Friedland and Peled [23] proved the rigorous estimates $0.6627989727 \pm 10^{-10}$ for $d = 2$ and the range $[0.7653, 0.7863]$ for $d = 3$. Here the upper bounds were obtained by the transfer matrix method, while the lower bounds relied on the Friedland-Tverberg inequality. The lower bound in the Friedland-Peled paper was subsequently improved by newer and newer results (see e.g. [21]) on Friedland's asymptotic matching conjecture which was finally proved by L. Gurvits [28]. Meanwhile, a non-rigorous estimate $[0.7833, 0.7861]$ was obtained via matrix permanents [33]. Concerning rigorous results, the most significant improvement was obtained recently by D. Gamarnik and D. Katz [24] via their new method which they called sequential cavity method. They obtained the range $[0.78595, 0.78599]$. More precise but non-rigorous estimates can be found in [10]. This paper uses Mayer-series with many coefficients computed in the expansion. The related paper [9] may lead to further development through the so-called Positivity conjecture of the authors.

Here we give some computational results arising from estimating certain integrals along matching measures.

Theorem 5.6 ([2]) *We have*

$$\tilde{\lambda}(\mathbb{Z}^3) = 0.7859659243 \pm 9.88 \cdot 10^{-7},$$

$$\tilde{\lambda}(\mathbb{Z}^4) = 0.8807178880 \pm 5.92 \cdot 10^{-6}.$$

$$\tilde{\lambda}(\mathbb{Z}^5) = 0.9581235802 \pm 4.02 \cdot 10^{-5}.$$

The bounds on the error terms are rigorous.

6 Dense Graphs and Matchings

It is possible to extend several ideas of this paper to dense graph limits. Here we assume some partial familiarity with the theory of dense graph limits. We only mention some simple results.

Theorem 6.1 *Suppose that (G_n) is a sequence of graphs convergent in the dense model. Let π_n be the uniform probability measure on roots of the matching polynomial $\mu(G_n, x)$. Then the rescaled measures $\frac{1}{\sqrt{v(G_n)}} \cdot \pi_n$ converge weakly.*

In particular, this allows us to associate "matching measures" to graphons. For instance, with this method one can prove the following result for the constant p graphon, in other words, the limiting distribution of Erdős–Rényi random graphs. This result was independently and prior proved in [11].

Theorem 6.2 ([11, 15]) *Let $p \in (0, 1)$, and let $(G_n)_n$ be a sequence of Erdős–Rényi random graphs $G_n \sim \mathbb{G}_{n,p}$. Let π_n be the uniform probability distribution on the roots of the matching polynomial of G_n. Then almost surely, the measures $\lambda_n := \frac{1}{\sqrt{n}}\pi_n$ converge weakly to the semicircle distribution SC_p whose density function is*

$$\rho_p(x) := \frac{1}{2\pi}\sqrt{4 - \frac{x^2}{p}}, \quad -2p \leq x \leq 2p.$$

Acknowledgements The author is grateful to László Márton Tóth, Ferenc Bencs, Endre Csóka and Viktor Harangi for various useful comments. The author is supported by the Marie Skłodowska-Curie Individual Fellowship grant no. 747430, by the Hungarian National Research, Development and Innovation Office, NKFIH grant K109684 and Slovenian-Hungarian grant NN114614, and by the ERC Consolidator Grant 648017.

References

1. Abért, M., Csikvári, P., Frenkel, P., Kun, G.: Matchings in Benjamini–Schramm convergent graph sequences. Transactions of the American Mathematical Society **368**(6), 4197–4218 (2016)
2. Abért, M., Csikvári, P., Hubai, T.: Matching measure, Benjamini–Schramm convergence and the monomer–dimer free energy. Journal of Statistical Physics **161**(1), 16–34 (2015)
3. Abért, M., Hubai, T.: Benjamini–Schramm convergence and the distribution of chromatic roots for sparse graphs. Combinatorica **35**(2), 127–151 (2015)
4. Alon, N., Friedland, S.: The maximum number of perfect matchings in graphs with a given degree sequence. The Electronic Journal of Combinatorics **15**(1) (2008)
5. Baxter, R.J.: Dimers on a rectangular lattice. Journal of Mathematical Physics **9**(4), 650–654 (1968)
6. Bollobás, B., McKay, B.D.: The number of matchings in random regular graphs and bipartite graphs. Journal of Combinatorial Theory, Series B **41**(1), 80–91 (1986)
7. Borgs, C., Chayes, J., Kahn, J., Lovász, L.: Left and right convergence of graphs with bounded degree. Random Structures & Algorithms **42**(1), 1–28 (2013)
8. Brégman, L.M.: Some properties of nonnegative matrices and their permanents. In: Soviet Math. Dokl, vol. 14, pp. 945–949 (1973)
9. Butera, P., Federbush, P., Pernici, M.: Higher-order expansions for the entropy of a dimer or a monomer-dimer system on d-dimensional lattices. Physical Review E **87**(6), 062113 (2013)
10. Butera, P., Pernici, M.: Yang-Lee edge singularities from extended activity expansions of the dimer density for bipartite lattices of dimensionality $2 \leq d \leq 7$. Physical Review E **86**(1), 011104 (2012)
11. Chen, X., Li, X., Lian, H.: The matching energy of random graphs. Discrete Applied Mathematics **193**, 102–109 (2015)
12. Csikvári, P.: Matchings in vertex-transitive bipartite graphs. Israel Journal of Mathematics **215**(1), 99–134 (2016)
13. Csikvári, P.: Lower matching conjecture, and a new proof of Schrijver's and Gurvits's theorems. Journal of European Mathematical Society **19**, 1811–1844 (2017)
14. Csikvári, P., Frenkel, P.E.: Benjamini–Schramm continuity of root moments of graph polynomials. European Journal of Combinatorics **52**, 302–320 (2016)
15. Csikvári, P., Frenkel, P.E., Hladký, J., Hubai, T.: Chromatic roots and limits of dense graphs. Discrete Mathematics **340**(5), 1129–1135 (2017)
16. Davies, E., Jenssen, M., Perkins, W., Roberts, B.: Independent sets, matchings, and occupancy fractions. Journal of the London Mathematical Society **96**(1), 47–66 (2017)
17. Dembo, A., Montanari, A.: Gibbs measures and phase transitions on sparse random graphs. Brazilian Journal of Probability and Statistics pp. 137–211 (2010)
18. Dembo, A., Montanari, A., Sly, A., Sun, N.: The replica symmetric solution for Potts models on d-regular graphs. Communications in Mathematical Physics **327**(2), 551–575 (2014)
19. Dembo, A., Montanari, A., Sun, N.: Factor models on locally tree-like graphs. The Annals of Probability **41**(6), 4162–4213 (2013)
20. Elkies, N., Kuperberg, G., Larsen, M., Propp, J.: Alternating-sign matrices and domino tilings (Part I). Journal of Algebraic Combinatorics **1**(2), 111–132 (1992)
21. Friedland, S., Gurvits, L.: Lower bounds for partial matchings in regular bipartite graphs and applications to the monomer–dimer entropy. Combinatorics, Probability and Computing **17**(03), 347–361 (2008)
22. Friedland, S., Krop, E., Markström, K.: On the number of matchings in regular graphs. The Electronic Journal of Combinatorics **15**(1), R110 (2008)
23. Friedland, S., Peled, U.N.: Theory of computation of multidimensional entropy with an application to the monomer–dimer problem. Advances in Applied Mathematics **34**(3), 486–522 (2005)
24. Gamarnik, D., Katz, D.: Sequential cavity method for computing free energy and surface pressure. Journal of Statistical Physics **137**(2), 205–232 (2009)

25. Godsil, C.D.: Algebraic combinatorics, vol. 6. CRC Press (1993)
26. Greenhill, C., Janson, S., Ruciński, A.: On the number of perfect matchings in random lifts. Combinatorics, Probability and Computing **19**(5-6), 791–817 (2010)
27. Gurvits, L.: Van der Waerden/Schrijver-Valiant like conjectures and stable (aka hyperbolic) homogeneous polynomials: one theorem for all. The Electronic Journal of Combinatorics **15**(1) (2008)
28. Gurvits, L.: Unleashing the power of Schrijver's permanental inequality with the help of the Bethe approximation. arXiv preprint arXiv:1106.2844 (2011)
29. Hammersley, J.: An improved lower bound for the multidimensional dimer problem. In: Mathematical Proceedings of the Cambridge Philosophical Society, vol. 64, pp. 455–463. Cambridge Univ Press (1968)
30. Hammersley, J.M.: Existence theorems and Monte Carlo methods for the monomer-dimer problem. Reseach papers in statistics: Festschrift for J. Neyman, edited by FN David, Wiley, London pp. 125–146 (1966)
31. Hammersley, J.M., Menon, V.V.: A lower bound for the monomer-dimer problem. IMA Journal of Applied Mathematics **6**(4), 341–364 (1970)
32. Heilmann, O.J., Lieb, E.H.: Theory of monomer-dimer systems. Communications in Mathematical Physics pp. 190–232 (1972)
33. Huo, Y., Liang, H., Liu, S.Q., Bai, F.: Computing monomer-dimer systems through matrix permanent. Physical Review E **77**(1), 016706 (2008)
34. Kasteleyn, P.W.: The statistics of dimers on a lattice: I. The number of dimer arrangements on a quadratic lattice. Physica **27**, 1209–1225 (1961)
35. Kenyon, R., Okounkov, A., Sheffield, S.: Dimers and amoebae. Annals of Mathematics pp. 1019–1056 (2006)
36. Kesten, H.: Symmetric random walks on groups. Transactions of the American Mathematical Society **92**(2), 336–354 (1959)
37. Laurent, M., Schrijver, A.: On Leonid Gurvits's proof for permanents. The American Mathematical Monthly **117**(10), 903–911 (2010)
38. Lelarge, M.: Counting matchings in irregular bipartite graphs and random lifts. In: Proceedings of the Twenty-Eighth Annual ACM-SIAM Symposium on Discrete Algorithms, pp. 2230–2237. Society for Industrial and Applied Mathematics (2017)
39. Linial, N.: Lifts of graphs (talk slides). http://www.cs.huji.ac.il/~nati/PAPERS/lifts_talk.pdf
40. Lovász, L.: Large networks and graph limits, vol. 60. American Mathematical Soc. (2012)
41. Lyons, R.: Asymptotic enumeration of spanning trees. Combinatorics, Probability and Computing **14**(04), 491–522 (2005)
42. McKay, B.D.: The expected eigenvalue distribution of a large regular graph. Linear Algebra and its Applications **40**, 203–216 (1981)
43. McKay, B.D.: Spanning trees in regular graphs. European Journal of Combinatorics **4**(2), 149–160 (1983)
44. Ruozzi, N.: The Bethe partition function of log-supermodular graphical models. In: Advances in Neural Information Processing Systems, pp. 117–125 (2012)
45. Schrijver, A.: Counting 1-factors in regular bipartite graphs. Journal of Combinatorial Theory, Series B **72**(1), 122–135 (1998)
46. Schrijver, A., Valiant, W.G.: On lower bounds for permanents. In: Indagationes Mathematicae (Proceedings), vol. 83, pp. 425–427. Elsevier (1980)
47. Sly, A., Sun, N.: The computational hardness of counting in two-spin models on d-regular graphs. In: Foundations of Computer Science (FOCS), 2012 IEEE 53rd Annual Symposium on, pp. 361–369. IEEE (2012)
48. Sly, A., Sun, N.: Counting in two-spin models on d–regular graphs. The Annals of Probability **42**(6), 2383–2416 (2014)
49. Straszak, D., Vishnoi, N.K.: Belief propagation, Bethe approximation and polynomials. arXiv preprint arXiv:1708.02581 (2017)
50. Temperley, H.N.V., Fisher, M.E.: Dimer problem in statistical mechanics-an exact result. Philosophical Magazine **6**(68), 1061–1063 (1961)

51. Vontobel, P.O.: Counting in graph covers: A combinatorial characterization of the Bethe entropy function. IEEE Transactions on Information Theory **59**(9), 6018–6048 (2013)
52. Voorhoeve, M.: A lower bound for the permanents of certain (0, 1)-matrices. In: Indagationes Mathematicae (Proceedings), vol. 82, pp. 83–86. Elsevier (1979)
53. Wilf, H.S.: On the permanent of a doubly stochastic matrix. Canadian Journal of Mathematics **18**, 758–761 (1966)

Sequential Importance Sampling for Estimating the Number of Perfect Matchings in Bipartite Graphs: An Ongoing Conversation with Laci

Persi Diaconis

Abstract Sequential importance sampling offers an alternative way to approximately evaluate the permanent. It is a stochastic algorithm which seems to work in practice but has eluded analysis. This paper offers examples where the analysis can be carried out and the first general bounds for the sample size required. This uses a novel importance sampling proof of Brégman's inequality due to Lovász.

Keywords Brégman's inequality · Permanent

MSC 2010 05A16 · 60C05

1 Introduction

Let $G = (V, W, E)$ be a bipartite graph with $|V| = |W| = n$ and E a set of edges from V to W. Let $\mathcal{M}$ be the set of perfect matchings. Assume throughout that $\mathcal{M}$ is non-empty. There is a large literature on computing and approximating $M = |\mathcal{M}|$. See [6] for background and applications in statistics. The magisterial [14] covers every aspect of matching theory. This paper studies an importance sampling algorithm for Monte Carlo approximation of M.

Algorithm 1 Let $v_1, v_2, \ldots, v_n$ be an enumeration of the vertices in V. Beginning at v_1 and proceeding in order:

- Check each edge coming out of v_1 to see if its removal, and the subsequent removal of the adjacent edges, leaves a graph allowing a perfect matching. Let J_1 be the set of available edges. Pick $e \in J_1$ uniformly and delete this edge.
- Repeat with v_2, forming J_2, and continue until a perfect matching is found. (The last step is forced.)

P. Diaconis (✉)
Departments of Mathematics and Statistics, Stanford University, Sequoia Hall, 390 Serra Mall, Stanford, CA 94305, USA
e-mail: diaconis@math.stanford.edu

- This generates a random matching π with probability

$$P(\pi) = \prod_{i=1}^{n} |J_i|^{-1}.$$

Let $T(\pi) = 1/P(\pi)$. Then $\sum_{\pi \in \mathcal{M}} P(\pi) = 1$ and

$$E(T(\pi)) = \sum_{\pi \in \mathcal{M}} T(\pi) P(\pi) = M.$$

This gives an unbiased estimate of M and one proceeds to generate $\pi_1, \pi_2, \ldots, \pi_N$ independently to give $T_1, T_2, \ldots, T_N$. The estimate is

$$\widehat{M} = \frac{1}{N} \sum_{j=1}^{N} T_i.$$

This is a sequential importance sampling algorithm: sequential because π is built up one step at a time, and importance sampling because $1/P(\pi)$ is used to weight π. A host of similar algorithms are in active use to estimate things like the number of graphs with given degree sequences [1], or the number of zero/one tables with given row and column sums [3], or the number of self-avoiding paths in a graph [10]. See [2] or [13] for surveys.

These importance sampling algorithms have notoriously large variability and it is natural to ask how large a sample N is required to assure $\widehat{M}$ is accurate, e.g., $P\{|\widehat{M} - M| > \epsilon\}$ is small. A fresh approach to this problem is suggested by [2]. We use their notation in the following.

If $\mathcal{X}$ is a measurable space, ν a probability on $\mathcal{X}$, $f : \mathcal{X} \to \mathbb{R}$, with $I(f) = \int f(x) \nu(dx) < \infty$. Let μ be a second probability on $\mathcal{X}$ which is "easy to sample from", and with $\mu \ll \nu$. Set $\rho = d\mu/d\nu$ so

$$\int f \rho \, d\mu = \int f \frac{d\nu}{d\mu} \, d\mu = I(f).$$

If $X_1, X_2, \ldots, X_N$ is an i.i.d. sample from μ, the importance sampling estimator is

$$\widehat{I}_N(f) = \frac{1}{N} \sum_{i=1}^{N} f(X_i) \rho(X_i).$$

Define the Kullback–Leibler divergence by

$$L = D(\nu \mid \mu) = \int \rho \log \rho \, d\mu = \int \log \rho \, d\nu = E_\nu \{\log \rho(Y)\}.$$

The main result of [2] shows that "e^L steps are necessary and sufficient for convergence" in the following sense:

(a) If $\|f\|_{2,L^2(\nu)} < \infty$ and $N = e^{L+t}$ for $t > 0$,

$$E\left|\widehat{I}_N(f) - I(f)\right| \leq \|f\|_{2,L^2(\nu)} \left[e^{-t/4} + 2P_\nu^{1/2}\left(\log \rho(Y) > L + \frac{t}{2}\right)\right].$$

(b) Conversely, if $f \equiv 1$ and $N = e^{L-t}$, $t > 0$, for any $\delta > 0$,

$$P\left\{\widehat{I}_N(f) > (1-\delta)\right\} \leq e^{-t/2} + P_\nu\left(\log \rho(Y) \leq L - t/2\right)\big/(1-\delta).$$

Remarks

1. To help think about this, suppose $\|f\|_{2,\nu} \leq 1$, e.g., f is the indicator function of a set. Part (a) says that if $N > e^{L+t}$ and $\log \rho(Y)$ is concentrated about its mean $(E_\nu\{\log \rho(Y)\} = L)$, then $\widehat{I}_N(f)$ is close to $I(f)$ with high probability. (Use Markov's inequality with (a).)
2. Conversely, part (b) shows that if $N = e^{L-t}$ and $\log \rho(Y)$ is concentrated about its mean, then of course $I(1) = 1$, but there is only a small probability that $\widehat{I}_N(1)$ is correct.

In the special case of perfect matchings, $\mathcal{X}$ is the set of perfect matchings

$$\nu(\pi) = 1/|\mathcal{X}|, \quad \mu(\pi) = \prod_{i=1}^n |J_i(\pi)|^{-1} \quad (\pi \in \mathcal{M} \text{ in the earlier notation}),$$

and

$$\rho(\pi) = \frac{d\nu}{d\mu}(\pi) = \prod_{i=1}^n |J_i(\pi)|/|\mathcal{X}|, \tag{1}$$

$$L = \frac{1}{|\mathcal{X}|} \sum_{\pi \in \mathcal{M}} \log \rho(\pi) = -E_\nu \log P(\pi) - \log |\mathcal{X}|. \tag{2}$$

The questions now become, "Given a graph G what is L, and is $\log \rho(Y)$ concentrated?"

In the next section, Lovász's proof of Brégman's inequality is used to show that if the ordering $v_1, v_2, \ldots, v_n$ is chosen randomly,

$$N = e^L \leq \prod (d_i!)^{1/d_i}\big/|\mathcal{X}|,$$

with d_i the degree of v_i in G. Section 3 works out a simple example using Fibonacci permutations where much can be computed. Section 4 suggests related research.

This paper began during a birthday conference for Donald Knuth in January 2018. Knuth had used sequential importance sampling to estimate the number of self-avoiding walks in an $n \times n$ grid [10] and often uses it to estimate the running time of backtracking algorithms [11]. At the conference I presented my work with Chatterjee applied to these problems. After my talk, Laci came up and said, "I once did some things that seem related. I hadn't heard about your topics but maybe they will be useful." He sent me some class notes using sequential importance sampling as a proof technique for Brégman's inequality which led to the bound outlined in this section. These sequential importance algorithms are widely used and this is the first case of a useful general theorem. Laci's book with Plummer is filled with a host of problems "equivalent to matchings." Thus Hall's marriage theorem (When does a graph have a perfect matching?), the Gale–Ryser theorem (When are two sets of numbers the row and column sums of a zero/one table?), and the *Erdős– Gallai* theorem (When is a given set of numbers the degree sequence of a graph?) are derivable from one another. All are used in the sequential sampling applications to guarantee that the algorithm doesn't "get stuck." Are there analogs of Brégman's inequality to these related problems? Does there exist a stochastic proof technique that allows parallel bounds on required sample size? I know that Laci has things to say about such questions and hope this paper will allow us to continue the conversation.

2 Brégman's Inequality

This section begins with Lovász's communication and then applies it to sequential importance sampling.

2.1 Lovász Communication

On January 10, 2018, Lovász sent the following communication (probably course notes). There are similar proofs of Brégman's inequality [15] but these gave just what was needed.

Branching Counting

Let F be a rooted tree of depth k, such that all N leaves of F are on the lowest level. For every leaf u let P_u be the path from the root to u. Let X be a random leaf, obtained by starting from the root, and wherever we are, selecting one child uniformly. Let Y be a uniformly selected leaf.

For every path P from the root to a leaf, let $d(P)$ be the product of the down-degrees of its node (except the leaf). Then

$$P(X = u) = \frac{1}{d(P_u)}.$$

Hence
$$\sum_u \frac{1}{d(P_u)} = 1$$

and
$$E(d(P_X)) = N.$$

The variance of the random variable $d(P_X)$ can be very large, so the following inequalities may be more useful:

Lemma
$$E(\log d(P_X)) \le \log N \le E(\log d(P_Y)).$$

Proof Since $\log x$ is a concave function, Jensen's inequality gives
$$E(\log d(P_Y)) = -\frac{1}{N} \sum_u \log \frac{1}{d(P_u)} \ge -\frac{1}{N} \cdot N \cdot \log \frac{1}{N} = \log N.$$

On the other hand, $x \log x$ is convex, so
$$E(\log d(P_X)) = -\sum_u \frac{1}{d(P_u)} \log \frac{1}{d(P_u)} \le N \left(-\frac{1}{N} \log \frac{1}{N}\right) = \log N.$$

Note: $E(\log d(P_X))$ is the entropy of X. □

Brégman's Theorem

Theorem 2.1 (Brégman) *Let G be a simple bipartite graph with bipartition $\{U, W\}$. Let $d_1, \ldots, d_n$ be the degrees of the nodes in W. Then the number of perfect matchings in G is at most*
$$(d_1!)^{1/d_1} \ldots (d_n!)^{1/d_n}.$$

Proof Let $\pi = (v_1, \ldots, v_n)$ be a permutation of W. Construct the following tree F_π: its nodes are those partial matchings that can be extended to a perfect matching, and match $v_1, \ldots, v_k$ for some k among the nodes of W. A node's parent is obtained by deleting the edge from v_k.

For a perfect matching M, permutation π, and $w \in W$, let $a(M, \pi, w)$ be the number of those neighbors of w that are matched by M to a node in W *not* preceding w in π. Let $M_{\pi,w}$ be the submatching of M covering exactly those nodes of W that precede w, then the degree in F_π of $M_{\pi,w}$ is at most $a(M, \pi, w)$. Hence for every perfect matching M,

$$\log d(P_M) = \sum_{w \in W} \log d_{F_\pi}(F(M_{pi,w})) \leq \sum_{w \in W} \log a(M, \pi, w).$$

Hence

$$\log N \leq E_Y(\log d(P_Y)) \leq \sum_{w \in W} E_Y(\log a(Y, \pi, w)).$$

Averaging over all permutations π,

$$\log N \leq \sum_{w \in W} E_\pi E_Y(\log a(Y, \pi, w)) = \sum_{w \in W} E_Y E_\pi(\log a(Y, \pi, w)).$$

In a random permutation π, the nodes in W matched with neighbors of w, the position of w is first, second, etc., with the same probability. Hence

$$E_\pi(\log a(Y, \pi, w)) = \frac{1}{d(w)} \sum_{j=1}^{d(w)} \log j = \frac{1}{d(w)} \log(d(w)!),$$

and so

$$\log N \leq \sum_{w \in W} E_Y \left(\frac{1}{d(w)} \log(d(w)!) \right) = \sum_{w \in W} \frac{1}{d(w)} \log(d(w)!). \qquad \square$$

2.2 Application to Importance Sampling

Consider the algorithm of Sect. 1 where the ordering $v_1, v_2, \ldots, v_n$ is chosen uniformly at random. Lovász's argument above shows

$$-E_v \log P(\pi) \leq \sum_{i=1}^{n} \frac{1}{d(v_i)} \log(d(v_i)!).$$

Using this in expression (2) gives

$$L \leq \sum_{i=1}^{n} \frac{1}{d(v_i)} \log(d(v_i)!) - \log |\mathcal{X}|, \qquad (3)$$

so

$$e^L \leq \prod (d_i!)^{1/d_i} / |\mathcal{X}|.$$

In a practical application of the algorithm one might use a deterministic ordering of the vertices, say by increasing degree. Also, there is no need to make the choices uniform; again, matching to low-degree edges makes some sense. As long as the chance of π is computable, $1/P(\pi)$ gives an unbiased estimator. In the parallel problems of graphs with given degree sequences and contingency tables, these practical choices can make a large difference. It seems difficult to give useful explicit bounds for the N required in these variations.

3 Example: Fibonacci Matchings

Here is a simple example where all the ingredients can be computed. Consider a tridiagonal restriction matrix A_n: ones on the main, super-, and subdiagonal, and zeros elsewhere.

$$A_4 = \begin{pmatrix} 1 & 1 & 0 & 0 \\ 1 & 1 & 1 & 0 \\ 0 & 1 & 1 & 1 \\ 0 & 0 & 1 & 1 \end{pmatrix}.$$

The number of perfect matchings is the Fibonacci number F_{n+1}. Indeed, 1 can only be matched to 1' or 2'. If it is matched to 1' the deleted graph is A_{n-1}. If it is matched to 2', then 2 must be matched to 1' and the deleted graph is A_{n-2}. These matchings are illustrated in Fig. 1. Thus there are five perfect matchings consistent with A_4; these are shown in Table 1 along with their associated probabilities if the vertices are tried in order 1, 2, 3, 4 for $P_1(\pi)$ and 2, 3, 4, 1 for $P_2(\pi)$. For larger n the possible matching probabilities can be quite different. In what follows the vertex order $1, 2, \ldots, n$ is studied.

Figure 2 shows a histogram of 1000 T_i when $n = 10$; then $F_{n+1} = 89$. The mean of 86.72 is reasonable but the minimum of 24, maximum of 288, and standard deviation of 45.3 give an indication of large variability.

Lemma 3.1 *The random matching algorithm of Sect. 1 for vertex order* $1, 2, \ldots, n$ *and the Fibonacci restriction matrix* A_n *results in choosing matching* π *with probability*

$$P_1(\pi) = \frac{1}{2^{n_{1-k}}},$$

Fig. 1 Fibonacci matchings

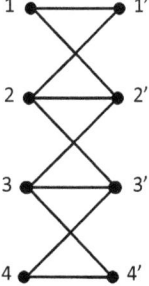

Table 1 Perfect matchings

π	1234	2134	1324	1243	2143
$P_1(\pi)$	1/8	1/4	1/4	1/8	1/4
$P_2(\pi)$	1/6	1/6	1/3	1/6	1/6

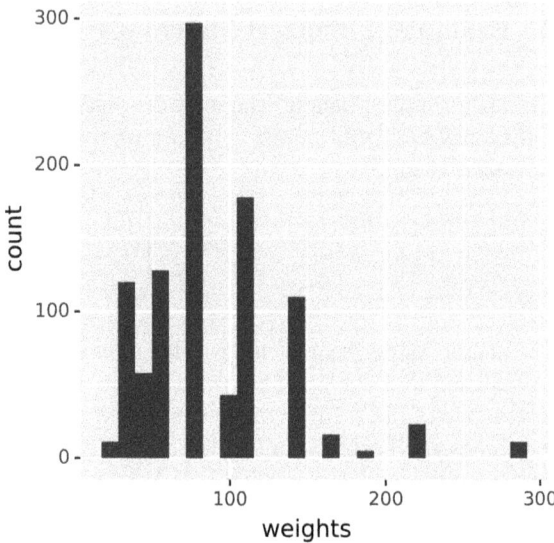

Fig. 2 Histogram of importance sampling weights

with $k_n = k(\pi)$ the number of transpositions in π not counting $(n, n-1)$.

Proof Working in order, if $1 \leftrightarrow 2$ is chosen then $2 \leftrightarrow 1$ is forced. If $1 \leftrightarrow 1$ is chosen then $2 \leftrightarrow 2$ or $2 \leftrightarrow 3$ is possible. This restarts after each transposition. The last move is always forced. □

Corollary 3.2 *For the ordering* $1, 2, \ldots, n$, *with* $\rho(\pi)$ *defined at* (1),

$$\log \rho(\pi) = \log 2\left[(n-1) - k(\pi)\right] - \log F_{n+1}.$$

The behavior of $k(\pi)$ follows from work of [6]. If $\kappa(\pi)$ is the number of transpositions in a Fibonacci permutation, under the uniform distribution $u(\pi) = 1/F_{n+1}$ they show, as n tends to infinity,

$$E_u(\kappa) = \frac{n(\sqrt{5}-1)}{2\sqrt{5}} + O(1),$$

$$\text{Var}_u(\kappa) = \frac{n}{5\sqrt{5}} + O(1),$$

$$P_u\left\{\frac{\kappa - \text{mean}}{\text{s.d.}} \leq x\right\} \to \frac{1}{\sqrt{2\pi}}\int_{-\infty}^{x} e^{-t^2/2}\, dt,$$

since $k(\pi) = \kappa(\pi) - \epsilon$ with $\epsilon = 1$ or 0. This proves:

Corollary 3.3 *Under the uniform distribution on perfect matchings, $\log \rho(\pi)$ is concentrated about its mean.*

It follows that the "$N = e^L$" sample size theorems of Sect. 1 are in force.

Corollary 3.4 *Let $\widehat{N}_1$ and $\widehat{N}_2$ be the sample sizes required from sequential importance sampling using the fixed order, Case 1, or a random order, Case 2. Then*

$$\widehat{N}_1 = \frac{2^{n\left(\frac{1}{2}+\frac{1}{2\sqrt{5}}\right)+O(n^{1/2})}}{F_{n+1}} = \frac{e^{n(0.5016\dots)+O(n^{1/2})}}{F_{n+1}} = e^{n(0.0204)+O(n^{1/2})},$$

$$\widehat{N}_2 = \frac{6^{n/3}+O(n^{1/2})}{F_{n+1}} = \frac{e^{n(0.5973\dots)+O(n^{1/2})}}{F_{n+1}} = e^{n(0.1161)+O(n^{1/2})}.$$

Remarks From these calculations, both algorithms require exponential sample sizes. However, the exponential constants are small, 0.02 and 0.11, and the deterministic ordering is slightly better. The Brégman is just an upper bound. Indeed, in joint work with Brett Kolesnik, we have computed $-E_r \log p(\pi) = x_n$ when a random ordering of the vertices is used in Algorithm 1. The result is

$$x_n = cn\,(1+o(1)), \qquad c = \left(\frac{13}{6} - \frac{2}{\sqrt{5}}\right)\frac{\log 2}{5} + \left(1 + \frac{1}{\sqrt{5}}\right)\frac{\log 3}{5} \doteq 0.4944.$$

Note that $c < (\log 2)(1 + 1/\sqrt{5})/2 \doteq 0.5016$ so random ordering is (slightly) better than deterministic, i.e., $\widehat{N} = e^{n(0.0132)+O(n^{1/2})}$.

4 Final Remarks

There are other methods for estimating the number of matchings in bipartite graphs, chief among them the Markov chain Monte Carlo algorithms. See [7, 9]. These come with provable guarantees but don't seem to work as well in practice. Extensions of the present approach with many further examples and theorems are in [4, 5]. A review of other algorithms and the need for random generation (as opposed to counting) is in [6].

Use of sequential importance sampling as a proof technique, as in Lovász's proof of Brégman's inequality, is interesting. [12] prove inequalities of Hoffman and Sidorenko for counting graph homomorphisms this way and [8] settle a graph coloring conjecture of Tomescu this way.

Finally, there are many classes of graphs where the number of perfect matchings is known. These include Ferris-type restrictions and dimer covering problems. These are grist for the mill of understanding sequential importance sampling. Fortunately, the wonderful book of [14] can lead us through this part of the maze.

Acknowledgements Support is acknowledged by National Science Foundation award DMS 1608182. My thanks to Paulo Ornstein for trying these estimators out; to Sourav Chatterjee for help with importance sampling; and to Joe Blitzstein whose thesis work suggested using sequential importance sampling for bipartite matchings, Leo Nagami and Don Knuth who gave detailed corrections added in proof. Thanks as well to Fan Chung, Ron Graham, and Brett Kolesnik, and to Laci Lovász for starting the conversation.

References

1. Blitzstein J, Diaconis P (2010) A sequential importance sampling algorithm for generating random graphs with prescribed degrees. Internet Math 6(4):489–522, https://doi.org/10.1080/15427951.2010.557277
2. Chatterjee S, Diaconis P (2018) The sample size required in importance sampling. Ann Appl Probab 28(2):1099–1135, https://doi.org/10.1214/17-AAP1326
3. Chen Y, Diaconis P, Holmes SP, Liu JS (2005) Sequential Monte Carlo methods for statistical analysis of tables. J Amer Statist Assoc 100(469):109–120
4. Chung F, Diaconis P, Graham R (2018) Permanental generating functions and sequential importance sampling, Adv. Applied Mathematics to appear
5. Diaconis P, Kolesnik B (2018) Randomized sequential importance sampling for estimating the number of perfect matchings in bipartite graphs, arxiv:1907.02333
6. Diaconis P, Graham R, Holmes SP (2001) Statistical problems involving permutations with restricted positions. In: State of the art in probability and statistics (Leiden, 1999), IMS Lecture Notes Monogr. Ser., vol 36, Inst. Math. Statist., Beachwood, OH, pp 195–222
7. Dyer M, Jerrum M, Müller H (2017) On the switch Markov chain for perfect matchings. J ACM 64(2):Art. 12, 33, https://doi.org/10.1145/2822322
8. Fox J, He X, Manners F (2017) A proof of Tomescu's graph coloring conjecture. ArXiv e-prints arxiv:1712.06067
9. Jerrum M, Sinclair A, Vigoda E (2004) A polynomial-time approximation algorithm for the permanent of a matrix with nonnegative entries. J ACM 51(4):671–697 (electronic)
10. Knuth DE (1976) Mathematics and computer science: Coping with finiteness. Science 194(4271):1235–1242, https://doi.org/10.1126/science.194.4271.1235
11. Knuth DE (2018) The Art of Computer Programming, Volume 4B, 1st edn. Addison-Wesley Professional, Fascicle 5: Mathematical Preliminaries Redux; Backtracking; Dancing Links
12. Levin DA, Peres Y (2017) Counting walks and graph homomorphisms via Markov chains and importance sampling. Amer Math Mon 124(7):637–641, https://doi.org/10.4169/amer.math.monthly.124.7.637
13. Liu JS (2001) Monte Carlo Strategies in Scientific Computing. Springer Series in Statistics, Springer-Verlag, New York
14. Lovász L, Plummer MD (2009) Matching Theory. AMS Chelsea Publishing, Providence, RI, https://doi.org/10.1090/chel/367

15. Spencer J (1990) The probabilistic lens: Sperner, Turán and Brégman revisited. In: A Tribute to Paul Erdős, Cambridge Univ. Press, Cambridge, pp 391–396

Tighter Bounds for Online Bipartite Matching

Uriel Feige

Abstract We study the online bipartite matching problem, introduced by Karp, Vazirani and Vazirani [1990]. For bipartite graphs with matchings of size n, it is known that the *Ranking* randomized algorithm matches at least $(1 - \frac{1}{e})n$ edges in expectation. It is also known that no online algorithm matches more than $(1 - \frac{1}{e})n + O(1)$ edges in expectation, when the input is chosen from a certain distribution that we refer to as D_n. This upper bound also applies to *fractional* matchings. We review the known proofs for this last statement. In passing we observe that the $O(1)$ additive term (in the upper bound for fractional matching) is $\frac{1}{2} - \frac{1}{2e} + O(\frac{1}{n})$, and that this term is tight: the online algorithm known as *Balance* indeed produces a fractional matching of this size. We provide a new proof that exactly characterizes the expected cardinality of the (integral) matching produced by *Ranking* when the input graph comes from the support of D_n. This expectation turns out to be $(1 - \frac{1}{e})n + 1 - \frac{2}{e} + O(\frac{1}{n!})$, and serves as an upper bound on the performance ratio of any online (integral) matching algorithm.

Keywords Online matching · Ranking algorithm · Fractional matching · Integer sequences

Subject Classifications 68W27 · 68Q25

1 Introduction

Given a bipartite graph $G(U, V; E)$, where U and V are the sets of vertices and $E \in U \times V$ is the set of edges, a matching $M \subset E$ is a set of edges such that every vertex is incident with at most one edge of M. Given a matching M, a vertex is referred to as either matched or exposed, depending on whether it is incident with

---

U. Feige (✉)
Department of Computer Science and Applied Mathematics,
The Weizmann Institute, Rehovot, Israel
e-mail: uriel.feige@weizmann.ac.il

an edge of M. A maximum matching in a graph is a matching of maximum cardinality, and a maximal matching is a matching that is not a proper subset of any other matching. Maximal matchings can easily be found by greedy algorithms, and maximum matchings can also be found by various polynomial time algorithms, using techniques such as alternating paths or linear programming (see [9] and references therein). In every graph, the cardinality of every maximal matching is at least half of that of the maximum matching, because every matched edge can exclude at most two edges from the maximum matching.

For simplicity of notation, for every n we shall only consider the following class of bipartite graphs, that we shall refer to as G_n. For every $G(U, V; E) \in G_n$ it holds that $|U| = |V| = n$ and that E contains a matching of size n (and hence G has a perfect matching). The vertices of U will be denoted by u_i (for $1 \leq i \leq n$) and the vertices of V will be denoted by v_i (for $1 \leq i \leq n$). All results that we will state for G_n hold without change for all bipartite graphs, provided that n denotes the size of the maximum matching in the graph.

Karp, Vazirani and Vazirani [7] introduced an online version of the maximum bipartite matching problem. This setting can be viewed as a game between two players: a maximizing player who wishes the resulting matching to be as large as possible, and a minimizing player who wishes the matching to be as small as possible. First, the minimizing player chooses $G(U, V; E)$ in private (without the maximizing player seeing E), subject to $G \in G_n$. Thereafter, the structure of G is revealed to the maximizing player in n steps, where at step j (for $1 \leq j \leq n$) the set $N(u_j) \subset V$ of vertices adjacent to u_j is revealed. At every step j, upon seeing $N(u_j)$ (and based on all edges previously seen and all previous matching decisions made), the maximizing player needs to irrevocably either match u_j to a currently exposed vertex in $N(u_j)$, or leave u_j exposed.

There is much recent interest in the online bipartite matching problem and variations and generalizations of it, as such models have applications for allocation problems in certain economic settings, in which buyers (vertices of U) arrive online and are interested in purchasing various items (vertices of V). For more details, see for example the survey by Metha [11].

An algorithm for the maximizing player in the online bipartite matching setting will be called *greedy* if the only vertices of U that it leaves unmatched are those vertices $u \in U$ that upon their arrival did not have an exposed neighbor (and hence could not be matched). It is not difficult to see that every non-greedy algorithm A can be replaced by a greedy algorithm A' that for every graph G matches at least as many vertices as A does. Hence we shall assume that the algorithm for the maximizing player is greedy, and this assumption is made without loss of generality, as far as the results in this manuscript are concerned.

Every greedy algorithm (for the maximizing player) produces a maximal matching, and hence matches at least half the vertices. For every deterministic algorithm, the minimizing player can select a bipartite graph G (that admits a perfect matching) that guarantees that the algorithm matches only half the vertices. (Sketch: The first $\frac{|U|}{2}$ arriving vertices have all of V as their neighbors, and the remaining $\frac{|U|}{2}$

are neighbors only of the $\frac{|V|}{2}$ vertices that the algorithm matched with the first $\frac{|U|}{2}$ vertices.)

To improve the size of the matching beyond $\frac{n}{2}$, Karp, Vazirani and Vazirani [7] considered randomized algorithms for the maximizing player. Specifically, they proposed an algorithm called *Ranking* that works as follows. It first selects uniformly at random a permutation π over the vertices V. Thereafter, upon arrival of a vertex u, it is matched to its earliest (according to π) exposed neighbor if there is one (and left unmatched otherwise). As the maximizing algorithm is randomized (due to the random choice of π), the number of vertices matched is a random variable, and we consider its expectation.

Let A be a randomized algorithm (such as *Ranking*) for the maximizing player. As such, for every bipartite graph G it produces a distribution over matchings. For a bipartite graph $G \in G_n$, we use the following notation:

- $\rho_n(A, G)$ is the expected cardinality of matching produced by A when the input graph is G.
- $\rho_n(A, -)$ is the minimum over all $G \in G_n$ of $\rho_n(A, G)$. Namely, $\rho_n(A, -) = \min_{G \in G_n}[\rho_n(A, G)]$.
- ρ_n is the maximum over all A (randomized online matching algorithms for the maximizing player) of $\rho_n(A, -)$. Namely, $\rho_n = \max_A[\rho_n(A, -)]$. (Showing that the maximum is attained is a technicality that we ignore here.)
- $\rho = \inf_n \frac{\rho_n}{n}$. Namely, ρ is the largest constant (independent of n) such that $\rho \cdot n \leq \rho_n$ for all n. (One might find a definition such as $\rho = \lim_{n \to \infty} \frac{\rho_n}{n}$ more natural, but it turns out that both definitions of ρ give the same value, which will be seen to be $1 - \frac{1}{e}$.)

Karp, Vazirani and Vazirani [7] showed that $\rho_n(Ranking, -) \geq (1 - \frac{1}{e})n - o(n)$, where e is the base of the natural logarithm (and $(1 - \frac{1}{e}) \simeq 0.632$). Unfortunately, that paper had only a conference version and not a journal version, and the proof presented in the conference version appears to have gaps. Later work (e.g., [2, 4, 12]), motivated by extensions of the online matching problem to other problems such as the *adwords* problem, presented alternative proofs, and also established that the $o(n)$ term is not required. There have also been expositions of simpler versions of these proofs. See [1, 3, 10], for example. Summarizing this earlier work, we have:

Theorem 1.1 *For every bipartite graph $G \in G_n$, the expected cardinality of the matching produced by* Ranking *is at least* $(1 - \frac{1}{e})n$. *Hence* $\rho_n(Ranking, -) \geq (1 - \frac{1}{e})n$, *and* $\rho \geq 1 - \frac{1}{e} \simeq 0.632$.

Karp, Vazirani and Vazirani [7] also presented a distribution over G_n, and showed that for every online algorithm, the expected size of the matching produced (expectation taken over random choice of graph from this distribution) is at most $(1 - 1/e)n + o(n)$. This distribution, that we shall refer to as D_n, is defined as follows. Select uniformly at random a permutation τ over V. For every j, the neighbors of vertex u_j are $\{v_{\tau(j)}, \ldots, v_{\tau(n)}\}$. The unique perfect matching M is the set of edges $(u_j, v_{\tau(j)})$ for $1 \leq j \leq n$.

To present the known results regarding D_n more accurately, let as extend previous notation.

- $\rho_n(A, D_n)$ is the expected cardinality of matching produced by A when the input graph G is selected according to distribution D_n. (Hence expectation is taken both over randomness of A and over selection from D_n.) By definition, for every algorithm A, $\rho_n(A, D_n)$ is an upper bound on $\rho_n(A, -)$.
- $\rho_n(-, D_n) \triangleq \max_A[\rho_n(A, D_n)]$ is the maximum over all A (randomized online algorithms for the maximizing player) of $\rho_n(A, D_n)$. By definition, for every n, $\rho_n(-, D_n)$ is an upper bound on ρ_n.

It is not hard to see (and was shown also in Lemma 13 of [7]) that for every two greedy online algorithms A and A' it holds that $\rho_n(A, D_n) = \rho_n(A', D_n)$. As greedy algorithms are optimal among online algorithms, and *Ranking* is a greedy algorithm, we have the following proposition.

Proposition 1.2 *For D_n defined as above,*

$$\rho_n(Ranking, D_n) = \rho_n(-, D_n) \geq \rho_n .$$

The result of [7] can be stated as showing that $\rho_n(-, D_n) \leq (1 - \frac{1}{e})n + o(n)$. Later analysis (see for example [12], or the lecture notes of Kleinberg [8] or Karlin [6]) replaced the $o(n)$ term by $O(1)$. Moreover, this upper bound holds not only for online randomized integral algorithms (that match edges as a whole), but also for online fractional algorithms (that match fractions of edges). Let us provide more details.

A fractional matching for a bipartite graph $G(U, V; E)$ is a nonnegative weight function w for the edges such that for every vertex $u \in U$ we have $\sum_{v \in N(u)} w(u, v) \leq 1$, and likewise, for every vertex $v \in V$ we have $\sum_{u \in N(v)} w(u, v) \leq 1$. The size of a fractional matching is $\sum_{e \in E} w(e)$. It is well known (see [9], for example) that in bipartite graphs, the size of the maximum fractional matching equals the cardinality of the maximum (integral) matching.

In the online bipartite fractional matching problem, as vertices of U arrive, the maximizing player can add arbitrary positive weights to their incident edges, provided that the result remains a fractional matching. We extend the ρ notation used for the integral case also to the fractional case, by adding a subscript f. Hence for example, $\rho_{f,n}(A, G)$ is the size of the fractional matching produced by an online algorithm A when $G \in G_n$ is the input graph.

It is not hard to see that in the fractional setting, randomization does not help the maximizing player, in the sense that any randomized online algorithm A for fractional matching can replaced by a deterministic algorithm A' that on every input graph produces a fractional matching of at least the same size. (Upon arrival of vertex u, the fractional weight that A' adds to edge (u, v) equals the expected weight that A adds to this edge, where expectation is taken over randomness of A.) Consequently, $\rho_{f,n} \geq \rho_n$, and every upper bound on $\rho_{f,n}$ is also an upper bound on ρ_n.

The following theorem summarizes the known upper bounds on $\rho_{f,n}$, which are also the strongest known upper bounds on ρ_n.

Tighter Bounds for Online Bipartite Matching

Theorem 1.3 *For D_n as defined above, $\rho_{f,n}(-, D_n) \leq (1 - \frac{1}{e})n + O(1)$. Consequently, $\rho_n(-, D_n) \leq (1 - \frac{1}{e})n + O(1)$.*

The combination of Theorems 1.1 and 1.3 implies the following corollary:

Corollary 1.4 *Using notation as above, $\rho = 1 - \frac{1}{e}$ and $\rho_f = 1 - \frac{1}{e}$. The Ranking algorithm (which produces an integral matching) is asymptotically optimal (for the maximizing player) for online bipartite matching both in the integral and in the fractional case. The distribution D_n is asymptotically optimal for the minimizing player, both in the integral and in the fractional case.*

In this manuscript, we shall be interested not only in the asymptotic ratios ρ and ρ_f, but also in the exact ratios ρ_n and $\rho_{f,n}$. Every (integral) matching is also a fractional matching, hence one may view *Ranking* also as an online algorithm for fractional matching. As such, *Ranking* is easily seen not to be optimal for some n. For example, when $n = 4$, tedious but straightforward analysis shows that a different known algorithm referred to as *Balance* (see Sect. 2) satisfies $\rho_{f,4}(Balance, -) > \rho_{f,4}(Ranking, -)$ (details omitted). However, for the integral case, it was conjectured in [7] that both *Ranking* and D_n are optimal for every n. Namely, the conjecture is:

Conjecture 1.5 $\rho_n = \rho_n(Ranking, D_n)$ *for every n.*

The above conjecture, though still open, adds motivation (beyond Proposition 1.2) to determine the exact value of $\rho_n(Ranking, D_n)$. This is done in the following theorem.

Theorem 1.6 *Let the function $a(n)$ be such that $\rho_n(Ranking, D_n) = \frac{a(n)}{n!}$ for all n. Then $a(n) = (n + 1)! - d(n + 1) - d(n)$, where $d(n)$ is the number of derangements (permutations with no fixed points) on the numbers $[1, n]$. Consequently, $\rho_n(-, D_n) = (1 - \frac{1}{e})n + 1 - \frac{2}{e} + O(\frac{1}{n!}) \simeq (1 - \frac{1}{e})n + 0.264$, and this is also an upper bound on ρ_n.*

The rest of this paper is organized as follows. In Sect. 2 we review a proof of Theorem 1.3. In doing so, we determine the value of the $O(1)$ term stated in the theorem, and also show that the upper bound is tight. Hence we end up proving the following theorem:

Theorem 1.7 *For every n, Balance is the fractional online algorithm with best approximation ratio, D_n is the distribution over graphs for which the approximation ratio is worst possible, and*

$$\rho_{f,n} = \rho_{f,n}(Balance, D_n) = (1 - \frac{1}{e})n + \frac{1}{2} - \frac{1}{2e} + O(\frac{1}{n}) \simeq (1 - \frac{1}{e})n + 0.316 .$$

In Sect. 3 we prove Theorem 1.6. The combination of Theorems 1.6 and 1.7 implies that $\rho_n < \rho_{f,n}$ for sufficiently large n. It also implies that $\rho_{f,n}(Balance, D_n) >$

$\rho_{f,n}(Ranking, D_n)$ for sufficiently large n. Hence Proposition 1.2 does not extend to online fractional matching.

In an appendix (Sect. 4) we review a proof (due to [3]) of Theorem 1.1, and derive from it an upper bound of $(1 - \frac{1}{e})n + \frac{1}{e}$ on $\rho_n(Ranking, D_n)$. This last upper bound is weaker than the upper bounds of Theorems 1.6 and 1.7, but its proof is different, and hence might turn out useful in attempts to resolve Conjecture 1.5.

1.1 Preliminaries—MonotoneG

When analyzing $\rho_n(Ranking, D_n)$ we shall use the following observation so as to simplify notation. Because *Ranking* is oblivious to names of vertices, the expected size of the matching produced by *Ranking* on every graph in the support of D_n is the same. Hence we shall consider one representative graph from D_n, that we refer to as the monotone graph *MonotoneG*, in which γ (in the definition of D_n) is the identity permutation. The monotone graph $G(U, V; E)$ satisfies $E = \{(u_i, v_j) \mid j \geq i\}$, and its unique perfect matching is $M = \{(u_i, v_i) \mid 1 \leq i \leq n\}$. Statements involving $\rho_n(Ranking, D_n)$ will be replaced by $\rho_n(Ranking, MonotoneG)$, as both expressions have the same value.

Likewise, the algorithm *Balance* is oblivious to names of vertices, and $\rho_{f,n}(Balance, D_n)$ will be replaced by $\rho_{f,n}(Balance, MonotoneG)$.

2 Online Fractional Matchings

Let us present a specific online fractional matching algorithm that is often referred to as *Balance*, which is the natural fractional analog of an algorithm by the same name introduced in [5]. *Balance* maintains a *load* $\ell(v)$ for every vertex $v \in V$, equal to the sum of weights of edges incident with v. Hence at all times, $0 \leq \ell(v) \leq 1$. Upon arrival of a vertex u with a set of neighbors $N(u)$, *Balance* distributes a weight of $\min[1, |N(u)| - \sum_{v \in N(u)} \ell(v)]$ among the edges incident with u, maintaining the resulting loads as balanced as possible. Namely, one computes a threshold t such that $\sum_{v \in N(u) \mid \ell(v) < t}(t - \ell(v)) = \min[1, |N(u)| - \sum_{v \in N(u)} \ell(v)]$, and then adds fractional value $t - \ell(v)$ to each edge (u, v) for those vertices $v \in N(u)$ that have load below t.

We first present a proof of Theorem 1.3 based on previous work. The theorem is restated below, with the additive $O(1)$ term instantiated. Previous work either did not specify the $O(1)$ additive term (e.g., in [6]), or derived an $O(1)$ term that is not tight (e.g., in [8]).

Theorem 2.1 *For every n it holds that*

$$\rho_{f,n}(-, D_n) = (1 - \frac{1}{e})n + \frac{1}{2} - \frac{1}{2e} + O(\frac{1}{n}) \simeq (1 - \frac{1}{e})n + 0.316 \ .$$

Moreover, $\rho_{f,n}(-, D_n) = \rho_{f,n}(Balance, D_n)$.

Proof For all graphs in the support of D_n, the size of the fractional matching produced by *Balance* is the same (by symmetry). Hence for simplicity of notation, consider the fractional matching produced by *Balance* when the input graph is the monotone graph *MonotoneG* (see Sect. 1.1). It is not hard to see that when vertex u_i arrives, *Balance* raises the load of each vertex in $\{v_i, \ldots, v_n\}$ by $\frac{1}{n-i+1}$. This can go on until the largest k satisfying $\sum_{i=1}^{k} \frac{1}{n-i+1} \leq 1$. Thereafter, when vertex u_{k+1} arrives, *Balance* can raise the load of its $n-k$ neighbors from $\sum_{i=1}^{k} \frac{1}{n-i+1}$ to 1. Hence altogether the size of the fractional matching is precisely $k + (n-k)(1 - \sum_{i=1}^{k} \frac{1}{n-i+1})$, for k as above.

The value of k can be determined as follows. It is known that the harmonic number $H_n = \sum_{i=1}^{n} \frac{1}{i}$ satisfies $H_n = \ln n + \gamma + \frac{1}{2n} + O(\frac{1}{n^2})$, where $\gamma \simeq 0.577$ is the Euler-Mascheroni constant. k is the largest integer such that $H_n - H_{n-k} \leq 1$. Defining $\alpha \triangleq \frac{n-k}{n}$, we have that

$$H_n - H_{n-k} = \ln n + \gamma + \frac{1}{2n} + O(\frac{1}{n^2}) - \ln \alpha n - \gamma - \frac{1}{2\alpha n} + O(\frac{1}{n^2})$$
$$= \ln \frac{1}{\alpha} - \frac{\frac{1}{\alpha} - 1}{2n} + O(\frac{1}{n^2}).$$

Choosing $\alpha = \frac{1}{e}$ (and temporarily ignoring the fact that in this case $k = (1 - \frac{1}{e})n$ is not an integer), we get that $H_n - H_{n-k} = 1 - \frac{e-1}{2n} + O(\frac{1}{n^2})$. The size of a matching is then

$$(1 - \frac{1}{e})n + \frac{n}{e}(\frac{e-1}{2n} + O(\frac{1}{n^2})) = (1 - \frac{1}{e})n + \frac{1}{2} + \frac{1}{2e} + O(\frac{1}{n})$$

as desired.

The fact that $k = (1 - \frac{1}{e})n$ above was not an integer requires that we round k down to the nearest integer. The effect of this rounding is bounded by the effect of changing the number of neighbors available to u_k and to u_{k+1} by one (compared to the computation without the rounding). Given that the number of neighbors is roughly $\frac{n}{e}$, the overall effect on the size of the fractional matching is at most $O(\frac{1}{n})$.

We conclude that $\rho_{f,n}(Balance, D_n) = (1 - \frac{1}{e})n + \frac{1}{2} + \frac{1}{2e} + O(\frac{1}{n})$, implying that $\rho_{f,n}(-, D_n) \geq (1 - \frac{1}{e})n + \frac{1}{2} + \frac{1}{2e} + O(\frac{1}{n})$. In remains to show that $\rho_{f,n}(-, D_n) \leq (1 - \frac{1}{e})n + \frac{1}{2} + \frac{1}{2e} + O(\frac{1}{n})$. This follows because *Balance* is the best possible online algorithm (for fractional bipartite matching) against D_n. Let us provide more details.

Given an input graph from the support of D_n, we shall say that a vertex $v \in V$ is *active* in round i if it is a neighbor of u_i. Initially all vertices are active, and after every round, one more vertex (chosen at random among the active vertices) becomes inactive, and remains inactive forever. Let $a(i)$ denote the number of active vertices at the beginning of round i, and note that $a(i) = n - i + 1$. Consider an arbitrary online algorithm. Let $L(i)$ denote the average load of the active vertices at the beginning of round i. Then in round i, the average load first increases by at most $\frac{1}{a(i)}$ (as long as it does not exceed 1) by raising weights of edges, and thereafter, making one vertex inactive keeps the average load unchanged in expectation (over choice of input from D_n). Hence in expectation, in every round, the average load does not exceed the value of the average load obtained by *Balance*. This means that in every round, in expectation, the amount of unused load of the vertex that became inactive is smallest when the online maximizing algorithm is *Balance*. Summing over all rounds and using the linearity of expectation, *Balance* suffers the smallest sum of unused load, meaning that it maximizes the final expected sum (over all V) of loads. The sum of loads equals the size of the fractional matching. □

We now prove Theorem 1.7.

Proof (*Theorem* 1.7) Given Theorem 2.1, it suffices to show that D_n is the worst possible distribution over input graphs for the algorithm *Balance*. Moreover, given that *Balance* is oblivious to the names of vertices, it suffices to show that *MonotoneG* is the worst possible graph for *Balance*.

Let $G \in G_n$ be a graph for which $\rho_{f,n}(Balance, G) = \rho_{f,n}(Balance, -)$. As *Balance* is oblivious to the names of vertices, we may assume that $\{(u_i, v_i) | 1 \leq i \leq n\}$ is a perfect matching in $G = G(U, V; E)$.

We use the notation $N(w)$ to denote the set of neighbors of a vertex w in the graph G. When running *Balance* on G, we use the notation $m(i, j)$ to denote the weight that the fractional matching places on edge (u_i, v_j) (and $m(i, j) = 0$ if $(u_i, v_j) \notin E$), and $m_i(j)$ to denote $\sum_{1 \leq \ell \leq i} m(u_\ell, v_j)$. Clearly, $m_i(j)$ is non-decreasing in i. The size of the final fractional matching is $m = \sum_{j=1}^{n} m_n(j)$. When referring to a graph G', we shall use the notation N' and m' instead of N and m.

An edge (u_i, v_j) with $j < i$ is referred to as a *backward* edge. □

Proposition 2.2 *Without loss of generality, we may assume that G has no backward edges. Hence $m_i(j) = m_j(j)$ for all $i > j$.* □

Proof Suppose otherwise, and let i be largest so that u_i has backward edges. Modify G by removing all backward edges incident with u_i, thus obtaining a graph G'. Compare the performance of *Balance* against the two graphs, G and G'. On vertices $u_1, \ldots, u_{i-1}$, both graphs produce the same fractional matching. The extent to which u_i is matched is at least as large in G as it is in G' (because also backward edges may participate in the fractional matching). Moreover, for every vertex v_j for $i < j \leq n$, it holds that $m'_i(j) \geq m_i(j)$. It follows that for every vertex u_ℓ for $\ell > i$, its marginal contribution to the fractional matching in G is at least as large as its marginal contribution in G'. Hence the fractional matching produced by *Balance* for G' is not larger than that produced for G. Repeating the above argument, all backward edges can be eliminated from G without increasing the size of the fractional matching. □

Lemma 2.3 *Without loss of generality we may assume that:*

1. $m_i(i) \leq m_j(j)$ (or equivalently, $m_n(i) \leq m_n(j)$) for all $i < j$.
2. $m_i(i) \geq m_i(j)$ for all i and j. □

Proof We first present some useful observations. For $1 \leq i < n$, consider the set $N(u_i)$ of neighbours of u_i in G (and recall that $v_i \in N(u_i)$, and that there are no backward edges). Then without loss of generality we may assume that $m_i(i) \geq m_i(j)$ for all $v_j \in N(u_i)$. This is because if there is some vertex $v_j \in N(u_i)$ with $m_i(j) > m_i(i)$, then it must hold (by the properties of *Balance*) that $m(i, j) = 0$. Hence the run of *Balance* would not change if the edge (u_i, v_j) is removed from G (and then $v_j \notin N(u_i)$).

Moreover, we may assume that $m_i(i) = m_i(j)$ for all $v_j \in N(u_i)$. Suppose otherwise. Then for $v_j \in N(u_i)$ with smallest $m_i(j)$, modify G to a graph G' as follows. For all $\ell < i$, make u_ℓ a neighbor of v_i if it was a neighbor of v_j, and make u_ℓ a neighbor of v_j iff it was a neighbor of v_i. The final size of the fractional matching in G' (which is $\sum_{j=1}^{n} m'_n(j)$) cannot be larger than in G. This is because $m'_i(i) < m_i(i)$, $m'_i(j) > m_i(j)$ and for $\ell \neq j$ satisfying $\ell > i$ it holds that $m'_i(\ell) = m_i(\ell)$. Moreover, as $m_i(i) < m_i(j) \leq 1$, u_i is fully matched in G and hence also in G', so the total size of fractional matching after step i is the same in both graphs. Thereafter, the marginal increase of the fractional matching at each step cannot be larger in G' than it is in G.

By the same arguments as above we may assume that $m_{i+1}(i+1) = m_{i+i}(j)$ for all $v_j \in N(u_{i+1})$.

Suppose now that item 1 fails to hold. Then for some $1 \leq i \leq n-1$ it holds that $m_i(i) > m_{i+1}(i+1)$. Vertices u_i and u_{i+1} cannot have a common neighbor because if they do (say, v_ℓ) it holds that $m_{i+1}(i+1) = m_{i+1}(\ell) \geq m_i(\ell) = m_i(i)$. Hence we may exchange the order of u_i and u_{i+1} (and likewise v_i and v_{i+1}) without affecting the size of the fractional matching produced by *Balance*.

Repeating the above argument whenever needed we prove item 1 of the lemma.

For $j < i$ item 2 holds because $m_i(j) = m_j(j) \leq m_i(i)$ (the last inequality follows from item 1). For $j > i$ item 2 holds because at the first point in time $\ell \leq i$ in which $m_\ell(j) = m_i(j)$ it must be that $m_\ell(j) = m_\ell(\ell)$, and item 1 implies that $m_\ell(\ell) \leq m_i(i)$. □

It is useful to note that Lemma 2.3 implies that there is some round number t such that for all $\ell \geq t$ vertex v_ℓ is fully matched (namely, $m_n(\ell) = 1$), and for every $\ell < t$ vertex v_ℓ is not fully matched (namely, $m_n(\ell) < 1$). As to vertices in u, for $\ell < t$ vertex u_ℓ is fully matched, for $\ell > t$ vertex u_ℓ contributes nothing to the fractional matching, and u_t is either partly matched or fully matched. Recalling that m denotes the size of the final fractional matching, we thus have (for t as above):

$$m = t - 1 + \sum_{j \geq t} m(t, j) . \qquad (1)$$

At every step i, the contribution of vertex v_i towards the fractional matching is finalized at that step, namely, $m_n(i) = m_i(i)$. Lemma 2.3 implies that for the worst graph G, this vertex v_i is the one with largest m_i value at this given step. Hence $m_i(i) = \max_{j \geq i}[m_i(j)]$ and we have:

$$m = \sum_{i=1}^{n} m_n(i) = \sum_{i=1}^{n} m_i(i) = \sum_{i=1}^{n} \max_{j \geq i}[m_i(j)] .$$

At this point it is intuitively clear why *MonotoneG* is the graph in G_n on which *Balance* produces the smallest fractional matching. This is because with *MonotoneG*, at each step i the fractional matching gets credited a value $m_i(i)$ that is the average of the values $m_i(j)$ for $j \geq i$, whereas for G its gets credited the maximum of these values. Below we make this argument rigorous.

Consider an alternative *averaging process* replacing algorithm *Balance*. It uses the same fractional matching as in *Balance* and the same $m(i, j)$ values, but maintains values $m'_i(i)$ that may differ from $m_i(i)$. At round 1, instead of being credited the maximum $m_1(1) = \max_{j \geq 1}[m_1(j)]$, the process is credited only the average $m'_1(1) = \frac{1}{n} \sum_{j=1}^{n} m_1(j)$. The remaining $\max_{j \geq 1}[m_1(j)] - \frac{1}{n} \sum_{j=1}^{n} m_1(j)$ is referred to as the *slackness* $s(1)$. More generally, at every round $i > 1$, instead of being credited by $\max_{j \geq i}[m_i(j)]$ at step i, the averaging process gets credit from two sources. One part of the credit is the average $\frac{1}{n-i+1} \sum_{j=i}^{n} m_i(j)$, where $s(i) = \max_{j \geq i}[m_i(j)] - \frac{1}{n-i+1} \sum_{j=i}^{n} m_i(j)$ is the slackness generated at round i. In addition, the process gets credit also for the slackness accumulated in previous rounds $\ell < i$, in such a way that each slackness variable $s(\ell)$ gets distributed evenly among the $n - \ell$ rounds that follow it. Hence we set

$$m'_i(i) = \frac{1}{n-i+1} \sum_{j=i}^{n} m_i(j) + \sum_{\ell=1}^{i-1} \frac{s(\ell)}{n-\ell} . \quad (2)$$

The averaging process continues until the first round t' at which $m'_{t'}(t') \geq 1$, at which point $m'_j(j)$ is set to 1 for all $j \geq t'$, and the process ends. The size of the fractional matching associated with the averaging process is $m' = \sum_{i=1}^{n} m'_i(i)$. Computing m' using the contributions of the vertices from U, for t' as above, we get that:

$$m' = t' - 1 + \sum_{j \geq t'} m(t', j) . \quad (3)$$

Proposition 2.4 *For the graph G, the size of the fractional matching produced by the averaging process is no larger than that produced by* Balance. *Namely, $m' \leq m$.*
□

Proof Compare Equations (1) and (3). If $t' = t$ then $m' = m$, and if $t' < t$ then $m' < m$. Hence it suffices to show that the assumption $t' \geq t$ implies that $t' = t$. This follows because $m_t(j) = 1$ for all $j \leq t$ (as noted above), and so:

$$m'_t(t) = \frac{1}{n-t+1} \sum_{j=t}^{n} m_t(j) + \sum_{\ell=1}^{t-1} \frac{s(\ell)}{n-\ell} = 1 + \sum_{\ell=1}^{t-1} \frac{s(\ell)}{n-\ell} \geq 1$$

where the last inequality holds because all slackness variables $s(\ell)$ are non-negative. □

Proposition 2.5 *For MonotoneG, running the averaging process and running Balance give the same process. Hence $m'(MonotoneG) = m(MonotoneG)$.* □

Proof This is because when running *Balance* on *MonotoneG*, at every round i we have that $m_i(i) = m_i(j)$ for all $j > i$. Hence there is no difference between the average and the maximum of the $m_i(j)$ for $j \geq i$. □

Proposition 2.6 *The size of the fractional matching produced by the averaging process for graph G is not smaller than the size it produces for MonotoneG. Namely, $m'(G) \geq m'(MonotoneG)$.* □

Proof Running the averaging process on graph G, we claim that for every round $i < t'$ we have that:

$$m'_i(i) = \sum_{k \leq i} \frac{1}{n-k+1} . \tag{4}$$

The equality can be proved by induction. For $i = 1$ both sides of the equality are $\frac{1}{n}$. For the inductive step, recalling Eq. 2 one can infer that

$$m'_{i+1}(i+1) = \frac{1}{n-i} \left((n-i+1)m'_i(i) - m'_i(i) + 1 \right)$$

where the $+1$ term is because $i < t'$. Likewise, the right hand side develops in the same way:

$$\sum_{k \leq i+1} \frac{1}{n-k+1} = \frac{1}{n-i} \left((n-i+1) \sum_{k \leq i} \frac{1}{n-k+1} - \sum_{k \leq i} \frac{1}{n-k+1} + 1 \right) .$$

The left hand side of Equation (4) concerns graph G. Observe that $m'_i(i)$ for *MonotoneG* exactly equals the right hand side of Equation (4). It follows that the averaging process ends at the same step t' both on the graph G and on *MonotoneG*, and up to step t' the accumulated fractional matching m' is identical. For rounds $j \geq t'$ we have that $m'_j(j) = 1$ for G and it cannot be larger than 1 for *MonotoneG*, proving the proposition. □

Combining the three propositions above we get that:

$$m(G) \geq m'(G) \geq m'(MonotoneG) = m(MonotoneG) .$$

This completes the proof of Theorem 1.7. □

3 Online Integral Matching

The first part of Theorem 1.6 is restated in the following theorem (recall the definition of the monotone graph *MonotoneG* in Sect. 1.1).

Theorem 3.1 *Let the function $a(n)$ be such that $\rho_n(Ranking, MonotoneG) = \frac{a(n)}{n!}$ for all n. Then $a(n) = (n+1)! - d(n+1) - d(n)$, where $d(n)$ is the number of derangements (permutations with no fixed points) on the numbers $[1, n]$.*

Proof When the input is *MonotoneG*, then for every permutation π used by *Ranking*, the matching M' produced satisfies the following two properties:

- All vertices in some prefix of U are matched, and then no vertices in the resulting suffix are matched. This is because all neighbors of u_{j+1} are also neighbors of u_j, so if u_{j+1} is matched then so is u_j.
- The order in which vertices of V are matched is consistent with the order π (for those vertices that are matched—some vertices of V may remain unmatched). In other words, if two vertices v_i and v_j are matched and $\pi(i) < \pi(j)$, then the vertex $u \in U$ matched with v_i arrived earlier (has smaller index) than the vertex $u' \in U$ matched with v_j.

Some arguments in the proof that follows make use of the above properties, without explicitly referring to them.

Fix n and *MonotoneG* as input. Let Π_n denote the set of all permutations over V. Hence $|\Pi_n| = n!$. *Ranking* picks one permutation $\pi \in \Pi_n$ uniformly at random. Recall our notation that $\pi(i)$ is the rank of v_i under π. We shall use π_i to denote the item of rank i in π (namely, $\pi_i = \pi^{-1}(i)$). For $i \le n$, let $a(n, i)$ denote the number of permutations $\pi \in \Pi_n$ under which π_i is matched.

Proposition 3.2 *For $a(n)$ as defined in Theorem 3.1 and $a(n, i)$ as defined above, it holds that $a(n) = \sum_{i=1}^{n} a(n, i)$.* □

Proof For a permutation $\pi \in \Pi_n$, let $x(\pi)$ denote the size of the greedy matching produced when *Ranking* uses π and the input graph in *MonotoneG*. Then by definition:

$$a(n) = \sum_{\pi \in \Pi_n} x(\pi).$$

By changing the order of summation:

$$\sum_{\pi \in \Pi_n} x(\pi) = \sum_{i=1}^{n} a(n, i).$$

Combining the above equalities proves the proposition. □

Proposition 3.2 motivates the study of the function $a(n, i)$.

Lemma 3.3 *The function $a(n, i)$ satisfies the following:*

1. $a(n, 1) = n!$ *for every* $n \geq 1$.
2. $a(n, i) = a(n, i + 1) + a(n - 1, i)$ *for every* $1 \leq i < n$. □

Proof The first statement in the lemma holds because in every permutation π, the item π_1 is matched with u_1. Hence it remains to prove the second statement.

Fixing $n > 1$ and $i < n$, consider the following bijection $B_i : \Pi_n \longrightarrow \Pi_n$, where given a permutation $\pi \in \Pi_n$, $B_i(\pi)$ flips the order between locations i and $i + 1$. Namely, $B_i(\pi)_i = \pi_{i+1}$ and $B_i(\pi)_{i+1} = \pi_i$ (we use $B_i(\pi)_i$ as shorthand notation for $(B_i(\pi))_i$). We compare the events that π_i is matched by the greedy matching when *Ranking* uses π with the event that $B_i(\pi)_{i+1}$ is matched by the greedy matching when *Ranking* uses $B_i(\pi)$.

There are four possible events:

1. Both π_i and $B_i(\pi)_{i+1}$ are matched.
2. Neither π_i nor $B_i(\pi)_{i+1}$ are matched.
3. π_i is matched but $B_i(\pi)_{i+1}$ is not matched.
4. π_i is not matched but $B_i(\pi)_{i+1}$ is matched.

Though any of the first three events may happen, the fourth event cannot possibly happen. This is because the item in location $i + 1$ in $B_i(\pi)$ is moved forward to location i in π, so if the greedy algorithm matches it (say to u_j) in $B_i(\pi)$, then the greedy algorithm must match it (either to the same u_j or to the earlier u_{j-1}) in π.

It follows that $a(n, i) - a(n, i + 1)$ exactly equals the number of permutations in which the third event happens. Hence we characterize the conditions under which the third event happens. Let u_j be the vertex matched with π_i in π. Up to the arrival of u_j, the behavior of *Ranking* on $B_i(\pi)$ and π is identical. Thereafter, for u_j not to be matched to $B_i(\pi)_{i+1} = \pi_i$, it must be matched to the earlier $B_i(\pi)_i$. Thereafter, for u_{j+1} not to be matched to $B_i(\pi)_{i+1}$, it must be that $B_i(\pi)_{i+1}$ is not a neighbor of u_{j+1}. But $B_i(\pi)_{i+1} = \pi_i$ is a neighbor of u_j (it was matched to u_j under π), and hence it must be that $\pi_i = v_j$. Summarizing, the third event happens if and only if the permutation $B_i(\pi)$ comes from the following class $\hat{\Pi}$, where permutations $\hat{\pi} \in \hat{\Pi}$ are those that have the property that $\hat{\pi}_i$ is matched, and $\hat{\pi}_{i+1} = v_j$, for the same j for which u_j is the vertex matched with $\hat{\pi}_i$. Consequently, $a(n, i) = a(n, i + 1) + |\hat{\Pi}|$.

To complete the proof of the lemma, it remains to show that $|\hat{\Pi}| = a(n - 1, i)$. Let $\Pi' \subset |\Pi_{n-1}|$ be the set of these permutations $\pi' \in \Pi_{n-1}$ under which *Ranking* (when $|U| = |V| = n - 1$) matches the item π'_i. □

Claim *For $\hat{\Pi}$ and Π' as defined above it holds that $|\hat{\Pi}| = |\Pi'|$.* □

Proof We first show a mapping from $\hat{\Pi}$ to Π'. Given $\hat{\pi} \in \hat{\Pi}$, let $v_j = \hat{\pi}_{i+1}$. To obtain permutation $\pi' \in \Pi_{n-1}$ from $\hat{\pi}$, remove v_j from $\hat{\pi}$, identify location k in $\hat{\pi}$ with location $k - 1$ in π' (for $i + 2 \leq k \leq n$), and identify item v_ℓ of $\hat{\pi}$ with item $v_{\ell-1}$ of π' (for $j + 1 \leq \ell \leq n$). We show now that $\pi' \in \Pi'$ (namely, π'_i is matched, when the input graph is *MonotoneG* with $|U| = |V| = n - 1$).

The vertices $u_1, \ldots, u_{j-1}$ are matched to exactly the same locations in π' and in $\hat{\pi}$, because the only vertices whose indices were decremented had index $\ell \geq j+1$, and are neighbors of $u_1, \ldots, u_{j-1}$ both before and after the decrement. Let $v_k = \hat{\pi}_i$ and note that $k > j$, because v_k is matched to u_j and it is not $v_j = \hat{\pi}_{i+1}$. Hence $\pi'_i = v_{k-1}$ and it too is a neighbor of u_j, because $j \leq k-1$. Hence π'_i will be matched to u_j.

Conversely, we have the following mapping from Π' to $\hat{\Pi}$. Given $\pi' \in \Pi'$, let u_j be the vertex matched π'_i. To obtain permutation $\hat{\pi} \in \hat{\Pi}$ from π', identify location k in $\hat{\pi}$ with location $k-1$ in π' (for $i+2 \leq k \leq n$), identify item v_ℓ of $\hat{\pi}$ with item $v_{\ell-1}$ of π' (for $j+1 \leq \ell \leq n$), and set $\hat{\pi}_{i+1} = v_j$. We show now that $\hat{\pi} \in \hat{\Pi}$.

As in the first mapping, the vertices $u_1, \ldots, u_{j-1}$ are matched to exactly the same locations in π' and in π. Let $v_k = \pi'_i$ and note that $k \geq j$, because v_k was matched to u_j. Hence $\hat{\pi}_i = v_{k+1}$ is neighbor of u_j, and will be matched to u_j. On the other hand, $\hat{\pi}_{i+1} = v_j$ will not be matched because it is not a neighbor of any of $[u_{j+1}, u_n]$. Hence $\hat{\pi} \in \hat{\Pi}$.

Given the two mappings described above (one is the inverse of the other) we have a bijection between Π' and $\hat{\Pi}$, proving the claim. □

The claim above implies that $|\hat{\Pi}| = |\Pi'| = a(n-1, i)$, and consequently that $a(n, i) = a(n, i+1) + a(n-1, i)$, proving the lemma. □

In passing, we note the following corollary.

Corollary 3.4 *For $a(n, i)$ and $a(n)$ as defined above, $a(n) = (n+1)! - a(n+1, n+1)$.*

Proof Using item 1 of Lemma 3.3 we have that $a(n+1, 1) = (n+1)!$. Applying item 2 of Lemma 3.3 iteratively for all $1 \leq i \leq n$ we have that $a(n+1, 1) - a(n+1, n+1) = \sum_{i=1}^{n} a(n, i)$. Proposition 3.2 shows that $\sum_{i=1}^{n} a(n, i) = a(n)$. Combining these three equalities we obtain $a(n) = (n+1)! - a(n+1, n+1)$, as desired. □

To obtain expressions for the values $a(n, i)$, let us introduce additional notation. A *fixpoint* (or *fixed point*) in a permutation π is an item that does not change its location under π (namely, $\pi(i) = i$). For $n \geq 1$ and $1 \leq i \leq n$ define $d(n, i)$ be the number of permutations over $[n]$ in which the only fixpoints (if any) are among the first i items. For example, $d(3, 1) = 3$ due to the permutations 132 (only 1 is a fixed point) 231 (no fixpoints) and 312 (no fixpoints).

Lemma 3.5 *The function $d(n, i)$ satisfies the following:*

1. $d(n, n) = n!$ *for every $n \geq 1$.*
2. $d(n, i+1) = d(n, i) + d(n-1, i)$ *for every $1 \leq i < n$.* □

Proof $d(n, n)$ denotes the number of permutations on $[n]$ with no restrictions, and hence $d(n, n) = n!$, which is the first statement of the lemma.

Consider now the second statement of the lemma. Let $\Pi_{n,i}$ denote the set of permutations in which the only fixpoints (if any) are among the first i items. Then the second statement asserts that $|\Pi_{n,i+1}| = |\Pi_{n,i}| + |\Pi_{n-1,i}|$. The set $\Pi_{n,i+1}$ can be partitioned in two. In one part $i+1$ is not a fixpoint. This part is precisely $\Pi_{n,i}$. In the second part, $i+1$ is a fixpoint. To specify a permutation in this part we need to specify the location of the remaining $n-1$ items, where the only fixpoints allowed are among the first i items. The number of permutations satisfying these constraints is $\Pi_{n-1,i}$, by definition. Hence indeed $|\Pi_{n,i+1}| = |\Pi_{n,i}| + |\Pi_{n-1,i}|$, proving the lemma. □

Corollary 3.6 *For every $n \geq 1$ and $1 \leq i \leq n$ it holds that $a(n, i) = d(n, n + 1 - i)$.* □

Proof The proof is by induction on n, and for every value of n, by induction on i.

For the base case $n = 1$, necessarily $i = 1$ (and hence also $n + 1 - i = 1$) and indeed we have $a(1, 1) = 1 = d(1, 1)$. Fixing $n > 1$, the base case for i is $i = 1$ (and $n + 1 - 1 = n$) and indeed we have that $a(n, 1) = n! = d(n, n)$. For the inductive step, consider $a(n, i)$ with $n > 1$ and $1 < i \leq n$, and assume the inductive hypothesis for $n' < n$ and the inductive hypothesis for n and $i' < i$. Then we have:

$$a(n, i) = a(n, i-1) - a(n-1, i-1)$$
$$= d(n, n-i+2) - d(n-1, n-i+1) = d(n, n-i+1).$$

The first equality is by Lemma 3.3, the second equality is by the inductive hypothesis, and the third equality is by Lemma 3.5. □

We can now complete the proof of Theorem 3.1. By Corollary 3.4 we have that $a(n) = (n+1)! - a(n+1, n+1)$. By Corollary 3.6 we have that $a(n+1, n+1) = d(n+1, 1)$. By definition, $d(n+1, 1)$ is the number of permutations on $[n+1]$ in which only item 1 is allowed to be a fixpoint. This number is precisely $d(n+1) + d(n)$ (where $d(j)$ are the derangement numbers), where the term $d(n+1)$ counts those permutations in which there is no fixpoint, and the term $d(n)$ counts those permutations in which item 1 is the only fixpoint. □

The second part of Theorem 1.6 is restated in the following Corollary.

Corollary 3.7 *For every n,*

$$\rho_n(Ranking, MonotoneG) = (1 + \frac{1}{e})n + (1 - \frac{2}{e}) + \nu(n)$$

where $|\nu(n)| < \frac{1}{n!}$.

Proof Theorem 3.1 shows that $a(n) = (n+1)! - d(n+1) - d(n)$, where $d(n)$ are the derangement numbers. It is known that $d(n) = \frac{n!}{e}$ rounded to the nearest integer. Hence $|d(n) - \frac{n!}{e}| < \frac{1}{2}$ and $|d(n+1) + d(n) - \frac{(n+1)!}{e} - \frac{n!}{e}| < 1$. Hence $|a(n) - (1 - \frac{1}{e})(n+1)! - \frac{n!}{e}| < 1$. Dividing by $n!$ and replacing $(1 - \frac{1}{e})(n+1)$ by $(1 - \frac{1}{e})n + 1 - \frac{1}{e}$ the corollary is proved. □

3.1 Some Related Sequences

To illustrate the values of some of the parameters involved in the proof of Theorem 3.1, consider a triangular table T where row n has n columns. The entries (for $1 \le i \le n$) are $d(n, i)$, as defined prior to Lemma 3.5. Recall that $d(n, i) = a(n, n + 1 - i)$, hence the table also provides the $a(n, i)$ values. We initialize the diagonal of the table by $d(n, n) = n!$. Thereafter we fill the remaining cells of table row by row, by using the relation $d(n, i) = d(n, i + 1) - d(n - 1, i)$, implied by Lemma 3.5. Finally, compute $a(n) = \sum_{i=1}^{n} a(n, i) = \sum_{i=1}^{n} d(n, i)$ by summing up each row. The table below shows the computation of $a(n)$ for $n \le 6$.

n	d(n,1)=a(n,n)	d(n,2)	d(n,3)	d(n,4)	d(n,5)	d(n,6)	a(n)
1	1						1
2	1	2					3
3	3	4	6				13
4	11	14	18	24			67
5	53	64	78	96	120		411
6	309	362	426	504	600	720	2921

The table T is identical in its definition to Sequence A116853, named *Difference triangle of factorial numbers read by upward diagonals*, in *The Online Encyclopedia of Integer Sequences* [13]. The row sums (and hence $a(n)$) in this table give Sequence A180191 (with an offset of 1 in the value of n), named *Number of permutations of [n] having at least one succession*. The first column (which equals $a(n, n)$) is the sequence A000255. These relations between $a(n)$ and the various sequences in [13] helped guide the statement and proof of Theorem 3.1.

The derangement numbers $d(n)$ (which form the sequence A000166) can be easily computed by the recurrence $d(n) = n \cdot d(n - 1) + (-1)^n$ (due to Euler). The table below shows the computation of $a(n) = (n + 1)! - d(n + 1) - d(n)$ for $n \le 7$.

n	n!	d(n)	a(n)
1	1	0	1
2	2	1	3
3	6	2	13
4	24	9	67
5	120	44	411
6	720	265	2921
7	5040	1854	23633
8	40320	14833	

Acknowledgements The work of the author is supported in part by the Israel Science Foundation (grant No. 1388/16). I thank several people whose input helped shape this work. Alon Eden and Michal Feldman directed me to the proof presented in [3], which is the one presented here (in the appendix) for Theorem 1.1. Thomas Kesselheim and Aranyak Mehta directed me to additional relevant references. The statement and proof of Theorem 3.1 were based on noting some numerical coincidences between the values of $a(n)$ for small n and sequences in *The Online Encyclopedia of Integer Sequences* [13]. Dror Feige wrote a computer program that computes $a(n)$, which made these numerical coincidences evident. Alois Heinz offered useful advice as to how to figure out proofs for various identities claimed in [13].

4 Appendix: A Performance Guarantee for *Ranking*

For completeness, we review here a proof of Theorem 1.1. The proof that we present uses essentially the same mathematical expressions as the proof presented in [2]. A simple presentation of the proof of [2] appeared in a blog post of Claire Mathieu [10] (with further slight simplifications made possible by a comment provided there by Pushkar Tripathi). We shall give an arguably even simpler presentation, due to Eden, Feldman, Fiat and Segal [3]. The proofs in [2, 10] make use of linear programming duality. The proof below is based on an economic interpretation, and a proof technique that splits *welfare* into the sum of *utility* and *revenue*. These last two terms turn out to be scaled versions of the dual variables used in [2, 10], but the proof does not need to make use of LP duality.

Proof (*Theorem* 1.1) Fix an arbitrary perfect matching M in G. Given a vertex $v \in V$, we use $M(v)$ to denote the vertex in U matched with v under M.

Recall that *Ranking* chooses a random permutation π over V. Equivalently, we may assume that every vertex $v_i \in V$ chooses independently uniformly at random a real valued weight $w_i \in [0, 1]$, and then the vertices of V are sorted in order of increasing weight (lowest weight first). This gives a random permutation π. The same permutation π is also obtained if each weight w_i is replaced by a "price" $p_i = e^{w_i - 1}$ and vertices are sorted by prices (because e^{x-1} is a monotonically increasing function in x). Observe that $p_i \in [\frac{1}{e}, 1]$, though it is not uniformly distributed in that range. The expected price that *Ranking* assigns to an item is:

$$E[p_i] = \int_0^1 e^{w_i - 1} dw_i = \frac{1}{e}(e - 1) = 1 - \frac{1}{e}. \tag{5}$$

It is convenient to think of the vertices of U as *buyers* and the vertices of V as *items*. Suppose that given $G(U, V; E)$, each vertex (buyer) $u \in U$ desires only items $v \in V$ that are neighbors of u (namely, u desires v iff $(u, v) \in E$), is willing to pay 1 for any such item, and wishes to buy exactly one item. The seller holding the items is offering to sell each item v_i for a price of p_i. Then given G, the matching produced by executing the *Ranking* algorithm is the same as the one that would be produced in a setting in which each buyer u_j, upon arrival, buys its cheapest exposed desired item, if there is any. If p_i is the price of the purchased item v_i, then the *revenue* that the seller extracts from the sale of v_i to u_j is $r(v_i) = p_i$, whereas the *utility* that the buyer extracts is $y(u_i) = 1 - p_i$. Consequently, the revenue plus utility extracted from a sale is 1, and the total revenue extracted from all sales plus the total utility sum up to exactly the cardinality of the matching.

To lower bound the expected cardinality of the matching, we consider each edge $(M(v_i), v_i) \in M$ separately, and consider the expectation $E[r(v_i) + y(M(v_i))]$, where expectation is taken over the choice of π. Using the linearity of the expectation, we will have that $\rho_n(Ranking, G) = \sum_{v_i \in V} E[r(v_i) + y(M(v_i))]$. □

Lemma 4.1 *For every $v_i \in V$ it holds that $E[r(v_i) + y(M(v_i))] \geq 1 - \frac{1}{e}$. Moreover, this holds even if expectation is taken only over the choice of random weight w_i (and hence of random price p_i) of item v_i, without need to consider other aspects of the random permutation π.* □

Proof Fix an arbitrary graph $G(U, V; E) \in G_n$, an arbitrary perfect matching M, and arbitrary prices $p_j \in [\frac{1}{e}, 1]$ for all items $v_j \neq v_i$. The price p_i for item v_i is set at random. Let M' denote the greedy matching produced by this realization of the *Ranking* algorithm (where each buyer upon its arrival is matched to the exposed vertex of lowest price among its neighbors, if there is any). Suppose as a thought experiment that item v_i is removed from V, and consider the greedy matching M'_{-i} that would have been produced in this setting. Let p denote the price of the item in V matched to $M(v_i)$ under M'_{-i}, and set $p = 1$ if $M(v_i)$ is left unmatched under M'_{-i}. Now we make two easy claims.

1. *If $p_i < p$, then v_i is matched in M'.* This follows because at the time that $M(v_i)$ arrived, either v_i was already matched (as desired), or it was available for matching with $M(v_i)$ and preferable (in terms of price) over all other items that $M(v_i)$ desires (as all have price at least $p > p_i$).
2. *The utility of $M(v_i)$ in M' satisfies $y(M(v_i)) \geq 1 - p$.* This follows because under M'_{-i} the utility of $M(v_i)$ is $1 - p$, and under the greedy algorithm considered, the introduction of an additional item (the item v_i when considering M') cannot decrease the utility of any agent. (At every step of the arrival process, the set of exposed vertices under M' contains the set of exposed vertices under M'_{-i}, and one more vertex.)

Using the above two claims and taking z to be the value satisfying $p = e^{z-1}$, we have:

$$E[y(M(v_i)) + r(v_i)] \geq 1 - p + Pr[p_i < p]p_i = 1 - e^{z-1} + \int_{w_i=0}^{z} e^{w_i-1} dw_i$$

$$= 1 - \frac{e^z}{e} + \frac{e^z - 1}{e} = 1 - \frac{1}{e}.$$

This completes the proof of Lemma 4.1. □

Using the linearity of the expectation, we have that

$$\rho_n(Ranking, G) = \sum_{v_i \in V} E[r(v_i) + y(M(v_i))] \geq (1 - \frac{1}{e})n.$$

Tighter Bounds for Online Bipartite Matching

This completes the proof of Theorem 1.1. □

One can adapt the proof presented above to the special case in which the input graph is *MonotoneG* (or more generally, comes from the distribution D_n). In this case one can upper bound the slackness involved in the proof of Theorem 1.1, and infer the following theorem. □

Theorem 4.2 *For every n it holds that $\rho_n(Ranking, MonotoneG) \leq (1 - \frac{1}{e})n + \frac{1}{e}$.*

Proof Recall the two properties mentioned in the beginning of the proof of Theorem 3.1. Recall also that the analysis of *Ranking* in the proof of Theorem 1.1 (within Lemma 4.1) involved the matching M' and other matchings M'_{-i}, and two claims. Let us analyse the slackness involved in these claims when the input is the monotone graph. The claims are restated with $M(v_i)$ replaced by u_i, because for the monotone graph $M(v_i) = u_i$.

The first claim stated that *if $p_i < p$, then v_i is matched in M'*. When the input is the monotone graph, then a converse also holds: if $p_i > p$, then v_i is not matched in M'. This follows because up to the time that u_i arrives and is matched, only vertices of V priced at most p are matched, and thereafter, no other vertex in U desires v_i. The event that $p_i = p$ has probability 0. Hence there is no slackness involved in the first claim—it is an *if and only if* statement.

The second claim stated that *the utility of u_i in M' is $y(u_i) \geq 1 - p$*. This inequality is not tight. Rather, the utility of u_i in M'_{-i} is $1 - p$, and $y(u_i)$ is not smaller. Let us quantify the slackness involved in this inequality by introducing slackness variables $s(u)$. For a vertex $u \in U$ we shall use the notation $y(u)$ to denote the utility of u under *Ranking*, and $y_{-v}(u)$ for the utility of u when vertex $v \in V$ is removed. The *slackness* $s(u_i)$ of vertex u_i is defined as $s(u_i) = y(u_i) - y_{-v_i}(u_i)$. □

Lemma 4.3 *For the monotone graph and an arbitrary vertex $u_j \in U$, the expected utility of u_j (expectation taken over choices of w_i for all $1 \leq i \leq n$ by the* Ranking *algorithm) is identical in the following two settings: when v_j is removed, and when v_n is removed. Namely, $E[y_{-v_j}(u_j)] = E[y_{-v_n}(u_j)]$.*

Proof Both v_j and v_n are neighbors of all vertices u_k arriving up to u_j (for $1 \leq k \leq j$). Hence whichever of the two vertices, v_j or v_n, is removed, the distributions of the outcomes of *Ranking* on the first j arriving vertices (including u_j) are the same. □

As a consequence of Lemma 4.3 we deduce that for the monotone graph, the expected slackness of every vertex $u \in U$ satisfies $E[s(u)] = E[y(u)] - E[y_{-v_n}(u)]$.

Lemma 4.4 *For the monotone graph and arbitrary setting of prices for the items (as chosen at random by* Ranking*), $\sum_{u \in U} s(u) \leq 1 - p_n$. Consequently, $\sum_{u \in U} E[s(u)] \leq \frac{1}{e}$, where expectation is taken over choice of weights w_i for vertices in V.* □

Proof Fix the prices p_i (hence π). Let $u_1, \ldots, u_k$ be the vertices of U matched under *Ranking*, and let $m(u_1), \ldots, m(u_k)$ be the vertices in V to which they are matched. Observe that the prices $p(m(u_i))$ (where $1 \le i \le k$) of these vertices form a monotonically increasing sequence. Necessarily, v_n is one of the matched vertices, because it is a neighbor of all vertices in U. Let j be such that $v_n = m(u_j)$.

Consider now what happens when v_n is removed. The vertices $u_1, \ldots, u_{j-1}$ are matched to $m(u_1), \ldots, m(u_j - 1)$ as before. As to the vertices $u_j, \ldots, u_{k-1}$, they can be matched to $m(u_{j+1}), \ldots, m(u_k)$, hence the algorithm will match them to vertices of no higher price. Specifically, for every i in the range $j \le i \le k-1$, vertex u_i will be matched either to $m(u_{i+1})$ or to an earlier vertex, though not earlier than $m(u_i)$. The vertex u_k may either be matched or be left unmatched. For simplicity of notation, we say that u_k is matched to either $m(u_{k+1})$ or to an earlier vertex, where $m(u_{k+1})$ is an auxiliary vertex of price 1 than indicates that u_k is left unmatched.

Note that:

$$\sum_{u \in U} y(u) = \sum_{i=1}^{k} y(u_i) = k - \sum_{i=1}^{k} p(m(u_i)),$$

and that:

$$\sum_{u \in U} y_{-v_n}(u) = \sum_{i=1}^{k} y_{-v_n}(u_i) \ge k - \sum_{i=1}^{j-1} p(m(u_i)) - \sum_{i=j+1}^{k+1} p(m(u_i)).$$

Hence we have that:

$$\sum_{u \in U} s(u) = \sum_{u \in U} y(u) - \sum_{u \in U} y_{-v_n}(u) \le p(m(u_{k+1})) - p(v_n).$$

Finally, noting that $p(m(u_{k+1})) \le 1$ and that $E[p(v_n)] = 1 - \frac{1}{e}$ (see Eq. (5)), the lemma is proved. □

As in the proof of Theorem 1.1 we have:

$$E[y(u_i) + r(v_i)] = 1 - p + s(u_i) + Pr[p_i < p]p_i = 1 - \frac{1}{e} + s(u_i).$$

Using the linearity of the expectation and Lemma 4.4 we have that:

$$\rho_n(Ranking, MonotoneG) = \sum_{v_i \in V} E[r(v_i) + y(u_i)]$$

$$= (1 - \frac{1}{e})n + \sum_{u \in U} E[s(u)] \le (1 - \frac{1}{e})n + \frac{1}{e}.$$

This completes the proof of Theorem 4.2. □

References

1. Benjamin E. Birnbaum, Claire Mathieu: On-line bipartite matching made simple. SIGACT News **39**(1), 80–87 (2008)
2. Nikhil R. Devanur, Kamal Jain, Robert D Kleinberg: Randomized Primal-Dual analysis of RANKING for Online BiPartite Matching. SODA 2013: 101–107
3. Alon Eden, Michal Feldman, Amos Fiat, Kineret Segal: An Economic-Based Analysis of RANKING for Online Bipartite Matching. Manuscripy, 2018. http://cs.tau.ac.il/~mfeldman/papers/EFFS18.pdf
4. Gagan Goel, Aranyak Mehta: Online budgeted matching in random input models with applications to Adwords. SODA 2008: 982–991.
5. Bala Kalyanasundaram, Kirk Pruhs: An optimal deterministic algorithm for online b-matching. Theor. Comput. Sci. 233(1-2): 319–325 (2000).
6. Anna Karlin. Online bipartite matching, lecture notes in course on Randomized Algorithms and Probabilistic Analysis, scribes Alex Polozav and Daryl Hansen, Spring 2013. https://courses.cs.washington.edu/courses/cse525/13sp/scribe/lec6.pdf
7. Richard M. Karp, Umesh V. Vazirani, Vijay V. Vazirani: An Optimal Algorithm for On-line Bipartite Matching. STOC 1990: 352–358.
8. Robert D. Kleinberg. Online bipartite matching algorithms, lecture note in course on Analysis of Algorithms, Fall 2012. http://www.cs.cornell.edu/courses/cs6820/2012fa/
9. L. Lovasz, M. D. Plummer. Matching Theory. Elsevier 1986.
10. Claire Mathieu. A CS's Professor Blog: A primal-dual analysis of the Ranking algorithm. June 25, 2011. http://teachingintrotocs.blogspot.co.il/2011/06/primal-dual-analysis-of-ranking.html
11. Aranyak Mehta. Online matching and ad allocation. *Foundations and Trends in Theoretical Computer Science*, 8(4):265–368, 2013.
12. Aranyak Mehta, Amin Saberi, Umesh V. Vazirani, Vijay V. Vazirani: AdWords and generalized online matching. J. ACM 54(5): 22 (2007).
13. N. J. A. Sloane, editor, The On-Line Encyclopedia of Integer Sequences, published electronically at https://oeis.org.

Minimum Cost Globally Rigid Subgraphs

Tibor Jordán and András Mihálykó

Abstract A d-dimensional framework is a pair (G, p), where $G = (V, E)$ is a graph and p is a map from V to $\mathbb{R}^d$. The length of an edge of G is equal to the distance between the points corresponding to its end-vertices. The framework is said to be globally rigid if its edge lengths uniquely determine all pairwise distances in the framework. A graph G is called globally rigid in $\mathbb{R}^d$ if every generic d-dimensional framework (G, p) is globally rigid. Global rigidity has applications in wireless sensor network localization, molecular conformation, formation control, CAD, and elsewhere. Motivated by these applications we consider the following optimization problem: given a graph $G = (V, E)$, a non-negative cost function $c : E \to \mathbb{R}_+$ on the edge set of G, and a positive integer d. Find a subgraph $H = (V, E')$ of G, on the same vertex set, which is globally rigid in $\mathbb{R}^d$ and for which the total cost $c(E') := \sum_{e \in E'} c(e)$ of the edges is as small as possible. This problem is NP-hard for all $d \geq 1$, even if c is uniform or G is complete and c is metric. We focus on the two-dimensional case, where we give $\frac{3}{2}$-approximation (resp. 2-approximation) algorithms for the uniform cost and metric versions. We also develop a constant factor approximation algorithm for the metric version of the d-dimensional problem, for every $d \geq 3$.

Keywords Global rigidity · Approximation algorithm · Redundant rigidity · Rigid framework

Subject Classifications 52C25 · 90C27 · 90C59

T. Jordán (✉)
Department of Operations Research, Eötvös University, and the MTA-ELTE Egerváry Research Group on Combinatorial Optimization, Pázmány Péter Sétány 1/C, Budapest 1117, Hungary
e-mail: jordan@cs.elte.hu

A. Mihálykó
Department of Operations Research, Eötvös University, Pázmány Péter Sétány 1/C, Budapest 1117, Hungary
e-mail: mihalyko@cs.elte.hu

© János Bolyai Mathematical Society and Springer-Verlag GmbH Germany, part of Springer Nature 2019
I. Bárány et al. (eds.), *Building Bridges II*, Bolyai Society Mathematical Studies 28, https://doi.org/10.1007/978-3-662-59204-5_8

1 Introduction

A d-dimensional *framework* is a pair (G, p), where $G = (V, E)$ is a graph and p is a map from V to $\mathbb{R}^d$. We also call (G, p) a *realization* of G in $\mathbb{R}^d$. Two realizations (G, p) and (G, q) are *equivalent* if $||p(u) - p(v)|| = ||q(u) - q(v)||$ holds for all pairs u, v with $uv \in E$, where $||.||$ denotes the Euclidean norm in $\mathbb{R}^d$. The frameworks (G, p) and (G, q) are *congruent* if $||p(u) - p(v)|| = ||q(u) - q(v)||$ holds for all pairs u, v with $u, v \in V$. This is the same as saying that (G, q) can be obtained from (G, p) by an isometry of $\mathbb{R}^d$.

We say that (G, p) is *globally rigid* in $\mathbb{R}^d$ if every d-dimensional realization (G, q) of G which is equivalent to (G, p), is congruent to (G, p). In other words, the framework is globally rigid if its edge lengths uniquely determine all pairwise distances. This property makes the notion of global rigidity a fundamental concept in problems where we are given partial information on the pairwise distances between pairs of a finite point set and our goal is to determine the configuration of the points, up to trivial transformations, see Sect. 1.3 below.

Saxe [24] showed that it is NP-hard to decide if even a 1-dimensional framework is globally rigid. The analysis and characterization of globally rigid frameworks become more tractable if we consider *generic frameworks*, i.e. frameworks (G, p) for which the set of coordinates of the points $p(v), v \in V(G)$, is algebraically independent over the rationals. Results of Connelly [6] and Gortler, Healy and Thurston [12] imply that the global rigidity of a generic framework (G, p) in $\mathbb{R}^d$ depends only on the graph G, for all $d \geq 1$. Hence we may define a graph G to be *globally rigid* in $\mathbb{R}^d$ if every (or equivalently, if some) generic realization of G in $\mathbb{R}^d$ is globally rigid. The problem of finding a polynomially verifiable characterization for graphs which are globally rigid in $\mathbb{R}^d$ has been solved for $d = 1, 2$, but is a major open problem when $d \geq 3$.

1.1 The Minimum Cost Globally Rigid Subgraph Problem

In this paper we consider the following algorithmic problem. The input is a graph $G = (V, E)$, a non-negative cost function $c : E \to \mathbb{R}_+$ on the edge set of G, and a positive integer d. The task is to find a subgraph $H = (V, E')$ of G, on the same vertex set, which is globally rigid in $\mathbb{R}^d$ and for which the total cost $c(E') := \sum_{e \in E'} c(e)$ of the edges is as small as possible.

We call this optimization problem the *Minimum cost globally rigid spanning subgraph problem* (or MCGRSS, for short). We shall focus on the following special cases of this problem: (i) if c is uniform, the goal is to find a *minimum size* globally rigid spanning subgraph (ii) in the *metric* MCGRSS problem the input graph G is complete and c satisfies the triangle inequality.

The Minimum cost globally rigid spanning subgraph problem (already in the special cases mentioned above) is NP-hard for all $d \geq 1$. The proof of this hardness result is given in Sect. 7. Therefore our aim is to design efficient approximation algorithms. We shall first consider the two-dimensional version of the problem and give a $\frac{3}{2}$-approximation algorithm for the minimum size globally rigid spanning subgraph problem as well as a 2-approximation algorithm for the metric version. We also show how the latter factor can be improved to 1.61 when the costs are defined by Euclidean distances in the plane.

In the second part of the paper we design constant factor approximation algorithms for the d-dimensional problem in the metric case, for all $d \geq 3$.

We can define—and we shall also consider—similar optimization problems by replacing global rigidity with redundant rigidity or rigidity (defined in Sect. 2 below) in the definition of MCGRSS. These problems are denoted by MCRRSS and MCRSS, respectively. It turns out that MCRRSS in $\mathbb{R}^d$ is also NP-hard for all $d \geq 1$. On the other hand, MCRSS is solvable in polynomial time in $\mathbb{R}^1$ and $\mathbb{R}^2$. The complexity status of MCRSS is open in $\mathbb{R}^d$ for $d \geq 3$.

1.2 Previous Work

It is a well-known folklore result in rigidity theory that a graph G is redundantly rigid (resp. globally rigid) in $\mathbb{R}^1$ if and only if it is 2-edge-connected (resp. 2-connected). Thus in the one-dimensional case of MCRRSS (resp. MCGRSS) we search for a minimum cost 2-edge-connected (resp. 2-connected) spanning subgraph. These problems, even in the uniform or metric version, are NP-hard, as they contain the Hamilton cycle problem as a special case. There are several constant factor approximation algorithms in the literature that deal with these problems, see e.g. [21]. In light of this connection the MCGRSS problem is a natural extension of these core problems from graph connectivity.

The only higher dimensional result we are aware of is due to García and Tejel [11]. They consider the minimum size redundantly rigid augmentation problem in the plane, which corresponds to MCRRSS in the special case when G is complete and $c(e) \in \{0, 1\}$ for all $e \in E(G)$. They show that this problem is NP-hard in general but can be solved in polynomial time if the graph to be augmented—that is, the graph of the edges of cost zero—is minimally rigid in $\mathbb{R}^2$. The minimum size globally rigid augmentation problem is briefly mentioned in [9, 17], along with some related results.

1.3 Motivation and Applications

One of the applications that inspired our research is the localization problem of two- and three-dimensional wireless sensor networks. In this problem the goal is to compute the locations of all sensors, when only a subset of the pairwise distances and locations is available. The network is localizable (that is, the localization problem has a unique solution) if and only if the corresponding framework is globally rigid [3]. In this framework the vertices correspond to the sensors and two vertices are adjacent if and only if the distance between them is known. Methods and results from rigidity theory have been used to solve a number of related problems. In particular, the characterization of localizability (assuming generic locations in the plane) and inductive constructions of localizable networks have been identified, see e.g. [1, 16]. Similar questions (concerning global rigidity or redundant rigidity) arise in molecular conformation, where the shape of a molecule is to be determined based on a subset of inter-atomic distances [26], in formation control [27], and elsewhere.

The minimum cost globally (or redundantly) rigid spanning subgraph problem may emerge in these applications when one wants to achieve, say, global rigidity by measuring (or recomputing, fixing, etc.) some pairwise distances in an optimal way. For example, it may happen that (i) certain distances are not computable, or more generally, the cost or time of computing pairwise distances may be different for different pairs, or preferences may be given to some pairs, or (ii) the level of noise in the distance data may be different, or (iii) the total length of the edges is a relevant factor, etc. These properties and parameters may be encodable in the cost and objective functions and then, assuming that the costs are uniform or metric, a near optimal solution can be obtained by using the approximation algorithms designed in this paper.

2 Rigid and Globally Rigid Graphs

In this section we collect the basic definitions and results from rigidity theory that we shall use. The framework (G, p) is *rigid* in $\mathbb{R}^d$ if there exists an $\epsilon > 0$ such that, if (G, q) is equivalent to (G, p) and $||p(v) - q(v)|| < \epsilon$ for all $v \in V$, then (G, q) is congruent to (G, p). It is known that, informally speaking, this is equivalent to saying that every continuous motion of the vertices of the framework in $\mathbb{R}^d$ which preserves all edge-lengths takes the framework to a congruent realization of G. It is clear that global rigidity implies rigidity.

As for global rigidity, the rigidity of frameworks in $\mathbb{R}^d$ is a generic property for all $d \geq 1$ [2]. We say that a graph G is *rigid* in $\mathbb{R}^d$ if every (or equivalently, if some) generic realization of G in $\mathbb{R}^d$ is rigid. See Fig. 1 for examples. A rigid graph $G = (V, E)$ in $\mathbb{R}^d$ is called *minimally rigid* if $G - e$ is not rigid for all $e \in E$.

It is known that the edge sets of the minimally rigid graphs on vertex set V correspond to the bases of the so-called d-dimensional *rigidity matroid*, defined on the edge set of a complete graph on V. Hence they have the same number of edges: for example, a minimally rigid graph in $\mathbb{R}^2$ on vertex set V has $2|V| - 3$ edges. The problem of finding a polynomially verifiable characterization for graphs which are rigid in $\mathbb{R}^d$ has been solved for $d = 1, 2$, but is a major open problem for $d \geq 3$. We refer the reader to [18, 19] for more details on rigid and globally rigid frameworks and graphs.

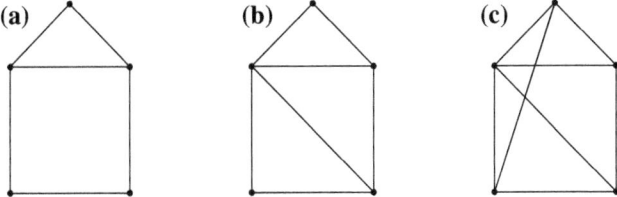

Fig. 1 Graphs which are **a** not rigid, **b** rigid but not globally rigid, **c** globally rigid in the plane

Fig. 2 The graphs obtained from K_3 (left) by a 0-extension operation (middle) followed by a 1-extension operation (right)

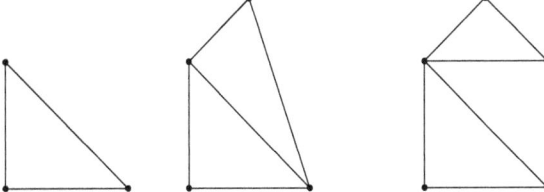

In the plane we have the following key result. Let $G = (V, E)$ be a graph. For a subset $X \subseteq V$ we use $i(X)$ to denote the number of edges induced by X. We say that G is *sparse* if

$$i(X) \leq 2|X| - 3 \text{ for all } X \subseteq V \text{ with } |X| \geq 2. \tag{1}$$

The operation *0-extension* adds a new vertex v to G and two new edges vx, vy for two distinct vertices x and y of G. The *1-extension* operation on edge uw and vertex z with $z \notin \{u, w\}$ adds a new vertex v, deletes uw, and adds three new edges vu, vw, vz. See Fig. 2.

The characterization of (minimally) rigid graphs is due to Laman.

Theorem 2.1 ([22]) *Let $G = (V, E)$ be a graph with $|E| = 2|V| - 3$. Then the following are equivalent:*
(i) G is minimally rigid in $\mathbb{R}^2$,
(ii) G is sparse,
(iii) G can be obtained from K_2 by a sequence of 0-extensions and 1-extensions.

As far as global rigidity is concerned, Hendrickson found the following necessary conditions for global rigidity in $\mathbb{R}^d$. We call a graph G *redundantly rigid* in $\mathbb{R}^d$ if $G - e$ is rigid in $\mathbb{R}^d$ for all $e \in E(G)$. A graph G is said to be *k-connected* if $G - X$ is connected for all $X \subset V(G)$ with $|X| \leq k - 1$.

Theorem 2.2 ([13]) *Let G be a globally rigid graph in $\mathbb{R}^d$ on at least $d + 2$ vertices. Then G is*
(i) $(d + 1)$-connected, and
(ii) redundantly rigid in $\mathbb{R}^d$.

These conditions together are also sufficient in $\mathbb{R}^1$ and $\mathbb{R}^2$. The one-dimensional result is folklore, see [14] for a proof. In the plane we have the following characterization.

Theorem 2.3 ([15]) *Let $G = (V, E)$ be a graph on at least four vertices. Then the following are equivalent:*
(i) G is globally rigid in $\mathbb{R}^2$,
(ii) G is 3-connected and redundantly rigid in $\mathbb{R}^2$,
(iii) G can be obtained from K_4 by a sequence of 1-extensions and edge additions.

We shall also use the following result of Nash-Williams [23]. Note that the graphs in the next theorem may have multiple edges.

Theorem 2.4 ([23]) *Let $G = (V, E)$ be a graph and let k be a positive integer. Then the edge set of G can be partitioned into k forests if and only if $i(X) \leq k|X| - k$ holds for all non-empty vertex sets $X \subseteq V$.*

2.1 Algorithms

The structural results presented in this section give rise to efficient combinatorial algorithms for testing whether a given graph $G = (V, E)$ is rigid, redundantly rigid, or globally rigid in the plane. These algorithms use the fact that the edge sets of the sparse subgraphs of a graph form the independent sets of the 2-dimensional rigidity matroid and boil down to the existence of an efficient subroutine for checking whether a graph is sparse or not. The matroidal property makes it possible to find a minimum cost rigid spanning subgraph of a rigid graph in $\mathbb{R}^2$ with respect to an arbitrary cost function on the edge set, in polynomial time. Each of these basic problems can be solved in $O(|V|^3)$ time or faster, see e.g. [4] for more details.

3 Minimum Size Globally Rigid Spanning Subgraphs

In this section we present two simple approximation algorithms for the minimum size globally (resp. redundantly) rigid spanning subgraph problems. We show that if we delete edges as long as possible, in a greedy fashion, maintaining the global (or redundant) rigidity of the graph, then we end up with a close-to-optimal solution.

A graph $G = (V, E)$ is called *minimally globally (resp. redundantly) rigid* in $\mathbb{R}^2$ if it is globally (resp. redundantly) rigid in $\mathbb{R}^2$ but $G - e$ is not globally (resp. redundantly) rigid in $\mathbb{R}^2$ for all $e \in E$.

Theorem 3.1 *Suppose that $G = (V, E)$ is minimally globally rigid in $\mathbb{R}^2$ with $|V| \geq 4$. Then $|E| \leq 3|V| - 6$.*

Proof Consider a sequence of graphs $G_1, G_2, \ldots, G_t$ for which $G_1 = K_4$, $G_t = G$, and G_i is obtained from G_{i-1} by an edge addition or 1-extension for all $2 \leq i \leq t$. Such a sequence exists by Theorem 2.3. Since G is minimally globally rigid, every edge addition operation used in this sequence adds an edge which will be split into two edges later by a 1-extension operation. This leads to a pairing, that is, a bijection between the added edges and a subset of the 1-extension operations. Each pair increases the number of vertices by one and the number of edges by three. A 1-extension operation alone increases the number of vertices by one and the number of edges by two. Thus, since K_4 satisfies $|E(K_4)| = 3|V(K_4)| - 6$, and the total number of edges added by the operations is not more than three times the number of added vertices, $G_t = G$ satisfies $|E| \leq 3|V| - 6$, as required. □

A globally (or redundantly) rigid graph G in $\mathbb{R}^2$ on vertex set V has at least $2|V| - 2$ edges by Theorems 2.2 and 2.1. Since testing global rigidity can be done in polynomial time, Theorem 3.1 leads to an efficient constant factor approximation algorithm.

Theorem 3.2 *There is a polynomial time $\frac{3}{2}$-approximation algorithm for the minimum size globally rigid spanning subgraph problem in $\mathbb{R}^2$.*

A similar situation holds for redundant rigidity. Here we use the following result (whose proof is substantially more complicated than that of Theorem 3.1).

Theorem 3.3 ([18]) *Suppose that $G = (V, E)$ is minimally redundantly rigid in $\mathbb{R}^2$ with $|V| \geq 7$. Then $|E| \leq 3|V| - 9$.*

As a corollary, we obtain:

Theorem 3.4 *There is a polynomial time $\frac{3}{2}$-approximation algorithm for the minimum size redundantly rigid spanning subgraph problem in $\mathbb{R}^2$.*

4 Structural Properties of Minimally Rigid Graphs

In the next two sections we consider the metric versions of the two-dimensional MCRRSS and MCGRSS problems. Our algorithms will first identify a minimum cost (minimally) rigid spanning subgraph of the input graph and then extend it to a feasible solution by adding new edges. In order to keep the total cost of these added edges low we need structural results on the minimally rigid subgraphs of a minimally rigid graph. We shall rely on some results of García and Tejel from [11] and also prove a number of new properties. In what follows a minimally rigid graph in the plane will be called a *Laman graph*.

4.1 The Extreme Classes of a Laman Graph

Let $G = (V, E)$ be a rigid graph. We say that an edge $e \in E$ is *redundant* in G if $G - e$ is rigid. Thus G is redundantly rigid if every edge of G is redundant. As we noted above, the Laman graphs on vertex set V are the bases of the two-dimensional rigidity matroid defined on the edge set of a complete graph on V. In particular, if G is Laman then $G + e$ has a unique (matroid) circuit, the *fundamental circuit of* e with respect to G. From this viewpoint the next lemma easily follows from some basic properties of matroids.

Lemma 4.1 *Let $G = (V, E)$ be a Laman graph and let $e = ij$ be an edge for some $i, j \in V$. Then*
(i) There is a unique fundamental circuit in $G + e$, denoted by $C(ij)$ or $C(e)$. This circuit contains e. $(V(C(e)), E(C(e)) - e)$ is a Laman subgraph of G, denoted by $L(ij) = (V(ij), E(ij))$ or simply $L(e)$.
(ii) For every edge $e' \in E(ij)$ the graph $(V, E + e - e')$ is a Laman graph, in which the fundamental circuit of e' is $C(ij)$. Moreover, if $e' \notin E(ij)$ then $(V, E + e - e')$ is not a Laman graph,
(iii) If G' is a Laman subgraph of G with $\{i, j\} \subseteq V(G')$ then $L(ij)$ is a subgraph of G'. Thus $L(ij)$ is equal to the intersection of all Laman subgraphs L_h of G with $\{i, j\} \subseteq V(L_h)$.

In other words $E(ij)$ is equal to the set of edges of G that become redundant in $G + e$. We may define $L(ij)$ even if $ij \in E(G)$. In this case $L(ij)$ is the single edge ij and $C(e)$ is a graph consisting of two parallel copies of ij.

For every $i, j \in V(G)$ we say that $L(ij)$ is a *generated* Laman subgraph of G whose *generator* is the edge ij. A Laman graph G is called *narrow* if $G = L(ij)$ for some i, j, that is, if it can be made redundantly rigid by adding one new edge. See Fig. 3. Otherwise it is said to be *wide*. We note that the authors in [11] use generated and non-generated, respectively, instead of narrow and wide. We feel the new terminology makes the statements and proofs more transparent.

Given a Laman graph G and a set $e_1, e_2, \ldots, e_k$ of new edges, let $L(e_1, e_2, \ldots, e_k)$ be the subgraph of G consisting of those edges of G that are redundant in $G + \{e_1, e_2, \ldots, e_k\}$.

Lemma 4.2 ([11, Lemma 4]) *Let G be a Laman graph. Then $L(e_1, e_2, \ldots, e_k) = L(e_1) \cup L(e_2) \cup \ldots \cup L(e_k)$.*

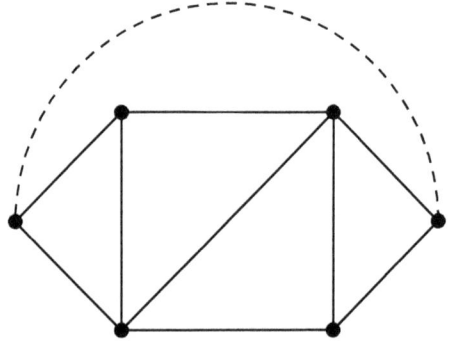

Fig. 3 A narrow Laman graph (solid edges). Adding the dotted edge makes it redundantly rigid

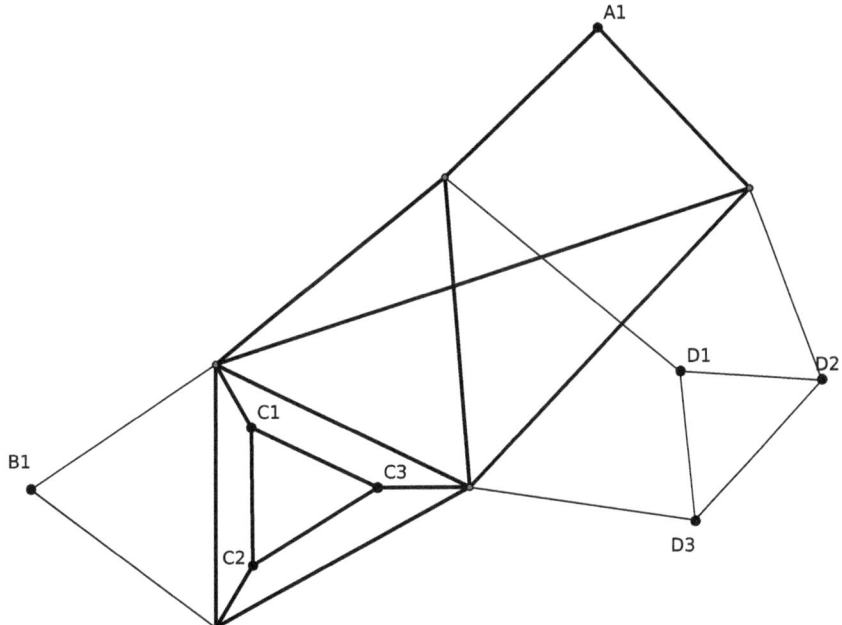

Fig. 4 The extreme classes in a Laman graph. This graph has four extreme classes: $\{A_1\}$, $\{B_1\}$, $\{C_1, C_2, C_3\}$, and $\{D_1, D_2.D_3\}$. The edges of the MGL subgraph generated by an edge connecting A_1 to some C_i ($1 \leq i \leq 3$) are thick

Thus adding a set of new edges $e_1, e_2, \ldots, e_k$ to a Laman graph G yields a redundantly rigid graph if and only if the union of the fundamental circuits of the edges e_i, $1 \leq i \leq k$, contains every edge of G. In a smallest redundantly rigid augmentation of G we may assume that for every new edge e_i the fundamental circuit of e_i is a *maximal (with respect to inclusion) generated Laman subgraph*, or simply an *MGL*. A vertex i of G is said to be *extreme* if there is a vertex j for which $L(ij)$ is an MGL of G.

Let $G = (V, E)$ be a Laman graph and let X be the set of its extreme vertices. We say that $i, i' \in X$ are *equivalent* if there exists a vertex $j \in X$ for which $L(ij)$ is an MGL and $L(ij) = L(i'j)$. García and Tejel verified that this is an equivalence relation on X, assuming that G is wide [11, Lemma 8]. We call the equivalence classes of X defined by this relation the *extreme classes* of G. See Fig. 4. The extreme vertices satisfy the following properties:

Lemma 4.3 ([11, Lemma 9]) *Let G be a wide Laman graph and let i_1, i_2 be extreme vertices of G. Then*
(i) if i_1 and i_2 are not equivalent then $L(i_1i_2)$ is an MGL,
(ii) if i_1 and i_2 are equivalent then $L(i_1i_2)$ is not an MGL,
(iii) if L' is MGL then L' contains extreme vertices from exactly two extreme classes of G. If i_1, i_2 are vertices from these two classes then $L' = L(i_1i_2)$.

The next result gives rise to an edge set whose addition makes every edge redundant.

Lemma 4.4 ([11, Lemma 10]) *Let G be a wide Laman graph. Suppose that G has h extreme classes with representative vertices $i_1, i_2, \ldots, i_h$. Then $G = \bigcup_{r=2}^{h} L(i_1 i_r)$.*

Thus G can be made redundantly rigid by adding $h - 1$ well chosen edges, based on the extreme classes. A more detailed analysis in [11] shows that in fact the optimum—the size of a smallest augmenting set—is equal to $\lceil \frac{h}{2} \rceil$, and that a set of representative vertices from the extreme classes as well as an optimal solution can be found in $O(n^2)$ time. We shall not use these facts concerning optimal augmentations but will rely on, and extend, some of the structural results on extreme classes from [11]. We shall use the following lemmas. The first one is well-known, see e.g. [15, Lemma 2.3].

Lemma 4.5 *Let $G = (V, E)$ be a Laman graph and let L_1, L_2 be Laman subgraphs of G with at least two vertices in common. Then their union as well as their intersection are also Laman subgraphs of G.*

Lemma 4.6 ([11, Lemma 5]) *Let G be a Laman graph on at least four vertices and let $L(ij)$ be an MGL subgraph of G. Then for every vertex $k \neq j$ the subgraphs $L(ij)$ and $L(jk)$ have at least one edge in common. In particular, $L(ij)$ contains all edges incident with i or j.*

A simple corollary is as follows.

Lemma 4.7 ([11]) *Let i, j, k be extreme vertices chosen from three different extreme classes. Then $L(ik) \subset L(ij) \cup L(jk)$.*

Proof Since $L(ij)$ and $L(jk)$ are MGL subgraphs of G, Lemma 4.6 implies that every edge incident with j belongs to both. Thus they have at least two vertices in common, which gives, by Lemma 4.5, that $L(ij) \cup L(jk)$ is a Laman subgraph of G. As it contains i and j, we must have $L(ik) \subset L(ij) \cup L(jk)$ by Lemma 4.1(iii).
□

4.2 Extreme Classes and Separating Pairs

Since the globally rigid graphs in the plane are 3-connected, a new set of edges whose addition to a Laman graph makes it globally rigid must eliminate all separating pairs. In order to handle this condition we next prove new structural results on the relation between extreme classes and separating pairs. We start with two preliminary lemmas about wide Laman graphs.

Lemma 4.8 *Let $i_1, i_2, \ldots, i_q$ be extreme vertices chosen from q different extreme classes. Then $L(i_1 i_q) \subset L(i_1 i_2) \cup L(i_2 i_3) \cdots \cup L(i_{q-1} i_q)$.*

Proof We apply induction on q. For $q = 3$ the lemma follows from Lemma 4.7. Now suppose that $q \geq 4$ and the lemma holds up to $q - 1$. Then $L(i_1 i_q) \subset L(i_1 i_{q-1}) \cup L(i_{q-1} i_q) \subset L(i_1 i_2) \cup \cdots \cup L(i_{q-2} i_{q-1}) \cup L(i_{q-1} i_q)$. □

Lemma 4.9 *Let T be a set of extreme vertices of G that contains exactly one vertex from each extreme class and let F be a set of edges for which (T, F) is connected. Then $G + F$ is redundantly rigid.*

Proof It follows from Lemma 4.4 that there exists an edge set J for which every edge of J is induced by T and $G + J$ is redundantly rigid. By Lemma 4.8 and the connectivity of (T, F) it follows that $G + F$ is redundantly rigid. □

Let $G = (V, E)$ be a 2-connected graph. We say that a pair $\{u, v\} \subset V$ is a *separating pair* in G if $G - \{u, v\}$ is disconnected. If X is the vertex set of a connected component of $G - \{u, v\}$, for some separating pair $\{u, v\}$, then X is called a *fragment*. For a vertex set $Z \subseteq V$ a vertex $w \in V - Z$ is called a *neighbour* of Z if there is an edge from w to some vertex of Z. The set of neighbours of Z is denoted by $N(Z)$. Thus $N(X)$ forms a separating pair for every fragment X. A minimal fragment of G (with respect to inclusion) is an *end*. A separating pair $\{u_1, v_1\}$ *crosses* another separating pair $\{u_2, v_2\}$ if u_1 and v_1 belong to different components of $G - \{u_2, v_2\}$. It is not hard to see that if $\{u_1, v_1\}$ crosses $\{u_2, v_2\}$ then $\{u_2, v_2\}$ crosses $\{u_1, v_1\}$. Hence these pairs are said to be *crossing* separating pairs. The next lemma is easy to verify.

Lemma 4.10 *Let G be 2-connected and suppose that v is a vertex of some end B of G. If $\{u, v\}$ is a separating pair for some vertex u then $N(B)$ and $\{u, v\}$ are crossing separating pairs.*

Lemma 4.11 ([15, Lemmas 2.6(a), 3.5(b)]) *Let G be a rigid graph on at least three vertices. Then*
(i) G is 2-connected, and
(ii) there are no crossing separating pairs in G.

Given two disjoint vertex sets $X, Y \subseteq V$ in a graph, the number of edges from X to Y is denoted by $d(X, Y)$.

Lemma 4.12 *Let $G = (V, E)$ be a Laman graph and let $X, Y \subset V$ with $|X \cap Y| = \{u, v\}$ and $d(X - Y, Y - X) = 0$. Then*
(i) if $uv \in E$ and $V = X \cup Y$ then $G[X]$ and $G[Y]$ are both Laman,
(ii) if $uv \notin E$ then at most one of $G[X]$ and $G[Y]$ is Laman. Furthermore, if $V = X \cup Y$ then exactly one of $G[X]$ and $G[Y]$ is Laman.

Proof First suppose that $uv \in E$ and $V = X \cup Y$. Then we have $2|V| - 3 = i(X) + i(Y) - 1 \leq 2|X| - 3 + 2|Y| - 3 - 1 = 2|V| - 3$. This implies (i). Next suppose that $uv \notin E$. If $G[X]$ and $G[Y]$ are both Laman then we have $2|X \cup Y| - 3 \geq i(X \cup Y) = i(X) + i(Y) = 2|X| - 3 + 2|Y| - 3 = 2|X \cup Y| - 2$, a contradiction. This proves the first part of (ii). By assuming that $V = X \cup Y$ and that neither of $G[X]$ or $G[Y]$ is Laman we have $2|V| - 3 = i(X) + i(Y) \leq 2|X| - 4 + 2|Y| - 4 = 2|V| - 4$, a contradiction. This completes (ii). □

Fig. 5 In this graph A_1 is an extreme vertex that belongs to a separating pair

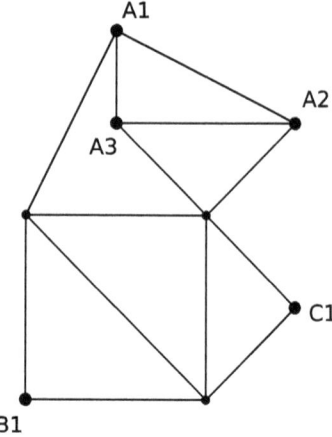

For a separating pair $\{u, v\}$ and a component C of $G - \{u, v\}$ let $\bar{C} = G[V(C) \cup \{u, v\}]$ be its *closure*.

Lemma 4.13 *Let G be a Laman graph and $\{u, v\}$ be a separating pair in G. Let the components of $G - \{u, v\}$ be denoted by $C_1, C_2, \ldots, C_t$. Then*
(i) if $uv \in E$ then $\bar{C}_i$ is Laman for all $1 \le i \le t$,
(ii) if $uv \notin E$ then there is a unique component, say C_1, for which $\bar{C}_1$ is Laman,
(iii) if $uv \notin E$ then $L(uv)$ intersects exactly one component of $G - \{u, v\}$.

Proof First observe that (i) follows by applying Lemma 4.12(i) to the sets $X = V(\bar{C}_i)$ and $Y = V - V(C_i)$. Next we assume $uv \notin E$. Then Lemma 4.12(ii) gives that at most one $\bar{C}_i$ is Laman. For a contradiction suppose that no $\bar{C}_i$ is Laman. Then $2|V| - 3 = |E| = \sum_1^t i(\bar{C}_j) = \sum_1^t (2|V(\bar{C}_j)| - 4) = 2|V| + 4(t - 1) - 4t = 2|V| - 4$, a contradiction. Finally, (iii) follows from (ii), since if $\bar{C}_1$ is Laman then it must contain the unique smallest Laman subgraph $L(uv)$ containing u, v. □

In a Laman graph an extreme vertex may belong to some separating pair, see Fig. 5. The next lemmas will show that it cannot happen to all vertices of an extreme class.

Lemma 4.14 *Let G be a Laman graph and let $\{u, v\}$ be a separating pair in G. Consider a pair x, y of vertices with $x \in A$, $y \in B$, where A, B are distinct connected components of $G - \{u, v\}$. Then $L(uv) \subseteq L(xy)$ and $L(ux) \subseteq L(xy)$.*

Proof By Lemma 4.1(iii), $L(xy)$ is a (smallest) Laman subgraph that contains x and y. Since $xy \notin E$, $L(xy)$ is 2-connected by Lemma 4.11(i). Thus we have $\{u, v\} \subseteq L(xy)$. Hence $L(uv) \subseteq L(xy)$. A similar argument gives $L(ux) \subseteq L(xy)$. □

Lemma 4.15 *Suppose that G is a wide Laman graph. Then every extreme class of G contains at least one vertex which is not part of any separating pair in G.*

Proof Consider an extreme class P of G and fix an extreme vertex $u \in P$. Suppose that $\{u, v\}$ is a separating pair for some $v \in V$. Since u is extreme, there exists an MGL $L(uj)$ for some extreme vertex j. Fix two components A, B of $G - \{u, v\}$ and a pair of vertices $x \in A$, $y \in B$.

We claim that $j \neq v$. To see this first note that if $uv \in E$, then $L(uv)$ is not an MGL, and hence $j \neq v$ follows. Next suppose that $uv \notin E$. Then we have $L(uv) \subseteq L(xy)$ by Lemma 4.14. Furthermore, the inclusion must be proper by Lemma 4.13(iii). This shows that that $L(uv)$ is not an MGL. Hence $j \neq v$ and the claim follows.

By symmetry we may assume that $j \notin B$. Then it follows from Lemma 4.14 that for every vertex $y \in B$ we have $L(uj) = L(yj)$ and hence y is also in P. By taking y to be a vertex of some end within B the lemma follows from Lemmas 4.10 and 4.11(ii). □

A similar result holds for narrow Laman graphs.

Lemma 4.16 *Let $G = (V, E)$ be a narrow Laman graph. Then there is a pair $u, v \in V$ which is disjoint from all separating pairs and for which $G + uv$ is redundantly rigid.*

Proof Since G is narrow, there is a pair $u_1, v_1 \in V$ for which $G + u_1 v_1$ is redundantly rigid. Suppose that $\{u_1, w\}$ is a separating pair for some vertex $w \in V$. By Lemma 4.13(iii) we must have $w \neq v_1$. Let A, B be two connected components of $G - \{u_1, w\}$ with $v_1 \in A$ and consider a vertex $u \in B$. By Lemma 4.14 we have $L(u_1 v_1) \subseteq L(uv_1)$. Since $L(u_1 v_1) = G$, it follows that $G + uv_1$ is redundantly rigid. $L(xv_1) = G$. By choosing u to be a vertex of some end within B, we may assume that u is disjoint from all separating pairs. Now applying a similar argument to the pair $\{u, v_1\}$ we obtain that there is a pair $\{u, v\}$ which is disjoint from all separating pairs and for which $L(uv) = G$. This completes the proof. □

5 Minimum Cost Globally Rigid Spanning Subgraphs

In this section we consider the metric MCRRSS and MCGRSS problems in the plane. To illustrate the main ideas, we start with the minimum cost redundantly rigid subgraph problem, for which we have a simpler approximation algorithm. Recall that the input of both problems is a complete graph $K = (V, E(K))$ on at least four vertices and a metric cost function $c : E(K) \to \mathbb{R}_+$.

Algorithm MinCostRedRig2

(i) Compute a minimum cost spanning Laman subgraph $G = (V, E)$ of K.
(ii) If G is a wide Laman graph, then find a set S of extreme vertices of G that contains exactly one vertex from each extreme class and compute a minimum cost spanning tree (S, F) of $K[S]$, where $K[S]$ is the subgraph of K induced by S. Output $(V, E + F)$.

(iii) If G is a narrow Laman graph, then find a new edge e for which $G + e$ is redundantly rigid. Output $(V, E + e)$.

Theorem 5.1 *Algorithm MinCostRedRig2 is a polynomial time 2-approximation algorithm for the metric MCRRSS in $\mathbb{R}^2$.*

Proof Consider an instance of MCRRSS. If G is wide, the output is a feasible solution by Lemma 4.9. If G is narrow, the output is feasible by construction. To verify the approximation ratio consider an optimal solution G^*. Let OPT denote the total cost of the edges of G^*. Since G^* is rigid, we have $c(E) \leq OPT$. We claim that G^* contains two edge-disjoint spanning trees. Indeed, since G^* is redundantly rigid, there exists a minimally rigid spanning subgraph H of $G^* - e$, for any fixed edge e of G^*: now Theorems 2.1 and 2.4 imply that $H + e$ is the union of two edge-disjoint spanning trees.

Suppose that G is wide and the output is obtained in step (ii). Since G^* contains two edge-disjoint spanning trees, a minimum cost spanning tree F^* of K satisfies $c(F^*) \leq \frac{OPT}{2}$. Furthermore, it is well-known that if c is metric and $S \subseteq V(G)$ then the cost of a minimum cost spanning tree in $K[S]$ has cost at most $2c(F^*)$. This follows by doubling the edges of F^* to obtain an Eulerian graph J and then shortcutting an Eulerian walk of J to obtain a spanning cycle C on S. Since c is metric and C contains a spanning tree of $K[S]$, it follows that the minimum cost spanning tree on S has cost at most $2c(F^*)$. Hence if G is wide then we have $c(E + F) \leq 2OPT$, as required.

Next suppose that G is narrow and the output is obtained in step (iii). Let $e = uv$ be the edge found for which $G + e$ is redundantly rigid. Since G^* contains two edge-disjoint uv-paths, there is a uv-path P with $c(P) \leq \frac{OPT}{2}$. By using that c is metric, we obtain $c(e) \leq c(P) \leq \frac{OPT}{2}$ and hence $c(E + e) \leq \frac{3}{2}OPT \leq 2OPT$, as claimed.

The polynomial running time of the algorithm follows by noting that a minimum cost spanning tree or a minimum cost spanning Laman subgraph can be found efficiently by a greedy algorithm. Moreover, as we remarked earlier, the extreme classes of G can also be found in polynomial time. □

Next we consider the metric MCGRSS in $\mathbb{R}^2$. The following algorithm is a refined version of Algorithm MinCostRedRig2.

Algorithm MinCostGlobRig2

(i) Compute a minimum cost spanning Laman subgraph $G = (V, E)$ of K.
(ii) If G is a wide Laman graph then find a set S of extreme vertices of G that contains exactly one vertex from each extreme class, so that the vertex belongs to no separating pair of G.
(iii) If G is a narrow Laman graph then find a pair $S = \{i, j\}$ of vertices, for which $G + ij$ is redundantly rigid and i, j belong to no separating pair of G.
(iv) Find a set T of vertices of G that contains exactly one vertex from each end W of G which is disjoint from S,

(v) Compute a minimum cost spanning tree (R, F) of $K[R]$, where $R = S \cup T$. Output $(V, E + F)$.

The steps of the algorithm are well-defined by Lemmas 4.15 and 4.16. We next show that the output is a feasible solution.

Lemma 5.2 *The output of Algorithm MinCostGlobRig2 is*
(i) 3-connected, and
(ii) redundantly rigid.

Proof First we prove (i). By the choice of the vertices in S (c. f. Lemmas 4.15, 4.16) and the vertices in T added from the ends (c.f. Lemmas 4.10, 4.11) no vertex in R belongs to a separating pair of G. Furthermore, for every end (and hence for every fragment) X we must have $X \cap R \neq \emptyset$. This implies that adding a tree on R eliminates every separating pair of G and hence makes it 3-connected.

Next we prove (ii) simultaneously for the two cases, that depend on whether G is wide or narrow. Let us fix two vertices $i, j \in S$ for which every internal vertex of the path P from i to j in (R, F) is a vertex in T. Let $P = i, t_1, t_2, \ldots, t_r, j$ and $L_i = L(t_i t_{i+1})$ for $1 \leq i \leq r - 1$. The key observation, which follows from Lemma 4.14, is that in the sequence $L(it_1), L_1, L_2, \ldots L_{r-1}, L(t_r j)$ each pair of consecutive Laman subgraphs have at least two vertices in common. By Lemma 4.5 this implies that their union is Laman. Hence $L(ij) \subseteq L(it_1) \cup L(t_1 t_2) \cup \ldots \cup L(t_{r-1} t_r) \cup L(t_r j)$. Then it follows that by adding the edges of P we make every edge of $L(ij)$ redundant. Therefore, by Lemma 4.9, adding F makes every edge of G redundant. □

An analysis similar to that of MinCostRedRig2, together with Lemma 5.2 above, gives our main result in $\mathbb{R}^2$.

Theorem 5.3 *Algorithm MinCostGlobRig2 is a polynomial time 2-approximation algorithm for the metric MCGRSS in $\mathbb{R}^2$.*

We have a family of instances showing that the approximation ratio of Algorithm MinCostRedRig2 (and of MinCostGlobRig2) is not better than $\frac{3}{2}$. Consider a complete graph K on $2s + 1$ vertices, for some integer $s \geq 2$. Fix a subset E of vertices of size s and define the costs of the edges of K so that the edges between vertices in E are of cost 2 while the cost of every other edge is equal to 1. The algorithm

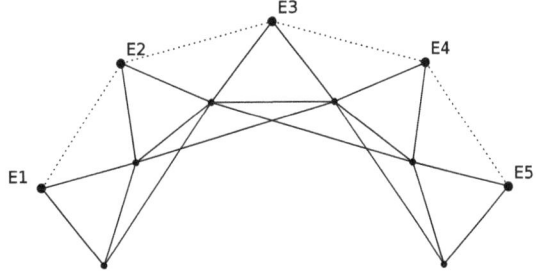

Fig. 6 The solid edges correspond to the spanning Laman subgraph. The dotted edges form a tree on its extreme vertices

may find, as the minimum cost spanning Laman subgraph, a graph in which each vertex in E is an extreme vertex of degree two. See Fig. 6 for the case $s = 5$. The minimum cost tree on these vertices has total cost $2s - 2$. Thus the output has cost $4s - 1 + 2s - 2 = 6s - 3$. On the other hand it is not hard to see that a feasible solution of cost $4s$ exists.

5.1 The Euclidean Case

In the Euclidean version of our problems the vertices correspond to points in $\mathbb{R}^2$ and the cost of an edge is the Euclidean distance of its endpoints. In this version, which may occur for example in the network localization problem, our algorithm has a better approximation ratio.

In order to show this, recall that in the Euclidean Steiner Tree Problem we are given a set S of points in the plane and the goal is to find a tree of minimum total length, which contains S. The tree may use points not in S. The ratio of the total length of a shortest spanning tree on S and the total length of a shortest Steiner tree with respect to S is the so called *Steiner ratio*. It was proved in [5] that the Steiner ratio is at most 1.22.

We can use this fact in the analysis of our algorithm and deduce that $c(F) \leq 1.22 c(F^*) \leq 0.61 OPT$, following the notation of Theorem 5.1. Thus the approximation ratio of the Euclidean version of MinCostRedRig2 (and MinCostGlobRig2) is 1.61.

6 Higher Dimensions

In this section we design an approximation algorithm for the d-dimensional metric MCGRSS problem, which works for every $d \geq 2$, with an approximation ratio that depends only on d.

The algorithm is rather simple and is based on the idea of graph powers. The kth power of graph G, denoted by G^k, is the graph on the same vertex set, in which two vertices are adjacent if and only if their distance in G is at most k. The input of the algorithm is an integer $d \geq 2$, a complete graph $K = (V, E(K))$ on at least $d + 2$ vertices and a metric cost function $c : E(K) \to \mathbb{R}_+$.

Algorithm MinCostGlobRigGen

(i) Compute a minimum cost spanning tree T of K.
(ii) By shortcutting $2T$ create a Hamilton cycle C on vertex set V.
(iii) Output C^d.

In step (ii) the graph $2T$ is obtained from T by replacing every edge of T by two parallel edges. The shortcutting operation is standard and we already used in

Fig. 7 A graph obtained in the process of constructing a globally rigid spanning subgraph of C_{13}^3 by 1-extensions. The last vertex added up to this point is v_i. The dotted edges have been deleted by the previous 1-extensions

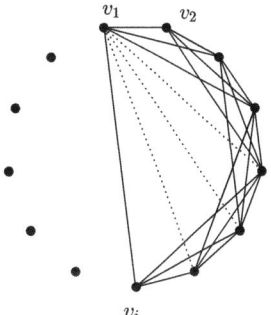

the analysis of Theorem 5.1: we find an Eulerian walk of $2T$ and by shortcutting repeated vertices we turn it into a Hamilton cycle.

The fact that the output is a feasible solution follows from the next lemma. Let C_n denote a cycle on n vertices.

Lemma 6.1 C_n^d *is globally rigid in* $\mathbb{R}^d$.

Proof If $n \leq 2d + 1$ then C_n^d is complete, and hence globally rigid in $\mathbb{R}^d$. So we may assume that $n \geq 2d + 2 \geq d + 2$. We shall prove that a spanning subgraph of C_n^d can be obtained from K_{d+2}, which is globally rigid, by a sequence of (d-dimensional) 1-extensions. This operation adds a new vertex v to the graph, deletes an edge uw, and adds $d + 1$ new edges incident with v, so that the set of new edges includes vu and vw. It is known that this operation preserves global rigidity in $\mathbb{R}^d$, see [6].

Label the vertices of C_n^d by $v_1, \ldots, v_n$ and start with a K_{d+2} on vertex set $v_1, \ldots, v_{d+2}$. In the first iteration perform a 1-extension which adds vertex v_{d+3}, deletes the edge $v_1 v_{d+2}$, and connects v_{d+3} to $v_{d+2}, v_{d+1}, \ldots, v_3$ and v_1. In the next iteration add v_{d+4} by a 1-extension on edge $v_1 v_{d+3}$ so that the new vertex is connected to the preceding d vertices and to v_1, and so on. After $n - d - 2$ iterations all vertices of C_n^d are included and the graph constructed is a globally rigid spanning subgraph of C_n^d. See Fig. 7. This completes the proof. □

The analysis of the algorithm will also use the following claim.

Lemma 6.2 *Suppose that* $G = (V, E)$ *is rigid in* $\mathbb{R}^d$ *with* $|E| = d|V| - d$ *and* $|V| \geq d + 1$, *for some* $d \geq 1$. *Then the edge set of* G *can be decomposed into* d *spanning trees.*

Proof Since $|E| = d(|V| - 1)$, it suffices to show that the edge set of G can be partitioned into d forests. We shall verify that G satisfies the condition in Theorem 2.4 for $k = d$. Before counting edges let us fix a minimally rigid spanning subgraph H of G. It is known that H has $d|V| - \binom{d+1}{2}$ edges and for every vertex set $X \subseteq V$ with $|X| \geq d + 1$ we have $i_H(X) \leq d|X| - \binom{d+1}{2}$. The number of edges of G which do not belong to H is equal to $\binom{d+1}{2} - d$.

Let $X \subseteq V$ be a non-empty vertex set. First suppose $|X| \geq d+1$. Then, by using the above bounds, we have $i_G(X) \leq i_H(X) + \binom{d+1}{2} - d \leq d|X| - d$, as required. Next suppose $|X| \leq d$. Then, since G has no parallel edges, we have $i_G(X) \leq \binom{|X|}{2} = \frac{|X|(|X|-1)}{2} \leq d(|X|-1) = d|X| - d$. This completes the proof. □

We are ready to analyse the algorithm. For simplicity we shall assume that $|V| \geq \binom{d}{2}$. We remark that if the input graph is smaller, a similar analysis gives the upper bound $2d + 2$ for the approximation ratio. Moreover, in this case enumerating all feasible solutions would also be an option for d fixed.

Theorem 6.3 *Algorithm MinCostGlobRigGen is a polynomial time $(d + \frac{2d}{d-1})$-approximation algorithm for the metric MCGRSS problem in $\mathbb{R}^d$, assuming that the size of the input graph is at least $\binom{d}{2}$.*

Proof The output is a feasible solution by Lemma 6.1. The polynomial running time is also clear. It remains to verify the approximation ratio.

Let $G^* = (V, E)$ be an optimal solution. Since it is globally rigid in $\mathbb{R}^d$, it is also redundantly rigid in $\mathbb{R}^d$ by Theorem 2.2. Thus we have $|E| \geq d|V| - \binom{d+1}{2} + 1$. Furthermore, $G^* = (V, E)$ is rigid in $\mathbb{R}^{d-1}$, too (an observation that follows easily from e.g. by the coning theorem of [7]). Also, since $|V| \geq \binom{d}{2}$, we have $|E| \geq d|V| - \binom{d+1}{2} + 1 = (d-1)|V| - (\binom{d}{2} + d) + 1 + |V| \geq (d-1)|V| - (d-1)$. Thus we can apply Lemma 6.2 to G^* and deduce that it contains $d - 1$ pairwise edge-disjoint spanning trees. Hence $c(F) \leq \frac{OPT}{d-1}$.

Therefore $c(C) \leq \frac{2OPT}{d-1}$. By using the metric property of c, the total cost of the edges of C^d that connect vertices which are of distance exactly k in C can be bounded by $kc(C)$. Thus

$$c(C^d) \leq (1 + 2 + \cdots + d)c(C) = \frac{d(d+1)}{2}c(C) \leq \frac{d(d+1)}{2}\frac{2}{d-1}OPT$$
$$= \frac{d(d+1)}{d-1}OPT = \left(d + \frac{2d}{d-1}\right)OPT,$$

as claimed. □

Note that the approximation ratio of MinCostGlobRigGen for $d = 2$ (and for $d = 3$) is equal to 6, which is substantially worse than that of algorithm MinCostGlobRig2. In the next subsection we show how to improve on this ratio in the three-dimensional case by using a more sophisticated analysis.

6.1 Improving the Ratio for $d = 3$

We start with a technical lemma.

Minimum Cost Globally Rigid Subgraphs

Lemma 6.4 *Let $K = (V, E)$ be a complete graph and let $c : E \to \mathbb{R}_+$ be a metric cost function. Suppose that $G = (V, F)$ is a 3-connected spanning subgraph of K which contains no subgraph isomorphic to K_6. Then for every $p > 0$ there is an N_p such that if $|V| \geq N_p$ then there exists a pair $\{e, f\} \subset E - F$ of edges with $c(e) + c(f) \leq c(G)p$.*

Proof First suppose that there is a vertex v with $d_G(v) \geq 3 + 4\lceil \frac{1}{p} \rceil$. Let $X = N_G(v)$. We claim that X induces at least $2\lceil \frac{1}{p} \rceil$ pairwise disjoint non-edges. Indeed, such a collection M of non-edges can be obtained in a greedy manner, using the fact that any subset of six vertices of X induces at least one non-edge. By using the metric property of c we can now deduce that

$$c(F) \geq \sum_{vu : u \in N_G(v)} c(vu) \geq \sum_{e \in M} c(e).$$

Thus the two edges e, f of M with the smallest cost satisfy $c(e) + c(f) \leq c(G)p$, as required.

Next suppose that $d_G(v) < 3 + 4\lceil \frac{1}{p} \rceil$ for all $v \in V$. Then, assuming $|V| > \sum_{i=0,\ldots,k} (3 + 4\lceil \frac{1}{p} \rceil)^i$ for some integer k, it follows that there exist two vertices $v_1, v_2 \in V$ for which the length of a shortest path from v_1 to v_2 in G is at least $k + 1$.

Take three internally disjoint chordless paths from v_1 to v_2. Let P be one of them with minimum total cost. Then we have $c(P) \leq \frac{1}{3}c(G)$. Furthermore, by taking a path of non-edges connecting every second vertex along P and using the fact that c is metric we obtain a set N of at least $\lfloor \frac{k+1}{2} \rfloor$ non-edges with $c(N) \leq c(P) \leq \frac{1}{3}c(G)$. Thus the two edges e, f of N with the smallest cost satisfy $c(e) + c(f) \leq \frac{1}{3} \frac{2}{\lfloor \frac{k+1}{2} \rfloor} c(G) \leq \frac{1}{3k} c(G)$. Hence by choosing $k \geq \frac{1}{3p}$ and $k \geq 4$, we have $c(e) + c(f) \leq c(G)p$, as required. $\square$

The next lemma leads to an improved bound by choosing arbitrary small $p < \frac{3}{2}$.

Lemma 6.5 *Let $K = (V, E)$ be a complete graph and let $c : E \to \mathbb{R}_+$ be a metric cost function. Suppose that $G = (V, F)$ is a globally rigid subgraph of K in $\mathbb{R}^3$ and let $H = (V, T)$ be a minimum cost spanning tree. Then for every $p > 0$ there is an N_p such that if $|V| \geq N_p$ then $c(T) \leq \frac{1}{3}(1 + p)c(G)$.*

Proof We shall use that, since G is globally rigid in $\mathbb{R}^3$, G is 3-connected and has a spanning proper subgraph G' with $3|V| - 6$ edges satisfying $i_{G'}(X) \leq 3|X| - 6$ for all $X \subseteq V$ with $|X| \geq 3$. Let's fix p.

First suppose $|F| \geq 3|V| - 3$. Then we can add three edges from F to G' and obtain a subgraph of G which contains three edge-disjoint spanning trees (by Nash-Williams' theorem). Hence $c(T) \leq \frac{2}{3}c(G)$.

Next suppose $3|V| - 5 \leq |F| \leq 3|V| - 4$. Then the sparsity property of G' implies that G contains no subgraph isomorphic to K_6. Now we may apply Lemma 6.4 to G and deduce that there is an N_p such that if $|V| \geq N_p$ then there exists a pair $\{e, f\} \subset E - F$ of edges with $c(e) + c(f) \leq c(G)p$. A similar argument gives that $G + e + f$ contains three edge-disjoint spanning trees, and hence $c(T) \leq \frac{1}{3}(1 + p)c(G)$. □

We can now deduce an upper bound on the approximation ratio of MinCostGlobRigGen for $d = 3$, at least for large enough graphs, which can be arbitrarily close to 4.

Theorem 6.6 *Let $K = (V, E)$ be a complete graph and let $c : E \to \mathbb{R}_+$ be a metric cost function. For every $p > 0$ there is an N_p such that if $|V| \geq N_p$ then the approximation ratio of MinCostGlobRigGen for $d = 3$ is at most $2\frac{1}{3}(1 + p)6 = 4(1 + p)$.*

7 Concluding Remarks

In this paper we introduced the Minimum cost globally rigid spanning subgraph problem in $\mathbb{R}^d$ and gave polynomial time approximation algorithms for the metric version. It remains an open problem to find similar results for general cost functions.

For Euclidean costs we obtained a somewhat better approximation ratio. It might be possible to find a polynomial time approximation scheme, like in the case of the k-connected spanning subgraph problem, see e.g. [8].

Finally we remark that a long list of similar problems can be obtained by replacing global rigidity in $\mathbb{R}^2$ (or equivalently, 3-connectivity and redundant rigidity) by other types of connectivity and sparsity requirements. The matroid on the edge set of a graph defined by the sparsity count of (1) happens to be a specific example of the so-called *count matroids*. These matroids can defined in a similar way by replacing $i(X) \leq 2|X| - 3$ by $i(X) \leq k|X| - l$ for some integers k, l with $l \leq 2k$, see [10, 25]. One can also define "redundant rigidity" with respect to these more general counts in a natural way. Partial results, extending the work in [11], have already been obtained by Király [20].

7.1 Hardness Results

For completeness we show that the problems considered in this paper are NP-hard. Since global rigidity is equivalent to 2-connectivity in $\mathbb{R}^1$, finding a smallest globally rigid spanning subgraph of a graph G on the line is more general than the Hamilton cycle problem. Hence MCGRSS is NP-hard in $\mathbb{R}^1$. By applying a sequence of $d - 1$ coning operations[1] to G, and assigning cost zero to each of the new edges, we can reduce the problem to the d-dimensional MCGRSS problem, for any given d,

---

[1] The *cone* of graph G is obtained from G by adding a new vertex v and new edges from v to all vertices of G. See Fig. 8. Connelly and Whiteley [7] proved that a graph G in globally rigid in $\mathbb{R}^d$ if and only if the cone of G is globally rigid in $\mathbb{R}^{d+1}$.

Fig. 8 The cone graph of a graph

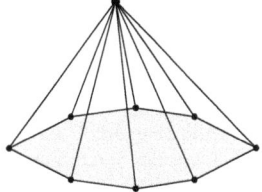

showing that MCGRSS is NP-hard in $\mathbb{R}^d$. A similar argument shows that MCRRSS is also NP-hard in $\mathbb{R}^d$ for all d.

A slightly more involved argument shows that these problems remain NP-hard in the metric case. Here we give the proof for MCGRSS in $\mathbb{R}^2$. Similar arguments can be used to extend the result to higher dimensions and to prove the hardness of metric MCRRSS in $\mathbb{R}^d$.

Theorem 7.1 *It is NP-hard to find a minimum cost globally rigid spanning subgraph in $\mathbb{R}^2$ of a given complete graph $G = (V, E)$ with respect to a metric cost function $c : E \to \mathbb{R}$.*

Proof We shall reduce the Hamilton cycle problem to our problem. Consider an instance $H = (V, E)$ of the Hamilton cycle problem. Let G be the cone of H, where the new vertex is denoted by v, and let K be the complete graph on vertex set $V \cup \{v\}$. We assign costs to the edges of K as follows.

For every edge $e = uv$ with $u, v \in V$ we let $c(e) = 1.1$ (resp. $c(e) = 1.9$) if $uv \in E$ (resp. if $uv \notin E$). For the remaining edges e of K, which are incident with v, we define $c(e) = 1$. We claim that H has a Hamilton cycle if and only if the minimum cost globally rigid spanning subgraph of K, with respect to c, has total cost $2.1|V|$.

To see this first suppose that H has a Hamilton cycle C. It is easy to see that the cone graph of C is globally rigid in $\mathbb{R}^2$. The total cost of the cone is $1.1|V| + |V| = 2.1|V|$. Next suppose that there is a globally rigid spanning subgraph F of K with cost at most $2.1|V|$. Since every globally rigid subgraph of K has at least $2|V|$ edges, the definition of c implies that F has exactly $2|V|$ edges and that it contains every edge incident with v.

Thus F is the cone graph of a 2-connected spanning subgraph C of H with $|V|$ edges. This shows that C is a Hamilton cycle in H. This completes the proof. □

Acknowledgements This work was supported by the Hungarian Scientific Research Fund grant no. K109240 and K115483, and the ÚNKP-18-3 New National Excellence Program of the Ministry of Human Capacities, Hungary.

References

1. B. D. O. Anderson, P. N. Belhumeur, T. Eren, D. K. Goldenberg, A. S. Morse, W. Whiteley, and Y. R. Yang. Graphical properties of easily localizable sensor networks. *Wireless Networks* 15 (2): 177–191, 2009.

2. L. Asimow and B. Roth. The rigidity of graphs, *Trans. Amer. Math. Soc.*, 245 (1978), pp. 279–289.
3. J. Aspnes, T. Eren, D. K. Goldenberg, A. S. Morse, W. Whiteley, Y. R. Yang, B. D. O. Anderson, and P. N. Belhumeur. A theory of network localization. *IEEE Transactions on Mobile Computing* Vol. 5 (2006) 1663–1678.
4. A.R. Berg and T. Jordán. Algorithms for graph rigidity and scene analysis. Proc. 11th Annual European Symposium on Algorithms (ESA) 2003, (G. Di Battista, U. Zwick, eds) Springer Lecture Notes in Computer Science 2832, pp. 78–89, 2003.
5. F.R.K. Chung and R.L. Graham. A new bound for Euclidean Steiner minimal trees, *Annals of the New York Academy of Sciences*, 440 (1985), pp. 328–346.
6. R. Connelly. Generic global rigidity, *Discrete Comput. Geom.* 33 (2005), pp 549–563.
7. R. Connelly and W. Whiteley. Global rigidity: the effect of coning, *Discrete Comput. Geom.* (2010) 43: 717–735.
8. A. Czumaj and A. Lingas. Approximation schemes for minimum-cost k-connectivity problems in geometric graphs, in: Handbook of approximation algorithms and metaheuristics, T.F. Gonzalez (ed.), CRC, 2007.
9. Z. Fekete, T. Jordán, Uniquely localizable networks with few anchors, Proc. Algosensors 2006, (S. Nikoletseas and J.D.P. Rolim, eds) Springer Lecture Notes in Computer Science 4240, pp. 176–183, 2006.
10. A. Frank. Connections in combinatorial optimization, Oxford University Press, 2011.
11. A. García and J. Tejel. Augmenting the rigidity of a graph in R^2, *Algorithmica*, February 2011, Volume 59, Issue 2, pp 145–168.
12. S. Gortler, A. Healy, and D. Thurston. Characterizing generic global rigidity, *American Journal of Mathematics*, Volume 132, Number 4, August 2010, pp. 897–939.
13. B. Hendrickson. Conditions for unique graph realizations, *SIAM J. Comput.* 21, 65–84, 1992.
14. B. Jackson. Notes on the rigidity of graphs. Lecture notes, Levico, 2007.
15. B. Jackson and T. Jordán. Connected rigidity matroids and unique realizations of graphs. *J. Combinatorial Theory Ser B*, 94:1–29, 2005.
16. B. Jackson and T. Jordán. Graph theoretic techniques in the analysis of uniquely localizable sensor networks, in: Localization Algorithms and Strategies for Wireless Sensor Networks (G. Mao and B. Fidan, eds.), IGI Global, 2009.
17. T. Jordán. Rigid and globally rigid graphs with pinned vertices, in: Bolyai Society Mathematical Studies, 20, G.O.H. Katona, A. Schrijver, T. Szőnyi, eds., Fete of Combinatorics and Computer Science, 2010, Springer, pp. 151–172.
18. T. Jordán. Combinatorial rigidity: graphs and matroids in the theory of rigid frameworks. In: Discrete Geometric Analysis, *MSJ Memoirs*, vol. 34, pp. 33–112, 2016.
19. T. Jordán and W. Whiteley. Global rigidity, *in* J. E. Goodman, J. O'Rourke, and Cs. D. Tóth (eds.), *Handbook of Discrete and Computational Geometry,* 3rd ed., CRC Press, Boca Raton, pp. 1661–1694.
20. C. Király. Rigid graphs and an augmentation problem, Tech. report 2015-03, Egerváry Research Group, Budapest, 2015.
21. G. Kortsarz and Z. Nutov. Approximating minimum-cost connectivity problems, in: Handbook of approximation algorithms and metaheuristics, T.F. Gonzalez (ed.), CRC, 2007.
22. G. Laman. On graphs and rigidity of plane skeletal structures, *J. Engineering Math.* 4 (1970), 331–340.
23. C.St.J.A. Nash-Williams. Decomposition of finite graphs into forests, *The Journal of the London Mathematical Society* 39 (1964) 12.
24. J.B. Saxe. Embeddability of weighted graphs in k-space is strongly NP-hard, Tech. Report, Computer Science Department, Carnegie-Mellon University, Pittsburgh, PA, 1979.
25. W. Whiteley. Some matroids from discrete applied geometry. In J. Bonin, J. Oxley, and B. Servatius, editors, *Matroid Theory*, volume 197 of *Contemp. Math.*, pages 171–311. Amer. Math. Soc., Providence, 1996.
26. W. Whiteley. Rigidity of molecular structures: generic and geometric analysis. In P.M. Duxbury and M.F. Thorpe, editors, *Rigidity Theory and Applications*, Kluwer/Plenum, New York, 1999.
27. C. Yu and B.D.O. Anderson. Development of redundant rigidity theory for formation control, *Int. J. Robust and Nonlinear Control*, Vol. 19, Issue 13, 2009, pp. 1427–1446.

Coloured and Directed Designs

Peter Keevash

To László Lovász on his seventieth birthday

Abstract We give some illustrative applications of our recent result on decompositions of labelled complexes, including some new results on decompositions of hypergraphs with coloured or directed edges. For example, we give fairly general conditions for decomposing an edge-coloured graph into rainbow triangles, and for decomposing an r-digraph into tight q-cycles.

Keywords Hypergraphs · Decompositions · Designs

Subject Classifications 05C65 · 05C70 · 05B05

1 Introduction

When can we decompose an object into copies of some other object? This vague question suggests a number of mathematical problems. Within graph theory, a fundamental instance of this question asks for a decomposition (i.e. partition of the edge set) of the complete graph K_n into copies of K_q. We require $n \geq q^2 - q + 1$ by Fisher's inequality (see e.g. [28, Theorem 19.6]). If q is one more than a prime power then the lines of a projective plane give a construction with $n = q^2 - q + 1$, but we do not know any construction with $n = q^2 - q + 1$ when q is not of this

---

Research supported in part by ERC Consolidator Grant 647678.

P. Keevash (✉)
Mathematical Institute, University of Oxford, Oxford, UK
e-mail: keevash@maths.ox.ac.uk

b	c	a	d
a	d	b	c
d	a	c	b
c	b	d	a

β	γ	α	δ
δ	α	γ	β
γ	β	δ	α
α	δ	β	γ

6	11	1	16
13	4	10	7
12	5	15	2
3	14	8	9

Fig. 1 Orthogonal and magic squares

form; the Prime Power Conjecture suggests that there are none. On the other hand, we may fix q and ask for conditions on n that guarantee a decomposition (perhaps only for large $n > n_0(q)$ so as to exclude the difficulties associated with the Prime Power Conjecture). The first such result, obtained by Kirkman in 1846 (see [30]), shows that K_n has a triangle decomposition iff n is 1 or 3 modulo 6.

These beginnings suggest several possible directions for further generalisation. From the combinatorial perspective (taken in this paper), one may ask for a decomposition of G by copies of H where G and H are any given graphs, or hypergraphs, or indeed other related structures (we will consider coloured and directed hypergraphs). On the other hand, the above questions also have natural interpretations in Design Theory, which suggests many further questions (some of which also have natural combinatorial interpretations). Perhaps the oldest topic in this area is that of Latin and Magic squares, which have their roots in antiquity (see [3, Chap. 2]); they were given prominence in the Western mathematical tradition by Euler in 1776, who posed the *36 officer's puzzle*, which was open until its solution by Tarry in 1900. In modern terminology, the result is that there is no pair of orthogonal Latin squares of order 6. A pair of orthogonal Latin squares of order 4 is illustrated in Fig. 1, together with an associated magic square (obtained by assigning values 1, 2, 3, 4 to a, b, c, d and 0, 4, 8, 12 to $\alpha, \beta, \gamma, \delta$).

In general, a Latin square of order n is a labelling of the cells of an n by n square with n symbols so that every symbol appears once in each row and once in each column. An equivalent combinatorial description is a triangle decomposition of $K_3(n)$, the complete tripartite graph with parts of size n. Indeed, we identify the three parts with the sets of rows, columns and symbols of the square, and then each cell corresponds to a triangle in the obvious way. For a pair of orthogonal Latin squares of order n we require two such squares with the extra condition that every pair of symbols appears together once; this is analogously equivalent to a K_4-decomposition of $K_4(n)$ (and similarly for larger numbers of mutually orthogonal Latin squares). We have chosen the pair in Fig. 1 with the extra property that both diagonals use all symbols in both squares, so as to obtain a magic square (all rows, columns and diagonals have the same sum). In Fig. 2 we illustrate the popular puzzle of completing a partially filled Sudoku square, which is a Latin square of order 9 partitioned into 3 by 3 subsquares each of which uses every symbol once.

Fig. 2 A completed Sudoku puzzle

1	8	4	9	6	3	7	2	5
5	6	2	7	4	8	3	1	9
3	9	7	5	1	2	8	6	4
2	3	9	6	5	7	1	4	8
7	5	6	1	8	4	2	9	3
4	1	8	2	3	9	6	5	7
9	4	1	3	7	6	5	8	2
6	2	3	8	9	5	4	7	1
8	7	5	4	2	1	9	3	6

We now consider the generalisations of the above problems from graphs to r-graphs (hypergraphs in which every edge has size r). When does an r-multigraph G have a decomposition into copies of some fixed r-graph H? The case that $H = K_q^r$ is the complete r-graph on q vertices is of particular interest, as a K_q^r-decomposition of K_n^r is equivalent to a Steiner (n, q, r) system, i.e. a collection of blocks of size q in a set of size n covering every set of size r exactly once. For example, if $(q, r) = (3, 2)$ a triangle decomposition of K_n is equivalent to a Steiner Triple System. More generally, giving each edge of K_n^r some fixed multiplicity λ, a K_q^r-decomposition of λK_n^r is equivalent to a (n, q, r, λ) design. Some necessary conditions for the existence of a K_q^r-decomposition of an r-multigraph G may be observed by considering the degrees. The degree of $e \subseteq V(G)$ is the number of edges of G containing e, i.e. the size of the neighbourhood $G(e) = \{f \subseteq V(G) \setminus e : e \cup f \in G\}$. We say G is K_q^r-divisible if $|G(e)|$ is divisible by $\binom{q-|e|}{r-|e|}$ for all $e \subseteq V(G)$; this is a necessary condition for a K_q^r-decomposition, as every copy of K_q^r containing e contains $\binom{q-|e|}{r-|e|}$ edges that contain e. For example, a necessary condition for the existence of a (n, q, r, λ) design is $\binom{q-i}{r-i} \mid \lambda \binom{n-i}{r-i}$ for all $0 \le i \le r - 1$. The Existence Conjecture, proved in [10], is that if $n > n_0(q, r, \lambda)$ is large and this divisibility condition holds then there is a (n, q, r, λ) design. More generally, we can find a K_q^r-decomposition in any K_q^r-divisible r-multigraph G that is sufficiently dense and quasirandom.

The Existence Conjecture has had a long history in Design Theory since 1853 when Steiner asked about the existence of Steiner (n, q, r) systems. Here we briefly mention a few highlights that are relevant to our discussion here. The case $r = 2$ was proved by Wilson [31–33] in the 1970s. Around the same time, Graver and Jurkat [6] and Wilson [34] showed that the divisibility condition suffices for an integral (n, q, r, λ) design, i.e. an assignment of integer weights w_Q to copies Q of K_q^r in K_n^r such that $\sum \{w_Q : e \in Q\} = \lambda$ for all $e \in K_n^r$. Rödl [24] showed the existence of approximate Steiner systems, i.e. that there are edge-disjoint copies of K_q^r in K_n^r such that only $o(n^r)$ edges are not covered; his semi-random (nibble) method is now an indispensable tool of modern Probabilistic Combinatorics. Teirlinck [26] was the first to show that there are *any* non-trivial (n, q, r, λ) designs for arbitrary r. Kuperberg, Lovett and Peled [14] gave an alternative probabilistic proof of this result (and the existence of many other regular combinatorial structures); their method was extended by Lovett, Rao and Vardy [18] to show the existence of 'large sets' of designs (for certain parameter sets). Glock, Kühn, Lo and Osthus [4] gave an alternative combinatorial proof of the Existence Conjecture (the proof in [10] used a randomised algebraic construction); they also weakened the typicality hypothesis of [10] (version 1) to an extendability hypothesis, similar to that subsequently used in [10] (version 2). Furthermore, in [5] they obtained analogous results on H-decompositions where H is any r-graph and G is an r-graph that is H-divisible, i.e. each degree $|G(e)|$ is divisible by the gcd of all degrees $|H(f)|$ with $|f| = |e|$.

Having discussed some hypergraph generalisations of Kirkman's result on triangle decompositions of K_n (Steiner Triple Systems), let us now consider such generalisations for triangle decompositions of $K_3(n)$ (Latin Squares). Besides being a combinatorially natural direction, this also has practical applications. For example, in software testing (see [9]), a K_q^r-decomposition of[1] $K_q^r(n)$ can be thought of as a sequence of tests to a program taking q inputs from $[n]$, so that for every r inputs all possible combinations are tested once (so an efficient K_q^r-covering of $K_q^r(n)$ suffices in this context). Another example is to a secret sharing scheme that distributes information to $q - 1$ bank clerks so that any r of them can open the safe but any $r - 1$ cannot: pick a random copy of K_q^r in the decomposition, give one vertex to each clerk, and make the final vertex the combination for the safe. High-dimensional permutations (also called Latin Hypercubes) are equivalent to K_{r+1}^r-decompositions of $K_{r+1}^r(n)$. In Sect. 2 we will show how the result of [11] implies an approximate formula for the number of such decompositions, thus confirming a conjecture of Linial and Luria [16]. The method applies in greater generality: as an other illustration we will give an approximate formula for the number of generalised Sudoku squares, via H-decompositions of $H(n)$ for an auxiliary 4-graph H.

In Sect. 3 we consider a common generalisation of the nonpartite and partite decompositions discussed above to a generalised partite setting in which the edges of H and G have the same intersection patterns with respect to some partitions of their vertex sets. This general setting encodes several further problems in Design Theory. For example, Kirkman's Schoolgirl Problem (a popular puzzle in the 19th century) asks for the construction of a Steiner Triple System that is resolvable, meaning that its blocks can be partitioned into perfect matchings (sets of triples covering every vertex exactly once). We will illustrate the generalisation to hypergraph decompositions given in [11]. We will also illustrate the construction in [11] of large sets of designs, i.e. decomposition of K_n^q into (n, q, r, λ) designs. An application of the latter (see [29]) is to the following 'Russian Cards' problem in information security. From a deck of n cards, we randomly deal cards so that Alice receives a cards, Eve $e < a$ cards and Bob $b = n - a - e$ cards. Alice wants to make a public announcement from which Bob can learn her cards (given the cards that he holds) while limiting the information that Eve receives (e.g. for any card that she does not hold she should not learn which of Alice or Bob holds it). A strategy for this problem can be identified with a partition of K_n^a, where edges represent the possible sets of cards for Alice, and

---

[1] For any hypergraph H we write $H(n)$ for its n-blowup.

Alice announces to which part her actual set belongs. An optimal (minimum number of parts) strategy such that Bob can learn Alice's hand corresponds to a partition of K_n^a into Steiner $(n, a, a - e)$ systems; furthermore, if $n > n_0(a, e)$ is large then it is secure against Eve, as for any card x that she does not hold, among the blocks disjoint from her hand in any of the Steiner systems, at least one contains x and at least one does not.

We will explain the statement of the result of [11] in Sect. 4, and illustrate it with two new applications in the subsequent two sections. In Sect. 5 we generalise the results on hypergraph decomposition discussed above to decompositions of hypergraphs where edges have colours which must be respected by the decomposition. As well as being combinatorially natural, such generalisations encode other problems of Design Theory (e.g. Whist Tournaments) and also fit within the large literature on rainbow versions of classical combinatorial results, which can encode seemingly unrelated questions (see e.g. [22]). In Sect. 6 we give a different generalisation, namely to decompositions of directed hypergraphs. This illustrates the following important feature of the result of [11]: it is fundamentally concerned with sets of functions (which we call labelled edges), so to apply it to sets of (unlabelled) edges (i.e. hypergraphs) we must encode an edge by a suitable set of labelled edges. This general setting has more applications, albeit at the expense of considerable effort required in setting up the theory in Sect. 4. However, this seems unavoidable, as there are divisibility phenomena even for unlabelled coloured hypergraphs that require labels to analyse (see [11, Sect. 1.5]). In Sect. 7 we give a common generalisation of the previous results for convenient use in applications. We conclude in Sect. 8 by discussing some directions for potential future research.

2 Partite Decompositions, Hypermutations, Sudoku

Over the next three sections we will gradually move from examples to the general setting. We start with this section by illustrating some results on hypergraph decompositions and some of their applications discussed in introduction. First we consider the nonpartite setting with the typicality condition from [10], which describes an r-graph where the common neighbourhood of small set of $(r - 1)$-sets behaves roughly as one would expect in a random r-graph of the same density.

Definition 2.1 Suppose G is an r-graph on $[n]$. The density of G is[2] $d(G) = |G|\binom{n}{r}^{-1}$. We say that G is (c, s)-*typical* if for any set A of $(r - 1)$-subsets of $V(G)$ with $|A| \leq s$ we have $\left|\bigcap_{f \in A} G(f)\right| = (1 \pm |A|c)d(G)^{|A|}n$.

The following result of [5] (see also [11, Theorem 1.5]) shows that any dense typical r-graph has an H-decomposition provided that it satisfies the necessary divisibility condition discussed above. Henceforth we fix parameters

---

[2] We identify any hypergraph with its edge-set, so $|G|$ is the number of edges.

$$h = 2^{50q^3} \text{ and } \delta = 2^{-10^3 q^5}.$$

Theorem 2.2 *Let H be an r-graph on $[q]$ and G be an H-divisible (c, h^q)-typical r-graph on $[n]$, where $n > n_0(q)$ is large, $d(G) > 2n^{-\delta/h^q}$, $c < c_0 d(G)^{h^{30q}}$ and $c_0 = c_0(q)$ is small. Then G has an H-decomposition.*

Next we set up some notation for stating the partite analogue of the previous result.

Definition 2.3 Let H be an r-graph. We call an r-graph G an H-blowup if $V(G)$ is partitioned as $(V_x : x \in V(H))$ and each $e \in G$ is f-partite for some $f \in H$, i.e. $f = \{x : e \cap V_x \neq \emptyset\}$.

We write G_f for the set of f-partite $e \in G$. For $f \in H$ let $d_f(G) = |G_f| \prod_{x \in f} |V_x|^{-1}$. We call G a (c, s)-typical H-blowup if for any $s' \leq s$ and distinct $e_1, \ldots, e_{s'}$ where each e_j is f_j-partite for some $f_j \in \binom{V(H)}{r-1}$, and any $x \in \cap_{j=1}^{s'} H(f_j)$ we have $\left| V_x \cap \bigcap_{j=1}^{s'} G(e_j) \right| = (1 \pm s'c) |V_x| \prod_{j=1}^{s'} d_{f_j \cup \{x\}}(G)$.

We say G has a partite H-decomposition if it has an H-decomposition using copies of H with one vertex in each part V_x.

We say G is H-balanced if for every $f \subseteq V(H)$ and f-partite $e \subseteq V(G)$ there is some n_e such that $|G_{f'}(e)| = n_e$ for all $f \subseteq f' \in H$.

Note in particular that the H-balance condition for $e = f = \emptyset$ implies equality of all $|G_{f'}|$ with $f' \in H$. If G has a partite H-decomposition then G must be H-balanced; the following result ([11, Theorem 1.7]) shows the converse for typical H-blowups.

Theorem 2.4 *Let H be an r-graph on $[q]$ and G be an H-balanced (c, h^q)-typical H-blowup on $(V_x : x \in V(H))$, where each $n/h \leq |V_x| \leq n$ for some large $n > n_0(q)$ and $d_f(G) > d > 2n^{-\delta/h^q}$ for all $f \in H$ and $c < c_0 d^{h^{30q}}$, where $c_0 = c_0(q)$ is small. Then G has a partite H-decomposition.*

In the previous result, we can not only show that G has a partite H-decomposition, but also give an approximate formula for the number of such decompositions. We will show some applications of this when G is a complete H-blowup. We start by considering the upper bound, which comes from the following result of Luria [20].

Theorem 2.5 *Let R be fixed and $D = D(N) \to \infty$ as $N \to \infty$. Suppose A is an R-graph on N vertices such that all vertex degrees are[3] $D + o(D)$ and all pair degrees are $o(D)$. Then the number of perfect matchings in A is at most $(De^{1-R} + o(D))^{N/R}$.*

When applying Theorem 2.5 to the setting of Theorem 2.4, we consider the auxiliary R-graph A on $V(A) = E(G)$ where edges correspond to copies of H, so $N = |G|$ and $R = |H|$. If we let $G = H(n)$ be the complete H-blowup of size n then $N = |H| n^r$ and the degree conditions of Theorem 2.5 hold with $D = n^{q-r}$. In fact, all

---
[3] The statement in [20] has D here, but the proof works with $D + o(D)$.

pair degrees are at most n^{q-r-1}. We deduce that the number of H-decompositions of $H(n)$ is at most $((e^{1-|H|} + o(1))n^{q-r})^{n^r}$. We will show below how a matching lower bound follows from Theorem 2.4. Before doing so, we discuss two applications.

First we consider the number $N_r(n)$ of r-dimensional permutations of order n, which is also the number of K^r_{r+1}-decompositions of $K^r_{r+1}(n)$. For $r = 2$ (Latin squares), Van Lint and Wilson [28, Theorem 17.3] obtained the approximate formula $N_2(n) = (n/e^2 + o(n))^{n^2}$; this was a short deduction from two celebrated breakthroughs on permanents (the proof of the Van der Waerden Conjecture by Falikman and by Egorychev and of the Minc Conjecture by Bregman). The upper bound can be obtained more simply by entropy inequalities, by which means Linial and Luria [20] showed $N_r(n) \leq (n/e^r + o(n))^{n^r}$, and Luria obtained the more general result in Theorem 2.5. However, the lower bound argument appeared not to generalise, even from Latin squares to Steiner Triple Systems, for which the approximate formula was a conjecture of Wilson [35], proved in [12]. In [11] we established the lower bound, thus giving the following approximate formula.

Theorem 2.6 *The number of r-dimensional permutations of order n is $(n/e^r + o(n))^{n^r}$.*

Our second application is to the number of generalised Sudoku squares, which are Latin squares of order n^2 partitioned into n by n subsquares each of which uses every symbol once (the usual Sudoku squares have $n = 3$). We encode these by the 4-graph H with $V(H) = \{x_1, x_2, y_1, y_2, z_1, z_2\}$ and $E(H) = \{x_1x_2y_1y_2, x_1x_2z_1z_2, y_1y_2z_1z_2, x_1y_1z_1z_2\}$. Then an H-decomposition of the complete n-blowup of H can be viewed as a Sudoku square, where we represent rows by pairs (a_1, a_2), columns by (b_1, b_2), symbols by (c_1, c_2) and boxes by (a_1, b_1); a copy of H with vertices $\{a_1, a_2, b_1, b_2, c_1, c_2\}$ represents a cell in row (a_1, a_2) and column (b_1, b_2) with symbol (c_1, c_2). The following estimate then follows from the estimate for general H given below.

Theorem 2.7 *The number of Sudoku squares with n^2 boxes of order n is $(n^2/e^3 + o(n^2))^{n^4}$.*

We conclude this section with the general formula that implies the two examples discussed above.

Theorem 2.8 *For any r-graph H on $[q]$, the number of H-decompositions of $H(n)$ is $((e^{1-|H|} + o(1))n^{q-r})^{n^r}$.*

Proof The upper bound comes from Theorem 2.5 applied to the auxiliary R-graph A described above (following the statement of Theorem 2.5). For the lower bound, we consider the random greedy matching process, in which we construct a sequence of vertex-disjoint edges $e_0, e_1, \ldots$ in A and subgraphs $A_0, A_1, \ldots$, where $A_0 = A$, e_i is a uniformly random edge of A_i, and A_{i+1} is obtained from A_i by deleting the vertices of e_i and all edges that intersect e_i. We will estimate the number of runnings of this process, stopped at some subgraph A_t which is quite sparse, but sufficiently dense and typical that Theorem 2.4 applies to show that A_t has a perfect matching. This will give a lower bound on the number of perfect matchings of A, i.e. H-decompositions of $H(n)$, which matches Luria's upper bound.

Bennett and Bohman [1] showed if A is a D-regular R-graph on N vertices with all pair degrees at most $L = o(D \log^{-5} N)$ then whp[4] the process persists until the proportion of uncovered vertices is at most $(L/D)^{1/2(R-1)+o(1)}$. (Their proof applies verbatim under the weaker assumption that all vertex degrees are $D \pm \sqrt{DL}$.) Here we have $L/D = n^{-1}$ and $R = |H|$, so we could run the process until the uncovered proportion is e.g. $n^{-1/2|H|}$, but we stop it when the remaining r-graph $G_t = V(A_t)$ has density $d = 3n^{-\delta/h^q}$. Furthermore, one can show that whp throughout the process the r-graphs $G_i = V(A_i)$ are (c, h^q)-typical H-blowups with $c < c_0 d^{h^{30q}}$ (similar lemmas in the nonpartite setting are well-known, see e.g. [2]). Then Theorem 2.4 can be applied to G_t, and we have a good estimate for the number of choices at each step of the process: at step i when all densities $d_f(G_i)$ with $f \in H$ are $d(i) = 1 - in^{-r}$ there are $(1 \pm 2|H|c)d(i)^{|H|}n^q$ edges of A_i (i.e. copies of H in G_i).

A simple counting argument will now give the required lower bound on the number of H-decompositions of $H(n)$. For $0 \le j \le j' \le t$, let us say that a running of the process from $A_0, \ldots, A_{j'}$ is j-good if G_i is (c, h^q)-typical for $1 \le i \le j$. Let $R_{j'}^j$ be the number of such runnings. Then $R_{j+1}^j/R_j^j = (1 \pm 2|H|c)d(j)^{|H|}n^q$ by typicality and $R_{j+1}^{j+1}/R_{j+1}^j = 1 \pm c$ (say) as whp typicality does not first fail at step $j+1$. Multiplying these estimates, the number of t-good runnings is $R_t^t = \prod_{j=0}^{t}((1 \pm 3|H|c)d(j)^{|H|}n^q)$. By Theorem 2.4, each t-good running can be completed to an H-decomposition of $H(n)$. We obtain a lower bound on the number of H-decompositions of $H(n)$ by dividing R_t^t by an upper bound of $\prod_{j=0}^{t}(n^r - j) = \prod_{j=0}^{t}(d(j)n^r)$ on the number of runnings giving rise to any fixed decomposition. A short calculation using Stirling's estimate on factorials gives the claimed lower bound $\prod_{j=0}^{t}((1 \pm 3|H|c)d(j)^{|H|-1}n^{q-r})$
$= ((e^{1-|H|} + o(1))n^{q-r})^{n^r}$. □

3 Generalised Partite Decompositions

In this section we state and give applications of a result that generalises both the nonpartite and partite decomposition results of the previous section to the generalised partite setting of the definition below (which is followed by some explanatory remarks).

Definition 3.1 Let H be an r-graph on $[q]$ and $\mathcal{P} = (P_1, \ldots, P_t)$ be a partition of $[q]$. Let G be an r-graph and $\mathcal{P}' = (P_1', \ldots, P_t')$ be a partition of $V(G)$. We say G has a $\mathcal{P}$-partite H-decomposition if it has an H-decomposition using copies $\phi(H)$ of H with all $\phi(P_i) \subseteq P_i'$.

---

[4]We say that an event E holds *with high probability* (whp) if $\mathbb{P}(E) = 1 - e^{-\Omega(n^c)}$ for some $c > 0$ as $n \to \infty$; by union bounds we can assume that any specified polynomial number of such events all occur.

For $S \subseteq [q]$ the $\mathcal{P}$-index of S is $i_\mathcal{P}(S) = (|S \cap P_1|, \ldots, |S \cap P_t|)$; similarly, we define the $\mathcal{P}'$-index of subsets of $V(G)$, and also refer to both as the 'index'.

For $\boldsymbol{i} \in \mathbb{N}^t$ we let $H_{\boldsymbol{i}}$ and $G_{\boldsymbol{i}}$ be the edges in H and G of index $\boldsymbol{i}$. Let $I = I(H) = \{\boldsymbol{i} : H_{\boldsymbol{i}} \neq \emptyset\}$. We call G an $(H, \mathcal{P})$-blowup if $G_{\boldsymbol{i}} \neq \emptyset \Rightarrow \boldsymbol{i} \in I$.

For $e \subseteq V(G)$ we define the degree vector $G_I(e) \in \mathbb{N}^I$ by $G_I(e)_{\boldsymbol{i}} = |G_{\boldsymbol{i}}(e)|$ for $\boldsymbol{i} \in I$. Similarly, for $f \subseteq [q]$ we define $H_I(f)$ by $H_I(f)_{\boldsymbol{i}} = |H_{\boldsymbol{i}}(f)|$. For $\boldsymbol{i}' \in \mathbb{N}^I$ let $H_{\boldsymbol{i}'}^I$ be the subgroup of $\mathbb{Z}^I$ generated by $\{H_I(f) : i_\mathcal{P}(f) = \boldsymbol{i}'\}$. We say G is $(H, \mathcal{P})$-divisible if $G_I(e) \in H_{\boldsymbol{i}'}^I$ whenever $i_{\mathcal{P}'}(e) = \boldsymbol{i}'$.

For $\boldsymbol{i} \in \mathbb{N}^t$ let $d_{\boldsymbol{i}}(G) = |G_{\boldsymbol{i}}| \prod_{j \in [t]} \binom{|P'_j|}{i_j}^{-1}$. We call G a (c, s)-typical $(H, \mathcal{P})$-blowup if for any $s' \leq s, \{f_1, \ldots, f_{s'}\} \subseteq \binom{V(G)}{r-1}, j \in [t]$ we have[5] $\left| P'_j \cap \bigcap_{k=1}^{s'} G(f_k) \right|$
$= (1 \pm s'c)|P'_j| \prod_{k=1}^{s'} d_{i(f_k)+e_j}(G)$.

The simplest examples of the previous definition are given by the trivial partitions with $t = 1$ (non-partite decompositions) or $t = q$ (partite decompositions). The latter is instructive for understanding the divisibility condition. We will illustrate it in the case that H is a (graph) triangle on [3], with parts $P_i = \{i\}$ for $i \in [3]$ and G is a tripartite graph with parts P'_i for $i \in [3]$. Then $I = \{\boldsymbol{i}^1, \boldsymbol{i}^2, \boldsymbol{i}^3\}$ with $\boldsymbol{i}^1 = (1, 1, 0)$, $\boldsymbol{i}^2 = (1, 0, 1)$, $\boldsymbol{i}^3 = (0, 1, 1)$. For each $\boldsymbol{i} \in I$ we have $G(\emptyset)_{\boldsymbol{i}} = |G_{\boldsymbol{i}}|$ and $H(\emptyset)_{\boldsymbol{i}} = |H_{\boldsymbol{i}}| = 1$, so the 0-divisibility condition is that the three bipartite pieces of G all have the same number of edges. For the 1-divisibility condition, we note that $H(1)_{\boldsymbol{i}^1} = H(1)_{\boldsymbol{i}^2} = 1$, $H(1)_{\boldsymbol{i}^3} = 0$ and $G(x_1)_{\boldsymbol{i}} = |G_{\boldsymbol{i}}(x_1)|$ for $x_1 \in P'_1$, so we require every vertex in P'_1 to have equal degrees into P'_2 and P'_3 (and similarly for each part). The 2-divisibility condition is trivially satisfied, so this completes the description. Our final remark on Definition 3.1 is that the typicality condition is a direct generalisation of that in Definition 2.3, allowing the possibility that both sides are zero if some $i(f_k) + e_j \notin I$.

Next we state a decomposition result in the generalised partite setting (a case of [11, Theorem 7.8]); the case $\mathcal{P} = ([q])$ implies Theorem 2.2 and the case $\mathcal{P} = (\{1\}, \ldots, \{q\})$ implies Theorem 2.4.

Theorem 3.2 *Let H be an r-graph on $[q]$ and $\mathcal{P} = (P_1, \ldots, P_t)$ be a partition of $[q]$. Let $n > n_0(q)$, $d > 2n^{-\delta/h^q}$ and $c < c_0 d^{h^{30q}}$, where $c_0 = c_0(q)$ is small. Suppose G is an $(H, \mathcal{P})$-divisible (c, h)-typical $(H, \mathcal{P})$-blowup wrt $\mathcal{P}' = (P'_1, \ldots, P'_t)$, such that each $n/h \leq |P'_i| \leq n$ and $d_{\boldsymbol{i}}(G) > d$ for all $\boldsymbol{i} \in I(H)$. Then G has a $\mathcal{P}$-partite H-decomposition.*

In the remainder of this section we give two applications of the following simplified version of the preceding result (the case that G is complete).

Theorem 3.3 *Let H be an r-graph on $[q]$ and $\mathcal{P} = (P_1, \ldots, P_t)$ be a partition of $[q]$. Suppose G is an $(H, \mathcal{P})$-divisible complete $(H, \mathcal{P})$-blowup wrt $\mathcal{P}' = (P'_1, \ldots, P'_t)$ such that each $n/h \leq |P'_i| \leq n$ with $n > n_0(q)$. Then G has a $\mathcal{P}$-partite H-decomposition.*

---

[5] Let $\{e_1, \ldots, e_t\}$ be the standard basis of $\mathbb{Z}^t$.

As our first application we reprove the result of [23] in the case that n is large on the existence of resolvable Steiner Triple Systems (for a hypergraph generalisation see [11, Theorem 7.9]).

Theorem 3.4 *Suppose $n = 6k + 3$ with $k \in \mathbb{N}$ is large. Then there is a resolvable Steiner Triple System of order n.*

Proof Let $H = K_4$ be the complete graph on 4 vertices, with $V(H) = [4]$ partitioned as $\mathcal{P} = (P_1, P_2)$, where $P_1 = [3]$ and $P_2 = \{4\}$. Let P_1' and P_2' be disjoint sets with $|P_1'| = n$ and $|P_2'| = (n-1)/2$. Let G be the graph with $V(G) = P_1' \cup P_2'$ whose edges are all pairs in $P_1' \cup P_2'$ not contained in P_2'. Then G is a complete $(H, \mathcal{P})$-blowup.

We claim that a resolvable Steiner Triple System of order n is equivalent to a $\mathcal{P}$-partite H-decomposition of G. To see this, suppose first that we have some $\mathcal{P}$-partite H-decomposition $\mathcal{H}$ of G. This means that $\mathcal{H}$ partitions $E(G)$, and each $\phi(H) \in \mathcal{H}$ has $\phi([3]) \subseteq P_1'$ and $\phi(4) \in P_2'$. Then $\mathcal{T} := \{\phi(H - 4) : \phi(H) \in \mathcal{H}\}$ is a triangle decomposition of the complete graph on P_1, i.e. a Steiner Triple System of order n. We can partition $\mathcal{H}$ as $(\mathcal{H}_y : y \in P_2')$, where each $\mathcal{H}_y = \{\phi(H) : \phi(4) = y\}$. Note that each $T_y = \{\phi([3]) : \phi(H) \in \mathcal{H}_y\}$ is a perfect matching on P_1; indeed, for each $x \in P_1$, as $\mathcal{H}$ partitions $E(G)$, there is a unique $\phi(H) \in \mathcal{H}$ containing xy, and then $\phi([3])$ is the unique triple in T_y containing x. Thus $\mathcal{T}$ is a resolvable Steiner Triple System. Conversely, the same construction shows that any resolvable Steiner Triple System gives rise to a $\mathcal{P}$-partite H-decomposition of G. Indeed, given a Steiner Triple System $\mathcal{T}$ on P_1 partitioned into perfect matchings, we arbitrarily label the perfect matchings as $(T_y : y \in P_2')$ and form a $\mathcal{P}$-partite H-decomposition of G by taking all $\phi(H)$ with $\phi([3]) \in T_y$ and $\phi(4) = y$ for some $y \in P_2'$. This proves the claim.

To complete the proof of the theorem, we show that Theorem 3.3 applies to give a $\mathcal{P}$-partite H-decomposition of G. In the notation of Definition 3.1, we have $I = I(H) = \{(2, 0), (1, 1)\}$ and need to show that $G_I(e) \in H_{i'}^I$ whenever $i_{\mathcal{P}'}(e) = i'$. First we consider $i_{\mathcal{P}'}(e) = (0, 0)$, i.e. $e = \emptyset$. We have $H_I(\emptyset) = (3, 3)$, as H contains 3 edges of each of the indices $(2, 0)$ and $(1, 1)$. Thus $H_{(0,0)}^I \leq \mathbb{Z}^2$ is generated by $(3, 3)$. We have $G_I(\emptyset) = (\binom{n}{2}, \binom{n}{2})$, as G contains $\binom{n}{2}$ edges inside P_1' and $\binom{n}{2}$ edges between P_1' and P_2'. As $3 \mid n$ we have $G_I(\emptyset) \in H_{(0,0)}^I$.

Next we consider $i_{\mathcal{P}'}(e) = (1, 0)$, i.e. $e \in P_1'$. We have $i_{\mathcal{P}}(f) = (1, 0)$ iff $f \in [3]$, and for any such f we have $H_I(f) = (2, 1)$, as f is contained in 2 edges of index $(2, 0)$ and 1 edge of index $(1, 1)$. Thus $H_{(1,0)}^I \leq \mathbb{Z}^2$ is generated by $(2, 1)$. We have $G_I(e) = (n - 1, (n - 1)/2)$, as e has degree $n - 1$ in P_1' and degree $(n - 1)/2$ in P_2'. As n is odd, $G_I(e) \in H_{(1,0)}^I$. The only remaining non-trivial case is that $i_{\mathcal{P}'}(e) = (0, 1)$, i.e. $e \in P_2'$. We have $i_{\mathcal{P}}(f) = (0, 1)$ iff $f = 4$, and $H_I(4) = (0, 3)$, as f is contained in no edges of index $(2, 0)$ and 3 edges of index $(1, 1)$. Thus $H_{(0,1)}^I \leq \mathbb{Z}^2$ is generated by $(0, 3)$. We have $G_I(e) = (0, n)$, as e has degree 0 in P_2' and degree n in P_1'. As $3 \mid n$ we have $G_I(e) \in H_{(1,0)}^I$. $\square$

Our second application is to reprove the existence of large sets of Steiner Triple Systems for large n (due to Lu, completed by Teirlinck, see [27]); see [11, Theorem 1.2] for the hypergraph version.

Theorem 3.5 *Suppose n is large and 1 or 3 mod 6. Then K_n^3 can be decomposed into Steiner Triple Systems.*

Proof Let $H = K_4$ be the complete 3-graph on 4 vertices, with $V(H) = [4]$ partitioned as $\mathcal{P} = (P_1, P_2)$, where $P_1 = [3]$ and $P_2 = \{4\}$. Let P_1' and P_2' be disjoint sets with $|P_1'| = n$ and $|P_2'| = n - 2$. Let G be the 3-graph with $V(G) = P_1' \cup P_2'$ whose edges are all triples $e \subseteq P_1' \cup P_2'$ with $|e \cap P_1'| \geq 2$. Then G is a complete $(H, \mathcal{P})$-blowup.

We claim that a decomposition of K_n^3 into Steiner Triple Systems is equivalent to a $\mathcal{P}$-partite H-decomposition of G. To see this, suppose we have some $\mathcal{P}$-partite H-decomposition $\mathcal{H}$ of G. We can partition $\mathcal{H}$ as $(\mathcal{H}_y : y \in P_2')$, where each $\mathcal{H}_y = \{\phi(H) : \phi(4) = y\}$. Note that each $T_y = \{\phi([3]) : \phi(H) \in \mathcal{H}_y\}$ is a Steiner Triple System on P_1; indeed, for each pair xx' in P_1, as $\mathcal{H}$ partitions $E(G)$, there is a unique $\phi(H) \in \mathcal{H}$ containing $xx'y$, and then $\phi([3])$ is the unique triple in T_y containing xx'. Furthermore, each triple in P_1' belongs to exactly one element of $\mathcal{H}$, and so to exactly one T_y. Thus $\{T_y : y \in P_2'\}$ is a decomposition of K_n^3 into Steiner Triple Systems. Conversely, the same construction converts any decomposition of K_n^3 into Steiner Triple Systems into a $\mathcal{P}$-partite H-decomposition of G.

To complete the proof of the theorem, we show that Theorem 3.3 applies to give a $\mathcal{P}$-partite H-decomposition of G. We have $I = I(H) = \{(3, 0), (2, 1)\}$ and need to show that $G_I(e) \in H_{i'}^I$ whenever $i_{\mathcal{P}'}(e) = i'$. First we consider $i_{\mathcal{P}'}(e) = (a, 0)$ with $0 \leq a \leq 2$. For any $f \subseteq V(H)$ with $i_{\mathcal{P}}(f) = (a, 0)$ we have $H_I(f) = (1, 3 - a)$, as f is contained in 1 edge of H with index $(3, 0)$ and $3 - a$ edges of H with index $(2, 1)$. Thus $H_{(a,0)}^I \leq \mathbb{Z}^2$ is generated by $(1, 3 - a)$. We have $G_I(e) = (\binom{n-a}{3-a}, (3 - a)\binom{n-a}{3-a})$, as e is contained in $\binom{n-a}{3-a}$ edges of G with index $(3, 0)$ and $\binom{n-a}{2-a}(n - 2) = (3 - a)\binom{n-a}{3-a}$ edges of G with index $(2, 1)$. Therefore $G_I(e) \in H_{(a,0)}^I$.

Next consider $i_{\mathcal{P}'}(e) = (0, 1)$, i.e. $e \in P_2'$. We have $i_{\mathcal{P}}(f) = (0, 1)$ iff $f = 4$, and $H_I(4) = (0, 3)$, as 4 is contained in 0 edges of index $(3, 0)$ and 3 edges of index $(2, 1)$. Thus $H_{(0,1)}^I \leq \mathbb{Z}^2$ is generated by $(0, 3)$. We have $G_I(e) = (0, \binom{n}{2})$, as e is contained in no edges of G with index $(3, 0)$ and $\binom{n}{2}$ edges of G with index $(2, 1)$. As $3 \mid \binom{n}{2}$ we have $G_I(e) \in H_{(0,1)}^I$.

The only remaining non-trivial case is $i_{\mathcal{P}'}(e) = (1, 1)$. We have $i_{\mathcal{P}}(f) = (1, 1)$ iff $f = a4$ for some $a \in [3]$. Then $H_I(f) = (0, 2)$, as f is contained in 0 edges of index $(3, 0)$ and 2 edges of index $(2, 1)$. Thus $H_{(1,1)}^I \leq \mathbb{Z}^2$ is generated by $(0, 2)$. We have $G_I(e) = (0, n - 1)$, as e is contained in no edges of G with index $(3, 0)$ and $n - 1$ edges of G with index $(2, 1)$. As n is odd, $G_I(e) \in H_{(1,1)}^I$. □

4 General Theory

In this section we state the main result of [11], from which all the other results in this paper follow. Most of the section will be occupied with preparatory definitions for the statement of the result, which we will illustrate with the following running example. Consider a graph G with $V(G) = [n]$ partitioned as (V_1, V_2), where there are no edges within V_2, edges within V_1 are red, and edges between V_1 and V_2 are blue or green. When does G have a decomposition into rainbow triangles?

4.1 Labelled Complexes and Embeddings

All decomposition problems that fit in our general framework are encoded by labelled complexes, which are sets of functions (which we think of as labelled edges) closed under taking restriction; this is analogous to (simplicial) complexes, which are sets of sets closed under taking subsets.

Definition 4.1 We call $\Phi = (\Phi_B : B \subseteq R)$ an R-system on V if $\phi : B \to V$ is injective for each $\phi \in \Phi_B$.

We call Φ an R-complex if whenever $\phi \in \Phi_B$ and $B' \subseteq B$ we have $\phi \mid_{B'} \in \Phi_{B'}$.

Let $\Phi_B^\circ = \{\phi(B) : \phi \in \Phi_B\}$, $\Phi_j^\circ = \bigcup\{\Phi_B^\circ : B \in \binom{R}{j}\}$, $\Phi_j = \bigcup\{\Phi_B : B \in \binom{R}{j}\}$, $V(\Phi) = \Phi_1^\circ$ and $\Phi^\circ = \bigcup\{\Phi_B^\circ : B \subseteq R\}$.

To apply Definition 4.1 in our example we take $V = V(G)$, $R = [3]$ and for each $B \subseteq [3]$ we let Φ_B consist of all injections $\phi : B \to V$ with $\phi(B \cap \{1, 2\}) \subseteq V_1$ and $\phi(B \cap \{3\}) \subseteq V_2$: we also call Φ the complete $(\{1, 2\}, 3)$-partite $[3]$-complex wrt (V_1, V_2). We think of $\phi \in \Phi_3$ as an embedding of the triangle on $[3]$ where 12 is red, 13 is blue and 23 is green. It is useful to consider all such embeddings, even though the only ones that can appear in a decomposition of G are those that are contained in G with $\phi(12)$ red, $\phi(13)$ blue and $\phi(23)$ green.

Next we consider the functional analogue of the subgraph notion for hypergraphs. Just as an embedding of a hypergraph H in a hypergraph G is an injection from $V(H)$ to $V(G)$ taking edges to edges, an embedding of labelled complexes is an injection taking labelled edges to labelled edges.

Definition 4.2 Let H and Φ be R-complexes. Suppose $\phi : V(H) \to V(\Phi)$ is injective. We call ϕ a Φ-embedding of H if $\phi \circ \psi \in \Phi$ for all $\psi \in H$.

In our example, Φ is as above, and H is the complete $(\{1, 2\}, 3)$-partite $[3]$-complex wrt $(\{1, 2\}, 3)$, i.e. each H_B with $B \subseteq [3]$ consists of all injections $\phi : B \to [3]$ with $\phi(B \cap \{1, 2\}) \subseteq \{1, 2\}$ and $\phi(B \cap \{3\}) \subseteq \{3\}$. We think of an edge e of the triangle on $[3]$ as being encoded by the set of labelled edges of H with image e, thus 12 is encoded by $\{(1 \mapsto 1, 2 \mapsto 2), (1 \mapsto 2, 2 \mapsto 1)\}$, 13 by $\{(1 \mapsto 1, 3 \mapsto 3), (2 \mapsto 1, 3 \mapsto 3)\}$, and 23 by $\{(2 \mapsto 2, 3 \mapsto 3), (1 \mapsto 2, 3 \mapsto 3)\}$. If ϕ is a Φ-embedding of H we encode the edges of the triangle on $\phi([3])$ by the corresponding sets of labelled edges: 12 by $\{(1 \mapsto \phi(1), 2 \mapsto \phi(2)), (1 \mapsto \phi(2), 2 \mapsto \phi(1))\}$, 13 by $\{(1 \mapsto \phi(1), 3 \mapsto \phi(3)), (2 \mapsto \phi(1), 3 \mapsto \phi(3))\}$, and 23 by $\{(2 \mapsto \phi(2), 3 \mapsto \phi(3)), (1 \mapsto \phi(2), 3 \mapsto \phi(3))\}$.

4.2 Extensions and Extendability

Next we will formulate our extendability condition.

Definition 4.3 Let $R(S)$ be the R-complex of all partite maps from R to $R \times S$, i.e. whenever $i \in B \subseteq R$ and $\psi \in R(S)_B$ we have $\psi(i) = (i, x)$ for some $x \in S$. If $S = [s]$ we write $R(S) = R(s)$.

Definition 4.4 Suppose $J \subseteq R(S)$ is an R-complex and $F \subseteq V(J)$. Define $J[F] \subseteq R(S)$ by $J[F] = \{\psi \in J : Im(\psi) \subseteq F\}$. Suppose ϕ is a Φ-embedding of $J[F]$. We call $E = (J, F, \phi)$ a Φ-extension of rank $s = |S|$. We write $X_E(\Phi)$ for the set or number of Φ-embeddings of J that restrict to ϕ on F. We say E is ω-dense (in Φ) if $X_E(\Phi) \geq \omega |V(\Phi)|^{v_E}$, where $v_E := |V(J) \setminus F|$. We say Φ is (ω, s)-extendable if all Φ-extensions of rank s are ω-dense.

In our example, we could consider extending some fixed rainbow triangle to an octahedron in which every triangle is rainbow. To implement this in the preceding two definitions, we let $J = [3](2)$ and $F = [3] \times \{1\}$. We identify F with $[3]$ by identifying each $(i, 1)$ with i. Then $J[F]_B = \{id_B\}$ for $B \subseteq [3]$ and ϕ is a Φ-embedding of $J[F]$ iff $\phi \in \Phi_3$. We think of $Im(\phi)$ as our fixed rainbow triangle, which has 2 vertices in V_1 and 1 vertex in V_2. Now consider any $\phi^+ \in X_E(\Phi)$ where $E = (J, F, \phi)$, i.e. ϕ^+ is a Φ-embedding of J that restricts to ϕ on F. For each $i \in [3]$ we have $(i \mapsto (i, 2)) \in J_1$, so $(i \mapsto \phi^+((i, 2))) \in \Phi_1$; thus $\phi^+((i, 2)) \in V_1$ if $i \in [2]$ or $\phi^+((i, 2)) \in V_2$ if $i = 3$. We think of $\{\phi^+((i, 1)), \phi^+((i, 2))\}$ for $i \in [3]$ as the opposite vertices of an octahedron extending the fixed triangle $Im(\phi)$. (We do not yet consider the colours; these will come into play when we consider Definition 4.5.) We have $X_E(\Phi) = (|V_1| - 2)(|V_1| - 3)(|V_2| - 1)$, as we choose 2 new vertices in V_1 and 1 in V_2, so E is $\Omega(1)$-dense if $|V_1|$ and $|V_2|$ are both $\Omega(n)$.

Next we augment our extendability condition to allow for various restrictions (coloured edges in our example).

Definition 4.5 Let Φ be an R-complex and $\Phi' = (\Phi^t : t \in T)$ with each $\Phi^t \subseteq \Phi$. Let $E = (J, F, \phi)$ be a Φ-extension and $J' = (J^t : t \in T)$ for some mutually disjoint $J^t \subseteq J \setminus J[F]$; we call (E, J') a (Φ, Φ')-extension. If $|T| = 1$ we identify $\Phi' \subseteq \Phi$ with (Φ').

We write $X_{E,J'}(\Phi, \Phi')$ for the set or number of $\phi^+ \in X_E(\Phi)$ with $\phi^+ \circ \psi \in \Phi^t_B$ whenever $\psi \in J^t_B$ and Φ^t_B is defined. We say (E, J') is ω-dense in (Φ, Φ') if $X_{E,J'}(\Phi, \Phi') \geq \omega |V(\Phi)|^{v_E}$.

We say (Φ, Φ') is (ω, s)-extendable if all (Φ, Φ')-extensions of rank s are ω-dense in (Φ, Φ').

For $G' = (G^t : t \in T)$ with each $G^t \subseteq \Phi^\circ$ and J' as above we write $X_{E,J'}(\Phi, G) = X_{E,J'}(\Phi, \Phi')$, where $\Phi' = (\Phi^t : t \in T)$ with each $\Phi^t = \{\phi \in \Phi : Im(\phi) \in G^t\}$. We say that (Φ, G') is (ω, s)-extendable if (Φ, Φ') is (ω, s)-extendable.

We continue the above example of extending a fixed rainbow triangle to an octahedron of rainbow triangles. We continue to ignore colours and first consider how the preceding definition can ensure that the octahedron is a subgraph of G. Indeed, if $\phi^+ \in X_{E,J\setminus J[F]}(\Phi, \Phi')$ with $\Phi' = \{\phi \in \Phi : Im(\phi) \in G\}$ then Φ'_B is only defined when $|B| = 2$, and for all $\psi \in J_2 \setminus J[F]$ we have $\phi^+ \circ \psi \in \Phi'$, i.e. $\phi^+(Im(\psi)) \in G$, as required. We also note for future reference that if for some r we have all $\Phi^t \subseteq \Phi_r$ then when checking extendability we can assume $J' \subseteq J_r \setminus J[F]$.

To implement colours, we let $T = \{12, 13, 23\}$, and for $t \in T$ let G^t be the set of edges of G of the appropriate colour (red if $t = 12$, blue if $t = 13$, green if $t = 23$), $\Phi^t = \{\phi \in \Phi : Im(\phi) \in G^t\}$ and $J^t = J_t \setminus J[F]$ for $t \in T$. If $\phi^+ \in X_{E,J'}(\Phi, \Phi')$ then for each $t \in T$, $\psi \in J_t \setminus J[F]$ we have $\phi^+(Im(\psi)) \in G^t$, as required. The extendability condition says that there are at least ωn^3 such octahedra of rainbow triangles containing ϕ (and similarly for any other extension of bounded size).

4.3 Adapted Complexes

A common feature of the decomposition results obtained from our main theorem is that they are implemented by a labelled complex equipped with a permutation group action, and the decomposition respects the orbits of the action, as in the following definitions.

Definition 4.6 Suppose Σ is a permutation group on R. For $B, B' \subseteq R$ we write $\Sigma_B^{B'} = \{\sigma \mid_B : \sigma \in \Sigma, \sigma(B) = B'\}$, $\Sigma^{B'} = \cup_B \Sigma_B^{B'}$ and $\Sigma^{\leq} = \cup_{B,B'} \Sigma_B^{B'}$.

Definition 4.7 Suppose Φ is an R-complex and Σ is a permutation group on R. For $\sigma \in \Sigma$ and $\phi \in \Phi_{\sigma(B)}$ let $\phi\sigma = \phi \circ \sigma \mid_B$. We say Φ is Σ-adapted if $\phi\sigma \in \Phi$ for any $\phi \in \Phi$, $\sigma \in \Sigma$.

Definition 4.8 For $\psi \in \Phi_B$ with $B \subseteq R$ we define the orbit of ψ by $\psi\Sigma := \psi\Sigma^B = \{\psi\sigma : \sigma \in \Sigma^B\}$. We denote the set of orbits by Φ/Σ. We write $Im(O) = Im(\psi)$ for $\psi \in O \in \Phi/\Sigma$.

Definition 4.9 Let Γ be an abelian group. For $J \in \Gamma^{\Phi_r}$ and $O \in \Phi_r/\Sigma$ we define J^O by $J^O_\psi = J_\psi 1_{\psi \in O}$. The orbit decomposition of J is $J = \sum_{O \in \Phi_r/\Sigma} J^O$.

The simplest example is when the permutation group is the entire symmetric group, e.g. if $R = [3]$ and $\Sigma = S_3$ then any $\phi \in \Phi_3$ has an orbit consisting of all six bijections from $[3]$ to $e = Im(\phi)$, which we would think of as encoding the edge e in a 3-graph. In our running example, we have $\Sigma = \{id, (12)\} \leq S_3$. We recall that if ϕ is a Φ-embedding of H then the edge $\phi(12)$ of Φ_2° is encoded by the labelled edges $(1 \mapsto \phi(1), 2 \mapsto \phi(2))$ and $(1 \mapsto \phi(2), 2 \mapsto \phi(1))$, and note that these form an orbit (and similarly for the other edges).

4.4 Decompositions

Now we set up the general framework for decompositions.

Definition 4.10 Let $\mathcal{A}$ be a set of R-complexes; we call $\mathcal{A}$ an R-complex family. If each $A \in \mathcal{A}$ is a copy of $\Sigma^{\le}$ we call $\mathcal{A}$ a $\Sigma^{\le}$-family. For $r \in \mathbb{N}$ we write $A_r = \bigcup \{A_B : B \in \binom{R}{r}\}$ and $\mathcal{A}_r = \cup_{A \in \mathcal{A}} A_r$.

Let Φ be an R-complex. We let $A(\Phi)$ denote the set of Φ-embeddings of A. We let $A(\Phi)^{\le}$ denote the $V(A)$-complex where each $A(\Phi)^{\le}_F$ for $F \subseteq V(A)$ is the set of Φ-embeddings of $A[F]$.

We let $\mathcal{A}(\Phi)^{\le}$ denote the $V(A)$-complex family $(A(\Phi)^{\le} : A \in \mathcal{A})$.

Let $\gamma \in \Gamma^{\mathcal{A}_r}$ for some abelian group Γ.

For $\phi \in A(\Phi)^{\le}$ with $A \in \mathcal{A}$ we define $\gamma(\phi) \in \Gamma^{\Phi_r}$ by $\gamma(\phi)_{\phi \circ \theta} = \gamma_\theta$ for $\theta \in A_r$ (zero otherwise). For $\phi \in A(\Phi)$ we call $\gamma(\phi)$ a γ-molecule. We let $\gamma(\Phi)$ be the set of γ-molecules.

Given $\Psi \in \mathbb{Z}^{\mathcal{A}(\Phi)}$ we define $\partial \Psi = \partial^\gamma \Psi = \sum_\phi \Psi_\phi \gamma(\phi) \in \Gamma^{\Phi_r}$. We also call Ψ an integral $\gamma(\Phi)$-decomposition of $\partial \Psi$ and call $\langle \gamma(\Phi) \rangle$ the decomposition lattice. If furthermore $\Psi \in \{0,1\}^{\mathcal{A}(\Phi)}$ (i.e. $\Psi \subseteq \mathcal{A}(\Phi)$) we call Ψ a $\gamma(\Phi)$-decomposition.

In our example, $\mathcal{A} = \{A\}$ consists of a single copy of the [3]-complex $\Sigma^{\le}$ on [3], which is identical with H as above, i.e. the complete $(\{1,2\}, 3)$-partite [3]-complex wrt $(\{1,2\}, 3)$. We let $\Gamma = \mathbb{Z}^3$ and denote the standard basis by e_{12}, e_{13}, e_{23}, which we think of as the colours red, blue and green. We define $\gamma \in \Gamma^{\mathcal{A}_2}$ by $\gamma_\theta = e_{Im(\theta)}$. The constituent parts of our decompositions are γ-molecules $\gamma(\phi)$, which encode rainbow triangles in Φ: we have $\phi \in A(\Phi)$ (which can be identified[6] with Φ_3), i.e. $\phi \circ \theta \in \Phi$ for all $\theta \in A = \Sigma^{\le}$, and e.g. the blue edge $\phi(1)\phi(3)$ is encoded by the coordinates $\gamma(\phi)_{\phi \circ \theta} = \gamma_\theta = e_{13}$ for $\theta \in A_2$ with $Im(\theta) = \{1,3\}$, i.e. $\theta = (1 \mapsto 1, 3 \mapsto 3)$ and $\theta = (2 \mapsto 1, 3 \mapsto 3)$. We encode any coloured graph G by $G^* \in (\mathbb{Z}^3)^{\Phi_2}$ defined by $G^*_\psi = e_{12}$ if $Im(\psi)$ is a red edge, $G^*_\psi = e_{13}$ if $Im(\psi)$ is a blue edge, $G^*_\psi = e_{23}$ if $Im(\psi)$ is a green edge. Then a $\gamma(\Phi)$-decomposition of G^* encodes a rainbow triangle decomposition of G.

Now we formalise in general the objects (atoms) that are being decomposed into molecules.

Definition 4.11 (*Atoms*) For any $\phi \in \mathcal{A}(\Phi)$ and $O \in \Phi_r / \Sigma$ such that $\gamma(\phi)^O \ne 0$ we call $\gamma(\phi)^O$ a γ-atom at O. We write $\gamma[O]$ for the set of γ-atoms at O. We say γ is elementary if all γ-atoms are linearly independent. We define a partial order $\le_\gamma$ on Γ^{Φ_r} where $H \le_\gamma G$ iff $G - H$ can be expressed as the sum of a multiset of γ-atoms.

In our example, atoms represent coloured edges. To see this, consider again the encoding of the blue edge $\phi(1)\phi(3)$ described above. The relevant orbit $O \in \Phi_2/\Sigma$ consists of the two labelled edges $(1 \mapsto \phi(1), 3 \mapsto \phi(3))$ and $(2 \mapsto \phi(1), 3 \mapsto$

---

[6]This identification is convenient but perhaps potentially confusing: depending on the context, we may identify the domain of ϕ with either the domain or the range of maps in Σ.

$\phi(3)$), and the relevant γ-atom at O is $\gamma(\phi)^O$ which is a vector supported on O with both coordinates equal to e_{13}. There are two other γ-atoms at O, which are vectors supported on O with both coordinates equal to e_{12} (meaning red edge), or both coordinates equal to e_{23} (meaning green edge). Thus γ is elementary, which is an important assumption in our main theorem, ensuring that our decomposition problems do not exhibit arithmetic peculiarities (as seen e.g. in the Frobenius coin problem).

4.5 Lattices

We conclude with a characterisation of the decomposition lattice $\langle \gamma(\Phi) \rangle$, with conditions that are somewhat analogous to the degree-based divisibility conditions considered above, but also account for the labels on the edges and the orbits of the group action. Throughout we let Φ be a Σ-adapted $[q]$-complex for some $\Sigma \leq S_q$, let $\mathcal{A}$ be a $\Sigma^\leq$-family and $\gamma \in \Gamma^{\mathcal{A}_r}$.

Definition 4.12 For $J \in \Gamma^{\Phi_r}$ we define $J^{\sharp} \in (\Gamma^Q)^\Phi$ by[7] $(J^{\sharp}_{\psi'})_B = \sum \{J_\psi : \psi' \subseteq \psi \in \Phi_B\}$ for $B \in Q := \binom{[q]}{r}$, $\psi' \in \Phi$. We define $\gamma^{\sharp} \in (\Gamma^Q)^{\cup \mathcal{A}}$ by $(\gamma^{\sharp}_{\theta'})_B = \sum \{\gamma_\theta : \theta' \subseteq \theta \in A_B\}$ for $B \in Q$, $\theta' \in A \in \mathcal{A}$. We let $\mathcal{L}_\gamma(\Phi)$ be the set of all $J \in \Gamma^{\Phi_r}$ such that $(J^{\sharp})^O \in \langle \gamma^{\sharp}[O] \rangle$ for any $O \in \Phi/\Sigma$.

We illustrate Definition 4.12 with our running example. We start with the orbit $O = \{\emptyset\}$, where $\emptyset$ denotes the unique function with domain $\emptyset$ (also denoting the empty set). Recall that we encode our coloured graph G by $G^* \in (\mathbb{Z}^3)^{\Phi_2}$ and write G^{ij} for the edges of G with colour corresponding to ij. Then $((G^*)^{\sharp}_{\emptyset})_{ij} = \sum_{\psi \in \Phi_{ij}} G^*_\psi$ equals $2|G^{12}|e_{12}$ if $ij = 12$ or $|G^{13}|e_{13} + |G^{23}|e_{23}$ otherwise. Similarly, $(\gamma^{\sharp}_{\emptyset})_{ij} = \sum_{\theta \in \Sigma^\leq_{ij}} \gamma_\theta$ equals $2e_{12}$ if $ij = 12$ or $e_{13} + e_{23}$ otherwise. The 0-divisibility condition is that $(2|G^{12}|e_{12}, |G^{13}|e_{13} + |G^{23}|e_{23}, |G^{13}|e_{13} + |G^{23}|e_{23})$ is an integer multiple of $(2e_{12}, e_{13} + e_{23}, e_{13} + e_{23})$, i.e. G has an equal number of edges of each colour.

Next consider the 1-divisibility condition for any orbit $O = \{(1 \to x), (2 \to x)\}$ with $x \in V_1$. For $i, i' \in [2]$, $j \neq i$ we have $((G^*)^{\sharp}_{i \to x})_{ij} = \sum \{G^*_\psi : \psi \in \Phi_{ij}, \psi(i) = x\}$, which equals $|G^{12}(x)|e_{12}$ if $j \in [2]$ or $|G^{13}(x)|e_{13} + |G^{23}(x)|e_{23}$ if $j = 3$. Also, $(\gamma^{\sharp}(i' \to x)_{i \to x})_{ij} = (\gamma^{\sharp}_{i \to i'})_{ij} = \sum \{\gamma_\theta : \theta \in \Sigma^\leq_{ij}, \theta(i) = i'\}$, which equals e_{12} if $j \in [2]$ or $e_{i'3}$ if $j = 3$. Thus we need $(|G^{12}(x)|e_{12}, |G^{13}(x)|e_{13} + |G^{23}(x)|e_{23}, |G^{13}(x)|e_{13} + |G^{23}(x)|e_{23})$ to lie in the group generated by $(e_{12}, e_{13}, 0)$, $(e_{12}, e_{23}, 0)$, $(e_{12}, 0, e_{13})$ and $(e_{12}, 0, e_{23})$, which holds iff $|G(x) \cap V_1| = |G(x) \cap V_2|$, i.e. each $x \in V_1$ has equal degrees in V_1 and in V_2.

The other 1-divisibility conditions are for orbits $O = \{3 \to x\}$ with $x \in V_2$. For $i \in [2]$ we have $((G^*)^{\sharp}_{3 \to x})_{i3} = \sum \{G^*_\psi : \psi \in \Phi_{i3}, \psi(3) = x\} = |G^{13}(x)|e_{13} + |G^{23}(x)|e_{23}$ and $(\gamma^{\sharp}(3 \to x)_{3 \to x})_{i3} = (\gamma^{\sharp}_{3 \to 3})_{i3} = \sum \{\gamma_\theta : \theta \in \Sigma^\leq_{i3}, \theta(3) = 3\} = e_{13}$

---

[7] The notation $\psi' \subseteq \psi$ means that ψ' is a restriction of ψ.

Coloured and Directed Designs   295

$+ e_{23}$, so we need $|G^{13}(x)| = |G^{23}(x)|$, i.e. each $x \in V_2$ has blue degree equal to green degree. There are no further conditions, as the 2-divisibility conditions hold trivially (we leave this verification to the reader).

Returning to the general setting, it is not hard to see $\langle \gamma(\Phi) \rangle \subseteq \mathcal{L}_\gamma(\Phi)$. The following result ([11, Lemma 5.19]) shows that the converse inclusion holds under an extendability assumption on Φ.

Lemma 4.13 *Let $\Sigma \leq S_q$, $\mathcal{A}$ be a $\Sigma^{\leq}$-family and $\gamma \in (\mathbb{Z}^D)^{\mathcal{A}_r}$. Let Φ be a Σ-adapted (ω, s)-extendable $[q]$-complex with $s = 3r^2$, $n = |V(\Phi)| > n_0(q, D)$ large and $\omega > n^{-1/2}$. Then $\langle \gamma(\Phi) \rangle = \mathcal{L}_\gamma(\Phi)$.*

4.6 Types and Regularity

Next we will formulate our regularity assumption, which can be thought of as robust fractional decomposition. First we give another notation for atoms.

Definition 4.14 For $\psi \in \Phi_B$ and $\theta \in \mathcal{A}_B$ we define $\gamma[\psi]^\theta \in \Gamma^{\psi\Sigma}$ by $\gamma[\psi]^\theta_{\psi\sigma} = \gamma_{\theta\sigma}$.

We will illustrate the various notations in our example for the atom $\gamma(\phi)^O$ representing a blue edge $\phi(1)\phi(3)$ as above. In the notation of Definition 4.10, we write $\gamma(\phi)^O = \gamma(\phi')$ where $\phi' = \phi \mid_{\{1,3\}}$ has domain $\{1, 3\}$, so if $\theta \in A_2$ with $Im(\theta) \subseteq Dom(\phi')$ then $\theta = (1 \mapsto 1, 3 \mapsto 3)$ or $\theta = (2 \mapsto 1, 3 \mapsto 3)$. In the notation of Definition 4.14, we write $\gamma(\phi)^O = \gamma[\phi']^\theta$ with $\theta = (1 \mapsto 1, 3 \mapsto 3)$, as $\gamma[\phi']^\theta$ is supported on $\phi' = (1 \mapsto \phi(1), 3 \mapsto \phi(3))$ with value $\gamma_\theta = e_{13}$ and on $\phi' \circ (12) = (2 \mapsto \phi(1), 3 \mapsto \phi(3))$ with value $\gamma_{\theta \circ (12)} = e_{13}$. We also think of this notation as 'an atom of type θ on ψ', where we define types in general as follows.

Definition 4.15 (*Types*) For $\theta \in \mathcal{A}_B$ we define $\gamma^\theta \in \Gamma^{\Sigma^B}$ by $\gamma^\theta_\sigma = \gamma_{\theta\sigma}$.

A type $t = [\theta]$ in γ is an equivalence class of the relation $\sim$ on any $\mathcal{A}_B$ with $B \in Q = \binom{[q]}{r}$ where $\theta \sim \theta'$ iff $\gamma^\theta = \gamma^{\theta'}$. We write T_B for the set of types in $\mathcal{A}_B$.
For $\theta \in t \in T_B$ and $\psi \in \Phi_B$ we write $\gamma^t = \gamma^\theta$ and $\gamma[\psi]^t = \gamma[\psi]^\theta$.
If $\gamma^t = 0$ call t a zero type and write $t = 0$.
If $\phi \in \mathcal{A}(\Phi)$ with $\gamma(\phi)^{\psi\Sigma} = \gamma[\psi]^t$ we write $t_\phi(\psi) = t$.

To illustrate the preceding definition on the above example of $\gamma[\phi']^\theta$ with $\theta = (1 \mapsto 1, 3 \mapsto 3)$ we think of $\{\theta\} \in T_{13}$ as the 'blue edge' type with $(\gamma[\phi']^\theta_{\phi'}, \gamma[\phi']^\theta_{\phi' \circ (12)}) = (\gamma^\theta_{id}, \gamma^\theta_{(12)}) = (\gamma_{1 \mapsto 1, 3 \mapsto 3}, \gamma_{2 \mapsto 1, 3 \mapsto 3}) = (e_{13}, e_{13})$. The possibility of a zero type is not relevant to our example, as it allows for non-edges when decomposing into copies of a non-complete graph. The 'red edge' type in T_{12} is $\{(1 \mapsto 1, 2 \mapsto 2), (1 \mapsto 2, 2 \mapsto 1)\}$, as $(\gamma^{1 \mapsto 1, 2 \mapsto 2}_{id}, \gamma^{1 \mapsto 1, 2 \mapsto 2}_{(12)}) = (\gamma_{1 \mapsto 1, 2 \mapsto 2}, \gamma_{1 \mapsto 2, 2 \mapsto 1}) = (e_{12}, e_{12})$ and $(\gamma^{1 \mapsto 2, 2 \mapsto 1}_{id}, \gamma^{1 \mapsto 2, 2 \mapsto 1}_{(12)}) = (\gamma_{1 \mapsto 2, 2 \mapsto 1}, \gamma_{1 \mapsto 1, 2 \mapsto 2}) = (e_{12}, e_{12})$.

Now we formulate our regularity assumption. The following definition can be roughly understood as saying that the vector J can be approximated by a non-negative linear combination of molecules, where all molecules that can be used (in that J contains all their atoms) are used with comparable weights (up constant factors).

Definition 4.16 (*Regularity*) Suppose γ is elementary and $J \in (\mathbb{Z}^D)^{\Phi_r}$ with $J^O \in \langle \gamma[O] \rangle$ for all $O \in \Phi_r/\Sigma$. For $\psi \in \Phi_B$ with $|B| = r$ we define integers J^t_ψ for all nonzero $t \in T_B$ by $J^{\psi\Sigma} = \sum_{0 \neq t \in T_B} J^t_\psi \gamma[\psi]^t$. Any choice of orbit representatives $\psi^O \in \Phi_{B^O}$ for each orbit $O \in \Phi_r/\Sigma$ defines an atom decomposition $J = \sum_{O \in \Phi_r/\Sigma} \sum_{0 \neq t \in T_{B^O}} J^t_{\psi^O} \gamma[\psi^O]^t$.

Let $\mathcal{A}(\Phi, J) = \{\phi \in \mathcal{A}(\Phi) : \gamma(\phi) \leq_\gamma J\}$. We say J is (γ, c, ω)-regular (in Φ) if there is $y \in [\omega n^{r-q}, \omega^{-1} n^{r-q}]^{\mathcal{A}(\Phi, J)}$ such that for all $B \in Q$, $\psi \in \Phi_B$, $0 \neq t \in T_B$ we have

$$\partial^t y_\psi := \sum \{y_\phi : t_\phi(\psi) = t\} = (1 \pm c) J^t_\psi.$$

For example, suppose $J = G^* \in (\mathbb{Z}^3)^{\Phi_2}$ encodes G as above. An atom decomposition expresses J as a sum where each summand encodes a coloured edge of G by some atom $\gamma[\psi^O]^t$ as discussed above. We have $\phi \in \mathcal{A}(\Phi, J)$ iff the molecule $\gamma(\phi)$ encodes a rainbow triangle in G. Then G^* is (γ, c, ω)-regular if we can assign each rainbow triangle in G a weight between ωn^{-1} and $\omega^{-1} n^{-1}$ so that the total weight of triangles on any edge is $1 \pm c$.

We require one further definition, used in the extendability hypothesis of Theorem 4.18 below.

Definition 4.17 For $L \in \Gamma^{\Phi_r}$ we let $\gamma[L] = (\gamma[L]^A : A \in \mathcal{A})$ where each $\gamma[L]^A$ is the set of $\psi \in A(\Phi)^{\leq}_r$ such that $\gamma(\psi) \leq_\gamma L$.

In our example, the extendability hypothesis says that for any Φ-extension $E = (J, F, \phi)$ of rank h there are many $\phi^+ \in X_E(\Phi)$ such that all edges $Im(\phi^+ \psi)$ with $\psi \in J_2 \setminus J[F]$ are edges of G with the correct colour (red if $\psi \in J_{12}$, blue if $\psi \in J_{13}$, green if $\psi \in J_{23}$). We illustrate this for extensions of some fixed rainbow triangle to an octahedron of rainbow triangles (recall $J = [3](2)$, $F = [3] \times \{1\} = [3]$ and let $J' = J_2 \setminus J[F]$). If $(\Phi, \gamma[G^*]^A)$ is $(\omega, 2)$-extendable we have at least $\omega |V_1|^2 |V_2|$ choices of $\phi^+ \in X_{E,J'}(\Phi, \gamma[G^*]^A)$. For each $\psi \in J_2 \setminus J[F]$ we have $\phi^+ \psi \in \gamma[G^*]^A$, i.e. $\gamma(\phi^+ \psi) \leq_\gamma G^*$. For example, if $\psi \in J_{13}$ with $\psi(1) = (1, 1)$ and $\psi(3) = (3, 2)$ then $\gamma(\phi^+ \psi)$ is the blue atom at $Im(\phi^+ \psi)$, i.e. the vector supported on the orbit with the two labelled edges $(1 \mapsto \phi^+((1,1)), 3 \mapsto \phi^+((3,2)))$ and $(2 \mapsto \phi^+((1,1)), 3 \mapsto \phi^+((3,2)))$, where both nonzero coordinates are e_{13}. For this ψ, the condition $\gamma(\phi^+ \psi) \leq_\gamma G^*$ says that G has a blue edge at $\phi^+((1,1))\phi^+((3,2))$. As ψ varies over J_2 we see that $Im(\phi^+)$ spans an octahedron of rainbow triangles.

Finally we can state the main result (Theorem 3.1) of [11] (recall $h = 2^{50 q^3}$ and $\delta = 2^{-10^3 q^5}$).

Theorem 4.18 For any $q \geq r$ and D there are ω_0 and n_0 such that the following holds for $n > n_0$, $n^{-\delta} < \omega < \omega_0$ and $c \leq \omega^{h^{20}}$. Let $\mathcal{A}$ be a $\Sigma^{\leq}$-family with $\Sigma \leq S_q$. Suppose $\gamma \in (\mathbb{Z}^D)^{\mathcal{A}_r}$ is elementary. Let Φ be a Σ-adapted $[q]$-complex on $[n]$. Let $G \in \langle \gamma(\Phi) \rangle$ be (γ, c, ω)-regular in Φ such that $(\Phi, \gamma[G]^A)$ is (ω, h)-extendable for each $A \in \mathcal{A}$. Then G has a $\gamma(\Phi)$-decomposition.

5 Coloured Hypergraphs

When can an edge-coloured graph be decomposed into rainbow triangles? In this section we illustrate the application of Theorem 4.18 to this question, and a hypergraph generalisation thereof. We start by formulating the general problem of decomposing an edge-coloured r-multigraph G by an edge-coloured r-graph H. For simplicity we assume that H is simple (one could allow multiple copies of edges in H provided they have distinct colours, but not multiple edges of a given colour, as then the associated γ in Definition 5.8 below is not elementary).

Definition 5.1 Suppose H is an r-graph on $[q]$, edge-coloured as $H = \cup_{d \in [D]} H^d$. We identify H with a vector $H \in (\mathbb{N}^D)^Q$, where each $(H_f)_d = 1_{f \in H^d}$ (indicator function) and $Q = \binom{[q]}{r}$.

Let Φ be an S_q-adapted $[q]$-complex on $[n]$. For $\phi \in \Phi_q$ we define $\phi(H) \in (\mathbb{N}^D)^{\Phi_r^\circ}$ by $\phi(H)_{\phi(f)} = H_f$. Let $\mathcal{H}$ be an family of $[D]$-edge-coloured r-graphs on $[q]$. Let $\mathcal{H}(\Phi) = \{\phi(H) : \phi \in \Phi_q, H \in \mathcal{H}\}$.

Let $G \in \mathbb{N}^{\Phi_r^\circ}$ be an r-multigraph $[D]$-edge-coloured as $G = \cup_{d \in [D]} G^d$, identified with $G \in (\mathbb{N}^D)^{\Phi_r^\circ}$. We call $\mathcal{H}' \subseteq \mathcal{H}(\Phi)$ with $\sum \mathcal{H}' = G$ an H-decomposition of G in Φ. We call $\Psi \in \mathbb{Z}^{\mathcal{H}(\Phi)}$ with $\sum_{H'} \Psi_{H'} H' = G$ an integral H-decomposition of G in Φ.

Note that copies of H in an integral H-decomposition of G can use edges $e \in \Phi_r^\circ$ with $G_e = 0$ or with the wrong colour, but all such terms must cancel. Before considering the general setting of the previous definition, we warm up by specialising to graphs ($r = 2$) and the case that Φ is the complete $[q]$-complex on $[n]$. We formulate a typicality condition for coloured graphs and a result on rainbow triangle decompositions analogous to that given in [12] for triangle decompositions of typical graphs.

Definition 5.2 Let G be a $[D]$-edge-coloured graph on $[n]$. For $\alpha \in [D]$, the α-density of G is $d(G^\alpha) = |G^\alpha| \binom{n}{2}^{-1}$. The density of G is $d(G) = |G| \binom{n}{2}^{-1}$. The density vector of G is $d(G)^* \in [0, 1]^D$ with $d(G)^*_\alpha = d(G^\alpha)$. Given vectors $\boldsymbol{x} \in [n]^t$ of vertices and $\boldsymbol{\alpha} \in [D]^t$ of colours we define the $\boldsymbol{\alpha}$-degree $d_G^{\boldsymbol{\alpha}}(\boldsymbol{x})$ of $\boldsymbol{x}$ in G as the number of vertices y such that $x_i y \in G^{\alpha_i}$ for all $i \in [t]$.

We say G is (c, h)-typical if $d_G^{\boldsymbol{\alpha}}(\boldsymbol{x}) = (1 \pm tc) n \prod_{i=1}^t d(G^{\alpha_i})$ for any such $\boldsymbol{x}$ and $\boldsymbol{\alpha}$ with $t \leq h$.

Theorem 5.3 Suppose G is a tridivisible (c, h)-typical $[D]$-edge-coloured graph on $[n]$, where $D \geq 4$, $n > n_0(D)$ is large, $h = 2^{10^3}$, $\delta = 2^{-10^6}$, $c < c_0 d(G)^{h^{90}}$ where $c_0 = c_0(D)$ is small, and each $n^{-\delta/2h^3} < d(G^\alpha) < (1/3 - n^{-\delta/2h^3}) d(G)$. Then G has a rainbow triangle decomposition.

Note that the tridivisibility condition (G has all degrees even and $3 \mid e(G)$) in Theorem 5.3 is necessary, as if we ignore the colours then we obtain a triangle decomposition of G; it is perhaps surprising that the colours do not impose any additional condition. We will deduce Theorem 5.3 from a more general result on typical r-multigraphs, as in the following definition.

Definition 5.4 Let G be a $[D]$-edge-coloured r-multigraph on $[n]$. For $\alpha \in [D]$, the α-density of G is $d(G^\alpha) = |G^\alpha|\binom{n}{r}^{-1}$. The density of G is $d(G) = |G|\binom{n}{r}^{-1}$. The density vector of G is $d(G)^* \in \mathbb{R}^D$ with $d(G)^*_\alpha = d(G^\alpha)$.

For $e \subseteq [n]$, the degree of e in G is $|G(e)|$; the degree vector is $G(e)^* \in \mathbb{N}^D$ with $G(e)^*_\alpha = |G^\alpha(e)|$.

Given vectors $\boldsymbol{f} \in \binom{[n]}{r-1}^t$ of $(r-1)$-sets and $\boldsymbol{\alpha} \in [D]^t$ of colours we define the α-degree of $\boldsymbol{f}$ in G as $d_G^{\boldsymbol{\alpha}}(\boldsymbol{f}) = \sum_{v \in [n]} \prod_{i=1}^t G^{\alpha_i}_{f_i \cup \{v\}}$.

We say G is (c, h)-typical if $d_G^{\boldsymbol{\alpha}}(\boldsymbol{f}) = (1 \pm tc)n \prod_{i=1}^t d(G^{\alpha_i})$ for any such $\boldsymbol{f}$ and $\boldsymbol{\alpha}$ with $t \leq h$.

Given a family $\mathcal{H}$ of $[D]$-edge-coloured r-graphs on $[q]$, we say G is (b, c)-balanced wrt $\mathcal{H}$ if there is $p \in [b, b^{-1}]^\mathcal{H}$ with $d(G)^* = (1 \pm c) \sum_H p_H d(H)^*$.

We say G is $\mathcal{H}$-divisible if each $G(e)^* \in \langle H(f)^* : f \in \binom{[q]}{|e|}, H \in \mathcal{H} \rangle$.

In the next lemma we show that in the case of rainbow triangles, the conditions in Definition 5.4 follow from the assumptions of Theorem 5.3.

Lemma 5.5 *Let $\mathcal{H}$ be the family of all $[D]$-edge-coloured rainbow triangles and G be a $[D]$-edge-coloured graph on $[n]$, with $D \geq 4$. Then*

i. *G is $\mathcal{H}$-divisible iff G is tridivisible, and*
ii. *If each $bD^2 < d(G^\alpha) < d(G)/3 - bD^3$ then G is $(b, 0)$-balanced wrt $\mathcal{H}$.*

Proof For (i), we need to know the integer span $Z(r, s)$ of the rows of a matrix $M(r, s)$ whose rows are indexed by $\binom{[s]}{r}$ and columns by $[s]$, with $M(r,s)_{e,i} = 1_{i \in e}$. It follows from [36, Theorem 2] (and is not hard to show directly) that $Z(r, s) = \{x \in \mathbb{Z}^s : r \mid \sum_i x_i\}$ for $s > r$. To apply this to the divisibility conditions, first consider $G(\emptyset)^* = (|G^1|, \ldots, |G^D|)$ and note that $H(\emptyset)^* = (|H^1|, \ldots, |H^D|)$ for $H \in \mathcal{H}$ are the rows of $M(3, D)$. We have $G(\emptyset)^* \in \langle H(\emptyset)^* : H \in \mathcal{H} \rangle$ iff $3 \mid \sum_\alpha |G^\alpha| = |G|$. Next, for any $v \in [n]$ we have $G(v)^* = (|G^1(v)|, \ldots, |G^D(v)|)$. As $H(x)^* = (|H^1(x)|, \ldots, |H^D(x)|)$ for $x \in [q]$, $H \in \mathcal{H}$ are the rows of $M(2, D)$ we have $G(v)^* \in \langle H(x)^* : x \in [q], H \in \mathcal{H} \rangle$ iff $2 \mid \sum_\alpha |G^\alpha(v)| = |G(v)|$. Finally, for any $uv \in \binom{[n]}{2}$ we have $G(uv)^* = (G^1_{uv}, \ldots, G^D_{uv})$ and $H(xy)^*$ for $xy \in \binom{[q]}{2}$, $H \in \mathcal{H}$ is the standard basis, so the 2-divisibility condition is trivial. Thus G is $\mathcal{H}$-divisible iff G is tridivisible.

For (ii), we note that the set of density vectors $d(H)^*$ for $H \in \mathcal{H}$ consists of all probability distributions on $[D]$ with 3 coordinates equal to $1/3$ and the rest zero. By [8, Theorem 46], any probability distribution x on $[D]$ is a convex combination of the vectors $d(H)^*$ iff $x_\alpha \leq 1/3$ for all $\alpha \in [D]$. Thus for any $x \in [0, 1]^D$ with each $3x_{\alpha'} \leq \sum_\alpha x_\alpha \leq 1$ there is some $p \in [0, 1]^\mathcal{H}$ with $x = \sum_H p_H d(H)^*$ and $\sum_H p_H = \sum_\alpha x_\alpha$. We apply this to $x = d(G)^* - b\sum_H d(H)^*$, noting that $\sum_\alpha x_\alpha = d(G) - b\binom{D}{3}$ and each $0 \leq x_\alpha = d(G^\alpha) - \frac{b}{3}\binom{D-1}{2} \leq \frac{1}{3}\sum_\alpha x_\alpha$. Then $p' = p + b\mathbf{1} \in [b, b^{-1}]^\mathcal{H}$ has $d(G)^* = \sum_H p'_H d(H)^*$. □

Next we consider how to encode decompositions of coloured multigraphs in the labelled edge setting of Theorem 4.18; this is similar to the running example used in the previous section.

Definition 5.6 Given a set e of size r, we write $e^{r \to q}$ for the set of all π^{-1} where $\pi : e \to [q]$ is injective. Given a $[D]$-edge-coloured r-multigraph $G = (G^d : d \in [D])$ we define $G^{r \to q} = ((G^{r \to q})^d : d \in [D])$ where each $(G^{r \to q})^d$ is the (disjoint) union of all $e^{r \to q}$ with $e \in G^d$.

Lemma 5.7 *Let H and G be $[D]$-edge-coloured r-multigraphs, $H^* = H^{r \to q}$ and $G^* = G^{r \to q}$. Then an (integral) H-decomposition of G is equivalent to an (integral) H^*-decomposition of G^*.*

Proof We associate any H-decomposition $\mathcal{D}$ of G with an H^*-decomposition $\mathcal{D}^*$ of G^*, associating each $\phi(H) \in \mathcal{D}$ with $\phi H^* := \{\phi \circ \theta : \theta \in H^*\} \in \mathcal{D}^*$. Then $e \in \phi(H^d)$ iff $e^{r \to q} \subseteq \phi H^{*d}$, as if $e = \phi(f)$ for some $f \in H^d$ and $\pi^{-1} \in e^{r \to q}$ then $\pi^{-1} = \phi \theta$, where $\theta = \phi^{-1} \pi^{-1} \in H^{*d}$, and conversely. The same proof applies to integral decompositions (defined in Definition 5.1). □

Definition 5.8 Given a family $\mathcal{H}$ of $[D]$-edge-coloured r-graphs on $[q]$, let $\mathcal{A} = \mathcal{A}^{\mathcal{H}} = \{A^H : H \in \mathcal{H}\}$ with each $A^H = S_q^{\le}$ and $\gamma = \gamma^{\mathcal{H}} \in (\mathbb{Z}^D)^{\mathcal{A}_r}$ with $\gamma_\theta = e_d$ (standard basis vector) if $\theta \in A_r^H$, $H \in \mathcal{H}$, $d \in [D]$ with $Im(\theta) \in H^d$ or $\gamma_\theta = 0$ otherwise.

Lemma 5.9 *In the notation of Definitions 5.1, 5.6 and 5.8, an (integral) $\mathcal{H}$-decomposition of G in Φ is equivalent to an (integral) $\gamma(\Phi)$-decomposition of G^*.*

Furthermore, if Φ is (ω, s)-extendable with $s = 3r^2$, $\omega > n^{-1/2}$ and $n > n_0(q)$ large then G has an integral $\mathcal{H}$-decomposition in Φ_q iff G is $\mathcal{H}$-divisible.

Proof For the first statement, the same argument as in Lemma 5.7 shows that an $\mathcal{H}$-decomposition of G in Φ is equivalent to an $\mathcal{H}^*$-decomposition of G^* in Φ (where $\mathcal{H}^* = \{H^* : H \in \mathcal{H}\}$), i.e. some $\mathcal{D} \subseteq \mathcal{H}^*(\Phi) = \{\phi H^* : H \in \mathcal{H}, \phi \in \Phi_q\}$ with $\sum \mathcal{D} = G^* \in (\mathbb{N}^D)^{\Phi_r}$. We can also view $\mathcal{D}$ as a $\gamma(\Phi)$-decomposition of G^* by identifying each $\phi H^* \in \mathcal{D}$ with the molecule $\gamma(\phi)$ where $\phi \in A^H(\Phi)$: indeed, if $\phi \pi^{-1} \in \phi H^{*d}$ with $d \in [D]$, where $e \in H^d$ and $\pi : e \to [q]$ is injective, then $\gamma(\phi)_{\phi \pi^{-1}} = \gamma_{\pi^{-1}} = e_d$. This proves the equivalence for decompositions, and the same argument applies to integral decompositions.

For the second statement, by Lemma 4.13 we have $\langle \gamma(\Phi) \rangle = L_\gamma(\Phi)$. By Definition 4.12 we need to show that G is $\mathcal{H}$-divisible iff $((G^*)^\sharp)^O \in \langle \gamma^\sharp[O] \rangle$ for any $O \in \Phi/S_q$.

Fix any $O \in \Phi/S_q$, write $e = Im(O) \in \Phi^\circ$ and $i = |e|$. Then $((G^*)^\sharp)^O \in ((\mathbb{Z}^D)^Q)^O = (\mathbb{Z}^D)^{Q \times O}$ is a vector supported on the coordinates (B, ψ') with $B' \subseteq B \in Q$ and $\psi' \in O \cap \Phi_{B'}$ with each $((G^*)^\sharp_{\psi'})_B) = \sum \{G^*_\psi : \psi' \subseteq \psi \in \Phi_B\} = (r - i)! G(e)^* \in \mathbb{N}^D$.

Also, $\langle \gamma^\sharp[O] \rangle$ is generated by $\gamma^\sharp$-atoms $\gamma^\sharp(\upsilon)$ at O, each of which is supported on the same coordinates (B, ψ') as $((G^*)^\sharp)^O$, with each $(\gamma^\sharp(\upsilon)_{\psi'})_B)$ equal to some $(r-i)! H(f)^*$ with $f \in \binom{[q]}{|e|}$, $H \in \mathcal{H}$. The lemma follows. □

Now we state our theorem on decompositions of typical coloured r-multigraphs. By Lemma 5.5 it implies Theorem 5.3. We will deduce it from Theorem 5.13 below.

Theorem 5.10 Let $\mathcal{H}$ be a family of $[D]$-edge-coloured r-graphs on $[q]$. Suppose G is a (c, h^q)-typical $[D]$-edge-coloured r-multigraph on $[n]$ with all $G_e^d < b^{-1}$ that is (b, c)-balanced wrt $\mathcal{H}$, where $n > n_0(q, D)$ is large, $d(G) > b := n^{-\delta/h^q}$, $c < c_0 d(G)^{h^{30q}}$ and $c_0 = c_0(q)$ is small. Then G has an $\mathcal{H}$-decomposition iff G is $\mathcal{H}$-divisible.

The next definition formulates the extendability and regularity conditions for coloured hypergraph decompositions; we will see below that they both follow from typicality. We remark that the extendability condition is stronger than simply requiring that each (Φ, G^i) is extendable (it is roughly equivalent to certain lower bounds on degree vectors $d_G^\gamma(x)$ as in Definition 5.2).

Definition 5.11 With notation as in Definition 5.1, we say $G \in (\mathbb{N}^D)^{\Phi_r^\circ}$ is $(\mathcal{H}, c, \omega)$-regular in Φ if there are $y_\phi^H \in [\omega n^{r-q}, \omega^{-1} n^{r-q}]$ for each $H \in \mathcal{H}$, $\phi \in \Phi_q$ with $\phi(H) \leq G$ (coordinate-wise) so that $\sum \{y_\phi^H \phi(H)\} = (1 \pm c)G$ (sum over all valid (H, ϕ), approximation coordinate-wise).

We say that (Φ, G) is (ω, h)-extendable if (Φ, G') is (ω, h)-extendable, where $G' = (G^1, \ldots, G^D)$.

The next theorem shows extendability and regularity suffice for the equivalence of decomposition and integral decomposition. For wider applicability we formulate it in the setting of exactly adapted complexes, as in the following definition, which allows for an S_q-adapted $[q]$-complex (such as the complete $[q]$-complex, suppressed in the statement of Theorem 5.10), or a generalised partite complex, which is exactly Σ-adapted for some subgroup Σ of S_q (such as that in the running example of the previous section).

Definition 5.12 We say that an R-complex Φ is exactly Σ-adapted if whenever $\phi \in \Phi_B$ and $\tau \in Bij(B', B)$ (set of bijections from B' to B) we have $\phi \circ \tau \in \Phi_{B'}$ iff $\sigma \in \Sigma_{B'}^B$.

We say Φ is exactly adapted if Φ is exactly Σ-adapted for some Σ.

Theorem 5.13 Let $\mathcal{H}$ be an family of $[D]$-edge-coloured r-graphs on $[q]$. Let Φ be an (ω, h)-extendable exactly adapted $[q]$-complex on $[n]$ where $n > n_0(q, D)$ is large, $n^{-\delta} < \omega < \omega_0(q, D)$ is small and $c = \omega^{h^{20}}$. Suppose $G \in (\mathbb{N}^D)^{\Phi_r^\circ}$ is $(\mathcal{H}, c, \omega)$-regular in Φ and (Φ, G) is (ω, h)-extendable. Then G has an $\mathcal{H}$-decomposition in Φ_q iff G has an integral $\mathcal{H}$-decomposition in Φ_q.

Proof By Lemma 5.9, it is equivalent to consider $\gamma(\Phi)$-decompositions of G^*, with notation as in Definitions 5.6 and 5.8. There are $D + 1$ types in γ for each $B \in Q$: the colour d type $\{\theta \in A_B^H : Im(\theta) \in H^d, H \in \mathcal{H}\}$ for each $d \in [D]$, and the nonedge type $\{\theta \in A_B^H : Im(\theta) \notin H \in \mathcal{H}\}$. Each γ^θ is e_d in all coordinates for θ in a colour d type or 0 in all coordinates for θ in a nonedge type, so γ is elementary. The atom decomposition of G^* is $G^* = \sum_{f \in \Phi_r^\circ} \sum_{d \in [D]} (G_f)_d f^d$, where $f_\psi^d = e_d f^{r \to q}$.

As G is $(\mathcal{H}, c, \omega)$-regular in Φ we have $\sum \{y_\phi^H \phi(H)\} = (1 \pm c)G$ for some $y_\phi^H \in [\omega n^{r-q}, \omega^{-1} n^{r-q}]$ for each $H \in \mathcal{H}$, $\phi \in \Phi_q$ with $\phi(H) \leq G$. As in the proof of the first part of Lemma 5.9, we can identify any such $\phi(H)$ with $\phi H^* \leq G^*$, and so (regarding $\phi \in A^H(\Phi)$) with $\gamma(\phi) \leq_\gamma G^*$, so $\phi \in \mathcal{A}(\Phi, G^*)$. Let $y_\phi = y_\phi^H$ for $\phi \in A^H(\Phi)$. For any $B \in Q$, $\psi \in \Phi_B$, $d \in [D]$, writing $t_d \in T_B$ for the colour d type, $\partial^{t_d} y_\psi = \sum\{y_\phi : t_\phi(\psi) = t_d\} = \sum\{y_\phi^H : Im(\psi) \in \phi(H^d), H \in \mathcal{H}\} = (1 \pm c)(G^*)_\psi^{t_d}$, so G^* is (γ, c, ω)-regular.

To apply Theorem 4.18, it remains to show that each $(\Phi, \gamma[G^*]^H)$ is (ω, h)-extendable. If $B \notin H$ then $\gamma[G^*]_B^H = \Phi_B$ and if $B \in H^d$ for $d \in [D]$ then $\gamma[G^*]_B^H = \{\psi \in \Phi_B : Im(\psi) \in G^d\}$. Consider any Φ-extension $E = (J, F, \phi)$ of rank h and $J' \subseteq J_r \setminus J[F]$. Let $J'' = (J^d : d \in [D])$ with each $J^d = \bigcup\{J_B' : B \in H^d\}$. As (Φ, G) is (ω, h)-extendable we have $X_{E,J''}(\Phi, G) > \omega n^{v_E}$. Consider any $\phi^+ \in X_{E,J''}(\Phi, G)$. For any $\psi \in J^d$ we have $\phi^+\psi \in \Phi$ and $Im(\phi^+\psi) \in G^d$, so $\phi^+\psi \in \gamma[G^*]^H$. Thus $\phi^+ \in X_{E,J'}(\Phi, \gamma[G^*]^H)$, so $(\Phi, \gamma[G^*]^H)$ is (ω, h)-extendable. □

Now we show that the extendability and regularity conditions follow from typicality, thus deducing our decomposition result for typical coloured r-multigraphs.

Proof of Theorem 5.14 Suppose G is an $\mathcal{H}$-divisible (c, h^q)-typical $[D]$-edge-coloured r-multigraph on $[n]$ that is (b, c)-balanced wrt $\mathcal{H}$, where $n > n_0(q, D)$ is large, $d(G) > b := 2n^{-\delta/h^q}$, $c < c_0 d(G)^{h^{30q}}$ and $c_0 = c_0(q)$ is small. We need to show that G has an $\mathcal{H}$-decomposition.

Let Φ be the complete $[q]$-complex on $[n]$. By Lemma 5.9 and $\mathcal{H}$-divisibility, G has an integral $\mathcal{H}$-decomposition in Φ_q. Let $p \in [b, b^{-1}]^{\mathcal{H}}$ with $d(G)^* = (1 \pm c) \sum_H p_H d(H)^*$. We can assume each colour $\alpha \in [D]$ is used at least once by $\mathcal{H}$, so $d(G^\alpha) \geq b/2Q$, where $Q = \binom{q}{r}$. To apply Theorem 5.13, it remains to check extendability and regularity.

We claim that (Φ, G) is (ω, h)-extendable with $\omega > n^{-\delta}$. To see this, consider any Φ-extension $E = (J, F, \psi)$ with $J \subseteq [q](h)$ and $J' = (J^d : d \in [D])$ for some mutually disjoint $J^d \subseteq J_r \setminus J[F]$. Let $V(J) \setminus F = \{x_1, \ldots, x_{v_E}\}$. For $i \in [v_E]$ we list the neighbourhood of x_i as $f^i = (f_1^i, \ldots, f_{t_i}^i)$ and let $\alpha^i \in [D]^{[t_i]}$ be such that each $f_j^i \cup \{x_i\}$ has colour α_j^i. Then the number of choices for x_i (weighted by edge-multiplicities) given any previous choices $\phi'|_{\{x_j: j<i\}}$ is $d_G^{\alpha^i}(\phi'(f^i)) = (1 \pm t_i c) n \prod_{j=1}^{t_i} d(G^{\alpha_j^i})$. As each $d(G^d) > b/2Q$ with $b = n^{-\delta/h^q}$, we deduce

$$X_{E,J'}(\Phi, G) = \sum_{\phi \in X_E(\Phi)} \prod_{d \in [D]} \prod_{f \in J^d} G_{\phi(f)}^d > n^{v_E - \delta}.$$

For regularity, taking $E = (J, F, \psi)$ as above with $J = [q](1)$, $J' = (H^d : d \in [D])$, $F = f \in H^\alpha$ with $H \in \mathcal{H}$, $\alpha \in [D]$, and $\psi \in Bij(f, e)$ with $e \in G^\alpha$, we obtain

$$X_{E,J'}(\Phi, G) = (1 \pm Qc) d(G^\alpha)^{-1} n^{q-r} \prod_{d \in [D]} d(G^d)^{|H^d|}.$$

Let $Z = n^{q-r} \prod_{d \in [D]} d(G^d)^{|H^d|}$ and $y_\phi = p_H(q)_r^{-1} Z^{-1} \prod_{d \in [D]} \prod_{f \in H^d} G_{\phi(f)}^d$ for each $\phi \in A^H(\Phi)$, $H \in \mathcal{H}$. Then each such $y_\phi \in [\omega n^{r-q}, \omega^{-1} n^{r-q}]$, as all $d(G^\delta) >$

$b/2Q$, $p_H < b^{-1}$ and $G^d_{\phi(f)} < b^{-1}$. Letting f vary over H^α, we have

$$\sum_H \sum_\phi y_\phi(\phi(H)_e)_\alpha = \sum_H p_H r!(q)_r^{-1} \sum_{f \in H^\alpha} z^{-1} \sum_{\phi \in X_E(\Phi)} \prod_{d \in [D]} \prod_{f \in H^d} G^d_{\phi(f)}$$
$$= \sum_H p_H Q^{-1} \sum_{f \in H^\alpha} (1 \pm 2Qc) d(G^\alpha)^{-1} G^\alpha_e = (1 \pm q^r c) G^\alpha_e.$$

Thus G is $(\mathcal{H}, q^r c, \omega)$-regular in Φ. □

We conclude with a theorem on coloured generalised partite decompositions, which can be used (we omit the details) to obtain a common generalisation of Theorems 3.2 and 5.10.

Definition 5.15 Let $\mathcal{H}$ be a family of $[D]$-edge-coloured r-graphs on $[q]$ and $\mathcal{P} = (P_1, \ldots, P_t)$ be a partition of $[q]$. Let $I^d = \{\mathbf{i} \in \mathbb{N}^t : \cup_H H^d_{\mathbf{i}} \neq \emptyset\}$ and $I = \cup_d I^d$.

Let Σ be the group of all $\sigma \in S_q$ with all $\sigma(P_i) = P_i$. Let Φ be an exactly Σ-adapted $[q]$-complex with parts $\mathcal{P}' = (P'_1, \ldots, P'_t)$, where each $P'_i = \{\psi(j) : j \in P_i, \psi \in \Phi_{\{j\}}\}$.

Let $G \in (\mathbb{N}^D)^{\Phi^\circ_r}$. We call G an $(\mathcal{H}, \mathcal{P})$-blowup if $G^d_{\mathbf{i}} \neq \emptyset \Rightarrow \mathbf{i} \in I^d$.

For $e \subseteq [n]$, $f \subseteq [q]$ we define $G(e)^*, H(f)^* \in (\mathbb{N}^D)^I$ by $(G(e)^*_{\mathbf{i}})_d = |G^d_{\mathbf{i}}(e)|$ and $(H(f)^*_{\mathbf{i}})_d = |H^d_{\mathbf{i}}(f)|$.

We say G is $(\mathcal{H}, \mathcal{P})$-divisible if each $G(e)^* \in \langle H(f)^* : f \in \binom{[q]}{|e|}, H \in \mathcal{H} \rangle$.

In the following extendability hypothesis we consider $G^d_{\mathbf{i}}$ undefined for $\mathbf{i} \notin I(H^d)$.

Theorem 5.16 *With notation as in Definition 5.15, suppose $n/h \leq |P'_i| \leq n$ with $n > n_0(q, D)$, G is an $(\mathcal{H}, \mathcal{P})$-divisible $(\mathcal{H}, \mathcal{P})$-blowup, G is $(\mathcal{H}, c, \omega)$-regular in Φ, and (Φ, G) is (ω, h)-extendable, where $n^{-\delta} < \omega < \omega_0(q, D)$ and $c = \omega^{h^{20}}$. Then G has a $\mathcal{P}$-partite $\mathcal{H}$-decomposition.*

Proof By Theorem 5.13 it suffices to show that G has an integral $\mathcal{H}$-decomposition in Φ_q, i.e. $G^* \in \langle \gamma(\Phi) \rangle = L_\gamma(\Phi)$ (by Lemmas 5.9 and 4.13). Consider any $\mathbf{i} \in I$ and $\mathbf{i}' \in \mathbb{N}^t$ with all $i'_j \leq i_j$. Let $m^{\mathbf{i}}_{\mathbf{i}'} = \prod_{j \in [t]} (i_j - i'_j)!$. For any $B' \subseteq B \in Q$ with $i_\mathcal{P}(B') = \mathbf{i}'$ and $i_\mathcal{P}(B) = \mathbf{i}$ and $\psi' \in \Phi_{B'}$ with $Im(\psi') = e$ we have $((G^*)^\sharp_{\psi'})_B) = \sum \{G^*_\psi : \psi' \subseteq \psi \in \Phi_B\} = m^{\mathbf{i}}_{\mathbf{i}'} G_{\mathbf{i}}(e)^* \in \mathbb{N}^D$. Writing $O = \psi'\Sigma$, for any $\psi \in O$ we have $((G^*)^\sharp_\psi)_B) = m^{\mathbf{i}}_{\mathbf{i}'} G_{\mathbf{i}}(e)^*$. Thus we obtain $((G^*)^\sharp)^O$ from $G(e)^*$ by copying coordinates and multiplying all copies of each $\mathbf{i}$-coordinate by $m^{\mathbf{i}}_{\mathbf{i}'}$. Similarly, for any $H \in \mathcal{H}$, $\theta' \in A^H_{B'}$, $f = Im(\theta')$ we have $(\gamma^\sharp_{\theta'})_B) = \sum \{\gamma_\theta : \theta' \subseteq \theta \in A^H_B\} = m^{\mathbf{i}}_{\mathbf{i}'} H_{\mathbf{i}}(f)^*$, so $\langle \gamma^\sharp [O] \rangle$ is generated by vectors $v^{Hf} \in (\mathbb{Z}^Q)^O$ where $H \in \mathcal{H}$, $f \subseteq [q]$ with $i_\mathcal{P}(f) = \mathbf{i}'$ and for each $\psi \in O$, $B \in Q$ we have $(v^{Hf}_\psi)_B = m^{\mathbf{i}}_{\mathbf{i}'} H_{\mathbf{i}}(f)^*$, where $\mathbf{i} = i_\mathcal{P}(B)$. Thus all vectors in $\langle \gamma^\sharp [O] \rangle$ are obtained from vectors $H(f)^*$ with $H \in \mathcal{H}$ and $i_\mathcal{P}(f) = i_{\mathcal{P}'}(e)$ by the same transformation that maps $G(e)^*$ to $((G^*)^\sharp)^O$. As G is $(\mathcal{H}, \mathcal{P})$-divisible we deduce $((G^*)^\sharp)^O \in \langle \gamma^\sharp [O] \rangle$ for any $O \in \Phi/\Sigma$, as required. □

6 Directed Hypergraphs

Our second illustration of Theorem 4.18 will be to decompositions of directed hypergraphs.

Definition 6.1 Let R be a set. An R-graph on V is a set G of injections from R to V. We call the elements of G arcs. If $R = [r]$ we call G an r-digraph. We say G is simple if $(Im(e) : e \in G)$ are all distinct. A copy of an R-graph H in an R-graph G is defined by an injection $\phi : V(H) \to V(G)$ such that $\phi H := \{\phi \circ e : e \in H\} \subseteq G$. An H-decomposition of G is a partition of G into copies of H.

Note that if $r = 2$ then a 2-digraph is equivalent to a digraph in the usual sense: we can think of an injection $f : [2] \to V$ as an arc directed from $f(1)$ to $f(2)$.

We will restrict our attention to H-decomposition problems in which H is simple; otherwise we obtain a non-elementary functional decomposition problem, which has arithmetic structure, and to which Theorem 4.18 does not apply.

Next we will state an example of our later theorem on r-digraph decompositions. Let KD_n^r denote the complete r-digraph on $[n]$, i.e. each of the $(n)_r = r!\binom{n}{r}$ injections from $[r]$ to $[n]$ is an arc. The r-digraph tight q-cycle $\circlearrowright_q^r$ has vertex set $[q]$ and arc set $\{\phi_j : j \in [q]\}$ with each $\phi_j(i) = i + j$, where addition wraps (we identify $q + i$ with i).

Theorem 6.2 *Suppose $q > r \geq 2$ and $n > n_0(q)$ with $q \mid (n)_r$. Then KD_n^r has a $\circlearrowright_q^r$-decomposition.*

Now we will describe the divisibility conditions in the general setting, and then illustrate them in the case $H = \circlearrowright_q^r$.

Definition 6.3 Let G be an r-digraph on $[n]$ and H be an r-digraph on $[q]$.

Given an injection $f : R' \to [n]$ with $R' \subseteq R$, we let $G \mid_f = \{e \in G : e \mid_{R'} = f\}$. The neighbourhood of f in G is the $(R \setminus R')$-graph $G(f) = \{e \mid_{R \setminus R'} : e \in G \mid_f\}$. The degree of f in G is $|G(f)|$.

We write I_t^s for the set of injections $\pi : [s] \to [t]$. For $\psi \in I_n^i$ we define the degree vector $G(\psi)^* \in \mathbb{N}^{I_r^i}$ by $G(\psi)_\pi^* = |G(\psi \pi^{-1})|$.

We say G is H-divisible if $G(\psi)^* \in \langle H(\theta)^* : \theta \in I_q^i \rangle$ for all $0 \leq i \leq r$, $\psi \subset I_n^i$.

Now we illustrate Definition 6.3 in the case $H = \circlearrowright_q^r$. For example, suppose $r = 2$, so H and G are digraphs. Writing $\emptyset$ for the element of I_n^0, we have $G(\emptyset)^* = (|G|)$ and $H(\emptyset)^* = (|H|) = (q)$, so the 0-divisibility condition is $q \mid |G|$. Next, for $\psi \in I_n^1$, writing $x = \psi(1) \in [n]$, we have $G(\psi)^* = (d_G^+(x), d_G^-(x))$, where $d_G^+(x) = |G(\psi)|$ is the number of arcs with $1 \mapsto x$ and $d_G^-(x) = |G(\psi \circ (1 \mapsto 2)^{-1})|$ is the number of arcs with $2 \mapsto x$. Also, for $\theta \in I_q^1$, writing $a = \theta(1) \in [q]$, we have $H(\theta)^* = (d_H^+(a), d_H^-(a)) = (1, 1)$, so the 1-divisibility condition is that G is vertex-regular, i.e. $d_G^+(x) = d_G^-(x)$ for all $x \in [n]$. Finally, for $\psi \in I_n^2$, $\theta \in I_q^2$ writing $x_i = \psi(i)$, $a_i = \theta(i)$, we have $G(\psi)^* = (1_{x_1 x_2 \in G}, 1_{x_2 x_1 \in G})$ and $H(\theta)^* = (1_{a_1 a_2 \in H}, 1_{a_2 a_1 \in H})$, so the 2-divisibility condition holds trivially. Next we describe the general $\circlearrowright_q^r$-divisibility conditions (proved in Lemma 6.5 below).

Definition 6.4 We define an equivalence relation $\sim$ on each I_r^i with $i \le r$ by $\theta \sim \theta'$ if for some $c \in \mathbb{Z}$ we have $\theta'(j) = \theta(j) + c$ for all $j \in [i]$ (where addition does not wrap). We say that G is shift regular if $G(\psi)_\theta^* = G(\psi)_{\theta'}^*$ whenever $\psi \in I_n^i$ and $\theta \sim \theta'$.

We note that $G = KD_n^r$ is shift regular, indeed $G(\psi)_\theta^* = (n)_r/(n)_i$ for any $\theta \in I_r^i$, $\psi \in I_n^i$. We also note that there is redundancy (symmetry) in the above definitions. Indeed, for $\psi \in I_n^i$, $\sigma \in S_i$, $\pi \in I_r^i$ we have $G(\psi\sigma)_\pi^* = |G(\psi\sigma\pi^{-1})| = G(\psi)_{\pi\sigma^{-1}}^*$, i.e. $G(\psi\sigma)^* = G(\psi)^*\sigma$, where S_i acts on I_n^i by $\psi \mapsto \psi\sigma = \psi \circ \sigma$ and on $\mathbb{N}^{I_r^i}$ by $(\nu\sigma)_\pi = \nu_{\pi\sigma^{-1}}$. Note that the latter is a right action as $(\nu(\sigma\tau))_\pi = \nu_{\pi(\sigma\tau)^{-1}} = \nu_{\pi\tau^{-1}\sigma^{-1}} = (\nu\sigma)_{\pi\tau^{-1}} = ((\nu\sigma)\tau)_\pi$. For any expression $G(\psi)^* = \sum_\theta n_\theta H(\theta)^*$ with $n \in \mathbb{Z}^{I_q^i}$ we have $G(\psi\sigma)^* = G(\psi)^*\sigma = \sum_\theta n_\theta H(\theta)^*\sigma = \sum_\theta n_\theta H(\theta\sigma)^*$, so it suffices to check H-divisibility on a system of coset representatives for the action of S_i on I_n^i. Furthermore, as $\theta \sim \theta'$ iff $\theta\sigma \sim \theta'\sigma$, and as $G(\psi)_{\theta\sigma}^* = |G(\psi(\theta\sigma)^{-1})| = G(\psi\sigma^{-1})_\theta^*$, it suffices to check shift regularity on a system of coset representatives for the action of S_i on I_q^i, e.g. all order-preserving elements.

Lemma 6.5 G is $\circlearrowright_q^r$-divisible iff G is shift regular and $q \mid |G|$.

Proof The 0-divisibility condition is $q \mid |G|$. Fix $0 < i \le r$. We classify the degree vectors $H(\theta)^*$ with $\theta \in I_q^i$. Note that $H(\theta)^*$ is the all-0 vector unless $Im(\theta)$ is contained in a cyclic interval of length r. By the cyclic symmetry of $\circlearrowright_q^r$ we have $H(\theta)^* = H(\theta + c)^*$ for any $c \in [q]$, defining $\theta + c \in I_q^i$ by $\theta(j) = \theta'(j) + c$ (where addition wraps). Thus we can assume $R := Im(\theta) \subseteq [r]$, i.e. $\theta \in I_r^i$. Note that $id_{[r]}$ is the unique arc of H containing id_R, so $1 = |H(id_R)| = H(\theta)_\theta^*$. Similarly, for each $c \in \mathbb{Z}$ such that $R + c \subseteq [r]$ (where addition does not wrap), $id_{[r]} - c$ is the unique arc of H containing $id_{R+c} - c$, so $1 = |H(id_{R+c} - c)| = H(\theta)_{\theta+c}^*$. All other coordinates of $H(\theta)^*$ are zero. We deduce that $H(\theta)^* = H(\theta')^*$ if $\theta \sim \theta'$, or otherwise $H(\theta)$ and $H(\theta')^*$ have disjoint support. Thus $G(\psi)^* \in \langle H(\theta)^* : \theta \in I_q^i \rangle$ iff G is constant on the support of each $H(\theta)^*$, i.e. G is shift regular. $\square$

Given Lemma 6.5, the case $H = \circlearrowright_q^r$ of the following result implies Theorem 6.2.

Theorem 6.6 *Suppose H is a simple r-digraph on $[q]$ and $n > n_0(q)$ is large. Then KD_n^r has an H-decomposition iff it is H-divisible.*

We will deduce Theorem 6.6 from a more general result in which we replace KD_n^r by any r-digraph G supported in a $[q]$-complex Φ that satisfies certain extendability and regularity conditions. The regularity condition is similar to those used earlier in the paper:

Definition 6.7 Let Φ be a $[q]$-complex on $[n]$, H be an r-digraph on $[q]$ and G be an r-digraph on $[n]$. We say G is (H, c, ω)-regular in Φ if there are $y_\phi \in [\omega n^{r-q}, \omega^{-1} n^{r-q}]$ for each $\phi \in \Phi_q$ with $\phi H \subseteq G$ so that $\sum_\phi y_\phi \phi H = (1 \pm c)G$.

Next we introduce some notation for the extendability condition and illustrate it for digraphs.

Coloured and Directed Designs   305

Definition 6.8 With notation as in Definition 6.7, let Q^H be the set of $B \in Q = \binom{[q]}{r}$ such that there is some $\theta_B \in H$ with $Im(\theta_B) = B$. Suppose H is simple, so that each θ_B is unique. Define $G^H \subseteq \Phi_r$ by $G_B^H = \{\psi \circ \theta_B^{-1} : \psi \in G\}$ if $B \in Q^H$ or $G_B^H = \Phi_B$ otherwise.

Examples Let $q = 3$, $r = 2$, G be a digraph on $[n]$ and Φ be the complete [3]-complex on $[n]$.

i. Let $H = \{(1 \mapsto 1, 2 \mapsto 2), (1 \mapsto 2, 2 \mapsto 3), (1 \mapsto 3, 2 \mapsto 1)\}$ be a cyclic triangle. For each $i \in [3]$ we have $G_{\{i,i+1\}}^H = \{(i \mapsto x, i+1 \mapsto y) : xy = (1 \mapsto x, 2 \mapsto y) \in G\}$ (interpreting $i+1 \mod 3$).
   If (Φ, G^H) is (ω, h)-extendable then for any disjoint sets $S_i \subseteq T_i$, $i \in [3]$ of size at most h and injection $\phi : S := \bigcup_{i=1}^3 S_i \to [n]$ there are at least $\omega n^{|T \setminus S|}$ injections $\phi^+ : T := \bigcup_{i=1}^3 T_i \to [n]$ extending ϕ such that for any $i \in [3]$, $x_i \in T_i$, $x_{i+1} \in T_{i+1}$ (addition mod 3) with $x_i x_{i+1} \not\subseteq S$ we have $(i \mapsto \phi^+(x_i), i+1 \mapsto \phi^+(x_{i+1})) \in G_{\{i,i+1\}}^H$, i.e. $\phi^+(x_i)\phi^+(x_{i+1}) \in G$.
   This is roughly equivalent to the following property: say that G is fully (ω, h)-extendable if for any disjoint $A, B \subseteq [n]$ of size at most h there are at least ωn vertices c such that $ca \in G$ for all $a \in A$ and $bc \in G$ for all $b \in B$. Indeed, if (Φ, G^H) is (ω, h)-extendable then G is fully (ω, h)-extendable (take $S_1 = T_1 = A$, $S_2 = T_2 = B$, $S_3 = \emptyset$, $|T_3| = 1$), and conversely, if G is fully (ω, h)-extendable then (Φ, G^H) is (ω^{3h}, h)-extendable (construct ϕ^+ one vertex at a time).

ii. Now let $H = \{(1 \mapsto 1, 2 \mapsto 2), (1 \mapsto 1, 2 \mapsto 3)\}$ be an outstar of degree two. For $i = 2, 3$ we have $G_{1i}^H = \{(1 \mapsto x, i \mapsto y) : xy \in G\}$, and $G_{23}^H = \Phi_{23}$ is complete. If (Φ, G^H) is (ω, h)-extendable then given S_i, T_i and ϕ as above, there are at least $\omega n^{|T \setminus S|}$ extensions ϕ^+ such that for any $i = 2, 3$, $x_1 \in T_1$, $x_i \in T_i$ with $x_1 x_i \not\subseteq S$ we have $\phi^+(x_1)\phi^+(x_i) \in G$.
    This is roughly equivalent to the following property: say that G is directedly (ω, h)-extendable if for any $A \subseteq [n]$ of size at most h there are at least ωn vertices c such that $ca \in G$ for all $a \in A$, and at least ωn vertices c such that $ac \in G$ for all $a \in A$.

The rough equivalence illustrated in the previous examples takes the following general form: if (Φ, G^H) is (ω, h)-extendable then (Φ, G) is (ω, h, H)-vertex-extendable (as in the next definition), and conversely, if G is (ω, h, H)-vertex-extendable then (Φ, G^H) is (ω^{qh}, h)-extendable.

Definition 6.9 With notation as in Definition 6.7, we say (Φ, G) is (ω, h, H)-vertex-extendable if for any $x \in [q]$ and disjoint sets A_i, $i \in [q] \setminus \{x\}$ of size at most h such that $(i \mapsto v_i : i \in [q] \setminus \{x\}) \in \Phi$ whenever each $v_i \in A_i$, there are at least ωn vertices $v \in \Phi_x^\circ$ such that

i. $(i \mapsto v_i : i \in [q]) \in \Phi$ whenever $v_x = v$ and $v_i \in A_i$ for each $i \neq x$,
ii. for each arc θ of H with $x \in Im(\theta)$, we have all arcs $(i \mapsto v_i : i \in [r])$ in G where $v_j = v$ for $j = \theta^{-1}(x)$ and $v_i \in A_{\theta(i)}$ for all $i \neq j$.

The following theorem when Φ and G are complete implies Theorem 6.6. Indeed, extendability is clear, and for regularity we let $y_\phi = |H|^{-1}(n)_r/(n)_q$ for each $\phi \in I_n^q$, so that for each $\psi \in I_n^r$ we have $\sum_\phi y_\phi(\phi H)_\psi = \sum_{\theta \in H} |H|^{-1}(n)_r(n)_q^{-1}|\{\phi : \psi = \phi\theta\}| = 1$.

Theorem 6.10 *Let H be a simple r-digraph on $[q]$, G be an r-digraph on $[n]$ and Φ be an (ω, h)-extendable S_q-adapted $[q]$-complex on $[n]$ where $n > n_0(q)$ is large, $n^{-\delta} < \omega < \omega_0(q)$ is small and $c = \omega^{h^{20}}$. Suppose G is (H, c, ω)-regular in Φ and (Φ, G^H) is (ω, h)-extendable. Then G has an H-decomposition in Φ_q iff G is H-divisible.*

To deduce this from Theorem 4.18 we will use the following equivalent encoding.

Definition 6.11 Given an injection $f : [r] \to X$, we write $f^{r \to q}$ for the set of all $f \circ \pi^{-1}$ where $\pi : [r] \to [q]$ is order-preserving. Given an r-digraph G, we let $G^{r \to q}$ be the (disjoint) union of all $f^{r \to q}$ with $f \in G$.

Lemma 6.12 *Let H and G be r-digraphs, $H^* = H^{r \to q}$ and $G^* = G^{r \to q}$. Then an (integral) H-decomposition of G is equivalent to an (integral) H^*-decomposition of G^*.*

Proof We associate any H-decomposition $\mathcal{H}$ of G with an H^*-decomposition $\mathcal{H}^*$ of G^*, associating each $\phi H \in \mathcal{H}$ with $\phi H^* \in \mathcal{H}^*$. Then $e \in \phi H$ iff $e^{r \to q} \subseteq \phi H^*$, as if $e = \phi\theta$ for some $\theta \in H$ and $e\pi^{-1} \in e^{r \to q}$ then $e\pi^{-1} = \phi\theta^*$, where $\theta^* = \theta\pi^{-1} \in H^*$, and conversely. The same proof applies to integral decompositions. □

Proof of Theorem 6.13 Let $H^* = H^{r \to q}$ and $G^* = G^{r \to q}$. Let $\mathcal{A} = \{A\}$ with $A = S_q^{\leq}$ and $\gamma \in \mathbb{Z}^{A_r}$ where each $\gamma_\theta = 1_{\theta \in H^*}$. Then a $\gamma(\Phi)$-decomposition of G^* is equivalent to an H^*-decomposition of G^*, and so (by Lemma 6.12) to an H-decomposition of G.

Next we claim that γ is elementary. To see this, we describe the type vectors $\gamma^\theta \in \{0, 1\}^{(S_q)^B}$ for $\theta \in A_B$, $B \in Q = \binom{[q]}{r}$. If $\gamma^\theta \neq 0$ then we can write $\theta = \theta_0\pi_0\sigma_0$ with $\theta_0 \in H$, $\pi_0 \in S_r$ and $\sigma_0 \in Bij(B, [r])$ order-preserving; this expression is unique, as θ_0 is determined by θ (as H is simple). For any $\sigma \in \Sigma^B$ we have $\gamma^\theta_\sigma = \gamma_{\theta\sigma}$ equal to 1 iff $\sigma = \sigma_0^{-1}\pi_0^{-1}\pi^{-1}$ where $\pi : [r] \to [q]$ is order-preserving. Thus there are $r! + 1$ types: the 0 type, and types t^{π_0} for each $\pi_0 \in S_r$, describing the $r!$ possible arcs with any given image. The supports of the t^{π_0} are mutually disjoint, so γ is elementary, as claimed.

The atom decomposition is $G^* = \sum_{e \in G} e^*$, where $e^* = e^{r \to q}$. As G is (H, c, ω)-regular in Φ, we have $\sum_\phi y_\phi \phi H = (1 \pm c)G$ (equivalently, $\sum_\phi y_\phi \phi H^* = (1 \pm c)G^*$) for some $y_\phi \in [\omega n^{r-q}, \omega^{-1} n^{r-q}]$ for each $\phi \in \Phi_q$ with $\phi H \subseteq G$ (equivalently, $\phi H^* \subseteq G^*$). For any such ϕ we have $\gamma(\phi) \leq_\gamma G^*$, so $\phi \in \mathcal{A}(\Phi, G^*)$. Also, for any $B \in Q$, $\psi \in \Phi_B$ and $0 \neq t \in T_B$, say with t supported on the set of all $\tau^{-1}\pi^{-1}$ where $\pi : [r] \to [q]$ is order-preserving, we have $\partial^t y_\psi = \sum \{y_\phi : t_\phi(\psi) = t\} = \sum \{y_\phi : \psi\tau \in \phi H^*\} = (1 \pm c)G^*_{\psi\tau} = (1 \pm c)(G^*)^t_\psi$, so G^* is (γ, c, ω)-regular.

Next we consider extendability. We have $\gamma[G^*] = \{\psi \in \Phi_r : \gamma(\psi) \leq_\gamma G^*\}$, so $\psi \in \Phi_B$ is in $\gamma[G^*]$ iff (a) no arc in H has image B, or (b) $\psi\theta_B \in G$ for the unique arc θ_B in H with $Im(\theta_B) = B$. Let $E = (J, F, \phi)$ be any Φ-extension of rank h and $J' \subseteq$

$J_r \setminus J[F]$. As (Φ, G^H) is (ω, h)-extendable we have $X_{E,J'}(\Phi, G^H) > \omega n^{v_E}$. Consider any $\phi^+ \in X_{E,J'}(\Phi, G^H)$. For any $\psi \in J'_B$ we have $\phi^+ \psi \in G^H_B$, so $\phi^+ \psi \theta_B \in G$, so $\phi^+ \psi \in \gamma[G^*]$. Thus $\phi^+ \in X_{E,J'}(\Phi, \gamma[G^*])$, so $(\Phi, \gamma[G^*])$ is (ω, h)-extendable.

To deduce the theorem from Theorem 4.18, it remains to consider divisibility. By Lemma 4.13 we have $\langle \gamma(\Phi) \rangle = \mathcal{L}_\gamma(\Phi)$. By Definition 4.12 we need to show that G is H-divisible iff $((G^*)^\sharp)^O \in \langle \gamma^\sharp[O] \rangle$ for any orbit $O \in \Phi/S_q$. To describe $((G^*)^\sharp)^O \in (\mathbb{Z}^Q)^O$, recall that if $\psi' \in O \cap \Phi_{B'}$ then $((G^*)^\sharp_{\psi'})_B)$ is the number of $\psi \in G^* \cap \Phi_B$ with $\psi \mid_{B'} = \psi'$. We can assume $B' \subseteq B$, otherwise this number is 0. Let $\pi_B : [r] \to B$ be order-preserving and $R = \pi_B^{-1}(B')$. Then $\psi \in G^* \cap \Phi_B$ iff $\psi \pi_B \in G$, and $\psi \mid_{B'} = \psi'$ iff $(\psi \pi_B) \mid_R = \psi' \pi_B$, so $((G^*)^\sharp_{\psi'})_B = |G(\psi' \pi_B)|$. Similarly, to describe $\langle \gamma^\sharp[O] \rangle$, recall that it is generated by vectors $\gamma^\sharp(\phi) \in (\mathbb{Z}^Q)^O$ where if $\psi' = \phi \theta'$ with $\theta' \in A_{B'}$ then $(\gamma^\sharp(\phi)_{\psi'})_B = (\gamma^\sharp_{\theta'})_B$ is the number of $\theta \in H_B^*$ with $\theta \mid_{B'} = \theta'$, which is $|H(\theta' \pi_B)|$.

Now fix $\psi \in O \cap \Phi_{[i]}$, where $O \in \Phi_i/S_q$. As G is H-divisible, there is $n \in \mathbb{Z}^{I^i_q}$ with $G(\psi)^* = \sum_\theta n_\theta H(\theta)^*$. Writing $\phi = \psi \theta^{-1}$, we claim that $((G^*)^\sharp)^O = \sum_\theta n_\theta \gamma^\sharp(\phi)$. To see this, note that it suffices to prove $((G^*)^\sharp)^O_{[r]} = \sum_\theta n_\theta \gamma^\sharp(\phi)_{[r]}$, as $((G^*)^\sharp_{\psi'})_B = |G(\psi' \pi_B)| = ((G^*)^\sharp_{\psi' \pi_B})_{[r]}$ and $(\gamma^\sharp(\phi)_{\psi'})_B = (\gamma^\sharp_{\phi^{-1}\psi'})_B = |H(\phi^{-1} \psi' \pi_B)| = (\gamma^\sharp(\phi)_{\psi' \pi_B})_{[r]}$. Now for any $\psi' \in O \cap \Phi_R$ with $R \subseteq [r]$, writing $\pi = (\psi')^{-1} \psi \in I^i_r$, we have $((G^*)^\sharp_{\psi'})_{[r]} = |G(\psi')| = G(\psi)^*_\pi = \sum n_\theta H(\theta)^*_\pi$, where each $H(\theta)^*_\pi = |H(\theta \pi^{-1})| = (\gamma^\sharp_{\theta \pi^{-1}})_{[r]} = (\gamma^\sharp(\phi)_{\psi'})_{[r]}$, so $((G^*)^\sharp_{\psi'})_{[r]} = \sum n_\theta (\gamma^\sharp(\phi)_{\psi'})_{[r]}$. □

7 All of the Above

For use in future applications (e.g. [13]), in this section we present a general theorem that simultaneously allows for the various flavours of decomposition considered in this paper (generalised partitions, colours and directions). We start with a definition that generalises our previous setting of simple r-digraphs to allow for colours, index vectors with respect to a partition, and different types of 'generalised arcs'; it is followed by some illustrative examples.

Definition 7.1 Let $\mathcal{P} = (P_1, \ldots, P_t)$ be a partition of $[q]$ such that if $x \in P_j$, $x' \in P_{j'}$, $j < j'$ then $x < x'$. Let $\mathcal{H}$ be a family of $[D]$-edge-coloured r-digraphs on $[q]$. For $i \in \mathbb{N}^t$ with $\sum_{j=1}^t i_j = r$ and $j \in [t]$ we define a partition $R(i) = (R(i)_1, \ldots, R(i)_t)$ of $[r]$ so that each $|R(i)_j| = i_j$ and $x < x'$ whenever $x \in R(i)_j$, $x' \in R(i)_{j'}$, $j < j'$. Suppose there are vectors $i^d \in \mathbb{N}^t$ and permutation groups Λ_j^d on $R(i)_j$ for all $d \in [D]$ and $j \in [t]$ such that if $H \in \mathcal{H}$ and $\theta \in H^d$ then

i. each $\theta(R(i)_j) \subseteq P_j$ (so[8] $i_\mathcal{P}(Im(\theta)) = i^d$), and
ii. for $\theta' \in Bij([r], Im(\theta))$ we have $\theta' \notin H \setminus H^d$, and $\theta' \in H^d$ iff $\theta^{-1}\theta' \in \Lambda^d := \prod_j \Lambda_j^d$.

We say that $\mathcal{H}$ is $(\mathcal{P}, \Lambda)$-canonical, where $\Lambda := (\Lambda^d : d \in D)$.

---

[8]Recall index vectors from Definition 3.1.

Examples

i. Let $q = 3$, $r = 2$ and $t = 1$, so $\mathcal{P} = ([3])$ and $R(2) = ([2])$. Let $D = 2$ and $\mathcal{H} = \{H\}$, where $H^1 = \{(1 \mapsto 1, 2 \mapsto 2), (1 \mapsto 1, 2 \mapsto 3)\}$ and $H^2 = \{(1 \mapsto 2, 2 \mapsto 3), (1 \mapsto 3, 2 \mapsto 2)\}$. Then $\mathcal{H}$ is canonical with $\Lambda_1^1 = \{id\}$ and $\Lambda_1^2 = S_2 = \{id, (12)\}$. One can interpret H as a mixed triangle, with arcs from 1 to 2 and 1 to 3 and an undirected edge between 2 and 3. In this interpretation, we are free to ignore the colours, as they do not affect whether a mixed graph G has an H-decomposition (the role of the colours is to ensure that under the encoding by arcs, an undirected edge encoded by two arcs cannot be decomposed into two actual arcs). In general, we think of an atom in some colour d as a 'generalised arc', which is encoded by some set of arcs invariant under the action of Λ^d on $[r]$. An actual arc corresponds to the case $\Lambda^d = \{id\}$ and an undirected edge to the case that each $\Lambda_j^d = Sym(R(i)_j)$.

ii. Let $q = 3$, $r = 2$, $t = 1$, $D = 2$ and $\mathcal{H} = \{H\}$, where $H^1 = \{(1 \mapsto 1, 2 \mapsto 2), (1 \mapsto 2, 2 \mapsto 3)\}$ and $H^2 = \{(1 \mapsto 3, 2 \mapsto 1)\}$. Then $\mathcal{H}$ is canonical with $\Lambda^1 = \Lambda^2 = \{id\}$. One can interpret H as a two-coloured cyclic directed triangle, with arcs of colour 1 from 1 to 2 and 2 to 3, and an arc of colour 2 from 3 to 1.

iii. Let $q = 3$, $r = 2$, $t = 2$, $\mathcal{P} = (\{1,2\}, \{3\})$, $D = 3$ and $\mathcal{H} = \{H\}$, where $H^1 = \{(1 \mapsto 1, 2 \mapsto 2)\}$, $H^2 = \{(1 \mapsto 1, 2 \mapsto 3)\}$ and $H^3 = \{(1 \mapsto 2, 2 \mapsto 3)\}$. We have $\boldsymbol{i}^1 = (2, 0)$, $R((2,0)) = ([2], \emptyset)$, $\boldsymbol{i}^2 = \boldsymbol{i}^3 = (1, 1)$, $R((1,1)) = (\{1\}, \{2\})$ and $\Lambda^1 = \Lambda^2 = \Lambda^3 = \{id\}$. One possible uncoloured interpretation of H is as a cyclic triangle $1 \to 2 \to 3 \to 1$ under the vertex partition $\mathcal{P}$. Here we are taking the natural interpretation of the colour 3 arc from 2 to 3 and the opposite interpretation of the colour 2 arc from 1 to 3, instead thinking of it as an arc from 3 to 1. Changing the direction of all arcs of colour 2 in both H and G has no effect on whether G has an H-decomposition, so this interpretation is equivalent to the natural interpretation in which we retain the given colours and directions. This illustrates the fact that in general there is no loss of generality from the assumption that the partitions $\mathcal{P}$ and $R(\boldsymbol{i})$ respect the orders of $[q]$ and $[r]$, as we are free to interpret different colours as encoding arcs with alternative partitions. We also note that there is no loss of generality in assuming that the index of an edge is determined by its colour (and indeed, we could have done so earlier in the paper).

For the main result of this section we adopt the setting of the following definition (see below for how it applies to the above examples).

Definition 7.2 Let $\mathcal{H}$ be a $(\mathcal{P}, \Lambda)$-canonical family of $[D]$-edge-coloured r-digraphs on $[q]$. We identify each $H \in \mathcal{H}$ with a vector $H \in (\mathbb{N}^D)^{I_q^r}$, where each $(H_f)_d = 1_{f \in H^d}$.

Let Σ be the group of all $\sigma \in S_q$ with all $\sigma(P_i) = P_i$. Let Φ be an exactly Σ-adapted $[q]$-complex with $V(\Phi) = [n]$ and parts $\mathcal{P}' = (P_1', \ldots, P_t')$, where each $P_i' = \{\psi(j) : j \in P_i, \psi \in \Phi_{\{j\}}\}$. For $\phi \in \Phi_q$ and $H \in \mathcal{H}$ we define $\phi H \in (\mathbb{N}^D)^{\Phi_{[r]}}$ by $(\phi H)_{\phi f} = H_f$. Let $\mathcal{H}(\Phi) = \{\phi H : \phi \in \Phi_q, H \in \mathcal{H}\}$.

Coloured and Directed Designs   309

Let $G \in (\mathbb{N}^D)^{\Phi_{[r]}}$ be an r-multidigraph $[D]$-edge-coloured as $G = \cup_{d \in [D]} G^d$.

We call $\mathcal{H}' \subseteq \mathcal{H}(\Phi)$ with $\sum \mathcal{H}' = G$ an H-decomposition of G in Φ, and $\Psi \in \mathbb{Z}^{\mathcal{H}(\Phi)}$ with $\sum_{H'} \Psi_{H'} H' = G$ an integral H-decomposition of G in Φ.

For $\psi \in I_n^i$ (injections $[i] \to [n]$) and $\theta \in I_q^i$ write $i_{\mathcal{P}'}(\psi) = i_{\mathcal{P}'}(Im(\psi))$ and $i_{\mathcal{P}}(\theta) = i_{\mathcal{P}}(Im(\theta))$. For $\psi \in I_n^i$ we define the degree vector $G(\psi)^* \in \mathbb{N}^{[D] \times I_r^i}$ by $G(\psi)^*_{d\pi} = |G^d(\psi \pi^{-1})|$. Similarly, for $\theta \in I_q^i$ we define $H(\theta)^* \in \mathbb{N}^{[D] \times I_r^i}$ by $H(\theta)^*_{d\pi} = |H^d(\theta \pi^{-1})|$.

For $i' \in \mathbb{N}^t$ we let $H\langle i' \rangle = \langle H(\theta)^* : i_{\mathcal{P}}(\theta) = i' \rangle$. We say G is $\mathcal{H}$-divisible (in Φ) if $G(\psi)^* \in H\langle i' \rangle$ whenever $i_{\mathcal{P}'}(\psi) = i'$.

We say G is $(\mathcal{H}, c, \omega)$-regular in Φ if there are $y^H_\phi \in [\omega n^{r-q}, \omega^{-1} n^{r-q}]$ for each $H \in \mathcal{H}, \phi \in \Phi_q$ with $\phi H \leq G$ so that $\sum \{y^H_\phi \phi H\} = (1 \pm c)G$.

For each $H \in \mathcal{H}$ and $B \in Q$ fix any $\theta_B \in H$ with $Im(\theta_B) = B$ if one exists.

For each $d \in [D]$ let $(G^H)^d = \bigcup \{\psi \circ \theta_B^{-1} : \theta_B \in H^d, G^d_\psi > 0\}$.

We say that (Φ, G^H) is (ω, h)-extendable if $(\Phi, ((G^H)^d : d \in [D]))$ is (ω, h)-extendable.

Examples

i. Recall the example of the mixed triangle: $q = 3, r = 2, t = 1, D = 2, \mathcal{H} = \{H\}$, $H^1 = \{(1 \mapsto 1, 2 \mapsto 2), (1 \mapsto 1, 2 \mapsto 3)\}$, $H^2 = \{(1 \mapsto 2, 2 \mapsto 3), (1 \mapsto 3, 2 \mapsto 2)\}$, $\Lambda_1^1 = \{id\}, \Lambda_1^2 = \{id, (12)\}$. Let Φ be the complete $[3]$-complex on $[n]$ and $G \in (\mathbb{N}^2)^{\Phi_2}$ be a $[2]$-edge-coloured 2-multidigraph. For the 2-divisibility condition we consider any $\psi \in I_n^2$, so that $G(\psi)^* \in \mathbb{N}^{[2] \times I_2^2}$. Ordering coordinates as $(1, id), (1, (12)), (2, id), (2, (12))$ we have $G(\psi)^* = (G^1_\psi, G^1_{\psi \circ (12)}, G^2_\psi, G^2_{\psi \circ (12)})$. The possible $H(\theta)^*$ with $\theta \in I_3^2$ are $(1, 0, 0, 0)$, $(0, 1, 0, 0)$ and $(0, 0, 1, 1)$. Thus the 2-divisibility condition is that $G^2_\psi = G^2_{\psi \circ (12)}$ for all $\psi \in I_n^2$, i.e. arcs of colour 2 always come in opposite pairs (which we interpret as an edge when we think of G as a mixed multigraph). As for 0-divisibility, writing $\emptyset$ for the function with empty domain, we have $G(\emptyset)^* = (|G^1|, |G^2|)$ and $H(\emptyset)^* = (|H^1|, |H^2|) = (2, 2)$, so we need $|G^1| = |G^2|$. In terms of mixed multigraphs, we need twice as many arcs as edges (each edge corresponds to a pair of arcs in G^2).

For the 1-divisibility conditions, consider any $\psi \in I_n^1$, so $G(\psi)^* \in \mathbb{N}^{[2] \times I_2^1}$. Let $x = Im(\psi) \in [n]$. Ordering coordinates as $(1, 1 \mapsto 1), (1, 1 \mapsto 2), (2, 1 \mapsto 1), (2, 1 \mapsto 2)$ we have $G(\psi)^* = (|G^1(1 \mapsto x)|, |G^1(2 \mapsto x)|, |G^2(1 \mapsto x)|, |G^2(2 \mapsto x)|) = (d_G^+(x), d_G^-(x), d_G(x), d_G(x))$, where in the mixed graph interpretation $d_G^\pm(x)$ denote in/outdegrees in arcs and $d_G(x)$ denotes degree in edges. We have $H(1 \mapsto 1)^* = (2, 0, 0, 0)$ and $H(1 \mapsto 2)^* = H(1 \mapsto 3)^* = (0, 1, 1, 1)$. Thus the 1-divisibility conditions are that each outdegree $d_G^+(x)$ is even and each $d_G(x) = d_G^-(x)$.

Now consider extendability. We have $(G^H)^1_{12} = G^1$, $(G^H)^1_{13} = G^1 \circ (1 \mapsto 1, 3 \mapsto 2) = \{(1 \mapsto x, 3 \mapsto y) : xy \in G^1\}$ and $(G^H)^2_{23} = G^2$ (for either choice of θ_{23} if arcs of colour 2 always come in opposite pairs). All other $(G^H)^d_B$ are undefined. If (Φ, G^H) is (ω, h)-extendable then for any sets $S_i \subseteq T_i, i \in [3]$ of size at most h and an injection $\phi : S := \bigcup_{i=1}^{3} S_i \to [n]$ there are at least $\omega n^{|T \setminus S|}$ injections $\phi^+ : T := \bigcup_{i=1}^{3} T_i \to [n]$ extending ϕ such that for any $1 \leq i < j \leq 3$, $x_i \in T_i$, $x_j \in T_j$ with $x_i x_j \not\subseteq S$ we have $(i \mapsto \phi^+(x_i), j \mapsto \phi^+(x_j)) \in (G^H)_{ij}^d$, i.e. $\phi^+(x_i)\phi^+(x_j) \in G^d$, where $d = 2$ if $ij = 23$ or $d = 1$ otherwise.

This is roughly equivalent to the following property: for any disjoint $A, B \subseteq [n]$ of size at most h there are at least ωn vertices c such that $ca \in G^2$ for all $a \in A$ and $bc \in G^1$ for all $b \in B$, and at least ωn vertices c such that $ca \in G^1$ for all $a \in A \cup B$.

ii. Recall the example of the two-coloured cyclic directed triangle: $q = 3$, $r = 2$, $t = 1$, $D = 2$, $\mathcal{H} = \{H\}$, $H^1 = \{(1 \mapsto 1, 2 \mapsto 2), (1 \mapsto 2, 2 \mapsto 3)\}$, $H^2 = \{(1 \mapsto 3, 2 \mapsto 1)\}$, $\Lambda^1 = \Lambda^2 = \{id\}$. Let Φ be the complete $[3]$-complex on $[n]$ and $G \in (\mathbb{N}^2)^{\Phi_2}$. The 2-divisibility condition is trivial. As $G(\emptyset)^* = (|G^1|, |G^2|)$ and $H(\emptyset)^* = (2, 1)$ the 0-divisibility condition is $|G^1| = 2|G^2|$. For $\psi \in I_n^1$, $x = Im(\psi) \in [n]$ we have $G(\psi)^* = (|G^1(1 \mapsto x)|, |G^1(2 \mapsto x)|, |G^2(1 \mapsto x)|, |G^2(2 \mapsto x)|) = (d_{G^1}^+(x), d_{G^1}^-(x), d_{G^2}^+(x), d_{G^2}^-(x))$. We have $H(1 \mapsto 1)^* = (1, 0, 0, 1)$, $H(1 \mapsto 2)^* = (1, 1, 0, 0)$ and $H(1 \mapsto 3)^* = (0, 1, 1, 0)$, which generate $H\langle 1 \rangle = \{v \in \mathbb{Z}^4 : v_1 + v_3 = v_2 + v_4\}$, so the 1-divisibility condition is $d_{G^1}^+(x) + d_{G^2}^+(x) = d_{G^1}^-(x) + d_{G^2}^-(x)$, i.e. the degree regularity condition $d_G^+(x) = d_G^-(x)$ needed for decomposition into cyclic triangles ignoring the colours.

As for extendability, we have $(G^H)_{12}^1 = G^1$, $(G^H)_{23}^1 = \{(2 \mapsto x, 3 \mapsto y) : xy \in G^1\}$ and $(G^H)_{13}^2 = \{(3 \mapsto x, 1 \mapsto y) : xy \in G^2\}$. If (Φ, G^H) is (ω, h)-extendable then for any S_i, T_i and ϕ as above there are at least $\omega n^{|T \setminus S|}$ extensions ϕ^+ such that for any $1 \leq i < j \leq 3$, $x_i \in T_i$, $x_j \in T_j$ with $x_i x_j \not\subseteq S$ we have $\phi^+(x_i)\phi^+(x_j) \in G^1$ if $ij \neq 13$ or $\phi^+(x_3)\phi^+(x_1) \in G^2$ if $ij = 13$.

This is roughly equivalent to: for any disjoint $A, B \subseteq [n]$ of size at most h there are

(1) at least ωn vertices c such that $ca \in G^1$ for all $a \in A$ and $bc \in G^1$ for all $b \in B$,

(2) at least ωn vertices c such that $ca \in G^1$ for all $a \in A$ and $bc \in G^2$ for all $b \in B$, and

(3) at least ωn vertices c such that $ca \in G^2$ for all $a \in A$ and $bc \in G^1$ for all $b \in B$.

iii. Recall the example of the cyclic triangle $1 \to 2 \to 3 \to 1$ with vertex partition $\mathcal{P} = (\{1, 2\}, \{3\})$: we have $q = 3$, $r = 2$, $t = 2$, $D = 3$, $\mathcal{H} = \{H\}$, $H^1 = \{(1 \mapsto 1, 2 \mapsto 2)\}$, $H^2 = \{(1 \mapsto 1, 2 \mapsto 3)\}$, $H^3 = \{(1 \mapsto 2, 2 \mapsto 3)\}$, $\mathbf{i}^1 = (2, 0)$, $R((2, 0)) = ([2], \emptyset)$, $\mathbf{i}^2 = \mathbf{i}^3 = (1, 1)$, $R((1, 1)) = (\{1\}, \{2\})$, $\Lambda^1 = \Lambda^2 = \Lambda^3 = \{id\}$. Let Φ be a complete $\mathcal{P}$-partite $[3]$-complex and $G \in (\mathbb{N}^3)^{\Phi_2}$. Note that $P_1' = \Phi_{\{1\}}^{\circ} = \Phi_{\{2\}}^{\circ}$ and $P_2' = \Phi_{\{3\}}^{\circ}$. The 2-divisibility condition is that arcs of G must respect the partition according to their colour, i.e. if $G_\theta^1 \neq 0$ then $Im(\theta) \subseteq P_1'$ and if $G_\theta^2 \neq 0$ or $G_\theta^3 \neq 0$ then $\theta(1) \in P_1'$ and $\theta(2) \in P_2'$. As $G(\emptyset)^* = (|G^1|, |G^2|, |G^3|)$ and $H(\emptyset)^* = (1, 1, 1)$ the 0-divisibility condition is $|G^1| = |G^2| = |G^3|$, i.e. in the uncoloured interpretation we have equal num-

bers of arcs (1) within P'_1, (2) from P'_1 to P'_2, and (3) from P'_2 to P'_1.
Now consider the 1-divisibility conditions. Let G' denote the arcs between P'_1 and P'_2 according to the uncoloured interpretation, where arcs from P'_1 to P'_2 correspond to G^3 and arcs from P'_2 to P'_1 correspond to G^2. Let $\psi \in I_n^1$ and $x = Im(\psi)$. Suppose first that $x \in P'_1$. Then $G(\psi)^* = (|G^1(1 \mapsto x)|, |G^1(2 \mapsto x)|, |G^2(1 \mapsto x)|, |G^2(2 \mapsto x)|, |G^3(1 \mapsto x)|, |G^3(2 \mapsto x)|) = (d^+_{G^1}(x), d^-_{G^1}(x), d^-_{G'}(x), 0, d^+_{G'}(x), 0)$. As $H(1 \mapsto 1)^* = (1, 0, 1, 0, 0, 0)$ and $H(1 \mapsto 2)^* = (0, 1, 0, 0, 1, 0)$, we obtain the conditions $d^+_{G^1}(x) = d^-_{G'}(x)$ and $d^-_{G^1}(x) = d^+_{G'}(x)$ for all $x \in P'_1$. Now suppose $x \in P'_2$. We have $G(\psi)^* = (0, 0, 0, d^+_{G'}(x), 0, d^-_{G'}(x))$ and $H(1 \mapsto 3)^* = (0, 0, 0, 1, 0, 1)$, so we need $d^+_{G'}(x) = d^-_{G'}(x)$ for all $x \in P'_2$.

As for extendability, we have $(G^H)^1_{12} = G^1$, $(G^H)^2_{13} = \{(1 \mapsto x, 3 \mapsto y) : xy \in G^2)\}$ and $(G^H)^3_{23} = \{(2 \mapsto x, 3 \mapsto y) : xy \in G^3)\}$. If (Φ, G^H) is (ω, h)-extendable then for any S_i, T_i and ϕ as above with $\phi(S_1), \phi(S_2) \subseteq P'_1$ and $\phi(S_3) \subseteq P'_2$ there are at least $\omega |P'_1|^{|T_1 \setminus S_1| + |T_2 \setminus S_2|} |P'_2|^{|T_3 \setminus S_3|}$ extensions ϕ^+ such that for any $1 \leq i < j \leq 3$, $x_i \in T_i$, $x_j \in T_j$ with $x_i x_j \not\subseteq S$ we have $\phi^+(x_i)\phi^+(x_j) \in G^{d_{ij}}$, where $d_{12} = 1$, $d_{13} = 2$, $d_{23} = 3$. This is roughly equivalent to:

(1) for any disjoint $A, B \subseteq P'_1$ of size at most h there are at least $\omega |P'_2|$ vertices $c \in P'_2$ such that $ca \in G'$ for all $a \in A$ and $bc \in G'$ for all $b \in B$, and
(2) for any disjoint $A \subseteq P'_1$, $B \subseteq P'_2$ of size at most h there are at least $\omega |P'_1|$ vertices $c \in P'_1$ such that $ca \in G'$ for all $a \in A$ and $bc \in G^1$ for all $b \in B$, and at least $\omega |P'_1|$ vertices $c \in P'_1$ such that $ca \in G^1$ for all $a \in A$ and $bc \in G'$ for all $b \in B$.

Similarly to Definition 6.9, we have the following general rough equivalence: if (Φ, G^H) is (ω, h)-extendable then (Φ, G) is (ω, h, H)-vertex-extendable (as in the next definition), and conversely, if G is (ω, h, H)-vertex-extendable then (Φ, G^H) is (ω^{qh}, h)-extendable.

Definition 7.3 With notation as in Definition 7.2, we say (Φ, G) is (ω, h, H)-vertex-extendable if for any $x \in [q]$ and disjoint sets A_i, $i \in [q] \setminus \{x\}$ of size at most h such that $(i \mapsto v_i : i \in [q] \setminus \{x\}) \in \Phi$ whenever each $v_i \in A_i$, there are at least ωn vertices $v \in \Phi^\circ_x$ such that

i. $(i \mapsto v_i : i \in [q]) \in \Phi$ whenever $v_x = v$ and $v_i \in A_i$ for each $i \neq x$,
ii. for each $d \in [D]$ and arc θ of H^d with $x \in Im(\theta)$, we have all arcs $(i \mapsto v_i : i \in [r])$ in G^d where $v_j = v$ for $j = \theta^{-1}(x)$ and $v_i \in A_{\theta(i)}$ for all $i \neq j$.

The main theorem of the section provides the above general setting with our usual conclusion (divisibility, regularity and extendability suffice for the existence of decompositions).

Theorem 7.4 With notation as in Definition 7.2, suppose all $n_1/h \leq |P'_i| \leq n_1$ with $n_1 > n_0(q, D)$, that G is $\mathcal{H}$-divisible and $(\mathcal{H}, c, \omega)$-regular in Φ, and all (Φ, G^H) are (ω, h)-extendable, where $n_1^{-\delta} < \omega < \omega_0(q, D)$ and $c = \omega^{h^{20}}$. Then G has an $\mathcal{H}$-decomposition in Φ.

Proof For $\psi \in \Phi_{[r]}$ we let ψ^* be the set of all $\psi \circ \pi^{-1}$ where $\pi : [r] \to [q]$ is order-preserving and $i_{\mathcal{P}}(\pi) = i_{\mathcal{P}'}(\psi)$. Similarly, for $\theta \in H_r$ we let θ^* be the set of all $\theta \circ \pi^{-1}$ where $\pi : [r] \to [q]$ is order-preserving and $i_{\mathcal{P}}(\pi) = i_{\mathcal{P}}(\theta)$. Let $G^* = \sum_{\psi \in \Phi_{[r]}} G_\psi \psi^*$ and $\mathcal{H}^* = \{H^* : H \in \mathcal{H}\}$ with each $(H^*)^d = (H^d)^* = \{\theta^* : \theta \in H^d\}$. Let $\mathcal{A} = \{A^H : H \in \mathcal{H}\}$ with each $A^H = \Sigma^\leq$ and $\gamma \in \mathbb{Z}^{\mathcal{A}_r}$ where each γ_θ is e_d if $\theta \in H^{d*}$ for some $H \in \mathcal{H}$, $d \in [D]$, otherwise zero. Then a $\gamma(\Phi)$-decomposition of G^* is equivalent to an $\mathcal{H}^*$-decomposition of G^*, and so, we claim, to an $\mathcal{H}$-decomposition of G.

For the latter equivalence, similarly to Lemma 6.12, we need to show for any $H \in \mathcal{H}, d \in [D], \phi \in \Phi_q$ that $\psi \in \phi H_r^d$ iff $\psi^* \subseteq \phi H_r^{d*}$. To see this, write $\psi = \phi\theta$, where $\theta \in H_r^d$ and let $i = i_{\mathcal{P}'}(\psi) = i_{\mathcal{P}}(\theta)$. For any $\psi' \in \psi^*$ we can write $\psi' = \psi\pi^{-1}$ where $\pi : [r] \to [q]$ is order-preserving with $i_{\mathcal{P}}(\pi) = i$, so $\psi' = \phi\theta'$ with $\theta' = \theta\pi^{-1} \in \theta^*$. Thus $\psi \in \phi H_r^d$ implies $\psi^* \subseteq \phi H_r^{d*}$. The converse is similar, so the claimed equivalence holds (and also for integral decompositions).

Next we claim that γ is elementary. To see this, we describe the type vectors γ^θ for $\theta \in A_B^H, B \in Q$. If $\gamma^\theta \neq 0$ then we can write $\theta = \theta_0 \tau_0 \pi_0^{-1}$ with $\theta_0 \in H, \tau_0 \in S_r$ and $\pi_0 \in Bij([r], B)$ order-preserving. Say $\theta_0 \in H^d$. As $\mathcal{H}$ is $(\mathcal{P}, \Lambda)$-canonical, for $\theta' \in Bij([r], Im(\theta_0))$ we have $\theta' \in H^d$ iff $\theta^{-1}\theta' \in \Lambda^d$. Fix a set X^d of representatives for the right cosets of Λ^d in S_r. Then we have a unique expression $\theta = \theta_0 \tau_0 \pi_0^{-1}$ with $\theta_0 \in H^d$ and $\tau_0 \in X^d$. For any $\sigma \in \Sigma^B$ we have $\gamma_\sigma^\theta = \gamma_{\theta\sigma} \in \{0, e_d\}$ equal to e_d iff $\sigma = \pi_0(\lambda\tau_0)^{-1}\pi^{-1}$ where $\lambda \in \Lambda^d$ and $\pi : [r] \to [q]$ is order-preserving with $i_{\mathcal{P}}(\pi) = i := i_{\mathcal{P}}(B)$. Thus, besides the 0 type, for each $B \in Q$ and $d \in [D]$ with $i^d = i$ we have $|X^d| = r!/|\Lambda^d|$ types $(t^{\tau_0} : \tau_0 \in X^d)$ describing all generalised arcs with any given image. Given d, the supports of the t^{τ_0} for $\tau_0 \in X^d$ are mutually disjoint, so γ is elementary, as claimed.

As G is $(\mathcal{H}, c, \omega)$-regular in Φ we have $y_\phi^H \in [\omega n^{r-q}, \omega^{-1} n^{r-q}]$ for each $H \in \mathcal{H}, \phi \in \Phi_q$ with $\phi H \leq G$ (equivalently, $\phi H^* \leq G^*$) so that $\sum\{y_\phi^H \phi H\} = (1 \pm c)G$. (equivalently, $\sum\{y_\phi^H \phi H^*\} = (1 \pm c)G^*$). We identify any such $\phi H^* \leq G^*$ with $\gamma(\phi) \leq_\gamma G^*$ (regarding $\phi \in A^H(\Phi)$), so $\phi \in \mathcal{A}(\Phi, G^*)$. Let $y_\phi = y_\phi^H$ for $\phi \in A^H(\Phi)$. For any $B \in Q, \psi \in \Phi_B, d \in [D]$ with $i^d = i := i_{\mathcal{P}'}(\psi)$ and $0 \neq t \in T_B$, say with t supported on the set of all $(\lambda\tau)^{-1}\pi^{-1}$ where $\lambda \in \Lambda^d$ and $\pi : [r] \to [q]$ is order-preserving with $i_{\mathcal{P}}(\pi) = i$, we have $\partial^t y_\psi = \sum\{y_\phi : t_\phi(\psi) = t\} = \sum\{y_\phi^H : \psi\tau \in \phi H^{d*}, H \in \mathcal{H}\} = (1 \pm c)G_{\psi\tau}^{d*} = (1 \pm c)(G^*)_\psi^t$, so G^* is (γ, c, ω)-regular.

Next we consider extendability. Fix $H \in \mathcal{H}$. We have $\gamma[G^*]^H = \{\psi \in A^H(\Phi)_r : \gamma(\psi) \leq_\gamma G^*\}$, so $\psi \in \Phi_B$ is in $\gamma[G^*]^H$ iff (a) no arc in H has image B, or (b) $\psi\theta_B \in G^d$ (i.e. $G_{\psi\theta_B}^d > 0$) for some (equivalently, all) $\theta_B \in H_r^d$ with $Im(\theta_B) = B$. Let $E = (J, F, \phi)$ be any Φ-extension of rank h and $J' \subseteq J_r \setminus J[F]$. Let $J'' = (J^d : d \in [D])$ with each $J^d = \bigcup\{J_B' : \theta_B \in H_r^d\}$. As (Φ, G^H) is (ω, h)-extendable we have $X_{E,J''}(\Phi, G^H) > \omega n^{v_E}$. Consider any $\phi^+ \in X_{E,J''}(\Phi, G^H)$. For any $\psi \in J_B^d$, $d \in [D]$ we have $\phi^+\psi \in (G^H)_B^d$, so $\phi^+\psi\theta_B \in G^d$, so $\phi^+\psi \in \gamma[G^*]^H$. Thus $\phi^+ \in X_{E,J'}(\Phi, \gamma[G^*]^H)$, so $(\Phi, \gamma[G^*])$ is (ω, h)-extendable.

To deduce the theorem from Theorem 4.18, it remains to show for any orbit $O \in \Phi/\Sigma$ that $((G^*)^\sharp)^O \in \langle \gamma^\sharp[O] \rangle$. Fix $\psi \in O \in \Phi_i/\Sigma$. Let $i' = i_{\mathcal{P}'}(\psi)$ and $I' = \{\theta \in$

$I_q^i : i_{\mathcal{P}}(\theta) = i'\}$. Write $\psi = \psi_0 \pi_0^{-1}$ with $\psi_0 \in I_n^i$ and $\pi_0 : [i] \to Dom(\psi)$ order-preserving. As G is $\mathcal{H}$-divisible, there is $n \in \mathbb{Z}^{\mathcal{H} \times I'}$ with $G(\psi_0)^* = \sum_{H,\theta} n_{H\theta} H(\theta)^*$, i.e. $|G^d(\psi_0 \pi^{-1})| = \sum_{H,\theta} n_{H\theta} |H^d(\theta \pi^{-1})|$ for all $d \in [D]$ and $\pi \in I_r^i$. Writing $\phi = \psi_0 \theta^{-1} \in A^H(\Phi)$, we claim $((G^*)^\sharp)^O = \sum_{H,\theta} n_{H,\theta} \gamma^\sharp(\phi)$. To see this, fix $\psi \sigma \in O$, $B \in Q$ and let $i = i_{\mathcal{P}}(B)$. We need to show for any $d \in [D]$ with $i^d = i$ that $|G_B^{d*}(\psi \sigma)| = \sum_{H,\theta} n_{H,\theta} |H_B^{d*}(\theta \pi_0^{-1} \sigma)|$, i.e. $|G^d(\psi_0 \pi^{-1})| = \sum_{H,\theta} n_{H\theta} |H^d(\theta \pi^{-1})|$, where $\pi^{-1} = \pi_0^{-1} \sigma \pi_B$ with $\pi_B : [r] \to B$ order-preserving; this is a case of the previous identity. □

8 Perspectives

The existence of designs established in [10] has seen several subsequent applications, some of which are particularly instructive as they require not only the existence but also that designs can be 'almost entirely random', in that the semi-random (nibble) construction of approximate designs by Rödl [24] can be completed to an actual design by an absorption process (Randomised Algebraic Construction in [10] or Iterative Absorption in [4]). In this vein, we mention the proof by Kwan [15] that almost all Steiner triple systems have perfect matchings, results on discrepancy of high-dimensional permutations by Linial and Luria [17], and the existence of bounded degree coboundary expanders of every dimension by Lubotzky, Luria and Rosenthal [19]. These results suggest that the new results in [11] may create more fruitful connections with the theory of high-dimensional expanders and other topics in high-dimensional combinatorics.

In Design Theory, the most fundamental problems that remain open are those concerning designs with large block sizes. Here we recall from the introduction the Prime Power Conjecture on projective planes, where we know that the divisibility conditions do not always suffice; the conjecture seems to reflect a philosophy that a combinatorial description of a sufficient rich structure somehow implies an algebraic characterisation. On the other hand, a conjecture that reflects the opposite philosophy is that Hadamard matrices (see [7]) of order n should exist whenever the trivially necessary conditions are satisfied (i.e. n is 1, 2 or divisible by 4). It is not clear how the methods of [4, 5, 10, 11] could apply to such problems, where a more fruitful direction may be the development of the approach of [14], which can allow for large block sizes. There are also many well-known open problems in Design Theory that do not involve large block sizes, and so may be more approachable by absorption techniques. Here we mention Ryser's Conjecture [25] that every Latin square of odd order should have a transversal; equivalently, any triangle decomposition of $K_3(n)$ for n odd should contain a triangle factor (perfect matching of triangles).

In Combinatorics, there are several natural directions in which one may seek to generalise the existence of various types of design, from extremal and/or probabilistic perspectives. A basic class of extremal questions is to determine the minimum degree threshold (which has various possible definitions) for decompositions (see e.g. [5, 21]). Natural probabilistic directions are thresholds for the existence of certain designs in random hypergraphs (e.g. Steiner Triple Systems in $G^3(n, p)$) or a theory of Random Designs analogous to the rich theory of Random Graphs.

Acknowledgements I would like to thank an anonymous referee for very detailed and helpful comments on the presentation of this paper.

References

1. P. Bennett and T. Bohman, A natural barrier in random greedy hypergraph matching, arXiv:1210.3581.
2. T. Bohman, A. Frieze and E. Lubetzky, Random triangle removal, *Adv. Math.* 280:379–438 (2015).
3. C. J. Colbourn and J. H. Dinitz, *Handbook of Combinatorial Designs*, 2nd ed. Chapman & Hall / CRC, Boca Raton, 2006.
4. S. Glock, D. Kühn, A. Lo and D. Osthus, The existence of designs via iterative absorption, arXiv:1611.06827.
5. S. Glock, D. Kühn, A. Lo and D. Osthus, Hypergraph F-designs for arbitrary F, arXiv:1706.01800.
6. J. E. Graver and W. B. Jurkat, The module structure of integral designs, *J. Combin. Theory Ser. A* 15:75–90, 1973.
7. J. Hadamard, Résolution d'une question relative aux déterminants, *Bull. des Sciences Math.* 17:240–246, 1893.
8. G. H. Hardy, J. E. Littlewood and G. Pólya, *Inequalities*, Cambridge University Press, 1952.
9. A. Hartman, Software and hardware testing using combinatorial covering suites, in: *Graph Theory, Combinatorics and Algorithms: Interdisciplinary Applications*, Springer, 237–266, 2005.
10. P. Keevash, The existence of designs, arXiv:1401.3665.
11. P. Keevash, The existence of designs II, arXiv:1802.05900.
12. P. Keevash, Counting designs, to appear in *J. Eur. Math. Soc.*
13. P. Keevash and K. Staden, The generalised Oberwolfach problem, manuscript.
14. G. Kuperberg, S. Lovett and R. Peled, Probabilistic existence of regular combinatorial objects, *Geom. Funct. Anal.* 27:919–972 (2017). Preliminary version in *Proc. 44th ACM STOC* (2012).
15. M. Kwan, Almost all Steiner triple systems have perfect matchings, arXiv:1611.02246.
16. N. Linial and Z. Luria, An upper bound on the number of high-dimensional permutations, *Combinatorica*, 34:471–486, 2014.
17. N. Linial and Z. Luria, Discrepancy of high-dimensional permutations, *Discrete Analysis* 2016:11, 8pp.
18. S. Lovett, S. Rao and A. Vardy, Probabilistic Existence of Large Sets of Designs, arXiv:1704.07964.
19. A. Lubotzky, Z. Luria and R. Rosenthal, Random Steiner systems and bounded degree coboundary expanders of every dimension, arXiv:1512.08331.
20. Z. Luria, New bounds on the number of n-queens configurations, arXiv:1705.05225.
21. R. Montgomery, Fractional clique decompositions of dense graphs, arXiv:1711.03382.
22. R. Montgomery, A. Pokrovskiy and B. Sudakov, Embedding rainbow trees with applications to graph labelling and decomposition, arXiv:1803.03316.
23. D.K. Ray-Chaudhuri and R.M. Wilson, Solution of Kirkman's schoolgirl problem, *Proc. Sympos. Pure Math.*, American Mathematical Society, XIX:187–203 (1971).
24. V. Rödl, On a packing and covering problem, *Europ. J. Combin.* 6:69–78 (1985).
25. H. Ryser, Neuere Probleme in der Kombinatorik, Vortrage über Kombinatorik, Oberwolfach, 69–91 (1967).

26. L. Teirlinck, Non-trivial t-designs without repeated blocks exist for all t, *Discrete Math.* 65:301–311 (1987).
27. L. Teirlinck, A completion of Lu's determination of the spectrum for large sets of disjoint Steiner triple systems, *J. Combin. Theory Ser. A* 57:302–305, 1991.
28. J. H. van Lint and R. M. Wilson, *A course in combinatorics*, Cambridge University Press, 2001.
29. C.M. Swanson and D.R. Stinson, Combinatorial solutions providing improved security for the generalized Russian cards problem, *Des. Codes Cryptogr.* 72:345–367, 2014.
30. R. Wilson, The early history of block designs, *Rend. del Sem. Mat. di Messina* 9:267–276 (2003).
31. R. M. Wilson, An existence theory for pairwise balanced designs I. Composition theorems and morphisms, *J. Combin. Theory Ser. A* 13:220–245 (1972).
32. R. M. Wilson, An existence theory for pairwise balanced designs II. The structure of PBD-closed sets and the existence conjectures, *J. Combin. Theory Ser. A* 13:246–273 (1972).
33. R. M. Wilson, An existence theory for pairwise balanced designs III. Proof of the existence conjectures, *J. Combin. Theory Ser. A* 18:71–79 (1975).
34. R. M. Wilson, The necessary conditions for t-designs are sufficient for something, *Utilitas Math.* 4:207–215 (1973).
35. R. M. Wilson, Nonisomorphic Steiner Triple Systems, *Math. Zeit.* 135:303–313 (1974).
36. R. M. Wilson, A diagonal form for the incidence matrices of t-subsets *vs.* k-subsets, *Europ. J. Combin* 11:609–615 (1990).

Efficient Convex Optimization with Oracles

Yin Tat Lee, Aaron Sidford and Santosh S. Vempala

Abstract Minimizing a convex function over a convex set is a basic algorithmic problem. We give a simple algorithm for the general setting in which the function is given by an evaluation oracle and the set by a membership oracle. The algorithm takes $\widetilde{O}(n^2)$ oracle calls and $\widetilde{O}(n^3)$ additional arithmetic operations. This results in more efficient reductions among the five basic oracles for convex sets and functions defined by Grötschel, Lovász and Schrijver (Algorithms Comb 2, (1988), [5]).

Keywords Convex optimization · Separation · Membership · Reductions

Subject Classifications 68W20 · 90C25

1 Introduction

Minimizing a convex function over a convex set is a basic algorithmic problem with special cases of independent interest motivated by a variety of applications. The classic book by Grötschel, Lovász and Schrijver [5] explored the problem and its applications in a very general setting, when the input instance is specified by one of a set of oracles. They show how convex optimization can be solved using variants of the Ellipsoid method, with only a polynomial number of calls to an oracle for the input set and a polynomial number of additional arithmetic operations. They use

---

Y. T. Lee
Paul G. Allen School of Computer Science and Engineering, University of Washington, 185 Stevens Way, CS 101, Seattle, WA 98195, USA
e-mail: yintat@uw.edu

A. Sidford
MS&E, Huang Engineering Center, 475 Via Ortega, Stanford, CA 94305, USA
e-mail: sidford@stanford.edu

S. S. Vempala (✉)
School of Computer Science, Georgia Tech, 266 Ferst Drive, Klaus 2222,
30306 Atlanta, GA, Georgia
e-mail: vempala@gatech.edu

this approach to establish polynomial-time algorithms for a range of combinatorial optimization problems. In the decades since then, the methods developed in this book have served as a foundation for the field of algorithmic complexity and continue to do so. Convex optimization today stands at the forefront of polynomial-time tractability. Recent algorithmic improvements to important problems (e.g., maxflow, submodular function minimization) have been closely tied to ideas and improvements for linear and convex optimization [4, 7–11, 13, 16, 17].

The motivation of this paper is to understand the best possible complexity of convex optimization in the setting of oracles. While [5] provided polynomial-time reductions, some reductions entail rather high degree polynomials, and our goal is to improve them. In particular, our focus is on the complexity of optimization using a *membership* oracle for the feasible set of solutions and an *evaluation* oracle for the objective function. The input is given by query access to the membership and evaluation oracles, along with bounds $0 < r < R$, and a point $x_0 \in K$ s.t. $B(x_0, r) \subseteq K \subseteq B(x_0, R)$, where $B(x_0, r)$ denotes the ball of radius r centered at $x_0 \in \mathbb{R}^n$. The reduction in [5] from linear optimization to membership appears to take $\Omega\left(n^{10}\right)$ calls to the membership oracle. This bound was improved using the random walk method and simulated annealing to $n^{4.5}$ [6, 12] (for structured convex sets, [1] provides improvements of up to a factor of $\sqrt{n}$.). On the other hand, it is well-known that with a *separation* oracle for the set (and a subgradient oracle for the function), this problem can be solved with $\tilde{O}(n)$ oracle queries using any of [2, 11, 19], which improve on $\tilde{O}(n^2)$ query complexity obtained via the Ellipsoid algorithm [5]. Nesterov and Spokoiny [14] have also given an algorithm with quadratic dependence on the dimension, but a polynomial dependence on the error parameter.

Our main result in this paper is a randomized algorithm that minimizes a convex function over a convex set using $\tilde{O}(n^2)$ membership and evaluation queries. We obtain this result implementing a separation oracle for a convex set (and a subgradient oracle for a convex function) using only $\tilde{O}(n)$ membership and evaluation queries (Sect. 3) and then using the known reduction from optimization to separation (Sect. 4). We state the result informally below. The formal statements, which allow an approximate membership oracle, are Theorems 3.6 and 4.1.

Theorem 1.1 *Let K be a convex set specified by a membership oracle, a point $x_0 \in \mathbb{R}^n$, and numbers $0 < r < R$ such that $B(x_0, r) \subseteq K \subseteq B(x_0, R)$. For any convex function f given by an evaluation oracle and any $\epsilon > 0$, there is a randomized algorithm that computes a point $z \in B(K, \epsilon)$ such that.*

$$f(z) \leq \min_{x \in K} f(x) + \epsilon \left(\max_{x \in K} f(x) - \min_{x \in K} f(x) \right)$$

with constant probability using $O\left(n^2 \log^2 \left(\frac{nR}{\epsilon r}\right)\right)$ calls to the membership oracle and evaluation oracle and $O(n^3 \log^{O(1)} \left(\frac{nR}{\epsilon r}\right))$ total arithmetic operations. □

Efficient Convex Optimization with Oracles

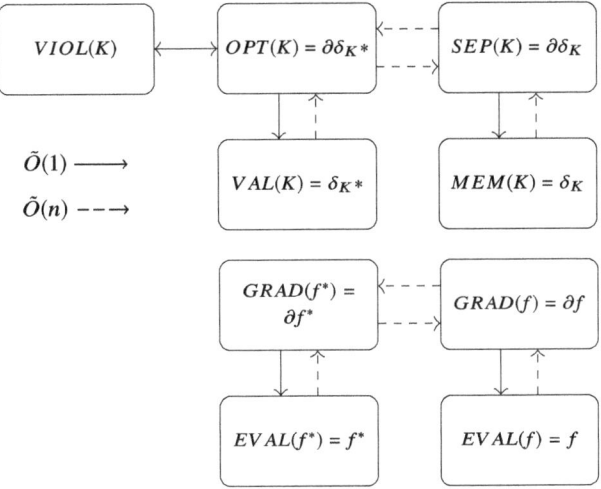

Fig. 1 The top diagram illustrates the relationships of the five oracles defined in [5]. The bottom diagram illustrates the relationships of oracles for a convex function f and its convex conjugate f^*

Protasov [15] gives an algorithm for approximately minimizing a convex function defined over an explicit convex body in $\mathbb{R}^n$, using $O(n^2 \log(n) \log(1/\varepsilon))$ function evaluations. Each iteration of his algorithm requires computing the convex hull, the John ellipsoid and the centroid of a set maintained by the algorithm, which would result in a very large number of calls to the membership oracle (in [15], the focus is on the number of function calls and it is assumed that the set is known to the algorithm). We remark that using the main idea from our algorithm, Protasov's method can be made more efficient, resulting in oracle complexity that is only a logarithmic factor higher, although still with a much higher arithmetic complexity than the results of this paper.

In Sect. 5 we develop two consequences of our main result. Grötschel, Lovász and Schrijver [5] define five basic problems over convex sets as oracles (OPTimization, SEParation, MEMbership, VIOLation and VALidity, see Sect. 2) and give polynomial-time reductions between them. With our new algorithm, several of these reductions become significantly more efficient, as summarized in Theorem 5.6. In discussing these reductions, it is natural to introduce oracles for convex functions. The relationships between set oracles and function oracles are described in Lemma 5.4 and those between function oracles in Lemma 5.5. Figure 1 illustrates these relationships and is an updated version of Fig. 4.1 from [5]. We suspect that the resulting complexities of reductions are all asymptotically optimal in terms of the dimension, up to logarithmic factors.

2 Preliminaries

Our notation and conventions are chosen for simplicity and consistency with Grötschel, Lov'asz and Schrijver [5]. We use $[n] \stackrel{\text{def}}{=} \{1, ..., n\}$. For a convex function $f : \mathbb{R}^n \to \mathbb{R}$ and $x \in \mathbb{R}^n$ we use $\partial f(x)$ to denote the set of subgradients of f at x:

$$\partial f(x) = \left\{ v : \quad \forall y, \ f(y) - f(x) \geq v^T(y - x) \right\}.$$

For $p \geq 1, \delta \geq 0$, and $K \subseteq \mathbb{R}^n$ we denote the set of points at distance at most δ from K in ℓ_p norm as follows:

$$B_p(K, \delta) \stackrel{\text{def}}{=} \left\{ x \in \mathbb{R}^n : \exists y \in K \text{ such that } \|x - y\|_p \leq \delta \right\}$$

For convenience, we overload notation by letting $B_p(x, \delta) \stackrel{\text{def}}{=} B_p(\{x\}, \delta)$ denote the ball of radius δ around x. We also define

$$B_p(K, -\delta) \stackrel{\text{def}}{=} \{x \in \mathbb{R}^n : B_p(x, \delta) \subseteq K\},$$

i.e., the set of points such that the δ radius balls around them are contained in K. Whenever p is omitted it is assumed that $p = 2$. Moreover, for any set $K \subseteq \mathbb{R}^n$, we let 1_K denote a function from $\mathbb{R}^n$ to $\mathbb{R} \cup \{+\infty\}$ such that $1_K(x) = 0$ if $x \in K$ and $1_K(x) = \infty$ otherwise.

We use $*$ to denote the convolution operator, i.e.

$$(f * g)(x) = \int_{\mathbb{R}^n} f(y) g(x - y) dy.$$

2.1 Oracles for Convex Sets

We recall the five basic oracles for a convex set, $K \subseteq \mathbb{R}^n$ defined in [5].

Definition 2.1 (*Optimization Oracle (OPT)*) Queried with a unit vector $c \in \mathbb{R}^n$ and real numbers $\delta, \delta' > 0$, with probability $1 - \delta'$, the oracle either

- finds a vector $y \in \mathbb{R}^n$ such that $y \in B(K, \delta)$ and $c^T x \leq c^T y + \delta$ for all $x \in B(K, -\delta)$, or
- asserts that $B(K, -\delta)$ is empty.

We let $\text{OPT}_{\delta, \delta'}(K)$ be the time complexity of this oracle. □

Definition 2.2 (*Violation Oracle (VIOL)*) Queried with a unit vector $c \in \mathbb{R}^n$, a real number γ and real numbers $\delta, \delta' > 0$, with probability $1 - \delta'$, the oracle either

- asserts that $c^T x \leq \gamma + \delta$ for all $x \in B(K, -\delta)$, or

- finds a vector $y \in B(K, \delta)$ with $c^T y \geq \gamma - \delta$.

We let $\text{VIOL}_{\delta,\delta'}(K)$ be the time complexity of this oracle. □

Definition 2.3 (*Validity Oracle (VAL)*) Queried with a unit vector $c \in \mathbb{R}^n$, a real number γ, and real numbers $\delta, \delta' > 0$, with probability $1 - \delta'$, the oracle either

- asserts that $c^T x \leq \gamma + \delta$ for all $x \in B(K, -\delta)$, or
- asserts that $c^T x \geq \gamma - \delta$ for some $x \in B(K, \delta)$.

We let $\text{VAL}_{\delta,\delta'}(K)$ be the time complexity of this oracle. □

Definition 2.4 (*Separation Oracle (SEP)*) Queried with a vector $y \in \mathbb{R}^n$ and real numbers $\delta, \delta' > 0$, with probability $1 - \delta'$, the oracle either

- assert that $y \in B(K, \delta)$, or
- find a unit vector $c \in \mathbb{R}^n$ such that $c^T x \leq c^T y + \delta$ for all $x \in B(K, -\delta)$.

We let $\text{SEP}_{\delta,\delta'}(K)$ be the time complexity of this oracle. □

Definition 2.5 (*Membership Oracle (MEM)*) Queried with a vector $y \in \mathbb{R}^n$ and real numbers $\delta, \delta' > 0$, with probability $1 - \delta'$, either

- assert that $y \in B(K, \delta)$, or
- assert that $y \notin B(K, -\delta)$.

We let $\text{MEM}_{\delta,\delta'}(K)$ be the time complexity of this oracle. □

In the main reductions, we simplify notation by using $\delta = \delta'$, i.e., the same parameter $\delta > 0$ to denote both the approximation error and the probability of failure.

2.2 Oracles for Convex Functions

Let f be a function from $\mathbb{R}^n$ to $\mathbb{R} \cup \{+\infty\}$. Recall that the dual function f^* is the convex (Fenchel) conjugate of f, defined as

$$f^*(y) = \sup_{x \in \mathbb{R}^n} \langle y, x \rangle - f(x).$$

In particular $f^*(0) = \inf f$. We will use the following two oracles for functions.

Definition 2.6 (*Evaluation Oracle (EVAL)*) Queried with a vector y with $\|y\|_2 \leq 1$ and real numbers $\delta, \delta' > 0$, with probability $1 - \delta'$, the oracle finds an extended real number α such that

$$\min_{x \in B(y,\delta)} f(x) - \delta \leq \alpha \leq \max_{x \in B(y,\delta)} f(x) + \delta. \tag{1}$$

We let $\text{EVAL}_\delta(f)$ be the time complexity of this oracle. □

Algorithm 1: SubgradConvexFunc(f, x, r_1, ε)

Require: $r_1 > 0$, $\|\partial f(z)\|_\infty \leq L$ for any $z \in B_\infty(x, 2r_1)$.

Set $r_2 = \sqrt{\frac{\varepsilon r_1}{\sqrt{n}L}}$.

Sample $y \in B_\infty(x, r_1)$ and $z \in B_\infty(y, r_2)$ independently and uniformly at random.

for $i = 1, 2, \cdots, n$ **do**
 Let α_i and β_i denote the end points of the interval $B_\infty(y, r_2) \cap \{z + se_i : s \in \mathbb{R}\}$.
 Set $\tilde{g}_i = \frac{f(\beta_i) - f(\alpha_i)}{2r_2}$ where we compute f to withing ε additive error.
end

Output $\tilde{g}$ as the approximate subgradient of f at x.

Definition 2.7 (*Subgradient Oracle (GRAD)*) Queried with a vector y with $\|y\|_2 \leq 1$ and real numbers $\delta, \delta' > 0$, with probability $1 - \delta'$, the oracle outputs an extended real number α satisfying (1) and a vector $c \in \mathbb{R}^n$ such that

$$\alpha + c^T(x - y) < \max_{z \in B(x, \delta)} f(z) + \delta \text{ for all } x \in \mathbb{R}^n \qquad (2)$$

We let $\text{GRAD}_\delta(f)$ be the time complexity of this oracle. □

3 From Membership to Separation

In this section, we show that how to implement a separation oracle for a convex set using a nearly linear number of queries to a membership oracle. We divide this into two steps. In the first step (Sect. 3.1), we compute an approximate subgradient of a given Lipshitz convex function. Using this, in the second step (Sect. 3.2), we compute an approximate separating hyperplane. The algorithms are stated in Algorithms 1 and 2. In the algorithms below when we refer to $\|\partial f(z)\|$, we mean the supremum over all elements of the set $\partial f(z)$. The parameter ρ below is a bound on the probability of failure.

The output of the algorithm for separation is a halfspace that approximately contains K, and the input point x is close to its bounding hyperplane. It uses a call to the subgradient function above.

3.1 Subgradient for a Lipschitz Convex Function

Here we show how to construct an approximate subgradient (or approximate separation) for a Lipschitz convex function given an evaluation oracle. Our construction is motivated by the following property of convex functions proved by Bubeck and Eldan [3, Lemma 6]: for any Lipschitz convex function f, there exists a small ball B such that f restricted to B is close to a linear function. By a small modification of their proof, we show this property in fact holds for almost every small ball (Lemma 3.1).

Efficient Convex Optimization with Oracles

Algorithm 2: Separate$_{\varepsilon,\rho}(K, x)$

Require: $B_2(0, r) \subset K \subset B_2(0, R)$.
if $MEM_\varepsilon(K)$ asserts that $x \in B(K, \epsilon)$ **then**
 | **Output:** "$x \in B(K, \varepsilon)$".
else if $x \notin B_2(0, R)$ **then**
 | **Output:** the halfspace $\{y : 0 \geq \langle y - x, x \rangle\}$.
end
Let $\kappa = R/r$, $\alpha_x(d) = \max_{d + \alpha x \in K} \alpha$ and $h_x(d) = -\alpha_x(d) \|x\|_2$.
The evaluation oracle of $\alpha_x(d)$ can be implemented via binary search and $MEM_\varepsilon(K)$.
Compute $\tilde{g} = \text{SubgradConvexFunc}(h_x, 0, r_1, 4\varepsilon)$ with $r_1 = n^{1/6} \varepsilon^{1/3} R^{2/3} \kappa^{-1}$ and the evaluation oracle of $\alpha_x(d)$.
Output: the halfspace
$$\left\{ y : \frac{56}{\rho} n^{7/6} R^{2/3} \kappa \varepsilon^{1/3} \geq \langle \tilde{g}, y - x \rangle \right\}$$

Leveraging this powerful fact, our algorithm is simple: we compute a random partial difference in each coordinate to get a subgradient (Algorithm 1). We prove that as long as the box we compute over is sufficiently small and the additive error of our evaluation oracle is sufficiently small, this yields an accurate separation/subgradient oracle in expectation (Lemma 3.2). We then obtain high probability bounds using Markov's inequality.

Lemma 3.1 *Let $0 < r_2 \leq r_1$ and f be a twice-differentiable convex function with $\|\nabla f(z)\|_\infty \leq L$ for any $z \in B_\infty(x, r_1 + r_2)$. For $y \in B_\infty(x, r_1)$, let $g(y) = \mathbb{E}_{w \sim B_\infty(y, r_2)} (\nabla f(w))$ be the average of the gradient of f over over $B_\infty(y, r_2)$. Then,*

$$\mathbb{E}_{y \in B_\infty(x, r_1)} \mathbb{E}_{z \in B_\infty(y, r_2)} \|\nabla f(z) - g(y)\|_1 \leq n^{3/2} \frac{r_2}{r_1} L.$$

Proof Let $h = \frac{1}{(2r_2)^n} f * \chi_{B_\infty(0, r_2)}$ where $\chi_{B_\infty(0, r_2)}$ is 1 on the set $B_\infty(0, r_2)$ and 0 outside. Using the divergence theorem, we have that

$$\int_{B_\infty(x, r_1)} \Delta h(y) dy = \int_{\partial B_\infty(x, r_1)} \langle \nabla h(y), n(y) \rangle \, dy$$

where $\Delta h(y) = \sum_i \frac{\partial^2}{\partial x_i^2} h(y)$ and $n(y)$ is the normal vector on $\partial B_\infty(x, r_1)$ the boundary of the box $B_\infty(x, r_1)$, i.e. standard basis vectors. Since f is L-Lipschitz with respect to $\|\cdot\|_\infty$ so is h, i.e. $\|\nabla h(z)\|_\infty \leq L$. Hence, we have that

$$\mathbb{E}_{y \in B_\infty(x, r_1)} \Delta h(y) \leq \frac{1}{(2r_1)^n} \int_{\partial B_\infty(x, r_1)} \|\nabla h(y)\|_\infty \|n(y)\|_1 \, dy$$
$$\leq \frac{1}{(2r_1)^n} \cdot 2n(2r_1)^{n-1} \cdot L$$
$$= \frac{nL}{r_1}.$$

By the definition of h, we have that

$$\mathbb{E}_{y \in B_\infty(x,r_1)} \mathbb{E}_{z \in B_\infty(y,r_2)} \Delta f(z) = \mathbb{E}_{y \in B_\infty(x,r_1)} \Delta h(y) \leq \frac{nL}{r_1}. \tag{3}$$

Let $\omega_i(z) = \langle \nabla f(z) - g(y), e_i \rangle$ for all $i \in [n]$. Since $\int_{B_\infty(y,r_2)} \omega_i(z) dz = 0$, the Poincare inequality for a box (see e.g. [18]) shows that

$$\int_{B_\infty(y,r_2)} |\omega_i(z)| \, dz \leq r_2 \int_{B_\infty(y,r_2)} \|\nabla \omega_i(z)\|_2 \, dz.$$

Since f is convex, we have that $\|\nabla^2 f(z)\|_F \leq \text{Tr} \nabla^2 f(z) = \Delta f(z)$ and hence

$$\sum_{i \in [n]} \|\nabla \omega_i(z)\|_2 = \sum_{i \in [n]} \|\nabla^2 f(z) e_i\|_2 \leq \sqrt{n} \|\nabla^2 f(z)\|_F \leq \sqrt{n} \Delta f(z).$$

Using this with $\|\nabla f(z) - g(y)\|_1 = \sum_i |\omega_i(z)|$, we have that

$$\int_{B_\infty(y,r_2)} \|\nabla f(z) - g(y)\|_1 \, dz \leq \sqrt{n} r_2 \int_{B_\infty(y,r_2)} \Delta f(z) dz.$$

Combining with the inequality (3) completes the proof. □

With the above fact in hand, asserting that on average, the gradient is approximated by its average in a small ball, we now proceed to construct an approximate subgradient.

Lemma 3.2 *Let $r_1 > 0$ and f be a convex function. Suppose that $\|\nabla f(z)\|_\infty \leq L$ for any $z \in B_\infty(x, 2r_1)$ and suppose that we can evaluate f to within ε additive error for $\varepsilon \leq r_1 \sqrt{n} L$. Let $\tilde{g} = \text{SubgradConvexFunc}(f, x, r_1, \varepsilon)$. Then, there is random variable $\zeta \geq 0$ with $\mathbb{E}\zeta \leq 2\sqrt{\frac{L\varepsilon}{r_1}} n^{5/4}$ such that for any q*

$$f(q) \geq f(x) + \langle \tilde{g}, q - x \rangle - \zeta \|q - x\|_\infty - 8nr_1 L.$$

Proof We assume that f is twice differentiable. For general f, we can reduce to this case by viewing it as a limit of twice-differentiable functions.

First, we assume that we can compute f exactly, namely $\varepsilon = 0$. Fix $i \in [n]$. Let $g(y)$ be the average of ∇f over $B_\infty(y, r_2)$. Then, for the function $\tilde{g}$ computed by the algorithm, we have that

$$\mathbb{E}_z |\tilde{g}_i - g(y)_i| = \mathbb{E}_z \left| \frac{f(\beta_i) - f(\alpha_i)}{2r_2} - g(y)_i \right|$$

$$\leq \mathbb{E}_z \frac{1}{2r_2} \int \left| \frac{df}{dx_i}(z + se_i) - g(y)_i \right| ds$$

$$= \mathbb{E}_z \left| \frac{df}{dx_i}(z) - g(y)_i \right|$$

where we used that both $z + se_i$ and z are uniform distribution on $B_\infty(y, r_2)$ in the last line. Hence, we have

$$\mathbb{E}_z \|\tilde{g} - \nabla f(z)\|_1 \leq \mathbb{E}_z \|\nabla f(z) - g(y)\|_1 + \mathbb{E}_z \|\tilde{g} - g(y)\|_1 \leq 2\mathbb{E}_z \|\nabla f(z) - g(y)\|_1.$$

Now, applying the convexity of f yields that

$$\begin{aligned}f(q) &\geq f(z) + \langle \nabla f(z), q - z \rangle \\ &= f(z) + \langle \tilde{g}, q - x \rangle + \langle \nabla f(z) - \tilde{g}, q - x \rangle + \langle \nabla f(z), x - z \rangle \\ &\geq f(z) + \langle \tilde{g}, q - x \rangle - \|\nabla f(z) - \tilde{g}\|_1 \|q - x\|_\infty - \|\nabla f(z)\|_\infty \|x - z\|_1.\end{aligned}$$

Now, using f is L-Lipschitz between x and z, we have that $f(z) \geq f(x) - L \cdot \|x - z\|_1$. Hence, we have

$$f(q) \geq f(x) + \langle \tilde{g}, q - x \rangle - \|\nabla f(z) - \tilde{g}\|_1 \|q - x\|_\infty - 2L \|x - z\|_1.$$

Note that $\|x - z\|_1 \leq n \cdot \|x - z\|_\infty \leq 2n(r_1 + r_2)$ by assumption. Moreover, we can apply Lemma 3.1 to bound $\|\nabla f(z) - \tilde{g}\|_1$ and use $r_2 = \sqrt{\frac{\varepsilon r_1}{\sqrt{n}L}} \leq r_1$ to get

$$f(q) \geq f(x) + \langle \tilde{g}, q - x \rangle - \zeta \|q - x\|_\infty - 8nr_1 L$$

with $\mathbb{E}\zeta \leq 2n^{3/2} \frac{r_2}{r_1} L$.

Since we only compute f up to ε additive error, it introduces $\frac{\varepsilon}{2r_2}$ additive error in $\tilde{g}_i$. Hence, we instead have that

$$\mathbb{E}\zeta \leq 2n^{3/2} \frac{r_2}{r_1} L + \frac{\varepsilon n}{2r_2}.$$

Setting $r_2 = \frac{1}{2}\sqrt{\frac{\varepsilon r_1}{\sqrt{n}L}}$ completes the proof. □

3.2 Separation for a Convex Set

Throughout this subsection, let $K \subseteq \mathbb{R}^n$ be a convex set that contains $B_2(0, r)$ and is contained in $B_2(0, R)$. Given some point $x \notin K$, we wish to separate x from K using a halfspace. To do this, we reduce this problem to computing an approximate subgradient of a Lipschitz convex function $h_x(d)$ defined for points in K. Roughly speaking, it is the "height" (or distance from the boundary) of a point d in the direction of x. Let

$$\alpha_x(d) = \max_{d + \alpha x \in K} \alpha \quad \text{and} \quad h_x(d) = -\alpha_x(d) \|x\|_2.$$

Fig. 2 The convex height function h_x

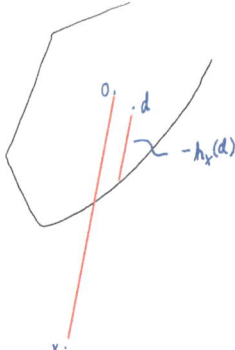

Note that $d + \alpha_x(d)x$ is the last point in K on the line through $d \in K$ in the direction of x, and $-h_x(d)$ is the ℓ_2 distance from this boundary point to d (Fig. 2).

Lemma 3.3 $h_x(d)$ *is convex over* K. □

Proof Let $d_1, d_2 \in K$ and $\lambda \in [0, 1]$. Now $d_1 + \alpha_x(d_1)x \in K$ and $d_2 + \alpha_x(d_2)x \in K$. Consequently,

$$[\lambda d_1 + (1 - \lambda)d_2] + [\lambda \cdot \alpha_x(d_1) + (1 - \lambda) \cdot \alpha_x(d_2)] x \in K.$$

Therefore, if we let $d \stackrel{\text{def}}{=} \lambda d_1 + (1 - \lambda)d_2$ we see that $\alpha_x(d) \geq \lambda \cdot \alpha_x(d_1) + (1 - \lambda) \cdot \alpha_x(d_2)$ and $h_x(\lambda d_1 + (1 - \lambda d_2) \leq \lambda h_x(d_1) + \lambda h_x(d_2)$ as claimed. □

Lemma 3.4 h_x *is* $\left(\frac{R+\delta}{r-\delta}\right)$-*Lipschitz over points in* $B_2(0, \delta)$ *for any* $\delta < r$. □

Proof Let d_1, d_2 be arbitrary points in $B(0, \delta)$. We wish to upper bound $|h_x(d_1) - h_x(d_2)|$ in terms of $\|d_1 - d_2\|_2$. We assume without loss of generality that $\alpha_x(d_1) \geq \alpha_x(d_2)$ and therefore

$$|h_x(d_1) - h_x(d_2)| = |\alpha_x(d_1) \|x\|_2 - \alpha_x(d_2) \|x\|_2| = (\alpha_x(d_1) - \alpha_x(d_2)) \|x\|_2.$$

Consequently, it suffices to lower bound $\alpha_x(d_2)$. We split the analysis into two cases.

Case 1: $\|d_2 - d_1\|_2 \geq r - \delta$. Since $0 \geq h_x(d_1), h_x(d_2) \geq -R - \delta$, we have that

$$|h_x(d_1) - h_x(d_2)| \leq R + \delta \leq \frac{R + \delta}{r - \delta} \|d_2 - d_1\|_2.$$

Case 2: $\|d_2 - d_1\|_2 \leq r - \delta$. We consider the point $d_3 = d_1 + \frac{d_2 - d_1}{\lambda}$ with $\lambda = \|d_2 - d_1\|_2 /(r - \delta)$. Note that

$$\|d_3\|_2 \leq \|d_1\|_2 + \frac{1}{\lambda} \|d_2 - d_1\|_2 \leq \delta + \frac{1}{\lambda} \|d_2 - d_1\|_2 \leq r.$$

Hence, $d_3 \in K$. Since $\lambda \in [0, 1]$ and K is convex, we have that $\lambda \cdot d_3 + (1 - \lambda) \cdot [d_1 + \alpha_x(d_1)x] \in K$. Now, we note that

$$\lambda \cdot d_3 + (1 - \lambda) \cdot [d_1 + \alpha_x(d_1)x] = d_2 + (1 - \lambda) \cdot \alpha_x(d_1)x$$

and this shows that

$$\alpha_x(d_2) \geq (1 - \lambda) \cdot \alpha_x(d_1) = \left(1 - \frac{\|d_2 - d_1\|_2}{r - \delta}\right) \cdot \alpha_x(d_1).$$

Since $d_1 + \alpha_x(d_1)x \in K \subset B_2(0, R)$, we have that $\alpha_x(d_1) \cdot \|x\|_2 \leq R + \delta$ and hence

$$\begin{aligned}
|h_x(d_1) - h_x(d_2)| &= (\alpha_x(d_1) - \alpha_x(d_2)) \cdot \|x\|_2 \\
&\leq \alpha_x(d_1) \cdot \|x\|_2 \frac{\|d_2 - d_1\|_2}{r - \delta} \\
&\leq \frac{R + \delta}{r - \delta} \|d_2 - d_1\|_2.
\end{aligned}$$

In either case, as claimed we have

$$|h_x(d_1) - h_x(d_2)| \leq \frac{R + \delta}{r - \delta} \|d_2 - d_1\|_2.$$

Lemma 3.5 *Let K be a convex set satisfying $B_2(0, r) \subset K \subset B_2(0, R)$. Given any $0 < \rho < 1$ and $0 \leq \varepsilon \leq r$. With probability $1 - \rho$, $\mathtt{Separate}_{\varepsilon,\rho}(K, x)$ outputs a half space that contains K.* □

Proof When $x \notin B_2(0, R)$, the algorithm outputs a valid separation for $B_2(0, R)$. For the rest of the proof, we assume $x \notin B(K, -\varepsilon)$ (due to the membership oracle) and $x \in B_2(0, R)$.

By Lemmas 3.3 and 3.4, h_x is convex with Lipschitz constant 3κ on $B_2(0, \frac{r}{2})$. By our assumption on ε and our choice of r_1, we have that $B_\infty(0, 2r_1) \subset B_2(0, \frac{r}{2})$. Hence, we can apply Lemma 3.2 to get that

$$h_x(y) \geq h_x(0) + \langle \tilde{g}, y \rangle - \zeta \|y\|_\infty - 24nr_1\kappa \qquad (4)$$

for any $y \in K$. Note that $-\frac{x}{\kappa} \in K$ and $h_x(-\frac{x}{\kappa}) \geq h_x(0) - \frac{1}{\kappa}\|x\|_2$. Hence, we have

$$h_x(0) - \frac{1}{\kappa}\|x\|_2 = h_x(-\frac{1}{\kappa}x) \geq h_x(0) + \left\langle \tilde{g}, -\frac{1}{\kappa}x \right\rangle - \frac{1}{\kappa}\zeta \|x\|_\infty - 24nr_1\kappa.$$

Therefore, we have

$$\langle \tilde{g}, x \rangle \geq \|x\|_2 - \zeta \|x\|_\infty - 24nr_1\kappa^2. \qquad (5)$$

Now, we note that $x \notin B(K, -\varepsilon)$. Using that $B(0, r) \subset K$, we have

$$\left(1 - \frac{\varepsilon}{r}\right) K \subset B(K, -\varepsilon).$$

Hence,

$$h_x(0) \geq -\left(1 - \frac{\varepsilon}{r}\right) \|x\|_2 \geq -\|x\|_2 + \varepsilon \kappa.$$

Therefore, we have

$$h_x(0) + \langle \tilde{g}, x \rangle \geq -\zeta \|x\|_\infty - 24 n r_1 \kappa^2 - \varepsilon \kappa$$

Combining this with (4), we have that

$$\begin{aligned} h_x(y) &\geq \langle \tilde{g}, y - x \rangle - \zeta \|y\|_\infty - \zeta \|x\|_\infty - 24 n r_1 \kappa - 24 n r_1 \kappa^2 - \varepsilon \kappa \\ &\geq \langle \tilde{g}, y - x \rangle - 2\zeta R - 48 n r_1 \kappa^2 - \varepsilon \kappa \end{aligned}$$

for any $y \in K$. Recall from Lemma 3.2 that ζ is a positive random scalar independent of y satisfying $\mathbb{E}\zeta \leq 2\sqrt{\frac{3\kappa\varepsilon}{r_1}} n^{5/4}$. For any $y \in K$, we have that $h_x(y) \leq 0$ and hence $\tilde{\zeta} \geq \langle \tilde{g}, y - x \rangle$ where $\tilde{\zeta}$ is a random scalar independent of y satisfying

$$\begin{aligned} \mathbb{E}\tilde{\zeta} &\leq 4\sqrt{\frac{3\kappa\varepsilon}{r_1}} n^{5/4} R + 48 n r_1 \kappa^2 + \varepsilon \kappa \\ &\leq 55 n^{7/6} R^{2/3} \varepsilon^{1/3} \kappa + \varepsilon \kappa \\ &\leq 56 n^{7/6} R^{2/3} \varepsilon^{1/3} \kappa \end{aligned}$$

where we used $r_1 = n^{1/6} \varepsilon^{1/3} R^{2/3}/\kappa$ and $0 \leq \varepsilon \leq r$. The result then follows using Markov's inequality. $\square$

Theorem 3.6 *Let K be a convex set satisfying $B_2(0, 1/\kappa) \subset K \subset B_2(0, 1)$. For any $0 \leq \eta < \frac{1}{2}$, we have that*

$$SEP_\eta(K) \leq O\left(n \log\left(\frac{n\kappa}{\eta}\right)\right) MEM_{(\eta/n\kappa)^{O(1)}}(K).$$

Proof First, we bound the running time. Note that the bottleneck is to compute h_x with ε additive error. Since $-O(1) \leq h_x(y) \leq 0$ for all $y \in B_2(0, O(1))$, one can compute $h_x(y)$ by binary search with $O(\log(1/\delta))$ calls to the membership oracle.

Next, we check that $\texttt{Separate}_{\delta,\rho}(K, x)$ is indeed a separation oracle. Note that $\tilde{g}$ may not be an unit vector and we need to re-normalize the $\tilde{g}$ by $1/\|\tilde{g}\|_2$. So, we need to lower bound $\|\tilde{g}\|_2$.

From (5) and our choice of r_1, if $\delta \leq \frac{\rho^3}{10^6 n^6 \kappa^6}$, then we have that

$$\langle \tilde{g}, x \rangle \geq \|x\|_2 - \zeta \|x\|_\infty - 24 n r_1 \kappa^2 \geq \frac{r}{4}.$$

Hence, we have that $\|\tilde{g}\|_2 \geq \frac{1}{4\kappa}$. Therefore, this algorithm is a separation oracle with error $\frac{400}{\rho} n^{7/6} \kappa^2 \delta^{1/3}$ and failure probability $O(\rho + \log(1/\delta)\delta)$.

$$\text{SEP}_{\Omega(\max(n^{7/6}\kappa^2 \delta^{1/3}/\rho + \rho + \log(1/\delta)\delta)}(K) \leq O(\log(1/\delta)) \text{MEM}_\delta(K).$$

Setting $\rho = \sqrt{n^{7/6} \kappa^2 \delta^{1/3}}$ and $\delta = \Theta\left(\frac{\eta^6}{n^{7/2} \kappa^6}\right)$, we have that

$$\text{SEP}_\eta(K) \leq O\left(\log\left(\frac{n\kappa}{\eta}\right)\right) \text{MEM}_{\eta^6/(n^{7/2}\kappa^6)}(K).$$

4 From Separation to Optimization

Once we have a separation oracle, our running times follow by applying a recent convex optimization algorithm by Lee, Sidford and Wong [11]. Previous algorithms [2, 19] also achieved $\tilde{O}(n)$ oracle complexity, but with a higher polynomial number of arithmetic operations. We remark that the theorem stated in [11] is slightly more general then the one we give below, but since we only need to minimize linear functions over convex sets, we state a simplified version here.

Theorem 4.1 (Theorem 42 of [11] Rephrased) *Let K be a convex set satisfying $B_2(0, r) \subset K \subset B_2(0, 1)$ and let $\kappa = 1/r$. For any $0 < \varepsilon < 1$, with probability $1 - \varepsilon$, we can compute $x \in B(K, \varepsilon)$ such that*

$$c^T x \leq \min_{x \in K} c^T x + \varepsilon \|c\|_2$$

with an expected running time of

$$O\left(n \log\left(\frac{n\kappa}{\varepsilon}\right)\right) \text{SEP}_\delta(K) + O\left(n^3 \log^{O(1)}\left(\frac{n\kappa}{\varepsilon}\right)\right),$$

where $\delta = (\frac{\varepsilon}{n\kappa})^{\Theta(1)}$. In other words, we have that

$$OPT_\varepsilon(K) = O\left(n \log\left(\frac{n\kappa}{\varepsilon}\right)\right) \text{SEP}_{(\frac{\varepsilon}{n\kappa})^{\Theta(1)}}(K) + O\left(n^3 \log^{O(1)}\left(\frac{n\kappa}{\varepsilon}\right)\right).$$

5 Reductions Between Oracles

In this section, we provide all other reductions among oracles defined in Sect. 2.1. To simplify notation we assume the convex set is contained in the unit ball and the convex function is defined on the unit ball. This can be done without loss of generality by scaling and shifting.

We remark that it is known that OPT and VIOL are equivalent up to the cost of a binary search.

Lemma 5.1 (Equivalence between OPT and VIOL) *Given a convex set K contained in the unit ball, we have that $VIOL_\delta(K) \leq OPT_\delta(K)$ and $OPT_\delta(K) \leq O\left(\log\left(1 + \frac{1}{\delta}\right)\right) \cdot VIOL_\delta(K)$ for any $\delta > 0$.* □

5.1 Relationships Between Set Oracles and Function Oracles

Next, to handle all these relationships efficiently, we find it convenient to instead look at oracles for convex functions and connect them to oracles for sets. For this purpose we note the following simple relationship between MEM(K) and EVAL(1_K) and between SEP(K) and GRAD(1_K).

Lemma 5.2 (MEM(K) and SEP(K)) *are membership and subgradient oracle of 1_K] For any convex set $K \subseteq \mathbb{R}^n$, we have that $MEM_\delta(K) = EVAL_\delta(1_K)$ and $SEP_\delta(K) = GRAD_\delta(1_K)$ for any $\delta > 0$.* □

Next, we note that the relationship between VAL(K) and EVAL(1_K^*) and between OPT(K) and GRAD(1_K^*).

Lemma 5.3 (VAL(K) and OPT(K)) *are membership and subgradient oracle of 1_K^*] Let K be a convex set such that $B(\mathbf{0}, r) \subset K \subset B(\mathbf{0}, 1)$ and let $\kappa = 1/r$. For any $\delta > 0$, we have that*

- $VAL_\delta(K) \leq EVAL_\delta(1_K^*)$ and $EVAL_\delta(1_K^*) \leq O(\log(\kappa/\delta)) \cdot VAL_{\Omega(\delta/(\kappa \log(1/\delta)))}(K)$.
- $OPT_\delta(K) \leq GRAD_{\delta/4}(1_K^*)$ and $GRAD_\delta(1_K^*) \leq OPT_{\delta/(3+\kappa)}(K)$.

where the oracle for 1_K^* is only defined on the unit ball. □

Proof For the first inequality, to implement the validity oracle, we need to compute β such that

$$\max_{x \in B(K, -\delta)} c^T x - \delta \leq \beta \leq \min_{x \in B(K, \delta)} c^T x + \delta \qquad (6)$$

for any unit vector c and $\delta > 0$. We note that

$$\min_{x \in B(c, \frac{\delta}{R})} 1_K^*(x) \geq 1_K^*(c) - \delta = \max_{x \in K} c^T x - \delta \geq \max_{x \in B(K, -\delta)} c^T x.$$

Therefore, (1) shows that the output α by $\text{EVAL}_\delta(1_K^*)$ with input $-c$ satisfies $\max_{x \in B(K,-\delta)} c^T x \leq \alpha + \delta$. Similarly, we have that $\min_{x \in B(K,\delta)} c^T x \geq \alpha - \delta$. Thus, the output of $\text{EVAL}_\delta(1_K^*)$ satisfies the condition (6). Hence, we have that $\text{VAL}_\delta(K) \leq \text{EVAL}_\delta(1_K^*)$.

For the second inequality, to implement the evaluation oracle of 1_K^*, we need to compute $1_K^*(c) = \max_{x \in K} c^T x$ for any vector c with $\|c\|_2 \leq 1$. Using that $B(0, r) \subset K$, we have $(1 - \frac{\delta}{r})K \subset B(K, -\delta)$. Hence, we have that

$$\max_{x \in B(K,-\delta)} c^T x \geq (1 - \frac{\delta}{r}) \max_{x \in K} c^T x \geq \max_{x \in K} c^T x - \kappa \delta.$$

On the other hand, we have that

$$\max_{x \in B(K,\delta)} c^T x \leq \max_{x \in K} c^T x + \delta.$$

Hence, by binary search on γ, $\text{VAL}_\delta(K)$ allows us to estimate $\max_{x \in K} c^T x$ up to $2(2 + \kappa)\delta$ additive error.

For the third inequality, to implement the optimization oracle, we let c be the vector we want to optimize. Let x be the output of $\text{GRAD}_\eta(1_K^*)$ on input c. Using (2) and (1), we have that

$$\min_{z \in B(c,\eta)} 1_K^*(z) + x^T(d - c) < \max_{z \in B(d,\eta)} 1_K^*(z) + 2\eta$$

for any vector d. Since 1_K^* is R-Lipschitz, we have that

$$1_K^*(c) + x^T(d - c) < 1_K^*(d) + 4\eta.$$

Putting $d = 0$, we have

$$\max_{x \in K} c^T x = 1_K^*(c) \leq c^T x + 4\eta.$$

Setting $\eta = \delta/4$, we see that x is a maximizer of $\max_{x \in K} c^T x$ up to δ additive error.

For the fourth inequality, to implement the subgradient oracle, we let c be the point we want to compute the subgradient such that $\|c\|_2 \leq 1$. Let y be the output of $\text{OPT}_\delta(K)$ with input c. Since $(1 - \frac{\delta}{r})K \subset B(K, -\delta)$, we have that

$$\max_{x \in K} c^T x \leq c^T y + \delta + \kappa \delta.$$

Therefore,

$$c^T y + (d - c)^T y \leq \max_{x \in K} c^T x + \delta + (d - c)^T y \leq d^T y + (2 + \kappa)\delta \leq \max_{x \in K} d^T x + (3 + \kappa)\delta.$$

Let $\alpha = c^T y$. Since $y \in B(K, \delta)$ and satisfies the guarantee of optimization oracle, α satisfies (1) with additive error δ. Furthermore, we note that

$$\alpha + y^T(d-c) \leq 1_K^*(d) + (3+\kappa)\delta.$$

Hence, it satisfies (2) with additive error $(3+\kappa)\delta$. □

Given a convex function $f : B_n \to [0, 1]$, define the convex set

$$K_f = \left\{ \left(\frac{x}{2}, \frac{t}{4}\right) \text{ such that } x \in B(0, 1) \text{ and } f(x) \leq t \leq 2 \right\}.$$

Lemma 5.4 • $MEM_\delta(K_f) \leq EVAL_{\delta/10}(f)$ and

$$EVAL_\delta(f) \leq O(\log(1/\delta))MEM_{\Omega(\delta/\log(1/\delta))}(K_f).$$

- $SEP_\delta(K_f) \leq GRAD_{\delta/10}(f)$ and $GRAD_\delta(f) \leq O(\log(1/\delta))SEP_{\Omega(\delta/\log(1/\delta))}(K_f)$.
- $GRAD_\delta(f^*) \leq OPT_{\delta/6}(K_f)$.

□

Proof The first two sets of reductions are clear.

For the last one, to implement the subgradient oracle, we let c be the point we want to compute the subgradient such that $\|c\|_2 \leq 1$. Let (y, t') be the output of $OPT_\delta(K_f)$ with input $(c, -1)$. Since $(1 - 4\delta)K_f \subset B(K_f, -\delta)$, we have that

$$\max_{(x,t) \in K_f} (c^T x - t) \leq c^T y - t' + 5\delta.$$

Since $(y, t') \in B(K_f, \delta)$, for any vector d, we have that

$$(c^T y - t') + (d-c)^T y \leq \max_{(x,t) \in K_f}(c^T x - t) + (d-c)^T y \leq d^T y - t' + 5\delta \leq \max_{(x,t) \in K_f}(d^T x - t) + 6\delta.$$

Let $\alpha = c^T y - t'$. Since $y \in B(K, \delta)$ and satisfies the guarantee of optimization oracle, α is a good enough approximation of $f^*(c)$. Furthermore, we note that

$$\alpha + y^T(d-c) \leq \max_{(x,t) \in K_f} d^T x + 5\delta = f^*(d) + 6\delta$$

Hence, it satisfies (2) with additive error 6δ. □

5.2 Relationships Between Convex Function Oracles

Due to the equivalences above, we can focus on the more general problem of the relationships between the following four function oracles:

- $EVAL_\delta(f)$, $GRAD_\delta(f)$, $EVAL_\delta(f^*)$, $GRAD_\delta(f^*)$.

Lemma 5.5 *Let f be a convex function defined on the unit ball with value between 0 and 1. For any $0 \leq \delta \leq \frac{1}{2}$, we have that*

- $EVAL_\delta(f) \leq GRAD_\delta(f) \leq O(n \log^2(\frac{n}{\delta})) MEM_{(\delta/n)^{O(1)}}(K_f)$

$$\leq O(n \log^2(\frac{n}{\delta})) EVAL_{(\delta/n)^{O(1)}}(f)$$

- $GRAD_\delta(f^*) \leq OPT_{\delta/6}(K_f)$ *and*

$$OPT_{\delta/6}(K_f) \leq O\left(n \log\left(\frac{n}{\delta}\right)\right) SEP_{(\delta/n)^{O(1)}}(K_f) + O\left(n^3 \log^{O(1)}\left(\frac{n}{\delta}\right)\right)$$
$$\leq O\left(n \log\left(\frac{n}{\delta}\right)\right) \cdot GRAD_{(\delta/n)^{O(1)}}(f) + O\left(n^3 \log^{O(1)}\left(\frac{n}{\delta}\right)\right).$$

Proof The bound $EVAL_{\delta,\eta}(f) \leq GRAD_{\delta,\eta}(f)$ is immediate from their definition.

To bound $GRAD(f)$ by $EVAL(f)$, we use Lemma 5.4 and get that

$$GRAD_\delta(f) \leq O(\log(\delta^{-1})) SEP_{\Omega(\delta/\log(\delta^{-1}))}(K_f).$$

Next, we note that for every $x \in B_n(0, 1)$, the point $(\frac{x}{2}, t) \in K_f$ for every $t \in \left[\frac{f(x)}{4}, \frac{1}{2}\right]$. Therefore, $B\left((0, \frac{3}{8}), 0.1\right) \subset K_f \subset B(0, 1)$. Hence (after a small shift to center the inner ball at zero, which doesn't affect the outcome), Theorem 3.6 shows that

$$SEP_\delta(K_f) \leq O(n \log(\frac{n}{\delta})) MEM_{(\delta/n)^{O(1)}}(K_f).$$

Hence, we have that

$$GRAD_\delta(f) \leq O(n \log^2(\frac{n}{\delta})) MEM_{(\delta/n)^{O(1)}}(K_f).$$

Applying Lemma 5.4 again, we have the result.

To bound $GRAD(f^*)$ by $GRAD(f)$, we again use Lemma 5.4 and Theorem 4.1 to get

$$GRAD_\delta(f^*) \leq OPT_{\delta/6}(K_f) \leq O\left(n \log\left(\frac{n}{\delta}\right)\right) SEP_{(\delta/n)^{O(1)}}(K_f) + O\left(n^3 \log^{O(1)}\left(\frac{n}{\delta}\right)\right)$$
$$\leq O\left(n \log\left(\frac{n}{\delta}\right)\right) \cdot GRAD_{(\delta/n)^{O(1)}}(f) + O\left(n^3 \log^{O(1)}\left(\frac{n}{\delta}\right)\right).$$

5.3 Relationships Between Convex Set Oracles

Theorem 5.6 *For any convex set K such that $B(0, 1/\kappa) \subset K \subset B(0, 1)$, for any $0 < \delta < \frac{1}{2}$, we have that*

1. $VIOL_\delta(K) \leq OPT_\delta(K)$ and $OPT_\delta(K) \leq O(\log(1 + \frac{1}{\delta})) \cdot VIOL_{\Theta(\delta/\log(1/\delta))}(K)$.
2. $MEM_\delta(K) \leq SEP_\delta(K)$ and $SEP_\delta(K) \leq O(n \log(\frac{n\kappa}{\delta})) \cdot MEM_{(\delta/n\kappa)^{O(1)}}(K)$.
3. $VAL_\delta(K) \leq OPT_\delta(K)$ and $OPT_\delta(K) \leq O(n \log^3(\frac{n\kappa}{\delta})) \cdot VAL_{(\delta/n\kappa)^{O(1)}}(K)$.
4. $OPT_\delta(K) = O\left(n \log\left(\frac{n\kappa}{\delta}\right) \cdot SEP_{(\delta/(n\kappa))^{O(1)}}(K) + n^3 \log^{O(1)}\left(\frac{n\kappa}{\delta}\right)\right)$.
5. $SEP_\delta(K) = O\left(n \log\left(\frac{n}{\delta}\right) \cdot OPT_{(\delta/(n\kappa))^{O(1)}}(K) + n^3 \log^{O(1)}\left(\frac{n}{\delta}\right)\right)$. □

Proof (1) follows from Lemma 5.1. (2) follows from Theorem 3.6. (4) follows from Theorem 4.1.

For (3), we use Lemmas 5.3, 5.5 and 5.3 to get

$$OPT_\delta(K) \leq GRAD_{\delta/4}(1_K^*) \leq O(n \log(\frac{n}{\delta})) EVAL_{(\delta/n)^{O(1)}}(1_K^*)$$
$$\leq O(n \log^2(\frac{n\kappa}{\delta})) VAL_{(\delta/n\kappa)^{O(1)}}(K)$$

where we used that 1_K^* is a function between 0 and 1.

For (5), we use Lemmas 5.2, 5.5 and 5.3

$$SEP_\delta(K) = GRAD_\delta(1_K) \leq O\left(n \log\left(\frac{n}{\delta}\right) \cdot GRAD_{(\delta/n)^{O(1)}}(1_K^*) + n^3 \log^{O(1)}\left(\frac{n}{\delta}\right)\right)$$
$$\leq O\left(n \log\left(\frac{n}{\delta}\right) \cdot OPT_{(\delta/(n\kappa))^{O(1)}}(K) + n^3 \log^{O(1)}\left(\frac{n}{\delta}\right)\right)$$

where we used that 1_K^* is a function between 0 and 1. □

Acknowledgements We thank Sebastien Bubeck, Ben Cousins, Sham Kakade and Ravi Kannan for helpful discussions, Yan Kit Chim for making the illustrations, and Xiaodi Wu for pointing out some typos in a previous version of the paper. This work was supported in part by NSF Awards CCF-1563838, CCF-1717349 and CCF1740551.

References

1. Jacob D. Abernethy and Elad Hazan. Faster convex optimization: Simulated annealing with an efficient universal barrier. In *Proceedings of the 33nd International Conference on Machine Learning, ICML 2016, New York City, NY, USA, June 19-24, 2016*, pages 2520–2528, 2016.
2. Dimitris Bertsimas and Santosh Vempala. Solving convex programs by random walks. *Journal of the ACM (JACM)*, 51(4):540–556, 2004.
3. Sébastien Bubeck and Ronen Eldan. Multi-scale exploration of convex functions and bandit convex optimization. *arXiv preprint* arXiv:1507.06580, 2015.
4. Paul Christiano, Jonathan A Kelner, Aleksander Madry, Daniel A Spielman, and Shang-Hua Teng. Electrical flows, laplacian systems, and faster approximation of maximum flow in undi-

rected graphs. In *Proceedings of the forty-third annual ACM symposium on Theory of computing*, pages 273–282. ACM, 2011.
5. Martin Grötschel, László Lovász, and Alexander Schrijver. *Geometric algorithms and combinatorial optimization*, volume 2. Algorithms and Combinatorics, 1988.
6. A. T. Kalai and S. Vempala. Simulated annealing for convex optimization. *Math. Oper. Res.*, 31(2):253–266, 2006.
7. Jonathan A Kelner, Yin Tat Lee, Lorenzo Orecchia, and Aaron Sidford. An almost-linear-time algorithm for approximate max flow in undirected graphs, and its multicommodity generalizations. In *Proceedings of the Twenty-Fifth Annual ACM-SIAM Symposium on Discrete Algorithms*, pages 217–226. SIAM, 2014.
8. Yin Tat Lee, Satish Rao, and Nikhil Srivastava. A new approach to computing maximum flows using electrical flows. In *Proceedings of the forty-fifth annual ACM symposium on Theory of computing*, pages 755–764. ACM, 2013.
9. Yin Tat Lee and Aaron Sidford. Path finding methods for linear programming: Solving linear programs in o(sqrt(rank)) iterations and faster algorithms for maximum flow. In *Foundations of Computer Science (FOCS), 2014 IEEE 55th Annual Symposium on*, pages 424–433. IEEE, 2014.
10. Yin Tat Lee and Aaron Sidford. Efficient inverse maintenance and faster algorithms for linear programming. In *Foundations of Computer Science (FOCS), 2015 IEEE 56th Annual Symposium on*, pages 230–249. IEEE, 2015.
11. Yin Tat Lee, Aaron Sidford, and Sam Chiu-wai Wong. A faster cutting plane method and its implications for combinatorial and convex optimization. In *Foundations of Computer Science (FOCS), 2015 IEEE 56th Annual Symposium on*, pages 1049–1065. IEEE, 2015.
12. L. Lovász and S. Vempala. Fast algorithms for logconcave functions: sampling, rounding, integration and optimization. In *FOCS*, pages 57–68, 2006.
13. Aleksander Madry. Navigating central path with electrical flows: From flows to matchings, and back. In *Foundations of Computer Science (FOCS), 2013 IEEE 54th Annual Symposium on*, pages 253–262. IEEE, 2013.
14. Yurii Nesterov and Vladimir Spokoiny. Random gradient-free minimization of convex functions. *Foundations of Computational Mathematics*, 17(2):527–566, Apr 2017.
15. V. Yu. Protasov. Algorithms for approximate calculation of the minimum of a convex function from its values. *Mathematical Notes*, 59(1):69–74, 1996.
16. Jonah Sherman. Nearly maximum flows in nearly linear time. In *Foundations of Computer Science (FOCS), 2013 IEEE 54th Annual Symposium on*, pages 263–269. IEEE, 2013.
17. Jonah Sherman. Area-convexity, l_∞ regularization, and undirected multicommodity flow. In *Proceedings of the 49th Annual ACM SIGACT Symposium on Theory of Computing, STOC 2017, Montreal, QC, Canada, June 19-23, 2017*, pages 452–460, 2017.
18. Stefan Steinerberger. Sharp l 1-poincaré inequalities correspond to optimal hypersurface cuts. *Archiv der Mathematik*, 105(2):179–188, 2015.
19. P. M. Vaidya. A new algorithm for minimizing convex functions over convex sets. *Math. Prog.*, 73:291–341, 1996.

Approximations of Mappings

Jaroslav Nešetřil and Patrice Ossona de Mendez

Abstract We consider mappings, which are structure consisting of a single function (and possibly some number of unary relations) and address the problem of approximating a continuous mapping by a finite mapping. This problem is the inverse problem of the construction of a continuous limit for first-order convergent sequences of finite mappings. We solve the approximation problem and, consequently, the full characterization of limit objects for mappings for first-order (i.e. FO) convergence and local (i.e. FO^{local}) convergence. This work can be seen both as a first step in the resolution of inverse problems (like Aldous–Lyons conjecture) and a strengthening of the classical decidability result for finite satisfiability in Rabin class (which consists of first-order logic with equality, one unary function, and an arbitrary number of monadic predicates). The proof involves model theory and analytic techniques.

Keywords Structural limit · Mappings · MSC: 03C13 (Finite structures)

1 Introduction

We consider the following *approximation problems*: Given an infinite structure with given first-order properties, as well as satisfaction probabilities for every first-order formula, can one find a finite structure with approximately similar properties and

---

Supported by grant ERC Synergy DYNASNET, and by the European Associated Laboratory "Structures in Combinatorics" (LEA STRUCO) P202/12/G061.

J. Nešetřil (✉) · P. Ossona de Mendez
Computer Science Institute of Charles University (IUUK), Malostranské nám.25,
11800 Prague 1, Czech Republic
e-mail: nesetril@kam.ms.mff.cuni.cz

P. Ossona de Mendez
e-mail: pom@ehess.fr

P. Ossona de Mendez
Centre d'Analyse et de Mathématiques Sociales (CNRS, UMR 8557),
190-198 avenue de France, 75013 Paris, France

© János Bolyai Mathematical Society and Springer-Verlag GmbH Germany,
part of Springer Nature 2019
I. Bárány et al. (eds.), *Building Bridges II*, Bolyai Society
Mathematical Studies 28, https://doi.org/10.1007/978-3-662-59204-5_11

satisfaction probabilities? What if we are not given the infinite structure, but only the satisfaction probability of first-order formulas?

These problems are in general intractable, as (even when considering no probabilities of satisfaction) it is known that deciding whether a sentence satisfied by an infinite structure is also satisfied by a finite structure is (in general) undecidable. Intensive studies have been conducted to determine decidable classes of structures and fragments of first-order logic. A maximal example is the *Rabin class*, which consists of all first-order sentences with arbitrary quantifier prefix and equality, one unary function symbol, and an arbitrary number of unary relation symbols (but no function or relation symbols of higher arity). The satisfiability problem and the finite satisfiability problem for this class are both decidable, but not elementary recursive [1].

Another particular case of our problem was considered extensively in the context of topological group theory, ergodic theory and graph limits, and concerns the class of bounded degree graphs (one binary symmetric symbol) and local first-order formulas with a single free variable. It can be formulated as follows: consider a unimodular probability measure μ defined on the set G^* of all countable rooted connected graphs endowed with the metric defined by the rooted neighborhood isomorphisms. Can μ be approximated by finite graphs? This question is known as the *Aldous–Lyons conjecture*. It is not just an isolated problem as a positive solution would have far-reaching consequences, by implying that all finitely generated groups are sofic (answering a question by Weiss [24]), the direct finiteness conjecture of Kaplansky [11] on group algebras, a conjecture of Gottschalk [8] on surjunctive groups in topological dynamics, the Determinant Conjecture on Fuglede–Kadison determinants, and Connes' Embedding Conjecture for group von Neumann algebras [3]. It is easily shown that Aldous-Lyons conjecture can be reduced to the approximation problem for quantifier-free formulas on structures with two functions f and g satisfying $f^2 = g^3 = \text{Id}$.

In this paper we solve the approximation problem for mappings, i.e. structures consisting of a set X and an (endo)function $f : X \to X$, and more generally we solve it for the whole Rabin class. At lest at first glance it is perhaps surprising that such a seemingly special case is quite difficult to handle.

Approximation problems recently appeared in the context of graph limits as so called *inverse problems*. In order to make the connection clear, we take time for a quick review of some of the fundamental notions and problems encountered in the domain of graph limits, and how they are related to the study of limits and approximations of algebras (that is of functional structures).

A sequence of (colored) graphs with maximum degree at most d converges if, for every integer r, the distribution of the isomorphism type of the ball of radius r rooted at a random vertex (drawn uniformly at random) converges. The limit object of a local convergent sequence of graphs is a *graphing*, that is a graph on a standard Borel space, which satisfies a *Mass Transport Principle*, which amounts to say that for every Borel subsets A, B it holds that

$$\int_A \deg_B(v)\,dv = \int_B \deg_A(v)\,dv.$$

An alternative description of graphings is as follows: a graphing is defined by a finite number of measure preserving involutions $f_1, \ldots, f_D$ on a standard Borel space, which define the edges of the graphing as the union of the orbits of size two of $f_1, \ldots, f_D$.

The idea to conceptualize limits of structures by means of convergence of the satisfaction probability of formulas in a fixed fragment of first-order logic has been introduced by the authors in [17]. In this setting, a sequence $(\mathbf{A}_n)_{n\in\mathbb{N}}$ of structures is *convergent* (or *X-convergent*) if, for every first-order formula ϕ in a fixed fragment X the probability $\langle \phi, \mathbf{A}_n\rangle$ that ϕ is satisfied in $\mathbf{A}_n$ for a random assignment of elements of $\mathbf{A}_n$ to the free variables of ϕ converges as n grows to infinity. If X is the set of all first-order formulas, then we speak about FO-convergence. This definition allowed us to consider limits of general combinatorial structures, and was applied to limits of sparse graphs with unbounded degrees [7, 18, 20, 21], matroids [12], and tree semi lattices [2].

The main result of [19] is the construction of a limit object for FO-convergent sequences of mappings (a *mapping* being an algebra with a single function symbol and—possibly—finitely many unary predicates).

Theorem 1.1 *Every FO-convergent sequence $(\mathbf{F}_n)_{n\in\mathbb{N}}$ of finite mappings such that $\lim_{n\to\infty} |F_n| = \infty$ has a modeling mapping limit $\mathbf{L}$, such that*

1. *the probability measure $\nu_\mathbf{L}$ is atomless;*
2. *the complete theory of $\mathbf{L}$ has the finite model property;*
3. *$\mathbf{L}$ satisfies the finitary mass transport principle.*

Let us explain the (undefined) notions appearing in this theorem:

(i) A *modeling* $\mathbf{L}$ is a *totally Borel structure*—that is a structure whose domain L is a standard Borel space, such that every definable set is Borel—endowed with a probability measure $\nu_\mathbf{L}$.
(ii) The measure $\nu_\mathbf{L}$ is *atomless* (or *continuous*, or *diffuse*) if for every $v \in L$ it holds $\nu_\mathbf{L}(\{v\}) = 0$. (As we consider only standard Borel spaces, this condition is equivalent to the condition that for every Borel subset A with $\nu_\mathbf{L}(A) > 0$ there exists a Borel subset B of A with $\nu_\mathbf{L}(A) > \nu_\mathbf{L}(B) > 0$.) The necessity of this condition is witnessed by the formula $x_1 = x_2$, as $\langle x_1 = x_2, \mathbf{F}\rangle = 1/|F|$ holds for every finite mapping $\mathbf{F}$. This conditions is thus required as soon as we consider QF-convergence.
(iii) the *finitary mass transport principle* (FMTP) means that for every Borel subsets X, Y of L and every positive integer k it holds

$$(\forall v \in Y)\,|f^{-1}(v) \cap X| = k \quad \Rightarrow \quad \nu_\mathbf{L}(f^{-1}(Y) \cap X) = k\nu_\mathbf{L}(Y)$$
$$(\forall v \in Y)\,|f^{-1}(v) \cap X| > k \quad \Rightarrow \quad \nu_\mathbf{L}(f^{-1}(Y) \cap X) > k\nu_\mathbf{L}(Y)$$

This condition can be reformulated as follows: the set of all y such that $f_\mathbf{F}^{-1}(y)$ is infinite has zero $\nu_\mathbf{F}$-measure, and for every Borel subsets X and Y of L (with $|f_\mathbf{F}^{-1}(y)| < \infty$ for all $y \in Y$) we have

$$\nu_\mathbf{F}(X \cap f_\mathbf{F}^{-1}(Y)) = \int_Y |f_\mathbf{F}^{-1}(y) \cap X| \, d\nu_\mathbf{F}(y). \tag{1}$$

When X and Y are definable subsets, the above condition is clearly required for being a limit.

(iv) the *finite model property* means that for every sentence θ satisfied by $\mathbf{L}$ there exists a finite mapping $\mathbf{F}$ that satisfies θ. This is indeed a necessary condition for $\mathbf{L}$ to be an elementary limit of finite mappings hence necessary as soon as we consider FO-convergence. As mentioned, the problem of existence of a finite mapping $\mathbf{F}$ satisfying a given sentence θ is decidable, though with huge time complexity.

Theorem 1.1 was proved as a combination of general results about limit distributions from [17] and methods developed in [18] for the purpose of graph-trees. This theorem has the following corollary.

Corollary 1.2 *Every* $\mathrm{FO}^{\mathrm{local}}$-*convergent sequence* $(\mathbf{F}_n)_{n \in \mathbb{N}}$ *of finite mappings (with* $\lim_{n \to \infty} |F_n| = \infty$*) has a modeling mapping* $\mathrm{FO}^{\mathrm{local}}$-*limit* $\mathbf{L}$*, such that*

1. *the probability measure* $\nu_\mathbf{L}$ *is atomless;*
2. $\mathbf{L}$ *satisfies the finitary mass transport principle.*

Proof Consider an FO-convergent subsequence. Such a subsequence exists by (sequential) compactness of FO-convergence. According to Theorem 1.1 this subsequence has a modeling mapping limit $\mathbf{L}$ satisfying all the requirements. This modeling limit is then a modeling $\mathrm{FO}^{\mathrm{local}}$-limit of $(\mathbf{F}_n)_{n \in \mathbb{N}}$.

The inverse problems aim to determine which objects are limits of finite mappings (for given types of convergence). The main contribution of this paper is the solution of such inverse problems. Namely, for FO and $\mathrm{FO}^{\mathrm{local}}$-convergence we show how to approximate a modeling mapping by a finite mapping. (For QF-convergence the inverse problem is much easier and was solved in [19].) The following are the main results of this paper.

Theorem 1.3 *Every atomless modeling mapping* $\mathbf{L}$ *that satisfies the finitary mass transport principle is the* $\mathrm{FO}^{\mathrm{local}}$-*limit of an* $\mathrm{FO}^{\mathrm{local}}$-*convergent sequence of finite mappings.*

Theorem 1.4 *Every atomless modeling mapping* $\mathbf{L}$ *with the finite model property that satisfies the finitary mass transport principle is the FO-limit of an FO-convergent sequence of finite mappings.*

Here is a rough outline of the proof of Theorem 1.4:

1. reduce to the case where the mapping modeling **L** has no connected component of measure greater than ϵ;
2. consider a derived modeling mapping **L'** obtained by removing all the elements with zero-measure rank-R local type;
3. cut all the short circuits by means of interpretation;
4. approximate the measure on the rank-R local types by a rational measure μ;
5. construct a finite mapping **F** such that the measure of each rank-r local type t is equal to what is derived from μ;
6. consider a finite mapping **M**, which is equivalent to **L** up to a huge quantifier rank, and merge it with a great number of copies of F to form an FO-approximation of **L**;
7. deduce, using interpretation, an FO-approximation of the original mapping modeling.

Theorem 1.3 is then proved by considering separately large and small connected components, and following a similar strategy as the proof of Theorem 1.4:

1. every connected modeling mapping is close (in the sense of local convergence) to a modeling mapping with finite height; such a modeling mapping has the finite model property, hence maybe FO-approximated thanks to Theorem 1.4;

(2–5) for a mapping modeling without connected components of measure greater than ϵ, construct a finite mapping **F** as in the steps (2)–(5) of the proof of Theorem 1.4;

6. then complete **F** by means of small models of missing necessary local types, merged with a great number of copies of **F**;
7. the FO^{local} approximation is obtained as the disjoint union of the FO^{local} approximations of large connected components and the $\text{FO}_1^{\text{local}}$ approximation of the remaining components (after careful tuning of the respective orders).

It should be noted that Theorems 1.3 and 1.4 allow to obtain approximations from a mapping modeling, which may have only finitely many unary predicates in its signature. The case where we allow infinitely many unary predicates easily restricts to this case, as (for given metrization of FO- and FO^{local}-convergence) for every $\epsilon > 0$ there exist $\epsilon' > 0$ and $C \in \mathbb{N}$ such that any ϵ'-approximation of the mapping considering only the first C unary predicates is an ϵ-approximation of the mapping when considering all the unary predicates. Hence Theorems 1.4 and 1.3 solves the approximation problem for the Rabin class modelings, too.

As a pleasing consequence of our general methods we believe that Theorems 1.3 and 1.4 can be formulated in a setting where we do not approximate a particular modeling **L** but rather consider the satisfaction probability of formulas. This would give a full solution of the second type approximation problem for the Rabin class.

2 Preliminaries

2.1 Facts from Finite Model Theory

We recall some basic definitions and facts from finite model theory. The interested reader is refereed to [4, 9, 10, 14–16].

A *signature* σ is a list function or relation symbols with their arities. A σ-*structure* $\mathbf{A}$ is defined by its *domain* A, its *signature* σ, and the interpretation in A of all the relations and functions in σ. The *Gaifman graph* of a σ-structure $\mathbf{A}$ is the graph with vertex set A, where two elements are adjacent if they belong to a same relation or are related by a function application. When we speak about the neighborhood of an element x in A or about the distance between two elements x and y in A, we mean the set of elements adjacent to x in the Gaifman graph of $\mathbf{A}$ or the graph distance between x and y in the Gaifman graph of $\mathbf{A}$. Also, for $u \in A$ and $r \in \mathbb{N}$ we denote by $B_r(\mathbf{A}, u)$ the r-*ball* of u in $\mathbf{A}$, that is the set of all elements of A at distance at most r from u.

We denote by $\mathrm{FO}(\sigma)$ the set of all first-order formulas (in the language defined by the signature σ). A formula ϕ (with p free variables) is *local* if its satisfaction only depends on a fixed r-neighborhood of its free variables, and we denote by $\mathrm{FO}^{\mathrm{local}}(\sigma)$ the set of all local formulas. Also, we denote by $\mathrm{QF}(\sigma)$ the set of all quantifier free formulas. When we consider sub-fragments where we restrict free variables to $x_1, \ldots, x_p$, we will add p as a subscript, as in $\mathrm{FO}_p(\sigma)$ or $\mathrm{FO}_p^{\mathrm{local}}(\sigma)$.

For a first-order formula ϕ with p free variables and a σ-structure $\mathbf{A}$ be define $\phi(\mathbf{A})$ as the set of all p-tuples of elements of $\mathbf{A}$ that satisfy the formula ϕ in $\mathbf{A}$, that is:

$$\phi(\mathbf{A}) = \{(v_1, \ldots, v_p) \in A^p : \mathbf{A} \models \phi(v_1, \ldots, v_p)\}.$$

In the following definition we consider signatures with a function symbol and finitely many unary predicates. Although Rabin class allows infinitely many unary predicates, this is not a real restriction in the context of approximation problems, but this assumption will make the definitions and notations simpler.

Definition 2.1 A *mapping* is a σ-structure, where the signature σ consists of a single unary function symbol f and (possibly) finitely many unary relation symbols $M_1, \ldots, M_c$.

Let $\mathbf{F}$ be a mapping. We denote by F the domain of $\mathbf{F}$ and by $f_\mathbf{F}$ the interpretation of the symbol f in $\mathbf{F}$ (thus $f_\mathbf{F} : F \to F$). Unary relations will be denoted by $M_i^\mathbf{F}$ (or simply just M_i). Note that the distance $\mathrm{dist}(u, v)$ between two elements u, v in a mapping $\mathbf{F}$ is the minimum value $a + b$ such that $a, b \geq 0$ and $f_\mathbf{F}^a(u) = f_\mathbf{F}^b(v)$.

Every formula $\phi \in \mathrm{FO}_1^{\mathrm{local}}(\sigma)$ is logically equivalent to a formula with no function composition. Such formulas we call *clean*.

Approximations of Mappings 343

> **Definition 2.2** The *quantifier rank* of a formula ϕ, denoted by $\mathrm{qrank}(\phi)$, is the minimum number of nested quantifiers in a clean formula equivalent to ϕ.
> The *local rank* of a local formula ϕ, denoted by $\mathrm{lrank}(\phi)$, is the minimum number of nested quantifiers in a clean formula equivalent to ϕ in which quantification is restricted to previously defined variables and their neighbors.

It is easily checked that for a given finite signature σ there exist only finitely many local formulas $\phi \in \mathrm{FO}_1^{\mathrm{local}}(\sigma)$ that have local rank at most r (up to logical equivalence).

A *local type* is any maximal consistent subset t of $\mathrm{FO}_1^{\mathrm{local}}(\sigma)$. The *local type* of an element v of a mapping $\mathbf{F}$ is the local type t such that $\mathbf{F} \models \phi(v)$ holds for every $\phi \in t$. A *rank r local type* is the subset of a all formulas with rank at most r in a local type. We denote by $\mathcal{T}_r(\sigma)$ the set of all rank r local types for signature σ. We denote by $\mathsf{Type}_r^{\mathbf{F}}(v)$ the rank r local type of an element v in a mapping $\mathbf{F}$.

Note that for every rank r local type $t \in \mathcal{T}_r$ there exists a clean formula $\varphi_t \in t$ (in which quantification is restricted to previously defined variables and their neighbors) such that φ_t is logically equivalent to the conjunction of all the formulas in t. (The formula φ_t will always have this meaning.) Thus for every σ-structure $\mathbf{F}$ and every $v \in F$ it holds that

$$\mathsf{Type}_r^{\mathbf{F}}(v) = t \quad \Longleftrightarrow \quad \mathbf{F} \models \varphi_t(v).$$

For $r < r'$, $t \in \mathcal{T}_r(\sigma)$ and $t' \in \mathcal{T}_{r'}(\sigma)$ we say that t' *refines* t, and write $t' \prec t$, if $\varphi_{t'} \vdash \varphi_t$ (i.e. if $t' \supseteq t$).

Given two mappings $\mathbf{F}$ and $\mathbf{F}'$, it is well known that $\mathbf{F}$ and $\mathbf{F}'$ satisfy the same sentences with quantifier rank at most r, what is denoted by $\mathbf{F} \equiv_r \mathbf{F}'$, if and only if Duplicator has a winning strategy for the r-rounds Ehrenfeucht–Fraïssé game.

Given two elements $v \in F$ and $v' \in F'$, testing whether $\mathsf{Type}_r^{\mathbf{F}}(v) = \mathsf{Type}_r^{\mathbf{F}'}(v')$ can be done using a variant of a Ehrenfeucht–Fraïssé game: We start by defining $u_0 = v$ and $u_0' = v'$. At each round $1 \leq k \leq r$, Spoiler chooses in F an element u_k adjacent to some of $u_0, \ldots, u_{k-1}$ (or in F' an element u_k' adjacent to some of $u_0', \ldots, u_{k-1}'$). Then Duplicator should choose $u_k' \in F'$ (or $u_k \in F$) so that for every $0 \leq i, j \leq k$ it holds

$$\mathbf{F} \models u_i = u_j \quad \Longleftrightarrow \quad \mathbf{F}' \models u_i' = u_j'$$
$$\mathbf{F} \models f(u_i) = u_j \quad \Longleftrightarrow \quad \mathbf{F}' \models f(u_i') = u_j'$$

Spoiler wins if Duplicator cannot make such a choice and $k \leq r$; otherwise, Duplicator wins. It is easily checked that $\mathsf{Type}_r^{\mathbf{F}}(v) = \mathsf{Type}_r^{\mathbf{F}'}(v')$ if and only if Duplicator has a winning strategy. We call this variant of Ehrenfeucht–Fraïssé game the *local Ehrenfeucht–Fraïssé game*.

For $r \leq r'$ we define the natural projection π_r mapping an r'-type t to the r-type $\pi_r(t)$, which is just the subset of all formulas in t with rank at most r. Obviously, if $r' > r$ then $\pi_r(\mathsf{Type}_{r'}^{\mathbf{F}}(v)) = \mathsf{Type}_r^{\mathbf{F}}(v)$.

Let σ, σ' be signatures of mappings. Let $M_1, \ldots, M_a$ be the symbols of the unary symbols in σ' (as usual f is the function symbol). The following is a standard definition.

Definition 2.3 A *basic interpretation* I of σ'-structures into σ-structures is defined by a formulas $\kappa_1, \ldots, \kappa_a$ with a single free variable, and a formula η with two free variables defining the graph of an endofunction, that is such that

$$\vdash \forall x \exists y \, \bigl(\eta(x, y) \wedge (\forall z)(\eta(x, z) \to (z = y))\bigr).$$

For every σ-structure **A**, the σ'-structure **B** = I(**A**) has same domain as **A** (i.e. $B = A$), its relations are defined by

$$\mathbf{B} \models M_i(v) \quad \Longleftrightarrow \quad \mathbf{A} \models \kappa_i(v)$$

and $f_\mathbf{B}$ is (implicitly) defined by

$$\mathbf{B} \models f(u) = v \quad \Longleftrightarrow \quad \mathbf{A} \models \eta(u, v).$$

The interpretation I is *trivial* if $\eta(x, y) := (f(x) = y)$ (hence $f_\mathbf{B} = f_\mathbf{A}$).

For every first order formula ϕ with p free variables (on the language of σ'-structures) the first-order formula $\mathsf{I}(\phi)$ is obtained by replacing (in a clean formula logically equivalent to ϕ) terms $M_i(x)$ by $\kappa_i(x)$ and terms $f(x) = y$ by $\eta(x, y)$. The formula $\mathsf{I}(\phi)$ is such that for every σ-structure **A** and every $v_1, \ldots, v_p \in B$ it holds

$$\mathbf{B} \models \phi(v_1, \ldots, v_p) \quad \Longleftrightarrow \quad \mathbf{A} \models \mathsf{I}(\phi)(v_1, \ldots, v_p).$$

Note that if ϕ and all the formulas defining I are local then $\mathsf{I}(\phi)$ is local and

$$\mathrm{lrank}(\mathsf{I}(\phi)) \leq \mathrm{lrank}(\phi) + \max(\mathrm{lrank}(\kappa_1), \ldots, \mathrm{lrank}(\kappa_a), \mathrm{lrank}(\eta)).$$

2.2 Structural Limits

We recall here some definitions and notations from [17].

Recall that a σ-structure is *Borel* if its domain is a standard Borel space, and all the relations and functions of the structure are Borel. For instance, the mapping **F** is *Borel* if the function $f_\mathbf{F} : F \to F$ and the subsets $M_i(\mathbf{F}) = \{v \in F : \mathbf{F} \models M_i(v)\}$ are Borel;

A stronger notion has been proposed in [17]:

Definition 2.4 A *σ-modeling* (or a *modeling* when σ is implied) is a σ-structure **M**, whose domain M is a standard Borel space endowed with a probability measure $\nu_{\mathbf{M}}$, and with the property that every definable subset of a power of M is Borel.

If **F** is a finite structure, it will be practical to implicitly consider a uniform probability measure $\nu_{\mathbf{F}}$ on F, for the sake of simplifying the notations.

Note that every modeling mapping is obviously Borel, but the converse does not hold true in general, as shown by the next example.

Remark 2.5 A counter-example of Lebesgue's belief that the projection to $\mathbb{R}$ of a Borel subset of $\mathbb{R}^2$ is Borel has been given by Suslin [22]. It follows that there exits a Borel subset $S \subseteq (0, 1] \times (0, 1]$, whose first projection (on $(0, 1]$) is not Borel. Consider the mapping **F** with domain $[0, 1] \times [0, 1]$, and signature $\sigma = (f, M)$ (where f is the function symbol and M is a unary relation), with $M(\mathbf{F}) = S$ and

$$f_{\mathbf{F}}(x, y) = \begin{cases} (x, 0) & \text{if } y \neq 0 \\ (0, 0) & \text{otherwise} \end{cases}$$

The mapping **M** is obviously Borel, but fails to be a modeling, as the set $f_{\mathbf{F}}(S)$ is first-order definable but not Borel.

Definition 2.6 Let **F** be a Borel σ-structure with associated probability measure $\nu_{\mathbf{F}}$, and let $\phi \in \mathrm{FO}(\sigma)$ be a formula with p free variables, such that $\phi(\mathbf{F})$ is a Borel subset of F^p.

The *Stone pairing* of ϕ and **F** is the satisfaction probability of ϕ in **F** for independent random assignments of elements of F to the free variables of ϕ with probability distribution $\nu_{\mathbf{F}}$, that is:

$$\langle \phi, \mathbf{F} \rangle = \nu_{\mathbf{F}}^{\otimes p}(\phi(\mathbf{F})), \tag{2}$$

where $\nu_{\mathbf{F}}^{\otimes p}$ stands for the product measure $\overbrace{\nu_{\mathbf{F}} \otimes \cdots \otimes \nu_{\mathbf{F}}}^{p \text{ times}}$ on F^p.

Note that if **F** is finite (meaning that F is finite) it holds that

$$\langle \phi, \mathbf{F} \rangle = \frac{|\phi(\mathbf{F})|}{|F|^p}.$$

Definition 2.7 Given a fragment X of $\mathrm{FO}(\sigma)$, a sequence $(\mathbf{F}_n)_{n\in\mathbb{N}}$ of finite σ-structures is X-*convergent* if, for every $\phi \in X$ the limit $\lim_{n\to\infty}\langle\phi,\mathbf{F}_n\rangle$ exists.

Moreover, a modeling $\mathbf{L}$ is a modeling X-*limit* of the sequence $(\mathbf{F}_n)_{n\in\mathbb{N}}$ and we note $\mathbf{F}_n \xrightarrow{X} \mathbf{L}$ if, for every first-order formula $\phi \in X$ it holds that

$$\langle\phi, \mathbf{L}\rangle = \lim_{n\to\infty}\langle\phi, \mathbf{F}_n\rangle.$$

Note that if $\mathbf{L}$ is a modeling X-limit of $(\mathbf{F}_n)_{n\in\mathbb{N}}$, the pairing $\langle\phi, \mathbf{L}\rangle$ is defined for every first-order formula ϕ, but its value is required to be equal to $\lim_{n\to\infty}\langle\phi, \mathbf{F}_n\rangle$ only when ϕ is in X.

Given a fragment X of $\mathrm{FO}(\sigma)$ (closed under $\vee$, $\wedge$, and $\neg$) the equivalence classes of formulas in X with respect to logical equivalence form an at most countable Boolean algebra, the Lindenbaum–Tarski algebra $\mathcal{L}_X$ of X. The Stone dual to this algebra is denoted by $S(\mathcal{L}_X)$. This is a Polish space, the clopen sets of which are in bijection with the elements of $\mathcal{L}_X$, the topology of which is generated by its clopen sets, and the points of which are the maximal consistent subsets of $\mathcal{L}_X$ (that is Boolean algebra homomorphisms from $\mathcal{L}_X$ to the 2 elements Boolean algebra). For instance, if $X = \mathrm{FO}_1^{\mathrm{local}}$ then $S(\mathcal{L}_X)$ is the space of local types. Considering the Borel σ-algebra gives $S(\mathcal{L}_X)$ the structure of a standard Borel space.

The following representation theorem was proved in [17] (for general σ and fragment X):

Theorem 2.8 *To every finite σ-structure or σ-modeling $\mathbf{F}$ corresponds a unique probability measure $\mu_{\mathbf{F}}$ on $S(\mathcal{L}_X)$, such that for every formula $\phi \in X$ it holds that*

$$\langle\phi, \mathbf{F}\rangle = \int_{S(\mathcal{L}_X)} I_{K(\phi)}(t)\,\mathrm{d}\mu_{\mathbf{F}}(t), \tag{3}$$

where $I_{K(\phi)}$ denotes the indicator function of the clopen subset $K(\phi)$ of $S(\mathcal{L}_X)$ dual to ϕ. Moreover, a sequence $(\mathbf{F}_n)_{n\in\mathbb{N}}$ is X-convergent if and only if the corresponding sequence of probability measures on $S(\mathcal{L}_X)$ is weakly convergent.

Note that if the fragment X includes all the fragment FO_0 of all first-order sentences the support of $\mu_{\mathbf{F}}$ projects into a single point $\mathrm{Th}(\mu_{\mathbf{F}})$ of $S(\mathcal{L}_{\mathrm{FO}_0})$, which is (equivalently) characterized by the property

$$\forall t \in \mathrm{Supp}(\mu_{\mathbf{F}}) \quad \mathrm{Th}(\mu_{\mathbf{F}}) = t \cap \mathrm{QF}_0. \tag{4}$$

We call $\mathrm{Th}(\mu_{\mathbf{F}})$ the *complete theory of* $\mu_{\mathbf{F}}$, as this is nothing but the complete theory of $\mathbf{F}$ retrieved from $\mu_{\mathbf{F}}$.

In this paper we shall be particularly interested by the probability measures $\mu_{\mathbf{F}}^{\text{loc}}$ defined by a σ-structure $\mathbf{F}$ on the space $\mathcal{T}_\infty(\sigma)$ of local types (which is dual to the Lindenbaum–Tarski algebra of local formulas with a single free variable) and $\mu_{\mathbf{F}}^{\text{loc}(r)}$ defined by a σ-structure $\mathbf{F}$ on the (finite) space $\mathcal{T}_r(\sigma)$ of rank r local types (which is dual to the Lindenbaum–Tarski algebra of local formulas with a single free variable and local rank at most r).

We denote by π_r the projection from the space of consistent subsets of $\text{FO}_1^{\text{local}}$ to the space of consistent subsets of $\text{FO}_1^{\text{local}}$ with maximum quantifier rank at most r.

$$\pi_r(t) = \{\phi \in t : \text{lrank}(\phi) \leq r\}.$$

Note that π_r maps local types to local types with local rank at most r.

The mapping $t \mapsto \pi_r(t)$ is measurable and it is immediate that $\mu_{\mathbf{F}}^{\text{loc}(r)}$ is the push-forward $\pi_r^*(\mu_{\mathbf{F}}^{\text{loc}})$ by π_r of the probability measure $\mu_{\mathbf{F}}^{\text{loc}}$ (and that a similar statement holds with any of the probability measures $\mu_{\mathbf{F}}^{\text{loc}(r')}$ with $r' > r$).

For an integer r and a σ-modeling $\mathbf{F}$, the following easy consequence of (2) and (3) will be helpfull: for every $t \in \mathcal{T}_r(\sigma)$ it holds that

$$\mu_{\mathbf{F}}^{\text{loc}(r)}(t) = \nu_{\mathbf{F}}(\varphi_t(\mathbf{F})) = \langle \varphi_t, \mathbf{F} \rangle. \tag{5}$$

2.3 Measuring Proximity

The topology of FO-convergence can be metrized by using the following ultrametric

$$d_{\text{FO}}(\mathbf{M}, \mathbf{N}) = \sum_{p \geq 0} \sum_{r \geq 0} 2^{-(p+r)} \text{Dist}_{p,r}(\mathbf{M}, \mathbf{N}), \tag{6}$$

where

$$\text{Dist}_{p,r}(\mathbf{M}, \mathbf{N}) = \sup\left\{ |\langle \phi, \mathbf{M} \rangle - \langle \phi, \mathbf{N} \rangle| : \phi \in \text{FO}_p, \text{qrank}(\phi) \leq r \right\}. \tag{7}$$

The following lemma is a direct consequence of [20, Theorem 13], which in turn follows from Gaifman locality theorem.

Lemma 2.9 *A mapping modeling $\mathbf{L}$ is the FO_p-limit of a sequence of finite mappings if and only if it is both the $\text{FO}_p^{\text{local}}$-limit of a sequence of finite mappings and the elementary limit of a sequence of finite mappings.*

For elementary convergence, the appropriate notion of proximity is the notion of r-equivalence, and it holds that $\text{Dist}_{0,r}(\mathbf{M}, \mathbf{N}) = 0$ if and only $\mathbf{M} \equiv_r \mathbf{N}$.

For local convergence, we define the following distances (for integers $p \geq 1$ and $r \geq 0$):

$$\text{Dist}^{\text{local}}_{p,r}(\mathbf{M}, \mathbf{N}) = \sup\left\{|\langle \phi, \mathbf{M}\rangle - \langle \phi, \mathbf{N}\rangle| \,:\, \phi \in \text{FO}^{\text{local}}_p, \text{lrank}(\phi) \leq r\right\}. \quad (8)$$

Note that (by Theorem 2.8) this is nothing but twice the total variation distance between the probability measures defined by $\mathbf{M}$ and $\mathbf{N}$ on the Stone dual of the algebra of local formulas with free variables within $x_1, \ldots, x_p$ and local rank at most r.

The following lemma is a direct consequence of Lemma 2.9.

Lemma 2.10 *For every fixed signature σ, every integers p, r, and every positive real $\epsilon > 0$ there exist an integer r' and a positive real $\epsilon' > 0$, such that for every σ-modelings $\mathbf{M}, \mathbf{N}$ it holds*

$$\mathbf{M} \equiv_{r'} \mathbf{N} \text{ and } \text{Dist}^{\text{local}}_{p,r'}(\mathbf{M}, \mathbf{N}) < \epsilon' \implies \text{Dist}_{p,r}(\mathbf{M}, \mathbf{N}) < \epsilon. \quad (9)$$

In sufficiently sparse structures, where the probability that two random elements are close is small, we can further reduce the computation of the local distance to the case of local formulas with a single free variable:

Lemma 2.11 *Let $\delta_r(x_1, x_2)$ be the formula $\text{dist}(x_1, x_2) \leq r$. Then for every integers p, r and every modelings $\mathbf{M}, \mathbf{N}$ it holds*

$$\text{Dist}^{\text{local}}_{p,r}(\mathbf{M}, \mathbf{N}) \leq 2p \, \text{Dist}^{\text{local}}_{1,r}(\mathbf{M}, \mathbf{N}) + \binom{p}{2}(\langle \delta_{2r}, \mathbf{M}\rangle + \langle \delta_{2r}, \mathbf{N}\rangle). \quad (10)$$

Proof Let ϕ be a local formula with local rank at most r. The satisfaction of ϕ only depends on the r-neighborhood of the free variables $x_1, \ldots, x_p$. It follows that there exists a finite family of types $\mathcal{F} \subseteq \mathcal{T}_r^p$ such that if $\text{dist}(v_i, v_j) > 2r)$ for every $1 \leq i < j \leq p$ then it holds that

$$\mathbf{M} \models \phi(v_1, \ldots, v_p) \iff \mathbf{M} \models \widehat{\phi}(x_1, \ldots, x_p),$$

where $\widehat{\phi}$ is the local formula

$$\widehat{\phi}(x_1, \ldots, x_p) := \bigvee_{(t_1, \ldots, t_p) \in \mathcal{F}} \bigwedge_{i=1}^{p} \varphi_{t_i}(x_i).$$

Moreover, $\widehat{\phi}(\mathbf{M})$ only differs from $\bigcup_{(t_1,\ldots,t_p) \in \mathcal{F}} \prod_{i=1}^{p} \varphi_{t_i}(\mathbf{M})$ on tuples $(v_1, \ldots, v_p)$ with $\text{dist}(v_i, v_j) \leq 2r$ for some $1 \leq i < j \leq p$. It follows that

Approximations of Mappings

$$\left|\langle\phi, \mathbf{M}\rangle - \sum_{(t_1,\ldots,t_p)\in\mathcal{F}} \prod_{i=1}^{p} \langle\phi_{t_i}, \mathbf{M}\rangle\right| < \binom{p}{2}\langle\delta_{2r}, \mathbf{M}\rangle,$$

as the probability that two random elements of M are at distance at most $2r$ is bounded (by union bound) by $\binom{p}{2}$ times the probability that two random elements are at distance at most $2r$, that is by the right hand side of the inequality.

Of course, the same holds for the modeling **N**.

Let $\mu_\mathbf{M}$ (resp. $\mu_\mathbf{N}$) be the probability measure defined by **M** (resp. **N**) on $\mathcal{T}_r(\sigma)$. As

$$\sum_{(t_1,\ldots,t_p)\in\mathcal{F}} \prod_{i=1}^{p} \langle\phi_{t_i}, \mathbf{M}\rangle = \mu_\mathbf{F}^{\otimes p}(\mathcal{F}),$$

and as it is well known that if ρ, λ are probability measures on a finite set it holds that

$$\|\rho^{\otimes p} - \lambda^{\otimes p}\|_{\mathrm{TV}} \leq p\|\rho - \lambda\|_{\mathrm{TV}}$$

we deduce

$$\frac{1}{2}\left|\sum_{(t_1,\ldots,t_p)\in\mathcal{F}} \prod_{i=1}^{p} \langle\phi_{t_i}, \mathbf{M}\rangle - \sum_{(t_1,\ldots,t_p)\in\mathcal{F}} \prod_{i=1}^{p} \langle\phi_{t_i}, \mathbf{N}\rangle\right| \leq \|\mu_\mathbf{M}^{\otimes p} - \mu_\mathbf{N}^{\otimes p}\|_{\mathrm{TV}}$$
$$\leq p\,\mathrm{Dist}_{1,r}^{\mathrm{local}}(\mathbf{M}, \mathbf{N}).$$

The statement of the lemma follows. □

2.4 The Finitary Mass Transport Principle

The domain of a mapping **F** is partitioned into countably many subsets

$$F_i = \{x \subset F : |f_\mathbf{F}^{-1}(x)| = i\}$$

for $i = 0, 1, \ldots$, and

$$F_\infty = \{x \in F : |f_\mathbf{F}^{-1}(x)| = \infty\}.$$

The mass transport principle for mappings takes the following form.

Definition 2.12 The *Finitary Mass Transport Principle* (FMTP) for **F** is the satisfaction of the following conditions:
- $\nu_\mathbf{F}(F_\infty) = 0$;

- for every measurable subsets A, B of $F \setminus F_\infty$ it holds that

$$\nu_{\mathbf{F}}(A \cap f_{\mathbf{F}}^{-1}(B)) = \int_B |f_{\mathbf{F}}^{-1}(y) \cap A|\, d\nu_{\mathbf{F}}(y) \tag{11}$$

Note that a direct consequence of the FMTP is that for every measurable subset A of F it holds that $\nu_{\mathbf{F}}(A) \geq \nu_{\mathbf{F}}(f_{\mathbf{F}}(A))$.

Intuitively, the FMTP describes the interplay of two measures: the probability measure $\nu_{\mathbf{F}}$ on F used to randomly select an element, and the counting measure (implicitly) used to count, for instance, the degree of an element. This principle ultimately relies of the fact that the local type of an element is (at least partly) determined by the local type of any of its neighbors.

Definition 2.13 The *transport operator* ξ is a mapping from the space of consistent subsets of $\mathrm{FO}_1^{\mathrm{local}}$ to itself, defined by

$$\xi(t) = \{\phi(x) \in \mathrm{FO}_1^{\mathrm{local}} : [(\exists z)\, (z = f(x) \wedge \phi(z))] \in t\}.$$

A fundamental property of the transport operator is that if $r' > r$ then for every σ-structure $\mathbf{F}$ it holds that

$$\mathsf{Type}_r^{\mathbf{F}} \circ f_{\mathbf{F}} = \pi_r \circ \xi \circ \mathsf{Type}_{r'}^{\mathbf{F}}, \tag{12}$$

what is depicted by the following diagram:

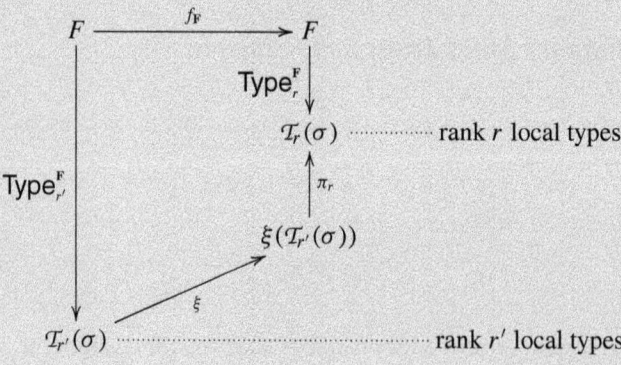

In other words, the rank r local type of the image by f of an element v is exactly the projection of the image by the transport operator of the rank $r + 1$ (or any rank $r' > r$) local type of v.

We now focus on another aspect of the FMTP.

Approximations of Mappings

Let $R > 2r$ be positive integers, and let ρ be a probability measure on $\mathcal{T}_R(\sigma)$ (and by extension on $\mathcal{T}_r(\sigma)$). Define

$$T_R(\rho) = \{\tau \in \mathcal{T}_r : \rho(\tau) > 0\}.$$

For $\tau \in T_R(\rho)$ and $t \in T_r(\rho)$ define

$$\operatorname{adm}^+(\tau, t) = \begin{cases} 1 & \text{if } \varphi_\tau(v) \vdash \varphi_t(f(v)) \\ 0 & \text{otherwise} \end{cases}$$

and let $\operatorname{adm}^-(\tau, t)$ be the maximum integer $a \in \{0, \ldots, r+1\}$ such that

$$\varphi_\tau(v) \vdash \exists x_1, \ldots, x_a \left(\bigwedge_{1 \le i \le a} (\varphi_t(x_i) \wedge f(x_i) = v) \wedge \bigwedge_{1 \le i < j \le a} (x_i \ne x_j) \right).$$

Definition 2.14 The probability measure ρ satisfies the (R, r)-restricted FMTP if there exists a function $s : T_R(\rho) \times T_r(\rho) \to \{0, 1, \ldots, r\} \cup (r, \infty)$, called *companion function* of ρ, such that for every $\tau \in T_R(\rho)$ and $t \in T_r(\rho)$ it holds

$$\min(r, \operatorname{adm}^-(\tau, t)) = \min(r, s(\tau, t)) \tag{13}$$

$$\sum_{\tau_1 \prec t_1} \operatorname{adm}^+(\tau_1, t_2)\mu(\tau_1) = \sum_{\tau_2 \prec t_2} s(\tau_2, t_1)\mu(\tau_2). \tag{14}$$

This notion is justified by the next lemma.

Lemma 2.15 *Let $R > 2r$ be positive integers.*
Let $\mathbf{L}$ be mapping modeling $\mathbf{L}$ that satisfies the FMTP and let μ be the probability measure on $\mathcal{T}_R$ defined by $\mu(\tau) = \nu_\mathbf{L}(\varphi_\tau(\mathbf{L}))$.
Then μ satisfies the (R, r)-restricted FMTP.

Proof For $\tau \in T_R(\mu)$ and $t \in T_r(\mu)$ define

$$w(\tau, t) = \frac{\nu_\mathbf{L}(f_\mathbf{L}^{-1}(\varphi_\tau(\mathbf{L})) \cap \varphi_t(\mathbf{L}))}{\nu_\mathbf{L}(\varphi_\tau(\mathbf{L}))}. \tag{15}$$

According to FMTP we have the following set of equations (where $\tau \in T_R(\mu)$ and $t \in T_r(\mu)$):

$$\min(\mathrm{adm}^-(\tau, t), r) = \min(w(\tau, t), r) \qquad (16)$$

$$\sum_{\tau_1 \prec t_1} \mathrm{adm}^+(\tau_1, t_2)\mu(\tau_1) = \sum_{\tau_2 \prec t_2} w(\tau_2, t_1)\mu(\tau_2). \qquad (17)$$

2.5 The Finite Model Property

An infinite σ-structure **M** has the *Finite Model Property* if every sentence θ satisfied by **M** has a finite model. In other words, **M** has the Finite Model Property if, for every integer r, there exists a finite σ-structure **F** with $\mathbf{F} \equiv_r \mathbf{M}$.

Deciding whether an infinite structure has the finite model property is extremely difficult, as deciding wether a sentence has a finite model is undecidable in general, see Trakhtenbrot [23].

However, it is clear from our definition that if a modeling **M** is an FO-limit of a sequence of finite structures then **M** does have the finite model property. When considering the problem of constructing an FO-approximation of a modeling **M**, we will not only assume that the modeling **M** has the finite model property, but that we can ask an oracle to provide us (for each integer r) with a finite structure **F** that is $\mathbf{F} \equiv_r \mathbf{M}$.

In some very particular cases, deciding whether a structure has the finite model property and constructing an elementary approximation can be easy. For instance, Lemma 4.3 below asserts that every mapping with finite height has the finite model property and describes how to construct an elementary approximation. The case of mappings is intermediate between the case of bounded height trees (which have the finite model property) and the case of relational structures with at least one relation symbol with arity at least two, for which satisfiability problem is undecidable. The *Rabin class* $[all, (\omega), (1)]_=$ of first-order logic with equality, one unary function and monadic predicates does not have the finite model property. (For instance, one can consider a sentence expressing that there exists a unique element which is not the image of another element, but that every other element is the image of exactly one element.) However, satisfiability problem and finite satisfiability problem for Rabin class are both decidable, though with huge complexity (the first-order theory of one unary function is not elementary recursive). For a general discussion on classical decision problems we refer the reader to [1].

2.6 Derived Modelings

Let **F** be a modeling mapping and let X be a non-zero measure first-order definable subset of F. We denote by $\mathbf{F} \upharpoonright_X$ the restriction of **F**, which is the modeling mapping with domain X, probability measure $\nu_{\mathbf{F} \upharpoonright_X} = \frac{1}{\nu_{\mathbf{F}}(X)} \nu_{\mathbf{F}}$ and

$$f_{\mathbf{F}\restriction_X}(v) = \begin{cases} f_{\mathbf{F}}(v) & \text{if } f_{\mathbf{F}}(v) \in X \\ v & \text{otherwise} \end{cases}$$

Remark 2.16 The condition that X is first-order definable ensures that $\mathbf{F} \restriction_X$ is a modeling. The condition that X is a Borel subset of F would not be sufficient. This may be seen analogously as in Example 2.5 above:

Consider the modeling mapping $\mathbf{F}$ with $F = [0, 1] \times [0, 1]$ and $f_{\mathbf{F}}$ maps (x, y) to $(x, 0)$, and $\nu_{\mathbf{F}}$ be the usual measure. Then $\mathbf{F}$ is clearly a modeling. Let X_0 be a Borel subset of $(0, 1) \times (0, 1)$ such that $f_{\mathbf{F}}(X)$ is not a Borel subset of $[0, 1] \times \{0\}$ (such a set can be derived from a standard example of non-Borel Σ_1^1 sets), and let $X = X_0 \cup [0, 1] \times \{0\}$. Then $\mathbf{F} \restriction_X$ is not a modeling as the definable subset $\{v : (\exists x)\, (x \neq v) \wedge (f(x) = v)\}$ is not Borel.

Lemma 2.17 *Let $\mathbf{F}$ be a mapping modeling and let X be a non zero-measure first-order definable subset of F. If $\mathbf{F}$ satisfies the FMTP then so does $\mathbf{F} \restriction_X$.*

Proof Let A, B be Borel subsets of X. Let $Z = \{v \in X : f_{\mathbf{F}}(v) \notin X\}$. As $\mathbf{F}$ satisfies the FTMP it holds

$$\nu_{\mathbf{F}\restriction_X}(A \cap f_{\mathbf{F}\restriction_X}^{-1}(B)) = \nu_{\mathbf{F}\restriction_X}(A \cap f_{\mathbf{F}\restriction_X}^{-1}(B \setminus Z)) + \nu_{\mathbf{F}\restriction_X}(A \cap f_{\mathbf{F}\restriction_X}^{-1}(B \cap Z))$$

$$= \frac{1}{\nu_{\mathbf{F}}(X)} \left(\nu_{\mathbf{F}}(A \cap f_{\mathbf{F}}^{-1}(B \setminus Z)) + \nu_{\mathbf{F}}(A \cap B \cap Z)\right)$$

$$= \frac{1}{\nu_{\mathbf{F}}(X)} \left(\int_{B \setminus Z} |f_{\mathbf{F}}^{-1}(y) \cap A|\, d\nu_{\mathbf{F}}(y) + \nu_{\mathbf{F}}(A \cap B \cap Z)\right)$$

$$= \int_B |f_{\mathbf{F}\restriction_X}^{-1}(y) \cap A|\, d\nu_{\mathbf{F}\restriction_X}(y)$$

Thus the FMTP holds for $\mathbf{F} \restriction_X$. □

We also note the following:

Lemma 2.18 *Let $\mathbf{M}$ be a modeling and let $\mathbf{M}^+$ be obtained from $\mathbf{M}$ by marking exactly one element of M with a new unary relation. Then*

1. *$\mathbf{M}^+$ is a modeling;*
2. *$\mathbf{M}^+$ satisfies the FMTP if and only if $\mathbf{M}$ satisfies the FMTP;*
3. *$\mathbf{M}^+$ has the finite model property if and only if $\mathbf{M}$ has the finite model property.*

Proof The first item was proved in [17]. The second item is obvious as $\mathbf{M}$ and $\mathbf{M}^+$ have the same Gaifman graph. As $\mathbf{M}$ is a trivial interpretation of $\mathbf{M}^+$, the finite model property for $\mathbf{M}^+$ implies the finite model property for $\mathbf{M}$. Conversely, assume $\mathbf{F} \equiv_{r+1} \mathbf{M}$ and start a Ehrenfeucht–Fraïssé game by choosing the element that is marked in $\mathbf{M}^+$. Assume Duplicator follows a winning strategy for the $(r + 1)$-rounds game, and mark the vertex chosen by Duplicator in $\mathbf{F}$. Then (continuing the game) we get that the marked structure is r-equivalent to $\mathbf{M}^+$. □

2.7 List of Symbols

For the benefit or the readers we include here a list of the main symbols used in this section.

Symbol	Signification
Introduced in Sect. 2.1	
σ	signature
F	mapping (Definition 2.1)
F	domain of structure **F**
$\phi(\mathbf{F})$	set of tuples satisfying ϕ in **F**
$B_r(\mathbf{F}, u)$	r-ball of u in **F**
FO	all first-order formulas
FO_p	first-order formulas with free variables within $x_1, \ldots, x_p$
FO_0	sentences
FO^{local}	local first-order formulas
$\text{FO}^{\text{local}}_p$	local first-order formulas with free variables within $x_1, \ldots, x_p$
QF	quantifier free first-order formulas
$\text{lrank}(\phi)$	local rank of formula ϕ (Definition 2.2)
t, τ	Local types
$\mathcal{T}_r(\sigma)$	set of all rank r local types
$\text{Type}_r^{\mathbf{F}}(v)$	rank r local type of v in **F**
$\varphi_t(x_1)$	characteristic formula of local type t
$\delta_r(x_1, x_2)$	formula expressing $\text{dist}(x_1, x_2) \leq r$
I	interpretation (Definition 2.3)
Introduced in Sect. 2.2	
$\nu_{\mathbf{F}}$	Probability measure on the domain F of **F** (Definition 2.4)
$\langle \phi, \mathbf{F} \rangle$	Stone pairing of ϕ and **F** (Definition 2.6)
$S(\mathcal{L}_X)$	Stone dual of Lindenbaum–Tarski algebra of X
$\mu_{\mathbf{F}}$	Representation measure of **F** (Theorem 2.8)
$\mu_{\mathbf{F}}^{\text{loc}}$	Representation measure of structure **F** for $\text{FO}_1^{\text{local}}$ fragment
$\text{Th}(\mu_{\mathbf{F}})$	Complete theory of $\mu_{\mathbf{F}}$
π_r	Projection to consistent subsets of $\text{FO}_1^{\text{local}}$ with quantifier rank at most r
$\mu_{\mathbf{F}}^{\text{loc}(r)}$	Pushforward of $\mu_{\mathbf{F}}^{\text{loc}}$ by π_r
Introduced in Sect. 2.4	
ζ	Transport operator (Definition 2.13)
$\text{adm}^+(\tau, t)$	Does $\varphi_\tau(v)$ imply $\varphi_t(f(v))$?
$\text{adm}^-(\tau, t)$	How many distinct u with $\varphi_t(u)$ and $f(u) = v$ if $\varphi_\tau(v)$?

3 First-Order Approximation

The aim of this section is to prove Theorem 1.4. The general strategy of the proof is depicted in Fig. 1:

Approximations of Mappings 355

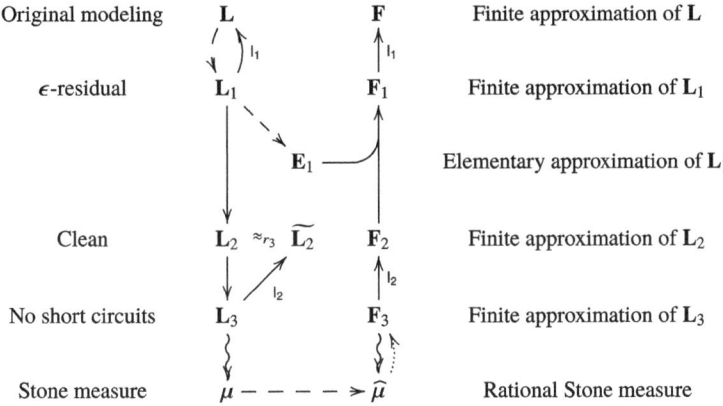

Fig. 1 Strategy for the proof of Theorem 1.4

1. Reduction $\mathbf{L} \to \mathbf{L}_1$, where $\mathbf{L}_1$ is ϵ_2-residual (i.e. has no connected component of measure greater than ϵ_2), with recovery interpretation I_1.
2. restriction $\mathbf{L}_1 \mapsto \mathbf{L}_2$ to no zero-measure rank-r_3 local types.
3. Transformation $\mathbf{L}_2 \mapsto \mathbf{L}_3$ killing all short circuits. Interpretation $\mathbf{L}_3 \overset{\mathsf{I}_2}{\to} \widetilde{\mathbf{L}}_2$, with local statistics close to $\mathbf{L}_2$.
4. Approximation of the rank-R local type measure μ of $\mathbf{L}_3$ by a rational measure $\widehat{\mu}$, still satisfying mass transport principle.
5. Construction of an exact model $\mathbf{F}_3$ of $\widehat{\mu}$, providing a finite approximation $\mathbf{F}_3$ of $\mathbf{L}_3$.
6. Rewiring the short cycles by means of interpretation I_2, leading to an approximation $\mathbf{F}_2$ of $\mathbf{L}_2$.
7. Construction of an elementary approximation $\mathbf{E}_1$ of $\mathbf{L}_1$.
8. Merge of $\mathbf{E}_1$ with a great number of copies of $\mathbf{F}_2$ to form an FO-approximation $\mathbf{F}_1$ of $\mathbf{L}_1$.
9. Interpretation $\mathbf{F}_1 \overset{\mathsf{I}_1}{\to} \mathbf{F}$ to get an FO-approximation of the original mapping modeling $\mathbf{L}$.

We shall reduce the complexity of the approximation problem by requiring more and more properties on the mapping modeling we want to approximate. The different properties we will consider for our mapping modeling are:

(P1) the modeling measure is atomless;
(P2) the modeling satisfies the FMTP;
(P3) the modeling has the finite model property;
(P4) the modeling is ϵ_2-residual;
(P5) the modeling is r_3-clean;
(P6) the modeling has no cycle of length smaller than r_4.

During the reduction process, we shall make use of additional unary relations to keep track of the properties of the original mapping. Therefore we shall consider larger and larger signatures:

- σ is the signature of both **L** and **F**. This signature contains a single unary function symbol f and (possibly) finitely many unary relation symbols.
- σ_1 is the signature of $\mathbf{L}_1$, $\mathbf{E}_1$, and $\mathbf{F}_1$. It is obtained by adding to σ the unary relation symbols $(A_i)_{1 \leq i \leq 2\lceil \epsilon_2^{-1}\rceil}$ and $(B_i)_{1 \leq i \leq 2\lceil \epsilon_2^{-1}\rceil}$.
- σ_2 is the signature of $\mathbf{L}_2$, $\tilde{\mathbf{L}}_2$ and $\mathbf{F}_2$. It is obtained by adding to σ_1 unary relations $(R_t)_{t \in \mathcal{T}_{r_3}(\sigma_1)}$.
- σ_3 is the signature of $\mathbf{L}_3$ and $\mathbf{F}_3$. It is obtained by adding to σ_2 unary relations $(U_i)_{1 \leq i \leq r_4}$ and unary relations $(T_t)_{t \in \mathcal{T}_{r_3}(\sigma_2)}$.

We fix integers p, r and a positive real $\epsilon > 0$. Our aim is to construct a finite mapping **F** such that $\mathrm{Dist}_{p,r}(\mathbf{L}, \mathbf{F}) < \epsilon$, that is such that for every first-order formula ϕ with at most p free variables and quantifier rank at most r, it holds that

$$|\langle \phi, \mathbf{L}\rangle - \langle \phi, \mathbf{F}\rangle| < \epsilon.$$

We first reduce the problem by separately considering local first-order formulas and sentences. It follows from Lemma 2.10 that there exist an integer r_1 and a positive real $\epsilon_1 > 0$ such that if $\mathbf{L} \equiv_r \mathbf{F}$ and $\mathrm{Dist}^{\mathrm{local}}_{p,r_1}(\mathbf{L}, \mathbf{F}) < \epsilon_1$ then it holds $\mathrm{Dist}_{p,r}(\mathbf{L}, \mathbf{F}) < \epsilon$. We further require $\epsilon_1 < 1/16$.

Let $r_2 = 4r_1^2$, $r_3 = 2r_2 + 1$, $r_4 = r_3!$, $\epsilon_2 = \epsilon_1/p^2$, $\epsilon_3 = \epsilon_1/4p$, $\epsilon_4 = \epsilon_5 = \epsilon_1/4p$, $N_1 = 2\lceil \epsilon_2^{-1}\rceil$, $r_5 = r_1 r_3 N_1 |\mathcal{T}_{r_3}(\sigma_3)|$.

3.1 From **L** to $\mathbf{L}_1$: Reduction to ϵ-Residual Case

We consider a signature augmented by $4\lceil \epsilon_2^{-1}\rceil$ marks $A_1, \ldots, A_{2\lceil \epsilon_2^{-1}\rceil}$ and $B_1, \ldots, B_{2\lceil \epsilon_2^{-1}\rceil}$, and the basic interpretation I_1 defined by

$$\eta(x, y) := \left[(f(x) = y)) \wedge \neg \bigvee_{i=1}^{2\lceil \epsilon_2^{-1}\rceil} A_i(x)\right] \vee \bigvee_{i=1}^{2\lceil \epsilon_2^{-1}\rceil} (A_i(x) \wedge B_i(y)).$$

We construct a mapping modeling $\mathbf{L}_1$ from **L** as follows.

We start by letting $\mathbf{L}_1$ be a copy of **L**, $j = \lceil \epsilon_2^{-1}\rceil + 1$, and we modify $\mathbf{L}_1$ as follows: We consider the connected component $\mathbf{C}_i$ ($1 \leq i \leq N \leq 1/\epsilon$) of $\mathbf{L}_1$ with measure $c_i = \nu_{\mathbf{L}}(\mathbf{C}_i) > \epsilon_2$. If $\mathbf{C}_i$ contains a non-trivial cycle, we arbitrarily select a vertex v on it, mark v with mark A_i, mark $f_{\mathbf{L}_1}(v)$ by mark B_i, and let $f_{\mathbf{L}_1}(v) = v$. For $u \in \mathbf{C}_i$ let

$$E(u) = \bigcup_{i \geq 1} f_{\mathbf{L}_1}^{-k}(u).$$

Suppose there exists $v \in C_i$ s.t. $\nu_{\mathbf{L}_1}(E(v)) > \epsilon_2$. As

$$\nu_{\mathbf{L}_1}(E(v)) = \lim_{k\to\infty} \nu_{\mathbf{L}_1}\left(\bigcup_{1\leq i\leq k} f_{\mathbf{L}_1}^{-k}(u)\right),$$

there exists some k s.t.

$$\sum_{u\in f^{-k}(v)} \nu_{\mathbf{L}_1}(E(u)) = \nu_{\mathbf{L}_1}\left(E(v) \setminus \bigcup_{1\leq i\leq k} f_{\mathbf{L}_1}^{-k}(v)\right) \leq \epsilon_2.$$

Therefore, there is some u s.t. $\nu_{\mathbf{L}_1}(E(u)) > \epsilon_2$ and $\nu_{\mathbf{L}_1}(E(x)) \leq \epsilon_2$ for all $x \in f_{\mathbf{L}_1}^{-1}(u)$.

Note that there exist at most c_i/ϵ_2 elements $u \in C_i$ such that $\nu_{\mathbf{L}_1}(E(u)) \geq \epsilon_2$ and $\nu_{\mathbf{L}_1}(E(x)) < \epsilon$ for every $x \in f_{\mathbf{L}_1}^{-1}(u)$. For each such element u, denoting $W = f_{\mathbf{L}_1}(u)$, we mark u by a mark B_j, every element in W by mark A_j, increase j by one, and redefine $f_{\mathbf{L}_1}(w) = w$ for every $w \in W$. As W is first-order definable with a parameter, the structure $\mathbf{L}_1$ is still a modeling. Doing this, the component $\mathbf{C}$ gives rise to (possibly uncountably many) small connected components of measure smaller than ϵ_2, and at most one connected component with measure ϵ_2. At the end of the day, we have used up to $2\lceil\epsilon_2^{-1}\rceil$ pairs of marks A_i and B_i, $\mathbf{L}_1$ is ϵ_2-residual, and $\mathbf{L} = \mathsf{l}_1(\mathbf{L}_1)$.

Lemma 3.1 $\mathbf{L}_1$ *satisfies the properties (P1) to (P4) and* $\mathbf{L} = l_1(\mathbf{L}_1)$.

Proof As $\nu_{\mathbf{L}_1} = \nu_{\mathbf{L}}$, (P1) holds for $\mathbf{L}_1$. The satisfaction of the FMTP for $\mathbf{L}$ obviously implies the satisfaction of the FMTP for $\mathbf{L}_1$ hence (P2) holds for $\mathbf{L}_1$.

The Finite Model Property for $\mathbf{L}$ implies the one for $\mathbf{L}_1$ (thus (P3) holds): For $r \in \mathbb{N}$, let $\mathbf{F}$ be a finite mapping such that $\mathbf{F} \equiv_{r+2\lceil\epsilon_1^{-1}\rceil} \mathbf{L}$. Start a Ehrenfeucht–Fraïssé game of length $r + 2\lceil\epsilon_1^{-1}\rceil$ by selecting in $\mathbf{L}$ the elements $v_1, \ldots, v_N$ marked $B_1, \ldots, B_N$ ($N \leq 2\lceil\epsilon_1^{-1}\rceil$) in $\mathbf{L}_1$, and let $z_1, \ldots, z_N$ be the corresponding elements of F chosen by Duplicator. We construct $\mathbf{F}_1$ from $\mathbf{F}$ by marking z_i by mark B_i, by marking every element in $Y_i = f_{\mathbf{F}}^{-1}(z_i)$ by mark A_i and letting $f_{\mathbf{F}_1}(y) = y$ for every $y \in Y_i$ (for $1 \leq i \leq N$). Then it is easily checked that Duplicator's winning strategy for the remaining r steps of the Ehrenfeucht–Fraïssé game between $\mathbf{L}$ and $\mathbf{F}$ defines a winning strategy for the r-step Ehrenfeucht–Fraïssé game between $\mathbf{L}_1$ and $\mathbf{F}_1$ hence $\mathbf{F}_1 \equiv_r \mathbf{L}_1$.

Property (P4) holds by construction, as well as the property that $\mathbf{L} = l_1(\mathbf{L}_1)$. □

3.2 From $\mathbf{L}_1$ to $\mathbf{L}_2$: Cleaning-Up

Definition 3.2 Let $r \in \mathbb{N}$. A mapping modeling $\mathbf{L}$ is *r-clean* if, for every formula $\phi \in \mathrm{FO}_1^{\mathrm{local}}$ with rank at most r it holds that

$$\mathbf{L} \models (\exists x)\phi(x) \iff \langle \phi, \mathbf{L} \rangle > 0.$$

In other words, a mapping modeling $\mathbf{L}$ is r-clean if every local type realized in $\mathbf{L}$ occurs with non zero probability.

We have proved that $\mathbf{L}_1$ satisfies (P1)–(P4). We now construct $\mathbf{L}_2$.

Define
$$T = \{t \in \mathcal{T}_{r_3}(\sigma_1) : \langle \varphi_t, \mathbf{L}_1 \rangle > 0\},$$

let $X = \bigvee_{t \in T} \varphi_t(\mathbf{L}_1)$—that is X is the subset of elements of $\mathbf{L}_1$ whose r_3-local type appears in $\mathbf{L}_1$ with no zero probability—and let $\mathbf{L}_2$ be obtained from $\mathbf{L}_1 \restriction_X$ by adding marks R_t ($t \in \mathcal{T}_{r_3}(\sigma_1)$), in such a way that for all $t \in \mathcal{T}_{r_3}(\sigma_1)$ and $v \in L_2$ it holds that

$$\mathbf{L}_2 \models R_t(v) \iff \mathbf{L}_1 \models \varphi_t(v) \iff \mathsf{Type}_{r_3}^{\mathbf{L}_1}(v) = t.$$

Lemma 3.3 *The mapping modeling $\mathbf{L}_2$ satisfies properties (P1)–(P5).*

Proof Let $\widehat{\mathbf{L}}_1$ be the σ_2-mapping obtained by the trivial interpretation adding marks R_t in such a way that $R_t(\widehat{\mathbf{L}}_1) = \varphi_t(\mathbf{L}_1)$. As we made use of a trivial interpretation, $\widehat{\mathbf{L}}_1$ is a modeling and properties (P1) to (P4) still hold. Note that $\mathbf{L}_2 = \widehat{\mathbf{L}}_1 \restriction_X$. It is immediate that (P1) and (P4) hold. According to Lemma 2.17, (P2) holds. If $\mathbf{F}$ is a finite elementary approximation of $\widehat{\mathbf{L}}_1$ then $\mathbf{F} \restriction_X$ is a finite elementary approximation of $\mathbf{L}_2$ hence $\mathbf{L}_2$ has the finite model property (P3). An easy r_3-step local Ehrenfeucht–Fraïssé game easily shows that if $u, v \in L_2$ have same rank ϵ_2 local type in $\mathbf{L}_1$ then they have the same rank r_3 local type in $\mathbf{L}_2$. It follows that $\mathbf{L}_2$ is r_3-clean thus (P5) holds. $\square$

3.3 From $\mathbf{L}_2$ to $\mathbf{L}_3$: Cutting the Short Cycles

Cutting the short cycles will allow to handle mapping modelings that are locally acyclic, which will strongly simplify the proofs. A natural procedure would be to consider a Borel transversal of all short cycles (which exists thanks to Borel selection theorem [13, p. 78]), to mark it, and to use an interpretation to kill the cycles at the mark. However, such an approach fails as marking a Borel subset of a modeling does not in general keep the property of being a modeling (see Remark 2.5). We shall use a different approach. Let Γ be the set $[r_4]$. We consider the σ_3-mapping modeling $\mathbf{L}_3$ with domain $L_3 = L_2 \times \Gamma$, measure $\nu_{\mathbf{L}_3} = \nu_{\mathbf{L}_2} \otimes \delta_\Gamma$ (where δ_Γ is the uniform measure on Γ), with (x, i) marked by $U_i, T_{\mathsf{Type}_{r_3}^{\mathbf{L}_2}(x)}$, and

$$f_{\mathbf{L}_3}(x, i) = (f_{\mathbf{L}_2}(x), i + 1 \bmod r_4).$$

An example of construction of $\mathbf{L}_3$ is shown on Fig. 2.

Lemma 3.4 *The mapping modeling $\mathbf{L}_3$ satisfies (P1)–(P6).*

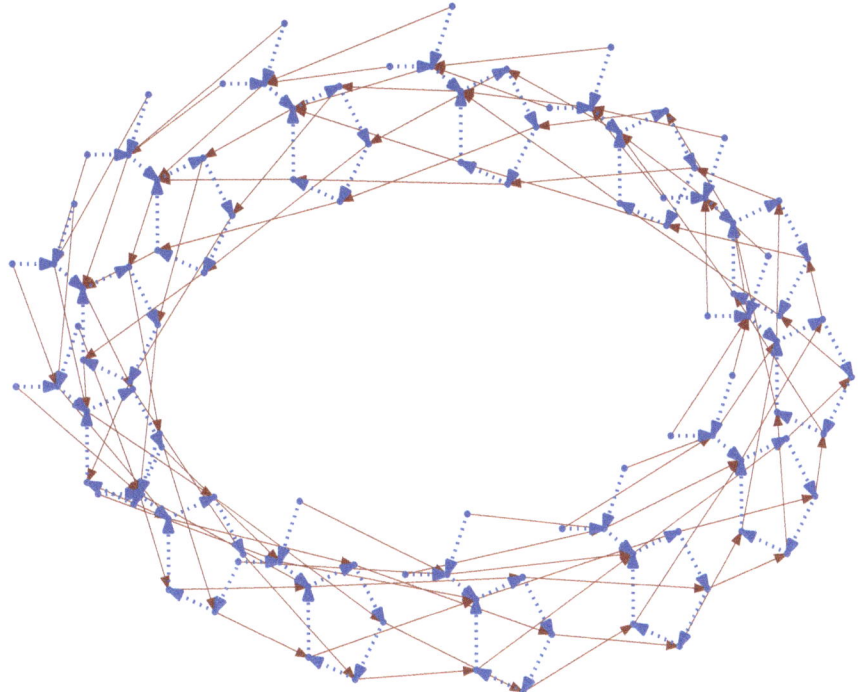

Fig. 2 Construction of L_3

Proof Property (P1) obviously holds.

As $\mathbf{L}_2$ satisfies the FMTP, so does $\mathbf{L}_3$. Indeed, let A, B be Borel subsets of L_3 such that $\deg_B^{\mathbf{L}_3}(v)$ is bounded for $v \in A$ and $\deg_A^{\mathbf{L}_3}(v)$ is bounded for $v \in B$. Then we can write $A = \bigcup_i A_i \times \{i\}$ and $B = \bigcup_j B_j \times \{j\}$, where the A_i's and the B_j's are Borel subsets of L_2. Then it holds that

$$\nu_{\mathbf{L}_3}(A \cap f_{\mathbf{L}_3}^{-1}(B)) = \frac{1}{r_4} \sum_i \nu_{\mathbf{L}_2}(A_i \cap f_{\mathbf{L}_2}^{-1}(B_{i+1 \bmod r_4}))$$

$$= \frac{1}{r_4} \sum_j \int_{B_j} |f_{\mathbf{L}_2}^{-1}(y) \cap A_{j-1 \bmod r_4}| \, d\nu_{\mathbf{L}_2}(y)$$

$$= \int_B |f_{\mathbf{L}_3}^{-1}(y) \cap A| \, d\nu_{\mathbf{L}_3}(y)$$

Hence (P2) holds.

It is immediate that if for some $R \in \mathbb{N}$ it holds that $\mathbf{F} \equiv_R \mathbf{L}_2$ then if $\mathbf{F}'$ is obtained from $\mathbf{F}$ in the same way that $\mathbf{L}_3$ is obtained from $\mathbf{L}_2$ it holds that $\mathbf{F}' \equiv_R \mathbf{L}_3$ (Duplicator's strategy immediately follows from its strategy in an Ehrenfeucht–Fraïssé game between $\mathbf{F}$ and $\mathbf{L}_2$). Thus (P3) holds.

It is easily checked that the measure of a connected component of $\mathbf{L}_3$ is at most the measure of its projection on $\mathbf{L}_2$. Thus (P4) holds.

As $r_4 > r_3$, an easy Ehrenfeucht–Fraïssé game shows that if two elements x, y of L_2 have the same r_3 local type in $\mathbf{L}_2$ and $1 \leq i, j \leq$ the (x, i) and (y, j) have the same r_3 local type in $\mathbf{L}_3$. Thus, as $\mathbf{L}_2$ is r_3-clean so is $\mathbf{L}_3$. Hence (P5) holds for $\mathbf{L}_3$.

By construction, $\mathbf{L}_3$ has no cycle of length smaller than r_4 thus (P6) holds. □

For $1 \leq \ell \leq r_3$ let Z_ℓ be the subset of all the $t \in \mathcal{T}_{r_3}(\sigma_2)$ that contain the formula $[(f^\ell(x) = x) \wedge \bigwedge_{i<\ell}(f^i(x) \neq x)]$ (which means that x belongs to a cycle of length ℓ).

Now we consider the basic interpretation I_2, with

$$\eta(x, y) := \left[\bigvee_{\ell=1}^{r_3}(\zeta_\ell(x) \wedge (y = f^{\ell-1}(x)))\right] \vee \left[(y = f(x)) \wedge \neg \bigvee_{\ell=1}^{r_3} U_i(x)\right],$$

where

$$\zeta_l(x) := U_l(x) \wedge \bigvee_{t \in Z_\ell} T_t(x),$$

which also forgets the marks U_i and T_t. Let $\widetilde{\mathbf{L}}_2 = \mathsf{I}_2(\mathbf{L}_3)$.

Lemma 3.5 *For every $\phi \in \mathrm{FO}_1^{\mathrm{local}}$ with rank at most r_3 it holds that*

$$\langle \phi, \widetilde{\mathbf{L}}_2 \rangle = \langle \phi, \mathbf{L}_2 \rangle.$$

Proof It is straightforward that for every $v \in L_2$ and every $i \in \Gamma$ it holds that

$$\mathsf{Type}_{r_3}^{\widetilde{\mathbf{L}}_2}(v, i) = \mathsf{Type}_{r_3}^{\mathbf{L}_2}(v).$$

Hence for every $\phi \in \mathrm{FO}_1^{\mathrm{local}}$ with rank at most r_3 it holds that

$$\langle \phi, \widetilde{\mathbf{L}}_2 \rangle = \langle \phi, \mathbf{L}_2 \rangle.$$

3.4 From μ to $\widehat{\mu}$: Approximating the Stone measure

Let $\mu = \mu_{\mathbf{L}_3}^{\mathrm{loc}(r_3)}$. As $\mathbf{L}_3$ satisfies the FMTP, according to Lemma 2.15, the probability measure μ satisfies the (r_3, r_2)-restricted FMTP.

Lemma 3.6 *There exists a rational probability measure $\widehat{\mu}$ on $\mathcal{T}_{r_3}(\sigma_3)$ with same support as μ, that satisfies the (r_3, r_2)-restricted MTP, and such that $\|\mu - \widehat{\mu}\|_{\mathrm{TV}} < \epsilon_5$.*

Proof Let w be a companion function for μ, and let

$$Q_1 = \{(\tau, t) \in T_{r_3}(\mu) \times T_{r_2}(\mu) : w(\tau, t) \leq r_2\}$$
$$Q_2 = \{(\tau, t) \in T_{r_3}(\mu) \times T_{r_2}(\mu) : w(\tau, t) > r_2\}$$

Consider the following set of Diophantine equations and inequalities with variables x_τ ($\tau \in T_{r_3}(\mu)$) and $y_{\tau,t}$ $((\tau, t) \in Q_2)$:

$$x_\tau > 0, \quad \sum_{\tau \in T_{r_3}} x_\tau = 1, \quad y_{\tau,t} \geq 0,$$

$$\sum_{\tau_1 \prec t_1} \mathrm{adm}^+(\tau_1, t_2) x_{\tau_1} = \sum_{\substack{\tau_2 \prec t_2 \\ (\tau_1, t_2) \in Q_1}} \mathrm{adm}^-(\tau_1, t_2) x_{\tau_2} + \sum_{\substack{\tau_2 \prec t_2 \\ (\tau_1, t_2) \in Q_2}} (r_2 x_{\tau_2} + y_{\tau_1, t_2})$$

Then this set defines a convex polytope containing a solution for $x_\tau = \mu(\tau)$ and $y_{\tau,t} = (w(\tau, t) - r_2)\mu(\tau)$.

Since this polytope has rational vertices, either the aforementioned solution is rational, or there is a strictly positive rational solution in any of its neighborhood. Let $(\widehat{x}_\tau, \widehat{y}_{\tau,t})$ be such a rational solution, such that $\sum_{\tau \in T_{r_3}(\sigma_3)} |x_\tau - \widehat{x}_\tau| < \epsilon_5$.

Define $\widehat{\mu}(\tau) = \widehat{x}_\tau$. Then $\widehat{\mu}$ has same support as μ and $\|\mu - \widehat{\mu}\|_{\mathrm{TV}} < \epsilon_5$, and $\widehat{\mu}$, with companion function $s(\tau, t) = r_2 + \widehat{y}_{\tau,t}/\widehat{x}_\tau$, satisfies the (r_3, r_2)-restricted FMTP. □

3.5 Constructing F_3

It is possible, by means of a (relatively low local rank) local formula, to specify that in the neighborhood of an element v, related in a given way (by means of a digraph D indicating which element is the image of which element), one finds an element u_1 with rank ρ_1 local type t_1, an element u_2 with rank ρ_2 local type t_2,..., and an element u_k with rank ρ_k local type t_k. This is the aim of the following definition.

Definition 3.7 Let σ be a mapping signature, let $k \in \mathbb{N}$, $\rho_1 > \rho_2 > \cdots > \rho_k \geq 0$, $t_1 \in \mathcal{T}_{\rho_1}(\sigma), \ldots, t_k \in \mathcal{T}_{\rho_k}(\sigma)$, and let $D \subseteq [k+1] \times [k+1]$ be the arc set of a digraph with outdegrees at most 1 and connected underlying graph. We define the *characteristic formula* $\theta \in \mathrm{FO}_1^{\mathrm{local}}(\sigma)$ of $((\rho_i)_{i \in [k]}, (t_i)_{i \in [k]}, D)$ inductively as follows:

$$\theta_{k+1}(x_1, \ldots, x_{k+1}) := \bigwedge_{1 \leq i < j \leq k+1} (x_i \neq x_j) \wedge \bigwedge_{(i,j) \in D} f(x_i) = x_j$$
$$\theta_i(x_1, \ldots, x_i) := \exists y_i \, [\varphi_{t_i}^{\rho_i}(y_i) \wedge \theta_{i+1}(x_1, \ldots, x_i, y_i)] \qquad (1 \leq i \leq k)$$
$$\theta(x) := \theta_1(x)$$

Note that the rank of θ is at most $\rho_1 + 1 = \max\{\rho_i + i : 1 \leq i \leq k\}$.

Lemma 3.8 *Let $\mathbf{F}$ be a σ_3-mapping with no cycle of length $1 < \ell \leq r_4$, and let $\Upsilon : F \to \mathcal{T}_{r_3}(\sigma)$ be such that*

1. *for every unary mark M in the signature and every $v \in F$, $M(v)$ holds in $\mathbf{F}$ if and only if $M(x) \in \Upsilon(v)$;*
2. *for every $1 \leq \ell \leq r_4$ and every $v \in F$ it holds that $[f^i(x) = x] \notin \Upsilon(v)$.*
3. *for every $v \in F$ it holds that*

$$\mathrm{adm}^+(\Upsilon(v), \pi_{r_2}(\Upsilon(f_\mathbf{F}(v)))) = 1;$$

4. *for every $v \in F$ and $t \in \mathcal{T}_{r_2}(\sigma_3)$ it holds that*

$$\min(r_2, \mathrm{adm}^-(\Upsilon(v), t)) = \min(r_2, |\{u \in f_\mathbf{F}^{-1}(v) : \pi_{r_2}(\Upsilon(u)) = t\}|).$$

Then for every $v \in \mathbf{F}$ it holds that $\mathsf{Type}_{r_2}^\mathbf{F}(v) = \pi_{r_2}(\Upsilon(v))$.

Proof First note that Property 3 implies that for every $0 \leq i < r_2$ and every $v \in F$ it holds that

$$(\pi_i \circ \Upsilon) \circ f_\mathbf{F} = \xi \circ (\pi_{i+1} \circ \Upsilon)x.$$

Note that this is analog to (12), which states that for every non-negative integer i and every mapping $\mathbf{M}$ it holds that

$$\mathsf{Type}_i^\mathbf{M} \circ f_\mathbf{M} = \xi \circ \mathsf{Type}_{i+1}^\mathbf{M}.$$

For $v \in F$, let $\mathbf{M}$ be a countable model of $(\exists x)\,\varphi_{\Upsilon(v)}(x)$, and let $z \in M$ be such that $\mathbf{M}_0 \models \varphi_{\Upsilon(v)}(z)$, that is $\mathsf{Type}_{r_3}^{\mathbf{M}_0}(z) = \Upsilon(v)$. By Property 3 it holds that $f^d(x) = x$ belongs to no $\Upsilon(u)$ at distance at most $r_3 - d$ from z. Considering the ball of radius $r_3 + 1$ around z we deduce that there exists a connected mapping $\mathbf{M}$ with a special element z, which has no cycle of length > 1 (hence the Gaifman graph of $\mathbf{M}$ is a tree), at most one fixed point at distance $r_3 + 1$ from z, and such that $\mathsf{Type}_{r_3}^\mathbf{M}(z) = \Upsilon(v)$.

In order to prove $\mathsf{Type}_{r_2}^\mathbf{F}(v) = \pi_{r_2} \circ \Upsilon(v) = \mathsf{Type}_{r_2}^\mathbf{M}(z)$ it is sufficient to prove that Duplicator has a winning strategy for the r_2 steps local Ehrenfeucht–Fraïssé game between $(\mathbf{F}, v)$ and $(\mathbf{M}, z)$.

Assume that for some $0 \leq k < r_2$ we have $v_0, \dots, v_k \in F$ and $z_0, \dots, z_k \in M$ with $v_0 = v$ and $z_0 = z$, such that $v_i \mapsto z_i$ is a partial isomorphism, and such that for every $0 \leq i \leq k$ it holds that

$$\mathsf{Type}_{r_2-i}^\mathbf{M}(z_i) = \pi_{r_2-i} \circ \Upsilon(v_i).$$

Now consider a Spoiler move. There are six cases:

(i) Spoiler chooses $v_{k+1} \in F$, and there exists $0 \leq a < k+1$ such that $f_\mathbf{F}(v_a) = v_{k+1}$.

In this case, $\mathsf{Type}^M_{r_2-a}(z_a) = \pi_{r_2-a} \circ \Upsilon(v_a)$ implies

$$\mathsf{Type}^M_{r_2-a-1} \circ f_M(z_a) = \xi \circ \mathsf{Type}^M_{r_2-a}(z_a)$$
$$= \xi \circ \pi_{r_2-a} \circ \Upsilon(v_a)$$
$$= \pi_{r_2-a-1} \circ \Upsilon \circ f_F(v_a)$$

Thus we can let $z_{k+1} = f_M(z_a)$.

(ii) Spoiler chooses $v_{k+1} \in F$, there exists $0 \le a < k+1$ such that $f_F(v_{k+1}) = v_a$, and for every $0 \le i < a$ it holds $f_F(v_i) \ne v_a$.
Let $b_1 < b_2 < \cdots < b_{\ell+1} = k+1$ be such that $f_F^{-1}(v_a) \cap \{v_0, \ldots, v_{k+1}\} = \{v_{b_1}, \ldots, v_{b_{\ell+1}}\}$. Note that $b_1 > a$ by assumption.
For $1 \le i \le \ell+1$, let $\rho_i = r_2 - b_i$ and $t_i = \pi_{\rho_i} \circ \Upsilon(v_{b_i})$. Let D be the set of pairs $(i, 1)$ for $2 \le i \le \ell+2$, and let $\theta(x)$ be the characteristic formula of $((\rho_i)_{i \in [\ell+1]}, (t_i)_{i \in [\ell+1]}, D)$. This formula has rank at most $\rho_1 + 1 \le r_2 - a$ so it holds that $\theta(x) \in \pi_{r_2-a} \circ \Upsilon(v_a) = \mathsf{Type}^M_{r_2-a}(z_a)$. Thus there exist $z'_{b_1}, \ldots, z'_{b_\ell}, z'_{k+1}$ in $f_M^{-1}(z_a) \setminus \{z_a\}$, such that

$$\mathsf{Type}^M_{\rho_i}(z'_{b_i}) = \pi_{r_2-b_i} \circ \Upsilon(v_{b_i}) \quad (1 \le i \le \ell+1). \tag{18}$$

If z'_{b_i} is not equal to z_{b_i} for every $1 \le i \le \ell$, let i be minimum such that $z'_{b_i} \ne z_{b_i}$.

- If $z_{b_i} = z'_{b_j}$ for some $j > i$ then it holds that

$$t_j = \mathsf{Type}^M_{\rho_j}(z'_{b_j}) = \mathsf{Type}^M_{\rho_j}(z_{b_i}) \subseteq \mathsf{Type}^M_{\rho_i}(z_{b_i}) = t_i$$

and we deduce that (19) still holds after exchange of z'_{b_i} and z'_{b_j}.
- Otherwise, we let $z'_{b_i} = z_{b_i}$ and remark that (19) still holds.

We repeat this process until we get $z'_{b_i} = z_{b_i}$ for every $1 \le i \le \ell$. Then we let $z_{k+1} = z'_{k+1}$.

(iii) Spoiler chooses $v_{k+1} \in F$, there exists $0 \le a < k+1$ such that $f_F(v_{k+1}) = v_a$, and there exists $0 \le i < a$ such that it holds that $f_F(v_i) = v_a$.
Let $b_1 < b_2 < \cdots < b_{\ell+1} = k+1$ be such that $f_F^{-1}(v_a) \cap \{v_0, \ldots, v_{k+1}\} = \{v_{b_1}, \ldots, v_{b_{\ell+1}}\}$.
Note that there can be only one $0 \le i < a$ s.t. $f_F(v_i) = v_a$, as otherwise the two vertices would not be connected before step a, so $b_1 < a < b_2$.
For $1 \le i \le \ell+1$, let $\rho_1 = r_2 - a$, $t_1 = \pi_{r_2-a} \circ \Upsilon(v_a)$, and $\rho_i = r_2 - b_i$, $t_i = \pi_{\rho_i} \circ \Upsilon(v_{b_i})$ for $2 \le i \le \ell+1$.
Let D be the set of pairs $(i, 2)$ for $i \in \{1, \ldots, \ell+2\} \setminus \{2\}$, and let $\theta(x)$ be the characteristic formula of $((\rho_i)_{i \in [\ell+1]}, (t_i)_{i \in \ell+1}, D)$. This formula has rank at most $\rho_1 + 1 \le r_2 - a + 1 \le r_2 - b_1$ so it holds that $\theta(x) \in \pi_{r_2-b_1} \circ \Upsilon(v_{b_1}) = \mathsf{Type}^M_{r_2-b_1}(z_{b_1})$. Thus, there exists $z'_a, z'_{b_2}, \ldots, z'_{b_\ell}, z'_{k+1} \in f_M^{-1} \circ f_M(z_{b_1})$, all distinct, such that $f_M(z_{b_1}) = z_a$ and

$$\mathsf{Type}^{\mathbf{M}}_{\rho_i}(z'_{b_i}) = \pi_{r_2-b_i} \circ \Upsilon(v_{b_i}) \qquad (2 \leq i \leq \ell+1). \tag{19}$$

As in the previous case, we can assume $z'_{b_i} = z_{b_i}$ for $2 \leq i \leq \ell$ and let $z_{k+1} = z'_{k+1}$.

(iv) Spoiler chooses $z_{k+1} \in M$, and there exists $0 \leq a < k+1$ such that $f_{\mathbf{M}}(z_a) = z_{k+1}$.
As in Case (i), $\mathsf{Type}^{\mathbf{M}}_{r_2-a}(z_a) = \pi_{r_2-a} \circ \Upsilon(v_a)$ implies

$$\mathsf{Type}^{\mathbf{M}}_{r_2-a-1} \circ f_{\mathbf{M}}(z_a) = \pi_{r_2-a-1} \circ \Upsilon \circ f_{\mathbf{F}}(v_a).$$

Thus we can let $v_{k+1} = f_{\mathbf{F}}(v_a)$.

(v) Spoiler chooses $z_{k+1} \in M$, there exists $0 \leq a < k+1$ such that $f_{\mathbf{M}}(z_{k+1}) = z_a$, and for every $0 \leq i < a$ it holds $f_{\mathbf{M}}(z_i) \neq z_a$.
Let $\tau = \Upsilon(v_a)$, let $t = \mathsf{Type}^{\mathbf{M}}_{r_2-(k+1)}(z_{k+1})$, and let p be the number of elements of $f_{\mathbf{M}}^{-1}(z_a) \cap \{z_0, \ldots, z_{k+1}\}$ with rank $(r_2 - (k+1))$ local type t.
By assumption, it holds that $\mathsf{Type}^{\mathbf{M}}_{r_2-a}(z_a) = \pi_{r_2-a}(\tau)$. Thus

$$\sum_{t' \prec t} \mathrm{adm}^-(\tau, t') \geq p,$$

where the sum is over local types $t' \in \mathcal{T}_{r_2}(\sigma_3)$ such that $t' \prec t$. According to Property 4, it holds that

$$\sum_{t' \prec t} \mathrm{adm}^-(\tau, t') = |\{u \in f_{\mathbf{F}}^{-1}(v_a) : \pi_{r_2-(k+1)}(\Upsilon(u)) = t\}|.$$

It follows that there exists $v_{k+1} \in f_{\mathbf{F}}^{-1}(v_a)$, distinct from $v_0, \ldots, v_k$, such that $\pi_{r_2-(k+1)}(\Upsilon(v_{k+1})) = \mathsf{Type}^{\mathbf{M}}_{r_2-(k+1)}(z_{k+1})$.

(vi) Spoiler chooses $z_{k+1} \in M$, there exists $0 \leq a < k+1$ such that $f_{\mathbf{M}}(z_{k+1}) = z_a$, and there exists $0 \leq i < a$ such that it holds that $f_{\mathbf{M}}(z_i) = z_a$.
This case is solved similarly, by considering the element z_i such that $f_{\mathbf{M}}(z_i) = z_a$, and showing that the number of elements of $f_{\mathbf{F}}^{-1}(v_i)$ with same rank $(r_2 - (k+1))$ local type as z_{k+1} is at least equal to the number of elements of $f_{\mathbf{F}}^{-1}(z_i) \cap \{z_0, \ldots, z_{k+1}\}$ with same rank $(r_2 - (k+1))$ local type as z_{k+1}.

Lemma 3.9 *Let $r_3 > 2r_2$ be positive integers, and let $\widehat{\mu}$ be a rational probability measure on $\mathcal{T}_{r_3}(\sigma_3)$, such that*

1. *$\widehat{\mu}$ is clean: for every $\tau \in \mathcal{T}_{r_3}(\sigma_3)$ with $\widehat{\mu}(\tau) > 0$ and for every $t \in \mathcal{T}_{r_3-1}(\sigma_3)$, if $\phi_t(f(x)) \in \tau$ then $\sum_{\tau' \prec t} \widehat{\mu}(\tau') > 0$;*
2. *for every $1 < i \leq r_3$ the formula $f^i(x) = x$ does not belong to any $\tau \in \mathcal{T}_{r_3}(\sigma_3)$ with positive $\widehat{\mu}$-measure;*
3. *the measure $\widehat{\mu}$ satisfies the (r_3, r_2)-restricted MTP.*

Then there exists a finite σ_3-mapping $\mathbf{F}_3$ such that for every local formula $\phi \in \mathrm{FO}_1^{\mathrm{local}}(\sigma_3)$ with local rank at most r_2 it holds that

$$\langle \phi, \mathbf{F}_3 \rangle = \sum_{\tau \ni \phi} \widehat{\mu}(\tau). \tag{20}$$

Proof Let $T_{r_3} = \{\tau \in \mathcal{T}_{r_3}(\sigma_3) : \widehat{\mu}(\tau) > 0\}$ and $T_{r_2} = \{\pi_{r_2}(\tau) : \tau \in T_{r_3}\}$.

Let $N_2 \in \mathbb{N}$ be such that $N_2 \widehat{\mu}$ is integral, and let $\zeta : [N_2] \to T_{r_3}$ be such that for every $\tau \in T_{r_3}$ it holds $|\zeta^{-1}(\tau)| = N_2 \widehat{\mu}(\tau)$.

We construct a (partial) mapping $g : [N_2] \to [N_2]$ inductively. We start with an empty domain. For each $i \in [N]$ (not yet in the domain), let $t = \pi_r(\zeta(i))$. We consider the elements of $[N_2]$ such that $\mathrm{adm}^-(\zeta(j), t)$ is either $r_2 + 1$, or greater than the number of $k \in g^{-1}(j)$ such that $\zeta(k) \prec t$. Among these elements, we choose one element j such that $\mathrm{adm}^-(t, \zeta(j))$ is minimal, and let $g(i) = j$.

Now we prove that the above construction never gets stuck and that, at the end of the day, for every $j \in [N_2]$ and every $t \in \mathcal{T}_{r_2}(\sigma_3)$ it holds that

$$\min(r_2, \mathrm{adm}^-(\zeta(j), t)) = \min(r_2, |\{k \in g^{-1}(j) : \zeta(k) \prec t\}|). \tag{21}$$

Assume for contradiction that the construction gets stuck when trying to extend the domain of g to some $i \in [N_2]$. Let $\tau = \zeta(i)$, let $t_1 = \pi_{r_2}(\tau)$, and let t_2 be the unique rank r_2 local type such that $\varphi_{t_2}(f(x)) \in \tau_1$. By assumption, for every $\tau_2 \in T_{r_3}$ with $\tau_2 \prec t_2$ it holds that $\mathrm{adm}^-(\tau_2, t_1) \leq r_2$. Hence, by the (r_3, r_2)-restricted MTP, it holds that

$$\sum_{\tau_1 \prec t_1} \mathrm{adm}^+(\tau_1, t_2) \mu(\tau_1) = \sum_{\tau_2 \prec t_2} \mathrm{adm}^-(\tau_2, t_1).$$

Thus

$$|\{i : \pi_{r_2}(\zeta(i)) = t_1\}| = \sum_j |\{k \in g^{-1}(j) : \zeta(k) \prec t_1\}|,$$

which contradicts the hypothesis that the construction gets stuck.

Now assume for contradiction that (21) does not hold. Then there exists t_1 and j_0 such that

$$|\{k \in g^{-1}(j_0) : \zeta(k) \prec t_1\}| < \min(r_2, \mathrm{adm}^-(\zeta(j_0), t_1)).$$

Let $t_2 = \pi_{r_2}(\zeta(j_0))$. According to the construction of g, it holds for every j such that $\zeta(j) \prec t_2$ that

$$|\{k \in g^{-1}(j) : \zeta(k) \prec t_1\}| \leq \min(r_2, \mathrm{adm}^-(\zeta(j), t_1)).$$

Hence we have

$$\sum_{\tau_2 \prec t_2} \min(r_2, \mathrm{adm}^-(\zeta(j), t)) \widehat{\mu}(\tau_2) > \frac{1}{N_2} \sum_{\tau_2 \prec t_2} \sum_{j:\zeta(j)=\tau_2} |\{k \in g^{-1}(j) : \zeta(k) \prec t_1\}|$$

$$= \frac{1}{N_2} \sum_{\tau_1 \prec t_1} \sum_{i:\zeta(i)=\tau_1} \mathrm{adm}^+(\zeta(i), t_2)$$

$$= \sum_{\tau_1 \prec t_1} \mathrm{adm}^+(\tau_1, t_2) \widehat{\mu}(\tau_1)$$

which contradicts the (r_3, r_2)-restricted MTP. Thus (21) holds.

The σ_3-mapping $\mathbf{F}_3$ has domain $[N_2]$. For every unary relation symbol $S \in \sigma_3$ we let $S(\mathbf{F}_3) = \{i \in \mathbf{F}_3 : S(x) \in \zeta(i)\}$, and define $f_{\mathbf{F}_3} = g$.

Note that $\mathbf{F}$ has no cycle of length ℓ with $1 \le \ell \le r_4$: as $f(x) \wedge U_i(x) \to U_{(i+1) \bmod r_4}(f(x))$ holds with probability 1. Hence, all the cycles have their length a multiple of r_4.

That $\mathsf{Type}^{\mathbf{F}}_{r_2}(v) = \pi_{r_2}(\zeta(v))$ holds for every $v \in F_3$ then follows from Lemma 3.8.

As a consequence of Lemma 3.6 and Eq. 20 it holds that

$$\mathrm{Dist}^{\mathrm{local}}_{1, r_2}(\mathbf{F}_3, \mathbf{L}_3) < \epsilon_5. \tag{22}$$

3.6 From $\mathbf{F}_3$ to $\mathbf{F}_2$: Rewiring Short Cycles

We now let $\mathbf{F}_2 = \mathsf{I}_2(\mathbf{F}_3)$. Every local formula $\phi \in \mathrm{FO}^{\mathrm{local}}_1(\sigma_2)$ with local rank at most $2r_1$ corresponds (for the I_2 interpretation) to a local formula $\widehat{\phi}$ with local rank at most $2r_1(2r_1 - 1) < r_2$. Then it holds that

$$|\langle \phi, \mathbf{F}_2 \rangle - \langle \phi, \widetilde{\mathbf{L}}_2 \rangle| = |\langle \widehat{\phi}, \mathbf{F}_3 \rangle - \langle \widehat{\phi}, \mathbf{L}_3 \rangle|.$$

Thus

$$\mathrm{Dist}^{\mathrm{local}}_{1, 2r_1}(\mathbf{L}_2, \mathbf{F}_2) \le \mathrm{Dist}^{\mathrm{local}}_{1, 2r_1}(\mathbf{L}_2, \widetilde{\mathbf{L}}_2) + \mathrm{Dist}^{\mathrm{local}}_{1, 2r_1}(\widetilde{\mathbf{L}}_2, \mathbf{F}_2) < \epsilon_5. \tag{23}$$

3.7 The Mapping $\mathbf{E}_1$: A Finite Model

A *terminal* of T_R is a type τ such that if t' is such that $\mathrm{adm}^+(\tau, t') = 1$ then $\sum_{\tau' \prec t'} \mu(\tau') = 0$. Importance of terminal types will be a consequence of the following useful fact:

Claim 3.10 *Let τ_1 be such that $\mu(\tau_1) > 0$, and let t_2 be such that $\mathrm{adm}^+(\tau_1, t_2) = 1$.*

Then at least one of the following holds:

1. *there exists $\tau_2 \prec t_2$ such that $\mu(\tau_2) > 0$;*
2. *there exists $\tau_2 \prec t_2$ such that $\mathrm{adm}^-(t_1, \tau_2) > r$.*

Proof Let t_1 be such that $\tau_1 \prec t_1$. Assume that for every $\tau_2 \prec t_2$ such that $\mathrm{adm}^+(\tau_1, t_2) = 1$ it holds $\mathrm{adm}^-(t_1, \tau_2) \leq r$. Then, according to the FMTP, it holds

$$r \sum_{\vartheta_2 \prec t_2} \mu(\vartheta_2) \geq \sum_{\vartheta_2 \prec t_2} \mathrm{adm}^-(\vartheta_2, t_1)\mu(\vartheta_2)$$

$$= \sum_{\vartheta_1 \prec t_1} \mathrm{adm}^+(\vartheta_1, t_2)\mu(\vartheta_1)$$

$$\geq \mu(\tau_1) > 0$$

Thus there exists $\tau_2 \prec t_2$ such that $\mu(\tau_2) > 0$. □

A type τ' is a *hub type* if there exists $\tau \prec t$ such that τ is a terminal and $\mathrm{adm}^-(\tau', t) > r$. Let $\tau_1, \ldots, \tau_k$ be the terminal types of **L**, and let $\tau'_1, \ldots, \tau'_k$ be associated hub types.

Lemma 3.11 *There exists a finite mapping **M** such that $\mathbf{M} \equiv_{r_s} \mathbf{L}$, and such that there are elements*

$$h_{1,1}, \ldots, h_{1,N_1}, \ldots, h_{k,1}, \ldots, h_{k,N_1} \in M,$$

pairwise at distance at least 2^r, such that $\mathsf{Type}^{\mathbf{M}}_{r_2}(h_{i,j}) = \tau'_i$.

Proof We consider the formula ζ with free variables

$$x_{1,1}, \ldots, x_{1,N_1}, \ldots, x_{k,1}, \ldots, x_{k,N_1},$$

defined by

$$\zeta := \left(\bigwedge_{(i,j) \neq (i',j')} \mathrm{dist}(x_{i,j}, x_{i',j'}) > 2r_1 \right) \wedge \left(\bigwedge_{1 \leq i \leq k} \bigwedge_{1 \leq j \leq N_1} \varphi_{\tau'_i}(x_{i,j}) \right)$$

and the sentence

$$\theta := (\exists x_{1,1}, \ldots, x_{1,N_2}, \ldots, x_{k,1}, \ldots, x_{k,N_1})\zeta$$

The hub types can be chosen in such a way that θ is satisfied in **L**. Indeed, for each $\tau' < t'$ such that $\mathrm{adm}(\tau, t') = 1$ the connected component of any $v \in \phi_{\tau'}(\mathbf{L})$ has measure 0 (as **L** is residual) hence it is possible, for each terminal τ to choose τ' in such a way that there are in **L** uncountably many connected components with an element in $\phi_{\tau'}(\mathbf{L})$.

3.8 From $\mathbf{E}_1$ and $\mathbf{F}_2$ to $\mathbf{F}_1$: Merging

Let $S = \{v_1, \ldots, v_k\}$ be the set of all terminal elements of $\mathbf{F}_2$, and let $\gamma(v_i)$ be the rank r-type corresponding to elements of $\mathbf{L}$ having type $\Upsilon(v_i)$ in $\mathbf{L}_1$.

Let $N_3 = \lceil \frac{|E_1|}{|F_2|\epsilon_2} \rceil$, and let F_1 be the disjoint union of E_1 and $F_2 \times [N_3] \times [N_1]$. If $v \in M$, we define $f_{\mathbf{F}_1}(v) = f_{\mathbf{E}_1}(v)$. Otherwise, if $(v, i, j) \in F_2 \times [N_3] \times [N_1]$ we define

$$f_{\mathbf{F}_1}(v, i, j) = \begin{cases} (f_{\mathbf{F}_2}(v), i, j) & \text{if } v \notin S \\ h_{a,i} & \text{if } v = v_a \in S \end{cases}$$

(See Fig. 3.)

We consider the finite mapping $\widetilde{\mathbf{E}}$ obtained from $\mathbf{E}_1$ as follows: For $1 \leq i \leq k$ and $1 \leq j \leq N_3$, and every $z \in E_1$ such that $f_{\mathbf{E}_1}(z) = h_{i,j}$ and $\mathsf{Type}_{r_1}^{\mathbf{E}_1}(z) = \gamma(v_i)$, we mark z by mark $A_{i,j}$ and let $f_{\widetilde{\mathbf{E}}}(z) = z$. For all other elements $z \in E_1$ we let $f_{\widetilde{\mathbf{E}}}(z) = f_{\mathbf{E}_1}(z)$. Moreover, each $h_{i,j}$ receives mark $B_{i,j}$. There is an easy basic quantifier-free interpretation I such that $\mathsf{I}(\widetilde{\mathbf{E}}) = \mathbf{E}_1$.

Now we consider the disjoint union $\widetilde{\mathbf{F}}$ of $\widetilde{\mathbf{E}}$ and $N_1 N_3$ copies of $\mathbf{F}_2$, such that terminal v_i in copy (j, k) is marked $A_{i,j}$, and we let $\mathbf{F}_1 = \mathsf{I}(\widetilde{\mathbf{F}})$.

Lemma 3.12 *The finite mappings $\mathbf{E}_1$ and $\mathbf{F}_1$ are r_1-equivalent.*

Proof It is a direct consequence of Hanf's locality theorem that $\widetilde{\mathbf{F}}$ is r_1-equivalent to $\widetilde{\mathbf{E}}$. It follows that $\mathbf{F}_1 = \mathsf{I}(\widetilde{\mathbf{F}})$ is r_1-equivalent to $\mathbf{E}_1 = \mathsf{I}(\widetilde{\mathbf{E}})$.

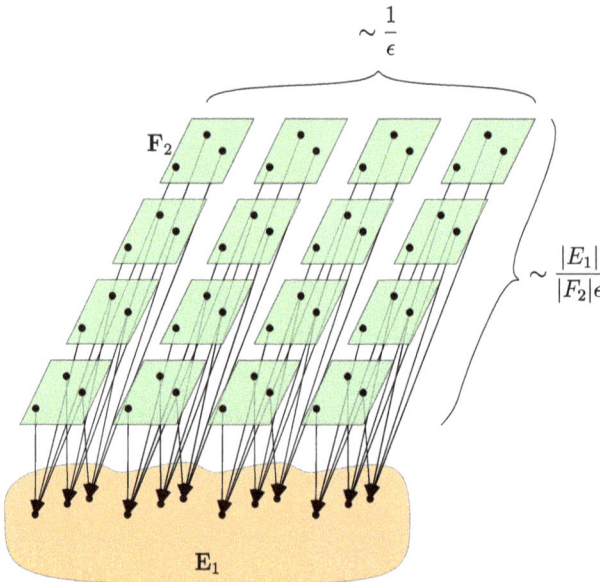

Fig. 3 Merging $\mathbf{M}$ with copies of $\mathbf{F}$

Approximations of Mappings 369

Lemma 3.13 *Each element (v, i, j) in a copy of $\mathbf{F}_2$ in $\mathbf{F}_1$ is such that*

$$\mathsf{Type}^{\mathbf{F}_1}_{r_1}(v, i, j) = \gamma(v).$$

Proof This follows easily from an Ehrefeucht-Fraïssé game.

Lemma 3.14
$$\langle [\mathrm{dist}(x_1, x_2) \leq 2r_1, \mathbf{F}_1 \rangle < \epsilon_2$$

Proof Every ball of radius $2r_1$ contains less than $|E_1| + N_3|F_2|$ elements. Thus the probability $\langle [\mathrm{dist}(x_1, x_2) \leq 2r_1, \mathbf{F}_1\rangle$ that two random elements in $\mathbf{F}_1$ are at distance at most $2r_1$ is less than $\frac{|E_1|+N_3|F_2|}{|E_1|+N_1 N_3|F_2|} < \epsilon_2$.

Lemma 3.15 *It holds that*
$$\mathrm{Dist}^{\mathrm{local}}_{1,r_1}(\mathbf{F}_1, \mathbf{L}_1) < \epsilon_3. \tag{24}$$

Proof Let $\phi \in \mathrm{FO}^{\mathrm{local}}_1$ be a formula with local rank at most r_1. Let $\psi = \bigvee_{t \ni \phi} R_t$, where the disjunction is over rank r_1-local types. Then $\langle\phi, \mathbf{L}_1\rangle = \rangle\psi, \mathbf{L}_2\rangle$. According to Lemma 3.13 it holds that

$$|\langle\phi, \mathbf{F}_1\rangle - \langle\psi, \mathbf{F}_2\rangle| \leq \frac{|E_1|}{|F_1|} \leq \frac{1}{1 + N_1 N_3 \frac{|F_2|}{|E_1|}} \leq \frac{1}{1 + \frac{2}{\epsilon_2^2}} < \frac{\epsilon_2^2}{2}.$$

Thus

$$|\langle\phi, \mathbf{F}_1\rangle - \langle\phi, \mathbf{L}_1\rangle| \leq |\langle\phi, \mathbf{F}_1\rangle - \langle\psi, \mathbf{F}_2\rangle| + |\langle\psi, \mathbf{F}_2\rangle - \langle\psi, \mathbf{L}_2\rangle|$$
$$< \frac{\epsilon_2^2}{2} + \epsilon_5 < \epsilon_3.$$

3.9 From $\mathbf{F}_1$ to $\mathbf{F}$: Approximation of the Original Mapping

At this stage, we have constructed a finite mapping $\mathbf{F}_1$ such that $\mathbf{L}_1 \equiv_{r_1} \mathbf{F}_1$ and $|\langle\psi, \mathbf{L}_1\rangle - \langle\psi, \mathbf{F}_1\rangle| < \epsilon_3$ for every $\psi \in \mathrm{FO}^{\mathrm{local}}_1$ with rank at most r_1.

Let $\mathbf{F} = \mathsf{I}(\mathbf{F}_1)$, where I_1 is the interpretation defined in Sect. 3.1. The following lemma ends the proof of Theorem 1.4.

Lemma 3.16 *For every formula ϕ with p free variables and rank at most r it holds that*
$$|\langle\varphi, \mathbf{L}\rangle - \langle\varphi, \mathbf{F}\rangle| < \epsilon.$$

Proof Let ϕ be a local formula with at most p free variables and rank at most r_1.

As $\mathbf{L}_1$ is ϵ_2-residual, according to Lemma 2.11, it holds that

$$\text{Dist}^{\text{local}}_{p,r_1}(\mathbf{L}_1, \mathbf{F}_1) \leq 2p\text{Dist}^{\text{local}}_{1,r_1}(\mathbf{L}_1, \mathbf{F}_1) + \binom{p}{2}(\langle \delta_{2r_1}, \mathbf{L}_1\rangle + \langle \delta_{2r_1}, \mathbf{F}_1\rangle)$$

$$< 2p\epsilon_3 + \binom{p}{2}\epsilon_2 < \epsilon_1.$$

We deduce from Lemma 2.9 and the definitions of r_1 and ϵ_1 that $|\langle \widetilde{\varphi}, \mathbf{L}_1\rangle - \langle \widetilde{\varphi}, \mathbf{F}_1\rangle| < \epsilon$ holds true for every first-order formula $\widetilde{\varphi}$ with at most p free variables and rank at most r.

Let φ be a first-order formula with at most p free variables and rank at most r. Then there exists a formula $\widetilde{\varphi}$ with at most p free variables and rank at most r such that $\langle \widetilde{\varphi}, \mathbf{L}_1\rangle = \langle \varphi, \mathbf{L}\rangle$ and $\langle \widetilde{\varphi}, \mathbf{F}_1\rangle = \langle \varphi, \mathbf{F}\rangle$. Hence $|\langle \varphi, \mathbf{L}\rangle - \langle \varphi, \mathbf{F}\rangle| < \epsilon$.

This ends the last reduction step in the proof of Theorem 1.4. As explained above (see Fig. 1 and comments preceding it) this finishes the proof of Theorem 1.4.

4 Local Approximation

The aim of this section is to prove Theorem 1.3 by following steps similar to those we followed to prove Theorem 1.4.

The first main difference is that we cannot use general first-order interpretations, but only local interpretations. Thus we cannot follow the first reduction step to reduce to the ϵ-residual case. Instead, we shall prove that every connected mapping modeling is close (for the topology of local convergence) to a connected mapping modeling with the finite model property, for which Theorem 1.4 applies. The strategy will be to consider first the connected components of $\mathbf{L}$ with non-negligible measures, and then the remaining components of the mapping modeling.

For ϵ-residual mapping modelings, we can follow the proof of Theorem 1.4 until Step 8. In this step, the model M will be replaced by the union of models of the hub local types.

4.1 Connected Mapping Modelings

Let $\mathbf{L}$ be a connected mapping modeling. We define a directed graph modeling $\widehat{\mathbf{L}}$ with countably many marks M and N as follows:

- The domain of $\widehat{\mathbf{L}}$ is L, with same probability measure;
- if $Z(\mathbf{L}) \neq \emptyset$, we arbitrarily mark a vertex $v \in Z(\mathbf{L}) \neq \emptyset$ with mark M and its image $f(v)$ with mark N;
- the arcs of $\widehat{\mathbf{L}}$ are the pairs $(v, f(v))$ for which v is not marked by M.

Approximations of Mappings

The following lemma is much stronger than what we need. It would be sufficient to say that for some d the ball of radius d around the root has measure at least $1 - \epsilon$. Now the idea is that the ball B of radius $d + r$ around the root of $\mathbf{L}$ not only has measure close to 1, but also has the property that less than ϵ measure of the elements have different rank r local type in $\mathbf{L}$ and $\mathbf{L} \mid B$. Now $\mathbf{L} \mid B$ has finite height hence enjoys the finite model property. An FO-approximation of $\mathbf{L} \mid B$ is then an FO^{local}-approximation of $\mathbf{L}$.

Lemma 4.1 *Let $\mathbf{L}$ be a connected mapping modeling with atomless measure $\nu_{\mathbf{L}}$ that satisfies the MTP, and let $\epsilon > 0$ be a positive real.*

Then for every $r \in L$ there exists $d \in \mathbb{N}$ such that the the subset $A \subseteq L$, defined as the union of the vertex sets of all the (undirected) paths of length at least $d + 1$ in $\widehat{\mathbf{L}}$ with endpoint r, has measure at most ϵ.

Proof There exists an even integer d such that the ball $B_{d/2}(\widehat{\mathbf{L}}, r)$ has measure at least $(1 - \epsilon/2)$. For $0 \leq i \leq d$, let S_i be the set of all vertices of A at distance exactly i from r. According to the MTP (and uniqueness of paths from a vertex v to r), and as $\nu_{\mathbf{L}}$ is atomless, it holds that

$$0 = \nu_{\mathbf{L}}(\{r\}) \leq \nu_{\mathbf{L}}(S_1) \leq \cdots \leq \nu_{\mathbf{L}}(S_d).$$

Thus it holds that

$$\nu_{\mathbf{L}}\left(\bigcup_{i=0}^{d/2} S_i\right) \leq \nu_{\mathbf{L}}\left(\bigcup_{i=d/2+1}^{d} S_i\right).$$

That is:

$$\nu_{\mathbf{L}}(A \cap B_{d/2}(\widehat{\mathbf{L}}, r)) \leq \nu_{\mathbf{L}}\left(\bigcup_{i=d/2}^{d} S_i\right)$$
$$\leq \nu_{\mathbf{L}}(L \setminus B_{d/2}(\widehat{\mathbf{L}}, r)).$$

Thus

$$\nu_{\mathbf{L}}(A) \leq \nu_{\mathbf{L}}(A \cap B_{d/2}(\widehat{\mathbf{L}}, r)) + \nu_{\mathbf{L}}(A \setminus B_{d/2}(\widehat{\mathbf{L}}, r))$$
$$\leq 2\nu_{\mathbf{L}}(L \setminus B_{d/2}(\widehat{\mathbf{L}}, r))$$
$$< \epsilon$$

Definition 4.2 Let $\mathbf{L}$ be a colored mapping modeling with finite height and let $r \in \mathbb{N}$. We define the *standard r-approximation* $\check{\mathbf{L}}$ of $\mathbf{L}$ as follows:

Let $C = Z(\mathbf{L})$ and $C_i = Z_i(\mathbf{L})$. For $x \in L$ let $h(x)$ be the minimum non-negative integer k such that $f_{\mathbf{L}}^k(x) \in C$. Note that $0 \leq h(x) \leq \text{height}(\mathbf{L})$. Let $p = \max_{x \in C \setminus C_1} h(x)$. We iteratively define sets X_i for $i = p, \ldots, 1$, together with an

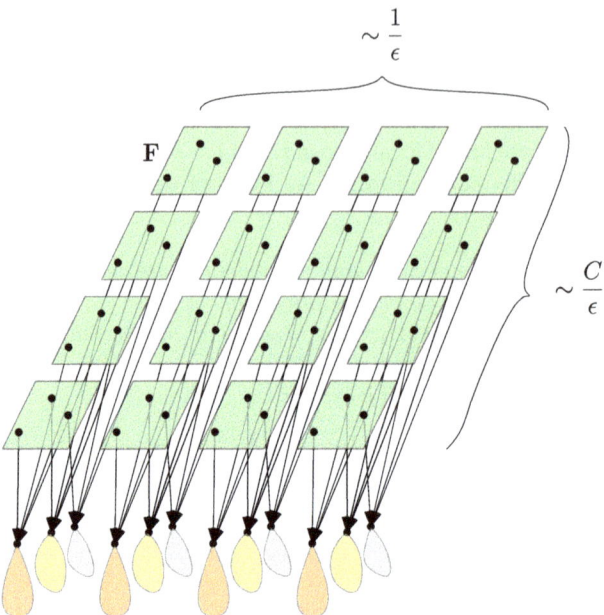

Fig. 4 Merging small models with many copies of **F**

equivalence relation $\sim_i$ on $h^{-1}(i) \cap f_{\mathbf{L}}(X_{i+1})$ (if $i < p$). We start with $i = p$ and define $\sim_p$ on L by $x \sim_{p+1} y$ if x and y have the same color. For every $y \in h^{-1}(i-1)$ we choose an inclusion maximal subset $I(y)$ of $f_{\mathbf{L}}^{-1}(y)$ containing no $r+1$ $\sim_i$-equivalent vertices. Then we define $X_i = \bigcup_{y \in h^{-1}(i-1)} I(y)$, and we define the equivalence relation $\sim_{i-1}$ on $h^{-1}(i-1) \cap f_{\mathbf{L}}(X_i)$ by $y_1 \sim_{i-1} y_2$ if for every $z \in f_{\mathbf{L}}^{-1}(y_1) \cup f_{\mathbf{L}}^{-1}(y_2)$ it holds that

$$|\{x_1 \in f_{\mathbf{L}}^{-1}(y_1) : x_1 \sim_i z\}| = |\{x_2 \in f_{\mathbf{L}}^{-1}(y_2) : x_2 \sim_i z\}|.$$

We now consider the restriction g of $f_{\mathbf{L}}$ to $C \cup \bigcup_{i=1}^{p} X_i$. Note that all the connected components have their size bounded by some fixed function of c and p. We consider an inclusion maximal union $\widehat{\mathbf{L}}$ of connected components of g containing no $r+1$ isomorphic connected components. The mapping $\widehat{\mathbf{L}}$ is then the restriction of $\mathbf{L}$ to $\widehat{\mathbf{L}}$. Note that $\widehat{\mathbf{L}}$ has its size bounded by some fixed function of c and p.

An alternate construction can be used, which is parametrized by a pair (r, R) of integers with $r \leq R$. The idea is as follows: we start from the standard R-approximation and then reduce every set of at least $k > r$ equivalent sons to r if either $k < R$ or some descendent of one of these sons as R equivalent sons. Then, according to MTP, the measure of the types of the vertices obtained by removing any R equivalent siblings and their descendants is at most $F(r, t)/R$. So one should require $R > F(r, t)/\epsilon$.

Lemma 4.3 *Every mapping* **L** *with finite height is r-equivalent to its standard r-approximation, hence has the finite model property.*

Proof An easy strategy for the r-round Ehrenfeucht–Fraïssé game shows that **L** is r-equivalent to its standard r-approximation. □

4.2 Merging with Hub Local Type Models

To each rank r hub local type τ_i' we associate a finite rooted mapping $(\mathbf{M}_i, h_i)$ such that $\mathsf{Type}_r^{\mathbf{M}_i}(h_i) = \tau_i'$. Let **M** be the disjoint union of the $\mathbf{M}_i$. We proceed to the merge of **M** with copies of **F** as in Step 8 of the proof of Theorem 1.4 (Fig. 4).

5 Concluding Remarks

In this paper we considered the approximation problem for mapping modelings. It would be interesting to consider the approximation problem where we have only the probability measure μ corresponding to the satisfaction probability of first-order formulas.

In such a setting, we shall consider probability measures μ on $S(\mathcal{L}_{\mathrm{FO}})$ that are invariant under the natural action of the infinite permutation group S_ω (acting by permuting the free variables in the formulas), whose support projects on a single point $\mathrm{Th}(\mu)$ of $S(\mathcal{L}_{\mathrm{FO}_0})$. The analogs of the property we required for modeling mappings are as follows: The condition for the modeling to be atomless corresponds to the property that the μ-measure of the clopen subset $K(x_1 = k_2)$ of $S(\mathcal{L}_{\mathrm{FO}})$ dual to the formula $x_1 = x_2$ is zero. The finite model property of the modeling corresponds to the property that every sentence in $\mathrm{Th}(\mu)$ has a finite model. The finitary mass transport principle for the modeling corresponds to the following property of μ: for every formulas $\phi, \psi \in \mathrm{FO}_1$ such that $\psi(x)$ entails that there exist exactly (resp. strictly more than) k elements $y_1, \ldots, y_k$ such that $\phi(y_i) \wedge f(y_i) = x$ we have $\mu(K(\phi)) = k\mu(K(\psi))$ (resp. $\mu(K(\phi)) > k\mu(K(\psi))$).

Admittedly the proofs presented in this paper are technical and complex. In a way this was expected as approximating modeling structures with two mappings seem to be fully out of reach.

An interesting question is to solve the inverse problem for acyclic modelings (the modeling equivalent of *treeings*). This problem has been solved in the bounded diameter case [20] by a complicated analysis, and in the bounded degree case by [5]. However the problem for general acyclic modelings remain open.

Another way to make the problem simpler is to assume that the acyclic modeling looks like a directed rooted tree. This is the motivation of the following problem stated in [18]: if a tree modeling is oriented in such a way that the root is a sink and non-roots have outdegree one and if any finite subset of the complete theory of the modeling has a connected finite model, is it true that the modeling is the FO-limit of a sequence of finite rooted trees?

Finally, we would like to mention that random mappings are not FO-convergent, as they do not satisfy a 0–1 law (the expected number of cycles of length r tend to $1/r$ [6]). However it might be possible that random mappings are FO^{local}-convergent.

References

1. Börger, E., Grädel, E., Gurevich, Y.: The classical decision problem. Springer Science & Business Media (2001)
2. Limits of structures and the example of tree-semilattices, Charbit, P., Hosseini, L., Ossona de Mendez, P. Discrete Mathematics **340**, 2589–2603 (2017). https://doi.org/10.1016/j.disc.2017.06.013
3. A. Connes, Classification of injective factors cases $II_1, II_\infty, III_\lambda, \lambda \neq 1$. Annals of Mathematics **104**(1), 73–115 (1976)
4. H.D. Ebbinghaus, J. Flum, *Finite Model Theory* (Springer-Verlag, 1996)
5. G. Elek, On the limit of large girth graph sequences. Combinatorica **30**(5), 553–563 (2010). https://doi.org/10.1007/s00493-010-2559-2
6. Flajolet, P., Odlyzko, A.: Random mapping statistics. In: Workshop on the Theory and Application of of Cryptographic Techniques, pp. 329–354. Springer (1989)
7. Gajarský, J., Hliněný, P., Kaiser, T., Král, D., Kupec, M., Obdržálek, J., Ordyniak, S., Tůma, V.: First order limits of sparse graphs: Plane trees and path-width. arXiv:1504.08122v1 [math.CO] (2015)
8. Gottschalk, W.: Some general dynamical notions. In: A. Beck (ed.) Recent Advances in Topological Dynamics, *Lecture Notes in Mathematics*, vol. 318, pp. 120–125. Springer Berlin Heidelberg (1973). https://doi.org/10.1007/BFb0061728
9. W. Hodges, *Model Theory* (Cambridge University Press, 1993)
10. W. Hodges, *A Shorter Model Theory* (Cambridge University Press, 1997)
11. I. Kaplansky, *Fields and rings* (University of Chicago Press, 1972)
12. Kardoš, F., Král, D., Liebenau, A., Mach, L.: First order convergence of matroids. arXiv:1501.06518v1 [math.CO] (2015)
13. A. Kechris, *Classical descriptive set theory* (Springer-Verlag, 1995)
14. Lascar, D.: La théorie des modèles en peu de maux. Cassini (2009)
15. L. Libkin, *Elements of Finite Model Theory* (Springer-Verlag, 2004)
16. D. Marker, *Model theory: an introduction, Graduate Texts in Mathematics*, vol. 217 (Springer Verlag, 2001)
17. Nešetřil, J., Ossona de Mendez, P.: A model theory approach to structural limits. Commentationes Mathematicæ Universitatis Carolinæ **53**(4), 581–603 (2012)
18. Nešetřil, J., Ossona de Mendez, P.: Modeling limits in hereditary classes: Reduction and application to trees. Electronic Journal of Combinatorics **23**(2), #P2.52 (2016)
19. Limits of mappings, Nešetřil, J., Ossona de Mendez, P. European Journal of Combinatorics **66**, 145–159 (2017). https://doi.org/10.1016/j.ejc.2017.06.021. SelectedpapersofEuroComb15(specialissue)
20. Nešetřil, J., Ossona de Mendez, P.: A unified approach to structural limits (with application to the study of limits of graphs with bounded tree-depth). Memoirs of the American Mathematical Society (2017). 117 pages; to appear
21. Nešetřil, J., Ossona de Mendez, P.: Existence of modeling limits for sequences of sparse structures. The Journal of Symbolic Logic (2018). To appear

22. M. Suslin, Sur une définition des ensembles mesurables b sans nombres transfinis. CR Acad. Sci. Paris **164**(2), 88–91 (1917)
23. B. Trakhtenbrot, The impossibility of an algorithm for the decision problem for finite domains. Doklady Akademii Nauk SSSR **70**, 569–572 (1950)
24. Weiss, B.: Sofic groups and dynamical systems. Sankhyā: The Indian Journal of Statistics, Series A (1961-2002) **62**(3), 350–359 (2000)

Subspace Arrangements, Graph Rigidity and Derandomization Through Submodular Optimization

Orit E. Raz and Avi Wigderson

*Dedicated with admiration to László Lovász,
on the occasion of his 70th birthday*

Abstract This paper presents a deterministic, strongly polynomial time algorithm for computing the matrix rank for a class of symbolic matrices (whose entries are polynomials over a field). This class was introduced, in a different language, by Lovász [16] in his study of flats in matroids, and proved a duality theorem putting this problem in $NP \cap coNP$. As such, our result is another demonstration where "good characterization" in the sense of Edmonds leads to an efficient algorithm. In a different paper Lovász [17] proved that all such symbolic rank problems have efficient probabilistic algorithms, namely are in BPP. As such, our algorithm may be interpreted as a derandomization result, in the long sequence special cases of the PIT (Polynomial Identity Testing) problem. Finally, Lovász and Yemini [20] showed how the same problem generalizes the *graph rigidity* problem in two dimensions. As such, our algorithm may be seen as a generalization of the well-known deterministic algorithm for the latter problem. There are two somewhat unusual technical features in this paper. The first is the translation of Lovász' flats problem into a symbolic rank one. The second is the use of submodular optimization for derandomization. We hope that the tools developed for both will be useful for related problems, in particular for better understanding of graph rigidity in higher dimensions.

Keywords Derandomization · Symbolic matrix rank · Graph rigidity

Subject Classifications 68Q15 · 68Q25

O. E. Raz (✉)
Department of Mathematics, University of British Columbia, Vancouver, BC, Canada
e-mail: oritraz@math.ubc.ca

A. Wigderson
School of Mathematics, Institute for Advanced Study, Princeton, NJ 08540, USA
e-mail: avi@ias.edu

1 Introduction

In this paper we provide a new *deterministic*, strongly polynomial time algorithm which can be viewed in two ways. The first is as solving a derandomization problem, providing a deterministic algorithm to a new special case of the PIT (Polynomial Identity Testing) problem. The second is as computing the dimension of the span a collection of subspaces in high dimensional space. Motivating and connecting the two is the problem of testing *graph rigidity*, to which an efficient deterministic algorithm is known only in the plane, and is open for higher dimensions. Accordingly, we will divide the introduction to explain these three problems.

1.1 Polynomial Identity Testing (PIT)

Let $\mathbb{K}$ be a field. Let $\mathbf{x} = (x_1, \ldots x_d)$ be a d-tuple of independent variables. The PIT problem is to determine, given a multivariate polynomial $p \in \mathbb{K}[\mathbf{x}]$, if $p \equiv 0$ (as a polynomial). Of course, the description of p as an input to this problem is central to its complexity, and many variants of this problem were considered. The most common formulation is when p is given by an arithmetic formula or circuit.[1]

The original version of this question was posed by Edmonds [5]. In his formulation, p is the determinant of a matrix whose entries are linear forms in $\mathbf{x}$ (we will refer such a matrix as a *symbolic* matrix). Lovász [17] proved that this problem is in BPP namely has a fast probabilistic algorithm (for fields $\mathbb{K}$ larger than the degree of p): indeed, the algorithm simply picks random elements from $\mathbb{K}$ and evaluates p (note that evaluating p is efficient in all three formulations above, and indeed in all formulations considered). This left open the problem of finding an efficient deterministic algorithm, namely derandomizing Lovász's algorithm for PIT.

? Open Problem

Is PIT $\in P$?

The importance of this seemingly specific open problem was revealed in an important result of Kabanets and Impagliazzo [13]. They showed that if the answer is positive (as everyone expects), this will imply non-trivial lower bounds on either arithmetic or Boolean circuits, well beyond current techniques.

The progress towards resolving this open problem has been by providing deterministic polynomial time algorithms for a large variety of special cases of it, with the idea of building up techniques. By far, in most of these results the special cases are

---

[1]When the input is a circuit, the degree of p is always assumed to be polynomial in the circuit's size, and in all cases considered in this paper this will be evident.

defined by restricting the input polynomial to lie in some complexity class. In these cases, progress in derandomization followed closely progress on lower bounds for the appropriate class (as is the case in the Boolean setting as well). There are literally dozens of such papers: many are mentioned and explained in the surveys [24, 26] and e.g. the recent paper [1].

In parallel, with motivation from algebra, geometry and other areas, a different collection of special cases of PIT was studied, of a structural nature. Here one works with Edmond's formulation, and develops an understanding (and often a polynomial time algorithm) for cases where the symbolic matrix has restricted structure. This includes for example the works [3, 4, 6, 9, 11, 21].

This paper contributes to the second line of research, providing new families of symbolic matrices for which PIT can be solved in deterministic polynomial time. To explain this structure we introduce some notation. We will work in a slightly more general setting, in two ways, as the results generalize to both. First, we will allow our symbolic matrices to have polynomial entries. In such cases, these polynomials will have simple formulas describing them. Second, we will be interested in computing the *rank* of the input symbolic matrix, not just whether its determinant vanishes. While seemingly a more general problem, this turns out to be equivalent to PIT (see e.g. [8, Appendix A][2]).

Let R be a family of polynomial maps $R = \{r : \mathbb{K}^d \to \mathbb{K}^n\}$. In all cases we assume the degree of all polynomials in all maps is at most n, and the number of variables d is at most polynomial in n, so we will think of n as the input size to the problem.

A family of maps R prescribes a family of symbolic matrices, so that each row is an image of the d-vector of variables $\mathbf{x}$ under some map in R. More formally, define PIT(R) to be the set of all symbolic matrices M (with n columns, and poly(n) rows) in which every row of the matrix is of the form $r(\mathbf{x})$, for some map $r \in R$. We will be interested in families R for which the ranks of matrices in PIT(R) can be computed in polynomial time.[3]

We first demonstrate the convenience of this notation. Call R *complete*, if a deterministic polynomial-time algorithm for PIT(R) implies a deterministic polynomial-time algorithm for PIT. Very simple maps are complete! It follows from Valiant's [30] hardness of the determinant for the class[4] *VP* that

---

[2]The proof in [8] is given for *non-commutative* rank, but the exact same proof works verbatim for our usual notion of rank over $\mathbb{K}(\mathbf{x})$.

[3]We identify the set of matrices and the computational problem of determining their ranks.

[4]The arithmetic analog of the Boolean class *P*.

Theorem 1.1 ([30]) *The class R_{affine} of affine linear maps is complete.*

Indeed, Valiant's original proof (see more detail here [15]) implies a stronger theorem. Even restricting the support of each row to have at most a single variable in some coordinate, is general enough to be complete.

Theorem 1.2 *The class R_{sparse} of affine linear maps, such that each map is non-constant in at most a single variable from $\{x_1, \ldots x_d\}$, is complete.*

We now turn to define the polynomial maps we will be interested in, and for which we will be able to provide efficient deterministic algorithms. Some motivation for interest in these maps will be given in the next two subsections.

Consider the following class R_2. Here $d = n$. Every $p \in R_2$ is of the form $\mathbf{x} \mapsto (A - A^T)\mathbf{x}$, where A is a rank-1 matrix. While this family may look very special, we note that the problem of graph rigidity in $\mathbb{R}^2$ (for which a polynomial time algorithm is known but far from trivial) is a very special case of PIT(R_2).[5]

Theorem 1.3 *PIT(R_2) can be solved in deterministic polynomial time, over a field $\mathbb{K}$ with sufficiently large characteristic (more precisely, when char($\mathbb{K}$) is larger than the number of rows of the input matrix or char($\mathbb{K}$) = 0).*

This construction can be generalized as follows. Here we will generate PIT instances whose entries are *polynomials*, rather than linear functions of the variables. For a k-dimensional tensor A of size n, denote by $\hat{A}$ its "anti-symmetric" version, namely where for every entry $(i_1, \ldots, i_k)$ we have

$$\hat{A}(i_1, \ldots, i_k) = \sum_{\sigma \in S_k} \text{sgn}(\sigma) A(i_{\sigma(1)}, \ldots, i_{\sigma(k)}).$$

Note that for $k = 2$ we have $\hat{A} = A - A^T$.

We now extend R_2, in which a matrix (namely a 2-dimensional tensor) acts on one vector of variables, to R_k, in which a k-dimensional tensor acts on $k - 1$ vectors of variables. Let R_k denote the following class of (degree $k - 1$) maps. Let $\mathbf{x}^1, \mathbf{x}^2, \ldots, \mathbf{x}^{k-1}$ be n-vectors of independent variables, so altogether $\mathbf{x} = (\mathbf{x}^1, \mathbf{x}^2, \ldots, \mathbf{x}^{k-1})$ is a vector of $(k - 1)n$ variables. A k-tensor of size n in each dimension acts on $\mathbf{x}$ simply with the i'th dimension acting on $\mathbf{x}^i$ for $i \in [k - 1]$. The output of this action is a vector (along dimension k) of length n of polynomials of degree $k - 1$, each linear in $\mathbf{x}^i$ for all i. Define R_k to be all maps defined by $\hat{A}$ for any rank-1 tensor A. Note that with this notation R_2 is precisely the class defined above.

Generalizing the above theorem we prove:

Theorem 1.4 *For every $k < n$, PIT(R_k) can be solved in deterministic polynomial time, over a field $\mathbb{K}$ with sufficiently large characteristic (more precisely, when char($\mathbb{K}$) is larger than the number of rows of the input matrix or char($\mathbb{K}$) = 0).*

---

[5]Moreover, the same family of rank-2, skew symmetric matrices is featured in a very different PIT problem: determining the maximum rank of a subspace generated by given such matrices. A deterministic polynomial time solution for this problem is given by Lovasz' celebrated matroid parity algorithm [18] (see also [19], Theorem 11.1.2).

1.2 Graph Rigidity

The problem of graph rigidity arises from several motivations, originally, mechanical engineering, where it was studied by Hilda Geiringer [22, 23] already in 1927, and later by Laman [14]. Rigidity theory is a fast-growing area, and we refer the interested reader to [27] for more background and recent approaches. Graph rigidiy has several versions, we describe perhaps the most common one, *generic* rigidity. It is supposed to capture the structural rigidity of a "bars and joints" framework described by a graph. We will not be formal here as precise definitions can be found e.g. in [2]. Here the relevant field for the geometric/physical interpretation is the Real numbers $\mathbb{R}$, and we use it in this subsection as in other papers on this problem (although the algebraic formulation is meaningful for every field $\mathbb{K}$).

Let $G(V, E)$ be an undirected graph on n vertices and m edges. An *embedding* of G in $\mathbb{R}^t$ is a map $\phi : V \to \mathbb{R}^t$. An embedding of G is called *rigid* if there is no perturbation of the vertex positions which preserves all edge lengths, other than the rigid motions of $\mathbb{R}^t$. The graph G is called *rigid* if every *generic* embedding of G is rigid (equivalently, if there exists an embedding of G which is rigid, see [2]). The main question is to determine if a given graph G is rigid (and more generally, compute the dimension of the non-rigid motions of a generic embedding, in case G is not rigid).

An extremely convenient formulation of the problem (as a PIT) is the following. Let $x_{v,j}$ be a set of variables indexed by $v \in V$ and $j \in [t]$. The intuition is that $(x_{v,1}, \ldots, x_{v,t})$ are the coordinates of a generic embedding of the vertex v in $\mathbb{R}^t$. Given G, construct a symbolic matrix $M_{G,t}$ of dimensions $m \times nt$, which may be viewed as a concatenation of t matrices, one for each dimension $j \in [t]$. Every row corresponds to an edge $\{u, v\} \in E$, and for each j, the column u, j contains the entry $x_{u,j} - x_{v,j}$, whereas the column v, j contains the the negation $x_{v,j} - x_{u,j}$.

It is not hard to prove that the rank (as usual, over $\mathbb{R}(x)$) of $M_{G,t}$ determines if G is rigid, and indeed the dimension of non-rigid motions (see [2] for the details). It is easy to see that for every graph G, the matrix $M_{G,2}$ is in the class $PIT(R_2)$ above. Indeed, let $e_1, \ldots, e_{2n}$ denote the standard basis vectors in $\mathbb{R}^{2n}$. For some $u < v \in [n]$, put $a = e_u - e_v$ and $b = e_{n+u} - e_{n+v}$. Consider the matrix $A = A_{u,v} := a^t b$. Then $(A - A^t)\mathbf{x}$, where $\mathbf{x} = (x_{21}, \ldots, x_{2n}, x_{11}, \ldots, x_{1n})$ is the $\{u, v\}$ row of $M_{G,2}$. Thus Theorem 1.3 yields as a corollary a polynomial time algorithm to determine whether a given graph G is rigid in $\mathbb{R}^2$. Such algorithms for rigidity in $\mathbb{R}^2$ are known (see [10, Section 2.2] and references therein). Note that the matrices $M_{G,t}$ make sense over any field $\mathbb{K}$, instead of $\mathbb{R}$, and Theorem 1.3 in fact provides a deterministic polynomial time algorithm to compute the rank of these matrices over any field $\mathbb{K}$ with large enough characteristic.

The symbolic matrix representation above shows that for every t, the problem of testing graph rigidity in $\mathbb{R}^t$ is in BPP, and it is a decades-old problem to whether it is also in P, even for the case $t = 3$.

Lovász and Yemini [20] have developed an alternative approach for studying graph rigidity in the plane, which obtains a somewhat finer characterization of rigidity than Geiringer-Laman's. What is even more interesting is their method. They show that the matrices $M_{G,2}$ can actually be obtained in the following way. First, with every edge $\{u, v\}$ associate a certain 2-dimensional subspace $f_{u,v} \subset \mathbb{R}^{2n}$. The intersection of this subspace $f_{u,v}$ with a *generic* hyperplane through the origin (of which the normal can be viewed essentially as the $2n$-vector of variables $x_{v,j}$) yields the $\{u, v\}$ row of $M_{G,2}$. In more detail, identify the vertices of G with the set $V = [n]$, and let $e_1, \ldots, e_{2n}$ denote the standard basis in $\mathbb{R}^{2n}$. Define $f_{u,v}$ to be the subspace of $\mathbb{R}^{2n}$ spanned by the pair of vectors $e_u - e_v$ and $e_{n+u} - e_{n+v}$ (note that the definition of $f_{u,v}$ is symmetric in u, v). Let $h(\mathbf{x})$ denote the subspace of $\mathbb{R}^{2n}$ orthogonal to the vector $\mathbf{x} = (y_1, \ldots, y_n, -x_1, \ldots, -x_n)$. It is not hard to verify (see [20] for the details) that $h(\mathbf{x}) \cap f_{u,v}$ is spanned by the $\{u, v\}$ row of $M_{G,2}$. Thus, for a generic $\mathbf{x}$, we have

$$\operatorname{rank} M_{G,2} = \dim \operatorname{span}\{h(\mathbf{x}) \cap f_{u,v} \mid \{u, v\} \in E\}.$$

Thus, the question of computing the rank of $M_{G,2}$ becomes the question of computing the dimension of the span of the resulting intersections (which here are simply lines) with a *generic* hyperplane. To analyze this, Lovász and Yemini use a theory developed by Lovász [16] which studies a similar problem for an arbitrary family of subspaces. The relevant part of Lovász's theory is introduced in the next subsection.

The idea of [20] can be applied also to rigidity in higher dimensions. For simplicity of the presentation, let us consider only the case $t = 3$. In this case we associate with each edge $\{u, v\} \in E$ a 3-dimensional subspace $g_{u,v}$ of $\mathbb{R}^{3n}$. Namely, the subspace spanned by the vectors $e_u - e_v$, $e_{n+u} - e_{n+v}$, $e_{2n+u} - e_{2n+v}$, where here $e_1, \ldots, e_{3n}$ stand for the standard basis of $\mathbb{R}^{3n}$. Let $\mathbf{x} = (x_1, \ldots, x_n, y_1, \ldots, y_n, z_1, \ldots, z_n)$ and define $\tilde{h}(\mathbf{x})$ to be the (codim 2) subspace of $\mathbb{R}^{3n}$ orthogonal to the pair of vectors

$$(y_1, \ldots, y_n, -x_1, \ldots, -x_n, 0, \ldots, 0)$$

$$(z_1, \ldots, z_n, 0, \ldots, 0, -x_1, \ldots, -x_n).$$

It is not hard to verify that $\tilde{h}(\mathbf{x}) \cap f_{u,v}$ is one dimensional and spanned by the $\{u, v\}$ row of $M_{G,3}$. Thus, for a generic choice of $\mathbf{x}$, we have

$$\operatorname{rank} M_{G,3} = \dim \operatorname{span}\{\tilde{h}(\mathbf{x}) \cap f_{u,v} \mid \{u, v\} \in E\}.$$

A crucial difference from the case $t = 2$ is that here a generic choice of $\mathbf{x}$ does not yield a generic codim 2 subspace $\tilde{h}(\mathbf{x})$ of $\mathbb{R}^{3n}$. From the perspective of this method and of our paper, this is "the reason" why rigidity in higher dimensions is more challenging.

1.3 Subspaces and Generic Hyperplanes

Let F be a collection of subspaces in $\mathbb{K}^d$. Let h be a generic hyperplane in $\mathbb{K}^d$, which without loss of generality can be taken to be all vectors perpendicular to

$\mathbf{x} = (x_1, \ldots x_d)$. For each subspace $f \in F$, let $f' = f \cap h$. Now consider the space spanned by the subspaces in $F' := \{f' \mid f \in F\}$ (note that the flats in F' are functions of $\mathbf{x}$). The question is, what is the dimension of span(F')?

One of the major results of Lovász' paper [16] is a formula, called $\rho(F)$ (which we redefine in Sect. 2), that determines this dimension for every family of subspaces, and for $\mathbf{x}$ satisfying a certain "general position" condition (see Definition 5.1). To show that a *generic* $\mathbf{x}$ satisfies Lovász's general position condition over any field (with large enough characteristic) is one main result of our paper (see Sect. 7). Note that this fact is mentioned (over the field $\mathbb{R}$) in [16] with no proof. This fact is again mentioned[6] and applied, again with no proof, in Tanigawa [28]. We see our paper as contributing to the completeness of these results.

When the subspaces F are derived from a graph in the manner described above to generate the rigidity matrix, Lovász and Yemini [20] write the explicit special case of the formula $\rho(F)$, which yields an elegant characterization. For the general case of an arbitrary family of subspaces F, the formula is given as the minimum, over all possible partitions of the family, of a certain easily computable function. As the number of partitions is exponential, there is no obvious efficient way of computing ρ. We have recently learned that the problem of computing ρ is a special case of minimizing, over all partitions of a set S, the *Dilworth truncation* of a given submodular function f defined over S; a strongly polynomial algorithm for this problem is given in Frank and Tardos [7, Chapters II.1 and IV.3]. In our paper we introduce an alternative[7] strongly polynomial algorithm for computing ρ, by reducing the original problem to a minimization problem of a certain submodular function. In fact, we prove our result to a more general quantity $\rho_c(F)$, introduced in Sect. 2. (Note that $\rho(F) = \rho_1(F)$ is the quantity from [16].)

Theorem 1.5 *There is a deterministic, strongly polynomial time algorithm to compute ρ_c for every real number c.*

Closing this circle, we will also prove that the problem of computing ρ_1 is *equivalent* to PIT(R_2). This will yield Theorem 1.3 as a corollary to Theorem 1.5.

1.4 Related Works and Applications

We see our result as a step towards better understanding of the algorithmic aspects of the notions and formulas introduced in Lovázs [16] and their applications.

Let us mention one related concept studied in Lovász [16] and discuss follow-up work by Tanigawa [28], which is related to Theorem 5.2 proved in this paper. It would

---

[6]In Tanigawa [28] an alternative general position condition is suggested, to supposedly correct a mistake in Lovász's paper. However we find the counter example in [28, footnote on p. 1416] false. We provide a full and detailed proof of Lovász's formula in Sect. 5.

[7]Our algorithm seems different than the one in [7], as it does not use duality.

be interesting to find efficient algorithms for the natural computational problem at hand. The reader may skip this subsection at first reading.

Let F be a finite family of subspace in $\mathbb{K}^d$ (where $\mathbb{K}$ is a field of characteristic 0). Let $X = \{x_f \mid f \in F\}$ be a collection of points in $\mathbb{K}^d$ such that $x_f \in f$ for each $f \in F$. The set X is said to be in *general position* with respect to F if, for every $f \in F$ fixed, the following holds: Any subspace spanned by members of F and points of $X \setminus \{x_f\}$ containing x_f must contain the whole flat f. Lovász shows that there exists a choice of a set X in general position with respect to any given family F. He then proves the following formula:

Theorem 1.6 (Lovász [16]) *Let F be a finite family of subspace in $\mathbb{K}^d$, and let $X = \{x_f \mid f \in F\}$ be in general position with respect to F. Then*

$$\operatorname{rank}(\operatorname{span} X) = \min_{G \subseteq F} \left\{ \operatorname{rank}\left(\operatorname{span} \bigcup G\right) + |F \setminus G| \right\}$$

An interesting application of Theorem 1.6 to the body-rod-bar rigidity problem is obtained by Tanigawa [28]. A *body-rod-bar framework* in $\mathbb{R}^d$ is defined as a structure consisting of d-dimensional subspaces (bodies) and $(d-2)$-dimensional flats (rods) mutually linked by one-dimensional lines (bars). (The term "rod" is appropriate for $d = 3$.) More formally, a d-dimensional body-rod-bar-framework is a triple (G, q, r), where $G = (V = B \cup R, E)$ is a graph, $r : R \to \operatorname{Gr}(d-1, \mathbb{R}^{d+1}) \subset \mathbb{P}(\bigwedge^{d-1}(\mathbb{R}^{d+1}))$ is the *rod-configuration* mapping a vertex $v \in R$ to a $(d-1)$-dimensional subspace r_v of $\mathbb{R}^{d+1}$, and $q : E \to \operatorname{Gr}(2, \mathbb{R}^{d+1}) \subset \mathbb{P}(\bigwedge^2(\mathbb{R}^{d+1}))$ is the *bar-configuration* mapping an edge $e \in E$ to a 2-dimensional subspace q_e in $\mathbb{R}^{d+1}$, such that

q_e and r_v have a nonzero intersection, whenever $v \in R$ is a vertex of e;

equivalently,
$$q_e \cdot r_v = 0, \quad \text{whenever } v \in R \text{ is a vertex of } e,$$

where here the dot product should be interpreted appropriately (see [28] for the details). Assume also that $r(u) \neq r(v)$ for every $u \neq v \in R$.

An *infinitesimal motion* of (G, q, r) is a mapping $m : B \cup R \to \bigwedge^{d-1}(\mathbb{R}^{d+1})$ such that
$$q_e \cdot (m(u) - m(v)) = 0, \quad \text{for every } e = \{u, v\} \in E. \tag{1}$$

An infinitesimal motion m is called *trivial* if either $m(u) = m(v)$ for all $u, v \in V$, or if, for some fixed $v_0 \in V$ we have $m(v_0) = r_{v_0}$ and $m(v) = 0$ for every $v \in V \setminus \{v_0\}$. Finally, a framework (G, q, r) is called *infinitesimally rigid* if every infinitesimal motion is trivial.

The body-rod-bar problem gives rise to a matroid $\operatorname{BR}(G, q, r)$ defined on the edge set E whose rank is the maximum size of independent linear equations in (1) (for unknown m). From the definition, (G, q, r) is infinitesimally rigid if and only if the rank of $\operatorname{BR}(G, q, r)$ is $\binom{d+1}{2}|V| - (\binom{d+1}{2} + |R|)$.

Theorem 1.7 (Tanigawa [28, Corollary 4.13]) *Let $G = (B \cup R, E)$ and suppose $d \geq 3$. Then, for almost all bar-configurations q and almost all rod-configurations r we have*

$$\text{rank}(E) = \min_{\Pi = \{F_0, \ldots, F_k\}} \left\{ |F_0| + \sum_{i=1}^{k} \left(\binom{d+1}{2} (V(F_i) - \binom{d+1}{2}) - R(F_i) \right) \right\},$$

where the minimum is taken over all partitions Π of E.

Tanigawa's proof is a nice combination of Theorem 1.6 with the other result of Lovász mentioned in the introduction, cited below as Theorem 5.2. Briefly, the first (simpler) step in the proof is to reduce the problem to the form of Theorem 1.6. That is, a family of flats F is introduced, and the question becomes to find the rank of a generic set of points $X = \{x_f \mid f \in F\}$. The family F resulted from the reduction can be described as follow: Each edge $e = \{u, v\}$ of G is associated with some fixed subspace f_e in $\left(\mathbb{P}(\bigwedge^2(\mathbb{R}^{d+1})) \right)^{|V|}$. Then $F = \{f_e \cap h(u) \cap h(v) \mid e = \{u, v\} \in E\}$, where $h_r(u), h_r(v)$ are subspaces depending on the choice of rod configuration r. Since r is taken generically, this imposes some genericity on the subspaces $h_r(v)$, but they are not exactly generic. The proof is then complete by proving a relaxed version of Theorem 5.2, and adding the subspaces $h_r(v)$ one after the other.

For more recent applications of [16, 20] see Tanigawa [28, 29].

1.5 Organization of This Paper

In Sect. 2 we introduce the function $\rho_c(F)$, which is the main object of this study. The rest of the paper has two separate parts. The first, in Sects. 3 and 4, describes the algorithm to compute ρ_c. In Sect. 3, we present and prove properties of the function ρ_c. Using these properties we describe, in Sect. 4, a deterministic strongly polynomial time algorithm that computes ρ_c over every field via submodular optimization. Note that, as there is an alternative algorithm [7] in the literature to efficiently compute functions like ρ_c, this part can be skipped.

The second part, in Sects. 5–7, describes the genericity proof of ρ. In Sect. 5, we state (and reprove) the result of Lovász [16] above, relating ρ_1 to the intersection of F with a hyperplane in "general position". A similar relation is obtained for ρ_c, for an integer $c > 0$ (see Theorem 5.5). In Sect. 6, we develop an explicit representation of a basis of the family F' resulting from this intersection, which give rise to the symbolic matrices $\text{PIT}(R_2)$ (and $\text{PIT}(R_k)$). Using this, we prove in Sect. 7 that most hyperplanes (and more generally, subspaces) satisfy the "general position" definition of Lovász, thus expressing the rank of a these symbolic matrices as appropriate $\rho(F)$. Using the algorithm above we can now compute these ranks deterministically and efficiently. This last section is the only one in which the size of the field $\mathbb{K}$ is important.

2 Subspaces, Partitions, and the Function ρ_c

We introduce the main objects of this study: Families of subspaces, their partitions, and the optimization problem we solve in this paper. We consider linear subspaces f of $\mathbb{K}^d$. Let $d(f)$ denote the dimension of a subspace f. For a family F of subspaces, we write $\operatorname{span} F := \operatorname{span} \bigcup_{f \in F} f$ and

$$d(F) := d(\operatorname{span} F).$$

A *partition* of F is a set $\Pi = \{P_1, \ldots, P_t\}$ of nonempty, pairwise disjoint subfamilies of F, such that $F = \bigcup_{i=1}^{t} P_i$. For a partition Π of F and a family of subspaces G, we define the *restriction* of Π to G by

$$\Pi \cap G := \{P \cap G \mid P \in \Pi, \ P \cap G \neq \emptyset\}. \tag{2}$$

If $G \subset F$, then $\Pi \cap G$ forms a partition of G.

Lovász [16] defined the following key function ρ of a family of subspaces, whose meaning will be revealed in Sect. 5. We actually generalize his definition to a family of functions ρ_c, for every $c > 0$ (his ρ is our ρ_1 for $c = 1$). Computing $\rho_c(F)$ in deterministic polynomial time given F, in Sect. 4, will be the key to our derandomization results.

Fix a constant $c > 0$. Let F be a finite family of subspaces in $\mathbb{K}^d$. For a partition Π of F, we define

$$\rho_c(F, \Pi) := \sum_{P \in \Pi} (d(P) - c).$$

$$\rho_c(F) := \min_{\Pi} \rho_c(F, \Pi), \tag{3}$$

where the minimum is taken over all partitions Π of F.

Definition 2.1 We say that Π is a *minimal* partition of F, with respect to the constant $c > 0$, if Π attains $\rho_c(F)$ and has the smallest possible number of parts.

Remark. In Corollary 3.2 we prove that, fixing $c > 0$, a minimal partition Π of a family F with respect to c is unique.

Notation.

We will use small letters f, g, h to denote subspaces in $\mathbb{K}^d$, capital letters F, G, P, Q to denote families of subspaces, and Π to denote partitions of a certain family F of subspaces. Note that the elements of a partition Π are themselves families of subspaces.

3 Properties of Minimal Partitions

In this and the next section we develop our algorithm in a fully self-contained manner. As mentioned in the introduction, the reader may skip these sections and apply the algorithm of [7] as a black box. In this section, we introduce some properties of minimal partitions, to be used in our algorithm. We find these properties interesting in their own right, but some may be known, indeed in more generality, for submosular functions.

3.1 Main Technical Lemma

We start with the following main technical lemma of this section.

Lemma 3.1 *Let F, G be families of subspaces in $\mathbb{K}^d$ with minimal partitions Π_F, Π_G, respectively. Assume that $Q \in \Pi_G$ and $Q \subset F$. Then Q is contained in one of the parts of Π_F.*

For the proof, the idea is to show that if, when considering a minimal partition for F, it "pays off" to put the elements of Q together, then it still "pays off" (or at least, harmless) to put these elements together, when this time considering a minimal partition for G.

Proof Consider the restriction $\Pi' := \Pi_F \cap Q$ of Π_F to Q (as defined in (2)). By assumption, $Q \subset F$, and thus Π' forms a partition of Q.

Our assumption that $Q \in \Pi_G$, and recalling that Π_G forms a minimal partition of G, implies that

$$\sum_{P \in \Pi'} (d(P) - c) \geq d(Q) - c. \tag{4}$$

Fixing some arbitrary order on the elements of Π', we write

$$\Pi' = (P'_1, \ldots, P'_t),$$

where $P'_i := P_i \cap Q$ is non-empty and $P_1, \ldots, P_t \in \Pi_F$ are distinct. Set $V'_0 := \{0\}$. For each $1 \leq i \leq t$, define

$$V'_i := \text{span}\left(\bigcup_{j=1}^{i} P'_j\right)$$

and put $r'_i := d(V'_i) - d(V'_{i-1})$ and $s'_i := d(P'_i) - r'_i$. Note that

$$d(Q) = \sum_{i=1}^{t} r'_i$$

and that
$$s'_i = d((\operatorname{span} P'_i) \cap V'_{i-1}). \tag{5}$$

With this notation, (4) can be rewritten as
$$\sum_{i=1}^{t}(r'_i + s'_i) - tc \geq \sum_{i=1}^{t} r'_i - c$$

which implies
$$\sum_{i=1}^{t} s'_i \geq c(t-1). \tag{6}$$

Next, we define
$$V_i := \operatorname{span}\left(\bigcup_{j=1}^{i} P_j\right)$$

and put $r_i := d(V_i) - d(V_{i-1})$ and $s_i := d(P_i) - r_i$. Similar to above, we have
$$d\left(\bigcup_{i=1}^{t} P_i\right) = \sum_{i=1}^{t} r_i$$

and
$$s_i = d((\operatorname{span} P_i) \cap V_{i-1}). \tag{7}$$

We claim that
$$\sum_{i=1}^{t}(d(P_i) - c) \geq d\left(\bigcup_{i=1}^{t} P_i\right) - c. \tag{8}$$

Indeed, the inequality (8) holds if and only if
$$\sum_{i=1}^{t}(r_i + s_i) - tc \geq \sum_{i=1}^{t} r_i - c$$

which holds if and only if
$$\sum_{i=1}^{t} s_i \geq c(t-1). \tag{9}$$

To prove the last inequality, notice that $V'_i \subset V_i$ and $\operatorname{span} P'_i \subset \operatorname{span} P_i$, for every i. Thus
$$d((\operatorname{span} P'_i) \cap V'_{i-1}) \leq d((\operatorname{span} P_i) \cap V_{i-1}).$$

Hence, by (5) and (7), we get $s_i' \leq s_i$. This fact combined with the inequality (6) implies (9) and hence also (8). Since Π_F is assumed to be minimal for F, we conclude that $t = 1$ and $Q \subset P_1$. This completes the proof. □

3.2 Uniqueness of Minimal Partitions

We prove uniqueness of minimal partitions.

Corollary 3.2 (Uniqueness) *Let F be a family of subspaces in $\mathbb{K}^d$ and let Π_1, Π_2 be minimal partitions of F. Then $\Pi_1 = \Pi_2$.*

Proof Let $\sim_1, \sim_2$ denote the equivalence relations on F induced by the partitions Π_1, Π_2, respectively. Let $f, g \in F$ and assume that $f \sim_1 g$. That is $f, g \in Q$, for some $Q \in \Pi_1$. Applying Lemma 3.1 (with F, $G := F$, and Q), we get that Q is contained in one of the parts in Π_2. Thus $f \sim_2 g$. By symmetry, we conclude that $f \sim_1 g$ if and only if $f \sim_2 g$. Thus $\Pi_1 = \Pi_2$, as claimed. □

Definition 3.3 Fix $c > 0$. Define $\Pi^*(F)$ to be *the* minimal partition of a family of subspaces F (with respect to c).

3.3 Monotonicity Properties

We prove the following "monotonicity" property of minimal partitions.

Corollary 3.4 (Monotonicity) *Let F, G be families of subspaces in $\mathbb{K}^d$ and assume that $G \subset F$. Then $\Pi^*(G)$ is a refinement of $\Pi^*(F) \cap G$.*

Proof Apply Lemma 3.1 to the families F and G. □

The following is another type of monotonicity property.

Lemma 3.5 *Let $F = \{f_1, \ldots, f_n\}$ be a family of n subspaces in $\mathbb{K}^d$. Let $f_i \subset f_i'$, for every $i = 1, \ldots, n$, and consider $F' := \{f_1', \ldots, f_n'\}$. For a partition Π of F, let Π' denote the partition of F' induced by Π, replacing each f_i by the corresponding f_i'. Then $(\Pi^*(F))'$ is a refinement of $\Pi^*(F')$.*

Proof Let $P \in \Pi^*(F)$ and assume without loss of generality that $P = \{f_1, \ldots, f_m\}$, for some $m \leq n$. It is easy to see, applying Lemma 3.1, that $\Pi^*(P) = \{P\}$.

Put $P' := \{f_1', \ldots, f_m'\}$. We claim that $\Pi^*(P') = \{P'\}$. First note that it suffices to prove the claim for the special case where $f_1 \subset f_1'$ and $f_i = f_i'$, for $i = 2, \ldots, m$, and then apply the same argument repeatedly to each i. To prove the calim for the special case, consider the family $Q = \{f_1, f_1'\}$. It is easy to see, by definition, that $\Pi^*(Q) = \{Q\}$. By Lemma 3.1, Q is contained in a part of $\Pi^*(G)$, for every family of subspaces G that contains Q. Moreover, since $f_1 \cup f_1' \subset f_1'$, we have

$$\rho_c(G) = \rho_c(G \setminus \{f_1\}) \text{ and } \Pi^*(G \setminus \{f_1\}) = \Pi^*(G) \cap (G \setminus \{f_1\})$$

for every such G (this follows directly from the definition of ρ_c and of Π^*). Define $G := \{f_1, f'_1, f_2, \ldots, f_m\}$. By what has just been argued, we have

$$\Pi^*(P') = \Pi^*(G) \cap P'. \tag{10}$$

Since $P, Q \subset G$, and applying Lemma 3.1, we get that each of P and Q is contained in a part of $\Pi^*(G)$. But $P \cap Q \neq \emptyset$, thus the set $P \cup Q$ must be contained in a part of $\Pi^*(G)$. Noting that $P \cup Q = G$, this implies that $\Pi^*(G) = \{G\}$. Combined with (10), this proves $\Pi^*(P') = P'$, as claimed.

Applying Lemma 3.1 to the families F', P', and with $P' \in \Pi^*(P')$, we conclude that P' is contained in one of the parts of $\Pi^*(F')$. Since this is true for every $P \in \Pi^*(F)$, the lemma follows. □

3.4 The Family $\hat{F}$

Let F be a family of subspaces in $\mathbb{K}^d$. We show that, in some sense, F can be replaced by a simpler family $\hat{F}$ defined next. With each $P \in \Pi^*(F)$ associate the subspace $f_P := \operatorname{span} P$. Then define the family

$$\hat{F} := \{f_P \mid P \in \Pi^*(F)\}.$$

Note that for $P \neq P'$ we have $f_P \neq f_{P'}$; otherwise, taking $P \cup P'$ yields a partition of F with strictly less parts and with smaller or equal value of ρ_c, contradicting the minimality of $\Pi^*(F)$.

The family F can be replaced by $\hat{F}$ in the sense of Lemma 3.6, and $\hat{F}$ is simpler in the sense of Lemma 3.7.

Lemma 3.6 *Let F, G be families of subspaces in $\mathbb{K}^d$. Then*

$$\rho_c(F \cup G) = \rho_c(\hat{F} \cup G) \text{ and } \Pi^*(F \cup G) \simeq \Pi^*(\hat{F} \cup G).$$

By the sign $\simeq$ we mean that the identity holds after identifying the partiton $\Pi^*(\hat{F} \cup G)$ of $\hat{F} \cup G$ with the partition of $F \cup G$ naturally induced by it. Concretely, the lemma asserts that

$$\Pi^*(F \cup G) = \{(\bigcup_{f_P \in \hat{Q}} P) \cup (G \cap \hat{Q}) \mid \hat{Q} \in \Pi^*(\hat{F} \cup G)\}.$$

Proof In the proof we often abuse notation and regard a partition of $\hat{F} \cup G$ as a one of $F \cup G$, as explained after the statement of the lemma. Let Π^* be the partition of $F \cup G$ induced by $\Pi^*(\hat{F} \cup G)$, given by

$$\Pi^* = \Big\{ (\bigcup_{f_P \in \hat{Q}} P) \cup (G \cap \hat{Q}) \mid \hat{Q} \in \Pi^*(\hat{F} \cup G) \Big\}.$$

We have $|\Pi^*| = |\Pi^*(\hat{F} \cup G)|$ and

$$\rho_c(F \cup G, \Pi^*) = \rho_c(\hat{F} \cup G, \Pi^*(\hat{F} \cup G)).$$

Thus
$$\rho_c(F \cup G) \leq \rho_c(\hat{F} \cup G).$$

To prove the inverse inequality, apply Lemma 3.1 to the families F and $F \cup G$. It follows that, for every $P \in \Pi^*(F)$, there exists $Q \in \Pi^*(F \cup G)$ such that $P \subset Q$. This means that $\Pi^*(F \cup G)$ induces a well-defined partition $\hat{\Pi}^*$ of $\hat{F} \cup G$ with $|\Pi^*(F \cup G)| = |\hat{\Pi}^*|$ and

$$\rho_c(F \cup G, \Pi^*(F \cup G)) = \rho_c(\hat{F} \cup G, \hat{\Pi}^*). \tag{11}$$

Concretely, $\hat{\Pi}^*$ is given by

$$\hat{\Pi}^* := \{\hat{Q} \mid Q \in \Pi^*(F \cup G)\},$$

where
$$\hat{Q} := \{f_P \mid P \subset Q, P \in \Pi^*(F)\} \cup (Q \cap G).$$

We have
$$\rho_c(F \cup G) = \rho_c(F \cup G, \Pi^*(F \cup G))$$
$$= \rho_c(\hat{F} \cup G, \hat{\Pi}^*)$$
$$\geq \rho_c(\hat{F} \cup G).$$

This proves that $\rho_c(F \cup G) = \rho_c(\hat{F} \cup G)$.

Next, we claim that $|\Pi^*(F \cup G)| = |\Pi^*(\hat{F} \cup G)|$. Indeed, by our argument above, the partition $\hat{\Pi}^*$ of $\hat{F} \cup G$ satisfies

$$\rho_c(\hat{F} \cup G, \hat{\Pi}^*) = \rho_c(\hat{F} \cup G) \text{ and } |\hat{\Pi}^*| = |\Pi^*(F \cup G)|.$$

Since $\Pi^*(\hat{F} \cup G)$ is taken to be the smallest that attains $\rho_c(\hat{F} \cup G)$, we get

$$|\Pi^*(\hat{F} \cup G)| \leq |\Pi^*(F \cup G)|.$$

Similarly, by our argument above, the partition Π^* of $F \cup G$ satisfies

$$\rho_c(F \cup G, \Pi^*) = \rho_c(F \cup G) \text{ and } |\Pi^*| = |\Pi^*(\hat{F} \cup G)|.$$

Thus,
$$|\Pi^*(F \cup G)| \leq |\Pi^*(\hat{F} \cup G)|.$$

This proves the claim.

By the uniqueness of minimal partition (see Corollary 3.2), we conclude that
$$\Pi^*(\hat{F} \cup G) = \hat{\Pi}^* \text{ and } \Pi^*(F \cup G) = \Pi^*.$$

This completes the proof of the lemma. □

Lemma 3.7 *Let F be a family of subspaces in $\mathbb{K}^d$. Then*
$$\Pi^*(\hat{F}) = \{\{\hat{f}\} \mid \hat{f} \in \hat{F}\}.$$

Proof Apply Lemma 3.6 with $G = \emptyset$. □

We introduce one more simple property that we need.

Lemma 3.8 $\widehat{F \cup G} = \hat{F} \cup G$.

Proof By Lemma 3.6, $\Pi^*(F \cup G) = \Pi^*(\hat{F} \cup G)$. The assertion then easily follows. □

4 An Algorithm for Computing $\rho_c(F)$

In this section we prove Theorem 1.5. That is, we introduce an algorithm to compute $\rho_c(F)$, for any number c and a given family F of n subspaces in $\mathbb{K}^d$, with polynomial running time in n (and in d). While we designed our algorithm for the class of functions ρ_c, it clearly works for a wider class of submodular functions. As it is different than the one in [7], we feel it would be interesting to explore its generality. Note that the problem is trivial for $c \leq 0$, which is why we consider only $c > 0$.

As mentioned in the introduction, the problem of computing ρ_c turns out to be an instance of a more general problem to which a strongly polynomial time algorithm is already known [7]. In more detail, the *Dilworth truncation* of a set function $b' : 2^S \to \mathbb{R} \cup \{\infty\}$ is defined as the function
$$b(X) = \min_{\Pi} \sum_{P \in \Pi} b'(P),$$

where the minimum is taken over all partitions Π of X.

Theorem 4.1 (Frank and Tardos [7, IV.3]) *Let $b' : 2^S \to \mathbb{R} \cup \{\infty\}$ be a submodular set function. Suppose that a minimizing oracle for b' is available. Then $b(S)$ can be* computed in a strongly polynomial time. The algorithm also constructs a partition Π of S for which $b(S) = \sum_{P \in \Pi} b'(P)$.

Remark. In [7], a more general result is proved.

4.1 High-Level Description of the Algorithm for ρ_c

The input to the algorithm is a number c and a family of subspaces $F = \{f_1, \ldots, f_n\}$ in $\mathbb{K}^d$ Write $F_i := \{f_1, \ldots, f_i\}$. The high-level scheme of the algorithm is the following:

1. $\hat{F}_1 \leftarrow \{f_1\}$.
2. **For** $i \leftarrow 2$ **to** n
    2.1. $\Pi \leftarrow$ **Compute** $\Pi^*(\hat{F}_{i-1} \cup \{f_i\})$
    2.2. $\hat{F}_i \leftarrow \{\text{span}(P) \mid P \in \Pi\}$
3. **Return** $\sum_{\hat{f} \in \hat{F}_n} (d(\hat{f}) - c)$

The heart of the algorithm is of course the missing description of Step 2.1, which computes, in the ith iteration, the minimal partition of the family $\hat{F}_{i-1} \cup \{f_i\}$ with respect to ρ.

Lemma 4.2 *The computation in Step 2.1 can be done in strongly-polynomial time.*

Recall that the minimal partition of $\hat{F}_{i-1}$ is the partition into singletons, by Lemma 3.7. So in this step we compute the effect on this partition of inserting one new subspace. We explain how to do so efficiently and prove Lemma 4.2 in Sect. 4.3 below. To describe and analyze step 2.1, we first need to recall submodular functions and optimization, which we do in Sect. 4.2. The proof of the lemma is then given in Sect. 4.3.

We are now ready to prove Theorem 1.5, assuming that Lemma 4.2 is true.

Proof *(Proof of Theorem 1.5) Correctness of the algorithm.* By Lemma 3.8, we have

$$\hat{F}_i = \widehat{\hat{F}_{i-1} \cup \{f_i\}}.$$

Thus the computation of $\hat{F}_i$ in Step 2.2 is correct. In view of Lemmas 3.6 and 3.7, the algorithm's output is $\rho_c(F)$, as needed.

Running time of the algorithm. We represent a k-dimensional subspace f in $\mathbb{K}^d$ by a $k \times d$ matrix whose rows form a basis for f. The dimension $d(f)$ of a subspace f is just the number of rows in the matrix representing the subspace, and hence can be computed in a constant time. Let P be a family of subspaces in $\mathbb{K}^d$. To compute span(P), we take the union of the rows of the matrices in P (representing subspaces) and apply Gauss elimination (using row operations only). If P has n subspaces, we will need to apply Gauss elimination to a matrix of dimensions at most $(nd) \times d$. The nonzero rows in the matrix received by this process will form a basis for span(P).

Now let F be a family of n subspaces in $\mathbb{K}^d$. Cleary, each line in the above description of the algorithm, when applied to F, is called at most n times. In each step, excluding Step 2.1, we are required to compute at most n times one of the operations just described (finding dimension or span) or simple operations such as addition. In view of Lemma 4.2, the proof is complete. □

4.2 A Submodular Set Function

Recall that a function s defined on the collection of subsets of a finite set A is called *submodular* if
$$s(X) + s(Y) \geq s(X \cup Y) + s(X \cap Y)$$
for all $X, Y \subset A$.

The following is proved by Schrijver in [25].

Theorem 4.3 (Schrijver [25]) *There exists a strongly polynomial-time algorithm minimizing a submodular function s, where s is given by an oracle. The number of oracle calls is bounded by a polynomial in the size of the underlying set. The algorithm also finds a minimizer X^* of s.*

In this section we consider a set function defined as follows. Let F be a family of subspaces in $\mathbb{K}^d$ and let $g \subset \mathbb{K}^d$ be a subspace not in F. Fix $c > 0$. Define $r_{F,g,c} : 2^F \to \mathbb{K}$ by
$$r_{F,g,c}(X) := d(X \cup \{g\}) - c + \sum_{f \in \overline{X}} (d(f) - c),$$

where $\overline{X} := F \setminus X$. We then put
$$r^*_{F,g,c} := \min_{X \subset F} r_{F,g,c}(X)$$

and we let $X^*_{F,g,c}$ denote a subset $X \subset F$ that attains $r^*_{F,g,c}$.

We show that $r_{F,g,c}$ is submodular.

Lemma 4.4 *Let F and g and c be as above. Then $r_{F,g,c}$ is submodular.*

Proof To simplify the notation, and as F, g, c are fixed, we write for short $r = r_{F,g,c}$. Let $X, Y \subset F$. We need to show
$$r(X) + r(Y) \geq r(X \cup Y) + r(X \cap Y).$$

Put $f_X := \text{span}(X \cup \{g\})$. By definition, we have

$$r(X) + r(Y) = d(X \cup \{g\}) + d(Y \cup \{g\}) + \sum_{f \in \tilde{X}} d(f) + \sum_{f \in \tilde{Y}} d(f) - c|\tilde{X}| - c|\tilde{Y}| - 2c$$

$$= d(f_X) + d(f_Y) + \sum_{f \in \tilde{X}} d(f) + \sum_{f \in \tilde{Y}} d(f) - c|\tilde{X}| - c|\tilde{Y}| - 2c.$$

By basic linear algebra, we have the identity

$$d(f_X) + d(f_Y) = d(\operatorname{span}(f_X \cup f_Y)) + d(f_X \cap f_Y).$$

Thus the last equality, after some rearranging, is

$$r(X) + r(Y) =$$
$$\left(d(\operatorname{span}(f_X \cup f_Y)) - c + \sum_{f \in \tilde{X} \cap \tilde{Y}} d(f) - c|\tilde{X} \cap \tilde{Y}| \right) + \left(d(f_X \cap f_Y) - c \right.$$
$$\left. + \sum_{f \in \tilde{X} \cup \tilde{Y}} d(f) - c|\tilde{X} \cup \tilde{Y}| \right)$$

Noting that $\operatorname{span}(f_X \cup f_Y) = \operatorname{span}(f_{X \cup Y})$ and that $\operatorname{span}(f_X \cap f_Y) \supset \operatorname{span}(f_{X \cap Y})$, we get

$$r(X) + r(Y) \geq \left(d(f_{X \cup Y}) - c + \sum_{f \in \overline{X \cup Y}} d(f) - c|\overline{X \cup Y}| \right)$$
$$+ \left(d(f_{X \cap Y}) - c + \sum_{f \in \overline{X \cap Y}} d(f) - c|\overline{X \cap Y}| \right)$$
$$= r(X \cup Y) + r(X \cap Y).$$

This proves the lemma. □

4.3 Inserting One Subspace

We are now ready to describe in detail Step 2.1 which computes $\widehat{F}_i$ given $\widehat{F}_{i-1}$ and f_i. More precisely, we describe a subroutine that receives as an input a family F with $F = \widehat{F}$ and a subspace g, and outputs $\Pi^*(F \cup \{g\})$.

We will need the following observation.

Lemma 4.5 *Let $G = F \cup \{g\}$ be a family of subspaces in $\mathbb{K}^d$. Let $Q_g \in \Pi^*(G)$ be the part that contains the subspace g. Then*

$$\Pi^*(G) \setminus \{Q_g\} \subset \Pi^*(F).$$

Proof For every $Q \in \Pi^*(G) \setminus \{Q_g\}$, we have $Q \subset F$. By Lemma 3.1, there exists $P \in \Pi^*(F)$ such that $Q \subset P$. Clearly, we also have $P \subset G$. Applying Lemma 3.1 once again, we get that also $P \subset Q$. Thus, $P = Q$ which means that $Q \in \Pi^*(F)$. □

Corollary 4.6 *Let F be a family of n subspaces in $\mathbb{K}^d$ with $\hat{F} = F$ and let g be another subspace in $\mathbb{K}^d$. Then $\rho_c(F \cup \{g\}) = r^*_{F,g,c}$ and*

$$\Pi^*(F \cup \{g\}) = \{X^*_{F,g,c} \cup \{g\}\} \cup \{\{f\} \mid f \in F \setminus X^*_{F,g,c}\},$$

where $X^*_{F,g,c}$ and $r^*_{F,g,c}$ are as defined in Sect. 4.2.

Proof This follows from the definitions of ρ_c and $r^*_{F,g,c}$, combined with Lemma 4.5. □

Proof (*Proof of Lemma* 4.2) Combinig Corollary 4.6 with Theorem 4.3, we get that the computation in Step 2.1 can be done in strongly-polynomial time. □

5 Intersecting Subspaces with a Hyperplane

In this section we state (and reprove) a result of Lovász [16], which explains the source of the function ρ (more precisely, taking ρ_c with $c = 1$) as the dimension of the intersections of a family of subspaces with a hyperplane in "general position". This connection has been used by Lovász to study certain questions about matroids in [16], and by Lovász and Yemini in [20] to study rigid structures in $\mathbb{R}^2$. We extend Lovász' treatment to arbitrary fields $\mathbb{K}$.

In Theorem 5.5 below, we further extend Lovász's result, in a straightforward manner, to apply to the intersection of a family of subspaces with an arbitrary subspace (of any co-dimension) in "general position", instead of only a (co-dimension 1) hyperplane.

Lovász [16] uses a very specific notion of genericity, which he calls *general position* defined below, and shows that ρ correctly computes the dimension of the intersection when the hyperplane is in general position with respect to the given family of subspaces. In Theorem 7.1 we will prove that indeed "general position" is a generic property, namely holds for almost all hyperplanes. This will complete the connection with the PIT problem solved in this paper.

A *hyperplane* in $\mathbb{K}^d$ is a subspace (subspace of $\mathbb{K}^d$) of codimension 1. Let F be a family of (nonzero) subspaces in $\mathbb{K}^d$ and let $h \subset \mathbb{K}^d$ be a hyperplane in $\mathbb{K}^d$. We denote by $F \cap h$ the family $\{f \cap h \mid f \in F\}$. Following Lovász, we have the following definition:

Definition 5.1 (*General Position*) We say that h is in *general position* with respect to F if, for every $A, B, C \subset F$, with A nonempty, we have:

(i) If $\text{span}(A) \subset h$, then $\text{span}(A) = \{0\}$.
(ii) If[8]

---

[8]Note that here one can take any of A, B, C to be the empty set, and we interpret $\text{span}(\emptyset) = \{0\}$.

$$\text{span}\,((A \cap h) \cup B) \cap \text{span}\,((A \cap h) \cup C) \subset h,$$

then

$$\text{span}\,((A \cap h) \cup B) \cap \text{span}\,((A \cap h) \cup C) = \text{span}(A \cap h).$$

Remark. In Sect. 6, we prove (in Theorem 7.1) that being in general position with respect to a given family F is a generic property; this fact is mentioned in [16] without a proof.

Theorem 5.2 (Lovász [16, Theorem 2.3]) *Let F be a family of subspaces in $\mathbb{K}^d$. Let h be a hyperplane in $\mathbb{K}^d$ in general position with respect to F. Then*

$$\rho_1(F) = d(F \cap h)$$

For completeness, we introduce a slightly more detailed proof, based on the line of argument from [16].

Proof (*Proof of Theorem* 5.2) Fix F and h as in the statement. Let $F' := F \cap h$. We need to show that $\rho_1(F) = d(F')$.

We first prove that $d(F') \leq \rho_1(F)$. That is, equivalently, we show that $d(F') \leq \rho_1(F, \Pi)$, for every partition Π of the family F. Let Π be a partition of F. For $P \in \Pi$, let $P' := P \cap h$. Then

$$\text{span}(F') = \text{span}\left(\bigcup_{P \in \Pi} \text{span}(P')\right)$$

and hence

$$d(F') \leq \sum_{P \in \Pi} d(P').$$

Note also that, for every $P \in \Pi$, we have $\text{span}(P') \subset \text{span}(P) \cap h$ and hence

$$d(P') \leq d(\text{span}(P) \cap h) = d(P) - 1,$$

where here we used property (i) of the general position assumption on h, namely, we used the fact that $\text{span}(P)$ is not contained in h. We conclude that

$$d(F') \leq \sum_{P \in \Pi}(d(P) - 1), \tag{12}$$

for every partition Π of F. This implies $d(F') \leq \rho_1(F)$.

To prove the reverse inequality, we show that, for a certain partition Π^* of F, the inequality (12) is in fact tight. We will construct Π^* explicitly subsequently refining a given partition. We describe the first step, which is indeed the general step (the proof will allow us to proceed recursively).

Define an equivalence relation on F as follows: For $f_1, f_2 \in F$, $f_1 \sim f_2$ if and only if
$$\mathrm{span}(F' \cup \{f_1\}) = \mathrm{span}(F' \cup \{f_2\}).$$

Let $\{P_1, \ldots, P_m\}$ be the partition (equivalence classes) of F induced by the relation $\sim$.

The main idea is to prove that after intersection with h, the spans of the parts P_i' become a direct sum decomposition of $\mathrm{span}(F')$. As we will see below, Π^* will be achieved by refining the partition $\{P_1, \ldots, P_m\}$ inductively.

Lemma 5.3 *We have*
$$\mathrm{span}(F') = \oplus_{i=1}^m \mathrm{span}(P_i'). \tag{13}$$

Before we prove Lemma 5.3, we establish some preliminary claims. Let $g_1, \ldots, g_m$ be the (distinct) subspaces $g_i := \mathrm{span}(F' \cup \{f\})$ for some $f \in P_i$ (note that by construction g_i is independent of the specific element $f \in P_i$ that we take).

We observe that, for every $1 \le i \le m$,
$$d(g_i) = d(F') + 1. \tag{14}$$

Indeed, by property (i) of general position, f is not contained in h and $\dim(f \cap h) = \dim(f) - 1$, for every $f \in F$. Hence, for every $f \in F$, one can choose a basis for f with all elements of the basis in h except for exactly one element b_f which is not in h. Thus, fixing any $f \in P_i$, we have
$$g_i = \mathrm{span}(F' \cup \{f\}) = \mathrm{span}(F' \cup \{b_f\}) = \mathrm{span}(F') \oplus \mathrm{span}\{b_f\}.$$

Thus, $d(g_i) = d(F') + 1$, as needed.

Next, we observe that, for $i \ne j$, we have
$$g_i \cap g_j = \mathrm{span}(F') \subset h. \tag{15}$$

Indeed, by construction $g_i \ne g_j$, and in particular $g_i \cap g_j \subsetneq g_i$. Combining this with (14), we get $d(g_i \cap g_j) \le d(g_i) - 1 = d(F')$. By the definition of g_i, g_j, we also have $\mathrm{span}(F') \subset g_i \cap g_j$. Hence $g_i \cap g_j = \mathrm{span}(F')$ and (15) follows.

Proof (*Proof of Lemma 5.3*) Here property (ii) of the general position definition will be crucial for the induction step. If $m = 1$ then (13) clearly holds. For $m \ge 2$, it suffices to show that, for every $2 \le k \le m$ and every distinct indices $1 \le i_1, \ldots, i_k \le m$, one has
$$\mathrm{span}(P_{i_1}' \cup \cdots \cup P_{i_{k-1}}') \cap \mathrm{span}(P_{i_k}') = \{0\}. \tag{16}$$

We prove (16) by induction on k. For $k = 2$, we need to show that $\mathrm{span}(P_{i_1}') \cap \mathrm{span}(P_{i_2}') = \{0\}$, for every distinct $1 \le i_1, i_2 \le m$. By the definition of the subspaces g_{i_1}, g_{i_2} and applying (15), we have
$$\mathrm{span}(P_{i_1}) \cap \mathrm{span}(P_{i_2}) \subset g_{i_1} \cap g_{i_2} \subset h.$$

Since h is in general position, using property (ii), this implies that $\text{span}(P_{i_1}) \cap \text{span}(P_{i_2}) = \{0\}$. This proves the induction base case $k = 2$.

Assume next that (16) holds for some $2 \leq k \leq m - 1$ fixed and for every distinct indices $1 \leq i_1, \ldots, i_k \leq m$. Let $1 \leq i_1, \ldots, i_{k+1} \leq m$ be some distinct indices. To establish the induction step we need to prove

$$\text{span}(P'_{i_1} \cup \cdots \cup P'_{i_k}) \cap \text{span}(P'_{i_{k+1}}) = \{0\}. \quad (17)$$

Observe that in order to prove (17) it suffices to show that

$$\text{span}(P'_{i_1} \cup \cdots \cup P'_{i_k}) \cap \text{span}(P'_{i_2} \cup \cdots \cup P'_{i_{k+1}}) \subset \text{span}(P'_{i_2} \cup \cdots \cup P'_{i_k}). \quad (18)$$

Indeed, assume that (18) holds. Then

$$\text{span}(P'_{i_1} \cup \cdots \cup P'_{i_k}) \cap \text{span}(P'_{i_{k+1}})$$
$$= \text{span}(P'_{i_1} \cup \cdots \cup P'_{i_k}) \cap \text{span}(P'_{i_2} \cup \cdots \cup P'_{i_{k+1}}) \cap \text{span}(P'_{i_{k+1}})$$
$$\subset \text{span}(P'_{i_2} \cup \cdots \cup P'_{i_k}) \cap \text{span}(P'_{i_{k+1}}),$$

where the first line uses the trivial fact that $\text{span}(P'_{i_{k+1}}) \subset \text{span}(P'_{i_2} \cup \cdots \cup P'_{i_{k+1}})$ and the second line is due to (18). By the induction hypothesis, we have

$$\text{span}(P'_{i_2} \cup \cdots \cup P'_{i_k}) \cap \text{span}(P'_{i_{k+1}}) = \{0\}.$$

Thus, assuming that (18) is true, (17) follows.

Finally, we now prove (18). Note that, by the definition of the subspaces g_i and using (15), we have

$$\text{span}(P_{i_1} \cup (P'_{i_2} \cup \cdots \cup P'_{i_k})) \cap \text{span}((P'_{i_2} \cup \cdots \cup P'_{i_k}) \cup P_{i_{k+1}}) \subset g_{i_1} \cap g_{i_{k+1}} \subset h.$$

Hence, our assumption that h is in general position with respect to F implies that in fact

$$\text{span}(P_{i_1} \cup (P'_{i_2} \cup \cdots \cup P'_{i_k})) \cap \text{span}((P'_{i_2} \cup \cdots \cup P'_{i_k}) \cup P_{i_{k+1}})$$
$$\subset \text{span}(P'_{i_2} \cup \cdots \cup P'_{i_k}).$$

This clearly implies (18). Thus we have established the inductive step and this completes the proof of Lemma 5.3. □

Recall that our goal is to show that (12) is tight for some partition Π^* of F. In view of Lemma 5.3, for the partition $\{P_1, \ldots, P_m\}$ defined above, one has

$$d(F') = \sum_{i=1}^{m} d(P'_i). \quad (19)$$

That is, we expressed the quantity $d(F')$ as the sum of the quantities $d(P_i')$ for certain subfamilies $P_1, \ldots, P_m$ of F. This allows to prove the existence of Π^* using induction on the size of F.

If $|F| = 1$, the unique partition on F clearly attains (12). For $|F| \geq 1$, let $\{P_1, \ldots, P_m\}$ be the partition of F given by Lemma 5.3, satisfying (19). If $m = 1$, the identity (19), combined with (14), gives

$$d(F') = d(P_1) - 1.$$

This means that (12) is tight, and thus $\Pi^* = \{P_1\}$. If $m > 1$, then each subfamily P_i has fewer elements than F. Applying the induction hypothesis, there exist subpartitions $\Pi_i^* = \{P_{i1}, \ldots, P_{im_i}\}$ of P_i, for each $1 \leq i \leq m$, satisfying

$$d(P_i) = \sum_{j=1}^{m_i} (d(P_{ij}) - 1).$$

Combined with (19), we get

$$d(F') = \sum_{i=1}^{m} \sum_{j=1}^{m_i} (d(P_{ij}) - 1).$$

So $\Pi^* := \bigcup_{i=1}^{m} \Pi_i^*$ forms a partition of F that attains (12). This completes the proof of the theorem. □

Remark 5.4 Note that in the inductive proof of Lemma 5.3, it was sufficient to consider not *all* k-subsets of the P_i in the given partition, but rather simply on intervals $P_2, P_3, \ldots, P_k$. The same induction on k works without change. Thus even after refinement, in the proof of this theorem we never need to apply the "general position" condition more than $|F|$ times. This will help us later bound the show that $\rho_1(F)$ correctly computes $\dim(F \cap h)$ for most (or generic) hyperplanes h even when $\mathbb{K}$ is finite and not too large.

We now generalize the theorem above to intersecting a family of subspaces with an arbitrary subspace. For this we need to extend the definition of "general position".

Let F be a family of subspaces in $\mathbb{K}^d$. Let $\{\mathbf{x}_1, \ldots, \mathbf{x}_k\}$ be a set of vectors, and define that the subspaces $h_i = \{\mathbf{x}_1, \ldots, \mathbf{x}_i\}^\perp$. Note that h_i is of codimension i in $\mathbb{K}^d$, and that $h_i' := h_i \cap h_{i-1}$ is a hyperplane in h_{i-1}, for $i = 1, \ldots, k$. We say that the subspace $h = h_k$ is in *general position* with respect to F if for all $i \in [k]$ we have that the hyperplane h_i' is in general position with respect to the family $F_i = F \cap h_{i-1}$.

Theorem 5.5 *Let F be a family of subspaces in $\mathbb{K}^d$. Let h be a subspace in $\mathbb{K}^d$ of codimension k in general position with respect to F. Then*

$$\rho_k(F) = d(F \cap h)$$

Proof We prove by induction on the codimension k. The case $k = 1$ is Theorem 5.2.

Let $\mathbf{x}_1, \ldots, \mathbf{x}_k \in \mathbb{K}^d$ be vectors such that $h = \{\mathbf{x}_1, \ldots, \mathbf{x}_k\}^\perp$ is in general position with respect to h. We know that h'_k is in general position with respect to the family $F_k := F \cap h_{k-1}$. By Theorem 5.2 again, we have

$$d(F \cap h) = d(F_k \cap h'_k) = \rho_1(F_k)$$
$$= \min_{\Pi_k} \sum_{P' \in \Pi_k} (d(P') - 1),$$

where the minimum ranges over all partitions Π_k of F_k. Note that Π_k induces a partition Π on F, in the obvious way. Moreover, for every $P' \in \Pi_k$ there exists $P \subset F$ such that $P' = P \cap h_{k-1}$. By induction, we get

$$d(P') = d(P \cap h_{k-1}) = \rho_{k-1}(P).$$

Thus,

$$d(F \cap h) = \min_{\Pi} \sum_{P \in \Pi} (\rho_{k-1}(P) - 1)$$
$$= \min_{\Pi} \sum_{P \in \Pi} \left(\left(\min_{\Pi_P} \sum_{Q \in \Pi_P} (d(Q) - k + 1) \right) - 1 \right),$$

where the first minimum (the outer one) in this exprssion is taken over all partitions Π of F, and, fixing Π and given $P \in \Pi$, the inner minimum is taken over all partitions Π_P of the family P.

Note that, for any partition Π of F, the partitions $\{\Pi_P \mid P \in \Pi\}$ induce a new partition Π' which is a refinement of Π. Namely, $\Pi' := \bigcup_{P \in \Pi} \Pi_P$. Note that taking $\Pi_P = \{P\}$ for each $P \in \Pi$, we get

$$d(F \cap h) \leq \min_{\Pi} \sum_{P \in \Pi} \left(\left(\sum_{Q \in \{P\}} (d(Q) - k + 1) \right) - 1 \right)$$
$$= \min_{\Pi} \sum_{P \in \Pi} (d(P) - k)$$
$$= \rho_k(F). \tag{20}$$

We now prove the inverse inequality. Fix a partition Π of F, and, for $P \in \Pi$, let Π_P^* be a partition of P that attains the minimum in

$$\min_{\Pi_P} \sum_{Q \in \Pi_P} (d(Q) - k + 1).$$

That is, the partitions Π_P^* satisfy

$$\sum_{P\in\Pi}\left(\left(\min_{\Pi_P}\sum_{Q\in\Pi_P}(d(Q)-k+1)\right)-1\right) = \sum_{P\in\Pi}\left(\left(\sum_{Q\in\Pi_P^*}(d(Q)-k+1)\right)-1\right)$$

Let $(\Pi')^*$ be the partition of F induced by $\bigcup\{\Pi_P^* \mid P \in \Pi\}$. Observe that

$$\begin{aligned}
d(F \cap h) &= \min_{\Pi}\sum_{P\in\Pi}\left(\left(\sum_{Q\in\Pi_P^*}(d(Q)-k+1)\right)-1\right) \\
&\geq \min_{\Pi}\sum_{P\in\Pi}\sum_{Q\in\Pi_P^*}((d(Q)-k+1)-1) \\
&= \min_{\Pi}\sum_{Q\in(\Pi')^*}(d(Q)-k) \\
&= \min_{(\Pi')^*}\sum_{Q\in(\Pi')^*}(d(Q)-k) \\
&\geq \min_{\Pi}\sum_{Q\in\Pi}(d(Q)-k) \\
&= \rho_k(F). \qquad (21)
\end{aligned}$$

Combining the inequalities (20) and (21), we get $d(F \cap h) = \rho_k(F)$. This completes the induction step, and therefore proves the theorem. □

6 Rank of Symbolic Matrices

In this section we show that the quantity $\rho_c(F)$ can be interpreted as the generic rank, defined as the rank over $\mathbb{K}(\mathbf{x})$, of a certain symbolic matrix associated with F. More concretely, for $\mathbf{x} \in \mathbb{K}^d$ let

$$h(\mathbf{x}) := (\text{span}\{\mathbf{x}\})^\perp.$$

We prove that $\rho_c(F)$ equals to the generic rank of a symbolic matrix whose entries are linear combinations of the coordinates of $\mathbf{x}$.

Our main result for the section is the following (note that this is Theorem 1.3 in the introduction).

Theorem 6.1 *Let $u_1, \ldots, u_n, v_1, \ldots, v_n \in \mathbb{K}^d$ be row vectors. Consider the symbolic matrix $A(\mathbf{x})$, with unknowns $\mathbf{x} = (x_1, \ldots, x_d)$, whose ith row is*

$$(v_i^t u_i - u_i^t v_i)\mathbf{x}$$

Then the (generic) rank of $A(\mathbf{x})$ can be computed in polynomial time.

To prove the theorem we use the property established in Theorem 5.2, interpreting the quantity $\rho_1(F)$ as the dimension of the space spanned by

$$F \cap h = \{f \cap h \mid f \in F\},$$

for any hyperplane h in general position with respect to F (see Definition 5.1). Taking $h = h(\mathbf{x})$ we prove, in Lemma 6.2, that the intersection $f \cap h(\mathbf{x})$ is the span of vectors with entries that are linear combinations of the coordinates of $\mathbf{x}$. We then prove, in Theorem 7.1, that, given a family F, $h(\mathbf{x})$ is in general position with respect to F, for every generic $\mathbf{x}$ (namely, for almost every $\mathbf{x} \in \mathbb{K}^d$). Finally, we use the algorithm for computing ρ_1 from Sect. 4.

Lemma 6.2 *Let f be an m-dimensional subspace in $\mathbb{K}^d$ and let $v_1, \ldots, v_m$ be a basis of f. Let $\mathbf{x} \in \mathbb{K}^d$ and assume that $f \not\subseteq h(\mathbf{x})$. Then $h(\mathbf{x}) \cap f$ is spanned by vectors of the form*

$$w_{ij} := (v_j \cdot \mathbf{x})v_i - (v_i \cdot \mathbf{x})v_j,$$

with $i \neq j$.
Moreover, if (wlog) $\mathbf{x} \cdot v_1 \neq 0$, then the set $\{w_{12}, \ldots, w_{1m}\}$ forms a basis of $f \cap h_{\mathbf{x}}$.

Proof We first observe that $w_{ij} \in f \cap h(\mathbf{x})$. Indeed, by definition, each w_{ij} is a linear combination of basis vectors for f, and thus $w_{ij} \in f$. We also have

$$\begin{aligned} w_{ij} \cdot \mathbf{x} &= ((v_j \cdot \mathbf{x})v_i - (v_i \cdot \mathbf{x})v_j) \cdot \mathbf{x} \\ &= (v_j \cdot \mathbf{x})(v_i \cdot \mathbf{x}) - (v_i \cdot \mathbf{x})(v_j \cdot \mathbf{x}) = 0. \end{aligned}$$

Thus $w_{ij} \in f \cap h(\mathbf{x})$.

We now show that w_{ij} also span $f \cap h(\mathbf{x})$. Indeed, we prove the stronger "moreover" statement.

Let $w \in f \cap h(\mathbf{x})$. Since $w \in f$ we may write $w = \sum_{i=1}^m a_i v_i$. Since $w \in h(\mathbf{x})$, we have $w \cdot \mathbf{x} = 0$ or

$$0 = \sum_{i=1}^m a_i v_i \cdot \mathbf{x}. \tag{22}$$

If $v_i \cdot \mathbf{x} = 0$ for every i, then $f \subseteq h(\mathbf{x})$, contradicting our assumption. We may therefore assume, without loss of generality, that $v_1 \cdot \mathbf{x} \neq 0$. In this case (22) can be rewritten as

$$a_1 = -\sum_{i=2}^m \frac{a_i v_i \cdot \mathbf{x}}{v_1 \cdot \mathbf{x}}.$$

We conclude that

$$w = \sum_{i=1}^m a_i v_i$$

$$= -\left(\sum_{i=2}^{m} \frac{a_i v_i \cdot \mathbf{x}}{v_1 \cdot \mathbf{x}}\right) v_1 + \sum_{i=2}^{m} a_i v_i$$

$$= \sum_{i=2}^{m} \frac{-a_i}{v_1 \cdot \mathbf{x}} ((v_i \cdot \mathbf{x}) v_1 - (v_1 \cdot \mathbf{x}) v_i)$$

$$= \sum_{i=2}^{m} \frac{-a_i}{v_1 \cdot \mathbf{x}} w_{1i}.$$

This completes the proof of the lemma. □

We observe an interesting consequence of Lemma 6.2, asserting that computing $\rho_1(F)$ for a family F can be reduced to computing $\rho_1(G)$, for a certain family G consisting only of planes (two-dimensional subspaces).

Corollary 6.3 *Let $F = \{f_1, \ldots, f_n\}$ be a family of subspaces in $\mathbb{K}^d$ and let $\{v_{i1}, \ldots, v_{im_i}\}$ be a basis of f_i, for $i = 1, \ldots, n$. Consider the family of two-dimensional subspaces*

$$G = \bigcup_{i=1}^{n} \{g_{ijk} \mid 1 \le j \ne k \le m_i\},$$

where $g_{ijk} = \text{span}\{v_{ij}, v_{ik}\}$. Then $\rho_1(F) = \rho_1(G)$.

Proof It follows easily from Theorem 7.1 that $h(\mathbf{x})$ is in general position with respect to both families F and G, for every generic $\mathbf{x} \in \mathbb{K}^d$. Fixing such $\mathbf{x} \in \mathbb{K}^d$ and applying Lemma 6.2, we see that $\text{span}(F \cap h(\mathbf{x})) = \text{span}(G \cap h(\mathbf{x}))$. By Theorem 5.2 this means that $\rho_1(F) = \rho_1(G)$, as needed. □

The following lemma is a natural extension of Lemma 6.2 to a similar description of the intersection of a given subspace with a generic one, where the latter is not necessarily of co-dimension 1. If the co-dimension is k, the basis elements of the intersection will be homogeneous polynomials of degree k in the entries of the generic vectors. This connection, together with our algorithm for computing ρ_k, will prove Theorem 1.4 from the introduction.

Lemma 6.4 *Let $k < m \le d$ be integers. Let f be an m-dimensional subspace in $\mathbb{K}^d$ and let $v_1, \ldots, v_m$ be a basis of f. Let $\mathbf{x}_1, \ldots, \mathbf{x}_k$ be vectors in $\mathbb{K}^d$ and define the subspace*

$$h := (\text{span}\{\mathbf{x}_1, \ldots, \mathbf{x}_k\})^{\perp}.$$

Assume that $\dim(f \cap h) = m - k$ (this extends the assumption $f \not\subseteq h(\mathbf{x})$ of the lemma above). Let X be the $k \times d$ matrix with $\mathbf{x}_i$ as its ith row. Let V denote the $d \times m$ matrix with v_j as its jth column. Put $M := XV$. So M is a $k \times m$ matrix with (i, j) entry being $\mathbf{x}_i \cdot v_j$. For every $I \subset [m]$ of cardinality k, let M_I denote the $k \times k$ matrix received by restricting to the columns of M with indices in I. Then $f \cap h$ is the span of vectors of the form

Subspace Arrangements, Graph Rigidity and Derandomization ...

$$w_S := \sum_{j=1}^{k+1} (-1)^j \det(M_{I_j}) v_{s_j},$$

where $S = \{s_1 < \ldots < s_{k+1}\} \subset [m]$ is of cardinality $k + 1$ and $I_j := S \setminus \{s_j\}$.

Moreover, if (wlog, given our assumption above), assuming that the last k columns of M are linearly independent, $f \cap h$ is spanned by the $m - k$ vectors w_S with S containing the last k columns.

Proof We first show that $w_S \in f \cap h$, for every $S \subset [m]$ of cardinality $k + 1$. For S fixed, we need to verify that w_S is orthogonal to each of $\mathbf{x}_1, \ldots, \mathbf{x}_k$. For every $1 \leq i \leq k$ we have

$$w_S \cdot \mathbf{x}_i = \sum_{j=1}^{k+1} (-1)^j \det(M_{I_j}) v_{s_j} \cdot \mathbf{x}_i.$$

Observe that the right-hand side is exactly the determinant of the matrix received by duplicating the ith row of M. Since the latter matrix is evidently singular, we conclude that $w_S \cdot \mathbf{x}_i = 0$, for every $i = 1, \ldots, k$. Thus $w_S \in h$. Clearly, we also have $w_S \in f$. Thus $w_S \in f \cap h$, as needed.

We now turn to prove that the vectors w_S generate $f \cap h$. Indeed we prove the stronger "moreover" statement that already the $m - k$ vectors w_S with S of size $k + 1$ that contain the last k columns span $f \cap h$. Recall that the last k columns of M are independent.

It will be convenient to add one more piece of (slightly informal) notation. Let M' be the matrix extending M with one more (say, 0'th) row, that contains in the jth coordinate the *vector* v_j. Note that, up to a sign, the determinant of any $k + 1$ minor of M' on columns S is precisely w_S.

Note also that column operations on M', and replacing w_S by the $k + 1$ minors of the resulting matrix, do not change the span of the vectors w_S. Moreover, note that column operations on the last k columns of M' do not change the vectors w_S, restricting to sets $S \subset I$ of size $k + 1$ that contain the indices of the last k columns. We may therefore assume, by performing such column operations, that the last k columns of M form the $k \times k$ identity matrix.

We will prove the lemma by induction on k. We already know that this statement holds for $k = 1$ (and any m) by Lemma 6.2. Assume it holds for $k - 1$ (and $m - 1$, this is all we need), and we will infer the statement for k. Consider the subspace h' orthogonal to the vectors $\mathbf{x}_1, \ldots, \mathbf{x}_{k-1}$, and the subspace f' spanned by the vectors $v_1, \ldots, v_{m-1}$, and form the associated $(k - 1) \times (m - 1)$ matrix, say N. Add to the matrix N the $0'th$ row to create N'. By induction, we know that the k-minors containing the last $k - 1$ columns of N' are vectors which span the $f' \cap h'$. For $i \in [m - k]$, let w_i' denote the basis vector that corresponds to the columns $\{i, m - k + 1, \ldots, m - 1\}$. Note that

$$f \cap h = \text{span}((f' \cap h') \cup \{v_m\}) \cap \{\mathbf{x}_k\}^\perp.$$

Now add to N' a last column for v_m and a last row for x_k to form M'. Fix $i \in [m-k]$, and write $w_i := w_{S_i}$, where $S_i = \{i, m-k+1, \ldots, m\}$. Due to the last k columns of M being the identity matrix, we have

$$w_i = (\mathbf{x}_k \cdot v_i) v_m - w_i'.$$

Moreover, one can check that in fact

$$\mathbf{x}_k \cdot v_i = \mathbf{x}_k \cdot w_i' \quad \text{and}$$
$$w_i' = (\mathbf{x}_k \cdot v_m) w_i'.$$

That is, $w_i = (\mathbf{x}_k \cdot w_i') v_m - (\mathbf{x}_k \cdot v_m) w_i'$. Applying Lemma 6.2, we get that the vectors w_i, for $i \in [m-k]$, form a basis for $f \cap h$, as needed. □

7 Generic Versus General Position

This section completes the cycle of connections, proving that most (namely, generic) hyperplanes, and indeed most subspaces, are in general position (in the Lovász sense of Sect. 5) with respect to any given family of subspaces. The proof will make use the explicit description we established in the previous section for a basis to the intersection of a family of subspaces and a hyperplane. Thus, computing the ranks of the symbolic matrices in Theorems 1.3 and 1.4 are equivalent to computing the functions ρ_1 and ρ_k respectively, which we can do efficiently by the algorithm of Sect. 4.

Theorem 7.1 *Let F be a family of subspaces in $\mathbb{K}^d$, and assume that either $\text{char}(\mathbb{K}) > |F|$ or $\text{char}(\mathbb{K}) = 0$. Then the hyperplane $h(\mathbf{x})$ is in general position (see Definition 5.1) with respect to F for almost every $\mathbf{x} \in \mathbb{K}^d$. More precisely, over finite fields all but $|F|/|\mathbb{K}|$- fraction of hyperplanes are not in general position, and for infinite fields they have measure zero.*

The proof of this theorem turns out to be more intricate than we imagined. We will give below a linear-algebraic proof that is valid for all fields $\mathbb{K}$. In the appendix we give an alternative, geometric proof which is valid for the field $\mathbb{R}$ of Real numbers.

Proof Fix subsets $A, B, C \subset F$. Our goal is to show that for

$$S := \text{span}_\mathbb{K}((A \cap h(\mathbf{x})) \cup B) \cap \text{span}_\mathbb{K}((A \cap h(\mathbf{x})) \cup C)$$

either $S \not\subset h(\mathbf{x})$ generically, or $S \subset A \cap h(\mathbf{x})$ generically. Indeed, we will prove that one of these alternative holds for every $\mathbf{x}$, except for those $\mathbf{x}$ that vanish on a certain nontrivial linear equation. Thus, if $\mathbb{K}$ is finite, the fraction of such exceptional values of $\mathbf{x}$ is $1/|\mathbb{K}|$. Since the number of choices of A, B, C is finite, we see that if $\mathbb{K}$ is large enough this probability remains negligible. Being a bit more careful, (see Remark 5.4 at the end of the proof of Theorem 5.2), there are at most $|F|$ applications of the "general position" definition, and so the fraction of "bad" $\mathbf{x}$ is at most $|F|/|\mathbb{K}|$ as stated.

It is easy to see that replacing B by $\mathrm{span}\, B$ and C by $\mathrm{span}\, C$ does not affect the subspace S. We may therefore assume that each of the families B, C contains a single subspace of $\mathbb{K}^d$.

Suppose that $B \cap C \neq \{0\}$, that is, that there exists $v \in B \cap C$, with $v \neq 0$. Clearly, we have $v \in S$ and the linear form $v \cdot \mathbf{x}$ not identically zero. Thus, for almost every $\mathbf{x}$, S is not contained in $h(\mathbf{x})$ and there is nothing to prove in this case. We may therefore assume that $B \cap C = \{0\}$. In this case, after a change of basis of $\mathbb{K}^d$, we may assume that $B = \mathrm{span}\{e_1, \ldots, e_k\}$ and $C = \{e_{k+1}, \ldots, e_{k+m}\}$, where $1 \leq k < k + m \leq d$ and $e_1, \ldots, e_d$ stand for the standard basis vectors in $\mathbb{K}^d$.

From now on we will regard $\mathbf{x}$ as a vector of variables, and work in the field of fractions $\mathbb{K}(\mathbf{x})$. In particular this makes all subspaces under consideration, A, B, C, $A \cap h(\mathbf{x})$ and of course $S = S(\mathbf{x})$ now subspaces of $\mathbb{K}(\mathbf{x})^d$ (by taking the span of their bases in $\mathbb{K}(\mathbf{x})^d$).

With this, our task becomes proving the following about these subspaces:

Claim Either $S \not\subseteq h(\mathbf{x})$, or $S \subset A \cap h(\mathbf{x})$. □

We will break this task to two. Clearly, it will suffice to prove the claim for any spanning set S' replacing S. So first we will prove that we can take S' to be the affine functions (of $\mathbf{x}$) in S, and then we will prove the claim for S'.

Lemma 7.2 *S is spanned by its elements which are affine functions of $\mathbf{x}$.*

Proof (*Proof of Lemma 7.2*) Recall that we showed, in Lemma 6.2, that $\mathrm{span}_{\mathbb{K}}(A \cap h)$ has a basis consisting of elements of the form $(u^t v - v^t u)\mathbf{x}$, for some $u, v \in \mathbb{K}^d$. Write $\{\mathbf{a}_1(\mathbf{x}), \ldots, \mathbf{a}_n(\mathbf{x})\}$ for a basis of $\mathrm{span}_{\mathbb{K}}(A \cap h)$ of this form.

Having bases for B, C and $A \cap h(\mathbf{x})$ we can express all elements of S as linear combinations of these bases. Thus, elements in S are described by solutions $\alpha, \alpha' \in \mathbb{K}^n, \beta \in \mathbb{K}^k, \gamma \in \mathbb{K}^m$ to the following system of linear equations.

$$\sum_{i=1}^{n} \alpha_i \mathbf{a}_i(\mathbf{x}) + \sum_{i=1}^{k} \beta_i e_i = \sum_{i=1}^{n} \alpha'_i \mathbf{a}_i(\mathbf{x}) + \sum_{i=1}^{m} \gamma_i e_{k+i} \qquad (23)$$

where $\alpha_i \in \mathbb{K}$ (resp., $\alpha'_i, \beta_i, \gamma_i \in \mathbb{K}$) is the ith entry of α (resp., α', β, γ).

By basic theory of linear algebra, there exists a set of solutions, each of the form

$$w = w(\mathbf{x}) = \sum_{i=1}^{n} \alpha_i(\mathbf{x})\mathbf{a}_i(\mathbf{x}) + \sum_{i=1}^{k} \beta_i(\mathbf{x})e_i = \sum_{i=1}^{n} \alpha'_i(\mathbf{x})\mathbf{a}_i(\mathbf{x}) + \sum_{i=1}^{m} \gamma_i(\mathbf{x})e_{k+i},$$
$$(24)$$

where $\alpha_i(\mathbf{x})$, $\alpha'_i(\mathbf{x})$, $\beta_i(\mathbf{x})$, $\gamma_i(\mathbf{x})$ are rational functions in the entries of $\mathbf{x}$, that together span the subspace S. Moreover, these rational functions are of degree at most $|F|$.

We will now strive to find a simpler spanning set S' for S, and then use it to prove Claim 19.

The first simplification is realizing (via common denominators) that without loss of generality we can assume that all $\alpha_i(\mathbf{x})$, $\alpha'_i(\mathbf{x})$, $\beta_i(\mathbf{x})$, $\gamma_i(\mathbf{x})$ are in fact *polynomials* in the entries of $\mathbf{x}$. These elements of S span the rest, after dividing by some fixed polynomial.

The next simplification (separating out homogeneous terms) shows that without loss of generality we can take all the polynomials in each of α, α', β, γ to be homogeneous of the same degree, which we may respectively call $\deg(\alpha)$, $\deg(\alpha')$, $\deg(\beta)$, $\deg(\gamma)$. These homogeneous solutions certainly span S, and now we refine their structure further.

Indeed, inspecting the system of equations we know more: since each entry of $\mathbf{a}_i(\mathbf{x})$, for every i is of degree one, we know that for some fixed integer $r \geq 0$, they must satisfy $\deg(\alpha) = \deg(\alpha') = r$ and $\deg(\beta) = \deg(\gamma) = r + 1$. We use this to stratify solutions w by degree, and say that the associated w has degree r. Let S_r be all solutions of degree r (note that each S_r is a subspace over $\mathbb{K}$, though we will not use this fact). We call solutions w of degree 0 *linear*. Our main simplification will come from showing that linear elements S_0 span S, which in this notation is a restatement of the lemma we are proving.

Claim span $S_0 = S$ □

We will prove this claim by induction on r, using our stratifications S_r of members of S. It is clearly true for $r = 0$. So assume S_0 spans S_r, and we need to prove that S_0 spans S_{r+1}. By induction, it suffices to prove that S_r spans S_{r+1}. The plan for this will be as follows. We will assume we have some $w \in S_{r+1}$. We will take all partial derivatives of its constituent polynomials with respect to each variable x_t, $t \in [d]$. From each of these we will generate an element $w_t \in S_r$, as the degree decreased by 1. Finally, we will show that w is a linear combination, indeed a very simple one, of the form : $(r + 1)w = \sum_{t=1}^{d} x_t w_t$. We now elaborate.

Fix $t \in [d]$. Let us take a derivative with respect to the variable x_t of $\mathbf{x}$, of both sides of the identity (24). We get

$$\sum_{i=1}^{n} \left(\frac{\partial \alpha_i(\mathbf{x})}{\partial x_t} \mathbf{a}_i(\mathbf{x}) + \alpha_i(\mathbf{x}) \frac{\partial \mathbf{a}_i(\mathbf{x})}{\partial x_t} \right) + \sum_{i=1}^{k} \frac{\partial \beta_i(\mathbf{x})}{\partial x_t} e_i =$$

$$\sum_{i=1}^{n} \left(\frac{\partial \alpha'_i(\mathbf{x})}{\partial x_t} \mathbf{a}_i(\mathbf{x}) + \alpha'_i(\mathbf{x}) \frac{\partial \mathbf{a}_i(\mathbf{x})}{\partial x_t} \right) + \sum_{i=1}^{m} \frac{\partial \gamma_i(\mathbf{x})}{\partial x_t} e_{k+i}$$

To define w_t we first define $\alpha(t)$, $\alpha'(t)$, $\beta(t)$, $\gamma(t)$ by appropriately collecting homogeneous terms, and making sure that $\alpha(t)$, $\alpha'(t) \in A \cap h$ are of degree r, and that $\beta(t) \in B$ and $\gamma(t) \in C$ are of degree $r + 1$:

- $\alpha(t)_i = \frac{\partial \alpha_i(\mathbf{x})}{\partial x_t}$
- $\alpha'(t)_i = \frac{\partial \alpha'_i(\mathbf{x})}{\partial x_t}$,
- For $i \in [k]$, $\beta(t)_i(\mathbf{x})$ is

$$\left[\sum_{s=1}^{n}(\alpha_s(\mathbf{x}) - \alpha'_s(\mathbf{x}))\frac{\partial \mathbf{a}_s(\mathbf{x})}{\partial x_t}\right]_i + \frac{\partial \beta_i(\mathbf{x})}{\partial x_t}$$

- For $i \in [m]$, $\gamma(t)_i(\mathbf{x})$ is

$$\left[\sum_{s=1}^{n}(\alpha'_s(\mathbf{x}) - \alpha_s(\mathbf{x}))\frac{\partial \mathbf{a}_s(\mathbf{x})}{\partial x_t}\right]_{k+i} + \frac{\partial \gamma_i(\mathbf{x})}{\partial x_t};$$

here we used $[v]_j$ to denote the jth entry of a vector v. Now we can formally define $w_t \in S_r$ as follows. We first observe that

$$\sum_{i=1}^{n}\alpha(t)_i(\mathbf{x})\mathbf{a}_i(\mathbf{x}) + \sum_{i=1}^{k}\beta(t)_i(\mathbf{x})e_i = \sum_{i=1}^{n}\alpha'(t)_i(\mathbf{x})\mathbf{a}_i(\mathbf{x}) + \sum_{i=1}^{m}\gamma(t)_i(\mathbf{x})e_{k+i}. \quad (25)$$

Indeed, note that (24), restricted to the jth component of the equation, implies that for every, $k + m < j \le n$, we have

$$\left[\sum_{i=1}^{n}(\alpha'_i(\mathbf{x}) - \alpha_i(\mathbf{x}))\mathbf{a}_i(\mathbf{x})\right]_j = 0.$$

From this it is straightforward to verify that the identity (25) indeed holds. Thus, letting

$$w_t := \sum_{i=1}^{n}\alpha(t)_i(\mathbf{x})\mathbf{a}_i(\mathbf{x}) + \sum_{i=1}^{k}\beta(t)_i(\mathbf{x})e_i,$$

for each t, the identity (25) implies that w_t is in S. Moreover, by our definition, w_t is of degree $r - 1$.

It remains to prove that w is spanned by the vectors w_t. For this, one basic fact we will need is that if $p(\mathbf{x})$ is any homogeneous polynomial of degree m, it satisfies

$$\sum_{t} x_t \cdot \frac{\partial p(\mathbf{x})}{\partial x_t} = mp(\mathbf{x}).$$

The second fact we will need follows from identity (24), when restricted to the jth component of the equation. For every $j \in [k]$,

$$\left[\sum_{i=1}^{n}(\alpha_i'(\mathbf{x}) - \alpha_i(\mathbf{x}))\mathbf{a}_i(\mathbf{x})\right]_j = \beta_j.$$

Combining these two properties, we get

- $\sum_t x_t \alpha(t) = r\alpha$
- $\sum_t x_t \beta(t) = r\beta$

and this implies that

$$rw = \sum_t x_t w_t.$$

Note that $r \neq 0$; indeed, for $\mathbb{K}$ with non-zero characteristic, we have $r < \text{char}(\mathbb{K})$. Thus the vectors w_t span w. This completes the induction step, and hence the proof of Lemma 7.2. □

To complete the proof of the theorem we now prove

Lemma 7.3 *Either S_0 is not contained in $h(\mathbf{x})$, or it is contained in $A \cap h(\mathbf{x})$.*

As the elements in S_0 are affine functions of $\mathbf{x}$, a violation of the first possibility will imply that $\mathbf{x}$ satisfy a linear equation, so the fraction of such vectors is at most $1/|\mathbb{K}|$ as requested.

Proof *(Proof of Lemma 7.3)* We first introduce some notation. Let $v(\mathbf{x})$ be a vector in $\mathbb{K}(\mathbf{x})^d$, such that each entry of $v(\mathbf{x})$ is some linear combination of $x_1, \ldots, x_d$, the coordinates of $\mathbf{x}$. Then $v(\mathbf{x})$ can be represented by a matrix $M \in \text{Mat}_{d \times d}(\mathbb{K})$, with constant entries, such that $M\mathbf{x} = v(\mathbf{x})$. Note that if M is skew-symmetric, this means that $(M\mathbf{x}) \cdot \mathbf{x} = (M^t \mathbf{x}) \cdot \mathbf{x} = -(M\mathbf{x}) \cdot \mathbf{x}$ or $2(M\mathbf{x}) \cdot \mathbf{x} = 0$, which means that $(M\mathbf{x}) \cdot \mathbf{x} = 0$, unless the characteristic of the field is 2. Conversely, if $M\mathbf{x} \cdot \mathbf{x} = 0$ for every $\mathbf{x} \in \mathbb{K}^d$ and so $M\mathbf{x} \cdot \mathbf{x}$ is the zero polynomial (in d variables), which implies that M is skew-symmetric.

Consider k such matrices $M_1, \ldots, M_k$, representing vectors $v_1(\mathbf{x}), \ldots, v_k(\mathbf{x})$, respectively. Then a linear combination $\sum_{i=1}^{k} \alpha_i M_i$ is a matrix that corresponds to a vector which is a linear combination of $v_1(\mathbf{x}), \ldots, v_k(\mathbf{x})$, namely, $v(\mathbf{x}) = \sum_i \alpha_i v_i(\mathbf{x})$. Thus $v(\mathbf{x})$ lies in the span of the vectors $v_i(\mathbf{x})$.

Assume first that $k + m = d$. We regard a $(k + m) \times (k + m)$ matrix M as a block matrix with $TL(M)$ (resp., $TR(M)$, $BL(M)$, $BR(M)$) denoting the top-left (resp., top-right, bottom-left, bottom-right) blocks. More precisely, $TL(M)$ (resp., $TR(M)$, $BL(M)$, $BR(M)$) stands for the submatrix induced by taking the first k (resp., first k, last m, last m) rows and first k (resp., last m, first k, last m) columns of M.

With some abuse of notation, we write $M \in Y$, for a subspace Y of $\mathbb{K}(\mathbf{x})^d$, if $M\mathbf{x} \in Y$. Recall that M is in h if and only if M is skew-symmetric. In particular, $TR(M) = -BL(M)^t$, for every $M \in \text{span}(A \cap h)$. Assume that for some $M \in \text{span}(A \cap h)$, we have $TR(M) \neq 0$ (and thus also $BL(M) \neq 0$). We claim that in this case there exists a matrix $\widetilde{M} \in S \setminus h$. To see this it is sufficient to show that there exist matrices $b \in B$ and $c \in C$ such that $M + b = c$ which is not skew-symmetric (and therefore not in h). Indeed, let b be defined by $TL(b) = -TL(M)$, $TR(b) = -TR(M)$, and $BL(b) = BR(b) = 0$. We define the matrix c by $TL(c) = TR(c) = 0$, $BL(c) = BL(M)$, $BR(c) = BR(M)$. Clearly, $b \in B$, $c \in C$ and $M + b = c$. If c is skew-symmetric, then we must have $BL(c) = BL(M) = 0$, contradicting our assumption on M. Thus $c = M + b$ is in $A \cap h$ but not in S. We conclude that in this case the general position requirement holds generically.

Assume next that for every $M \in \text{span}(A \cap h)$, we have $TR(M) = BL(M) = 0$. Recall that $\text{span}(A \cap h)$ is spanned by matrices of the form $v^t u - u^t v$ for some $u, v \in \mathbb{K}^d$. Assume that $TR(v^t u - u^t v) = BL(v^t u - u^t v) = 0$ for such a matrix. We claim that in this case at least one of $TL(v^t u - u^t v)$ or $BR(v^t u - u^t v)$ is the zero matrix. Indeed, put $M = v^t u - u^t v$, and assume that $TL(M) \neq 0$. The for some $1 \leq i_0 \neq j_0 \leq k$ we have $u_{i_0} v_{j_0} \neq u_{j_0} v_{i_0}$. In particular, not both $u_{i_0} v_{j_0}$ and $u_{j_0} v_{i_0}$ are zero. Assume, without loss of generality, that $u_{i_0} v_{j_0} \neq 0$. That is, $u_{i_0}, v_{j_0} \neq 0$. Suppose that $u_\ell = 0$ for every $\ell > k$. In this case it is clear that $BR(M) = 0$ and the claim is proved. Therefore, we may assume that for some $\ell > k$ we have $u_\ell \neq 0$. Since we $BL(M) = 0$, we have in particular $u_\ell v_j = u_j v_\ell$, for every $j = 1, \ldots, k$. In particular, $u_\ell v_{j_0} = u_{j_0} v_\ell$. Note that since $v_{j_0} \neq 0$ and $u_\ell \neq 0$, we must have that also $v_\ell, u_{j_0} \neq 0$. Thus, we get $\frac{v_{i_0}}{u_{i_0}} = \frac{v_\ell}{u_\ell}$ and $\frac{v_{j_0}}{u_{j_0}} = \frac{v_\ell}{u_\ell}$. Combining these equalities, we get that $u_{i_0} v_{j_0} = u_{j_0} v_{i_0}$, contradicting our assumption. This proves the claim.

This implies that $\text{span}(A \cap h)$ is a direct sum $U \oplus V$ of matrices with entries supported only on $TL(M)$ for $M \in U$ and matrices supported by $BR(M)$ for $M \in V$.

Now let $w \in S$. By the definition of S, w can be written as $w = a + b = a' + c$ for some $a, a' \in \text{span}(A \cap h)$, $b \in B$, $c \in C$. Write $a = a_U + a_V$, where $a_U \in U$ and $a_V \in V$. Similarly, write $a' = a'_U + a'_V$. Then $a_U + a_V + b = a'_U + a'_V + c$, or $a_U - a'_U + b = a'_V - a_V + c$. But then, we must have $b = a'_U - a_U$ and $c = a_V - a'_V$, which in particular implies that $b, c \in \text{span}(A \cap h)$.

Since $a_U - a'_U \in U$ and $a'_V - a_V \in V$, this implies that, without loss of generality, we may assume $a \in U$ and $a' \in V$. Thus also $w = a + b = a' + c \in \text{span}(A \cap h)$. We conclude that $w \in \text{span}(A \cap h)$ for every $w \in S$. Thus the general position requirement holds in this case.

We now prove the remaining case where $k + m < d$, by reducing it to the case $k + m = d$ just discussed. Write $k + m = d - z$, for some $z > 0$. Repeat the above argument ignoring the last z rows and last z columns of every matrix used along the proof. Note that for $a \in A \cap h$, a is skew-symmetric, and adding a matrix $b \in B$ or $c \in C$ will result with a matrix which is either in h or not in h, independent of the last z rows and columns of a. Indeed, for $b \in B$ and $c \in C$ these rows and columns are zero, and therefore they cannot affect the skew-symmetry of $a + b$ or $a' + c$. □

This completes the proof of Theorem 7.1. □

Having established the connection between genericity and general position, we can now complete the proof of Theorem 6.1.

Proof *(Proof of Theorem 6.1)*. Consider the family of subspaces $F = \{f_1, \ldots, f_n\}$, where $f_i := \text{span}\{u_i, v_i\}$, for each $i = 1, \ldots, n$. Let $\mathbf{x} = (x_1, \ldots, x_d)$ and consider $h := (\text{span}\{\mathbf{x}\})^\perp$. In view of Lemma 6.2, we have

$$\text{rank} A(\mathbf{x}) = d(\{f \cap h \mid f \in F\}).$$

On the other hand, by Theorem 5.2, we have $d(\{f \cap h \mid f \in F\}) = \rho_1(F)$. Thus there exists a deterministic strongly-polynomial time algorithm to compute $\text{rank} A(\mathbf{x})$. □

We note that in the exact same way, our ability to efficiently compute ρ_k for every integer k by Theorem 1.5, and the characterization above, completes the proof of Theorem 1.4 from the introduction.

Acknowledgements We would like to thank Ze'ev Dvir for many illuminating discussions. We thank Amir Shpilka and Roy Meshulam for useful comments on an earlier version of the paper. We also thank Jan Vondrak for telling us about Dilworth truncation. The first author was partially supported from NSF grant DMS-1128155. The second author was partially supported from NSF grant CCF-1412958.

Appendix: Proof of Theorem 7.1 over $\mathbb{R}$

Here we provide an alternative proof of Theorem 7.1 which works over the field of Real numbers. One advantage of working over $\mathbb{R}$ is that we have the notions of a manifold and of the dimension of a manifold available. In the proof below, we use the fact that the set of linear subspaces of $\mathbb{R}^d$ can be viewed as a manifold. Then, to show that a certain set has measure zero, it is sufficient to show that this set has lower dimension. This allows us to obtain a more straightforward proof for the case $\mathbb{K} = \mathbb{R}$.

Proof *(Proof over $\mathbb{R}$)* We first prove that property (i) in Definition 5.1 is a generic propery. Fix $A \subset F$ and put $g = \text{span}(A)$. For $\mathbf{x} \in \mathbb{S}^{d-1}$ with $g \subset h(\mathbf{x})$, we have $\mathbf{x} \in \mathbb{S}^{d-1} \cap g^\perp$. If $d(g) \geq 1$, this means that $\mathbf{x}$ lies in a lower-dimensional sphere, which is a measure-zero subset of $\mathbb{S}^{d-1}$. Since F is finite (and so the number of different sub-families A is finite), we conclude that for every $\mathbf{x} \in \mathbb{S}^{d-1}$, excluding a finite union of certain lower-dimensional sub-spheres of $\mathbb{S}^{d-1}$, $h(\mathbf{x})$ satisfies property (i) in Definition 5.1.

We now prove that property (ii) in Definition 5.1 is a generic property. Fix some subfamilies $A, B, C \subset F$. We first handle certain degenerate cases. Note that if

$$\text{span}(A \cap h(\mathbf{x})) = \text{span}((A \cap h(\mathbf{x})) \cup B) \cap \text{span}((A \cap h(\mathbf{x})) \cup C), \quad (26)$$

for some $\mathbf{x} \in \mathbb{S}^{d-1}$, then $h(\mathbf{x})$ clearly satisfies property (ii). Using Lemma 6.2, condition (26) defines an algebraic subvariety of $\mathbb{S}^{d-1}$. In particular, (26) either holds for every $\mathbf{x} \in \mathbb{S}^{d-1}$ or holds only for $\mathbf{x}$ taken from a subset of $\mathbb{S}^{d-1}$ of measure zero. In the former case this means that, with respect to the subfamilies A, B, C, property (ii) in Definition 5.1 holds for $h(\mathbf{x})$ for every $\mathbf{x} \in \mathbb{S}^{d-1}$ and there is nothing to prove. Therefore we can assume that we are in the complementary case. Namely, we assume that for almost every $\mathbf{x} \in \mathbb{S}^{d-1}$ we have

$$\operatorname{span}(A \cap h(\mathbf{x})) \subsetneq \operatorname{span}((A \cap h(\mathbf{x})) \cup B) \cap \operatorname{span}((A \cap h(\mathbf{x})) \cup C). \tag{27}$$

Our next step is to identify the set of subspaces g of the form $g = \operatorname{span}(A \cap h(\mathbf{x}))$, for some $\mathbf{x} \in \mathbb{S}^{d-1}$, and determine its dimension as a subset of the Grassmannian.

We need the following observation. Let

$$r := \max_{\mathbf{x} \in \mathbb{S}^{d-1}} d(A \cap h(\mathbf{x}))$$

We claim that $d(A \cap h(\mathbf{x})) = r$, for almost every $\mathbf{x} \in \mathbb{S}^{d-1}$. Indeed, by Lemma 6.2, one can write a basis for $\operatorname{span}(A \cap h(\mathbf{x}))$ with entries that are linear combinations in the coordinates of $\mathbf{x}$. In particular, $d(A \cap h(\mathbf{x}))$ can be expressed as the rank of a certain symbolic matrix, with entries depending linearly in the coordinates of $\mathbf{x}$. This implies that $d(A \cap h(\mathbf{x})) = r$ for every $\mathbf{x} \in \mathbb{S}^{d-1}$, excluding some subset of $\mathbb{S}^{d-1}$ of measure zero, which proves our claim. (Here we used the fact that the maximal rank of a given symbolic matrix is the same as the *generic* rank of the matrix.)

Let S_0 denote the subset of $\mathbf{x} \in \mathbb{S}^{d-1}$ such that either $d(A \cap h(\mathbf{x})) < r$ or (26) holds for $h(\mathbf{x})$. As argued above $S_0 \subset \mathbb{S}^{d-1}$ has measure zero. Let $\operatorname{Gr}(r, d)$ denote the Grassmannian of r-dimensional subspaces of $\mathbb{R}^d$, regarded as an affine variety. We define a map $\phi : \mathbb{S}^{d-1} \setminus S_0 \to \operatorname{Gr}(r, d)$ by

$$\mathbf{x} \mapsto \operatorname{span}(A \cap h(\mathbf{x})).$$

We claim that the image of ϕ is r-dimensional. Indeed, let $g \in \operatorname{Im}(\phi)$ and let $\mathbf{x} \in \phi^{-1}(g)$. By definition of the domain of ϕ, we have $\mathbf{x} \notin S_0$ and thus $d(g) = r$. This means g has maximal dimension. Observe that this guarantees that, for every $\mathbf{x} \in g^\perp$, we have $\operatorname{span}(A \cap h(\mathbf{x})) = g$. (Indeed, $\mathbf{x} \in g^\perp$ certainly implies that $g \subset \operatorname{span}(A \cap h(\mathbf{x}))$ and since $d(A \cap h(\mathbf{x})) \leq r = d(g)$, we have equality.) That is, $\phi^{-1}(g) = (\mathbb{S}^{d-1} \setminus S_0) \cap g^\perp$ and, in paticular,

$$\dim(\phi^{-1}(g)) = d - 1 - r$$

(dimension here is as a manifold). We conclude that

$$\dim \operatorname{Im}(\phi) = d - 1 - (d - 1 - r) = r, \tag{28}$$

as claimed.

Next, define

$$S'_1 = \{\mathbf{x} \in \mathbb{S}^{d-1} \mid \text{span}((A \cap h(\mathbf{x})) \cup B) \cap \text{span}((A \cap h(\mathbf{x})) \cup C) \subset h(\mathbf{x})\}.$$

Our goal is to show that S'_1 has measure zero, as a subset of the sphere. For this, it suffices to show that $S_1 := S'_1 \setminus S_0$ has measure zero (since S_0 is of measure zero). Consider the restriction of ϕ to S_1. Let $g \in \text{Im}(\phi|_{S_1})$ and let $\mathbf{x} \in \phi|_{S_1}^{-1}(g)$. Set

$$g' := \text{span}((A \cap h(\mathbf{x})) \cup B) \cap \text{span}((A \cap h(\mathbf{x})) \cup C).$$

Since $\mathbf{x} \notin S_0$, we have (27) which means

$$d(g') \geq r + 1.$$

Since we assume also that $\mathbf{x} \in S'_1$, we have $\mathbf{x} \in (g')^\perp$. So

$$\dim(\phi|_{S_1}^{-1}(g)) \leq d(g')^\perp \leq d - 1 - (r + 1) = d - r - 2. \qquad (29)$$

Clearly we also have $\text{Im}(\phi|_{S_1}) \subset \text{Im}(\phi)$, and thus, using (28),

$$\dim(\text{Im}(\phi|_{S_1})) \leq r. \qquad (30)$$

Combining (29) and (30), we get that

$$\dim S_1 = \dim(\text{Im}(\phi|_{S_1})) + \dim(\phi|_{S_1}^{-1}(g)) \leq d - 2.$$

This completes the proof of the lemma. $\square$

References

1. Agrawal, M., Saha, C., Saptharishi, R., and Saxena, N.: Jacobian hits circuits: Hitting sets, lower bounds for depth-d occur-k formulas and depth-3 transcendence degree-k circuits. SIAM J. Comput. **45.4**, 1533–1562 (2016).
2. Asimow, L., and Roth, B.: The rigidity of graphs. Trans. Amer. Math. Soc. **245**, 279–289 (1978)
3. Brooksbank, P. M., and Luks, E. M.: Testing isomorphism of modules. J. Algebra **320.11**, 4020–4029 (2008)
4. Chistov, A., Ivanyos, G., and Karpinski, M.: Polynomial time algorithms for modules over finite dimensional algebras. In *Proceedings of the 1997 ACM International Symposium on Symbolic and Algebraic Computation (ISSAC)* (1997), 68–74.
5. Edmonds, J.: Systems of distinct representatives and linear algebra. J. Res. Natl. Bur. Stand. **71**, 241–245 (1967)
6. Fenner, S., Gurjar, R., and Thierauf, T.: Bipartite perfect matching is in quasi-NC. In *Proceedings of the 48th Annual ACM SIGACT Symposium on Theory of Computing (STOC)* (2016), 754–763.
7. Frank, A., and Tardos, É. : Generalized polymatroids and submodular flows. Mathematicl Programming **42**, 489–563 (1988)

8. Garg, A., Gurvits, L., Oliveira, R., and Wigderson, A.: Operator scaling: theory and applications. In arXiv:1511.03730v3
9. Geelen, J. F.: Maximum rank matrix completion. Linear Algebra Appl. **288**, 211–217 (1999)
10. Hendrickson, B.: Conditions for unique graph realizations. SIAM J. Comput. **21**, 65–84 (1992)
11. Ivanyos, G., Karpinski, M., and Saxena, N.: Deterministic polynomial time algorithms for matrix completion problems. SIAM J. Comput. **39.8**, 3736–3751 (2010)
12. Jordan, T.: Combinatorial rigidity: Graphs and matroids in the theory of rigid frameworks. MSJ Memoirs **34**, 33–112 (2016)
13. Kabanets, V., and Impagliazzo, R.: Derandomizing polynomial identity tests means proving circuit lower bounds. Computational Complexity **13**, 1–46 (2004)
14. Laman, G.: On graphs and rigidity of plane skeletal structures. J. Engrg. Math. **4**, 333–338 (1970)
15. Liu, H., and Regan, K.: Improved construction for universality of determinant and permanent. Inf. Process. Lett. **100.6**, 233–237 (2006)
16. Lovász, L.: Flats in matroids and geometric graphs. In *Combinatorial Surveys*, Proc. 6th British Combinatorial Conf., P. Cameron, Ed., Academic Press, New York, 1977, pp. 45–89.
17. Lovász, L.: On determinants, matchings, and random algorithms. In *International Symposium on Fundamentals of Computation Theory*, 565–574 (1979)
18. Lovász, L.: Matroid matching and some applications. J. Combin. Theory Ser. B **28.2**, 208–236 (1980)
19. Lovász, L., and Plummer, M. D.: *Matching Theory*, American Mathematical Soc., Providence, Rhode Island, 2009.
20. Lovász, L., and Yemini, Y.: On generic rigidity in the plane. SIAM J. Alg. Disc. Math. **3**, 91–98 (1982)
21. Meshulam, R.: On the maximal rank in a subspace of matrices. Q. J. Math. **36.2**, 225–229 (1985)
22. Pollaczek-Geiringer, H.: Über die Gliederung ebener Fachwerke. Z. Angew. Math. Mech. (ZAMM) 7: **1**, 58–72 (1927).
23. Pollaczek-Geiringer, H.: âŁœZur Gliederungstheorie räumlicher Fachwerke. Z. Angew. Math. Mech. (ZAMM) **12.6**, 369–376 (1932).
24. Saxena, N.: Progress on polynomial identity testing-II. In *Perspectives in Computational Complexity*, Springer International Publishing, 2014, 131–146.
25. Schrijver, A.: A combinatorial algorithm minimizing submodular functions in strongly polynomial time. J. Combinat. Theory, Ser. B **80.2**, 346–355 (2000)
26. Shpilka, A., and Yehudayoff, A.: Arithmetic circuits: A survey of recent results and open questions. Found. Trends Theor. Comput. Sci. **5**, 207–388 (2010)
27. Sitharam, M., St. John, A., and Sidman, J. (editors), *Handbook of Geometric Constraint Systems Principles*, CRC press, Taylor& Francis group (2019).
28. Tanigawa, S.: Generic Rigidity Matroids with Dilworth Truncations. SIAM J. Discrete Math. **26.3**, 1412–1439 (2012)
29. Tanigawa, S.: Matroids of gain graphs in applied discrete geometry. Trans. Amer. Math. Soc. **367**, 8597–8641 (2015)
30. Valiant, L.: The complexity of computing the permanent. Theor. Comput. Sci. **8**, 189–201 (1979)

Finding k Partially Disjoint Paths in a Directed Planar Graph

Alexander Schrijver

Abstract The *partially disjoint paths problem* is: *given:* a directed graph, vertices $r_1, s_1, \ldots, r_k, s_k$, and a set F of pairs $\{i, j\}$ from $\{1, \ldots, k\}$, *find:* for each $i = 1, \ldots, k$ a directed $r_i - s_i$ path P_i such that if $\{i, j\} \in F$ then P_i and P_j are disjoint. We show that for fixed k, this problem is solvable in polynomial time if the directed graph is planar. More generally, the problem is solvable in polynomial time for directed graphs embedded on a fixed compact surface. Moreover, one may specify for each edge a subset of $\{1, \ldots, k\}$ prescribing which of the $r_i - s_i$ paths are allowed to traverse this edge.

Keywords Disjoint paths · Partially disjoint paths · Directed graph · Planar graph · Free partially commutative group · Graph group

MSC codes 05C10 · 05C85 · 68R15 · 68W32 · 90C27

1 Introduction

In this paper we show that the following problem, the k *partially disjoint paths problem*, is solvable in polynomial time for directed planar graphs, for each fixed k:

(1) *given:* a directed graph $D = (V, E)$, vertices $r_1, s_1, \ldots, r_k, s_k$ of D, and a set F of pairs $\{i, j\}$ from $\{1, \ldots, k\}$,

---

The research leading to these results has received funding from the European Research Council under the European Union's Seventh Framework Programme (FP7/2007-2013)/ERC grant agreement n° 339109.

A. Schrijver (✉)
Korteweg-de Vries Institute for Mathematics, University of Amsterdam, P.O. Box 94248, 1090 Amsterdam, The Netherlands
e-mail: lex@cwi.nl

CWI, Amsterdam, The Netherlands

find: for each $i = 1, \ldots, k$, a directed $r_i - s_i$ path P_i in D such that if $\{i, j\} \in F$ then P_i and P_j are disjoint.

Here 'disjoint' means vertex-disjoint. So F prescribes the set of pairs of paths that are forbidden to intersect.

This paper extends [15], where *all* pairs of paths are prescribed to be disjoint (so F is the set of all pairs from $\{1, \ldots, k\}$). Also the method of [15] based on free groups and cohomology is extended to free partially commutative groups (but also some simplifications of the method in [15] have been included in the present paper).

The partially disjoint paths problem comes up in multi-commodity routing where certain commodities are forbidden to use the same facility, to avoid clashes of conflicting commodities (radio frequencies, soccer fan gangs, chemicals (including gases) through a pipeline network, or time slots in routing on a VLSI-chip).

The disjoint paths problem is well-studied, and generally NP-complete, implying a fortiori that the partially disjoint paths problem is generally NP-complete. The problem is NP-complete if we do not fix k, even in the undirected case (Lynch [10]). Moreover, it is NP-complete for $k = 2$ for directed graphs (Fortune, Hopcroft, and Wyllie [8]). This is in contrast to the undirected case (if NP$\neq$P), where Robertson and Seymour [14] showed that, for any fixed k, the k disjoint paths problem is polynomial-time solvable for any graph (not necessarily planar).

The edge-disjoint and vertex-disjoint versions of these problems can be reduced to each other in case of general (undirected or directed) graphs, so that the complexity status for the edge-disjoint versions follows. However, when restricted to planar graphs, the complexity is less clear. Even for the following problem it is not known whether it is polynomial-time solvable or NP-complete: given a directed planar graph $D = (V, E)$ and vertices r and s, find a directed $r - s$ path P and a directed $s - r$ path Q such that P and Q are edge-disjoint. (For a survey of results on disjoint paths till 2003 we refer to Chap. 70 of [16].)

Our method for the partially disjoint paths problem (1) for directed planar graphs consists of a number of layers and reductions:

(2)   (i) The top layer is to select a homology type for the solution. The number of potentially feasible homology types can be bounded by $(2|E(G)| + 1)^{4k^2}$. This is the only level where the 'fixed k' comes in.

(ii) For each homology type, one can find in polynomial time a solution of that type, if it exists. The formalism to keep track of homology is that of flows over a 'graph group': the group given by generators $g_1, \ldots, g_k$ and relations $g_i g_j = g_j g_i$ for all i, j with $\{i, j\} \notin F$.

(iii) Finding such a solution of the prescribed homology type is done by reduction to a 'cohomology feasibility problem' in a (generally nonplanar) *extension* of the planar dual of the input graph. (This is why we need cohomology—homology in the original, planar graph seems not enough, mainly because disjoint paths should not only be edge-disjoint, but also vertex-disjoint.)

(iv) This cohomology feasibility problem is reduced to a 2-satisfiability problem, whose input is based on a (polynomial) number of 'pre-feasible' solutions for the cohomology feasibility problem.

(v) Finding these pre-feasible solutions forms the bottom layer of the algorithm. It consists of a rather brute-force, but yet polynomial-time, constraint satisfaction method (adapting an instance as long as it is not pre-feasible).

In our description, we start at the bottom and work our way up to the top layer.

The method rests on quite basic combinatorial group theory. The approach allows application of the algorithm where the embedding of the graph in the plane is given in an implicit way, viz. by a list of the cycles that bound the faces of the graph, or alternatively by the clockwise order of edges incident with v, for each vertex v.

Our method directly extends to directed graphs on any fixed compact surface and to inputs where for each edge e a subset K_e of $\{1, \ldots, k\}$ is given that prescribes which of the $r_i - s_i$ paths may traverse e. We did not see if the method would extend to a polynomial-time algorithm if, instead of fixing k, we fix the number of faces by which $r_1, s_1, \ldots, r_k, s_k$ can be covered.

Our algorithm is a 'brute force' polynomial-time algorithm. We did not aim at obtaining the best possible running time bound, as we presume that there are much faster (but possibly more complicated) methods for problem (1) for directed planar graphs than the one we describe in this paper.

We could not avoid that k pops up in the degree of the polynomial in (2)(i). In fact, Cygan, Marx, Pilipczuk, and Pilipczuk [3] recently showed that there exists a constant t, independent of k, such that the k (fully) disjoint paths problem for directed planar graphs is solvable in $O(n^t)$ time, for any fixed k. So k only shows up in the coefficient of the polynomial. In other words, the problem is 'fixed parameter tractable'. This raises the question if also the *partially* disjoint paths problem is fixed parameter tractable for directed planar graphs.

In the case of undirected planar graphs, it was shown by Reed, Robertson, Schrijver, and Seymour [13] that the k disjoint paths problem can be solved in *linear* time, for any fixed k. This algorithm utilizes methods from Robertson and Seymour's graph minors theory. For general undirected graphs, the k disjoint paths problem is solvable in time $O(n^2)$ for any fixed k [9, 14].

2 Graph Groups

Our method uses the framework of combinatorial group theory, viz. groups defined by generators and relations. For background literature on combinatorial group theory we refer to Magnus, Karrass, and Solitar [12] and Lyndon and Schupp [11].

In particular we utilize 'graph groups'. These groups are studied *inter alia* by Baudisch [2], Droms [6], Servatius [17], Wrathall [18], and Esyp, Kazachkov, and Remeslennikov [7]. Specific properties of graph groups that we will use are given in [2, 7], but we will also need several other properties that seem not to have been considered before, in particular concerning phenomena like convexity and periodicity emanating in graph groups.

We first give some standard terminology. Let $g_1, \ldots, g_k$ form an abstract set of *generators*. Call $g_1, g_1^{-1}, \ldots, g_k, g_k^{-1}$ *symbols*. A *word* (of *size* t) is a sequence $\alpha_1 \cdots \alpha_t$ where each α_j is a symbol. The empty word (of size 0) is denoted by $\emptyset$. Define $(g_i^{-1})^{-1} := g_i$, and $(\alpha_1 \cdots \alpha_t)^{-1} := \alpha_t^{-1} \cdots \alpha_1^{-1}$.

Let $g_1, \ldots, g_k$ be generators, and let F be a set of unordered pairs $\{i, j\}$ from $[k]$ with $i \neq j$. So $([k], F)$ is an undirected graph. (Throughout this paper: $[k] := \{1, \ldots, k\}$.)

Then the group $G = G_F$ is generated by the generators $g_1, \ldots, g_k$, with relations

(3) $\qquad g_i g_j = g_j g_i$ for each pair $\{i, j\} \notin F$.

Let 1 denote the unit element of G_F. So $1 = \emptyset$.

If F consists of *all* pairs, the group G_F is the *free group* generated by $g_1, \ldots, g_k$. If $F = \emptyset$ G_F is the *free commutative group* generated by $g_1, \ldots, g_k$. In this case, G_F is isomorphic to $\mathbb{Z}^k$.

The group G_F is called a *graph group*, or a *free partially commutative group*, or a *right-angled Artin group*, or a *semifree group*. (Our definition (3) of graph group differs in a nonessential way from that generally used, where the graph describes the pairs of commuting generators, rather than the pairs of noncommuting generators. Definition (3) is more convenient for our purposes. For instance, it implies that the group G_F is equal to the direct product of the groups obtained from each component of the graph $([k], F)$.)

2.1 Independent Symbols, Commuting, Reduced Words

We review the basics of graph groups, referring to Baudisch [2] and Esyp, Kazachkov, and Remeslennikov [7] for the elaboration of some details.

To describe G_F, call symbols α and β *independent* if $\alpha \in \{g_i, g_i^{-1}\}$ and $\beta \in \{g_j, g_j^{-1}\}$ for some $\{i, j\} \notin F$ with $i \neq j$. So if α and β are independent then $\alpha\beta = \beta\alpha$ and $\beta \neq \alpha^{\pm 1}$. (It follows from (5) below that also the converse implication holds.)

Call words w and v *equivalent* if v arises from w by iteratively:

(4) (i) replacing $x\alpha\alpha^{-1}y$ by xy or vice versa, where α is a symbol,
(ii) replacing $x\alpha\beta y$ by $x\beta\alpha y$ where α and β are independent symbols.

By *commuting* we will mean applying (ii) iteratively.

Then the elements of G_F are equivalence classes of words, which we can indicate by words, although one should obviously keep in mind that different words can indicate one group element. We will write $w \equiv v$ if we want to stress that w and v are equal as words. We denote G_F by G if F is clear from the context.

A word w is called *reduced* if it is not equal (as a word) to $x\alpha y\alpha^{-1}z$ where α is a symbol independent of all symbols occurring in y. Note that reducedness is a property of words, and that it is invariant under commuting. We say that a symbol α *occurs in* an element x of G, or that x *contains* α, if α occurs in any reduced word representing x. Two elements x and y of G are called *independent* if any symbol in x and any symbol in y are independent. (In particular, $\beta \neq \alpha^{\pm 1}$ for any symbols α in x and β in y.)

The following is basic—see Lemma 2.3 in [7]:

(5) Let w and x be reduced words with $w = x$ as group elements. Then word x can be obtained from w by a series of commutings.

Define, for $x \in G$, $|x|$ as the size of any reduced word representing x. So $|xy| \leq |x| + |y|$ for all $x, y \in G$.

(5) implies that testing if $w = 1$ is easy: just replace (iteratively) any contiguous subword $\alpha y \alpha^{-1}$ by y where α is a symbol and y is a word independent of α. The final word is empty if and only if $w = 1$. This gives a test for equivalence of words w and x: just test if $wx^{-1} = 1$. So the 'word problem' for graph groups is solvable in polynomial time. (In fact it can be solved in linear time—see Wrathall [18].)

It will be convenient to emphasize when the concatenation of two reduced words x and y gives again a reduced word (without cancellation as in (4)(i)). In other words, when $|xy| = |x| + |y|$.

To this end, we add an abstract new element $*$ to G and define a multiplication $\cdot$ on $G \cup \{*\}$ as follows. Let $x, y \in G$. Then $x \cdot y := xy$ if $|xy| = |x| + |y|$, and $x \cdot y := *$ if $|xy| < |x| + |y|$. So $x \cdot y$ belongs to G if for any reduced words x' and y' representing x and y one has that the concatenation of x' and y' is reduced. So no symbol in x' cancels out any symbol in y'. If we moreover set $* \cdot x := *$ and $x \cdot * := *$ for all $x \in G \cup \{*\}$, we obtain an associative multiplication $\cdot$ on $G \cup \{*\}$.

The only purpose of introducing $*$ is to have a convenient and formally correct tool to write, for $x, y, z \in G$, $x = y \cdot z$, which is equivalent to $x = yz$ and $|x| = |y| + |z|$. By extension, for $x, y_1, \ldots, y_n \in G$, $x = y_1 \cdot \ldots \cdot y_n$ is equivalent to $x = y_1 y_2 \ldots y_n$ and $|x| = |y_1| + |y_2| + \cdots + |y_n|$. That is, in the concatenation of reduced words $y_1, \ldots, y_n$ there is no cancelation. The element $*$ will not occur anymore below.

While the multiplication $\cdot$ is associative, it is generally not the case that if $xy = x \cdot y$ and $yz = y \cdot z$ then $xyz = x \cdot y \cdot z$, because in xyz, symbols in x might cancel out symbols in z. Nevertheless, the following holds. Call $y \in G$ a *segment* of $a \in G$ if there exist $x, z \in G$ with $a = x \cdot y \cdot z$. Then:

(6) If $xy = x \cdot y$ and $yz = y \cdot z$, then y is a segment of xyz.

To see this, let x, y, and z be reduced words, and consider the concatenation of x, y, z. In the cancellation to obtain a reduced word, only symbols in x and symbols in z can cancel each other out (since the concatenations x, y and y, z are reduced). So y survives as a segment of xyz.

2.2 The Partial Order $\leq$

Most of what follows in this section is, explicitly or implicitly, in [7]. Let $x, y \in G$. We write $x \leq y$ if $y = x \cdot a$ for some a (namely, $a = x^{-1}y$). (If $x \leq y$, x is called in [7] a *left divisor* of y.) So if y is given as reduced word, it means that y can be commuted so that the first $|x|$ symbols in y form x (by (5)). Also, $x \leq y$ if and only if $|y| = |x| + |x^{-1}y|$.

It is easy to derive from the norm properties of $|.|$ that $\leq$ is a partial order. In fact, the partial order $\leq$ is a lattice if we add to G an element ∞ at infinity (Propositions 3.10 and 3.12 in [7]). This follows from the existence of the meet $x \wedge y$ for all $x, y \in G$. Then the join $\vee$ exists for all x, y for which there exists $z \in G$ with $x, y \leq z$ (then $x \vee y$ is the meet of all such z). So adding an element ∞ with $\infty \geq x$ for all x, makes $(G \cup \{\infty\}, \leq)$ to be a lattice. Then $x \vee y = \infty$ if there is no $z \in G$ with $z \geq x, y$.

If finite, the join $x \vee y$ can be described as follows (Proposition 3.18 in [7]):

(7) Let $x, y \in G$ and define $x' := (x \wedge y)^{-1}x$ and $y' := (x \wedge y)^{-1}y$. Then $x \vee y < \infty$ if and only if x' and y' are independent.

Moreover:

(8) If $x \vee y < \infty$ then $x \vee y = (x \wedge y) \cdot x' \cdot y' = x(x \wedge y)^{-1}y$.

For any $a \in G$ define:

(9) $a^{\downarrow} := \{x \in G \mid x \leq a\}$ and $a^{\uparrow} := \{x \in G \mid x \geq a\}$.

The norm characterization of $\leq$ implies for all x, y, z with $x \leq y, z$:

(10) $y \leq z$ if and only if $x^{-1}y \leq x^{-1}z$.

Hence for any $a \in G$, the function $a^{\uparrow} \to a^{-1}a^{\uparrow}$ with $b \mapsto a^{-1}b$ for $b \in a^{\uparrow}$ is an order isomorphism, and therefore:

(11) If $b, c \geq a$ then $a^{-1}(b \wedge c) = a^{-1}b \wedge a^{-1}c$, and if moreover $b \vee c < \infty$, then $a^{-1}(b \vee c) = a^{-1}b \vee a^{-1}c$.

Proposition 2.1 *Let $x_1, \ldots, x_t$ be such that $x_i \vee x_j < \infty$ for all i, j. Then $x_1 \vee \cdots \vee x_t < \infty$.*

Proof It suffices to show this for $t = 3$, since for $t \geq 4$ we can apply induction. Let $x, y, z \in G$ with $x \vee y, x \vee z, y \vee z < \infty$. Define $a := x \wedge (y \vee z)$.

As $x \vee y < \infty$ we know by (7) that $(x \wedge y)^{-1}x$ and $(x \wedge y)^{-1}y$ are independent. As $x \wedge y \leq a \leq x$, this implies that $a^{-1}x$ and $(x \wedge y)^{-1}y$ are independent. Now (using $x \wedge y = x \wedge y \wedge (y \vee z) = a \wedge y$ and (8))

(12) $(x \wedge y)^{-1}y = (a \wedge y)^{-1}a = a^{-1}a(a \wedge y)^{-1}y = a^{-1}(a \vee y)$.

So $a^{-1}x$ and $a^{-1}(a \vee y)$ are independent. Similarly, $a^{-1}x$ and $a^{-1}(a \vee z)$ are independent. So $a^{-1}x$ and $a^{-1}(a \vee y) \vee a^{-1}(a \vee z)$ are independent. The latter is equal to, by (11),

(13) $\quad a^{-1}(a \vee y) \vee a^{-1}(a \vee z) = a^{-1}(a \vee y \vee z) = a^{-1}(y \vee z).$

So $a^{-1}x$ and $a^{-1}(y \vee z)$ are independent. Hence by (7), $x \vee (y \vee z) < \infty$. ∎

We say that a symbol α is a *minimal symbol* of $x \in G$ if $\alpha \leq x$ (so α can be commuted so as to become the first symbol). Similarly, α is a *maximal symbol* if $\alpha^{-1} \leq x^{-1}$, i.e., if $x\alpha^{-1} \leq x$ (so α can be commuted so as to become the last symbol).

We will need that for all $x, y, z \in G$:

(14) $\quad$ If $y \leq xyz = x \cdot y \cdot z$, then $y \leq xy$.

This can be seen by induction on $|z|$, the case $|z| = 0$, that is $z = 1$, being trivial. Suppose $z \neq 1$, and let α be a maximal element of z, let $z' := z\alpha^{-1}$, and suppose that $y \not\leq xyz'$. Let α occur m times in y. As $y \leq xyz$ and $y \not\leq xyz'$, α occurs $m - 1$ times in xyz'. This contradicts the fact that α occurs m times in y, hence in $x \cdot y \cdot z'$. This proves (14).

We also will need that for all $x, y, z \in G$:

(15) $\quad$ If $x, y \leq z$ then $(x \wedge y)^{-1}x \leq y^{-1}z$.

To see this, let $b := x \vee y = y(x \wedge y)^{-1}x$. As $y \leq b \leq z$, we have $(x \wedge y)^{-1}x = y^{-1}b \leq y^{-1}z$ by (10), proving (15).

Moreover, for all $x, y, z \in G$:

(16) $\quad$ If $y^{-1}x \wedge y^{-1}z = 1$ then $(x \wedge z) \vee (x \wedge y) \vee (y \wedge z) = y$.

To prove this, we can assume (by (11)) that $x \wedge y \wedge z = 1$. Let $a := y \wedge z$, $b := x \wedge z$, and $c := x \wedge y$. Then $a \wedge b = a \wedge c = b \wedge c = 1$. Hence, by (7), $a, b,$ and c are pairwise independent, and $bc \leq x$, $ac \leq y$, $ab \leq z$. Let $x', y',$ and z' satisfy $x = bcx'$, $y = acy'$, and $z = abz'$. Since $ac = ca$ and $bc = cb$, we know $y^{-1}x = y'^{-1} \cdot a^{-1} \cdot b \cdot x'$. Hence, as a and b are independent, $y'^{-1} \cdot b \leq y^{-1}x$. Similarly, $y'^{-1} \cdot b \leq y^{-1}z$. As $y^{-1}x \wedge y^{-1}z = 1$, we have $y' = 1$ and $b = 1$. Hence $(x \wedge z) \vee (x \wedge y) \vee (y \wedge z) = a \vee b \vee c = ac = y$.

(We finally remark, but will not need, that the lattice $G \cup \{\infty\}$ is not distributive (if $F \neq \emptyset$), while for each $a \in G$, the sublattice $a^{\downarrow}$ is distributive.)

2.3 Convex Sets

The function $\text{dist}(x, y) := |x^{-1}y|$ is a metric, since, for all $x, y \in G$, (i) $|x| = 0 \iff x = 1$, (ii) $|x^{-1}| = |x|$, and (iii) $|xy| \leq |x| + |y|$. Note that this distance is left-invariant: $\text{dist}(zx, zy) = \text{dist}(x, y)$ for all x, y, z.

We call a subset L of G *convex* if L is nonempty and if $x, z \in L$ and $\text{dist}(x, y) + \text{dist}(y, z) = \text{dist}(x, z)$ imply $y \in L$. Since the distance function is left-invariant, if L is convex also yL is convex, for each $y \in G$.

Proposition 2.2 *A nonempty subset L of G is convex if and only if*

(17) (i) *if $x \leq y \leq z$ and $x, z \in L$ then $y \in L$,*
(ii) *if $x, y \in L$ then $x \wedge y \in L$ and, if $x \vee y$ is finite, $x \vee y \in L$.*

Proof Necessity follows from the facts that $x \leq y \leq z$ implies $\text{dist}(x, y) + \text{dist}(y, z) = \text{dist}(x, z)$, that $\text{dist}(x, y) = \text{dist}(x, x \wedge y) + \text{dist}(x \wedge y, y)$ and that, if $x \vee y$ is finite then $\text{dist}(x, y) = \text{dist}(x, x \vee y) + \text{dist}(x \vee y, y)$.

To see sufficiency, let $\text{dist}(x, y) + \text{dist}(y, z) = \text{dist}(x, z)$ with $x, z \in L$. We must show $y \in L$. So $|x^{-1}y| + |y^{-1}z| = |x^{-1}z|$, hence $y^{-1}x \wedge y^{-1}z = 1$. Hence by (16), $(x \wedge z) \vee (x \wedge y) \vee (y \wedge z) = y$. So $x \wedge z \leq x \wedge y \leq x$ and $x \wedge z \leq y \wedge z \leq z$. Therefore, $x \wedge y$ and $y \wedge z$ belong to L and hence y belongs to L. ∎

This implies that $a^\uparrow$ and $a^\downarrow$ are convex. Moreover, each convex set L has a unique minimal element $\min L$. This in fact characterizes convex sets:

(18) *A nonempty subset L of G is convex if and only if for each $a \in G$, aL has a unique minimal element.*

Here necessity follows from (17)(ii). To see sufficiency, let $x, z \in L$ and $y \in G$ such that $\text{dist}(x, y) + \text{dist}(y, z) = \text{dist}(x, z)$. We prove $y \in L$. We may assume $y = 1$ (as the condition is invariant under resetting $L \to y^{-1}L$). Let a be the unique minimal element in L. So $x \geq a$ and $z \geq a$. On the other hand, $|x| + |z| = |x^{-1}z|$, and hence $x \wedge z = 1$. So $a = 1$, proving (18).

Clearly, the intersection of any number of convex sets is convex again. Moreover, convex sets satisfy the following 'Helly-property':

(19) *Let L_1, L_2, L_3 be convex sets with $L_i \cap L_j \neq \emptyset$ for all $i, j = 1, 2, 3$. Then $L_1 \cap L_2 \cap L_3 \neq \emptyset$.*

For choose $x \in L_1 \cap L_2$, $y \in L_1 \cap L_3$, $z \in L_2 \cap L_3$. Without loss of generality, $z = 1$ (as we can replace L_1, L_2, L_3 by $z^{-1}L_1, z^{-1}L_2, z^{-1}L_3$). Now $x \wedge y \in L_1 \cap L_2 \cap L_3$.

This proves (19), which implies the following. As usual, define $XY := \{xy \mid x \in X, y \in Y\}$ and $X^{-1} := \{x^{-1} \mid x \in X\}$, for $X, Y \subseteq G$. Then:

(20) *Let $L_1, L_2,$ and L_3 be convex, with $L_1 \cap L_2 \neq \emptyset$. Then $L_1 L_3^{-1} \cap L_2 L_3^{-1} = (L_1 \cap L_2) L_3^{-1}$.*

Indeed, trivially, $L_1 L_3^{-1} \cap L_2 L_3^{-1} \supseteq (L_1 \cap L_2) L_3^{-1}$. To see the reverse inclusion, let $x \in L_1 L_3^{-1} \cap L_2 L_3^{-1}$. Since $x \in L_1 L_3^{-1}$, we know $x^{-1} L_1 \cap L_3 \neq \emptyset$. Similarly, $x^{-1} L_2 \cap L_3 \neq \emptyset$. Since also $L_1 \cap L_2 \neq \emptyset$, (19) gives $x^{-1} L_1 \cap x^{-1} L_2 \cap L_3 \neq \emptyset$. Hence $x \in (L_1 \cap L_2) L_3^{-1}$.

2.4 Ideals and Closed Sets

A subset I of G is an *ideal* if I is nonempty and

(21) (i) if $y \leq x$ and $x \in I$ then $y \in I$,
 (ii) if $x, y \in I$ and $x \vee y$ is finite then $x \vee y \in I$.

Since 1 belongs to any ideal, by comparing (17) and (21) one sees that each ideal is convex. Moreover, for any $L \subseteq G$ and $x \in L$:

(22) L is convex if and only if $x^{-1}L$ is an ideal.

Call H *closed* if both H and H^{-1} are ideals. In particular, if H is closed and $x \in H$, then any segment of x belongs to H.

Proposition 2.3 *If H is closed and $x, y \in G$, then $x \leq y$ implies $\min(xH) \leq \min(yH)$.*

Proof We can assume that $y = x\alpha$ for some symbol α (as we can assume $x \neq y$, so that $x \leq x\alpha \leq y$ for some symbol α, which implies inductively $\min(xH) \leq \min(x\alpha H) \leq \min(yH)$). Let $c \in H$ with $yc = \min(yH)$. It suffices to show that there exists $d \in H$ with $xd \leq yc$.

If $\alpha^{-1} \leq c$, let $d := \alpha c$, in which case $d \in H$ (as H^{-1} is an ideal) and $xd = yc$. If $\alpha^{-1} \not\leq c$, then let $d := c$, in which case $xd \leq yc$. Indeed, as $yc = \min(yH)$ and H is an ideal, we know $c \leq y^{-1}$. Since $c \leq y^{-1}$ and $\alpha^{-1} \leq y^{-1}$, α^{-1} and c are independent. As moreover $\alpha^{-1} \leq y^{-1}$, we know $\alpha^{-1} \leq c^{-1}y^{-1}$. Therefore, $yc\alpha^{-1} \leq yc$, and hence $xc = y\alpha^{-1}c = yc\alpha^{-1} \leq yc$. ∎

This is used in showing:

(23) *If L is convex and H is closed, then LH is convex. Moreover, $\min(LH) = \min(\min(L)H)$.*

To show that LH is convex, by (18) it suffices to show that LH has a unique minimal element (as each xL is again convex). Let $a = \min(L)$ and choose $c \in H$ with $ac := \min(aH)$. Then for each $x \in L$ and $y \in H$, we have by Proposition 2.3, as $a \leq x$, $ac = \min(aH) \leq \min(xH) \leq xy$. So ac is the unique minimal element in LH. Hence $\min(LH) = \min(aH) = \min(\min(L)H)$, and we have (23).

This implies:

(24) *If H and H' are closed, then HH' is closed.*

Indeed, H is an ideal, hence convex, hence by (24), HH' is convex. As $1 \in HH'$, it follows that HH' is an ideal. Similarly, $(HH')^{-1}$ is an ideal. So HH' is closed.

This gives for any closed H and $x, z \in G$:

(25) $xHz^{-1} = x^{\downarrow}H(z^{\downarrow})^{-1} \cap x^{\uparrow}H(z^{\downarrow})^{-1} \cap x^{\downarrow}H(z^{\uparrow})^{-1} \cap x^{\uparrow}H(z^{\uparrow})^{-1}.$

This follows from (20) and (23), as $x^{\uparrow}, x^{\downarrow}, z^{\uparrow}$, and $z^{\downarrow}$ all are convex, hence $z^{\downarrow}H^{-1}$ and $z^{\uparrow}H^{-1}$ are convex. Then (20) gives that the right-hand side in (25) is equal to $xH(z^{\downarrow})^{-1} \cap xH(z^{\uparrow})^{-1}$. Applying (20) to the inverse of this set, we obtain (25).

2.5 Peaks

An element of G is called a *peak* if it has precisely one maximal symbol. The peaks are precisely the join-irreducible elements of G with respect to $\vee$. If the maximal symbol equals α, then p is called an α-peak. For each $x \in G$ and symbol α, all α-peaks $p \leq x$ are totally ordered by $\leq$. Moreover:

(26)   Each $x \in G$ is the join of all peaks $p \leq x$.

To see this, let y be the join of all peaks $p \leq x$. If $y \neq x$, choose a maximal symbol α of $y^{-1}x$. Then α is also a maximal symbol of x. Write $x = \xi_1 \cdot \ldots \cdot \xi_n$ with symbols $\xi_1, \ldots, \xi_n$, in such a way that the maximum j for which $\xi_j = \alpha$ is minimized. Then $q := \xi_1 \cdot \ldots \cdot \xi_j$ is an α-peak with $q \leq x$. So $q \leq y$. This however contradicts the fact that $y^{-1}x$ has maximal symbol α, thus showing (26).

For each $x \in G$ and symbol α, let $|x|_\alpha$ be the number of occurrences of symbol α in x (not considering α^{-1}).

Proposition 2.4 *Let $x, y \in G$ and let $p \leq x$ and $q \leq y$ be α-peaks satisfying $|p^{-1}x|_\alpha = |q^{-1}y|_\alpha$. Then $|pq^{-1}| \leq |xy^{-1}|$.*

Proof By induction on $|x| + |y|$. If x is not an α-peak, let β be a maximal symbol of x with $\beta \neq \alpha$. Let $x' := x\beta^{-1}$. If β is also a maximal symbol of y, then we can apply induction to x' and $y' := y\beta^{-1}$, since $x'(y')^{-1} = xy^{-1}$. If β is not a maximal symbol of y then $|x'y^{-1}| \leq |xy^{-1}|$ (as β is not canceled in the concatenation of x and y^{-1}), and hence we can apply induction to x' and y.

So we can assume that x is an α-peak, and similarly that y is an α-peak. If $|p^{-1}x|_\alpha = 0$, then $x = p$ and $y = q$, and we are done. If $|p^{-1}x|_\alpha > 0$, then $x > p$ and $y > q$ and we can apply induction to $x' := x\alpha^{-1}$ and $y' := y\alpha^{-1}$. Note that $|p^{-1}x'|_\alpha = |p^{-1}x|_\alpha - 1 = |q^{-1}y|_\alpha - 1 = |q^{-1}y'|_\alpha$. ∎

We use this proposition only in obtaining the following:

(27)   Let $p \leq r \leq ar$ with p an α-peak. Then there exists a' with $p \leq a'p$, $|p^{-1}a'p|_\alpha = |r^{-1}ar|_\alpha$, and $|a'| \leq |a|$.

This follows by applying Proposition 2.4 to $x := r$ and $y := ar$, taking for q the (unique) α-peak satisfying $p \leq q \leq ar$ with $|q^{-1}ar|_\alpha = |p^{-1}r|_\alpha$, which shows that we can take $a' := qp^{-1}$. (Note that $|q^{-1}ar|_\alpha = |p^{-1}r|_\alpha$ is equivalent to $|p^{-1}q|_\alpha = |r^{-1}ar|_\alpha$, since $|p^{-1}q|_\alpha + |q^{-1}ar|_\alpha = |p^{-1}ar|_\alpha = |p^{-1}r|_\alpha + |r^{-1}ar|_\alpha$.)

We also need:

(28)   If $x \leq y$ and α is a symbol not occurring in $x^{-1}y$, then $|xp^{-1}| \leq |yp^{-1}|$ for each α-peak p.

Indeed, by induction we can assume that $x^{-1}y = \beta$ for some symbol $\beta \neq \alpha$. Then in yp^{-1}, the maximal symbol β of y is not cancelled, since otherwise β would be maximal symbol also of p, hence $\beta = \alpha$, contradicting our assumption. Hence $|xp^{-1}| \leq |x\beta p^{-1}| = |yp^{-1}|$.

2.6 Connectedness and Cyclic Reducedness

We study periodicity of symbols in elements of G in order to obtain control on 'stalling' in the algorithm. For this we need Proposition 2.8 below—the other results in Sects. 2.6–2.9 are only needed to prove Proposition 2.8.

Call an element b of G *connected* if the generators occurring in b induce a connected subgraph of $([k], F)$. So b is connected if and only if there are no $a, c \in G$ with $b = ac$, $a \neq 1 \neq c$, and a and c independent. Each peak is connected.

Call an element b of G *cyclically reduced* if $b \wedge b^{-1} = 1$. So b is cyclically reduced if and only if $b^2 = b \cdot b$. Also, if b is cyclically reduced, then for each $s \geq 0$: $b^s = b \cdot b \cdot \ldots \cdot b$ (cf. [2]).

The following proposition will be used in proving Propositions 2.6 and 2.7.

Proposition 2.5 *Let $c, d \in G$ satisfy $d \leq dc$, $c \leq dc$, and $d \not\leq c$. Suppose that d is connected and that all minimal symbols of c occur in d. Then $|c| \leq |c \wedge d|^2$. If d is moreover cyclically reduced, then $c \leq d^{|c \wedge d|}$.*

Proof The proof is by induction on $|c|$. Let $c' := (c \wedge d)^{-1} c$. Then $c' \leq c$, by (15), since $c, d \leq dc$. Also, $d \leq dc'$ (as $d \leq dc$ and $c' \leq c$) and $c' \leq dc'$ (by (16), as $c' \leq c \leq dc$ and $dc' \leq dc$). As $d \not\leq c$ and $c' \leq c$, we know $d \not\leq c'$. Moreover, as $c' \leq c$, all minimal symbols of c' occur in c, hence in d.

If $c' \wedge d < c \wedge d$, then by induction $|c'| \leq |c' \wedge d|^2$ and, if d is cyclically reduced, $c' \leq d^{|c' \wedge d|}$. Hence $|c| = |c \wedge d| + |c'| \leq |c \wedge d| + |c' \wedge d|^2 \leq |c \wedge d|^2$ and $c \leq c \vee d = d(c \wedge d)^{-1} c = dc' \leq d^{|c' \wedge d|+1} \leq d^{|c \wedge d|}$, as required.

So we can assume $c' \wedge d = c \wedge d$. As $c, d \leq dc$, $c \vee d < \infty$. So $(c \wedge d)^{-1} c = c'$ and $(c \wedge d)^{-1} d$ are independent. Hence $c' \wedge d = c \wedge d$ and $(c \wedge d)^{-1} d$ are independent. As $c \wedge d \neq d$ and as d is connected, we know $c \wedge d = 1$. So c and d are independent. Since all minimal symbols of c occur in d, this implies $c = 1$ and the bounds are trivial. ∎

2.7 Conjugates

An element c of G is called a *conjugate* of $a \in G$ if $c = x^{-1} a x$ for some $x \in G$. Then:

(29) For each $a \in G$, each conjugate c of a contains a segment $x^{-1} a x$ with x using only generators occurring in a.

Indeed, let $c = y^{-1} a y$. Then (29) can be proved by induction on $|y|$. If $z := y \wedge a \neq 1$, replace y by $z^{-1} y$ and a by $z^{-1} a z$, and apply induction (this resetting does not change $y^{-1} a y$). So we can assume that $y^{-1} a = y^{-1} \cdot a$ and similarly that $ay = a \cdot y$. Hence a is a segment of $y^{-1} a y$ by (6), proving (29).

We use this in proving:

(30) If a and b are independent, then, each conjugate c of ab contains a segment which is a conjugate of a.

Indeed, by (29) c has a segment $x^{-1}abx$ with x only using generators occurring in ab. As a and b are independent we can write $x = yz$ with y only using generators occurring in a and z only using generators occurring in b. Hence $y^{-1}ay$ and $z^{-1}bz$ are independent, and so $y^{-1}ay$ is a segment of $x^{-1}abx$.

Proposition 2.5 implies:

Proposition 2.6 *Let d be connected and cyclically reduced. Then for each $n \geq 0$, each conjugate c of $d^{n+2|d|}$ contains d^n as segment.*

Proof Choose $x \in G$ with $c = x^{-1}d^{n+2|d|}x$ and $|x|$ as small as possible. Then $d \not\leq x$, otherwise replacing x by $d^{-1}x$ contradicts the minimality of x. Let $y := x \wedge d^{n+|d|}$. Then $y \leq d^{n+|d|+1}$ and $d \leq dy \leq d^{n+|d|+1}$, hence by (14), $y \leq dy$. Since $y \leq d^{n+|d|}$, all minimal symbols of y occur in d. Hence by Proposition 2.5, $y \leq d^{|d|}$. So $d^{|d|} = y \cdot a$ for some a. Hence for $z := y^{-1}x$ one has $x^{-1}d^{n+|d|} = z^{-1} \cdot a \cdot d^n$, implying $x^{-1}d^{n+|d|} = x^{-1}d^{|d|} \cdot d^n$.

By symmetry, $d^{n+|d|}x = d^n \cdot d^{|d|}x$. So by (6), d^n is a segment of

(31) $\quad x^{-1}d^{|d|}d^n d^{|d|}x = c.$ ∎

2.8 Periodicity

We give conditions for the eventual periodicity of a peak:

Proposition 2.7 *Let q be connected and contain symbol α, and let p be an α-peak with $p \leq pq$. Then there exists an α-peak r and $t \geq 0$ with $p = r \cdot q^t$ and $|r| \leq 2|pqp^{-1}|^2$.*

Proof Let $a := pqp^{-1}$. Then $|q| \leq |a|$, as $|p| + |q| = |pq| = |ap| \leq |p| + |a|$. If $|p| \leq 2|a|^2$, we can take $r := p$ and $t := 0$. So we can assume that $|p| > 2|a|^2$.

Let $m := |q|_\alpha$, and let p' be minimal with the properties that $p' \leq p$ and $|p'|_\alpha \geq |p|_\alpha - m$. Then $p' = 1$ or p' is an α-peak. By showing that $(p')^{-1}p = q$ we are done, since then we can apply induction, as $p'q(p')^{-1} = pqp^{-1}$.

Define $c := p^{-1}$, $d := q^{-1}$, and $u := c \wedge dc$. As $u \leq c$, $u \leq dc$, and $d \leq dc$, we know $u \leq du = d \cdot u$ (by (14)). Since $a = (dc)^{-1}c$, we have $|a| = |c| + |dc| - 2|u|$, and so, as $|c| = |p| > 2|a|^2$ and $|a| \geq |d|$:

(32) $\quad 2|u| = |c| + |dc| - |a| = 2|c| + |d| - |a| > 4|a|^2 + |d| - |a| \geq 2|d|^2 + 2|d|.$

Hence $|u| > |d|^2 + |d| \geq |u \wedge d|^2$. Therefore, $d \leq u$ by Proposition 2.5.

Let $b := p^{-1}p'$. So we must show $b = d$. Since $q^{-1} = d \leq u \leq c = p^{-1}$ and $m = |q|_\alpha$, we know $p' \leq pq^{-1}$, that is, $d \leq b$. On the other hand, $b \wedge u \leq d$, since $b \wedge u \leq u \leq dc = q^{-1}p^{-1}$, $|b \wedge u|_\alpha \leq m = |q|_\alpha$, and p is a peak. So $b \wedge u = d$, and hence we must show $b \leq u$.

Let $u' := d^{-1}u$. Since $d \leq u \leq du$, we have $u' \leq u$ (by (10)). Moreover, as $b, u \leq c$, $b \vee u < \infty$. Hence $u' \leq u \leq b \vee u = bd^{-1}u = b \cdot u'$. Since $|u'| = |u| - |d| > |d|^2 = |u \wedge b|^2 \geq |u' \wedge b|^2$, Proposition 2.5 gives $b \leq u'$, implying $b \leq u$. ∎

A g.c.d argument shows:

(33) Let p, r, r', q, q' be α-peaks, and let $t, t' \geq 0$ with $p = r \cdot q^t = r' \cdot (q')^{t'}$ and $\frac{1}{3}|p|_\alpha \geq |r|, |r'|, |q|, |q'|$. Then there exists d such that q and q' are powers of d.

Indeed, define $g := \max\{|r|, |r'|, |q|, |q'|\}$, $m := |p|_\alpha$, $u := |q|_\alpha$, and $u' := |q'|_\alpha$. So $m \geq 3g$. As p is an α-peak, we can uniquely write $p = p_1 \cdot p_2 \cdot \ldots \cdot p_m$, with each p_i being an α-peak. Since r is an α-peak with $|r|_\alpha \leq g$, and since $p = r \cdot q^t$, the sequence $z := (p_{g+1}, \ldots, p_m)$ is periodic with period u. Similarly, z is periodic with period u'. Moreover, z has at least $u + u'$ terms, since $m - g \geq 2g \geq u + u'$.

This implies[1] that z is periodic with period $v := \gcd\{u, u'\}$. Let $d := z_{m-v+1} \cdot \ldots \cdot z_m$. Then $q = z_{m-u+1} \cdot \ldots \cdot z_m = d^{u/v}$ and similarly $q' = d^{u'/v}$.

2.9 A Main Tool

We now come to a main tool for bounding the complexity of our algorithm (which we will use in Sect. 3.3).

Proposition 2.8 *Let p be an α-peak, and let $a, a' \in G$ be such that $p \leq ap$ and $p \leq a'p$ and such that α occurs in $p^{-1}a'p$. If $|p| \geq 8|a|^3$, then each conjugate of a has a segment s satisfying $|p^{-1}aps^{-1}|_\alpha \leq 2|a'|^2$.*

Proof We can assume $|p^{-1}ap|_\alpha > 2|a'|^2$, as otherwise we can take $s := 1$. This implies $|a| \geq |ap| - |p| = |p^{-1}ap| \geq |a'|$.

Let q be the component of $p^{-1}ap$ that contains α; that is, q is the element such that $p^{-1}ap = qu$ for some u independent of q, with q connected and containing α. Similarly, let q' be the component of $p^{-1}a'p$ that contains α. Note that $|q| \leq |a|$ and $|q'| \leq |a'|$.

By (28) applied to $x := pq$ and $y := ap$ we have $|pqp^{-1}| \leq |a|$. Hence, by Proposition 2.7, $p = r \cdot q^t$ for some α-peak r with $|r| \leq 2|pqp^{-1}|^2 \leq 2|a|^2$. Similarly, $p = r' \cdot (q')^{t'}$ for some α-peak r' with $|r'| \leq 2|a'|^2 \leq 2|a|^2$. Now

(34) $|p|_\alpha \geq t = (|p| - |r|)/|q| \geq (8|a|^3 - 2|a|^2)/|a| \geq 6|a|^2 \geq 3\max\{|r|, |r'|, |q|, |q'|\}$,

---

[1]If a and b are periods of $x = (x_1, \ldots, x_n)$ with $a < b$ and $a + b \leq n$, then $b - a$ is a period of x. For let $1 \leq i \leq n - (b - a)$. We show $x_{i+(b-a)} = x_i$. If $i \leq n - b$ then $x_{i+(b-a)} = x_{(i+b)-a} = x_{i+b} = x_i$. If $i > n - b$ then $i > a$ (as $i > n - b \geq a$, since $a + b \leq n$ by assumption), hence $x_{i+(b-a)} = x_{(i-a)+b} = x_{i-a} = x_i$.

Hence by (33), $q = d^n$ for some d and $n \geq 0$, with $|d| \leq |q'| \leq |a'|$. As $t \geq 1$ by (34), q is cyclically reduced, hence also d is cyclically reduced. As q is connected, also d is connected.

Let c be a conjugate of a. Then c is a conjugate of $p^{-1}ap = qu$, with u independent of q. Hence by (30), c contains a segment c' which is a conjugate of $q = d^n$. Now $n = |q|/|d| \geq |p^{-1}ap|_\alpha/|a'| > 2|a'| \geq 2|d|$. Hence, by Proposition 2.6, c' contains $s := d^{n-2|d|}$ as segment. Then $qs^{-1} = d^{2|d|}$, hence $|p^{-1}aps^{-1}|_\alpha = |qs^{-1}|_\alpha \leq |qs^{-1}| = 2|d|^2 \leq 2|a'|^2$. ∎

2.10 The Function $\mu_{a,H} : x \mapsto \min(a^{-1}x\uparrow H)$

The following function $\mu_{a,H} : G \to G$ forms an important ingredient in our algorithm. Fixing $a \in G$ and a closed set $H \subseteq G$, it is defined by

(35) $\mu_{a,H}(x) := \min(a^{-1}x\uparrow H)$

for $x \in G$, which is well-defined as $a^{-1}xH$ is convex by (23). So for each $x \in G$:

(36) $\mu_{a,H}(x) \leq a^{-1}x$.

Moreover, for all $x, y \in G$:

(37) $\mu_{a,H}(x) \leq y$ if and only if $a \in x\uparrow H(y\downarrow)^{-1}$,

since $\min(a^{-1}x\uparrow H) \leq y$ if and only if $a^{-1}bh = c$ for some $b \in x\uparrow$, $h \in H$ and $c \in y\downarrow$.

Proposition 2.9 *Let $a \in G$ and $H \subseteq G$ be closed, and set $\mu := \mu_{a,H}$. Then for all $x, y \in G$:*

(38)
- (i) *if $x \leq y$ then $\mu(x) \leq \mu(y)$,*
- (ii) $\mu(x \wedge y) \leq \mu(x) \wedge \mu(y)$,
- (iii) *if $x \vee y$ is finite, then $\mu(x \vee y) = \mu(x) \vee \mu(y)$.*

Proof Since $x \leq y$ implies $x\uparrow \supseteq y\uparrow$, we have (i). Then (ii) follows from (i). To see (iii), we have $\mu(x) \vee \mu(y) \leq \mu(x \vee y)$ by (i). In particular, $\mu(x) \vee \mu(y)$ is finite. To see the reverse inequality, set $d := \mu(x) = \min(a^{-1}x\uparrow H)$ and $e := \mu(y) = \min(a^{-1}y\uparrow H)$. Then, by (37) and as both d^{-1} and e^{-1} belong to $((d \vee e)\downarrow)^{-1}$,

(39) $a \in x\uparrow Hd^{-1} \cap y\uparrow He^{-1} \subseteq x\uparrow H((d \vee e)\downarrow)^{-1} \cap y\uparrow H((d \vee e)\downarrow)^{-1} = (x \vee y)\uparrow H((d \vee e)\downarrow)^{-1}$,

where the equality follows from (20) (as $x\uparrow \cap y\uparrow = (x \vee y)\uparrow$). So by (37), $\mu(x \vee y) = \min(a^{-1}(x \vee y)\uparrow H) \leq d \vee e$. ∎

The composition of functions $\mu_{a,H}$ have the following property. Let $x, a, a' \in G$ and let H and H' be closed. Then

(40) $\mu_{a',H'}(\mu_{a,H}(x)) \leq \mu_{aa',HH'}(x)$.

Indeed, by the definitions of HH' and $\mu_{aa',HH'}(x)$, there exist $x' \geq x$ and $c \in H$, $c' \in H'$ with $\mu_{aa',HH'}(x) = (aa')^{-1}x'cc'$. Now $\mu_{a,H}(x) \leq a^{-1}x'c$. Hence, using (38)(i),

(41) $\mu_{a',H'} \circ \mu_{a,H}(x) \leq \mu_{a',H'}(a^{-1}x'c) \leq (a')^{-1}(a^{-1}x'c)c' = \mu_{aa',HH'}(x)$.

2.11 Polynomial-Time Algorithms

Let I be an ideal and $x \in G$. Then there is a unique largest element $y \leq x$ with $y \in I$. We can find it in polynomial time if membership of I can be tested in polynomial time:

(42) Let I be an ideal of which we can test membership in polynomial time. Then for any $x \in G$, we can find the maximal element $y \leq x$ with $y \in I$ in polynomial time.

To see this, grow a word $y \leq x$ with $y \in I$, starting with $y = 1$. If there is a minimal symbol in $y^{-1}x$ with $y\alpha \in I$, replace y by $y\alpha$. If no such α exists, y is as required.

Note that y is the closest (with respect to dist) element in I to x. Hence, by the left-invariance of the distance function, $x^{-1}y$ is the closest element in $x^{-1}I$ to 1. That is: $x^{-1}y = \min(x^{-1}I)$. Therefore:

(43) Let I be an ideal of which we can test membership in polynomial time. Then for any $z \in G$, we can find $\min(zI)$ in polynomial time.

Note that we can test membership of $x^\uparrow$ and of $x^\downarrow$ in polynomial time. Hence:

(44) For any $y, x \in G$, we can find $\min(y^{-1}x^\uparrow)$ in polynomial time.

This follows from (43) setting $I := x^{-1}x^\uparrow$ and $z := y^{-1}x$.

If H is closed, then for any $y \in G$, $y^{-1}x^\uparrow H$ is convex (by (23)). Its minimum can be found in polynomial time:

(45) Let H be a closed set of which we can test membership in polynomial time. Then for any $x, a \in G$, $\mu_{a,H}(x) = \min(a^{-1}x^\uparrow H)$ can be found in polynomial time.

Indeed, by (23), $\min(a^{-1}x^\uparrow H) = \min(\min(a^{-1}x^\uparrow)H)$. Hence (45) follows from (44) and (43).

Finally,

(46) Let H be a closed set of which we can test membership in polynomial time. Then for any $x, y, z \in G$, we can test in polynomial time if y belongs to $x^\uparrow H(z^\uparrow)^{-1}$.

Indeed, $y \in x^{\uparrow}H(z^{\uparrow})^{-1}$ if and only if $y^{-1}x^{\uparrow}H \cap z^{\uparrow} \neq \emptyset$. This is the case if and only if $z^{-1}y^{-1}x^{\uparrow}H \cap z^{-1}z^{\uparrow} \neq \emptyset$. The latter statement is equivalent to: $\min(z^{-1}y^{-1}x^{\uparrow}H)$ belongs to the ideal $z^{-1}z^{\uparrow}$. As this minimum can be determined in polynomial time by (45) and as membership of $z^{-1}z^{\uparrow}$ can be tested in polynomial time, we have proved (46).

3 The Cohomology Feasibility Problem

Let $D = (V, E)$ be a directed graph and let G be a group. Two functions $\varphi, \psi : E \to G$ are called *cohomologous* if there exists a function $f : V \to G$ such that $\psi(e) = f(u)^{-1}\varphi(e)f(w)$ for each edge $e = (u, w)$. One directly checks that this gives an equivalence relation.

We give a polynomial-time algorithm for the following *cohomology feasibility problem* for graph groups:

(47) *given:* a directed graph $D = (V, E)$, an undirected graph $([k], F)$, a function $\varphi : E \to G_F$, and for each edge e a closed set $H(e) \subseteq G_F$,
find: a function $\psi : E \to G_F$ such that ψ is cohomologous to φ and such that $\psi(e) \in H(e)$ for each $e \in E$.

The running time of the algorithm for this problem is bounded by a polynomial in $n := |V|$, $m := |E|$, $\sigma := \max\{|\varphi(e)| \mid e \in E\}$, and the maximum time needed to test if any word of polynomial size belongs to $H(e)$ (over all edges e). (The number k of generators can be bounded by $m\sigma$, since we may assume that all generators occur among the $\varphi(e)$.) More precisely, there exist polynomials p_1 and p_2 such that the problem takes time $p_1(n, m, \sigma)\tau_H(p_2(n, m, \sigma))$, where $\tau_H(x)$ is the time needed to test membership of the $H(e)$ for words of size at most x.

Note that, by the definition of cohomologous, equivalent to finding a function ψ as in (47), is finding a function $f : V \to G_F$ satisfying:

(48) $\quad f(u)^{-1}\varphi(e)f(w) \in H(e)$ for each edge $e = (u, w)$.

We call such a function f *feasible*.

We can assume that

(49) $\quad |\varphi(e)| \leq 1$ for each edge e.

Indeed, if $\varphi(e) = xy$ for edge $e = (u, w)$, we can split the edge into two edges $(u, v), (v, w)$, where v is a new vertex, and define $\varphi(u, v) := x$, $\varphi(v, w) := y$, $H(u, v) := H(e)$, and $H(v, w) := \{1\}$. The new problem is equivalent to the original problem: if f is a solution to the original problem, we can set $f(v) := yf(w)$, and obtain a solution for the new problem; conversely, if f is a solution to the original problem, forgetting the value of f on v, we obtain a solution to the original problem.

3.1 Pre-feasible Functions

Given the input of the cohomology feasibility problem (47), we call a function $f : V \to G_F$ *pre-feasible* if for each edge $e = (u, w)$ of D there exist $x \geq f(u)$ and $z \leq f(w)$ such that $x^{-1}\varphi(e)z \in H(e)$. Clearly, each feasible function is pre-feasible. There is a trivial pre-feasible function f, defined by $f(v) := 1$ for each $v \in V$. Note that f is pre-feasible if and only if

(50) $\quad \mu_{\varphi(e), H(e)}(f(u)) \leq f(w)$ for each edge $e = (u, w)$.

The collection G_F^V of all functions $f : V \to G_F$ can be partially ordered by: $f \leq g$ if and only if $f(v) \leq g(v)$ for each $v \in V$. Then G_F^V forms a lattice if we add an element ∞ at infinity. Let $\wedge$ and $\vee$ denote meet and join. Then (38) (ii) and (iii) directly give:

(51) Let f_1 and f_2 be *pre-feasible functions*. Then $f_1 \wedge f_2$ and, if $f_1 \vee f_2 < \infty$, $f_1 \vee f_2$ are pre-feasible again.

It follows that for each function $f : V \to G_F$ there is a unique smallest pre-feasible function $\bar{f} \geq f$, provided that there exists at least one pre-feasible function $g \geq f$. If no such g exists we set $\bar{f} := \infty$. By (51), $\overline{f \vee g} = \bar{f} \vee \bar{g}$ for any two functions f, g with $f \vee g$ finite.

3.2 A Subroutine Finding $\bar{f}$

Condition (50) suggests a 'constraint satisfaction' algorithm to find $\bar{f}$ for a given function f. Let input $D = (V, E)$, F, φ, H for the cohomology feasibility problem be given. For any edge e, we can determine $\mu_{\varphi(e), H(e)}(f(u))$ in polynomial time, by (45).

Subroutine to find $\bar{f}$: Find an edge $e = (u, w)$ for which

(52) $\quad m_e := \mu_{\varphi(e), H(e)}(f(u)) \not\leq f(w)$.

If $m_e \vee f(w)$ is finite, reset $f(w) := m_e \vee f(w)$ and iterate. If $m_e \vee f(w) = \infty$, output $\bar{f} := \infty$. If no such edge e exists, output $\bar{f} := f$.

Then the output of the subroutine (if any) is correct. For let f' be the reset function. If $\bar{f}$ is finite, then $f \leq f' \leq \bar{f}$, since $f'(w) = \mu_{\varphi(e), H(e)}(f(u)) \vee f(w) \leq \mu_{\varphi(e), H(e)}(\bar{f}(u)) \vee \bar{f}(w) = \bar{f}(w)$, since $\bar{f}$ is pre-feasible. So in this case $\bar{f}' = \bar{f}$. This moreover implies that if $m_e \vee f(w) = \infty$ then $\bar{f} = \infty$.

3.3 Running Time of the Subroutine

For each walk $P = e_1 e_2 \ldots e_t$, where $e_1, \ldots, e_t$ are consecutive edges of D, we set $\varphi(P) := \varphi(e_1)\varphi(e_2)\ldots\varphi(e_t)$ and $H(P) := H(e_1)H(e_2)\ldots H(e_t)$.

We will study the running time of the subroutine under the condition that the cohomology feasibility problem has a solution, or more weakly, that

(53)   for each closed walk C, $H(C)$ contains a conjugate of $\varphi(C)$.

For any function $f : V \to G_F$, let $|f| := \max\{|f(v)| \mid v \in V\}$.

Proposition 3.1 *If (49) and (53) hold, then for any $f : V \to G_F$ with $\bar{f}$ finite:*

(54) $|\bar{f}| \leq k|f| + 220k(2nk)^9$.

Proof Define $c := 2nk$ and $m := 3c^2$, and suppose $|\bar{f}| > k|f| + 220kc^9$. Consider the (infinite) directed graph $\mathcal{D}$ with vertex set $\mathcal{V} := \{(v, p) \mid v \in V, p \text{ peak}\}$ and edge set all pairs $((v, p), (w, q)) \in \mathcal{V} \times \mathcal{V}$ with $e = (v, w) \in E$ and $q \leq \mu_{\varphi(e), H(e)}(p)$. Let $\mathcal{S}$ be the set of all vertices (v, p) of $\mathcal{D}$ with $p \leq f(v)$. Then for each $v \in V$ and peak p:

(55) $\bar{f}(v) = \bigvee\{p \mid \text{there exists a walk in } \mathcal{D} \text{ from } \mathcal{S} \text{ to } (v, p)\}$.

This follows with (26) and Proposition (9)(iii). Hence:

(56)   if $p \leq \bar{f}(v)$, then there exists a walk in $\mathcal{D}$ from $\mathcal{S}$ to (v, p).

Since $|\bar{f}| > k|f| + 220kc^9$, there exists $w \in V$ with $|\bar{f}(w)| > k|f| + 220kc^9$. Hence there exists a peak $q \leq \bar{f}(w)$ with $|q| > |f| + 220c^9$ (since $\bar{f}(w)$ is a join of at most k peaks). By (56), there is a walk $\mathcal{P}$ in $\mathcal{D}$ from $\mathcal{S}$ to (w, q). Choose a shortest such walk $\mathcal{P}$; let it have length ℓ. Since $|\varphi(e)| \leq 1$ by assumption, $|p| \leq |\varphi(e)^{-1}p'| \leq |p'| + 1$ for each edge $((v, p), (v', p'))$ of $\mathcal{D}$, where $e = (v, v')$. Hence $\ell \geq |q| - |f| > 220c^9$.

Let $\mathcal{P}$ traverse vertices $(v_0, q_0), (v_1, q_1)\ldots, (v_\ell, q_\ell)$ of $\mathcal{D}$, in this order. So $v_\ell = w$ and $q_\ell = q$. For each $u \in V$ and each symbol α let $I_{u,\alpha}$ denote the set of indices $j \in \{\ell - 3c^3, \ldots, \ell\}$ such that $v_j = u$ and q_j is an α-peak. Then there exist $u \in V$ and a symbol α such that $|I_{u,\alpha}| > 3c^3/c = m$. Choose $j_0 < j_1 < \cdots < j_m$ in $I_{u,\alpha}$. Set $p_i := q_{j_i}$ for $i = 0, 1, \ldots, m$. So each p_i is an α-peak and $p_0 < p_1 < \cdots < p_m$. We will apply Proposition 2.8 to $p := p_0$.

Let C_i be the $u - u$ walk $v_{j_{i-1}}, v_{j_{i-1}+1}, \ldots, v_{j_i-1}, v_{j_i}$ in D, and let $C := C_1 C_2 \ldots C_m$ (the concatenation of $C_1, \ldots, C_m$). Then $|C| \leq 3c^3$ and so there exists $i \in \{1, \ldots, m\}$ with $|C_i| \leq |C|/m \leq 3c^3/m = c$.

Let $a := \varphi(C)^{-1}$ and $a'' := \varphi(C_i)^{-1}$. Then $|a| = |\varphi(C)| \leq |C| \leq 3c^3$ and $|a''| = |\varphi(C_i)| \leq |C_i| \leq c$. This gives

(57) $|p_0| \geq |q| - 3c^3 \geq 220c^9 - 3c^3 \geq 216c^9 = 8(3c^3)^3 \geq 8|a|^3$.

Moreover, $p_0 \leq p_m \leq \mu_{\varphi(C), H(C)}(p_0) \leq ap_0$ and similarly $p_i \leq p_{i+1} \leq a'' p_i$. As α occurs in $p_i^{-1} p_{i+1}$, α also occurs in $p_i^{-1} a'' p_i$. By (27), $p_0 \leq a' p_0$ for some a' with $|a'| \leq |a''|$ and $|p_0^{-1} a' p_0|_\alpha = |p_i^{-1} a'' p_i|_\alpha$.

As we assume that (53) holds, some conjugate x of $\varphi(C)$ belongs to $H(C)$. So x^{-1} is a conjugate of a. Hence, by Proposition 2.8, x^{-1} has a segment s such that $|p_0^{-1} a p_0 s^{-1}|_\alpha \leq 2|a'|^2 \leq 2c^2 < 3c^2 = m$. Now $p_m \leq \min(\varphi(C)^{-1} p_0^\uparrow H(C)) \leq a p_0 s^{-1}$, since $s^{-1} \in H(C)$, as $H(C)$ is closed and $x \in H(C)$. So $|p_0^{-1} p_m|_\alpha < m$, contradicting the fact that p_m contains at least m α's more than p_0. ∎

This implies, where $n := |V(D)|$, and where, for any $\sigma \in \mathbb{Z}_+$, $\tau_H(\sigma)$ is the maximum time needed to test if any word of size $\leq \sigma$ belongs to $H(e)$, for any given edge e.

Theorem 3.2 *There exist polynomials p_1 and p_2 such that, if (49) and (53) hold, then the running time of the subroutine is bounded by*

(58) $\quad p_1(n, k, |f|) \tau_H(p_2(n, k, |f|))$.

Proof At each iteration, we increase $|f(v)|$ for some vertex v. Hence Proposition 3.1 implies that, if $\bar{f}$ is finite, the number of iterations is bounded by some polynomial p_1 in $n, k, |f|$. If the subroutine exceeds this number of iterations, we conclude that $\bar{f} = \infty$.

Since in each iteration, the reset f' satisfies $|f'| \leq |\bar{f}|$, and since $|\bar{f}|$ is bounded by a polynomial p_2 in $n, k, |f|$, in each iteration we only need to test membership of words of size at most $p_2(n, k, |f|)$. ∎

3.4 A Polynomial-Time Algorithm for the Cohomology Feasibility Problem for Graph Groups

We now describe the algorithm for the cohomology feasibility problem for graph groups. Let input $D = (V, E)$, F, φ, H of (47) be given.

Let $\mathcal{F}$ be the collection of all functions $f : V \to G_F$ such that for each edge $e = (u, w)$ of D there exist $x \geq f(u)$ and $z \geq f(w)$ satisfying $x^{-1}\varphi(e)z \in H(e)$; equivalently:

(59) $\quad \varphi(e) \in f(u)^\uparrow H(e)(f(w)^\uparrow)^{-1}$.

This can be tested in polynomial time by (46). So for any given function f one can check in polynomial time whether f belongs to $\mathcal{F}$. Trivially, if $f \in \mathcal{F}$ and $g \leq f$ then $g \in \mathcal{F}$. Moreover:

(60) Let $f_1, \ldots, f_t$ be functions such that $f_i \vee f_j \in \mathcal{F}$ for all i, j. Then $f := f_1 \vee \cdots \vee f_t \in \mathcal{F}$.

Proof We must show that for each edge $e = (u, w)$, $\varphi(e)$ belongs to

(61) $\quad f(u)^\uparrow H(e)(f(w)^\uparrow)^{-1}$.

Since $f_i \vee f_j \in \mathcal{F}$ for all i, j, we know

(62) $\quad \varphi(e) \in f_i(u)^\uparrow H(e) f_j(w)^{\uparrow -1}$

for all i, j. Hence by (20),

(63) $\quad \varphi(e) \in \bigcap_i \bigcap_j f_i(u)^\uparrow H(e)(f_j(w)^\uparrow)^{-1} =$

$$\left(\bigcap_i f_i(u)^\uparrow\right) H(e) \left(\bigcap_j (f_j(w)^\uparrow)^{-1}\right) = f(u)^\uparrow H(e)(f(w)^\uparrow)^{-1}.$$

Here i and j range over $1, \ldots, t$. ∎

In the following theorem, 'solvable in polynomial time' means as before that there exist polynomials p_1 and p_2 such that the problem is solvable in time $p_1(n + m, k, \rho)\tau_H(p_2(n + m, k, \rho))$, where $n := |V(D)|$, $m := |E(D)|$, ρ is the maximum of $|\varphi(e)|$ over all $e \in E$, and where $\tau_H(\sigma)$ again is the maximum time needed to test if any word of size $\leq \sigma$ belongs to $H(e)$, for any given edge e.

Theorem 3.3 *The cohomology feasibility problem for graph groups is solvable in polynomial time.*

Proof We can assume again that $|\varphi(e)| \leq 1$ for each edge e. Moreover, we can assume that with each edge $e = (u, w)$ also $e^{-1} = (w, u)$ is an edge, with $\varphi(e^{-1}) = \varphi(e)^{-1}$ and $H(e^{-1}) = H(e)^{-1}$.

For any $e = (u, w) \in E$, let f_e be the function defined by

(64) $\quad f_e(v) := \begin{cases} \varphi(e) & \text{if } v = u, \\ 1 & \text{if } v \neq u. \end{cases}$

Let L be the set of pairs $\{e, e^{-1}\}$ from E such that $\varphi(e) \notin H(e)$. Let N be the collection of all pairs $\{e, d\}$ from E such that the function $\bar{f}_e \vee \bar{f}_d = \infty$, or is finite and does not belong to $\mathcal{F}$ (possibly $e = d$).

Choose a subset B of E such that B intersects each pair in L and such that B contains no pair in N. This is a special case of the 2-satisfiability problem, and hence can be solved in polynomial time. Assuming that there exists a feasible function f, then B exists, as $B := \{e = (u, v) \in E \mid \varphi(e) \leq f(u)\}$ would have the required properties.

If we find B, define f by:

(65) $\quad f(v) := \bigvee_{e \in B} \bar{f}_e.$

We are done by proving that f is feasible. Since $\bar{f}_e \vee \bar{f}_b < \infty$ for each pair $\{e, d\} \subseteq B$, we know $f < \infty$. Moreover, f is the join of a finite number of pre-feasible functions, and hence f is pre-feasible. So by (25) it suffices to prove that for each edge $e = (u, w)$:

(66) (i) there exist $x \geq f(u)$ and $z \geq f(w)$ such that $x^{-1}\varphi(e)z \in H(e)$,
 (ii) there exist $x \leq f(u)$ and $z \leq f(w)$ such that $x^{-1}\varphi(e)z \in H(e)$.

To prove (66)(i), note that it is equivalent to: $f \in \mathcal{F}$. As $\bar{f}_e \vee \bar{f}_b \in \mathcal{F}$ for all $a, b \in B$, (60) gives $f \in \mathcal{F}$.

To prove (66)(ii), if it does not hold then $\varphi(e) \notin H(e)$, hence $\{e, e^{-1}\} \in L$. So e or e^{-1} belongs to B. By symmetry, we can assume that $e \in B$. So $f_e \leq f$, and therefore $\varphi(e) \leq f(u)$. So we can take $x := \varphi(e)$ and $z := 1$ in (66)(ii). ∎

An analysis of this algorithm shows that the cohomology feasibility problem has a solution if and only if for each vertex u and each pair C, C' of (undirected) $u - u$ walks in D there exists $x \in G$ such that $x^{-1}\varphi(C)x \in H(C)$ and $x^{-1}\varphi(C')x \in H(C')$. This condition is trivially necessary.

4 Planar Graphs

We repeat the *partially disjoint paths problem* for directed planar graphs:

(67) *given:* a directed planar graph $D = (V, E)$, vertices $r_1, s_1, \ldots, r_k, s_k$ of D, and a set F of pairs $\{i, j\}$ with $i, j \in [k]$,
find: a k-tuple $\mathcal{P} = (P_1, \ldots, P_k)$, where P_i is a directed $r_i - s_i$ path P_i, for $i = 1, \ldots, k$, such that P_i and P_j are disjoint whenever $\{i, j\} \in F$.

We can assume without loss of generality:

(68) $r_1, s_1, \ldots, r_k, s_k$ are distinct, each r_i has outdegree 1 and indegree 0, and each s_i has indegree 1 and outdegree 0.

Again, let G_F be the graph group generated by $g_1, \ldots, g_k$ and relations $g_i g_j = g_j g_i$ whenever $\{i, j\} \notin F$. For each solution $\mathcal{P}$ of (67), let $\chi_\mathcal{P} : E \to G_F$ be defined by:

(69) $\chi_\mathcal{P}(e) := \prod_{P_i \text{ traverses } e}^{i} g_i$.

The order in which we take this product is irrelevant, since if both P_i and P_j traverse e, then $\{i, j\} \notin F$ and hence $g_i g_j = g_j g_i$.

Let $\mathcal{F}$ be the collection of faces of D. Call $\varphi, \psi : E \to G_F$ *homologous* if there exists $f : \mathcal{F} \to G_F$ such that for each edge e: $\psi(e) = f(F)^{-1}\varphi(e)f(F')$, where F and F' are the left-hand and the right-hand face at e, respectively (seen when traversing e in forward direction).

4.1 Finding Partially Disjoint Paths of Prescribed Homology

We first consider the homology version of the partially disjoint paths problem:

(70) *given:* a directed planar graph $D = (V, E)$, vertices $r_1, s_1, \ldots, r_k, s_k$ of D satisfying (68), a set F of pairs $\{i, j\}$ with $i, j \in [k]$, and a function $\varphi : E \to G_F$,

find: a solution $\mathcal{P}$ of (67) such that $\chi_{\mathcal{P}}$ is homologous to φ.

Proposition 4.1 *Problem* (70) *is solvable in polynomial time.*

Proof We can assume that problem (70) has a solution[2] — that is, φ is homologous to $\chi_{\mathcal{P}}$ for some solution $\mathcal{P}$ of (67).

Let $\mathcal{F}$ be the collection of faces of D. Consider the dual directed graph $D^* = (\mathcal{F}, E^*)$, where for each edge e of D there is a directed edge $e^* \in E^*$ from the face at the left-hand side of e to the face at the right-hand side of e. We define $\hat{\varphi}(e^*) := \varphi(e)$ and

(71) $\quad H(e^*) := \{\prod_{i \in I} g_i \mid I \subseteq [k], I \text{ stable set in } ([k], F)\}$,

where I is *stable* if it contains no pair in F as subset. Note that $H(e^*)$ is a closed subset of G_F.

We extend the planar graph D^* by a number of further 'nonplanar' edges, as follows. Consider any vertex $v \notin \{r_1, s_1, \ldots, r_k, s_k\}$ of D and two faces F and F' of D incident with v. Let $e_1, \ldots, e_t$ be the edges incident with v that are crossed when going clockwise from F to F' around v. Then add to D^* an edge $e_{v,F,F'}$ from F to F', and define

(72) $\quad \hat{\varphi}(e_{v,F,F'}) := \varphi(e_1)^{\sigma_1} \ldots \varphi(e_t)^{\sigma_t}$,

where, for $j = 1, \ldots, t$, $\sigma_j := 1$ if e_j is oriented away from v and $\sigma_j := -1$ if e_j is oriented towards v. Note that, as by assumption φ is homologous to $\chi_{\mathcal{P}}$ for some solution $\mathcal{P}$ of (67), we necessarily have $\hat{\varphi}(e_{v,F',F}) = \hat{\varphi}(e_{v,F,F'})^{-1}$.

Moreover, define

(73) $\quad H(e_{v,F,F'}) := \{\prod_{i \in I} g_i^{\tau(i)} \mid I \subseteq [k], I \text{ stable set in } ([k], F), \tau : I \to \{+1, -1\}\}$.

Also $H(e_{v,F,F'})$ is a closed subset of G_F.

Let $\hat{D}$ be the extended directed graph, and consider the cohomology feasibility problem Π for $(\hat{D}, \hat{\varphi})$, in which we require that the output is only weakly allowed on the edges in E'. As, by assumption, φ is homologous to $\chi_{\mathcal{P}}$ for some $\mathcal{P}$, problem Π has a solution, namely $\chi_{\mathcal{P}}$. Conversely, let $\psi : E(\hat{D}) \to G_F$ be any solution of Π. Define $\check{\psi} : E(D) \to G_F$ by $\check{\psi}(e) := \psi(e^*)$ for $e \in E(D)$. Then $\check{\psi}$ is equal to $\chi_{\mathcal{P}}$ for some $\mathcal{P}$. This because, by our assumption, φ is homologous to $\chi_{\mathcal{P}}$ for some $\mathcal{P}$. Hence, for the edge e incident with r_i, $\varphi(e)$ is conjugate to g_i (as by (68) e is incident at both sides to the same face), therefore $\check{\psi}(e)$ is conjugate to g_i. Since $\check{\psi}(e)$ is allowed, it follows that in fact $\check{\psi}(e) = g_i$. So only P_i traverses e, and in the forward direction. Similarly for the edge incident with s_i.

Therefore, the proposition follows from Theorem 3.3. ∎

---

[2] Let A a polynomial-time algorithm that finds a solution for feasible instances. When we apply A to any instance, then if feasible, we find a feasible solution, and if infeasible, A gets stuck or has not delivered a solution in polynomial time.

4.2 Enumerating Homologies of Disjoint Paths

We finally describe an algorithm that finds, for any input of (67), a collection Φ of functions $\varphi : E \to G_F$ with the property that

(74) for each solution $\mathcal{P}$ of (67), $\chi_\mathcal{P}$ is homologous to at least one $\varphi \in \Phi$.

So, although there exist infinitely many homology classes (if $F \neq \emptyset$), in our algorithm we can restrict ourselves to a number of homology classes that, for fixed k, is bounded by a polynomial in the size of the graph.

Proposition 4.2 *Fixing k, a collection Φ satisfying (74) can be found in polynomial time.*

Proof Again, we can assume (68). Moreover, we can assume that D is weakly connected and that (for the convenience of the exposition) each $i \in [k]$ is contained in at least one pair $\{i, j\}$ in F (otherwise we can easily reduce the problem). We also can assume that each vertex $v \neq r_1, s_1, \ldots, r_k, s_k$ has total degree $\deg(v)$ equal to 3: replace v by a directed circuit of length $\deg(v)$ and attach the edges incident with v to the different vertices of the circuit (in a planar manner of course). Any φ found for the modified graph can be 'shrunk' to the smaller graph.

Choose a spanning tree T in D. We will consider graphs T' obtained from T by replacing each edge e by a number (possibly 0) of parallel edges. These edges form a *parallel class*, denoted by π_e. Each such graph T' is trivially planar, by drawing the edges properly parallel in the plane.

We moreover consider undirected walks in such graphs T'. (An *undirected walk* may traverse edges in any direction.) Call undirected walks W and W' in T' *crossing* if there is a vertex v and distinct edges e_1, e_2, e_3, e_4 of T' incident with v, in clockwise or counterclockwise order, such that W traverses e_1 and e_3 consecutively, and W' traverses e_2 and e_4 consecutively. If $W = W'$, we say that W is *self-crossing*.

In particular, we consider k-tuples $\mathcal{W} = (W_1, \ldots, W_k)$ of undirected walks in T' such that

(75) (i) W_i runs from r_i to s_i and is not self-crossing, for each $i = 1, \ldots, k$,
   (ii) W_i and W_j are not crossing, for each $\{i, j\} \in F$,
   (iii) each edge of T' is traversed by precisely one W_i.

The last condition implies that T' is determined by $W_1, \ldots, W_k$.

For each k-tuple $\mathcal{W}$ satisfying (75), define $\varphi_\mathcal{W} : E \to G_F$ as follows. If e is an edge of D not in T, set $\varphi_\mathcal{W}(e) := 1$. If $e = (u, w)$ is an edge of T, let $e_1, \ldots, e_t$ be the edges in π_e, from left to right with respect to the orientation (u, w) of e. Let $\alpha_j := g_i$ if e_j is traversed by W_i in the direction from u to w, and let $\alpha_j := g_i^{-1}$ if e_j is traversed by W_i in the direction from w to u. Define $\varphi_\mathcal{W}(e) := \alpha_1 \ldots \alpha_t$. Then:

(76) For each solution $\mathcal{P}$ of (67), there exists T' and $\mathcal{W} = (W_1, \ldots, W_k)$ satisfying (75) such that $\chi_\mathcal{P}$ and $\varphi_\mathcal{W}$ are homologous and such that, for each $i = 1, \ldots, k$, each parallel class in T' is traversed at most $2|E|$ times by W_i.

To see this, reroute $P_1, \ldots, P_k$ along T as follows. For each e in $E(D) \setminus E(T)$, let Q_e be the path in T connecting the ends of e. Order the edges in $E(D) \setminus E(T)$ as $e_1, e_2, \ldots, e_m$ such that if Q_{e_j} is longer than Q_{e_i} then $j > i$. Then for $j = 1, \ldots, m$, if P_i traverses e_j, reroute P_i along Q_{e_j}; that is, add edges parallel to the edges in Q_{e_j}, in the disk enclosed by e_j and Q_{e_j}, and replace e_j in P_i by the new edges (in order). This gives T' and $W_1, \ldots, W_k$ as required, proving (76).

So to cover all homology classes of solutions of problem (67), it suffices to enumerate all T' and $W_1, \ldots, W_k$ satisfying (75).

In fact, we can assume that each W_i is *non-returning* in the following sense. Let W_i traverse edges e, vertex v, and edges e' consecutively.

(77) (i) If $v \notin \{r_1, s_1, \ldots, r_k, s_k\}$, then e and e' belong to different parallel classes incident with v.
(ii) If $v \in \{r_j, s_j\}$ for some $j \in [k]$, then e and e' enclose the starting or ending edge of W_j.

This can be attained as follows. Suppose W_i, e, v, e' violate (77). Fixing v, choose W_i, e, e' such that the number of edges inbetween of (that is, enclosed by) e and e' is as small as possible. Then each edge inbetween of e and e' is traversed by some W_j with $j \neq i$ (as W_i is not self-crossing) and $\{i, j\} \notin F$ (as W_i and W_j cross). So deleting e and e' from W_i and from T', gives a walk system $\mathcal{W}'$ again satisfying (77), with $\varphi_{\mathcal{W}'} = \varphi_\mathcal{W}$, and with a smaller total length. Iterating this, we end up with each W_i non-returning.

This implies that if $v \in V \setminus \{r_1, s_1, \ldots, r_k, s_k\}$ has degree 1 in T, it is incident with no edges in T'. Delete such vertices from T repeatedly. Let T_0 be the final graph. It is a tree with maximum degree 3 and with $2k$ vertices of degree 1 (namely, $r_1, s_1, \ldots, r_k, s_k$). Hence T_0 has $2k - 2$ vertices of degree 3. The vertices of degree 1 and 3 are connected by $4k - 3$ internally vertex-disjoint paths, together forming T_0.

Let $\mathcal{W} = (W_1, \ldots, W_k)$ be a k-tuple of walks satisfying (75) and (77). Consider a vertex v of degree 2 in T_0, say incident with edges e and e' of T_0. By (77), W_i traverses edges in π_e as often as it traverses edges in $\pi_{e'}$.

For each $i = 1, \ldots, k$, define $h_i : E(T_0) \to \{0, 1, \ldots, 2|E|\}$ by: $h_i(e)$ is the number of times that W_i traverses π_e, in any direction (for $e \in E(T_0)$). Then we can derive $\varphi_\mathcal{W}$ from $h_1, \ldots, h_k$, without knowing $\mathcal{W}$:

Claim. Given $h_1, \ldots, h_k$, one can find $\varphi_\mathcal{W}$ in polynomial time.

Proof Consider any $\{i, j\} \in F$. Let T'' be the subgraph of T' consisting of the edges traversed by W_i and W_j. We know T'' since we know h_i and h_j. We determine an undirected graph H with vertex set $E(T'')$, calling two edges in $E(T'')$ *associated* if they form an edge of H.

First, consider any vertex v of T_0 of degree 2. Let e and e' be the edges of T_0 incident with v, and consider the parallel classes π_e and $\pi_{e'}$ in T''. As $|\pi_e| = h_i(e) + h_j(e) = h_i(e') + h_j(e') = |\pi_{e'}|$, we can order the edges in π_e as $e_1, \ldots, e_m$ from left to right when going towards v, and similarly, the edges in $\pi_{e'}$ as $e'_1, \ldots, e'_m$, from left to right when going away from v. For each $t = 1, \ldots, m$, we 'associate' e_t and e'_t.

Next, consider any vertex v of T_0 of degree 3. Let e, e', and e'' be the edges of T_0 incident with v. Consider the parallel classes π_e, $\pi_{e'}$, and $\pi_{e''}$ in T''. As W_i and W_j are non-returning (that is, satisfy (77)(i)), we know that there exist nonnegative integers a, b, and c such that $|\pi_e| = b + c$, $|\pi_{e'}| = a + c$, and $|\pi_{e''}| = a + b$. These numbers are unique and can be directly calculated from $|\pi(e)|$, $|\pi(e')|$, and $|\pi(e'')|$. This implies that the edges in $\pi_e \cup \pi_{e'} \cup \pi_{e''}$ can uniquely be pairwise 'associated' in such a fashion that any two associated pairs of edges are noncrossing at v and such that no two edges in the same parallel class are associated.

Finally, consider any vertex v of T_0 of degree 1. So v belongs to

(78) $\{r_1, s_1, \ldots, r_k, s_k\}$.

Let e be the edge of T_0 incident with v. Let $e_1, \ldots, e_t$ be the edges in the parallel class π_e of T'', in order. 'Associate' e_i with e_{t+1-i} for each $i = 1, \ldots, \lfloor \frac{1}{2}t \rfloor$. So if t is odd (which is the case if and only if $v \in \{r_i, s_i, r_j, s_j\}$), one edge in π_e remains unassociated at v (namely the middle edge).

Then the graph H with $E(T'')$ as vertex set and all pairs of associated edges of T'' as edges of H, consists of two paths, corresponding to W_i and W_j in T'. These sets of edges form two walks that we can orient, one from r_i to s_i, the other from r_j to s_j. Then for each edge e of T_0 we know the order, from left to right, in which W_i and W_j traverse the parallel class π_e of T'', and we can derive the direction. Concluding, we can derive the subword of $\varphi_{\mathcal{W}}(e)$ made up by the symbols g_i, g_i^{-1}, g_j, and g_j^{-1}. (It is important here that we know that H comes from an $r_i - s_i$ walk W_i and an $r_j - s_j$ walk W_j. So H contains no circuit, for which we would not know whether it belongs to W_i or to W_j.)

As we can do this for each $\{i, j\} \in F$, we can derive $\varphi_{\mathcal{W}}(e)$. This follows from the fact that for any word w with symbols $g_1, g_1^{-1}, \ldots, g_k, g_k^{-1}$, if we know for each $\{i, j\} \in F$ the subword $w_{i,j}$ of w made up by g_i, g_i^{-1}, g_j, g_j^{-1}, we can determine w as word up to transposition of commuting symbols (but without cancellation): Start by finding an $i \in [k]$ and $\alpha \in \{g_i, g_i^{-1}\}$ that occurs first in $w_{i,j}$ for each j with $\{i, j\} \in F$. By transposition we can assume that α is the first symbol of w. Then delete the first α from each $w_{i,j}$ with $\{i, j\} \in F$, and iterate.

Thus, temporarily, we do not cancel g_i with g_i^{-1} or g_j with g_j^{-1}, but work in the semigroup generated by $g_1, g_1^{-1}, \ldots, g_k, g_k^{-1}$ with relations $g_i g_j = g_j g_i$, $g_i g_j^{-1} = g_j^{-1} g_i$, and $g_i^{-1} g_j^{-1} = g_j^{-1} g_i^{-1}$ for all distinct i, j with $\{i, j\} \notin F$. At the end we factor out to the group G_F.

Concluding, we can find the element of G_F represented by word w. (Here we use the assumption that each i is contained in some pair in F.) □

We finally describe the required algorithm. Enumerate all k-tuples of functions $h_1, \ldots, h_k : E(T_0) \to \{0, 1, \ldots, 2|E|\}$ with the property that if e and e' are the edges of T_0 incident with a vertex of T_0 of degree 2, then $h_i(e) = h_i(e')$ for each i. Determine, if possible, $\varphi_{\mathcal{W}}$. All such $\varphi_{\mathcal{W}}$ form Φ.

Since T_0 consists of vertices of degree 1 and 3 together with $4k - 3$ internally vertex-disjoint paths connecting these vertices, there are $((2|E| + 1)^{4k-3})^k$ such k-tuples $h_1, \ldots, h_k$. For fixed k, this is polynomially bounded. ∎

4.3 Finding Partially Disjoint Paths

Concluding, we have:

Theorem 4.3 *For each fixed k, the partially disjoint paths problem in directed planar graphs is solvable in polynomial time.*

Proof Directly from Propositions 4.1 and 4.2. ∎

5 Some Extensions and Open Questions

The theorem can be extended to the case where for each edge e of D a subset K_e of $[k]$ is given, prescribing that e may be traversed only by paths P_i with $i \in K_e$. This amounts to restricting I in (71) to subsets of K_e. Instead of requiring disjointness of certain pairs of paths, one may relax this to requiring that certain pairs of paths are noncrossing: so they are allowed to 'touch' each other in a vertex, but not to cross. This amounts to deleting the 'nonplanar' edges $e_{v,F,F'}$.

One may impose further conditions of the following kind. Choose an (undirected) path Q in the dual graph D^*, connecting two faces F and F' of D. Then one may restrict the total 'flow' of paths P_i in D that intersect Q: as long as the restriction can be described by a closed subset of G_F, this requirement translates into an extra nonplanar edge added to the dual graph D^*, like before we did for paths in D^* connecting two faces incident with a vertex v.

Moreover, the theorem extends to directed graphs D on any fixed compact surface instead of planar graphs. Then, instead of considering the spanning tree T in Sect. 4.2, one considers a minimal connected spanning subgraph T that is cellularly embedded, i.e., each face is a disk (assuming without loss of generality that D is cellularly embedded). Fixing the surface, the number of edges in T is only a fixed amount more than in a spanning tree, and the enumeration of homology classes can be bounded accordingly.

We finally mention some open questions. The running time of the algorithm above is bounded by a polynomial with exponent depending on k (in fact, $O(k^2)$). This raises the question if the problem is 'fixed parameter tractable'; that is, can the degree of the polynomial be fixed independently of k, while the dependence of k is only in the coefficient. As mentioned, this question was answered confirmatively by Cygan, Marx, Pilipczuk, and Pilipczuk [3] for the *k fully* disjoint paths problem in directed planar graphs.

Another open question is if the condition of fixing k can be relaxed to fixing other parameters of the graph $\Gamma = ([k], F)$. One may think of fixing the maximum degree of Γ, or (more weakly) fixing the chromatic number of Γ, or (even more weakly) fixing the clique number of Γ. A different open question is if instead of fixing k, it suffices to fix the number of faces that can cover all terminals (by the face boundaries). Moreover, as mentioned in the introduction, the complexity of the corresponding edge-disjoint version is unknown.

Let us finally ask whether the polynomial-time solvability of the cohomology feasibility problem for graph groups (polynomial-time even for *unfixed k*) has other applications, for instance to free partially commutative semigroups as studied for inhomogeneous sorting and scheduling of concurrent processes (cf. Anisimov and Knuth [1], Diekert [4, 5]).

Acknowledgements The author thanks the referee for helpful corrective remarks and suggestions.

References

1. A.V. Anisimov, D.E. Knuth, Inhomogeneous sorting, *International Journal of Computer and Information Sciences* 8 (1979) 255–260.
2. A. Baudisch, Kommutationsgleichungen in semifreien Gruppen, *Acta Mathematica Academiae Scientiarum Hungaricae* 29 (1977) 235–249.
3. M. Cygan, D. Marx, M. Pilipczuk, M. Pilipczuk, The planar directed k-vertex-disjoint paths problem is fixed-parameter tractable, in: *2013 IEEE 54th Annual Symposium on Foundations of Computer Science*, IEEE, 2013, pp. 197–206.
4. V. Diekert, On the Knuth-Bendix completion for concurrent processes, *Theoretical Computer Science* 66 (1989) 117–136.
5. V. Diekert, *Combinatorics on Traces*, Lecture Notes in Computer Science 454, Springer, Berlin, 1990.
6. C. Droms, Isomorphisms of graph groups, *Proceedings of the American Mathematical Society* 100 (1987) 407–408.
7. E.S. Esyp, I.V. Kazachkov, V.N. Remeslennikov, Divisibility theory and complexity of algorithms for free partially commutative groups, in: *Groups, Languages, Algorithms*, Contemporary Mathematics 378, American Mathematical Society, Providence, R.I., 2005, pp. 319–348.
8. S. Fortune, J. Hopcroft, J. Wyllie, The directed subgraph homeomorphism problem, *Theoretical Computer Science* 10 (1980) 111–121.
9. K. Kawarabayashi, Y. Kobayashi, B. Reed, The disjoint paths problem in quadratic time, *Journal of Combinatorial Theory, Series B* 102 (2012) 424–435.
10. J.F. Lynch, The equivalence of theorem proving and the interconnection problem, *(ACM) SIGDA Newsletter* 5:3 (1975) 31–36.
11. R.C. Lyndon, P.E. Schupp, *Combinatorial Group Theory*, Springer, Berlin, 1977.
12. W. Magnus, A. Karrass, D. Solitar, *Combinatorial Group Theory*, Wiley-Interscience, New York, 1966.
13. B.A. Reed, N. Robertson, A. Schrijver, P.D. Seymour, Finding disjoint trees in planar graphs in linear time, in: *Graph Structure Theory* (N. Robertson, P. Seymour, eds.), *Contemporary Mathematics* 147, American Mathematical Society, Providence, R.I., 1993, pp. 295–301.

14. N. Robertson, P.D. Seymour, Graph minors. XIII. The disjoint paths problem, *Journal of Combinatorial Theory, Series B* 63 (1995) 65–110.
15. A. Schrijver, Finding k disjoint paths in a directed planar graph, *SIAM Journal on Computing* 23 (1994) 780–788.
16. A. Schrijver, *Combinatorial Optimization – Polyhedra and Efficiency*, Springer, Berlin, 2003.
17. H. Servatius, Automorphisms of graph groups, *Journal of Algebra* 126 (1989) 34–60.
18. C. Wrathall, The word problem for free partially commutative groups, *Journal of Symbolic Computation* 6 (1988) 99–104.

Embedding Graphs into Larger Graphs: Results, Methods, and Problems

Miklós Simonovits and Endre Szemerédi

Dedicated to Laci Lovász on his 70th birthday

Abstract Extremal Graph Theory is a very deep and wide area of modern combinatorics. It is very fast developing, and in this long but relatively short survey we select some of those results which either we feel very important in this field or which are *new breakthrough* results, or which—for some other reasons—are very close to us. Some results discussed here got stronger emphasis, since they are connected to Lovász (and sometimes to us).

Keywords Extremal graphs · Stability · Regularity Lemma · Semi-random methods · Embedding

Subject Classification Primary 05C35 · Secondary 05D10, 05D40, 05C45, 05C65, 05C80

1 Introduction

We dedicate this survey paper to Laci Lovász, our close friend, on the occasion of his 70th birthday. Beside learning mathematics from our professors, we learned a lot

---

This research was supported by the grants NKFIH 116769 and NKFIH 119528.

M. Simonovits (✉) · E. Szemerédi
Alfréd Rényi Institute of Mathematics, Hungarian Academy of Sciences,
Budapest, Hungary
e-mail: miki@renyi.hu

E. Szemerédi
e-mail: szemered@rutgers.edu

of mathematics from each other, and we emphasise that we learned a lot of beautiful mathematics from Laci.

Extremal Graph Theory is a very deep and wide area of modern combinatorics. It is very fast developing, and in this long but relatively short survey we select some of those results which either we feel very important in this field or which are *new breakthrough* results, or which—for some other reasons—are very close to us. Some results discussed here got stronger emphasis, since they are connected to Lovász (and sometimes to us).[1] The same time, we shall have to skip several very important results; often we just start describing some areas but then we refer the reader to other surveys or research papers, or to some survey-like parts of research papers. (Fortunately, nowadays many research papers start with excellent survey-like introduction.)

Extremal graph theory became a very large and important part of Graph Theory, and there are so many excellent surveys on parts of it that we could say that this one is a "survey of surveys". Of course, we shall not try to cover the whole area, that would require a much longer survey or a book.

Also we could say that there are many subareas, "rooms" in this area, and occasionally we just enter a "door", or "open a window" on a new area, point out a few theorems/phenomena/problems, explain their essence, refer to some more detailed surveys, and move on, to the next "door".

It is like being in our favourite museum, having a very limited time, where we must skip many outstanding paintings. The big difference is that here we shall see many very new works as well.

One interesting feature of Extremal Graph Theory is its very strong connection and interaction with several other parts of Discrete Mathematics, and more generally, with other fields of Mathematics. It is connected to Algebra, Commutative Algebra, Eigenvalues, Geometry, Finite Geometries, Graph Limits (and through this to Mathematical Logic, e.g., to Ultra Product, to Undecidability), to Probability Theory, application of Probabilistic Methods, to the evolution of Random Structures, and to many other topics. It is also strongly connected to Theoretical Computer Science through its methods, (e.g., using random and pseudo-random structures), expanders, property testing, and also it is strongly connected to algorithmic questions. This connection is fruitful for both sides.

One reason of this fascinating, strong interaction is that in Extremal Graph Theory often seemingly simple problems required the invention of very deep new methods (or their improvements). Another one can be that combinatorial methods start becoming more and more important in other areas of Mathematics. A third reason is, perhaps,— as Turán thought and used to emphasise,—that Ramsey's theorem and *his theorem* are applicable because they are generalizations of the Pigeon Hole Principle. Erdős wrote several papers on how to apply these theorems in Combinatorial Number Theory, (we shall discuss [258], but see also, e.g., [272, 273, 275, 276, 282, 284],

---

[1] We shall indicate the given names mostly in case of ambiguity, in cases where there are two mathematicians with the same family name, (often, but not always, father and son). We shall ignore this "convention" for Erdős, Lovász and Turán.

Erdős, Sárközy, and Sós [311], Alon and Erdős [31]).[2] Beside putting emphasis on the results we shall emphasise the development of methods also very strongly. We shall skip discussing our results on the Erdős–Sós and Loebl–Komlós–Sós conjectures on tree embedding, since they are well described in [386],[3] however, we shall discuss some other tree-embedding results. We are writing up the proof of the Erdős–Sós Conjecture [8].[4]

Since there are several excellent books and surveys, like Bollobás [119, 120], Füredi [370, 372], Füredi–Simonovits [386], Lovász [589, 591], Simonovits [757, 761–763], Simonovits and Sós [767], about Extremal Graph Theory, or some parts of it, here we shall often shift the discussion into those directions which are less covered by the "standard" sources.[5] Also, we shall emphasise/discuss the surprising connections of Extremal Graph Theory and other areas of Mathematics. We had to leave out several important new methods, e.g.,

- We shall mostly leave out results on Random Graphs;
- Razborov's Flag Algebras [669], (see also Razborov [670], Pikhurko and Razborov [650], Grzesik [423], Hatami, Hladký, Král', Norine, and Razborov, [453], and many others);
- the Hypergraph Container Method, "recently" developed, independently, by Balogh, Morris, and Samotij [77], and by Saxton and Thomason [731]. Very recently Balogh, Morris, and Samotij published a survey on the Container method in the ICM volumes [78];
- the method of Dependent Random Choice, see e.g., Alon, Krivelevich, and Sudakov [38] or, for a survey on this topic see Fox and Sudakov [353], or many other sources…
- and we have to emphasise, we had to leave out among others almost everything connected to the Universe of Integers, e.g., Sum-free sets, the Cameron-Erdős conjecture, the Sum-Product problems, …

Further, we skipped several areas, referring the readers to more authentic sources. Thus, e.g., one of the latest developments in Extremal Graph Theory is the surprising, strong development in the area of Graph Limits, coming from several sources. A group of researchers meeting originally at Microsoft Research, started investigating problems connected to graph limits, for various reasons. We mention here only Laci Lovász and Balázs Szegedy [595], Borgs, Chayes, Lovász, Sós, and Vesztergombi

---

[2]Sometimes we list papers in their time-order, in some other cases in alphabetical order.

[3]The proof of the approximate Loebl–Komlós–Sós conjecture was first attacked in [12] and then proved in a sequence of papers, from Yi Zhao [824], Cooley [202], … Hladký, Komlós, Piguet, Simonovits, Stein, and Szemerédi [466–469].

[4]The first result on the Loebl Conjecture was an "approximate" solution of Ajtai, Komlós, and Szemerédi [12].

[5]Essential parts of this survey are connected to Regularity Lemmas, Blow-up lemmas, applications of Absorbing techniques, where again, there are several very important and nice surveys, covering those parts, e.g., Alon [22], Gerke and Steger [392], Komlós and Simonovits [546], Kühn and Osthus [563, 565], Rödl and Ruciński [689], Steger [782], and many others.

[137–141], and Freedman, Lovász, and Schrijver [360]. Lovász has published a 500pp book [591] about this area.

How do we refer to papers? We felt that in a survey like this it is impossible to refer to all the good papers. Also, we tried to "introduce" many authors. So if a paper is missing from this survey that does not mean that we felt it was not worth including it. Further, wherever we referred to a paper, we mentioned (all) its authors, unless the paper and its authors had been mentioned a few lines earlier. (We never use *at al!*) In the "References" we mentioned the given names as well, mostly twice: (i) first occurrence and (ii) first occurrence as the first author.

— · —

Extremal Graph Theory could have started from a number theoretical question of Erdős [258], however, he missed to observe that this is a starting point of a whole new theory. Next came Turán's theorem [808], with his questions, which somewhat later triggered a fast development of this area. In between, starting from a topological question, Erdős and Stone [321] proved their theorem (here Theorem 2.8), which later turned out to be very important in this area. Among others, this easily implies the Erdős–Simonovits Limit Theorem [312], strengthened to the Erdős–Simonovits Stability theory [271, 751], and much later led to Szemerédi's Regularity Lemma [792]. All these things will be explained in more details.

We shall also discuss several important methods: Stability, Regularity Lemma, Blow-up Lemma, Semi-random Methods, Absorbing Lemma, and also some further, sporadic methods.

In this survey we mention hypergraph extremal problems only when this does not become *too technical* for most of the readers: thus we shall not discuss, among others, the very important Hypergraph Regularity Lemmas.[6]

Repetitions. This paper "covers" a huge area, with a very involved structure. So we shall occasionally repeat certain assertions, to make the paper more readable.

Notation. Below, for a while we shall consider simple graphs, (hypergraphs, loops and multiple edges are excluded) and for graphs (or hypergraphs) the first subscript almost always denotes the number of vertices: G_n, S_n, $T_{n,p}$ ...are graphs on n vertices. There will be just two exceptions: $K_d(n_1, \ldots, n_d)$ means a d-partite complete graph with n_i vertices in its ith class; if we list a family of graphs $\mathcal{L} = \{L_1, \ldots, L_t\}$, then again, the subscript is not necessarily the number of vertices. The maximum and minimum degrees will be denoted by $d_{\max}(G)$ and $d_{\min}(G)$, respectively. (In the hypergraph section $\delta_1(\mathbb{H}^{(k)})$ is also the minimum degree.) C_ℓ, P_ℓ, K_ℓ denote the cycle, path, and the complete graph of ℓ vertices, respectively. Further, $e(G)$, $v(G)$, $\chi(G)$ and $\alpha(G)$ denote the number of edges, vertices, the chromatic number, and the independence number,[7] respectively. For a graph G, $V(G)$ and $E(G)$ denote the sets of vertices and edges. If $U \subseteq V(G)$ then $G[U]$ is the subgraph induced by U.

---

[6]For a concise "description" of this topic, see Rödl, Nagle, Skokan, Schacht, and Kohayakawa [687], and also Solymosi [770].
[7]Often called stability number.

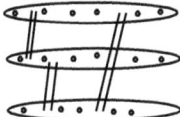

Fig. 1 Turán graph

Given a family $\mathcal{L}$ of excluded graphs, **ex**$(n, \mathcal{L})$ denotes the maximum of $e(G_n)$ for a graph G_n not containing excluded subgraphs, and **EX**$(n, \mathcal{L})$ denotes the family of extremal graphs for $\mathcal{L}$: the graphs not containing excluded subgraphs but attaining the maximum number of edges. Let $T_{n,p}$ be the Turán graph: the graph obtained by partitioning n vertices into p classes as equally as possible and joining two vertices if and only if they are in distinct classes (Fig. 1).

Given some graphs $L_1, \ldots, L_r$, $R(L_1, \ldots, L_r)$ is the Ramsey number: the smallest integer R for which all r-edge-coloured K_R have a (monochromatic) subgraph L_i in some colour i. If all these graphs are complete graphs, $L_i = K_{p_i}$, then we use the abbreviation $R(p_1, \ldots, p_r)$ for this Ramsey number.

Remark 1.1 (Graph Sequences) Speaking of $o(n^2)$ edges, or $o(n)$ vertices, ..., we cannot speak of an individual graph, only about a sequence of graphs. In this paper, using $o(.)$, we always assume that $n \to \infty$. Also, since $A(n) + o(n)$ and $A(n) - o(n)$ mathematically are exactly the same, we shall be cautious with formulating some of our theorems.

2 The Beginnings

2.1 Very Early Results

The first, simplest extremal graph result goes back to Mantel.

Theorem 2.1 (Mantel (1907), [606]) *If a graph G_n does not contain a K_3 then it has at most $\lfloor \frac{n^2}{4} \rfloor$ edges.*

Turán, not knowing of this theorem,[8] proved the following generalization:

Theorem 2.2 (Turán (1941, 1954, 1989), [808, 809, 811])[9] *If a graph G_n does not contain a K_{p+1} then $e(G_n) \leq e(T_{n,p})$, and the equality is achieved only for $G_n = T_{n,p}$.*

---

[8]The last paragraph of Turán's original paper is as follows: "...Further on, I learned from the kind communication of Mr. József Krausz that the value of $d_k(n)$ given on p 438 for $k = 3$ was found already in 1907 by W. Mantel (Wiskundige Opgaven, vol 10, pp. 60–61). I know this paper only from the reference Fortschritte d. Math. vol 38, p. 270."

[9]Turán's papers originally written in Hungarian were translated into English after his death, thus [811] contains the English translation of [808].

The above theorems are sharp, since $T_{n,p}$ is p-chromatic and therefore it does not contain K_{p+1}. These theorems have very simple proofs and there was also an earlier extremal graph theorem discovered and proved by Erdős. He set out from the following "combinatorial number theory" problem.

Problem 2.3 (*Erdős* (1938), [258]) Assume that $\mathcal{A} := \{a_1, \ldots, a_m\} \subseteq [1, n]$ satisfies the "multiplicative Sidon property" that all the pairwise products are different. More precisely, assume that

$$\text{if} \quad a_i a_j = a_k a_\ell \quad \text{then} \quad \{i, j\} = \{k, \ell\}. \tag{1}$$

How large can $|\mathcal{A}|$ be?

The primes in $[2, n]$ satisfy (1). Can one find many more integers with this "multiplicative Sidon property"? To solve Problem 2.3, Erdős proved

Theorem 2.4 (Erdős [258]) *If $G \subseteq K(n, n)$ and $C_4 \nsubseteq G$ then $e(G) \leq 3n\sqrt{n}$.*

This may have been the first non-trivial extremal graph result. It is interesting to remark that this paper contained the first *finite geometric construction*, due to Eszter Klein, to prove the lower bound $\mathbf{ex}(n, C_4) \geq cn\sqrt{n}$. Let $\pi(n)$ be the number of primes in $[2, n]$. Using Theorem 2.4, Erdős proved

Theorem 2.5 *Under condition* (1),

$$|\mathcal{A}| \leq \pi(n) + O(n^{3/4}).$$

As to the sharpness of this theorem, Erdős wrote:
"Now we prove that the error term cannot be better than $O(n^{3/4}/(\log n)^{3/2})$. First we prove the following lemma communicated to me by Miss E. Klein:
Lemma. *Given $p(p + 1) + 1$ elements, (p a prime), we can construct $p(p + 1)$ $+1$ combinations taken $p + 1$ at a time, having no two elements in common.*"

Though the language is somewhat archaic, it states the existence of a finite geometry on $p^2 + p + 1$ points: this seems to be the first application of finite geometric constructions in this area. Later Erdős applied graph theory in number theory several times. Among others, he returned to the above question in [272] and proved that the lower bound is sharp.

Much later a whole theory developed around these types of questions. Here we mention only some results of Sárközy, Erdős, and Sós [311], the conjecture of Cameron and Erdős [174], and the papers of Ben Green [413], and of Alon, Balogh, Morris, and Samotij [25].

Remark 2.6 (a) One could ask, what is the connection between $\mathbf{ex}(n, C_4)$ and Theorem 2.5. Erdős wrote each non-prime integer $a_i \in \mathcal{A}$ as $a_i = b_{j(i)} d_{j(i)}$, where b_j are the primes in $[n^{2/3}, n]$ and all the integers in $[1, n^{2/3}]$, and d_j are the integers in $[1, n^{2/3}]$. So the non-prime numbers defined a bipartite graph $G[\mathcal{B}, \mathcal{D}]$, and each a_i defined an edge in it. If this graph contained a C_4, the corresponding four integers would have violated (1). Consider those integers a_i for which $b_{j(i)} < 10\sqrt{n}$ and $d_{j(i)} < 10\sqrt{n}$. Not having C_4 in the corresponding subgraph, we can have at most $c\sqrt{n}^{3/2} = cn^{3/4}$ such integers. This does not cover all the cases, however, we can partition all the integers into a "few" similar subclasses.[10] This proves that

$$|\mathcal{A}| \leq \pi(n) + O(n^{3/4}).$$

(b) There is another problem solved in [258], also by reducing it to an extremal graph problem. It is much simpler, and we skip it.

(c) The "ADDITIVE SIDON" condition assumes that $a_i + a_j = a_k + a_\ell$ implies $\{i, j\} = \{k, \ell\}$. In case of SUM-FREE sets we exclude $a_i + a_j = a_k$. See e.g., Cameron [173], Bilu [109] and many other papers on integers, or groups.

(d) Some related results can be found, e.g., in Chan, Győri, and Sárközy [176].

Remarks 2.7 (a) Many extremal problems formulated for integers automatically extend to finite Abelian groups, or sometimes to any finite group.

Thus, e.g., a *Sidon sequence* can be (and is) defined in any Abelian group as a subset for which $a_i + a_j \neq a_k + a_\ell$, unless $\{i, j\} = \{k, \ell\}$. A paper of Erdős and Turán [324] estimates the maximum size Sidon subset of $[1, n]$. Several papers of Erdős investigate Sidon problems, e.g., [259, 262]. This problem was extended to groups, first by Babai [55], Babai and Sós [57]. (Surprisingly, for integers there are sharp differences in analysing the density of finite and of infinite Sidon sequences, see Sect. 4.6.) Problems on "sum-free sets" were generalized to groups by Babai, Sós, and then by Gowers [405], by Balogh, Morris, and Samotij [76], and by Alon, Balogh, Morris, and Samotij [25].[11]

(b) A few more citations from this area are: Alon [21], Green and Ruzsa [415], Lev, Łuczak, and Schoen [574], Sapozhenko [719–721], …

— · —

The next extremal graph result was motivated by topology. It is

Theorem 2.8 (Erdős–Stone (1946), Theorem [321]) *For any fixed* $t \geq 1$,

$$\mathbf{ex}(n, K_{p+1}(t, \ldots, t)) = \left(1 - \frac{1}{p}\right)\binom{n}{2} + o(n^2).$$

We mention here one more "early extremal graph theorem", strongly connected to the Erdős–Stone Theorem:

---

[10] E.g., we can put a_i into $\mathcal{A}_i$ if the smallest prime divisor of a_i is in $(2^t, 2^{t+1}]$ and use a slight generalization of Theorem 2.4 to $K(m, n)$.

[11] A longer annotated bibliography of O'Bryant can be downloaded from the Electronic Journal of Combinatorics [160] on Sidon sets.

Theorem 2.9 (Erdős, Kővári, Sós-Turán (1954), [554])[12,13]

$$\mathbf{ex}(n, K(a,b)) \leq \frac{1}{2}\sqrt[a]{b-1} \cdot n^{2-\frac{1}{a}} + O(n). \tag{2}$$

Below let $a \leq b$. A very important conjecture is the sharpness of (2):

Conjecture 2.10 (KST is sharp) *For every pair of positive integers $a \leq b$, there exists a constant $c_{a,b} > 0$ for which*

$$\mathbf{ex}(n, K(a,b)) > c_{a,b} \, n^{2-\frac{1}{a}}. \tag{3}$$

We can replace $c_{a,b}$ by $c_{a,a}$. The conjecture follows from Erdős [258] if $a = 2$. A sharp construction can be found in [554] for the bipartite case,[14] (and later an even sharper one from Reiman [675]), and a sharp construction is given by Erdős, Rényi, and Sós [308] and by Brown for $C_4 = K(2,2)$ and not necessarily bipartite G_n. Brown also found a sharp construction [150], for $K(3,3)$.[15] Much later, Conjecture 2.10 was proved for $a \geq 4$ and $b > a!$ by a Kollár–Rónyai–Szabó construction [532]. This was slightly improved by Alon–Rónyai–Szabó [42]: the weaker condition $b > (a-1)!$ is also enough for (3).

— · —

Theorems of Erdős and Stone and of Kővári, T. Sós, and Turán have a very important consequence.

Corollary 2.11 $\mathbf{ex}(n, \mathcal{L}) = o(n^2)$ *if and only if $\mathcal{L}$ contains a bipartite graph.*

Actually, the theorem of Kővári, T. Sós, and Turán and the fact that $T_{n,2}$ is bipartite imply the stronger dichotomy:

Corollary 2.12 *If $L \in \mathcal{L}$ is bipartite, then $\mathbf{ex}(n, \mathcal{L}) = O(n^{2-(2/v(L))}) = o(n^2)$ and if $\mathcal{L}$ contains no bipartite graphs, then $\mathbf{ex}(n, \mathcal{L}) \geq \lfloor \frac{n^2}{4} \rfloor$.*

The case when $\mathcal{L}$ contains at least one bipartite graph will be called DEGENERATE. It is one of the most important and fascinating areas of Extremal Graph Theory, and Füredi and Simonovits have a long survey [386] on this. Here we shall deal only very shortly with DEGENERATE extremal graph problems.

— · —

---

[12] Mostly we call this result the Kővári-T. Sós-Turán theorem. Here we added the name of Erdős, since [554] starts with a footnote according to which "As we learned after giving the manuscript to the Redaction, from a letter of P. Erdős, he has found most of the results of this paper." Erdős himself quoted this result as Kővári-T. Sós-Turán theorem.

[13] For Hungarian authors we shall mostly use the Hungarian spelling of their names, though occasionally this may differ from the way their name was printed in the actual publications.

[14] Here "sharp" means that not only the exponent $2 - \frac{1}{p}$ but the value of $c_{a,b}$ is also sharp.

[15] The sharpness of the multiplicative constant followed from a later result of Füredi.

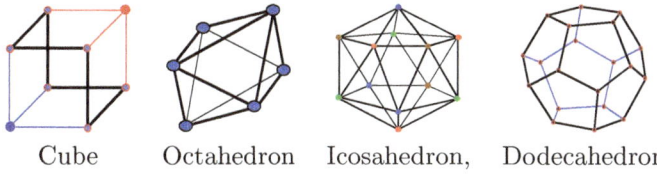

Cube   Octahedron   Icosahedron,   Dodecahedron

Fig. 2 Excluded platonic graphs

The main contribution of Turán was that he had not stopped at proving Theorem 2.2 but continued, asking the "right" questions:

What is the maximum of $e(G_n)$ if instead of excluding K_{p+1} we exclude some arbitrary other subgraph L?[16]

To provide a starting point, Turán asked for determining the extremal numbers and graphs for the Platonic bodies: Cube, Octahedron, Dodecahedron, and Icosahedron,[17] for paths, and for lassos [267],[18] see Fig. 2. Here the extremal graph problem of the Cube was (partly) solved by Erdős and Simonovits [313]. We return to this problem in Subsection 2.16. The problems of the Dodecahedron and Icosahedron were solved by Simonovits, [755, 756], using the Stability method, see Sect. 2.12.

The general question can be formulated as follows:

Given a family $\mathcal{L}$ of forbidden graphs, what is the maximum of $e(G_n)$ if G_n does not contain subgraphs $L \in \mathcal{L}$?

2.2 Constructions

Mostly in an extremal graph problem first we try to find out how do the extremal structures look like. In the nice cases this is equivalent to finding a construction providing the lower bound in our extremal problems, and then we try to find the matching upper bound. Here we shall not go into more details, however, we shall return to this question in Sect. 9, on Matchings, 1-factors, and the Hamiltonicity of Hypergraphs.

2.3 Some Historical Remarks

Here we make two remarks.

---

[16] Later this question was generalized to excluding an arbitrary family of subgraphs, however, that was only a small extension.

[17] The tetrahedron is K_4, covered by Turán theorem.

[18] Lasso is a graph where we attach a path to a cycle. Perhaps nobody considered the lasso-problem carefully, however, very recently Sidorenko solved a very similar problem of the keyrings [750].

(A) When Turán died in 1976, several papers appeared in his memory, analysing, among others, his influence on Mathematics. Erdős himself wrote several such papers, e.g., [281, 283]. In [280] he wrote, on Turán's influence on Graph Theory:

"In this short note, I will restrict myself to Turán's work in Graph Theory, even though his main work was in analytic number theory and various other branches of real and complex analysis. Turán had the remarkable ability to write perhaps only one paper or to state one problem in various fields distant from his own; later others would pursue his idea and a new subject would be born. In this way Turán initiated the field of Extremal Graph Theory. ...

...Turán also formulated several other extremal problems on graphs, some of which were solved by Gallai and myself [289]. I began a systematic study of extremal problems in Graph Theory in 1958 on the boat from Athens to Haifa and have worked on it since then. The subject has grown enormously and has a very large literature;..."

Observe that Erdős implicitly stated here that until the early 60's most of the results in this area were sporadic.

(B) Here we write about Extremal Graph Theory at length, still, if one wants to tell what Extremal Graph Theory is, and what it is not, that is rather difficult. We shall avoid answering this question, however, we remark that since the Goodman paper [400] and the Moon–Moser paper [617] an alternative answer was the following. Consider some "excluded subgraphs" $L_1, \ldots, L_t$, count the multiplicities of their copies, $m(L_i, G_n)$, in G_n, and Extremal Graph Theory consists of results asserting some inequalities among them. Since the emergence of Graph Limits this approach became stronger and stronger. One early "counting" example is

Theorem 2.13 (Moon and Moser (1962), [617]) *If t_k is the number of K_k in G_n, then*

$$k(k-2)t_k \geq t_{k-1}\left(\frac{(k-1)^2 t_{k-1}}{t_{k-2}} - n\right).$$

2.4 Early Results

If we restrict ourselves to simple graphs, *some central theorems* assert that for ordinary graphs the general situation is almost the same as for K_{p+1}: the extremal graphs S_n and the *almost extremal* graphs G_n are *very similar* to $T_{n,p}$. The similarity of two graph sequences (G_n) and (H_n) means that one can delete $o(n^2)$ edges of G_n and add $o(n^2)$ edges to obtain H_n.

The general asymptotics of $\mathbf{ex}(n, \mathcal{L})$ and the asymptotic structure of the extremal graphs are described by

Theorem 2.14 (Erdős–Simonovits (1967, 1968), [269, 271, 751]) *Let*

$$p := \min_{L \in \mathcal{L}} \chi(L) - 1. \tag{4}$$

If $S_n \in \mathbf{EX}(n, \mathcal{L})$, then one can delete from and add to S_n $o(n^2)$ edges to obtain $T_{n,p}$.

A much weaker form of this result immediately follows from Erdős–Stone Theorem.

Theorem 2.15 (Erdős–Simonovits (1966), [312]) *Defining p by* (4),

$$\mathbf{ex}(n, \mathcal{L}) = \left(1 - \frac{1}{p}\right)\binom{n}{2} + o(n^2).$$

One important message of these theorems is that for simple graphs the extremal number and the extremal structure are determined up to $o(n^2)$ by the minimum chromatic number of the excluded subgraphs. In some sense this is a great luck: there are many generalizations of the Turán-type extremal graph problems, but we almost never get so nice answers for the natural questions in other areas.

Speaking about the origins of Extremal Graph Theory, we have to mention the *dichotomy* that occasionally we have very nice extremal structures but in the DEGENERATE case the extremal graphs seem to have much more complicated structures (unless $\mathcal{L}$ contains a tree or a forest). Actually this may explain why Erdős missed to observe the importance of his Theorem 2.4, on $\mathbf{ex}(n, C_4)$.

To solve some extremal graph problems, Simonovits defined the *Decomposition Family* (see, e.g., [758]).

Definition 2.16 (*Decomposition,* $\mathcal{M} = \mathcal{M}(\mathcal{L})$) Given a family $\mathcal{L}$ with $p := \min\{\chi(L) : L \in \mathcal{L}\} - 1$, then $M \in \mathcal{M}$ if $L \subseteq M \otimes K_{p-1}(t, \ldots, t)$ for $t = v(M)$, where $L \otimes H$ denotes the graph obtained by joining each vertex of L to each vertex of H.

The meaning of this is that $M \in \mathcal{M}$ if we cannot embed M into one class of a $T_{n,p}$ without obtaining an excluded $L \in \mathcal{L}$. Thus, e.g., if we put a C_4 into the first class of $T_{n,p}$, then the resulting graph contains a $K_{p+1}(2, 2, \ldots, 2)$. Therefore C_4 is in the decomposition class of $K_{p+1}(2, \ldots, 2)$.

…Given a family $\mathcal{L}$ of excluded subgraphs, the decomposition family $\mathcal{M} = \mathcal{M}(\mathcal{L})$ determines (in some sense) the error terms and the finer structure of the $\mathcal{L}$-extremal graphs. Namely, the error terms depend on $\mathbf{ex}(n, \mathcal{M})$, see [751].

The next theorem "explains", why is $T_{n,p}$ extremal for K_{p+1}.

Definition 2.17 (*Colour-critical edge*) The edge $e \in E(L)$ is called critical, if $\chi(L - e) < \chi(L)$.

Of course, in such cases, $\chi(L - e) = \chi(L) - 1$. Each edge of an odd cycle is critical. In Fig. 2.4, e.g., one can see the Grötzsch graph (often incorrectly called Mycielski graph). It is 4-chromatic but all its edges are critical. On the other hand, in the Petersen, or the Dodecahedron graphs, there are no critical edges.[19,20] The next theorem solves all cases when L has a critical edge.

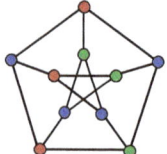
Petersen graph,
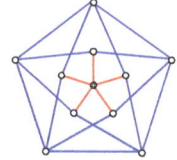
Grötzsch graph,
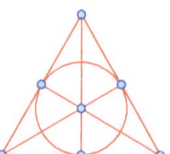
Fano hypergraph

Theorem 2.18 (Simonovits)[21] *Define $p = p(\mathcal{L})$ by (4). The following statements are equivalent:*
  (a) *Some $(p+1)$-chromatic $L \in \mathcal{L}$ has a critical edge.*
  (b) *There exists an n_0 such that for $n > n_0$, $T_{n,p}$ is extremal for $\mathcal{L}$.*
  (c) *There exists an n_1 such that for $n > n_1$, $T_{n,p}$ is the only extremal graph for $\mathcal{L}$.*[22]

Here (a) is equivalent to that adding an edge to $K_p(t, \ldots, t)$, for $t = v(L)$, we get a graph containing L, and equivalent to that $K_2 \in \mathcal{M}$. The extremal results on the Dodecahedron D_{20} and Icosahedron I_{12} follow from [756], and [755]. We skip the details referring the reader to the survey of Simonovits [762].

Meta-Theorem 2.19 *(Simonovits)* *"Whatever" we can prove for $L = K_{p+1}$, with high probability, we can also prove it for any L having a critical edge.*

2.5 Which Universe?

Extremal problems exist in a much more general setting: Theorem 2.5 is, e.g., an extremal theorem on sets of integers. In general, we fix the family of some objects, e.g., integers, graphs, hypergraphs, r-multigraphs, where some r is fixed and the edge-multiplicity is bounded by r. We exclude some substructures, and try to optimize some (natural) parameters. More generally, putting some bounds on the number of one type of substructures, we try to maximize (or minimize) the number of some other substructures. This approach can be found in the paper of Moon and Moser [617], or in Lovász and Simonovits, [593].[23] The paper of Alon and Shikhelman [47] is also about this question, in a more general setting. (We should also mention here the famous conjecture of Erdős, on the number of pentagons in a triangle-free

---

[19] If the automorphism group of G is edge-transitive, then either all the edges are critical, or none of them. By the way, in [762], Simonovits discusses these questions in more details, among others, the extremal problems of generalized Petersen graphs.

[20] The definition applies to hypergraphs as well, the triples of the Fano hypergraph are also critical.

[21] For $p = 2$ this was also known (at least, implicitly) by Erdős.

[22] Here (c)→(b) is trivial, and one can prove that (b) implies (c) with $n_1 = n_0 + 3p$.

[23] Bollobás [117] also contains similar, strongly related results.

graph, well approximated by Győri [440], (and by Füredi, unpublished) and then solved by Grzesik [423] and by Hamed Hatami, Hladký, Král, Norine, and Razborov [453].) With the development of the theory of graph limits this viewpoint became more and more important. Below we list some of the most common Universes, and some related papers/surveys.

(a) INTEGERS, as we have seen above, in Problem 2.3 or Theorem 2.5. Among many other references, here we should mention the book of Tao and Vu on Additive Combinatorics [801], the Geroldinger–Ruzsa book [393], and also a new book of Bajnok [59].
(b) ABELIAN GROUPS, see e.g., Babai–Sós, [57] Gowers [403, 405] and also [393], a survey of Tao and Vu [802] and [59].[24]
(c) GRAPHS: this is the main topic of this survey;
(d) DIGRAPHS and MULTIGRAPHS, with bounded arc/edge multiplicity,[25] see Brown–Harary [156], Brown–Erdős–Simonovits [152, 153]; Sidorenko [748], for longer surveys see Brown and Simonovits [158] and [159], Bermond and Thomassen [105], Thomassen [805], Bang-Jensen and Gutin [88], Jackson and Ordaz [474];
(e) HYPERGRAPHS, see e.g., de Caen [171], Füredi [370], Sidorenko [749], Keevash [498][26];
(f) EXTREMAL SUBGRAPHS OF RANDOM GRAPHS: Babai, Simonovits, and Spencer [56], Brightwell, Panagiotou, and Steger [148], Rödl-Schacht [703], Rödl [686], Schacht [732], ...DeMarco and Kahn [234–236], or
(g) EXTREMAL SUBGRAPHS OF PSEUDO-RANDOM GRAPHS, see e.g., Krivelevich and Sudakov [559] Aigner-Horev, Hàn, and Schacht [5], Conlon and Gowers [193, 199], or Conlon, Fox, and Zhao [198], or e.g., Allen, Böttcher, Kohayakawa, and Person [16, 17].

Perhaps one of those who first tried to compare various universes and analyze their connections was Vera Sós [774, 775]. In [774] she considered the connections between Graph Theory, Finite Geometries, and Block Designs. The emphasis in these papers was on the fact that basically the same problems occur in these areas in various settings, and these areas are in very strong connection, interaction, with each other.[27] Most of the above universes we shall skip here, to keep this survey relatively short, however, below we consider some extremal problems on integers.

---

[24]There are several earlier results on similar questions, e.g., Yap [819, 820], Diananda and Yap [240], yet they are slightly different, or several papers of A. Street, see [785].
[25]One has to assume that the edge-multiplicity is bounded, otherwise even for the excluded K_3 in the Universe of multigraphs we would get arbitrary many edges. As an exception, in the Füredi-Kündgen theorem [378] no such bound is assumed.
[26]And many others, see e.g., Rödl and Rucinski [689], or the much earlier Bermond, Germa, Heydemann, and Sotteau [104], and the corresponding Sects. 8 and 9.
[27]Vera Sós did not call these areas Universes.

Integers

We do not intend to describe this very wide area in detail, yet we start with some typical, important questions in this area, i.e., in the theory of extremal problems on subsets of integers. As we have stated in Problem 2.3, Erdős considered the multiplicative Sidon problem in [258]. Even earlier Erdős and Turán formulated the following conjecture for subsets of integers:

Conjecture 2.20 (Erdős–Turán (1936), [323])[28] *If $\mathcal{A} \subseteq [1, n]$ does not contain a k-term arithmetic progression, then $|\mathcal{A}| = o(n)$ (as $n \to \infty$).*[29]

The proof seemed those days very difficult. Even the simplest case $k = 3$ is highly non-trivial: it was first proved by K.F. Roth, in 1953 [679], and for $k = 4$ those days the conjecture seemed even more difficult.[30] The conjecture was proved by Szemerédi, first for $k = 4$ [789], and then for any k:

Theorem 2.21 (Szemerédi (1975), [791]) *Let k be a fixed integer. If $r_k(n)$ is the maximum number of integers, $a_1, \ldots, a_m \in [1, n]$ not containing a k-term arithmetic progression, then $r_k(n) = o(n)$.*

Ergodic theory and Szemerédi Theorem. Not much later that Szemerédi proved this theorem, Fürstenberg gave an alternative proof, in 1977, using ergodic theory [387]. This again is an example where seemingly simple combinatorial problems led to very deep theories. One advantage of Fürstenberg's approach was that it made possible for him and Katznelson and their school to prove several important generalizations, e.g., the high dimensional version [388], Bergelson and Leibman [102] proved some polynomial versions of the original theorem, and later the density version of Hales–Jewett theorem [389].

Polymath on Hales–Jewett theorem. As we just stated, one of the important generalizations of Theorem 2.21 is the *density version of* Hales–Jewett theorem, obtained by Fürstenberg and Katznelson [389] which earlier seemed hopeless.

There is a big difference between the Hales–Jewett Theorem and Szemerédi's theorem: just to explain the meaning of the Hales–Jewett Theorem or its density version, is more difficult than to explain the earlier ones. It is one of the most important Ramsey-type theorems, asserting, that—fixing the parameters appropriately—a high dimensional r-coloured structure will contain a (small) monochromatic substructure, a so called "combinatorial line". We remark that there was a similar result of Graham and Rothschild (on n-parameter sets) earlier, [409].

---

[28] Actually, here they formulated this for $r_3(n)$.

[29] Speaking of arithmetic progressions we always assume that its terms are distinct.

[30] The conjectures on $r_k(n)$ were not always correct. Vera Sós wrote a paper [776] on the letters between Erdős and Turán during the war, where one can read that Szekeres e.g., conjectured that for $n = \frac{1}{2}(3^\ell + 1)$ $r_k(n) \le 2^\ell$. This was later disproved by Behrend [96]. (This conjecture is also mentioned in [323].)

A simplified version of this was given by Austin [52].[31] The Polymath project [656] provided a completely elementary proof of this theorem. A nice description of this is the MathSciNet description of the "PolyMath" proof of this (see MathSciNet MR2912706, or the original paper, [656]).[32]

Erdős conjecture on the sum of reciprocals. One of the central questions in this area was if there are arbitrary long arithmetic progressions consisting of primes. In 1993, Erdős wrote a paper on his favourite theorems [287], where he wrote that in those days the longest arithmetic progression of primes had 17 integers (and was obtained with the help of computers). Now we all know the celebrated result:

Theorem 2.22 (Green and Tao (2008), [416]) *The set of primes contains arbitrary long arithmetic progressions.*

We close this part with the related famous open problem of Erdős which would imply Theorem 2.22. Then we list some estimates on $r_k(n)$.

Problem 2.23 (*Erdős* [287]) Let $\mathcal{A} = \{a_1, \ldots, a_n, \ldots\} \subseteq \mathbb{Z}$ be a set of positive integers. Is it true that if $\sum \frac{1}{a_i} = \infty$, then for each integer $k > 2$, $\mathcal{A}$ contains a k-term arithmetic progression?

Estimates on $r_k(n)$. First Roth proved that $r_3(n) = O(\frac{n}{\log \log n})$.[33] Many researchers worked on improving the estimates on $r_3(n)$, or more generally, on $r_k(n)$. Roth's estimate was followed by the works of Heath-Brown [462] and Szemerédi [793], and then Bourgain [145]. One of the last breakthroughs was

Theorem 2.24 (Sanders [717]) *Suppose that $\mathcal{A} \subseteq \{1, \ldots, N\}$ contains no 3-term arithmetic progressions. Then*

$$|\mathcal{A}| = O\left(\frac{(\log \log N)^5}{\log N} N\right).$$

The exponent of $\log \log n$ was brought down to 4 by T. Bloom [112].

Remark 2.25 (Lower bounds) Clearly, for $k \geq 3$, $r_k(n) \geq r_3(n)$. Behrend [96] proved that there exists a $c > 0$ for which

$$r_3(n) \geq \frac{n}{e^{c\sqrt{\log n}}}. \tag{5}$$

This was improved by Elkin [254], and then, in a much more compact way, by Green and Wolf [420].

---

[31] See also Austin [51], Beigleböck [97], Bergelson and Leibman [103] Gowers, [406], Polymath [655],...

[32] Similarly to the proof of $r_3(n) = o(n)$ from the Ruzsa–Szemerédi Triangle Removal Lemma, (see Theorem 5.26) Rödl, Schacht, Tengen and Tokushige proved $r_k(n) = o(n)$ and several of its generalizations in [700] "elementarily", i.e. not using ergodic theoretical tools. On the other hand, they remarked that those days no elementary proof was known on the Density Hales–Jewett theorem.

[33] Using $\log \log n$ we always assume that $n \geq 100$, and therefore $\log \log n > 3/2$.

One could ask, what do we know about $r_4(n)$. A major breakthrough was due to Gowers [402], according to which for every $k \geq 3$ there exists a $c_k > 0$ for which

$$r_k(n) = O\left(\frac{n}{\log\log^{c_k} n}\right). \quad \text{(Actually, } c_k = 2^{-(2^{k+9})} \text{ works.)}$$

Green and Tao improved this for $k = 4$ to

$$r_4(n) = O\left(\frac{n}{\log^c n}\right) \quad \text{for some constant} \quad c > 0.$$

See also Green and Tao [417, 419] and the survey of Sanders [718].

Other important problems on Integers: We start with the following remark. A property $\mathcal{P}$ is always a family of subsets of some fixed set. It is *monotone decreasing* if $X \in \mathcal{P}$ and $Y \subseteq X$ implies that $Y \in \mathcal{P}$. (Two examples of this are (i) the sets of integers not containing solutions of some given equations, and (ii) the family of graphs not containing an L.) When we fix a Universe and a "monotone" property $\mathcal{P}$, then, beside asking for the size of the extremal sets X for $\mathcal{P}$, we may also ask, e.g., how many $X \in \mathcal{P}_n$ are there, where $\mathcal{P}_n \subseteq \mathcal{P}$ is defined by some parameter n of these objects. We may also ask, what is their typical structure. For graphs these are the Erdős-Kleitman-Rothschild-type problems [297], discussed in Sect. 2.13. The same questions can also be asked for extremal problems on integers, and we shall not return to them later, therefore we list some of them here, together with "their extremal problems".

- We have started with the multiplicative Sidon problem (Thm 2.5), and there is also the problem of additive Sidon sets.
- We wrote about excluding the k-term arithmetic progressions, in this subsection.[34]
- Another important area is the problem of sum-free sets, see, e.g., Cameron and Erdős [174], Alon, Balogh, Morris, and Samotij [26], Łuczak and Schoen [603, 604], Sapozhenko [719],…Balogh, Liu, Sharifzadeh, and Treglown [72], Balogh, Morris, and Samotij [76].
- Very active research characterises the Sum-Product problems, introduced by Erdős and Szemerédi [322], see also Gy. Elekes [253], Bourgain, Gambourd, and Sarnak [146], Solymosi [771] among the very many related papers.
- Important and deep questions can be listed in connection with Freiman–Ruzsa-type results (see e.g., [361]).

Recommended surveys, papers: Ruzsa [708, 709], Solymosi [771], Granville and Solymosi [412], Pomerance and Sárközy [657], …and the survey of Shkredov [742].

---

[34] Subsubsections will also be called Subsections.

Groups

We have considered some problems about the Universe of Integers. It is, of course, natural to ask the analogous questions for groups. There are many results of this type. Here we mention only a few papers on groups. Most of the definitions immediately generalize from integers to any Abelian group. We have already mentioned that Sidon sets in groups were investigated by Babai and Sós [57]. From among the many-many further similar extensions we mention only Alon, Balogh, Morris, and Samotij [26, 76], Gowers [405], on quasi-random groups, Green [414], Green and Ruzsa [415], Lev, Łuczak, and Schoen [574], Sapozhenko [719, 721], and B. Szegedy on Gowers norms and groups [788].

There are also results on non-Abelian groups, e.g., Sanders [716], and the paper of Babai and Sós [57] considers both Abelian and non-Abelian groups (and try to determine the maximum size of a Sidon set in them), and we refer the reader to these papers.

Remark 2.26 For each property $\mathcal{P}$, one can also investigate the $\mathcal{P}$-maximal, or the $\mathcal{P}$-minimal structures. Here, e.g., one can try to count the maximal subsets of property $\mathcal{P}$, in $[1, n]$, or in a group $\mathcal{G}$, ... Thus e.g., Balogh, Liu, Sharifzadeh and Treglown [72] count the maximal sum-free subsets, while Balogh, Bushaw, Collares, Liu, Morris, and Sharifzadeh [65] describe the typical structure of graphs with no large cliques.[35]

Remark 2.27 There are several results on subsets of $\mathcal{G}$ without non-trivial arithmetic progressions where $\mathcal{G}$ is a group, or a linear vector space. Here we mention only the paper of Croot, Lev, and Péter Pach [210] on the linear vector space $\mathbb{Z}_4^n$. J. Wolf provides a very clear and detailed description of this paper and related results, in the MathSciNet-MR3583357. See also the related post of T. C. Tao, and the paper of Ellenberg and Gijswijt [255], building on [210].

2.6 Ramsey or Density?

One difference between Ramsey and Turán theories is that in the Turán case we have density statements, while in the Ramsey case the densities are not enough to ensure the occurrence of a monochromatic substructure. A trivial example of this is the problem of $R(4, 4)$, yet, instead of this we consider another trivial example: connectedness. If we RED-BLUE-edge-colour a K_n, then either we have a RED connected spanning subgraph, or a BLUE one. However, we may have $\approx \frac{1}{2}\binom{n}{2}$ edges in both colours, not enough for a connected spanning subgraph.

More generally, if we r-colour the edges of a graph G, and consider its subgraphs G^i defined by the ith colour, and we assert that—under some conditions,—G has a monochromatic L, because it has at least $\frac{1}{r}e(G)$ edges, that is a Turán-type, density theorem.

---

[35]The description of the typical structure is a stronger result than just counting them.

The Erdős–Turán conjecture, and its proof, the Szemerédi theorem, came from the van der Waerden theorem, [813] according to which, if r and k are fixed, and we r-colour the integers in an arbitrary way, then there will be a monochromatic arithmetic progression of length k. The Erdős–Turán conjecture was the corresponding density conjecture: any infinite sequence of integers of positive lower density contains an arithmetic progression of k terms.

Many Ramsey problems are very different from density problems, however, in some other cases a Ramsey problem may be basically a density problem. Sometimes a density theorem generalizes a Ramsey-type result in a very non-trivial way. In this survey mostly we are interested in density problems.

Example 2.28 For any tree T_k, trivially, $\mathbf{ex}(n, T_k) < (k-2)n$. This implies that $R(T_k, T_k) < 4k$, and for r colours $R_r(T_k, \ldots, T_k) \leq 2kr$.

So, for trees the Ramsey problem is a density problem, up to a constant. For more details, see e.g., the paper of Faudree and Simonovits [336].

2.7 Why Are the Extremal Problems Interesting?

Extremal graph problems are interesting on their own, they emerge in several branches of Discrete Mathematics, e.g., in some parts of Graph Theory not directly connected to Extremal Graph Theory, in Combinatorial Number Theory, and also they are strongly connected to Ramsey Theory.

Erdős wrote several papers on how can Graph Theory be applied in Combinatorial Number Theory, or in Geometry, see e.g., [272]. András Sárközy returned to the investigation and generalization of Erdős' results discussed in Problem 2.3: the next step was to analyze the case when no product of six distinct numbers from $\mathcal{A}$ was a square.[36] The corresponding graph theoretical lemmas were connected to $\mathbf{ex}(n, m, C_6)$[37] and were established by Erdős, A. Sárközy, and Sós [311] by G. N. Sárközy [722], and by E. Győri [441]. (Similar cycle-extremal results and similar methods were used also in a paper of Dietmann, Elsholz, Gyarmati and Simonovits [241], but for somewhat different problems.)

Extremal graph theory is strongly connected to many other parts of Mathematics, among others, to Number Theory, Geometry, the theory of Finite Geometries, Random Graphs, Quasi-Randomness, Linear Algebra, Coding Theory.

The application of constructions based on finite geometry became important and interesting research problems, we mention here just a few, such as Reiman [675], Hoffman–Singleton [472], Benson [101], Brown [150]. Erdős, Rényi, and Sós [308] …and refer the reader again to the surveys of Vera Sós [774], Füredi and Simonovits [386] or the papers of Lazebnik, Ustimenko, and Woldar, e.g., [567] and others.

---

[36]If (1) is violated then $a_i a_j a_k a_\ell$ is a square.

[37]$\mathbf{ex}(n, m, \mathcal{L})$ is the maximum number of edges an $\mathcal{L}$-free graph $G \subset K(n, m)$ can have. This problem may produce surprising phenomena when $n = o(m)$.

These constructions are connected to the construction of expander graphs (Ramanujan graphs) by Margulis [609–611], and Lubotzky, Phillips, and Sarnak [599], which use highly non-trivial mathematics, and in some sense are strongly connected to Extremal Graph Theory.[38] Here we recommend the survey of Alon in the Handbook of Combinatorics [22] and several of his results on the eigenvalues of graphs, e.g., [20] or the Alon–Milman paper [39]. Another surprise was that in [532] deeper results from algebra also turned out to be very useful. In some other cases (e.g., Bukh and Conlon [161, 162]) randomly chosen polynomial equations were used for constructions in extremal graph theory. We return to these questions in Sect. 2.16.

Remark 2.29 There are cases, when important methods came from that part of Discrete Mathematics which is not directly Extremal Graph Theory, however, is very strongly connected to it. One example is (perhaps) the Lovász Local Lemma [299], originally invented for problems very strongly connected to Extremal Graph Theory.[39]

2.8 Ramsey Theory and the Birth of the Random Graph Method

There are many cases when some Ramsey-type theorem is very near to a density theorem. In graph theory perhaps one of the first such results was that of Chvátal [186]. Faudree, Schelp and others also proved many results on the Ramsey topic, where in one colour we exclude a large tree. Faudree and Simonovits discussed in [336] this connection.

Erdős Magic. Turán thought that two-colouring K_n, say, in RED and BLUE, we shall always have a monochromatic K_m with $m \geq \sqrt{n}$. The reason he thought this was (most probably) that for $n = m^2$, $T_{n,m}$ yields a 2-colouring of K_n (where the edges of $T_{n,m}$ are RED, the others are BLUE), and thus K_n does not contain any RED K_{m+1}, neither a BLUE K_{m+1} and this construction seemed to be very nice. So Turán thought this maybe the best. When Turán asked Erdős about this, right after the war, Erdős answered that in a random colouring of K_n the largest monochromatic K_p has order at most $(2 + o(1)) \log_2 n$ (for sharper results see e.g., Bollobás and Erdős [125]). In some sense, this was the beginning of the Theory of Random graphs. Joel Spencer calls this "the Erdős Magic" and discusses this story in detail, e.g., in [779], or in [780], where he describes this and also the whole story of $R(3, k)$, its estimate

---

[38]The Margulis–Lubotzky–Phillips–Sarnak papers are eigenvalue-extremal, however, as Alon pointed out, (see the last pages of [599]), these constructions are "extremal" for many other graph problems as well.

[39]The Lovász Local Lemma is one of the most important tools in Probabilistic Combinatorics (including the application of probabilistic methods). Its proof is very short, and it is described, among others, in the Alon–Spencer book [48], in Spencer [778], or in the original paper, available at the "Erdős homepage" [827].

by Erdős [261], by Ajtai, Komlós, and Szemerédi [9], the application of the Lovász Local Lemma [299] by Spencer[777], and finally the matching deep result of Jeong Han Kim [515], using the Rödl nibble and many other deep tools.

So we see that most of the graphs G_n are counterexamples to this conjecture of Turán, however, we cannot construct graph sequences (G_n) without complete graphs on $\lfloor c \log n \rfloor$ vertices and independent set of vertices of size $\lfloor c \log n \rfloor$. Actually, to construct such graphs is a famous open problem of Paul Erdős, weakly approximated, but still unsolved.[40]

One of the beautiful conjectures is

Conjecture 2.30 (Vera Sós) *A Ramsey graph is quasi-random.*

Of course, here we should know what is a Ramsey graph and when is a graph quasi-random. We formulate this only in the simplest case. Given an integer m, let $N = R(m, m)$ be the smallest integer for which 2-colouring K_N we must have a monochromatic K_m. A Ramsey graph is a graph on $R(m, m) - 1$ vertices not containing K_m, nor m independent vertices. The notion of quasi-randomness came originally, in a slightly hidden form from the works of Andrew Thomason [803] (connected to some Ramsey problems). Next it was formulated in a more streamlined form by Chung, Graham, and Wilson [185] and here, without going into details, we "define" it as follows.

Definition 2.31 For $p > 0$ fixed, a sequence (G_n) of graphs is p-quasi-random if $e(G_n) = p\binom{n}{2} + o(n^2)$, and the number of C_4's in G_n is $6\binom{n}{4}p^4 + o(n^4)$, as in the random Binomial graph $R_{n,p}$, with edge-probability p.

The following beautiful theorem also supports the Sós Conjecture.

Theorem 2.32 (Prömel and Rödl [661]) *For any $c > 0$ there exists a $c^* > 0$ such that if neither G_n nor its complementary graph contains a $K_{\lceil c \log n \rceil}$, then G_n contains all the graphs H_ℓ of $\ell = \lceil c^* \log n \rceil$ vertices.*

— · —

LOVÁSZ META-THEOREM. *Many years ago Lovász formulated the principle that the easier is to obtain a "construction" for a problem by Random Methods, the more complicated it is to obtain it by "real construction".*

Supporting examples were those days, among others,
- the above "missing" Ramsey Construction,
- the Expander Graph problem, and
- also some good codes from Information Theory.

The random graphs have good expander properties. Expander graphs are important in several areas, among others, in Theoretical Computer Science. The fact that

---

[40]One problem with this sentence is that the notion of "construction" is not well defined, one of us witnessed a discussion between Erdős and another excellent mathematician about this, but they strongly disagreed. As to the constructions, we mention the Frankl–Wilson construction of Ramsey graphs [359], or some papers of Barak, Rao, Shaltiel, and Wigderson [89] and others.

Embedding Graphs into Larger Graphs: Results, Methods, and Problems 465

the random graphs are expanders were there in several early papers implicitly or explicitly, see e.g., Erdős and Rényi, [307], Pósa [660].

Remark 2.33 Pinsker [651] proved the existence of bounded degree expanders, using Random Methods, and the construction of Margulis [607] was a breakthrough in this area. They were used e.g., in the AKS Sorting networks [11].

2.9 Dichotomy, Randomness and Matrix Graphs

In the areas considered here, there are two extreme cases: (a) sometimes for some small constant ν we partition the n vertices into ν classes $U_1, \ldots, U_\nu$ and join the vertices according to the partition classes they belong to: if some vertices $x \in U_i$ and $y \in U_j$, $x \neq y$ are joined then all the pairs x', y' are joined for which $x' \in U_i$ and $y' \in U_j$, $x' \neq y'$. These graphs can be described by a $\nu \times \nu$ matrix, and therefore can be called MATRIX- GRAPHS.[41] (Similar approach can be used in connection with edge-coloured graphs, multigraphs and digraphs.)

Often such structures are the extremal ones, in some other cases the random graphs. We could say that a dichotomy can be observed: sometimes the extremal structures are very simple, in some other cases they are very complicated, randomlike, fuzzy. (One very important feature of the random graphs is that they are expanders. This is why in a random graph much fewer edges ensure Hamiltonicity than in an arbitrary graph.)

2.10 Ramsey Problems Similar to Extremal Problems

In some cases the Ramsey graphs are chaotic, see above, in some other (mainly off-diagonal) cases they are very similar to $T_{n,k}$. Below we shall discuss only those cases of off-diagonal Ramsey Numbers that are strongly connected to Extremal Graph Theory.[42] The area where the required monochromatic subgraphs are not complete graphs started with a paper of Gerencsér and Gyárfás [390]. Given two graphs, L and M, the Ramsey number $N = R(L, M)$ is the minimum integer N for which any RED-BLUE-colouring of K_N contains a RED L or a BLUE M. Gyárfás and Gerencsér started investigating these problems, Bondy and Erdős [135] discussed the case when both L and M are cycles. Chvátal [186] proved that

Theorem 2.34 *If T_m is any fixed m-vertex tree, then*

---

[41] A generalization of these graphs is the generalized random graph, where we join the two vertices with probability p_{ij}, independently.
[42] The Ramsey numbers $R(L, M)$ form a twice infinite matrix whose rows and columns are indexed by the graphs L and M. If $L \neq M$, then $R(L, M)$ is called "off-diagonal".

$$R(T_m, K_\ell) = (\ell - 1)(m - 1) + 1.$$

A construction yielding a lower bound is obvious: consider the Turán graph $T_{(\ell-1)(m-1), m-1}$. Colour its edges in RED and the complementary graph in BLUE. Observe that the construction yielding a lower bound in this theorem is the Turán graph $K_{\ell-1}(m - 1, \ldots, m - 1)$. Faudree, Schelp and others proved many results on these types of off-diagonal Ramsey problems.

Remark 2.35 (Simple Ramsey extremal structures) In several cases the Ramsey-extremal, or at least the Ramsey-almost extremal structures can be obtained from partitioning n vertices into a bounded number $\mathcal{C}_1, \ldots, \mathcal{C}_m$ of classes of vertices and colouring an edge xy according to the classes of their endpoints: all the edges where $x \in \mathcal{C}_i$ and $y \in \mathcal{C}_j$ have the same colour, for all $1 \le i, j \le m$. This applies, e.g., to the path Ramsey numbers described by Gerencsér and Gyárfás [390].

Occasionally some slight perturbation of such structures also provides Ramsey-extremal colourings. Such examples occur in connection with the cycle Ramsey numbers, e.g., the ones in the Bondy–Erdős conjecture on the Ramsey numbers on odd cycles [135], and in many other cases.

2.11 Applications in Continuous Mathematics

Toward the end of his life Turán wrote a series of papers, starting perhaps with [810], the last ones with Erdős, Meir, and Sós, [300–303] on the application of his theorem in estimating the number of short distances in various metric spaces, or in estimating some integrals, potentials.[43] He also liked mentioning a similar result of Katona [488] where Katona applied Turán's theorem to distributions of random variables. Perhaps the first result of Katona in this area was

Theorem 2.36 *Let $a_1, \ldots, a_n$ be d-dimensional vectors, $(d \ge 1)$, with $|a_i| \ge 1$ for $i = 1, \ldots, n$. Then the number of pairs (a_i, a_j) $(i \ne j)$ satisfying $|a_i + a_j| \ge 1$ is at least*

$$\begin{cases} t(t-1) & \text{if} \quad n = 2t \quad \text{(even)} \\ t^2 & \text{if} \quad n = 2t+1 \quad \text{(odd)}. \end{cases}$$

Somewhat later A. Sidorenko (under the influence of Katona) also joined this research [743, 745]. They proved continuous versions of discrete (extremal graph) theorems, mostly to apply it in analysis and probability theory.

Remark 2.37 Sidorenko also reformulated the Erdős–Simonovits conjecture [760] in the language of integrals, [746, 747]. The original conjecture had various forms,

---

[43] We mention just a few related papers, for a more detailed description of this area see the remarks of Simonovits in [811], and the surveys of Katona [490, 491].

but all these forms asserted that if L is bipartite, and E is noticeably larger than $\mathbf{ex}(n, L)$, then among all the graphs G_n with E edges, the Random Graph has the fewest copies of L. The weakest form of this conjecture is that

Conjecture 2.38 (Erdős–Simonovits *[760]*) *For any bipartite L, there exist two constants, $C = C_L > 0$ and $\gamma = \gamma_L > 0$, such that if $e(G_n) > C\mathbf{ex}(n, L)$, then G_n contains at least*

$$\gamma \cdot n^v \left(\frac{E}{n^2}\right)^e$$

copies of L, for $e = e(L)$ and $v = v(L)$.

These forms were primarily referring to the sparse case, when E is slightly above $\mathbf{ex}(n, L)$. On the other hand, Sidorenko's form becomes meaningful only for dense graph sequences. For some more details on this, see Füredi and Simonovits, [386] or Simonovits [760], or Sidorenko [746].

2.12 The Stability Method

In this section we shall describe the Stability method in a somewhat abstract form, but not in its most general form. Stability in these cases mostly means that for a property $\mathcal{P}$ we conjecture that the optimal objects have some simple structure, and the almost optimal structures are very similar to the (conjectured) optimal ones, in some mathematically well defined sense. There are various forms of the stability methods, here we restrict ourselves to one of them. A "property" below is always a subset of the Universe. Generally we have two properties, $\mathcal{P}$ and a much simpler property/subset $\mathcal{Q} \subseteq \mathcal{P}$. If a family $\mathcal{L}$ of excluded graphs is given, then $\mathcal{P}_n := \mathcal{P}(n, \mathcal{L})$ is the family of n-vertex $\mathcal{L}$-free graphs.

The STABILITY METHOD means that the optimization is easy on $\mathcal{Q}_n$ and we reduce the "optimization on $\mathcal{P}_n$" to "optimization on $\mathcal{Q}_n$", e.g.,—when we try to maximize the number of edges,—by considering a conjectured extremal graph S_n and showing that if $G_n \in \mathcal{P}_n - \mathcal{Q}_n$, then $e(G_n) < e(S_n)$. So, since the maximum is at least $e(S_n)$ it must be attained in $\mathcal{Q}_n$.

We start with three examples. We wish to maximize some function $e(G)$ on the n-element objects of $\mathcal{P}$, denoted by $\mathcal{P}_n$.

Examples: (where $e(G_n)$ is the number of edges).

(a) $\mathcal{P}$ means that $K_{p+1} \not\subseteq G$ and $\mathcal{Q}$ is the family of p-chromatic graphs. It is easy to maximize $e(G_n)$ for $\leq p$-chromatic graphs.
(b) $\mathcal{P}$ is the family of Dodecahedron-free graphs:

$$\mathcal{P} := \{G \, : \, D_{12} \not\subseteq G\},$$

$\mathcal{P}_n := \mathcal{P}(n, D_{12})$, and $\mathcal{Q}_n = \mathcal{Q}(n, p, s)$ is the family of n-vertex graphs from which one can delete $\le s - 1$ vertices to get a $\le p$-chromatic graph. It is easy to prove that $\mathcal{Q}(n, 6, 2) \subseteq \mathcal{P}$.[44]

Simonovits conjectured that the extremum is attained by a graph $H(n, 2, 6)$, where $H(n, p, s)$ is the generalization of $T_{n,p}$: the n-vertex graph having the maximum number of edges in $\mathcal{Q}(n, p, s)$. Next he proved that if $G_n \in \mathcal{P} - \mathcal{Q}(n, 2, 6)$ then $e(G_n) < e(H(n, 2, 6)) - \frac{1}{2}n + O(1)$. So, to maximize $e(G_n)$ was reduced to maximizing it in $\mathcal{Q}(n, 2, 6)$, which is easy.

(c) $\mathcal{P}$ is the family of Octahedron-free graphs and $\mathcal{Q}_n$ is the family of those graphs G_n where $V(G_n)$ can be partitioned into V_1 and V_2 so that $G[V_1]$ does not contain C_4 and $G[V_2]$ does not contain P_3. Again, it is not too difficult to prove that $\mathcal{Q}_n \subseteq \mathcal{P}$. Using this, Erdős and Simonovits, applying a stability argument [314], determined the exact extremal graphs for large n. (Actually, they proved a much more general theorem on $\mathbf{EX}(n, K_{p+1}(a_1, \ldots, a_p))$.)

It is worth mentioning the simplest case of (c):

Theorem 2.39 (Erdős–Simonovits [314]) *There exists an n_0 such that if $n > n_0$ and S_n is extremal for the Octahedron graph $K(2, 2, 2)$, then $V(S_n)$ can be partitioned into into two parts, A and B so that A spans a C_4-extremal graph in S_n, B spans a P_3-extremal graph and each $x \in A$ is joined to each $y \in B$.*

The product conjecture asserts that

Conjecture 2.40 (Simonovits, see [759]) *If the decomposition class $\mathcal{M}$ of a finite $\mathcal{L}$ does not contain trees or forests, then each $S_n \in \mathbf{EX}(n, \mathcal{L})$ is a product of p subgraphs of $(n/p) + o(n)$ vertices, where p is defined by* (4).

— · —

Let us fix a Universe, for the sake of simplicity, the universe of graphs or hypergraphs. Now we repeat what we said above, in a slightly more detailed form. The method of stability means that

(a) We consider an extremal graph problem, where some property $\mathcal{P}$, and two parameters n and e are given (mostly n is the number of vertices and e is the number of the edges) and we try to optimize, say maximize e for fixed n, on $\mathcal{P}_n \subseteq \mathcal{P}$, where $\mathcal{P}_n$ is the family of objects in $\mathcal{P}$ having the parameter n.
(b) We have a property $\mathcal{Q} \subseteq \mathcal{P}$ "strongly" connected to the considered extremal problem. $\mathcal{Q}_n$ is the corresponding subfamily of $\mathcal{Q}$ with parameter n. We assume that the maximization is difficult on $\mathcal{P}_n$ but easy on $\mathcal{Q}_n$.
(c) We prove that the maximum is smaller on $\mathcal{P}_n - \mathcal{Q}_n$ than on $\mathcal{P}_n$, therefore the extremal objects in $\mathcal{P}_n$, (i.e. the ones achieving the maximum) must be also in $\mathcal{Q}_n$, where it is easy to find them.

---

[44]This is equivalent to that deleting any 5 vertices of D_{12} one gets $\ge$ 3-chromatic graphs.

This approach, introduced by Simonovits [751], turned out to be very fruitful for many problems, e.g., for the extremal problems of the icosahedron, dodecahedron, and octahedron, and for several other graph problems, and in several hard hypergraph problems, e.g., in case of the Fano hypergraph extremal problem [385, 504], or the results of Füredi, Pikhurko, and Simonovits [380, 381], (and many similar hypergraph results) see Sect. 8. We can say that in the last twenty-thirty years it became widely used. Below we list some papers connected to the stability method, from a much longer list. See e.g., Balogh, Mousset, Skokan, [80], Ellis, [256], Friedgut [362], Füredi, Kostochka, and Luo [375], Füredi, Kostochka, Luo, and Verstraëte [376, 377] Gowers and Hatami [408], Gyárfás, Sárközy, and Szemerédi, [438], Keevash [495], Keevash and Mubayi [503] Nikiforov and Schelp [633], Mubayi [623], Patkós [642], Samotij [715], Tyomkyn and Uzzell [812].

The stability method is used, e.g., in Sect. 6.3, more precisely, in the corresponding paper [446] of P. Hajnal, S. Herdade, and Szemerédi,—however, in a much more complicated form,—to provide a new proof of the Pósa–Seymour conjecture, without using the Regularity Lemma, or the Blow-up Lemma.

These papers were selected from many-many more and below we add to them some on the stability of the Erdős–Ko–Rado [298] which started with the paper of Hilton and Milner [463], Balogh, Bollobás, and Narayanan [61] and continued with several further works, like Das and Tran [226], Bollobás, Narayanan, and A. Raigorodskii [129], Devlin and Jeff Kahn [239], Ellis, Keller, and Lifshitz, [257], …

Remark 2.41 In [751], where Simonovits introduced this Stability Method, another stability proof method was also introduced, the method of PROGRESSIVE INDUCTION. That meant that the extremal graphs became more and more similar to the conjectured extremal graphs as n increased, and finally they coincided. This approach was useful when the conjectured theorem could have been proved easily by induction, but it was difficult to prove the Induction Basis.

2.13 The "Typical Structure"

The results considered here are related to the situation described in the previous section, on the $\mathcal{P} - \mathcal{Q}$-stability, and have the following form: we have two properties, a complicated one, $\mathcal{P}$, and a simpler one, $\mathcal{Q} \subset \mathcal{P}$ and we assert:

Almost all n-vertex $\mathcal{P}$-graphs have property $\mathcal{Q}$.

Among the simplest ones we have already mentioned or will discuss the following ones.

(a) Almost all K_{p+1}-free graphs are p-chromatic [297, 531].
(b) Almost all Berge graphs are perfect [664] (cf Remark 8.11).
(c) Almost all $K(2, 2, 2)$-free graphs have a vertex-partition, where the first class is C_4-free and the second one is P_3-free, [64].

Erdős conjectured that, given a family $\mathcal{L}$ of forbidden graphs, it may happen that a large part of the $\mathcal{L}$-free graphs are subgraphs of some extremal graphs $S_n \in \mathbf{EX}(n, \mathcal{L})$, in the following sense. Denote by $\mathcal{P}(n, \mathcal{L})$ the family of n-vertex $\mathcal{L}$-free graphs. The subgraphs of an $\mathcal{L}$-extremal graph S_n provide $2^{\mathbf{ex}(n, \mathcal{L})}$ $\mathcal{L}$-free graphs.

Conjecture 2.42 (Erdős) *If $\mathcal{L}$ contains no bipartite L, then*

$$|\mathcal{P}(n, \mathcal{L})| = O(2^{(1+o(1))\mathbf{ex}(n,\mathcal{L})}).$$

The first such result, by Erdős, Kleitman, and Rothschild [297] asserted that for $\mathcal{L} = \{K_{p+1}\}$,

$$\log_2 |\mathcal{P}(n, \mathcal{L})| = (1 + o(1))\mathbf{ex}(n, \mathcal{L}).$$

Later Erdős, Frankl, and Rödl [288] proved Erdős' conjecture, for any non-degenerate case. Kolaitis, Prömel, and Rothschild [531] extended some related results to the case of L with critical edges, see Meta-Theorem 2.19. An important related result is that of Prömel and Steger [665]. Slowly a whole theory was built up around this question. Here we mention in details just a few results, and then list a few related papers.

Theorem 2.43 (Erdős, Frankl, and Rödl [288]) *If $\mathcal{L}$ does not contain bipartite graphs, and $\mathcal{P}(n, \mathcal{L})$ denotes the family of n-vertex $\mathcal{L}$-free graphs, then*

$$|\mathcal{P}(n, \mathcal{L})| < 2^{\mathbf{ex}(n,\mathcal{L})+o(n^2)}.$$

For sharper results, see Balogh, Bollobás, and Simonovits [62–64]. The cases (a) and (c) mentioned above are also connected to this Erdős-Frankl-Rödl theory.

If one counts the number of L-free graphs G_n for a bipartite L, then one faces several difficulties. Formally Theorem 2.43 remains valid, but becomes trivial. We recommend the papers of Kleitman and Winston [523], Kleitman and Wilson [522] and of Morris and Saxton [618].

We get another "theory" if we exclude *induced subgraphs*, see e.g., Prömel and Steger [662, 665], Alekseev [13], Bollobás and Thomason [130, 131]. The theory is similar, however, the minimum chromatic number of $L \in \mathcal{L}$ must be replaced by another, similar colouring number. For some further, related results see Alon, Balogh, Bollobás, and Morris [24]. ...

A more general and sharper question is when the considered family of graphs is $\mathcal{P}$, and the property $\mathcal{Q}$ is strongly connected to $\mathcal{P}$, then one can ask: is it true that almost all graphs $G_n \in \mathcal{P}$ are also in $\mathcal{Q}$. This often holds, e.g., almost all K_3-free graphs are bipartite. Again, a finer result, explained below, is

Theorem 2.44 (Osthus, Prömel, and Taraz [636]) *Let $T_p(n, \Gamma)$ denote the family of K_p-free graphs with Γ edges. If*

$$t_3 = t_3(n) := \frac{\sqrt{3}}{4} n^{3/2} \sqrt{\log n},$$

then for any fixed $\varepsilon > 0$, the probability that a random K_3-free graph on n vertices and Γ edges is bipartite,

$$\mathbb{P}(G_n \in \mathcal{T}_3(n, \Gamma) \implies \chi(G_n) = 2) \to \begin{cases} 1 & \text{if } \Gamma = o(n), \\ 0 & \text{if } \frac{1}{2}n \leq \Gamma \leq (1 - \varepsilon)t_3(n); \\ 1 & \text{if } \Gamma \geq (1 + \varepsilon)t_3(n). \end{cases} \quad (6)$$

The most important line of (6) is the third line. Actually, this "story" started with a result of Prömel and Steger [666], where the threshold-estimate was around $n^{7/4}$ for the third line of (6). Prömel and Steger conjectured that the right exponent is $3/2$. Łuczak [601] proved a slightly weaker related result, where the exponent was $3/2$, however, instead of asking for bipartite graphs, he asked only for "almost bipartite" graphs.

The meaning of this theorem is that for very small $\Gamma = e(G_n)$ most of the graphs will have no cycles, therefore they will be bipartite. For slightly larger $e(G_n)$ odd cycles will (also) appear, so there will be a "slightly irregular" interval, and then, somewhat above $t_3(n)$ everything becomes nice: almost all triangle-free graphs are bipartite.

Remark 2.45 (a) One could ask why $cn\sqrt{n \log n}$ is the threshold for our problem. As [636] explains, this is connected to the fact that this is the threshold where the diameter of a random graph becomes 2.

(b) A nice result of this paper extends the theorems from K_3-free graphs to C_{2h+1}-free random graphs.

(c) Another important generalization of this result is due to Balogh, Morris, Samotij, and Warnke [79] to any complete graph K_p.

FURTHER INFORMATION can be found on these questions in the paper of Balogh, Morris, Samotij, and Warnke [79], which, besides formulating the main results of [79], i.e., extending Theorem 2.44 to any K_p,[45] provides an excellent survey of this area and its connection to several other areas, among them to the problems on extremal subgraphs of Random Graphs, investigated also by Conlon and Gowers [199], ..., see Sect. 2.5/§(f).

As we have mentioned, here the situation for the degenerate (bipartite) case is completely different. Related results are, e.g., Balogh and Samotij [85–87], or Morris and Saxton [618].

Historical Remarks 2.46 This whole story started (perhaps) outside of Graph Theory, with some works of Kleitman and Rothschild, see [519–521].

A hypergraph analog of these results was proved by Nagle and Rödl [628].

Among the newer results we have mentioned or should mention several results of Prömel and Steger, e.g., [663, 664], Alon, Balogh, Bollobás, and Morris [24] and, on hypergraphs, Person and Schacht [645], Balogh and Mubayi [81, 82], ...

---

[45] The Master Thesis of Warnke contained results on K_4.

2.14 Supersaturated Graphs

When Turán proved his theorem, Rademacher immediately improved it:

Theorem 2.47 (Rademacher, unpublished) *If $e(G_n) > \lfloor \frac{n^2}{4} \rfloor$ then G_n contains at least $\lfloor \frac{n}{2} \rfloor$ copies of K_3.*

This is sharp: putting an edge into a larger class of $T_{n,2}$ we get $\lfloor \frac{n}{2} \rfloor$ triangles. More generally, putting k edges into the larger class of $T_{n,p}$ we get $\approx k(\frac{n}{p})^{p-1}$ copies of K_{p+1}, and in particular, for $p = 2$ we get $k \lfloor \frac{n}{2} \rfloor$ triangles. So Erdős generalized Rademacher Theorem:

Theorem 2.48 (Erdős (1962), [265]) *There exists a $c > 0$ such that for any $0 < k < cn$, if $e(G_n) \geq \lfloor \frac{n^2}{4} \rfloor + k$ then G_n contains at least $k \lfloor \frac{n}{2} \rfloor$ copies of K_3.*

Erdős conjectured that his result holds for any $c \leq \frac{1}{2}$.[46] He also generalized his result to K_{p+1} in [274].[47] Lovász and Simonovits proved the Erdős conjecture, in [592], and a much more general theorem in [593].

Let $F(n, L, E)$ be the minimum number of copies of $L \subseteq G_n$ with $e(G_n) = E > \text{ex}(n, L)$ edges. Lovász and Simonovits determined $F(n, K_{p+1}, E)$, for $e(T_{n,p}) < E < e(T_{n,p}) + c_p n^2$, for an appropriately small $c_p > 0$, using the stability method, and, more generally, for any $q \geq p$, and $e(T_{n,q}) \leq E < e(T_{n,q}) + c_q n^2$. In Sect. 2.15 we formulate the related, widely applicable Lovász–Simonovits Stability Theorem.

Remark 2.49 The Lovász–Simonovits method did not work in the general case, farther away from the Turán numbers. Their Supersaturated Graph result, on the number of complete subgraphs, was extended by Fisher and Ryan [343], by Razborov [670], then by Nikiforov [632], and finally, by Reiher [673]. For a related structural stability theorem see also the paper of Pikhurko and Razborov [650].

We complete this section with a C_5-Supersaturated theorem:

Theorem 2.50 (Erdős (1969), [274]) *If $e(G_{2n}) = n^2 + 1$ then G_{2n} contains at least $n(n-1)(n-2)$ pentagons.*

As Erdős remarks, a $K(n, n)$ with an extra edge added shows that his theorem is sharp. For some generalizations see Mubayi [624]. For early results on supersaturated graph results see e.g., the survey of Simonovits [760], explaining how the proofs of some extremal theorems depend on supersaturated graph results, the papers of Blakley and Roy [111], of Erdős [265, 274], Erdős and Simonovits [316, 317], Brown and Simonovits [158], the next subsection, and many further results.

---

[46] Again, there is some difference between the cases of even and odd n.
[47] Erdős' paper contains many further interesting and important results.

2.15 Lovász–Simonovits Stability Theorem

To prove and generalize Erdős' conjecture on K_3-supersaturated graphs, Lovász and Simonovits proved a "sieve", the simplest form of which is the following:

Theorem 2.51 *For any constant $C > 0$, there exists an $\varepsilon > 0$ such that if $|k| < \varepsilon n^2$ and G_n has $\lfloor \frac{n^2}{4} \rfloor + k$ edges and fewer than $C|k|n$ triangles K_3, then one can change $O(|k|)$ edges in G_n to get a bipartite graph.*

Here mostly we use $k > 0$, but if we have the theorem for $k > 0$, that immediately implies its extension for $k \leq 0$ as well. Let $m(L, G)$ denote the number of labeled copies of L in G. The more general form is related to any K_p and not only around $\mathbf{ex}(n, K_p)$, but more generally, when we wish to estimate $m(K_p, G)$ and $e(G_n)$ is around any $\mathbf{ex}(n, K_q)$, for $q \geq p$.

In the next, more general theorem t and d are defined by

$$e(G_n) = \left(1 - \frac{1}{t}\right)\frac{n^2}{2} \quad \text{and} \quad d = \lfloor t \rfloor.$$

Theorem 2.52 (Lovász–Simonovits [593]) *Let $C \geq 0$ be an arbitrary constant. There exist positive constants $\delta > 0$ and a $C' > 0$ such that if $-\delta n^2 < k < \delta n^2$ and G_n is a graph with*

$$e(G_n) = e(T_{n,p}) + k$$

edges and

$$m(K_p, G_n) < \binom{t}{p}\left(\frac{n}{p}\right)^p + Ckn^{p-2},$$

then there exists a $K_d(n_1, \ldots, n_d)$ such that $\sum n_i = n$, $|n_i - \frac{n}{d}| < C'\sqrt{k}$ and G_n can be obtained from $K_d(n_1, \ldots, n_d)$ by changing at most $C'k$ edges.

Here t can be regarded as a "fractional Turán-class-number". To explain the meaning of this theorem, remember that if one puts k edges into the first class of a $T_{n,p}$, that creates $\approx c_1 k n^{p-2}$ copies of K_p. This theorem asserts that in a graph G_n with $e(T_{n,p}) + k$ edges, either we get much more copies of K_p, or G_n must be very similar in structure to $T_{n,p}$.[48]

Theorem 2.52 can be used in many cases, e.g., it provides a clean and simple proof of the Erdős–Simonovits Stability Theorem.

---

[48] This theorem may remind us of the Removal Lemma, (see Sect. 5.4) yet, it is different in several aspects. Both they assert that either we have many copies of L in G_n, or we can get an L-free graph from G_n by deleting a few edges. However, the Removal Lemma has no condition on $e(G_n)$ and the Lovász–Simonovits theorem provides a much stricter structure.

This result can also be used for negative values of k, (and sometimes we need this), however, then we should replace k by $|k|$ in some of the formulas.

Remark 2.53 Assume that $p \geq 3$ and $k = \gamma n^2$, for some constant $\gamma > 0$. If one knows Theorem 2.52 for K_p, then one has it for any p-chromatic L, by applying the Erdős Hypergraph Theorem 8.1 to the v-uniform hypergraph on $V(G_n)$, for $v := v(L)$, whose hyperedges are the vertices of the copies of L in G_n. (For the details see, e.g., Brown and Simonovits [158].) Hence Theorem 2.52 is more general than the Erdős–Simonovits Stability theorem, since it does not completely exclude $L \subset G_n$, only assumes that G_n does not contain too many copies of L.

Remark 2.54 One could ask why do we call Theorem 2.52 a "sieve". Without answering this question, we make two remarks.

(a) The methods used here could be considered in some sense "primitive" predecessors of what today is called Razborov's Flag algebras.

(b) This whole story started with a "survey" paper of Lovász [585], written in Hungarian, the title of which was "Sieve methods".

Remark 2.55 Often the Lovász–Simonovits sieve can be replaced by the Removal Lemma.

Remark 2.56 The original proofs of the Erdős–Simonovits Limit theorem could have used the Regularity Lemma, (described in Sect. 5), or this theorem, but they had not: actually they were proved earlier. An alternative approach to prove the Erdős–Simonovits Stability theorem is to use the Regularity Lemma, however, then one needs the stability theorem itself for complete graphs. A simple, beautiful proof of that the stability holds for K_p was found by Füredi [374], who used for this purpose the Zykov symmetrization [826].

2.16 Degenerate Versus Non-degenerate Problems

We remind the reader that an extremal graph problem is DEGENERATE if $\mathbf{ex}(n, \mathcal{L}) = o(n^2)$, or in case of r-uniform hypergraphs, $\mathbf{ex}(n, \mathcal{L}) = o(n^r)$. Another way to describe a DEGENERATE extremal problem is that $\mathcal{L}$ contains a bipartite L (and for hypergraphs, $\mathcal{L}$ contains an L with strong chromatic number r, see Claim Corollary 8.2). The survey of Füredi and Simonovits [386] describes the details. Here we shall be very brief, describe just a few results and formulate three conjectures and describe some connections to Geometry, Finite Geometry, and Commutative Algebra.

So let us restrict ourselves to simple graphs. The simplest questions are when $L = K(a, b)$ and when $L = C_{2k}$. (The extremal problem of the paths P_k is a theorem of Erdős and Gallai [289].)

Assume that $a \leq b$. In Conjecture 2.10, $\mathbf{ex}(n, K(a, a)) > c_a n^{2-(\frac{1}{a})}$ is conjectured. For $a = 2$ and $a = 3$ the sharp lower bounds came from finite geometric constructions of Erdős, Rényi, Sós, [308] and Brown [150]. The random methods gave only much weaker lower bounds. By [306],[49]

$$\mathbf{ex}(n, K(a, b)) > c_a n^{2 - \frac{1}{a} - \frac{1}{b}}.$$

The Kollár–Rónyai–Szabó construction [532] and its improvement, the Alon–Rónyai–Szabó [42] construction, (proving the sharpness of Theorem 2.9 when a is much smaller than b) used Commutative Algebra, and Lazebnik, Ustimenko, and Woldar have several more involved algebraic constructions.

Remark 2.57 One of Erdős' favourite geometry problem was the following: Given n points in $\mathbb{E}^d$, how many equal (e.g., unit) distances can occur among them. Among others, he observed that if in $\mathbb{E}^3$ we join two points iff their distance is 1, then the resulting graph does not contain $K(3, 3)$. In Brown's construction this is turned around: the vertices of a graph G_n are the $n = p^3$ points of a finite 3-dimensional affine space $AG(p, 3)$. W.G. Brown joined two points (x, y, z) and (x', y', z') if their "distance" was "appropriate":

$$(x - x')^2 + (y - y')^2 + (z - z')^2 = \alpha, \quad (\bmod p).$$

The appropriate choice meant e.g., that if $p = 4k - 1$ then α was any non-zero quadratic residue.[50] Then Brown proved that (for some primes p and some α) the resulting graph contains no $K(3, 3)$ and has $\approx \frac{1}{2} n^{2-(1/3)}$ edges. (Surprisingly, as Füredi proved in [373],—as to the multiplicative constant $\frac{1}{2}$,—for $K(3, 3)$ the Brown construction is the sharp one, not the upper bound of Theorem 2.9.) The Commutative Algebra constructions [42, 532] can be regarded as extensions of Brown's construction, however, with deeper mathematics in the background.

Question 2.58 Sometimes in our constructions we use commutative structures, sometimes non-commutative ones. One could ask, what is the advantage of using non-commutative structures.

The same question can also be asked in connection with the so called Ramanujan graphs, see e.g., [598, 599], or [609]. The answer is simple: the Cayley graphs of commutative groups are full of short even cycles. We illustrate this through the girth problem.

Theorem 2.59 (Bondy–Simonovits, Even Cycle: C_{2k} [136])

$$\mathbf{ex}(n, C_{2k}) \leq c_1 k n^{1+(1/k)}. \tag{7}$$

Conjecture 2.60 (Sharpness) *The exponent $1 + (1/k)$ is sharp, i.e.*

---

[49] Actually, the First Moment Method yields a little better estimate, but far from being satisfactory. One can also see that as a is fixed and b gets larger, the "random construction" exponents converge to the optimal one. This motivates, among others, [42, 532].
[50] This choice ensured that the neighbourhoods did not contain three collinear points.

$$\mathbf{ex}(n, C_{2k}) \geq c_k n^{1+(1/k)} \quad \text{for some} \quad c_k > 0.$$

The first unknown case is $k = 4$, $\mathbf{ex}(n, C_8)$. Theorem 2.59 is sharp for C_4, C_6, C_{10}, see [150, 258, 308] and [101]. For some related constructions see also Wenger [814].

Now, to answer Question 2.58, observe, that we often use in our algebraic constructions Cayley graphs, where in the commutative cases we have many "coincidences", leading to many $C_{2k} \subset G_n$, which can be avoided in the non-commutative cases. A very elegant (and important) example of this is:

Construction 2.61 (Margulis, [608]) There exist infinite Cayley graph sequences $(G_{n,d})$ of degree $d = 2\ell$ with girth greater than $c \frac{\log n}{\log(d-1)}$.

Remark 2.62 We have emphasized that Extremal Graph Theory is connected to many other areas in Mathematics. The Margulis constructions [608] are connected to Coding Theory. In somewhat different ways, several papers of Füredi and Ruszinkó are also extremal hypergraph results strongly connected to (or motivated by) Coding Theory, e.g., [383].

A more detailed analysis of these questions can be found, e.g., in Alon's survey [22], or in the Füredi–Simonovits survey [386].

Remark 2.63 (a) It was a longstanding open question if one can improve the coefficient of $n^{1+(1/k)}$ in the Bondy–Simonovits theorem, from ck to $o(k)$. After several "constant"-improvements, Boris Bukh and Zilin Jiang [163][51] proved that

$$\mathbf{ex}(n, C_{2k}) \leq 80\sqrt{k} \log k \cdot n^{1+(1/k)} + O(n). \tag{8}$$

According to [163], Bukh thinks that Conjecture 2.60 does not hold: he conjectures that for sufficiently large, but fixed k,

$$\mathbf{ex}(n, C_{2k}) = o(n^{1+(1/k)}).$$

It is very "annoying" that we cannot decide this, not even for C_8.

(b) Related constructions were provided by Lazebnik, Ustimenko, and Woldar [567–569] and by Imrich [473].

(c) For some ordered versions of the C_{2k} problem see, e.g., the (very new) results of Győri, Korándi, Methuku, Tomon, Tompkins, and Vizer [443].

Historical Remarks 2.64 (a) Actually, before the Erdős–Simonovits paper [313] Erdős conjectured that the exponents can be only either $(2 - \frac{1}{k})$ or $(1 + \frac{1}{k})$. This conjecture was "killed" in [313], by some "blow-up" of the cube.

(b) Later Erdős and Simonovits conjectured that (i) for any rational exponent $\alpha \in [1, 2)$ there exist degenerate extremal problems $\mathbf{ex}(n, \mathcal{L})$ for which

---

[51] The original version claimed a slightly better estimate.

$$\mathbf{ex}(n, \mathcal{L})/n^\alpha \to c_\mathcal{L} > 0, \tag{9}$$

and (ii) for any degenerate problem $\mathbf{ex}(n, \mathcal{L})$ there exists a rational $\alpha \in [1, 2)$ for which (9) holds. Recently, Bukh and Conlon [162] proved (i). (For hypergraphs Frankl [354] has some earlier, corresponding results, see also Fitch [344].)

2.17 Dirac Theorem: Introduction

Difficult problems always played a central role in the development of Graph Theory. We shall mention here two important problems, the Dirac theorem on Hamiltonian cycles (which is not so difficult) and the Hajnal–Szemerédi theorem on equitable partitions.

Theorem 2.65 (Dirac (1952), [242]) *If $n \geq 3$ and $d_{\min}(G_n) \geq n/2$, then G_n contains a Hamiltonian cycle.*

If $n = 2h$, then $K(h-1, h+1)$ has no Hamiltonian cycle, showing that Theorem 2.65 is sharp.[52] One beautiful feature of this theorem is that as soon as we can guarantee a 1-factor, we get a Hamiltonian cycle.

This theorem triggered a wide research, see, e.g., Ore's theorem [635], Pósa's theorem on Hamiltonian graphs [658], and related results. We shall return to discuss generalizations of Dirac's Theorem, above all, the Pósa–Seymour conjecture and the hypergraph generalizations in Sects. 6.3 and 9.

2.18 Equitable Partition

We close this introductory part with a famous conjecture of Erdős, proved by András Hajnal and Szemerédi.

Definition 2.66 (*Equitable colouring*) A proper vertex-colouring of a graph G is EQUITABLE if the sizes of any two colour classes differ by at most one.

Theorem 2.67 (A. Hajnal and E. Szemerédi (1970), [444]) *For every positive integer r, every graph with maximum degree at most r has an equitable colouring with $r + 1$ colours.*

The theorem is often quoted in its complementary form. The sharpness is shown by the complementary graph of an almost-Turán graph, i.e. the union of complete graphs K_r and K_{r+1}.

---

[52]The union of two complete graphs of $n/2$ vertices having at most one common vertex in common also show the sharpness. For $= 2\ell - 1$ one can use $K(\ell, \ell - 1)$ for sharpness.

The theorem was proved first only for K_3, by Corrádi and A. Hajnal [209]. Then came the proof of Hajnal and Szemerédi. Much shorter and simpler proofs of Theorem 2.67 were found independently by Kierstead and Kostochka, [510] and Mydlarz and Szemerédi [627]. The paper of Kierstead, Kostochka, Mydlarz, and Szemerédi [513] provides a faster (polynomial) algorithm to obtain the equitable colouring.[53]

Multipartite case. This theorem was generalized in several different ways, also, considered for multipartite graphs G_n by Martin and Szemerédi [613], Csaba and Mydlarz [216], Extensions towards multipartite hypergraphs can be found, e.g., in Lo-Markström [576]. It is applied to prove some other graph theorems, e.g., in Komlós–Sárközy–Szemerédi [540, 542, 543], and many other cases. We shall return to these questions in Sect. 7.

The theorem was extended also to directed graphs, by Czygrinow, DeBiasio, Kierstead, and Molla [219].

Random Graphs. The problem of equitable partitions in Random Graphs was also discussed (and in some sense solved) in works of Bohman, Frieze, Ruszinkó, and Thoma [114]. Their result was improved by Johansson, Jeff Kahn, and Van Vu [480].

2.19 Packing, Covering, Tiling, L-Factors

When speaking of "packing", sometimes we mean edge-disjoint embedding of just two graphs (this is connected to some complexity questions from Theoretical Computer Science) and sometimes we try to cover the whole graph with some vertex-disjoint copies of a graph.[54]

There was a period, when—because of Theoretical Computer Science,—packing a graph into the complementary graph of another (i.e. the above problem for two *edge-disjoint* graphs) was a very actively investigated topic. This was connected to EVASIVENESS, i.e. to the problem, how many "questions" are needed to decide if a graph G_n has property $\mathcal{P}$.

The whole area is described in a separate chapter of Bollobás' "Extremal Graph Theory" [119]. For some further related details see also the papers of Bollobás and Eldridge, e.g., [123] with the title "Packing of Graphs and Applications to Computational Complexity". Here we mention also the result of P. Hajnal [445], (improving some important earlier results). He proves that the randomized decision tree complexity of any nontrivial monotone graph property of a graph with n vertices is $\Omega(n^{4/3})$. See also Bollobás [118], and [517, 518]. This is again a nice example where Combinatorics and Theoretical Computer Science are in a very strong interaction.

---

[53] See the introduction of [513]. They also point out the applications of this theorem, e.g., in [36, 478] for "deviation bounds" for sums of weekly dependent random variables, and in the Rödl-Ruciński proof of the Blow-up Lemma [688].

[54] In the Gyárfás Conjecture we try to pack many different trees into a complete graph.

Given a graph G_n, and a sample graph L, a PERFECT TILING is a covering of $V(G_n)$ with *vertex-independent* copies of L, and an ALMOST-TILING is covering at least $n - o(n)$ vertices of it. Tiling is sometimes a tool, a method, in other cases it is the aim. The Corrádi–Hajnal and the Hajnal–Szemerédi theorems, in Sect. 2.18, were also about packing=tiling of graphs.

Perhaps the first case when tiling was used in a proof was the Rademacher–Turán theorem of Erdős [265], Here Erdős covered as many vertices of G_n by vertex-independent triangles as he could, to prove the theorem.

We can consider the problem of TILINGS in a more general way. Here we have a (small) sample graph L and wish to embed into G_n as many vertex-independent copies[55] of L as possible. The question is, given an integer t, which conditions on G_n ensure, e.g., t vertex-independent copies of L. We shall return to this question later, here we mention just a few related results, as illustrations. As we mentioned, Theorem 2.67 is an example of such results. Below we formulate some more general results.

Definition 2.68 Given two graphs L and G, where $v(L)$ divides $v(G)$, an L-factor of G is a family of $v(G)/v(L)$ vertex-disjoint copies of L,

Theorem 2.69 (Alon and Yuster (1996), [49]) *For every $\varepsilon > 0$, and for every positive integer h, there exists an $n_0 = n_0(\varepsilon, h)$, such that for every "sample" graph L_h, every "host" graph G_n with $n = h\ell > n_0$ vertices and minimum degree*

$$d_{\min}(G_n) \geq \left(1 - \frac{1}{\chi(L)} + \varepsilon\right) n \qquad (10)$$

contains an L-factor.

As to the usage of names, tiling, packing, and perfect L-factor are almost the same: given a graph G and a sample graph L, we wish to embed into G as many vertex-independent copies of L as possible, and if they (almost) cover $V(G)$, then we speak about an (almost) perfect tiling/packing.

Komlós extended the notion of L-factor by saying that G has an L-factor, if it contains $\lfloor v(G)/v(L) \rfloor$ vertex-independent copies of L. Alon and Yuster conjectured and Komlós, Sárközy and Szemerédi [543] proved the following[56]

Theorem 2.70 (Komlós, Sárközy, and Szemerédi) *For every L there is a constant $K = K_L$ such that if*

$$d_{\min}(G_n) \geq \left(1 - \frac{1}{\chi(L)}\right) n + K_L,$$

then G_n has an L-factor.

---

[55]Or, in other cases edge-disjoint copies. Here "vertex-independent" and "vertex-disjoint" are the same.

[56]Actually, they formulated two "similar" conjectures, we consider only one of them.

Komlós considered the tiling situation in [534] in a more general way. He considered degree conditions for finding many disjoint copies of a fixed graph L in a large graph G. Let $\tau(n, L, M)$ be the minimum m for which if $d_{\min}(G_n) \geq m$, then there is an L-matching covering at least M vertices in G. For any fixed $x \in (0, 1)$, Komlós determined

$$f_L(x) = \lim_{n \to \infty} \frac{1}{n} \tau(n, L, xn).$$

Thus, e.g., Theorem 8 of [534] determines (for a fixed but arbitrary L) a sharp min-degree condition on G_n to enable us to cover $\approx xn$ vertices of G_n by copies of L. Among others, Komlós analyzed the strange and surprising differences between the cases when we try to cover G_n with vertex-independent copies of L completely, and when we try only to cover it almost completely, or when we try to cover only a large part, $\lfloor xn \rfloor$ vertices of it. Several further details and a careful analysis of this situation can be found in the paper of Komlós [534].

— · —

Kawarabayashi conjectured that Theorem 2.70 can be improved by taking into account the particular "colouring structure" of L. This was proved by Cooley, Kühn, and Osthus [203], and then by Kühn and Osthus [562]. Komlós, in [534] defined a chromatic number $\chi_{cr}(L)$, improving the earlier results, by using this χ_{cr}. Kühn and Osthus defined another "colouring parameter" $\mathbf{hc}_\chi$ depending on the sizes of the colour-classes in the optimal colourings of L. Using $\mathbf{hc}_\chi$, they defined a $\chi^*(L) \in [\chi(L) - 1, \chi(L)]$ and proved

Theorem 2.71 (Kühn and Osthus (2009), [562]) *If $\delta(L, n)$ is the smallest integer m for which every G_n with $d_{\min}(G_n) \geq m$ has a perfect L-packing then*

$$\delta(L, n) = \left(1 - \frac{1}{\chi^*(L)}\right) n + O(1).$$

Bipartite Packing. The packing problems, as many similar problems, have also bipartite versions. Hladký and Schacht—extending some results of Yi Zhao [823]—proved

Theorem 2.72 (Hladký and Schacht (2010), [471]) *Let $1 \leq s < t$, $n = k(s + t)$. If*

$$\Phi_2(n, L) := \begin{cases} \frac{1}{2}n + s - 1 & \text{if } k \text{ is even} \\ \frac{1}{2}(n + t + s) - 1 & \text{if } k \text{ is odd,} \end{cases}$$

then each subgraph $G \subseteq K(n, n)$ with minimum degree at least $\Phi_2(n, L)$ contains a $K(s, t)$-factor, and this is sharp, except if $t \geq 2s + 1$ and k is odd.

For the "missing case" (when $t \geq 2s + 1$ and k is odd) see Czygrinow and DeBiasio [218].

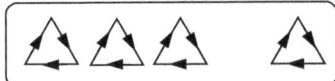

Fig. 3 Cyclic triangles

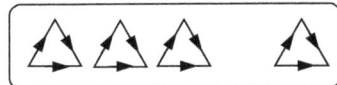

Fig. 4 Transitive triangles

Remark 2.73 Here we have to be careful with using the word "factor": Lovász, in his early papers [583, 584] called a subgraph $H_n \subseteq G_n$ an f-factor if

$$f : V(G_n) \to \mathbb{N}, \quad \text{and} \quad \deg_H(x) = f(x) \quad \text{for each} \quad x \in V(G_n).$$

Directed graphs. There are many further results connected to this area. We shall return to some of them later, e.g., in Sects. 6.7 and 9.6.

Here we close this area with two analogous results on ORIENTED GRAPHS: directed graphs without loops, where between any two vertices there is at most one arc. $\delta^0(G)$ is the minimum of all the in- and out-degrees.

Theorem 2.74 (Keevash and Sudakov [505]) *There exist a fixed $c > 0$, and an n_0 such that if G_n is an oriented graph on $n > n_0$ vertices and $\delta^0(G_n) > (\frac{1}{2} - c)n$, then G_n contains a "cyclic triangle" tiling which leaves out at most 3 vertices. This is sharp (Fig. 3).*

Actually, Keevash and Sudakov [505] describe the history of this theorem in detail, explain several related results, and prove that the theorem is sharp in the sense that if $n \equiv 3 \pmod{18}$ then one cannot guarantee the covering of the whole graph with cyclic triangles, even under a stronger degree-condition. Further, they prove some other packing theorems where the lengths of some cycles are prescribed but they need not be triangles. We close with

Theorem 2.75 (Balogh, Lo, and Molla [73]) *There exists an n_0 such that for every $n \equiv 0 \pmod{3}$, if $n > n_0$, then any oriented graph G_n on n vertices with $\delta^0(G_n) \geq \frac{7n}{18}$ contains a TT_3-tiling, where TT_3 is the transitively oriented triangle (Fig. 4).*

This area is fairly active nowadays, we refer to several papers on equitable colouring of Kostochka and others, e.g., [509, 550, 551], and we mention just some extensions to bipartite tiling, e.g., Czygrinow and DeBiasio [218], and to oriented graphs, by Czygrinow, DeBiasio, Kierstead, and Molla [219, 223], Yuster on tournaments [821] or [218, 221], or to hypergraphs, e.g., Pikhurko [647] or Czygrinow, DeBiasio, and Nagle [221], and several papers of Yi Zhao and Jie Han, e.g., [451], and many-many others. Some of these results are good examples of the application of the ABSORBING METHOD, discussed in Sect. 6.5. This is, e.g., the case with the Balogh-Lo-Molla theorem or the Czygrinow-DeBiasio-Nagle result just mentioned.

3 "Classical Methods"

Here, (giving some preference to results connected to Laci Lovász) we still skip the area of graph limits, we skipped the applications of Lovász Local Lemma (though Lovász Local Lemma is among the most important tools in this area and it came from research strongly connected to extremal graph problems [299]), but we mention two other, less known results of Lovász, strongly connected to extremal graph theory.

3.1 Detour I: Induction?

Speaking of methods in Extremal Graph theory, we mostly avoided speaking of hypergraph results, partly because they are often much more involved than the corresponding results on ordinary graphs. One is the construction of graphs with high girth and high chromatic number. Erdős used Random Graphs to prove the existence of such graphs [263, 264] and there was a long-standing challenge to *construct* them.[57] Lovász often solved "such problems" by trying to use induction, and when this did not work directly, to look for and find a stronger/more general assertion where the induction was already applicable. In *this* case Lovász generalized the problem to hypergraphs and used induction [582]. His proof was a tour de force, rather involved but worked. It was the first "construction" to get graphs of high girth and high chromatic number.

How is this problem connected to Extremal Graph Theory?

Theorem 3.1 *If $\mathcal{L}$ is a finite class of excluded graphs, then $\mathbf{ex}(n, \mathcal{L}) = O(n)$ iff $\mathcal{L}$ contains some tree (or forest).*

To prove this, one needs the Erdős result [264] about high girth graphs G_n with $e(G_n) > n^{1+\alpha}$ edges, or the Erdős-Sachs theorems [310], ..., or some corresponding Lubotzky-Phillips-Sarnak-Margulis graphs, see e.g., [598, 599]. (Lovász' tour de force construction was a big breakthrough into this direction though it was not quantitative, which is needed above.)

Remark 3.2 Since then several alternative constructions were found. We mention here the construction of Nešetřil and Rödl [630]. Perhaps one of the best is the construction of Ramanujan graphs by Lubotzky, Phillips, and Sarnak [599] and Margulis [611]: it is very direct and elegant. It has only one "disadvantage": to verify that it is a good construction, one has to use deep Number Theory. (For more detailed description of the topic, see [599], or the books of Lubotzky [597] and of Davidov, Sarnak, and Valette [227].)

---

[57] We must repeat that the meaning of "to construct them" is not quite well defined. Let us agree for now that the primary aim was to eliminate the randomness.

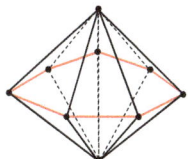

Fig. 5 Double cone

3.2 Detour II: Applications of Linear Algebra

We start with repeating a definition.

Definition 3.3 H is a colour-critical graph[58] if for any edge e of H, $\chi(H - e) < \chi(H)$. It is k-colour-critical if, in addition, it is k-chromatic.

The theory of colour-critical graphs is a fascinating area. Erdős writes about its beginnings, e.g., in his article in the memory of Gabriel Dirac [286]. Gallai also had several interesting conjectures on colour critical graphs. One of them, on the independence number of a 4-chromatic colour-critical graph, was disproved by a construction of Brown and Moon [157], and then by Simonovits [753] and Toft [807]. The disproof was strongly connected to a hypergraph extremal problem discussed also by Brown, Erdős, and Sós [154]. Lovász improved the corresponding results, proving the following sharp and much more general result (Fig. 5).

Theorem 3.4 (Lovász (1973, 1978), [586, 587]) *Let $\alpha_k(n)$ denote the maximum number of independent vertices in a k-critical graph on n vertices. Then*

$$n - 2kn^{1/(k-2)} \le \alpha_k(n) \le n - (k/6)n^{1/(k-2)}.$$

The lower bound is based on generalizing the Brown–Moon construction and the upper bound improves the result of Simonovits, $\alpha_4(n) \le n - c_2 n^{2/5}$. Simonovits in [753, 754] used a hypergraph extremal problem, where the excluded hypergraphs were 3-uniform double-cones.[59] One of his results was basically equivalent to a results of Brown, Erdős, Sós [154], where the excluded graphs were all the triangulations of a 3-dimensional sphere (the double cones are among these hypergraphs). Actually, Simonovits proved

Theorem 3.5 (Simonovits [753]) *If $\mathcal{C}_r^3$ denotes the (infinite) family of 3-uniform r-cones, then*

$$\mathbf{ex}_3(n, \mathcal{C}_r^3) = O(n^{3-\frac{1}{r}}).$$

---

[58]More precisely, edge-colour-critical graph.

[59]An r-cone is a 3-uniform hypergraph obtained from a cycle $x_1, \ldots, x_k$ by adding r new vertices $y_1, \ldots, y_r$ and all the triples $y_j x_i x_{i+1}$ (where $x_{k+1} = x_1$).

(Watch out, here the subscript r is not the number of vertices!) Simonovits, and then Toft, and Lovász, reduced the general case of the Gallai problem to the case when each $x \in X$ had degree $d(x) = k - 1$, and these vertices x defined a $(k-1)$-uniform hypergraph $\mathcal{H}_m$ on the remaining vertex-set $\mathcal{M} := V(G_n) - X$. Simonovits proved (for $k = 4$) that this hypergraph $\mathcal{H}_m$ does not contain any double-cone. Therefore $e(\mathcal{H}_m) = O(m^{5/2})$. This led to his estimate. Lovász—starting with the same approach—excluded many more $(k-1)$-uniform hypergraphs. To make his argument easier to follow, we restrict ourselves to the case $k = 4$. Let $\mathcal{M} := \{u_1, \ldots, u_m\}$. The neighbourhoods of u_i's defined a 3-uniform hypergraph on $[1, m]$, and Lovász attached to each u_i a 0-1 vector of length $\binom{m}{2}$, where, for a neighbourhood $N(u_i) = \{a, b, c\}$, Lovász put 1's into the places (a, b), (b, c), and (c, a), thus obtaining a 0-1 vector $\mathbf{x}_i$ of length $\binom{m}{2}$, with three coordinates equal to 1's (and all the others were equal to 0). The 4-criticality implied that these vectors were linearly independent.[60] Therefore, their number was at most $\binom{m}{2}$, i.e., we obtained that $e(\mathcal{H}_m) = |X| \leq \binom{m}{2}$. This gave the upper bound in Theorem 3.4.

So he used the Linear Algebra method, basically unknown those days, to get the sharp result in his extremal graph problem. That gave a sharp result also in the Gallai problem.

Remark 3.6 For any monotone graph property $\mathcal{P}$ we may define the critical structures: G_n belonging to $\mathcal{P}$ but after the deletion of any edge (or, in other cases, any vertex) we get a graph outside of $\mathcal{P}$. If we have a graph-parameter on graphs, then criticality mostly means increasing/decreasing this parameter by deleting any edge. Criticality was discussed for the stability number, chromatic numbers, diameter, and can be investigated for many other monotone properties. Among criticality problems colour-criticality seems to be one of the most interesting ones.

The fascinating area of colour-critical graphs was started by G. Dirac, see e.g., Dirac [243, 245] and Erdős [286]. There are several results on it in the Lovász book [589]. We skip here the topic of colour-critical hypergraphs, listing just a few fascinating results on them: e.g., Abbott and Liu [2] and results of Krivelevich [556]. Deuber, Kostochka, and Sachs [238], Rödl and Siggers [704] and Anstee, Fleming, Füredi, and Sali [50, 384]. Toft [807], and Simonovits [753], Rödl and Tuza [707], Sachs and Stiebitz [712–714] Stiebitz, [783] Kostochka and Stiebitz [552], Stiebitz, Tuza, and Voigt [784].

See also the survey of Sachs and Stiebitz [714].

We conclude this part with two open problems.

Problem 3.7 (*Erdős*) Is it true that if (G_n) is a sequence of 4-critical graphs, then $d_{\min}(G_n) = o(n)$?

Simonovits [753] and Toft [807] constructed 4-colour-critical graph sequences (G_n) with $d_{\min}(G_n) > c\sqrt[3]{n}$.

---

[60]In graph-theoretical language, Lovász excluded all the 3-graphs for which the Sperner Lemma holds: for which each pair was contained by an even number of triples.

Problem 3.8 (*Linial*) How many triples can a 3-uniform $\mathcal{H}_n$ have without containing a triangulation of the torus.[61]

Remark 3.9 In some sense this geometric/linear algebraic approach helped Lovász to solve the famous Shannon conjecture on the capacity of C_5, in [588].

4 Methods: Randomness and the Semi-random Method

Of course, the title of this chapter may seem too pretentious. We shall skip several important methods and related results here, or just touch on them, primarily since they are well described in several other places, and partly since the bounded length of this survey does not allow us to go into details.

Thus we shall skip most of the results *directly* connected to random graphs or random graph constructions, while we shall touch on the pseudo-random graphs. For random methods, the readers are referred to the books of Alon and Spencer [48], Bollobás [121], or that of Janson, Łuczak, and Ruciński [476], the survey of Molloy [614], or the book of Molloy and Reed [615].

Here of course most of these sources are well known, like e.g., the books [48] or [476]; we mention only Molloy's excellent chapter [614], which is a survey on this topic, and perhaps got less attention than it deserves. It contains many important details and explanations, and perhaps would fit the best to our topic, with the exception that it concentrates more on colouring and we concentrate more on independent sets in particular graphs.[62]

Listing the methods left out here, we should mention the extremely important works on the Hypergraph Regularity Lemma, primarily that of Frankl and Rödl [358] on 3-uniform hypergraphs, (This was among the first ones). Of course, we should mention the whole school of Rödl, among others, e.g., the papers of Rödl, Nagle, Skokan, Schacht, and Kohayakawa, [527, 687][63] Rödl and Schacht [701], Rödl and Skokan [705, 706], and Kohayakawa, and many further results, and, on the other hand, related to this, several works of Tim Gowers [403, 404], Green and Tao [418] Terry Tao [798] …

4.1 Various Ways to Use Randomness in Extremal Graph Theory

Random graphs are used in the area discussed here in several different ways.

---

[61] Observe that this is motivated by [154], and that we formulated it in its simplest case, however, we (more precisely, Nati Linial) meant a whole family of problems. He spoke about them in his talk in the Lovász Conference, 2018.

[62] The same applies to the book of Molloy and Reed [615].

[63] This is a PNAS "survey", with an accompanying paper of Solymosi [770].

(a) There are many cases where constructions seem unavailable but random graphs may be used to replace them. We have mentioned the Erdős Magic [261]. Another similar direct approach of Erdős and Rényi [306] was that

$$\mathbf{ex}(n, L) \geq c_L n^{2-v(L)/e(L)}. \tag{11}$$

(b) In other cases we use modified random graphs: (11) is useless for cycles, but the modified random graph (in the simplest case the First Moment Method) worked in the papers of Erdős [263, 264], and in many similar cases. We mention here the Erdős-Spencer book [320], and its follower, the very popular book of Alon and Spencer [48].
(c) Since the very important paper of Erdős and Rényi [306]—on the evolution of Random Graphs,—the investigation of the changes/phase-transitions in the structure of Random Graphs, as the number of edges is slowly increasing— became a central topic of Combinatorics. Probably the first profound book on this was that of Bollobás [121]. Also we should definitely mention here the newer book of Janson, Łuczak, and Ruciński [476], and the Molloy–Reed book [615].
(d) Extremal graph theory optimizes on a Universe, and this Universe may be the family of Random Graphs.[64] Since a question of Erdős was answered by Babai, Simonovits, and Spencer [56], (here Theorem 5.43) this also became a very interesting and popular topic. We shall return to this question in Sect. 5.7.
(e) The Semi-random method was introduced by Ajtai, Komlós, and Szemerédi, for graphs in [10] (to be applied in Combinatorial Number Theory, to Sidon Sequences), and later Rödl, in his famous solution of the Erdős-Hanani problem [296] about block designs developed the absolutely important method, now called the Rödl Nibble [684].

In this short part we describe the Semi-Random method and the Rödl Nibble very superficially. According to their importance we should provide here a longer description, but the Rödl Nibble has an excellent description, a whole chapter, in the Alon–Spencer book [48], and it is more complicated and technical than that we could easily provide a sufficiently detailed description of it. (Beside referring to the Rödl Nibble, described in [48], we also mention the original Komlós–Pintz–Szemerédi paper [536], to see the origins, and also the Pippenger–Spencer paper [652], and the Jeff Kahn paper [482] proving the asymptotic weakening of the very famous Erdős–Faber–Lovász conjecture. (The Jeff Kahn paper also contains a fairly detailed description of the method.) The more recent paper of Alon, Kim, and Spencer [37] is also related to the previous topic.

---

[64] Again, this case differs from the others: if we try to optimize some parameter on all n-vertex graphs, or on the subgraphs if the d-dimensional cube,..., that problem is well defined for individual graphs, while the assertions on the subgraphs of a random graphs make sense only in some asymptotic sense, the assertions always contain the expression "almost surely as $n \to \infty$".

4.2 The Semi-random Method

The semi-random method was introduced for graphs by Ajtai, Komlós, and Szemerédi [10], in connection with infinite Sidon sequences. Later it was extended by Komlós, Pintz, and Szemerédi to 3-uniform hypergraphs in [536], to disprove a famous conjecture of Heilbronn, discussed in Sects. 4.7–4.9. The method was further extended by Ajtai, Komlós, Pintz, Spencer, and Szemerédi in [7] and by Duke, Lefmann, and Rödl [248].

Here we are concerned with three topics, strongly connected to each other and to the estimates of the independence number of K_3-free graphs, or analogous hypergraph results. The semi-random method had the form where the independence number of a graph or a hypergraph had to be estimated from below, under the condition that the graph had no short cycles. The topics are

(a) **Sidon's problem on infinite sequences.** A Sidon sequence is a sequence of integers in which no two (distinct pairs) have the same sum. What is the maximum density of such a sequence?
(b) **Heilbronn problem** for the "minimum areas" in geometric situations.
(c) **Ramsey problem** $R(3, k)$. This will be obtained as a byproduct, for free.

4.3 Independent Sets in Uncrowded Graphs

A graph, hypergraph $\mathcal{H}$ is called UNCROWDED if it does not contain short cycles. For graphs we excluded triangles, for hypergraphs cycles of length 2, 3, or 4. In a hypergraph a cycle can be defined in several ways. Here a k-cycle ($k \geq 2$)[65] is a sequence of k different vertices: $x_1, \ldots, x_{k-1}, x_k = x_0$, and a sequence of k different edges: $E_1, \ldots, E_k$ such that $x_{i-1}, x_i \in E_i$ for $i = 1, 2 \ldots, k$. The cycle above is called SIMPLE if $E_i \cap \left(\cup_{j \neq i} E_j \right) = \{x_{i-1}, x_i\}$ for $i = 1, 2 \ldots, k$. In a hypergraph $\mathcal{H}$, a 2-cycle is a pair of two hyperedges intersecting in at least two vertices; a vertex set $A \subset \mathcal{H}$ is INDEPENDENT if it does not contain hyperedges. The maximum size of an independent set in $\mathcal{H}$ is denoted by $\alpha(\mathcal{H})$. There are several lower bounds concerning independent sets in k-uniform uncrowded hypergraphs, mostly having the following form:

Hypergraphs having no short cycles have large independent sets.

For ordinary graphs the following theorem, connected to infinite Sidon sequences, was the starting point.

Theorem 4.1 (Ajtai, Komlós, Szemerédi (1980, 1981), [9, 10]) *If in a triangle-free graph G_n the average degree is $t := \frac{2e(G_n)}{n}$, then the independence number*

$$\alpha(G_n) \geq \frac{1}{100} \cdot \frac{n}{t} \log t. \tag{12}$$

---

[65] Often called a Berge k-cycle: in Fig. 14(b) the edges of a C_6 are covered by 3-tuples.

Turán's theorem, or the greedy algorithm would give only n/t and in the random graphs we have the extra $\log t$ factor. The meaning of this theorem is, perhaps, that excluding K_3 forces a much larger independent set and a randomlike behaviour.

Ajtai, Erdős, Komlós, and Szemerédi also proved

Theorem 4.2 ([6]) *If G_n is K_p-free then $\alpha(G_n) \geq c'(n/t) \ln A$ where $A = (\ln t)/p$ and c' is an absolute constant.*

See also Shearer [738–741] and Denley [237].

Proof-Sketch

Originally Theorem 4.1 had two different proofs. One of them was an induction on the number of vertices, [9], (and a similar, perhaps more elegant proof—also using induction on n—was given by Shearer for its sharpening, see [738]). The other proof, from [10] used an iterated random construction which later developed into the Rödl Nibble. This approach turned out to be fairly important, so we "sketch" its main idea, suppressing most of the technical details, and following the description from p10 of [10].

(a) Since G_n is triangle-free, $\alpha(G_n) > d_{\max}(G_n) \geq t$. So we may assume that the degrees are smaller than $\frac{1}{100} \cdot \frac{n}{t} \log t$. Similarly, we may assume that $t \geq \sqrt{n \log n}$, since otherwise $t > \frac{1}{100} \frac{\log t}{t} n$, implying (12): $F(t) = t^{-2} \log t$ is monotone decreasing; for $t = \sqrt{n \log n}$ we have $F(t) \approx \frac{1}{2n}$. This proves the assertion.

(b) We select a subset $\mathcal{K} \subset V(G_n)$ of $\frac{1}{110}(n/t)$ independent vertices.

(c) Next we consider a vertex-set $\mathcal{M} \subseteq V(G_n) - \mathcal{K}$ of $\approx n/2$ vertices not joined to $\mathcal{K}$: we need a lemma about the existence of such an $\mathcal{M}$.

(d) Another lemma assures us that the crucial quantity n/t does not drop too much when we move from G_n to $G_m = G[\mathcal{M}]$. (It drops only by a factor $\vartheta := 1 - \frac{1}{t} - c_{10}\sqrt{t/n} > 1 - t^{1/3}$.)

(e) If we are lucky, then we can iterate this step $\approx \frac{1}{2}\log t$ times: we gain a $\frac{1}{2}\log t$ factor and get Theorem 4.1.

(f) On the other hand, if we are "unlucky" and get stuck in the rth step, then for the corresponding t_r we have that it is too large: $t_r^{-1/3} > 1/\log t$. But then we get in this last step alone enough independent vertices:

$$\frac{n_r}{t_r + 1} > (\log t)^{-3} \frac{n}{2^r} > \frac{n(\log t)^{-3}}{\sqrt{t}} > n\frac{\log t}{t}.$$

Summarizing: In the typical case we can choose small independent subsets in $V(G_n)$ basically $\log t$ times to gain a $\log t$ factor. It is important that discarding these small vertex sets, we do not ruin the structure of the remaining part.

4.4 Uncrowded Hypergraphs

Most probably, the earliest result on hypergraphs using the semi-random method was the following one.

Theorem 4.3 (Komlós, Pintz, and Szemerédi [536], Lemma 1) *There exists a threshold t_0 and a constant $c > 0$ such that if $\mathbb{H}_n^{(3)}$ is a 3-uniform hypergraph on n vertices, with average degree $t^{(3)}(\mathbb{H}_n^{(3)})$, and not containing simple cycles of length at most 4, then for $t := \sqrt{t^{(3)}(\mathbb{H}_n^{(3)})} > t_0$, $t = O(n^{1/10})$, we have*

$$\alpha(\mathbb{H}_n^{(3)}) > c\frac{n}{t}\sqrt{\log t}.$$

The problems we discuss here were reduced to finding lower bounds on the independence number $\alpha(\mathbb{H})$ of a graph or hypergraph $\mathbb{H}$ under the assumption that $\mathbb{H}$ has no short cycles.[66] The above theorem and its versions were enough for the early applications, in 1980's, however, as it turned out in [248], only hypergraph cycles of length 2 had to be excluded: the following much newer generalization can be very useful in some new applications.

Theorem 4.4 (Duke, Lefmann, and Rödl [248], Theorem 2) *Let $\mathbb{H}$ be a k-uniform hypergraph on n vertices. Let Δ be the maximum degree of $\mathbb{H}$. Assume that $\Delta \leq t^{k-1}$ and $t > t_0$. If $\mathbb{H}$ doesn't contain 2-cycles (two hyperedges with at least two common vertices), then*

$$\alpha(\mathcal{H}) = \Omega\left(\frac{n}{t}(\log t)^{\frac{1}{k-1}}\right).$$

Theorem 4.4 for $k = 3$ implies Theorem 4.3. One advantage of it is that in our applications we may have many simple cycles of length 3 and 4, but Theorem 4.4 still can be applied.

— · —

There are many results in this field. We mention here only (i) Duke, Lefmann, and Rödl [248] on Uncrowded Hypergraphs, (ii) Bertram-Kretzberg, Hofmeister, and Lefmann [106, 107], some generalizations and results on the algorithmic aspects of the Heilbronn Problem: finding efficiently the large independent set in an uncrowded hypergraph, and (iii) Lefmann and Schmitt [573] and Lefmann [571], on the higher dimensional Heilbronn problem.

In [10] it is remarked that it is enough to assume that the number of triangles is small, instead of assuming that there are no K_3's at all. Shearer [738, 739] improved the constant in Theorem 4.1 in an ingenious way:

---

[66] Actually, in [536] one needs to exclude only cycles of length 2, 3, and 4, where a cycle of length 2 is a pair of hyperedges intersecting in at least two vertices. Even this is improved in the next theorem.

Theorem 4.5 (Shearer [738]) *Let* $f(t) = \frac{t \log t - t + 1}{(t-1)^2}$. *Then for any triangle-free* G_n, *with average degree* t,

$$\alpha(G_n) \geq f(t) \cdot n. \tag{13}$$

This improves (12), and (in some sense) this is sharp. Related extensions were found by Denley [237], and by Shearer in cases when we assume that the odd girth of the graph is large [740], and also when we wish to use finer information on the degree distribution. Kostochka, Mubayi, and Verstraëte [549] proved some hypergraph versions of this theorem, giving lower bounds on the stability number under the condition that certain cycles are excluded.

4.5 Ramsey Estimates

Observe that—as a byproduct,—Theorem 4.1 immediately yields that

Theorem 4.6 (Ajtai, Komlós, Szemerédi, on Ramsey theorem [9]) *There exists a constant* $c > 0$ *such that*

$$R(3, m) \leq \frac{m^2}{c \log m}. \tag{14}$$

Proof Indeed, if $n > \frac{m^2}{c \log m}$, and $K_3 \not\subseteq G_n$, then either G_n has a vertex x of degree $\delta_G(x) \geq m$, yielding an independent m-set $N_G(x)$, or by Theorem 4.1,

$$\alpha(G_n) > c \frac{\frac{m^2}{c \log m}}{m} \log m = m$$

proving Theorem 4.6. □

Theorem 4.6 improves Erdős' old result [263], by a $\log m$ factor.[67] For many years it was open if (14) is just an improvement of the Erdős result or it is a breakthrough. Jeong Han Kim [515] proved much later, (using among others the Rödl Nibble) that this bound is sharp.

Theorem 4.7 (Kim, Ramsey (1995), [515])[68]

$$R(3, m) \geq \tilde{c} \frac{m^2}{\log m}. \tag{15}$$

So the $r(3, m)$-problem is one of the very few nontrivial infinite cases on Ramsey numbers where the order of magnitude is determined. Bohman and Keevash [115]

---

[67] Of course, Shearer's improvement yields an improvement of c in (14).
[68] Actually, the proof works with $\tilde{c} = \frac{1}{162} - o(1)$.

and Fiz Pontiveros, Griffiths, and Morris [345] independently proved that $R(3, k) \geq (\frac{1}{4} + o(1))\frac{k^2}{\log k}$. Recently Shearer's estimate was "strengthened" as follows.

Theorem 4.8 (Davies, Jenssen, Perkins, and Roberts [228]) *If G_n is a triangle-free graph with maximum degree t, and $\mathcal{I}(G_n)$ is the family of independent sets in G_n, then*

$$\frac{1}{|\mathcal{I}(G_n)|} \sum_{I \in \mathcal{I}(G_n)} |I| \geq (1 + o(1))\frac{\log t}{t} n.$$

This is a strengthening in the sense that here the average size is large, but it is a weakening: it uses the maximum degree, instead of the average degree. For further details see, e.g., the introduction of [228].

Remark 4.9 The above questions are connected to another important question: under some condition, what can be said about the number of independent sets in a graph or a hypergraph? Without going into details, we remark that these questions are connected to the container method, (for references see the Introduction).

Remark 4.10 Further related results can be found, e.g., in papers of Cooper and Mubayi, and Dutta [206–208] which count the number of independent sets in triangle-free graphs and hypergraphs.

4.6 Infinite Sidon Sequences

Finite Sidon sequences are well understood, the maximum size of a Sidon subset of $[1, n]$ is around $\sqrt{n}$, [178, 324]. However, infinite Sidon sequences seem much more involved. The greedy algorithm provides an infinite Sidon sequence (a_n) with $a_n > cn^{1/3}$. This was slightly improved by using Theorem 4.1, but only by $\sqrt[3]{\log n}$:

Theorem 4.11 (Ajtai, Komlós, and Szemerédi (1981), [10]) *There exists an infinite Sidon sequence B for which, if $B(n)$ denotes the number of elements of it in $[1, n]$, then*

$$B(n) \geq c(n \log n)^{1/3}.$$

As it is remarked in [10], Erdős conjectured that $B(n) > n^{(1/2)-\varepsilon}$ is possible. As to Sidon sets, later Theorem 4.11 was improved "dramatically":

Theorem 4.12 (Ruzsa (1998), [710]) *There exists an infinite Sidon sequence B for which, if $B(n)$ denotes the number of elements in $[1, n]$, then*

$$B(n) \geq n^{\sqrt{2}-1+o(1)}.$$

So the importance of this Ajtai–Komlós–Szemerédi result [10] was that this was the beginning of the Semi-Random method. The following generalization was proved by Cilleruelo [189]. Call a sequence $\mathcal{A} = \{a_i\}_{i=1}^{\infty}$ h-Sidon if all the sums $a_{i_1} + \cdots + a_{i_h}$ are distinct for $a_{i_1} \leq \cdots \leq a_{i_h}$.

Theorem 4.13 (Cilleruelo (2014), [189]) *For any $h \geq 2$ there exists an infinite h-Sidon sequence $\mathcal{A}$ with*

$$A(n) = n^{\sqrt{(h-1)^2+1}-(h-1)+o(1)},$$

where $A(n)$ counts the number of elements of $\mathcal{A}$ not exceeding n.

For $h = 2$ Cilleruelo provides an explicit construction. We remark also that, by a "random construction" of Erdős and Rényi [305], for any $\delta > 0$, there exists an infinite sequence $Q := (a_1, \ldots, a_n, \ldots)$ for which the number of solutions of $a_i + a_j = h$ is bounded, for all h, and $a_k = O(k^{2+\delta})$.

4.7 The Heilbronn Problem, Old Results

Problem 4.14 (*Heilbronn's problem on the area of small triangles*) Consider n points in the unit square (or in the unit disk) no three of which are collinear. What is the maximum of the minimum area of triangles, defined by these points where the maximum is taken for all n-element point-sets?[69]

This maximum of the minimum will be denoted by Δ_n. Erdős gave a simple construction where this minimum area was at least $\frac{1}{2n^2}$: for a prime $p \approx n$, consider all the points $(\frac{1}{p}(i, f(i))$, where $f(i) = i^2 \pmod{p}$. So

$$\frac{1}{2n^2} < \Delta_n \leq \frac{1}{n-1}.$$

Heilbronn conjectured that $\Delta = O(\frac{1}{n^2})$. This was disproved by

Theorem 4.15 (Komlós, Pintz, and Szemerédi (1981), [535]) $\Delta_n = \Omega(\frac{\log n}{n^2})$: *For some constant $c > 0$, for infinitely many n, there exist n points in the unit square for which the minimum area is at least $c\frac{\log n}{n^2}$.*

The proof of Theorem 4.15 is based on Theorem 4.3.

4.8 Generalizations of Heilbronn's Problem, New Results

Péter Hajnal and Szemerédi [448] used the Duke-Lefmann-Rödl lower bound (Theorem 4.4) to prove two new geometrical results. The first one [448] is closely related to Heilbronn's triangle problem, discussed in [535, 678, 680, 682, 683, 734].

---

[69] If three of them are collinear that provides 0.

Consider an n-element point set $\mathbb{P} \subset \mathbb{E}^2$. Instead of triangles we can take k-tuples from $\mathbb{P}$ and consider the area of the convex hull of the k chosen points. Denote the minimum area by $H_k(\mathbb{P})$, its maximum for the n-element sets $\mathbb{P}$ by $H_k(n)$. So $\Delta_n = H_3(n)$. The best lower bound on $H_3(n)$ from [536], and some trivial observations are summarized in the next line: there exists a constant $c > 0$ such that

$$c \frac{\sqrt{\log n}}{n^2} \leq \Delta_n = H_3(n) \leq H_4(n) \leq H_5(n) \leq \ldots = O\left(\frac{1}{n}\right).$$

We mention two major open problems:

Problem 4.16 Is it true that for any $\varepsilon > 0$, $H_3(n) = O(1/n^{2-\varepsilon})$?
Is it true that $H_4(n) = o(1/n)$?

One is also interested in finding good lower bounds on $H_4(n)$. Schmidt [734] proved that $H_4(n) = \Omega(n^{-3/2})$. The proof is a construction of a point set $\mathbb{P}$ by a simple greedy algorithm. In [106], Bertram-Kretzberg, Hofmeister, and Lefmann provided a new proof, and extensions of this result. They also asked whether Schmidt's bound can be improved by a logarithmic factor. Using the semi-random method, Péter Hajnal and Szemerédi improved Schmidt's bound and settled the problem of [106].

Theorem 4.17 (P. Hajnal and E. Szemerédi [448]) *For some appropriate constant $c > 0$, for any $n > 3$, there exist n points in the unit square for which the convex hull of any 4 points has area at least $cn^{-3/2}(\log n)^{1/2}$.*

4.9 The Heilbronn Problem, an Upper Bound

The first upper bound was Roth's fundamental result that $\Delta_n \ll \frac{1}{n\sqrt{\log \log n}}$.[70] Schmidt [734] improved this to $\Delta_n \ll \frac{1}{n\sqrt{\log n}}$. Roth returned to the problem and improved the earlier results to $\Delta_n \ll 1/n^{1.117}$. Not only his bound was much better, his method was also groundbreaking. He combined methods from analysis, geometry, and functional analysis. On these results see the survey of Roth [683]. Roth's result was improved by Pintz, Komlós, and Szemerédi:

Theorem 4.18 (Pintz, Komlós, Szemerédi [535]) $\Delta_n \ll e^{c\sqrt{\log n}} n^{-8/7}$.

---

[70] Here $f \ll g$ is the same as $f \leq cg$, for some absolute constant $c > 0$.

4.10 The Gowers Problem

P. Hajnal and E. Szemerédi considered the following related but "projective" question[71]:

Problem 4.19 (*Gowers* [407]) Given a planar point set $\mathbb{P}$, what is the minimum size of $\mathbb{P}$ that guarantees that one can find n collinear points (points on a line) or n independent points (no three on a line) in $\mathbb{P}$?

Gowers noted that in the grid at least $\Omega(n^2)$ points are needed, and if we have $2n^3$ points without n points on a line, then a simple greedy algorithm finds n independent points. Payne and Wood [643] improved the upper bound $O(n^3)$ to $O(n^2 \log n)$. They considered an arbitrary point set with much fewer than n^3 points, and without n points on a line, but they replaced the greedy algorithm by a Spencer lemma, based on a simple probabilistic sparsification.[72]

Péter Hajnal and Szemerédi improved the Payne-Wood upper bounds by improving the methods. They also started with a random sparsification, but after some additional preparation (getting rid of 2-cycles) they could use the semi-random method (see [248]) to find a large independent set.

Theorem 4.20 (Hajnal and Szemerédi [448]) *There exists a constant $C > 0$ such that in any planar point set $\mathbb{P}$ of size $C \cdot \frac{n^2 \log n}{\log \log n}$, there are n points that are collinear or independent.*

4.11 Pippenger–Spencer Theorem

In the theory developing around the semi-random methods, one should mention the papers of Rödl [684], and of Frankl and Rödl [357], ...

One important step was the Pippenger–Spencer result [652], asserting that if the degrees in a k-uniform hypergraph are large and the CODEGREES are relatively small, then the hypergraph has an almost-1-factor.

Definition 4.21 A MATCHING of a hypergraph $\mathcal{H}$ is a collection of pairwise disjoint hyperedges. The CHROMATIC INDEX $\chi'(\mathcal{H})$ of $\mathcal{H}$ is the least number of matchings whose union covers the edge set of $\mathcal{H}$. The CODEGREE $\delta_\ell(X)$ is the number of hyperedges containing the ℓ-tuple $X \subseteq V(\mathcal{H})$, and $\delta_\ell(\mathcal{H})$ is the minimum of this, taken over all the ℓ-tuples of vertices in $\mathcal{H}$.

We formulate the theorem, in a slightly simplified form.

---

[71] Actually, Hajnal and Szemerédi found this problem on Gowers' homepage, but, as it turned out, from [643], originally the problem was asked by Paul Erdős, [285].

[72] Actually, above we spoke about the "diagonal case" but [643] covers some off-diagonal cases too.

Theorem 4.22 (Pippenger and Spencer [652]) *For every $k \geq 2$ and $\delta > 0$ there exists a $\delta' > 0$ and an n_0 such that if $\mathcal{H}_n$ is a k-uniform hypergraph with $n > n_0$ vertices, and*

$$d_{\min}(\mathcal{H}_n) > (1 - \delta) d_{\max}(\mathcal{H}_n),$$

and the codegrees are small:

$$\delta_2(\mathcal{H}_n) < \delta' d_{\min}(\mathcal{H}_n)$$

then the chromatic index

$$\chi'(\mathcal{H}_n) \leq (1 + \delta) d_{\max}(\mathcal{H}_n).$$

The meaning of the conclusion is that the set of hyperedges can be partitioned into packings (or matchings), almost all of which are almost perfect. Also the edges can be partitioned into coverings, almost all of which are almost perfect. This theorem strengthens and generalizes a result of Frankl and Rödl [357].

4.12 Erdős–Faber–Lovász Conjecture

We close this part with the beautiful result of Jeff Kahn on the famous Erdős–Faber–Lovász conjecture. Faber writes in [326][73]:

"In 1972, Paul Erdős, László Lovász and I got together at a tea party in my apartment in Boulder, Colorado. This was a continuation of the discussions we had had a few weeks before in Columbus, Ohio, at a conference on hypergraphs. We talked about various conjectures for linear hypergraphs analogous to Vizing's theorem for graphs (see [327]). Finding tight bounds in general seemed difficult so we created an elementary conjecture that we thought would be easy to prove. We called this the *n* sets problem: given *n* sets, no two of which meet more than once and each with *n* elements, color the elements with *n* colors so that each set contains all the colors. In fact, we agreed to meet the next day to write down the solution. Thirty-eight years later, this problem is still unsolved in general." (See [676] for a survey of what is known.)

The original conjecture says:

Conjecture 4.23 *If G_n is the union of k complete graphs K_k, any two of which has at most one common vertex, then $\chi(G_n) \leq k$.*

As we stated, originally the conjecture was formulated using Linear Hypergraphs.[74] A weakened asymptotic form of this was proved by Jeff Kahn:

---

[73] In citations we use our numbers, not the original ones.
[74] Perhaps the expression "linear hypergraph" was unknown those days.

Theorem 4.24 (Jeff Kahn (1991), [482]) *If $A_1, \ldots, A_k \subseteq [n]$ are nearly disjoint, in the sense that $|A_i \cap A_j| \leq 1$ for all pairs $i \neq j$, then $\chi'(H) \leq (1 + o(1))n$.*

Jeff Kahn gave an elegant proof of this assertion, and also a clear description of the sketch of his proof, which is also a nice description of the Semi-Random method. The proof is based on a technical generalization of the Pippenger–Spencer Theorem [652].

Remark 4.25 (a) Originally the Erdős–Faber–Lovász conjecture had a slightly different form, see above, or, e.g., Erdős [279]. Jeff Kahn refers to Hindman [464] who rephrased it in this form. Kahn also remarks that Komlós suggested to prove an asymptotic weakening.

(b) We also remark that the fractional form of this problem was solved by Kahn and Seymour [483].

(c) Vance Faber proved [326] that if there are some counter-examples to the Erdős–Faber–Lovász conjecture, they should be in some sense in the "middle range", according to their densities.

Remark 4.26 (On Keevash' existence-proof of Block designs) One of the recent results that is considered a very important breakthrough is that of Peter Keevash [500, 501], according to which, if we fix some parameters for some block-designs, and the corresponding trivial divisibility conditions are also satisfied, then the corresponding block designs do exist, assumed that the set of elements is sufficiently large. Important but simpler results in the field were obtained by Richard Wilson [815–818], Vojta Rödl [684], using—among others—the methods described above, above all, the Rödl Nibble. The excellent paper of Gil Kalai [484] explains this area: what one tries to prove and how the semi-random methods are used. As a very important contribution, approach, see also the papers of Glock, Kühn, Lo, and Osthus [396, 397]. We do not go into details but again, refer the interested readers to the paper of Kalai written for the general audience, or, at least, for most of the combinatorics.

5 Methods: Regularity Lemma for Graphs

As we have already stated, Regularity Lemma is applicable in many areas of Discrete Mathematics. A weaker, "bipartite" version was used in the proof of Szemerédi Theorem [791] according to which $r_k(n) = o(n)$. Also weaker versions were used in the first applications in Graph Theory [711, 790]. The first case when its standard form (Theorem 5.3) was needed was the Chvátal–Szemerédi theorem [188] on the parametrized form of the Erdős–Stone theorem.[75] The Regularity Lemma is so successful, perhaps because of the following.

---

[75]The question was that if $e(G_n) = e(T_{n,p}) + \varepsilon n^2$, how large $K_{p+1}(m, \ldots, m)$ can be guaranteed in G_n? This maximum $m = m(p, \varepsilon)$ had a very weak estimate in [321]. This was improved to $c(p, \varepsilon) \log n$ by Bollobás and Erdős [124], which was improved by Bollobás, Erdős, and Simonovits [127]. Chvátal and Szemerédi needed the Regularity Lemma to get the "final" result, sharp up to a multiplicative absolute constant.

To embed a graph H into G is much easier if G is a random graph than if it is an arbitrary graph. The Regularity Lemma asserts that every graph G can be approximated by a "generalized random graph", more precisely, by a "generalized quasi-random graph". But then we may embed H into an almost random graph which is much easier. (Also, many other things are easier for Random Graphs.)

Lovász and Szegedy [595] wrote a beautiful paper on the Regularity Lemma, where they wrote[76]:

"Roughly speaking, the Szemerédi Regularity Lemma says that the node set of every (large) graph can be partitioned into a small number of parts so that the subgraphs between the parts are random-like. There are several ways to make this precise, some equivalent to the original version, some not...."

To formulate the Regularity Lemma, we define (i) the ε-regular pairs of vertex-sets in a graph, (ii) the generalized random graphs, and (iii) the generalized quasi-random graphs.

Given a graph G, with the disjoint vertex sets $X, Y \subseteq V(G)$, the edge-density between X and Y is

$$d(X, Y) := \frac{e(X, Y)}{|X||Y|}.$$

Regular pairs are highly uniform bipartite graphs, namely ones in which the density of any "large" induced subgraph is about the same as the overall density of the whole graph.

Definition 5.1 (*Regular pairs*) Let $\varepsilon > 0$. Given a graph G and two disjoint vertex sets $A \subseteq V(G)$, $B \subseteq V(G)$, we say that the pair (A, B) is ε-regular if for every $X \subseteq A$ and $Y \subseteq B$ satisfying

$$|X| > \varepsilon |A| \quad \text{and} \quad |Y| > \varepsilon |B|$$

we have

$$|d(X, Y) - d(A, B)| < \varepsilon.^{77} \tag{16}$$

We can also describe the Regularity Lemma as a statement asserting that any graph G_n can be approximated well by generalized random graphs. However, first we have to define the generalized Erdős–Rényi random graph, then the generalized quasi-random graph sequences.

Definition 5.2 (*Generalized Random Graphs*) Given a matrix of probabilities, $A := (p_{ij})_{r \times r}$ and integers $n_1, \ldots, n_r$. We choose the subsets $U_1, \ldots, U_r$ with $|U_i| = n_i$ and join $x \in U_i$ to $y \in U_j$ with probability p_{ij}, independently.

The GENERALIZED QUASI-RANDOM graphs are obtained when we also fix an $\varepsilon > 0$ and instead of joining U_i to U_j using random independent edges, we join them with ε-regular bipartite graphs of the given density p_{ij}.

---

[76] This was formulated by many researchers.
[77] In random graphs this holds for sufficiently large disjoint vertex sets.

Regularity Lemma asserts that the large graphs can be approximated by generalized quasi-random graphs well.

Theorem 5.3 (Szemerédi (1978), [792]) *For every $\varepsilon > 0$ and M_0 there are two constants $M(\varepsilon, M_0)$ and $N(\varepsilon, M_0)$ with the following property: for every graph G_n with $n \geq N(\varepsilon, M_0)$ vertices there is a partition of the vertex set into ν classes*

$$V = V_1 \cup V_2 \cup \ldots \cup V_\nu$$

such that

- $M_0 \leq \nu \leq M(\varepsilon, M_0)$,
- $||V_i| - |V_j|| \leq 1$, $(1 \leq i < j \leq \nu)$
- *all but at most $\varepsilon \nu^2$ of the pairs (V_i, V_j) are ε-regular.*

The lower bound on the number of classes is needed mostly to make the number of edges within the clusters small. If e.g., $M_0 > 1/\varepsilon$, then $\sum e(V_i) < \frac{1}{2}\varepsilon n^2$. So mostly we can choose $M_0 = 1/\varepsilon$. We do not really need that $||V_i| - |V_j|| \leq 1$, however, we do need that the V_i's are not too large. In the applications we very often use the *Reduced graph* or *Cluster graph* H_ν defined as follows[78]:

Definition 5.4 (*Cluster Graph $H_\nu = H_\nu(G_n)$*) Given a graph G_n, and two constants $\varepsilon, \tau > 0$, the corresponding Cluster Graphs are obtained as follows. We apply the Regularity Lemma with ε, obtaining the partition $V_1, \ldots, V_\nu$. The vertices of H_ν are the vertex-sets V_i (called Clusters) and we connect the Cluster V_i to V_j if (V_i, V_j) is ε-regular in G_n and $d(V_i, V_j) > \tau$.

Remark 5.5 Mostly we use the cluster graph as follows: (i) We set out with a graph sequence (G_n), satisfying some conditions $\mathcal{P}$, (ii) take the cluster graphs H_ν, (iii) derive that (H_ν) must have some properties $\mathcal{P}^*$ (because of the combinatorial conditions on G_n), (iv) therefore we can apply some "classical" graph theorem to H_ν, (v) this helps to describe H_ν[79] and (vi) having this information on H_ν helps us to prove what we wanted.

...Or, in case (v-vi), instead of, say, estimating $e(G_n)$, we embed some given graph U_μ into H_ν, and using this, we embed a (much larger) W_m into G_n.

Remark 5.6 Given a graph sequence (G_n), it may happen that we have very different regular partitions, e.g., in one of them the densities are around 0 and 1, in the other they are around $\frac{1}{2}$.

The natural question if (ε, τ) and G_n determine H_ν is considered in the paper of Alon, Shapira, and Stav [46]. The answer is "No, but under some conditions YES". However, here the reader should be cautions, we have not defined when do we consider two regular partitions near to each other.

---

[78] Perhaps the name "Reduced Graph" comes from Simonovits, the "Cluster Graph" from Komlós, and the theorem itself was originally called "Uniformity Lemma": the name "Regularity Lemma" became popular only later.

[79] Estimate $e(H_\nu)$ or prove some structural property of H_ν.

So the cluster graph is not determined by these parameters, (neither the ε-regular partition), and τ is much larger than ε, however, in the applications $\tau \to 0$ as $\varepsilon \to 0$. (If e.g., we try to embed a relatively small graph (?) with bounded degree Δ, then $\tau \approx \varepsilon^{1/(2\Delta)}$ is mostly enough for our proofs, see also, e.g., Sárközy, [725].).

There are many surveys on Regularity Lemmas and its applications, e.g., Komlós and Simonovits [546], Komlós, Shokoufandeh, Simonovits, and Szemerédi [545]; the survey of Komlós [533] (formally on the Blow-up Lemma) is also very informative, and we also recommend a more recent survey of Szemerédi [795].

There are several results on sparse graphs, see e.g., surveys of Kohayakawa and Rödl [528] or of Gerke and Steger on the Sparse Regularity Lemma [392].

There are several works on some other aspects of the Regularity Lemma (or Regularity Lemmas), e.g., whether one needs exceptional classes, or how many clusters are needed (see e.g., Gowers [401]), or how can it be viewed from other points of views, e.g., Tao [798], however, we skip most of them.

Regularity Lemma and Parameters. The Regularity Lemma is very inefficient in the sense that it works only for very-very large graphs.

There are three natural questions concerning this: (a) do we need the exceptional cluster-pairs, and (b) how large the threshold $n_0(M_0, \varepsilon)$ must be, (c) how many clusters are needed. The answer to the first question was given by the "half-graph" where one must use exceptional pairs. It is not quite clear how many exceptional pairs are needed in general.

Several results are known according to which if we fix $\varepsilon > 0$ and M_0, then we have to use many-many clusters. The first such result was due to Gowers [401]. (See also Moshkovitz and Shapira, [621].) For sharper results we refer the reader to Fox, László Miklós Lovász, and Yufei Zhao [350], ...

5.1 Ramsey Problems, Cycles

As to the Ramsey Theory, we shall not introduce the standard notation here. For an excellent source, see the book of Graham, Rothschild, and Spencer [410].

The Ramsey problems were extended to arbitrary graphs first by Gerencsér and Gyárfás [390]. Several results were proved in this area by Faudree and Schelp [333–335], (and some parallel to Faudree and Schelp, by Vera Rosta [677]) and more generally, by Erdős, Faudree, Rousseau, and Schelp, and others. Bondy and Erdős [135] formulated several important conjectures for the case when the excluded graphs are paths, or cycles. In case of cycles, it turned out that the parity of the length of cycles is also very important. Again, we shall mention only a few related results, and then mention a few papers: this area is too large to describe it here in more detail.

We have to start with the remark that in most cases considered below the (conjectured) extremal structures come from some "matrix-graph-sequence": n vertices of a K_N are partitioned into a few classes and then we colour $E(K_N)$ so that the colours depend only on the classes of the endvertices. These structures are very simple and nice, not chaotic/random like at all, so the proofs are also similar to the proofs of some extremal problems. Often we have some stability of the Ramsey-extremal structures (Fig. 6).

One of the many questions Bondy and Erdős [135] asked was whether for an odd $n \geq 3$, is it true that $R_r(C_n, C_n, \ldots, C_n) = 2^{r-1}(n-1) + 1$? The construction suggesting/motivating this conjecture is recursive: for two colours we take two BLUE complete K_{n-1} and join them completely by GREEN edges. If we have already constructions on $r - 1$ colours, take two such constructions, and a new colour, say, RED, and join the two constructions completely with this new colour. (Watch out, if we use different sets of colours, there are other, similar but more complicated constructions, colour-connections between them. As r increases, the number of non-equivalent constructions increases. For $r = 3$ we have two similar, yet different constructions.)

Conjecture 5.7 (Bondy and Erdős) *Let n be odd. If we r-colour K_N for $N = 2^{r-1}(n-1) + 1$, then we shall have a monochromatic C_n.*

The following approximation of the Bondy–Erdős conjecture, for three colours, was a breakthrough in this area.

Theorem 5.8 (Łuczak (1999), [600]) *If n is odd, then*

$$R(C_n, C_n, C_n) = 4n + o(n).$$

The proof was highly non-trivial. Applying stability methods, the $o(n)$ was eliminated:

Theorem 5.9 (Kohayakawa, Simonovits, and Skokan [530]) *If n is a sufficiently large odd integer, then*[80]

$$R(C_n, C_n, C_n) = 4n - 3.$$

Recently Jenssen and Skokan proved the corresponding general conjecture, [479], for odd cycles, at least if n is large. They use both the Regularity Lemma and the Stability method, adding some non-linear optimization tools to the usual methods, too.

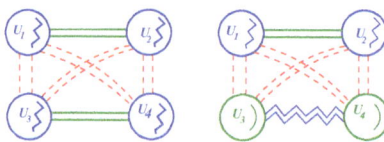

Fig. 6 Cycle-Ramsey with three colours

---

[80] Watch out, mostly it does not matter, but here, in case of sharp Ramsey results one has to distinguish the lower and upper Ramsey numbers. The upper one is the smallest one for which there is no good colouring, here $4k - 3$. The lower Ramsey number is $R(L_1, \ldots, L_r) - 1$.

The situation with the paths and the even cycles is different. We could say that the reason is that the above colouring contains many monochromatic even cycles. This is why for three colours the Ramsey number is half of the previous one. Figaj and Łuczak proved

Theorem 5.10 (Figaj, Łuczak [341]) *If $\alpha_1 \geq \alpha_2, \alpha_3 > 0$, then for $m_i := 2\lfloor \alpha_i n \rfloor$ ($i = 1, 2, 3$)*
$$R(C_{m_1}, C_{m_2}, C_{m_3}) = (2\alpha_1 + \alpha_2 + \alpha_3)n + o(n).$$

Corollary 5.11 *If $m_1 \geq m_2 \geq m_3$ are even, then*
$$R(P_{m_1}, P_{m_2}, P_{m_3}) = m_1 + \frac{1}{2}(m_2 + m_3) + o(m_1).$$

In a recent paper Figaj and Łuczak [342] determined the asymptotic value of $R(C_\ell, C_m, C_n)$, when the parities are arbitrary and the asymptotic values of $\ell/n = \alpha$ and $m/n = \beta$ are given, as $n \to \infty$. We mention here just the following theorem, that proves a conjecture of Faudree and Schelp.

Theorem 5.12 (Gyárfás, Ruszinkó, Sárközy, and Szemerédi [432]) *There exists an n_0 such that if $n > n_0$ is even then $R(P_n, P_n, P_n) = 2n - 2$, if n is odd then $R(P_n, P_n, P_n) = 2n - 1$.*

The results listed above use the Regularity Lemma, and one thinks they could be proved without it, too. The difference between paths and even cycles is not that large:

Theorem 5.13 (Benevides and Skokan (2009), [99]) *If n is even, and sufficiently large, then $R(C_n, C_n, C_n) = 2n$.*

These proofs use (among others) the Coloured Regularity Lemma. Hence all the assertions not using $o(..)$ are proved only for very large values of n. This area is again large and ramified. We shall return to some corresponding Ramsey hypergraph results in Sect. 9.1.

5.2 Ramsey Theory, General Case

Burr and Erdős conjectured [164] that the 2-colour Ramsey number $R(H_n, H_n)$ is linear in n for bounded degree graphs.[81] First some weaker bound was found by József Beck, but then the conjecture was proved by

Theorem 5.14 (Chvátal–Rödl–Szemerédi–Trotter (1983), [187]) *For any $\Delta > 0$ there exists a constant $\Gamma = \Gamma(\Delta)$ such that for any H_n with $d_{\max}(H_n) < \Delta$, we have $R(H_n, H_n) < \Gamma n$.*

---

[81] Here by linear we mean $O(n)$.

The proof was based on the Regularity Lemma, applied to a graph G_n defined by the RED edges of a RED-BLUE-coloured K_N. Then Turán's theorem and Ramsey's theorem were applied to the Coloured Cluster Graph H_ν. The RED-BLUE colouring of K_N defines a RED-BLUE colouring of H_ν: the cluster-edge $V_i V_j$ gets the colour which has the majority. Then the proof is completed by building up a RED copy of H_n in the RED-BLUE K_N, if H_ν contained a sufficiently large RED complete subgraph. The next, stronger conjecture is still open.

Conjecture 5.15 (Burr–Erdős) *If condition $d_{\max}(H_n) < \Delta$ is replaced by the weaker condition that for any $H^* \subset H_n$ we have $d_{\min}(H^*) < \Delta$, then $R(H_n, H_n) = O(n)$.*

Remark 5.16 The actual bound of [187] on the multiplicative constant in Theorem 5.14 was fairly weak. This was improved in several steps, first by Eaton [250], then by Graham, Rödl, and Ruciński [411], and finally by Conlon [192], and Fox and Sudakov [352].

5.3 Ramsey–Turán Theory

When Turán, around 1969 [810], started a series of applications of his graph theorem in other fields of Mathematics, in some sense it turned out that $T_{n,p}$ is too regular, has too simple structure: from the point of view of applications perhaps it would be better to have a version of his theorem for graphs with less regular, more complicated structure, but providing better estimates.[82] This led Vera Sós, in [773] to ask more general questions than the original Turán question, and that led her to the Turán-Ramsey theory. One of her questions was

Problem 5.17 (*Sós* (1969), [773]) Fix r sample graphs $L_1, \ldots, L_r$. What is the maximum of $e(G_n)$ if the edges of G_n can be r-coloured so that the ith colour does not contain L_i.

The answer was given by

Theorem 5.18 (Burr, Erdős, and Lovász (1976), [167]) *For a family $\{L_1, \ldots, L_r\}$ of fixed graphs, let R be the smallest integer for which there exists an m such that if $T_{m,R}$ is arbitrarily r-coloured, then for some i the ith colour contains an L_i. Then*

$$\mathbf{RT}(n; L_1, \ldots, L_r) = \mathbf{ex}(n, K_{R-1}) + o(n^2).$$

Actually, this easily follows from the Erdős–Stone theorem. (Further, obviously, we can choose $m = \sum v(L_i)$.) Later it turned out that the really interesting problem is to determine $\mathbf{RT}(n; L_1, \ldots, L_{r-1}, o(n))$, which in the simplest case $r = 2$ means the following.[83]

---

[82] Actually, Turán's corresponding results, or the Erdős–Sós-type Ramsey–Turán theorems were not used in "applications", however the Ajtai–Komlós–Szemerédi-type results are also in this category and, as we have seen in Sects. 4.2–4.10, they were used in several beautiful and important results.

[83] One has to be cautious with this notation, when we write $o(n)$ instead of a function $f(n)$.

Problem 5.19 (*Erdős–Sós* (1970), [319]) Consider a graph sequence (G_n). Estimate $e(G_n)$ if $L \not\subset G_n$, and the independence number, $\alpha(G_n) = o(n)$.

The case of odd complete graphs was solved by

Theorem 5.20 (Erdős and Sós (1970), [319]) *Let (G_n) be a graph sequence. If $K_{2\ell+1} \not\subset G_n$ and $\alpha(G_n) = o(n)$, then*

$$e(G_n) \leq \left(1 - \frac{1}{\ell}\right)\binom{n}{2} + o(n^2),$$

and this is sharp.

Here the sharpness, i.e. the lower bound is easy: let $m = \lfloor n/\ell \rfloor$ and embed into each class of a $T_{n,\ell}$ a graph G_m not containing K_3, with $\alpha(G_m) = o(m)$. By the "probabilistic constructions" of Erdős [264] we know the existence of such graphs.[84] The obtained graph S_n provides the lower bound in Theorem 5.20: it does not contain $K_{2\ell+1}$ and $\alpha(S_n) = o(n)$.

The problem of estimating $\mathbf{RT}(n, K_4, o(n))$ turned out to be much more difficult and this was among the first ones solved (basically!) by the Regularity Lemma, where the upper bound was given by Szemerédi [790] and the sharpness of this upper bound was proved by Bollobás and Erdős [126]. It turned out that

Theorem 5.21 (Szemerédi (1972), [790], Bollobás and Erdős (1976), [126])

$$\mathbf{RT}(n, K_4, o(n)) = \left(\frac{1}{8} + o(1)\right)n^2.$$

The proof of the upper bound used (a simpler (?) form of) the Regularity Lemma. The first graph theoretical successes of the Regularity Lemma were on the $f(n, 6, 3)$-problem, see Theorem 5.26, and the Ramsey–Turán theory: Theorem 5.21, and its generalization by Erdős, Hajnal, Sós, and Szemerédi [295], see also [292], and [294] by Erdős, Hajnal, Simonovits, Sós, and Szemerédi. Since those days this area has gone through a very fast development, Simonovits and Sós have a longer survey on Ramsey–Turán problems [767]. However, even since the publication of [767], many beautiful new results have been proved, e.g., by Mubayi and Sós [626], Schelp [733], Fox, Loh, Zhao [349], Sudakov [786], Balogh and Lenz [70, 71], and many further ones.

In the proof of the upper bound of Theorem 5.21 one needed something more than what was used in the earlier arguments: Until this point in most applications of the Regularity Lemmas the densities were near to 0 or near to 1. Here the densities in the Regular Partitions turned out to be near 0 or $\frac{1}{2}$. Fixing the appropriate ε and $\tau = \sqrt[4]{\varepsilon}$, one important step was that the Cluster Graph H_ν was triangle-free: $K_3 \not\subset H_\nu$, and

---

[84] These are the graphs we considered in connection with $R(n, 3)$ in Sect. 2.16. There are many such graphs obtained by various, more involved constructions.

another was that in the Regular Partition, $d(U_i, U_j) \leq \frac{1}{2} + o(1)$. This implies the upper bound in Theorem 5.21, namely

$$e(G_n) \leq \mathbf{ex}(\nu, K_3) \cdot \left(\frac{1}{2} + o(1)\right) \left(\frac{n}{\nu}\right)^2 + \text{negligible terms} = \left(\frac{1}{8} + o(1)\right) n^2.$$

As to the lower bound in Theorem 5.21, the Erdős-Bollobás construction used a high dimensional isoperimetric inequality, similar to a related construction of Erdős and Rogers [309]. One remaining important open question is

Problem 5.22 (*Erdős*) Is $\mathbf{RT}(n, K(2,2,2), o(n)) = o(n^2)$ or not?

Rödl proved that

Theorem 5.23 (Rödl [685]) *There exist a graph sequence* (G_n) *with* $\alpha(G_n) = o(n)$ *and* $K_4 \not\subset G_n$, $K(3,3,3) \not\subset G_n$ *for which* $e(G_n) \geq \frac{1}{8}n^2 + o(n^2)$.[85]

Phase transition. It can happen that $f(n)$ is a "small" function, much smaller than "just" $o(n)$, say $f(n) = o(\sqrt{n})$, and then we see kind of a phase transition: one can prove better estimates on $\mathbf{RT}(n, L, f(n))$. In other words, sometimes when $f(n)$ is replaced by a slightly smaller $g(n)$, the Ramsey–Turán function noticeably drops. Such result can be found, e.g., in Sudakov [786], in Balogh, Hu, and Simonovits [69], or in Bennett and Dudek [100],...Bennett and Dudek also list several related results and papers. Some roots of this phenomenon go back to [293, 295].

We close this part with a two remarks on PHASE TRANSITION results. Perhaps the simplest approach is to investigate $RT(L_1, \ldots, L_r, f(n))$, when we know of $f(n)$ only that $f(n) = o(n)$, however, $f(n)/n \to 0$ can be arbitrary slow. There are two directions from here:

(a) when we know of $f(n)$ that $f(n)/n \to 0$ relatively fast, so fast that $\lim \frac{1}{n^2} RT(L_1, \ldots, L_r, f(n))$ becomes smaller than for a slightly larger $g(n)$.

(b) The other direction is when we investigate the δ-dependence of

$$\liminf \frac{1}{n^2} RT(L_1, \ldots, L_r, \delta n).$$

There are some very new interesting results in both areas, we refer the reader to the above mentioned [786] and [69], and also to papers of Lüders and Reiher [605] and of Kim, Kim, and Liu [516] settling several phase-transition problems earlier unsolved. They determined $\mathbf{RT}(n, K_3, K_s, \delta n)$ for $s = 3, 4, 5$ and sufficiently small δ, confirming—among others—a conjecture of Erdős and Sós from 1979, and settling some conjectures of Balogh, Hu, and Simonovits, according to which $\mathbf{RT}(n, K_8, o(\sqrt{n \log n})) = \frac{n^2}{4} + o(n^2)$.

---

[85] Actually, Rödl proved a slightly stronger theorem, answering a question of Erdős, but the original one, Problem 5.22, is still open.

5.4 Ruzsa–Szemerédi Theorem, Removal Lemma

We start with a slightly simplified version of a result of G. Dirac [244], a generalization of Turán's theorem:

Theorem 5.24 (Dirac (1963), [244]) *Let $p \geq 2$, $q \in [1, p]$ and $n \geq p + q$ be integers. If G_n is a graph with $e(G_n) > e(T_{n,p})$ edges, then G_n contains a subgraph of $p + q$ vertices and $\binom{p+q}{2} - (q - 1)$ edges.*

Some related results were also independently obtained by Erdős, see [286], and also [267]. Erdős asked the following problem.

Problem 5.25 ([267]) Let $\mathcal{L}_{k,\ell}$ be the family of graphs L with k vertices and ℓ edges. Determine (or estimate) $f(n, k, \ell) := \mathbf{ex}(n, \mathcal{L}_{k,\ell})$.

Several results were proved on this problem by Erdős [267], Simonovits [752], Griggs, Simonovits, and Thomas [421], Simonovits [762] and others. Extending these types of results to 3-uniform hypergraphs in [154] and then to r-uniform hypergraphs in [155], Brown, Erdős and Sós started a systematic investigation of $f_r(n, k, \ell)$ defined as the maximum number of hyperedges an r-uniform n-vertex hypergraph $\mathcal{H}_n^{(r)}$ can have without containing some k-vertex subgraphs with at least ℓ hyperedges.

Below mostly we shall restrict ourselves to the case of 3-uniform hypergraphs, i.e. $r = 3$. Several subcases were solved in [154], but the problem of $f(n, 6, 3)$, i.e. the case when no 6 vertices contain 3 triplets, turned out to be very difficult. One can easily show that $f(n, 6, 3) \leq \frac{1}{6}n^2$. Ruzsa and Szemerédi proved that

Theorem 5.26 (Ruzsa–Szemerédi (1978), [711])

$$cn \cdot r_3(n) < f(n, 6, 3) = o(n^2). \tag{17}$$

The crucial tool was a consequence of the Regularity Lemma:

Theorem 5.27 (Ruzsa–Szemerédi Triangle Removal Lemma) *If (G_n) is a graph sequence with $o(n^3)$ triangles, then we can delete $o(n^2)$ edges from G_n to get a triangle-free graph.*

Clearly, (17) implies Roth theorem that $r_3(n) = o(n)$. There is a more general form of Theorem 5.27, the so called REMOVAL LEMMA:

Theorem 5.28 (Removal Lemma) *If $v(L) = h$, then for every $\varepsilon > 0$ there is a $\delta > 0$ for which, if a G_n contains at most δn^h copies of L, then one can delete εn^2 edges to destroy all the copies of L in G_n.*

Remark 5.29 (Induced matchings) Let us call some edges of a graph G_n STRONGLY Independent, if there are no edges joining two of them. The Ruzsa–Szemerédi theorem can be formulated also without using hypergraphs. Indeed, for each $x \in V(\mathcal{H})$, the LINK of x, i.e. the pairs uv forming a hyperedge in $\mathcal{H}$ with x, form a so-called

INDUCED MATCHING: not only its edges are independent but they are pairwise STRONGLY INDEPENDENT.[86] So the Ruzsa–Szemerédi theorem has the following alternative form:

Theorem 5.30 (Ruzsa–Szemerédi [711]) *If $E(G_n)$ can be covered by n induced matchings, then $e(G_n) = o(n^2)$.*

FOR SOME FURTHER APPLICATIONS[87] and results connected to Induced Matchings see also the papers of Alon, Moitra, and Sudakov [40], and also of Birk, Linial, and Meshulam [110]. Alon, Moitra, and Sudakov describe several constructions and their applications (in theoretical computer science) connected to the Ruzsa–Szemerédi theorem, more precisely, to dense graphs that can be covered by a given number of induced matchings.

Remark 5.31 Among the original questions of Brown, Erdős, and Sós, the estimate of $f(n, 7, 4)$ is still unsolved. For some special graphs connected to groups, Solymosi [772] has a solution. Actually, very recently Nenadov, Schreiber, and Sudakov [631] extended this result.

Remark 5.32 (a) Though Theorem 5.28 was not explicitly formulated in [711], implicitly it was there. Later it was explicitly formulated, e.g., in the paper of Erdős, Frankl, and Rödl [288] and Füredi [371, 372], and soon it became a central research topic, partly because it is often applicable and partly because—though for ordinary graphs the Removal Lemma easily follows from the Regularity Lemma, for hypergraphs everything is much more involved.

(b) The Ruzsa–Szemerédi theorem was extended by Erdős, Frankl, and Rödl [288] to r-uniform hypergraphs. They proved that

$$f_r(n, 3r - 3, 3) = o(n^2).$$

(c) Actually, $f(n, 6, 3)$ is a fixed function of n, and—though [711] asserts only that $f(n, 6, 3) = o(n^2)$,—it has some better asymptotics and they are very interesting and important, since these results have many applications. J. Fox gave a new, more effective proof of the Removal Lemma [347], not using the Regularity Lemma, leading also to better estimates on $f(n, 6, 3)$, according to which

$$f(n, 6, 3) < \frac{n^2}{2^{\log^* n}}. \tag{18}$$

(d) A more detailed description of this topic can be found in two surveys of Conlon and Fox [194, 195]. We skip all the details connected to "Tower" functions and "Wowzer" functions,[88] (which show that the Regularity-Lemma methods are very

---

[86] Here we assume that the mindegree is at least 3.
[87] Some applications of the Ruzsa–Szemerédi theorem are given in Sect. 5.5.
[88] With the exception of the next theorem.

inefficient). Conlon and Fox discuss the estimates connecting the various parameters, in details.

(e) Some related results and generalizations were proved by Sárközy and Selkow [727, 728], and by Alon and Shapira [45]. The emphasis in [45] is on obtaining a new lower bound to generalize Theorem 5.26.[89]

(f) The Ruzsa–Szemerédi removal lemma was applied also by Balogh and Petříčková [84], to give a sharp estimate on the number of maximal triangle-free graphs G_n.

We close this part with quoting a result of Jacob Fox. Let $TF(t)$ be the "tower" of 2's of height t: $TF(1) = 2$ and $TF(t+1) = TF(2^{TF(t)})$.

Theorem 5.33 (Fox [347]) *Fix a sample graph L_h (on h vertices) and an $\varepsilon > 0$. Let $\delta := 1/TF(\lceil 5h^4 \log 1/\varepsilon \rceil)$. Every G_n with at most δn^h copies of L_h can be made L_h-free by removing at most εn^2 edges.*

Remark 5.34 The surveys of Conlon and Fox, e.g., [195] also discuss the connection of property testing if G_n contains an induced copy of H, and its connection to the "Induced Removal Lemma".

The interested reader is recommended to read [347] or the survey of Conlon and Fox [195].

— · —

Remark 5.35 (a) Mostly we skip the references to graph limits, however, here we should mention the Elek-Szegedy Ultra-product approach to graph limits, and within that to the Removal Lemma, see [251, 252].

(b) T. Tao also has a variant of the Hypergraph removal lemma [799] which he uses to prove a Szemerédi-type theorem on the Gaussian primes [800].

Remark 5.36 (Szemerédi theorem and the Clique-union Lemma) V. Rödl, as an invited speaker of the ICM 2014, in Seoul, spoke about the relations described above. Generalizing the above approach, Peter Frankl and he tried to prove $r_k(n) = o(n)$, using this combinatorial approach. The reader is suggested to look up the movie about his lecture (ICM 2014 Video Series, Aug 21).

Some generalizations. The Ruzsa–Szemerédi theorem has two parts, and both have several important and interesting generalizations, yet, in this paper we mostly skip the results connected to the lower bounds. Several related results can be found in the papers of Erdős, Frankl, and Rödl [288], or e.g., in the paper of Füredi and Ruszinkó on excluding the grids [383], and also in Sárközy and Selkow [728], and Alon and Shapira [45].

Remark 5.37 Solymosi tried to formulate a non-trivial Removal Lemma for bipartite excluded graphs, however Timmons and Verstraëte [806] provided infinitely many "counterexamples".

---

[89] Actually, this assertion is somewhat more involved, see the introduction of [45].

5.5 Applications of Ruzsa–Szemerédi Theorem

As we have mentioned, one of the early successes of a simpler form of the Regularity Lemma was the answer to the question of Brown, Erdős, and Sós, according to which $f(n, 6, 3) = o(n^2)$, (and the proof of the triangle removal lemma, implying this). Fox writes in [347]: "The graph removal lemma has many applications in graph theory, additive combinatorics, discrete geometry, and theoretical computer science." Here we mention two graph theoretical applications, one of which is strongly connected to Turing Machines.

Füredi Theorem on Diameter Critical Graphs

Call a graph G DIAMETER-d-CRITICAL, if it has diameter d and deleting any edge the diameter becomes at least $d + 1$ (or G gets disconnected). Such graphs are, e.g., C_k or $T_{n,p}$. Restrict ourselves to $d = 2$: consider diameter-critical graphs of diameter 2. As Füredi describes in [371], Plesník observed [653] that for all known diameter-2-critical graphs G_n, $e(G_n) \leq \lfloor \frac{n^2}{4} \rfloor$. Independently, Simon and Murty conjectured [170] that

Conjecture 5.38 *If G_n is a diameter-2-critical graph, then $e(G_n) \leq \lfloor \frac{n^2}{4} \rfloor$.*

This seemed to be a beautiful but very difficult conjecture. Füredi (1992), [371] proved it for large n, using the Ruzsa–Szemerédi theorem:

Theorem 5.39 (Füredi (1992), [371]) *There exists an n_0 such that if $n > n_0$, then the Murty-Simon conjecture holds.*

Remark 5.40 In the proof of Theorem 5.39 we get a very large n_0. In some sense Theorem 5.39 settles the conjecture, at least for most of us. Yet a lot of work has been done to prove it for reasonable values of n, too. Plesník proved, instead of $e(G_n) \leq \frac{1}{4}n^2$, that $e(G_n) \leq \frac{3}{8}n^2$, Caccetta and Häggkvist [170] improved this to $e(G_n) \leq 0.27n^2$, and G. Fan [328] proved for $n > 25$ that $e(G_n) \leq 0.2532n^2$.[90]

There are many related results proved and some conjectures on the diameter-d-critical graphs, see e.g., a survey of Haynes, Henning, van der Merwe, and Yeo, [460], or Po-Shen Loh and Jie Ma [580]. This later one disproves a Caccetta-Häggkvist conjecture on the average edge-degree of diameter-critical graphs (and contains some further nice results as well).

Triangle Removal Lemma in Dual Anti-Ramsey Problems

Burr, Erdős, Graham and Sós started investigating the following "dual Anti-Ramsey" problem [166], (see also [165]):

---

[90] Watch out, some of these papers, e.g., [328] are from before Füredi's result, some others are from after it.

Given a sample graph L, n, and E, what is the minimum number of colours $t = \chi_S(n, E, L)$ such that any graph G_n with E edges can be edge-coloured with t colours so that all the copies $L \subset G_n$ are Rainbow coloured (i.e. the edges are coloured without colour repetition)?[91]

They observed that for $L = P_4$ this question can be solved using $f(n, 6, 3) = o(n^2)$. Actually, their solution for $L = P_4$ is reducing the problem to the problem of estimating $f(n, 6, 3)$. For $L = C_5$ (which seemed one of the most interesting cases), they proved:

Theorem 4.1 of [166]. *There exists an n_0 such that if $n > n_0$ and $e = \lfloor \frac{n^2}{4} \rfloor + 1$, then*

$$c_1 n \leq \chi_S(n, e, C_5) \leq \left\lfloor \frac{n}{2} \right\rfloor + 3. \tag{19}$$

Erdős and Simonovits [318], using the Lovász–Simonovits Stability theorem, (see Subsection 2.15) proved that the upper bound (obtained from a simple construction) is sharp.

Theorem 5.41 (Erdős and Simonovits) *There exists a threshold n_0 such that if $n > n_0$, and a graph G_n has $E = \lfloor \frac{n^2}{4} \rfloor + 1$ edges and we colour its edges so that every C_5 is 5-coloured, then we have to use at least $\lfloor \frac{n}{2} \rfloor + 3$ colours.*

To apply the Lovász–Simonovits Stability, they needed the result of [166] on P_4. So, again, the removal lemma was one of the crucial tools. (The application of the Lovász–Simonovits Stability can be replaced here by a second application of the Removal Lemma and the Erdős–Simonovits Stability, however that approach would be less elementary and effective.) They also proved several further results, however, we skip them. Simonovits has also some further results in this area, e.g., on the problem of C_5 when $e(G_n) = \lfloor \frac{n^2}{4} \rfloor + k$, and k is any fixed number, or tending to ∞ very slowly, and also on the problem of C_7. Another conjecture of Burr, Erdős, Graham, and Sós was (almost completely) proved by Sárközy and Selkow [729].

5.6 Hypergraph Removal Lemma?

There are several ways to approach the Removal Lemma and the Hypergraph Removal Lemma. Rödl and Schacht [702] describe a hypergraph generalization of the Removal Lemma. In the nice introduction of [702] they also write

"…the result of Alon and Shapira [44] is a generalization which extends all the previous results of this type where the triangle is replaced by a possibly infinite family $\mathcal{F}$ of graphs and the containment is induced…".

We have promised to avoid some areas that are much more difficult/technical than the others. Unfortunately, it is not easy to decide what is "technical". Komlós and

---

[91] The corresponding extremal value will be denoted by $\chi_S(n, E, L)$. Here S stands for "strong" in χ_S. It is the STRONG CHROMATIC NUMBER of the $v(L)$-uniform hypergraph whose hyperedges are the $v(L)$-sets of vertices of the copies of $L \subset G_n$.

Simonovits [546] tried to provide an easy introduction to the Regularity Lemma. If one knows enough non-discrete mathematics, e.g., compactness, metric spaces, then the Lovász-Szegedy paper [595] is an easy introduction to the area of connections of Regularity Lemma and other, related areas.

The situation with hypergraphs is different. That area seems inherently difficult to "learn". The paper of Rödl, Nagle, Skokan, Schacht, and Kohayakawa [687] and the companion paper of Solymosi [770] helps a lot to bridge this difficulty. The difficulties come from two sources. The first one is that there are useful and not so useful versions of the Hypergraph Regularity Lemma, and the useful ones are difficult to formulate. Layers come in, e.g., for 3-uniform hypergraphs: beside considering the vertices and hyperedges, we have to consider some auxiliary graphs.

In [687] twelve theorems are formulated, the last two are the hypergraph regularity and the hypergraph counting lemmas. It is nice that the hypergraph removal lemma keeps its simple form.

Theorem 5.42 (See Theorem 5 of [687]) *For any fixed integers $\ell \geq k \geq 2$ and $\varepsilon > 0$, there exists a $\xi = \xi(\ell, k, \varepsilon)$ and an $n_0(\ell, k, \varepsilon)$, such that if $\mathbb{F}_\ell^{(k)}$ is a k-uniform hypergraph on ℓ vertices and $\mathbb{H}_n^{(k)}$ in a k-uniform hypergraph of n vertices, with fewer than ξn^ℓ copies of $\mathbb{F}_\ell^{(k)}$, then it may be transformed into an $\mathbb{F}_\ell^{(k)}$-free hypergraph by deleting εn^k hyperedges.*

Solymosi remarked that the appropriate hypergraph removal lemma implies the higher dimensional version of the Szemerédi theorem. We close this part by quoting Solymosi [770]:

"There is a test to decide whether a hypergraph regularity is useful or not. Does it imply the Removal Lemma? If the answer is yes, then it is a correct concept of regularity indeed. On the contrary, applications of the hypergraph regularity could go beyond the Removal Lemma. There are already examples for which the hypergraph regularity method, combined with ergodic theory, analysis, and number theory, are used efficiently to solve difficult problems in mathematics."

The hypergraph removal lemma was approached from several directions. Among others, Tao considered it in [799], Elek and B. Szegedy [251] approached it from the direction of Non-Standard Analysis, Rödl and Schacht from their general hypergraph regularity theory.

5.7 The Universe of Random Graphs

The following result answers a question of Erdős:

Theorem 5.43 (Babai, Simonovits, and Spencer (1990), [56]) *There exists a $p_0 < \frac{1}{2}$ such that a random graph $R_{n,p}$ with edge-probability $p > p_0$ almost surely has the following property $\mathcal{B}_L$, for $L = K_3$: all its triangle-free subgraphs with maximum number of edges are bipartite.*[92]

---

[92] Here "almost surely" means that its probability tends to 1 as $n \to \infty$.

Actually, they proved much more general results. Consider the following assertion, depending on L and p.

$(\mathcal{B}_{L,p})$ *All the L-free subgraphs $F_n \subset R_{n,p}$ having maximum number of edges are $\chi(L)$-chromatic, almost surely, if $n > n_0(L, p)$.*

They proved $(\mathcal{B}_{L,p})$ for all cases when L has a critical edge[93] and p is nearly $\frac{1}{2}$. They also proved several related weaker results in the general case, when L was arbitrary, and $p > 0$. They could not extend their results to sparse graphs, primarily because that time the sparse Regularity Lemma did not exist. Soon it was "invented" by Kohayakawa.[94] The results of Babai, Simonovits, and Spencer were generalized, first by Brightwell, Panagiotou, and Steger [148], and then, in various ways, by others. So the first breakthrough towards sparse graphs was

Theorem 5.44 (Brightwell, Panagiotou, and Steger (2012), [148]) *There exists a constant $c > 0$ for which choosing a random graph $R_{n,p}$ where each edge is taken independently, with probability $p = n^{-(1/2)+\varepsilon}$, the largest triangle-free subgraph F_n of $R_{n,p}$ is bipartite, with probability tending to 1.*

They remarked that the conclusion cannot hold when $p = \frac{1}{10} \frac{\log n}{\sqrt{n}}$, since these $R_{n,p}$ contain, almost surely, an induced C_5 whose edges are not contained in triangles of this $R_{n,p}$. All the edges of $R_{n,p}$ not covered by some $K_3 \subset R_{n,p}$ must belong to F_n. Therefore now $C_5 \subset F_n$: F_n is not bipartite. The proof of [148], similarly to that of the original proof of Babai, Simonovits, and Spencer, uses a stability argument, however, instead of the original Regularity Lemma it uses the Sparse Regularity Lemma. In some sense the "final" result was found by DeMarco and Jeff Kahn [235, 236]. They proved (among others) that

Theorem 5.45 (DeMarco and Kahn (2015), [236]) *For each r there exists a constant $C = C_r > 0$ for which choosing a random graph $R_{n,p}$ where each edge is taken independently, with probability*

$$p > Cn^{-\frac{2}{(r+1)}} \log^{\frac{2}{(r+1)(r-2)}} n,$$

the largest K_r-free subgraph T_n of $R_{n,p}$ is almost surely $r-1$-partite.

A hypergraph analog was proved by Balogh, Butterfield, Hu, and Lenz [66]. (They again used the stability approach.)

5.8 Embedding Large Trees

There are many results where one fixes a sample graph L and tries to embed it into a Random Graph $R_{n,p}$. (See e.g., Erdős–Rényi [306].) Here we try to embed a fixed

---

[93] See Meta-Theorem 2.19.
[94] Rödl also knew it, but it seems that he had not published it.

tree T_m of $m \approx (1-\alpha)n$ vertices into a (random) graph R_n. However, we use a slightly different language.

A relatively new notion of RESILIENCE was introduced by Sudakov and Vu [787]. Fix a graph property $\mathcal{P}$. The resilience of G_n is its "edit" distance[95] to graphs not having property $\mathcal{P}$.[96] Balogh, Csaba, and Samotij [68] proved

Theorem 5.46 *Let α and γ be (small) positive constants and assume that $\Delta \geq 2$. There exists a constant $C > 0$ (depending on α, γ, and Δ) such that for all $p = p(n) \geq C/n$, the local resilience of $R_{n,p}$ with respect to the property of containing all trees T_m of $m := \lfloor (1-\alpha)n \rfloor$ vertices and maximum degree $d_{\max}(T_m) \leq \Delta$ is almost surely greater than $\frac{1}{2} - \gamma$.*

As a subcase, this contains

Theorem 5.47 (Balogh, Csaba, and Samotij [68]) *Let α be a positive constant, and assume that $\Delta \geq 2$. There exists a constant $C > 0$ (depending on α, and Δ) such that for all $p = p(n) \geq C/n$, $R_{n,p}$ contains all trees T_m of $m \leq (1-\alpha)n$ vertices and maximum degree $d_{\max}(T_m) \leq \Delta$, almost surely, as $n \to \infty$.*

The proof uses a sparse Regularity Lemma and a theorem of Penny Haxell [457] on embedding bounded degree trees into "expanding" graphs.

5.9 Extremal Subgraphs of Pseudo-random Graphs

Another direction of research is when the Universe (instead of Random Graphs) consists of more general objects, say of pseudo-random graphs. These are natural directions:

(a) whenever we can prove a result for complete graphs K_p, we have a hope to extend it to any L with critical edges, and

(b) whenever we know something for Random Graphs, there is a chance that it can be extended to random-looking objects (say to quasi-random graphs, or Pseudo-Random graphs, or to expanders graphs.[97])

We mention here a few such papers: Thomason [803, 804], Krivelevich and Sudakov [559] are nice and detailed surveys on Pseudorandom graphs. Aigner-Horev, Hàn, and Schacht [5], and D. Conlon, Fox, and Yufei Zhao [198] also are two more recent nice papers (surveys?) on this topic. We recommend these papers, and also Kühn and Osthus [564]. For hypergraphs see, e.g., Haviland and Thomason, [454, 455], or Kohayakawa, Mota, and Schacht [526].

---

[95]The "edit" distance is the same used in [751]: the minimum number of edges to be changed to get from G_n a graph isomorphic to H_n.

[96]Though we formulate a theorem on the local resilience of graphs for some graph property, we shall not define here the notion of local and global resilience: we refer the reader to the papers of Sudakov and Vu [787], or Balogh, Csaba, and Samotij [68], or suggest to read only Theorem 5.47.

[97]We have defined only the quasi-random graphs here, for pseudo-random graphs see e.g., [559, 803], for expanders see e.g., [20].

5.10 Extremal Results on "Slightly Randomized" Graphs

Bohman, Frieze, and Martin [113] proved Hamiltonicity for graphs R_n obtained from a non-random graph G_n^0 whose minimum degree was $d < n/2$ where R_n is obtained by adding to G_n^0 a random binomial graph $R_{n,p}$ whose edge-probability is a small $p = p_n > 0$. Dudek, Reiher, Ruciński, and Schacht continued this line, using this model, and proving in [247] that the stronger conclusion of Pósa–Seymour conjecture holds almost surely for the obtained graph, under their conditions: instead of a Hamiltonian cycle they obtained a power of it.

5.11 Algorithmic Aspects

Whenever we use an existence-proof in combinatorics, it is natural to ask if we can turn it into an algorithm. This was the case with the Lovász Local Lemma, and this is the case with the Regularity Lemma, and also with the Blow-up Lemma.

The algorithmic aspects of the Lovász Local Lemma were investigated first by József Beck [95], and later by Moser and Tardos [619, 620], and many others.

— · —

Alon, Duke, Lefmann, Rödl, and Yuster in [30] showed that one can find the Regular Partition, and the Cluster Graph of a G_n fairly efficiently. Actually, they proved two theorems: (a) to decide if a partition is ε-regular is difficult, but (b) to find an ε-regular partition of a given graph is easy. More precisely,

Theorem 5.48 ((1992), [30]) *Given a graph G_n and an $\varepsilon > 0$, and a partition $(V_0, \ldots, V_k)$,[98] it is Co-NP-complete to decide if this partition is ε-regular.*

Theorem 5.49 ([30]) *For every $\varepsilon > 0$ and every $t > 0$, there exists a $Q(\varepsilon, t)$ such that for every G_n with $n > Q(\varepsilon, n)$ one can find an ε-Regular Partition (described in Thm 5.3) in $O(M(n))$ steps, where $M(n)$ is the "number of steps" needed to multiply two $0 - 1$ matrices over the integers.*[99,100]

The algorithmic problem with the Regularity Lemma is that we may have too many clusters. Therefore a direct way to transform it into an efficient algorithm may be hopeless. The Frieze–Kannan version often solves this problem, see Sect. 5.13, or [364, 365, 485], or Lovász-Szegedy [595].

The above methods were needed and extended to hypergraphs, see e.g., Nagle, Rödl, and Schacht [629].

---

[98] Here we have $k + 1$ classes, since originally there was also an exceptional class V_0, different from the others. This V_0 can be forgotten: its vertices can be distributed in the other classes.

[99] The theorem also has a version on parallel computation.

[100] Here we do not define the "steps" and ignore again the difference caused by neglecting V_0 in Theorem 5.3.

Remark 5.50 We needed this short section *here*, since it helps to understand the next part better: otherwise it would come later.

5.12 Regularity Lemma for the Analyst

As soon as the theory of Graph Limits turned into a fast-developing research area, it became interesting, what happens with the regularity lemmas in this area. As we have mentioned, Lovász wrote first a long survey [590], then a thick book on graph limits [591], so we shall not discuss it here, but mention just one aspect. The paper of Lovász and Szegedy [595] with the title "Szemerédi Lemma for the analyst" not only described the Regularity Lemma in terms of Mathematical Analysis, but also described the Weak Regularity Lemma (i.e. the Frieze–Kannan version [365]), and the Strong Regularity Lemma [33] of Alon, Fischer, Krivelevich, and M. Szegedy, and connected the Regularity Lemma to ε-nets in metric spaces, and to Compactness. We quote part of the Introduction of their paper.

"Szemerédi's regularity lemma was first used in his celebrated proof of the Erdős–Turán Conjecture on arithmetic progressions in dense sets of integers.[101] Since then, the lemma has emerged as a fundamental tool in Graph Theory: it has many applications in Extremal Graph Theory, in the area of 'Property Testing' in computer science, combinatorial Number Theory, etc. ...Tao described the lemma as a result in probability. Our goal is to point out that Szemerédi's lemma can be thought of as a result in analysis. We show three different analytic interpretations. The first one is a general statement about approximating elements in Hilbert spaces which implies many different versions of the Regularity Lemma, and also potentially other approximation results. The second one presents the Regularity Lemma as the compactness of an important metric space of 2-variable functions. ...The third analytic interpretation shows the connection between a weak version of the regularity lemma and covering by small diameter sets, i.e., dimensionality. ...We describe two applications of this third version: ...and an algorithm that constructs the weak Szemerédi partition as Voronoi cells in a metric space."

Actually, it is surprising that such a short paper can describe such an involved situation in such a compact way, also including the proofs.

---

[101] As we have mentioned, this is not quite true. It was invented to prove a conjecture of Bollobás, Erdős, and Simonovits on the parametrized Erdős–Stone theorem, and was first used in the paper of Chvátal and Szemerédi [188]. A weaker, bipartite, asymmetric version of it was used to prove that $r_k(n) = o(n)$.

5.13 Versions of the Regularity Lemma

There are many versions of the Regularity Lemma. Below we list some of them, with very short descriptions. Many of them are difficult to invent but have you invented them, their proofs are often very similar to the original proof. (In case of Hypergraph Regularity Lemmas the situation is completely different.)

We have already described the Regularity Lemma.

Coloured version. There is an easy generalization of the Regularity Lemma in which the edges of a graph G_n are r-coloured, for some fixed r, so we have r edge-disjoint graphs, and we wish to find a vertex-partition which satisfies the Regularity Lemma in all the colours, simultaneously. This is possible and often needed, e.g., in the Erdős–Hajnal–Sós–Szemerédi extension of Theorem 5.21 (in [295]), and more generally, this was used in Ramsey-type theorems and Ramsey–Turán-type theorems, and later in many similar cases.

Weak Regularity Lemma. There is an important weakening of the Regularity Lemma, namely the Frieze–Kannan Weak Regularity Lemma [365], see also [364, 366] and the Frieze–Kannan–Vempala approach [367]. The difference between the Regularity Lemma and the Weak Regularity Lemma is that the later one ensures ε-regularity only for "much larger subsets", and (therefore) needs much fewer clusters. Here the Weak ε-regularity means that given a partition $U_1, \ldots, U_k$, for a subset $X \subset V(G_n)$ we hope to have

$$\mathcal{E}(X) := \sum d(U_i, U_j) \cdot |U_i \cap X||U_j \cap X|$$

edges in $G[X]$, so we conclude that $\mathcal{E}(X)$ is close to $e(X)$.

— · —

While the original algorithm of Frieze and Kannan is randomized, Dellamonica, Kalyanasundaram, Martin, Rödl, and Shapira [233] provided a deterministic $O(n^2)$ algorithm, analogous to Theorem 5.49, to find the Frieze–Kannan Partition.

This regularity lemma is more connected to Statistics,[102] and in many cases, where one can apply a regularity lemma for an existence proof, the algorithmic versions of the regularity lemmas provide algorithms in these applications too.

As we wrote, these algorithms are slow because of the very large number of clusters,[103] but in many such cases the **Algorithmic Weak Regularity Lemma** can also be used, and then it provides a much faster algorithm, basically because it requires essentially fewer classes in the weak ε-regular partition.

One should remark that the Frieze–Kannan Regularity Lemma can be iterated and then it provides a proof of the original Szemerédi Regularity Lemma.

---

[102] Principal component analysis, see e.g., Frieze, Kannan, Vempala, and Drineas [246, 367].

[103] There are many results showing that the number of clusters must be very large. The first such result is due to Gowers [401].

On the other end, there is the **Strong Regularity Lemma** of Alon, Fischer, Krivelevich, and M. Szegedy [33]. The advantage of the Strong Regularity Lemma is that it can be applied in several cases where the original Regularity Lemma is not enough, primarily when we are interested in induced subgraphs.

We shall not formulate this strong lemma but include an explanation of it, from Alon–Shapira [43] (with a slight simplification). Alon and Shapira write:

"…This lemma can be considered a variant of the standard Regularity Lemma, where one can use a function that defines $\varepsilon > 0$ as a function of the size of the partition, rather then having to use a fixed ε as in Lemma 2.2."

Large part of this is described in the paper of Lovász and Szegedy [595].

5.14 Regularity Lemma for Sparse Graphs

The Regularity Lemma can be used in many cases but has several important limitations.

(a) Because of the large threshold n_0, one cannot combine it with computer programs, checking the small cases. In other words, it is a theoretical result but it cannot be used in practice.

(b) The most serious limitation is that we can apply it for embedding H into G only if the degrees in H are (basically) bounded.

(c) Another one is that it can be applied only to dense graphs G_n. This problem is partly solved by the Sparse Regularity Lemma, established by Kohayakawa [524], and Rödl, see also [528].

For sparse graph sequences, i.e. when $e(G_n) = o(n^2)$, the original Regularity Lemma is trivial but does not give any information. Having a bipartite subgraph $H[U, V] \subseteq G_n$, consider the following "rescaled" density:

$$d_{H,p}(U, V) := \frac{e(U, V)}{p \cdot |U||V|}. \tag{20}$$

If $p > 0$ is very small, e.g., $p := n^{-2/3}$, then the condition

$$|d_{H,p}(X, Y) - d_{H,p}(U, V)| < \varepsilon$$

does not say anything without $1/p$ in (20), but with $1/p$ it is a reasonable and strong restriction for sparse graphs. The Sparse Regularity Lemma says that for "nice" graphs H_n there is a partition $V_1, \ldots, V_\nu$, described in the Regularity Lemma, even if we use this stronger regularity requirement (20). Which are the "nice" graphs?

(d) Sparse regularity lemmas are well applicable when the graph G_n to be approximated by generalized quasi-random graphs does not contain subgraphs whose density is much above the edge-density of G_n, e.g., for some bound b and the edge density $p = e(G_n)/\binom{n}{2}$,

$$E(V_1, V_2) < bp|V_1||V_2| \quad \text{if} \quad |V_1|, |V_2| > \eta n. \tag{21}$$

This is the situation, e.g., when we consider (non-random) subgraphs of random graphs, see Sect. 5.7.

The Sparse Regularity Lemma sounds the same as the original Regularity Lemma, with two differences: we use the modified uniformity: (20), instead of (16), and the extra condition (21), for some fixed b.

One of the latest developments in this area is a sparse version of the Regularity Lemma due to Alex Scott [735]. Scott succeeded in eliminating the extra condition on the sparse graph that it had no (relatively) high density subgraphs. However, this had some price, discussed by Scott in Sect. 4 of [735]. This new sparse Regularity Lemma was used in several cases, e.g., by P. Allen, P. Keevash, B. Sudakov, and J. Verstraëte, in [19].

Remark 5.51 As we wrote, the Sparse Regularity Lemma can be used basically if G_n does not contain subgraphs much denser than the whole graph. Gerke and Steger wrote an important survey about it and about its applicability [392], see also [391]. Its applicability is discussed (among others) in the paper of Conlon, Gowers, Samotij, and Schacht [200], with its connection to the Kohayakawa-Łuczak-Rödl conjecture [525]. We also warmly recommend the paper of Conlon, Fox, and Yufei Zhao [198].

5.15 Quasi-random Graph Sequences

Quasi-random sequences are very important, e.g., in Theoretical Computer Science, and also very interesting, for their own sake. In Graph Theory they emerged in connection with some Ramsey problems. The first detailed, pioneering results in this direction are due to A. Thomason, see e.g., [803, 804]. He was motivated by some Ramsey Problems.

Chung, Graham, and Wilson [185] developed a theory in which *six properties of random graphs* were formulated which are equivalent for any infinite sequence (G_n) of graphs. The graphs having these properties are called *quasi-random*.

Quasi-randomness exists in other universes as well, e.g., there exist quasi-random subsets of integers, groups, see Gowers [405], tournaments, see Chung and Graham [183, 184], of real numbers, digraphs [338], of hypergraphs, e.g., Chung [180–182], Rödl and Kohayakawa [528], permutations, see Cooper [205], Král' and Pikhurko [555], and in many other settings...

Quasi-randomness and the Regularity Lemmas are very strongly connected. This was first established in a paper of Simonovits and Sós [765]:

Theorem 5.52 (Simonovits, Sós (1991), [765]) *A sequence of graphs (G_n) is p-quasi-random iff for every $\kappa > 0$ and $\varepsilon > 0$, there exist two thresholds $k_0(\varepsilon, \kappa)$ and $n_0(\varepsilon, \kappa)$ such that for $n > n_0(\varepsilon, \kappa)$ G_n has an ε-Regular Partition where all the pairs (V_i, V_j) are ε-regular with densities between $p - \varepsilon < d(V_i, V_j) < p + \varepsilon$ and $\kappa < k < k_0(\varepsilon, \kappa)$.*

Remark 5.53 Actually, a slightly stronger theorem holds. On the one hand, we may allow $\varepsilon\binom{k}{2}$ exceptional pairs (V_i, V_j) to ensure p-quasi-randomness. On the other hand, if (G_n) is p-quasi-random, we can find a partition where there are no exceptional pairs.

The corresponding generalization for sparse graphs was proved by Kohayakawa and Rödl. For a longer and detailed survey see their paper [528].

FOR FURTHER RELATED RESULTS see, e.g., Simonovits and Sós [766, 768], Skokan and Thoma [769], Yuster [822], Shapira and Yuster [737], Gowers on Counting Lemma [403], on quasi-random groups [405], and also Janson [475], Janson and Sós [477] on the connections to Graph Limits.

A generalization of the notion of Quasi-random graphs was investigated by Lovász and Sós [594] which corresponds to generalized random matrix-graphs.

5.16 Blow-Up Lemma

Several results exist about embedding *spanning* subgraphs into dense graphs. Many of the proofs use a relatively new and very powerful tool, called BLOW-UP Lemma. Here we describe it in a fairly concise way. The Blow-up Lemma is mostly used to embed a *bounded degree* graph H into a graph G as a *spanning subgraph*. The reader is also referred to the excellent "early" survey of Komlós [533] (explaining a lot of important background details about the Regularity Lemma and the Blow-Up lemma, and how to use them) or to the surveys of Komlós and Simonovits, [546], Komlós, Simonovits, Shokoufandeh, and Szemerédi, [545]. The "Doctor of Sciences" Thesis of Sárközy [723] is also an excellent source in this area. For some newer results on the topic see e.g., Rödl and Ruciński [688], Keevash [497] who extended the method to hypergraphs, Sárközy [725], Böttcher, Kohayakawa, and Taraz, and Würfl [143] who extended it to d-degenerate graphs.[104] A long survey of Allen, Böttcher, Hàn, Kohayakawa, and Person [18] discusses several features of the Blow-up Lemma applied to random and pseudo-random graphs. Recently Allen, Böttcher, Hàn, Kohayakawa, and Person extended the Blow-up lemma to sparse graphs [18].

— · —

We start with a definition.

Definition 5.54 $((\varepsilon, \delta)$-*super-regular pair*) Let G be a graph, $U, W \subseteq V(G)$ be two disjoint vertex sets, $|U| = |W|$. The vertex-set pair (U, W) is (ε, δ)-super-regular if it is ε-regular and $d_{\min}(G[U, W]) \geq \delta|U|$.

The Blow-up Lemma asserts that (ε, δ)-regular pairs behave as complete bipartite graphs from the point of view of embedding *bounded degree* subgraphs. In other

---

[104] They call it d-arrangeable.

words, for every large $\Delta > 0$ and small $\delta > 0$ there exist an $\varepsilon > 0$ such that if in the Cluster graph H_ν the min-degree condition also holds and we replace the (ε, δ)-regular pairs (U_i, U_j) by complete bipartite graphs, and then we can embed (the bounded degree) H_n into this new graph $\widetilde{G}_n$, then we can embed it into the original G_n as well. The low degree vertices of G_n could cause problems. Therefore, for embedding spanning subgraphs, one needs all degrees of the host graph large. That's why using regular pairs is not sufficient any more, we need *super-regular pairs*. Again, the Blow-up Lemma plays a crucial role in embedding *spanning* graphs H_n into G_n.

The difficulty is in embedding the "last few" vertices. The original proof of the Blow-up Lemma starts with a probabilistic greedy algorithm, and then uses a König-Hall argument to complete the embedding. The proof is quite involved.

Theorem 5.55 (Blow-up Lemma, Komlós–Sárközy–Szemerédi (1997), [539]) *Given a graph H_ν of order ν and two positive parameters δ, Δ, there exists an $\varepsilon > 0$ such that if $n_1, n_2, \ldots, n_r$ are arbitrary positive integers and we replace the vertices of H_ν with pairwise disjoint sets $V_1, V_2, \ldots, V_\nu$ of sizes $n_1, n_2, \ldots, n_\nu$, and construct two graphs on the same vertex-set $V = \bigcup V_i$ so that*

(i) the first graph $H_\nu(n_1, \ldots, n_\nu)$ is obtained by replacing each edge $\{v_i, v_j\}$ of H_ν with the complete bipartite graph $K(n_i, n_j)$ between the corresponding vertex-sets V_i and V_j,

(ii) and second, much sparser graph $H_\nu^*(n_1, \ldots, n_\nu)$ is obtained by replacing each $\{v_i, v_j\}$ with an (ε, δ)-super-regular pair between V_i and V_j,
then if a graph L with $d_{\max}(L) \leq \Delta$ is embeddable into $H_\nu(n_1, \ldots, n_\nu)$ then it is embeddable into the much sparser $H_\nu^*(n_1, \ldots, n_\nu)$ as well.

The Blow-up Lemma has several different proofs, e.g., Komlós, Sárközy, and Szemerédi first gave a randomized embedding [539], and then they gave a derandomized version [541] as well. Other proofs were given, e.g., by Rödl, Ruciński [688], Rödl, Ruciński, and Taraz [699], and Rödl, Ruciński, and Wagner [699].

Remark 5.56 G.N. Sárközy gave a very detailed version of the proof of the Blow-up lemma, [725], where he calculated all the related details very carefully in order that he and Grinshpun could use it in their later work [422], Theorems 7.22, 7.23: without this they could prove only a weaker result.

5.17 Regularity Lemma in Geometry

Until now we restricted our consideration to applications of the Regularity Lemma to embed sparse graphs into dense graphs. The Regularity Lemma has several applications in other fields as well, and beside this, versions tailored to these other fields. Here we mention only a few such versions, very briefly.

A theorem of Green and Tao [418] is a good example of this. They prove a so-called ARITHMETIC REGULARITY LEMMA that can be applied in ADDITIVE COMBI-

NATORICS, in several problems similar to Szemerédi theorem on arithmetic progressions. However, here we wish to discuss Geometry.

There are several cases where we restrict our consideration to some particular graphs, e.g., to graphs coming from geometry. In this case one may hope for much better estimates in some cases than for arbitrary graphs. A whole theory was built up around such problems, see, e.g., Erdős [260, 270], a survey of Szemerédi [796], Szemerédi and Trotter [797], Pach and Sharir [640] ...or the book of Pach and Agarwal [639]. We remark here only that the Regularity Lemma also has some strengthened forms, see, e.g., the improvement of some results of Fox, Gromov, Lafforgue, Naor, and Pach [348] by Fox, Pach, and Suk [351]...

So the Regularity Lemma can be applied in Geometry in many cases and the fact that we apply it to a geometric situation implies that in many cases the connection between the clusters will be a (basically) complete connection, or very few edges, instead of having randomlike connections. This is not so surprising, since many of the geometric relations are described by polynomials, or analytic functions, (?) and in these cases, if we have "many" 0's of a polynomial (of several variables) then the corresponding polynomial must vanish everywhere.[105] It seems that in most cases where the Regularity Lemma is used in Geometry, its use can be eliminated.

The interested reader is suggested to read, e.g., the survey of J. Pach [638], or his book with Agarwal [639], the papers of Alon, Pach, Pinchasi, Radoičić, and Sharir [41], on SEMI-ALGEBRAIC sets, or the paper of Fox, Gromov, Lafforgue, Naor, and Pach [348], or the paper of Pach and Solymosi, [641], or of J. Pach [637]. The result of Karasev, Kynčl, Paták, Patáková, and Martin Tancer [487] is also related to this topic.

6 With or Without Regularity Lemma?

The Regularity Lemma is one of the most effective, most efficient lemmas in Extremal Graph Theory. In several cases we can prove a result with the Regularity Lemma, but later we find out that it can easily (or not so easily?) be proven without the Regularity Lemma as well. So it is natural to ask if it is worth getting rid of the application of Regularity Lemma, if we can. The answer is not so simple. We should mention a disadvantage and two advantages of using the Regularity Lemma.

The *disadvantage* is that it can be applied only to very large graphs. This means that "in practice it is of no use". This bothers some people, e.g., those who wish to find out,—sometimes with the help of computers—the truth for the smaller values as well. Many of us feel this unimportant, some others definitely prefer eliminating the use of Regularity Lemma if it is possible. First we quote an opinion against the usage of the Regularity Lemma. In Sect. 5.5 we discussed the beautiful conjecture of Murty and Simon on the maximum number of edges a diameter-2-critical graph can have. We have mentioned that Füredi [371] solved this problem in 1992, using

---

[105] One form of this is expressed in the Combinatorial Nullstellensatz of Alon [23].

the Ruzsa–Szemerédi theorem that $f(n, 6, 3) = o(n^2)$. Much later, in 2015, a survey [460] on the topic described this area very active and wrote:

"The most significant contribution to date is an astonishing asymptotic result due to Füredi (1992) who proved that the conjecture is true for large n, that is, for $n > n_0$ where n_0 is a tower of 2's of height about 10^{14}. As remarked by Madden (1999), 'n_0 is an inconceivably (and inconveniently) large number: it is a tower of 2's of height approximately 10^{14}.' Since, for practical purposes, we are usually interested in graphs which are smaller than this, further investigation is warranted...."

First of all, it is not clear if it is correct to call Füredi theorem an "asymptotic result". The *advantage* of applying the Regularity Lemma is that it often provides a proof where we have no other proofs, and, in other cases, a much more transparent proof than the proof without it. (Thus for example, the beautiful theorem of Erdős, Kleitman, and Rothschild [297] was proved originally without the Regularity Lemma, but for many of us the Regularity Lemma provides a more transparent proof.)

Perhaps one of the first cases where we met a situation where the Regularity Lemma could be eliminated was a paper of Bollobás, Erdős, Simonovits, and Szemerédi, [128] discussing several distinct extremal problems, and one of them was which could be considered as one germ of PROPERTY TESTING:

Problem 6.1 (*Erdős*) Is it true that there exists a constant M_ε such that if one cannot delete εn^2 edges from G_n to make it bipartite then it has an odd cycle C_ℓ with $\ell < M_\varepsilon$?

The answer was

Theorem 6.2 (Bollobás, Erdős, Simonovits, and Szemerédi (1977), [128]) *YES, if one cannot delete εn^2 edges from G_n to make it bipartite then one can find an odd cycle C_ℓ with $\ell < \frac{1}{\varepsilon}$.*

In [128] we gave two distinct proofs of this theorem, one using the Regularity Lemma and another one, without the Regularity Lemma. Interestingly enough, these questions later became very central and important in the theory of Property Testing, but there, in the works of Alon, Krivelevich, Shapira, and others, (see e.g., [33]) it turned out that such property testing results depend primarily on whether one can apply the Regularity Lemma or not. Just to illustrate this, we mention from the many similar results the paper of Alon, Fischer, Newman, and Shapira [34], the title of which is "A combinatorial characterization of the testable graph properties: It's all about regularity".[106]

Remark 6.3 Duke and Rödl [249] extended Theorem 6.2 to higher chromatic number, answering another question of Erdős, see also the ICM lecture of Rödl [686], and also the result of Alon and Shapira on property testing [43]. (Rödl: "Further refinement was given by Austin and Tao" [53].)

---

[106]Two remarks should be made here: (a) Originally Property Testing was somewhat different, see e.g., Goldreich, Goldwasser, and Ron [399]. (b) The theory of graph limits also has a part investigating property testing, see e.g., [138, 140],..., [596].

So we emphasise again that there are several cases where certain results can be proved with and without the Regularity Lemma, and the proof with the Regularity Lemma may be much more transparent, however, the obtained constants are much worse. Before proceeding we formulate a meta-conjecture about the "elimination".

Meta-Conjecture 6.4 *(Simonovits) The use of the Regularity Lemma can be eliminated from those proofs where*
(i) the conjectured extremal structures are "generalized random graphs" with a fixed number of classes and densities 0 and 1, and
(ii) at least one of the densities is 0 and one of them is 1.

One has to be very careful with this—otherwise informative—Meta-Conjecture:
(a) First of all, mathematically it is not quite well defined, what do we mean by "eliminating the Regularity Lemma".
(b) Further, without (ii) the Ruzsa–Szemerédi Theorem could be regarded as a "counter-example".[107]
(c) It is not well defined if using graph limits we regard as elimination of the Regularity Lemma or not?
(d) The first two-three results which we like proving nowadays to illustrate the usage of the Regularity Lemma, e.g., the Ramsey–Turán estimate for K_4 (Thm 5.21) and Ruzsa–Szemerédi Theorem, originally were proved using some weaker forms of the Regularity Lemma.

In several cases originally the Regularity Lemma was used to obtain some results, but then it was *easily* eliminated. Such examples are Erdős and Simonovits [315], or Pach and Solymosi [641], on Geometric Graphs, or results in the paper of Erdős, S.B. Rao, Simonovits, and Sós, [304]. Often in the published versions we do not even find the traces of the original proof with Regularity Lemma, anymore ...

One interesting case of this discussion is the proof of

Conjecture 6.5 (Lehel's conjecture [54]) *If we 2-colour the edges of K_N, then $V(K_N)$ can be covered by two vertex-disjoint monochromatic cycles of distinct colours.*

Remark 6.6 The first reference to Conjecture 6.5 can be found in the Ph.D. thesis of Ayel [54]. The conjecture was first proved by Łuczak, Rödl, and Szemerédi [602], using the Regularity Lemma. Of course, this worked only for very large values of n. The Regularity Lemma type arguments were eliminated by Peter Allen [14]. The difference between the two proofs is that Allen covers K_N by monochromatic cliques, using Ramsey's theorem, instead of using Regularity type arguments. Hence the threshold in the first proof is very large, and in Allen's proof it is "only" 2^{18000}. Finally, surprisingly, Bessy and Thomassé [108] found a simple and short proof of the conjecture, without using the Regularity Lemma, or any deep tool, and which worked for all n.

---

[107] Actually, J. Fox eliminated the application of the Regularity Lemma from the proof of the Triangle Removal Lemma, see Sect. 5.4 or [347].

Conjecture 6.5 was extended by Gyárfás to any number of colours:

Conjecture 6.7 (Gyárfás, [291, 424]) *If the edges of K_N are r-coloured, then $V(K_N)$ can be covered by $p(r) = r$ vertex-disjoint monochromatic cycles (where K_1, K_2 are also considered as cycles).*

Remark 6.8 This was "slightly" disproved for $k \geq 3$ by Pokrovskiy [654]. Here "slightly" means that in his counterexample there is one vertex which could not be covered, however, $p(r) = r + 1$ is still possible. We shall return to the Gyárfás conjecture (often called Gyárfás-Lehel conjecture) in Sect. 7.4.

Remark 6.9 The results of Bessy and Thomassé and of Erdős, Gyárfás, and Pyber, (from Sect. 6.5) were extended to LOCAL r- COLOURING by Conlon and Maya Stein [201], where local r-colouring means that each vertex is adjacent only to at most r distinct colours, but the total number of colours may be much larger.

6.1 Without Regularity Lemma

There are several cases where eliminating the use of the Regularity Lemma from the proof is or would be important. Above we discussed this and listed such cases. Here we mention two further cases. Fox gave a proof of the Removal Lemma without using the Regularity Lemma, in [347] (slightly simplified in the beautiful survey of Conlon and Fox [194]). This improves several estimates in some related cases. Conlon, Fox and Sudakov [197] recently removed using the Regularity Lemma from the proof of a theorem of Simonovits and Sós [766], which helped to understand the situation better.

6.2 Embedding Spanning or Almost-Spanning Trees

Originally we planned to write—among others,—about our results on tree embeddings: about the solutions of the Erdős–Sós conjecture and the Komlós–Sós conjecture. However, they are described in [386] and in [465–470], respectively, and we shall return to these topics elsewhere.

There are many results where we try to embed large or actually spanning trees into a graph G_n. We mention a few of them. Some of them describe a case when G_n is a random or random like graph, e.g., pseudo-random, expanding,... If we know something for random graphs, that often can (easily?) be extended to these cases: quasi-random, pseudo-random, or expander graphs. The "Resilience results" of Sects. 5.8 and 5.9 were of this type. One of the early results on expander graphs was

Theorem 6.10 (Friedman and Pippenger [363]) *If for every $X \subseteq V(G_n)$, with $|X| \leq 2k - 2$,*

$$|\Gamma(X)| \geq (d+1)|X|,$$

then G_n contains all trees T_k with $d_{\max}(T_k) \leq d$.

Remark 6.11 The Friedman-Pippenger theorem "embeds" only relatively small trees. It was extended by Balogh, Csaba, Pei, and Samotij [67], where a result of Haxell [456] was simplified, and then used. This guaranteed embedding almost spanning trees into "expanding graphs". We skip the precise formulation and the details, because they may look technical at first sight.

For some earlier related works see Alon–Chung [29], and Beck [94].

— · —

Now we consider a conjecture of Bollobás [119], on embedding bounded degree trees, proved by

Theorem 6.12 (Komlós, Sárközy, and Szemerédi [537, 544]) *For every $\varepsilon > 0$ and $\Delta > 0$, there exists an n_0 for which, if T_n is a tree on n vertices with $d_{\max}(T_n) \leq \Delta$, and G_n is a graph on n vertices with*

$$d_{\min}(G_n) \geq \frac{n}{2} + \varepsilon n, \tag{22}$$

then $T_n \subseteq G_n$, assuming that $n > n_0(\Delta)$.

Komlós, G. Sárközy, and Szemerédi [544] gave an improvement of Theorem 6.12, where they proved that a tree T_n can be embedded into G_n even if its maximum degree is allowed to be as large as large as $c\frac{n}{\log n}$.

Another improvement was

Theorem 6.13 (Csaba, Levitt, Nagy-György, and Szemerédi [215]) *(a) For any constant $\Delta > 0$ there exists a constant $c_\Delta > 0$ such that if T_n is a tree on n vertices with $d_{\max}(T_n) \leq \Delta$, and G_n is a graph on n vertices with*

$$d_{\min}(G_n) \geq \frac{n}{2} + c_\Delta \log n, \tag{23}$$

then $T_n \subseteq G_n$, assuming that $n > n_0(\Delta)$.

(b) There exist infinitely many graphs G_n with $d_{\min}(G_n) \geq \frac{1}{2}n + \frac{1}{17} \log n$ not containing the complete ternary tree $T_n^{[3]}$.

So the bound in (23) is tight. The proofs in [537, 544] used the Regularity Lemma and the Blow-up lemma, while [215] did not use them, and provided smaller n_0 and a sharper theorem. B. Csaba also extended the above results to WELL- SEPARABLE graphs:

Definition 6.14 An infinite graph sequence (H_n) is WELL- SEPARABLE, if one can delete $o(n)$ vertices of H_n so that each connected component of the remaining graph has $o(n)$ vertices, as $n \to \infty$.

Theorem 6.15 (Csaba [214]) *For every $\varepsilon, \Delta > 0$, there exists an $n_0 = n_0(\varepsilon, \Delta)$ such that if (H_n) is well-separable, $n > n_0$, and $d_{\max}(H_n) \leq \Delta$, and*

$$d_{\min}(G_n) > \left(1 - \frac{1}{2(\chi(H) - 1)} + \varepsilon\right) n, \qquad (24)$$

then H_n can be embedded into G_n.

For trees (or, more generally, for bipartite graphs H_n) (24) reduces to $d_{\min}(G_n) \geq \frac{1}{2}n + \varepsilon n$: Theorem 6.15 is a generalization of Theorem 6.12. Another version is where we assume that G_n is bipartite, see Csaba [213].

6.3 Pósa–Seymour Conjecture

Speaking of extremal problems, we could consider problems where we ask one of the following questions:

(a) how large $e(G_n)$ ensures a property $\mathcal{P}$?
(b) how large $d_{\min}(G_n)$ ensures $\mathcal{P}$?
(c) which (Ore-type) degree sum conditions $d(x) + d(y) \geq f_O(n, \mathcal{P})$ ensure $\mathcal{P}$, where we assume this only for independent vertices x, y;
(d) Given a graph G_n with the degree sequence $d_1, d_2, \ldots, d_n$, does it ensure $\mathcal{P}$?

In the next part we consider two of these questions: (b), called Dirac-type problems, and (c) called Ore-type problems.

— · —

Hamiltonicity of graphs is a central problem in graph theory. There are many results of type (d), where some conditions on the degrees ensure the Hamiltonicity. In the Introduction we formulated one of the first such results, Theorem 2.65:
DIRAC THEOREM.

If $d_{\min}(G_n) \geq n/2$, and $n \geq 3$, then G_n contains a Hamiltonian cycle.

Fig. 7 Square of a cycle

As we have mentioned, this is sharp. We shall go into two distinct directions from Dirac's Theorem: here we shall consider some generalizations for simple graphs, while in Sect. 9 we shall discuss some hypergraph extensions. A natural question analogous to Dirac's theorem was asked by Pósa (see Erdős [267] in 1965). The reader is reminded that the kth power $L := M^k$ of a graph M is obtained from M by joining all the pairs of vertices $x \ne y$ having distance $\rho_M(x, y) \le k$ (Fig. 7).

Conjecture 6.16 (Pósa) *If for a graph G_n $d_{\min}(G_n) \ge \frac{2}{3}n$, then G_n contains the square of a Hamiltonian cycle.*

This was generalized by Seymour in 1973:

Conjecture 6.17 (Seymour [736]) *Let G_n be a graph on n vertices. If $d_{\min}(G_n) \ge \frac{k}{k+1}n$, then G_n contains the kth power of a Hamiltonian cycle.*

For $k = 1$, this is just Dirac's theorem, for $k = 2$ the Pósa conjecture. The validity of the general conjecture implies the notoriously hard Hajnal–Szemerédi theorem (i.e. Theorem 2.67).[108]

Remark 6.18 Observe that for $\ell \ge k+1$, we have $K_{k+1} = P_{k+1}^k \subseteq P_\ell^k$. Hence $T_{n,k}$ does not contain P_ℓ^k. On the other hand, $P_n^k \subset T_{n,k+1}$. This provides some further motivation for the above conjectures.

In the earlier parts we mostly considered embedding problems where the graph H_m to be embedded into the "host graph" G_n had noticeably fewer vertices than G_n, and thus one could use the Regularity Lemma. As we mentioned, when we embed *spanning* subgraphs, the embedding of the last few vertices may create serious difficulties and *this difficulty* was overcome by using the Regularity Lemma—Blow-up Lemma method. First in [540] Komlós, Sárközy and Szemerédi proved Conjecture 6.17 in its weaker, asymptotic form:

Theorem 6.19 (Pósa–Seymour conjecture - approximate form (1998), [540]) *For any $\varepsilon > 0$ and positive integer k there is an $n_k(\varepsilon)$ such that if $n > n_k(\varepsilon)$ and*

$$d_{\min}(G_n) > \left(1 - \frac{1}{k+1} + \varepsilon\right)n,$$

then G_n contains the kth power of a Hamilton cycle.

Next they got rid of ε in [538, 542]: they proved both conjectures for $n \ge n_k$, without the extra $\varepsilon > 0$.

Theorem 6.20 (Komlós, Sárközy, and Szemerédi [542] *For every integer $k > 0$ there exists an n_k such that if $n \ge n_k$, and*

---

[108] Actually, this was the motivation for Pósa. (Later the Hajnal-Szemerédi theorem was proved in simpler ways.)

$$d_{\min}(G_n) \geq \left(1 - \frac{1}{k+1}\right)n, \qquad (25)$$

then the graph G_n contains the kth power of a Hamilton cycle.

The proofs used the Regularity Lemma [792], the Blow-up Lemma [539, 541] and the Hajnal–Szemerédi Theorem [444]. Since the proofs used the Regularity Lemma, the resulting n_k was very large (it involved a tower function). The use of the Regularity Lemma was removed by Levitt, Sárközy and Szemerédi in a new proof of Pósa's conjecture in [575]. Much later, finally, Péter Hajnal, Simao Herdade, and Szemerédi found a new proof of the Seymour conjecture [446] that avoids the use of the Regularity Lemma, thus resulting in a "completely elementary" proof and a much smaller n_k.

Historical Remarks. Partial results were obtained earlier on the Pósa–Seymour conjecture, e.g., by Jacobson (unpublished), Faudree, Gould, Jacobson, and Schelp [331], Häggkvist (unpublished), Genghua Fan and Häggkvist [329], and Fan and Kierstead [330]. Fan and Kierstead also announced a proof of the Pósa conjecture if the Hamilton cycle is replaced by Hamilton path. (Noga Alon observed that this already implies the Alon–Fischer theorem mentioned in Sect. 2.19, since the square of a Hamilton path contains all unions of cycles.) We skip the exact statements of these papers, but mention that Châu, DeBiasio, and Kierstead [177] proved Pósa Conjecture for all $n > 8 \times 10^9$.

Stability Remark. A crucial lemma of the proof in [446] is a "structural stability" assertion that for some constant $\gamma > 0$, either G_n contains an "almost independent" set of size $\frac{n}{k+1}$ or $d_{\min}(G_n) < (\frac{k}{k+1} - \gamma)n$.

6.4 Ore-Type Results/Pancyclic Graphs

As we have mentioned, Hamiltonian problems, above all, Dirac Theorem on Hamiltonicity of graphs, led to several important research directions. Here there is a significant difference between the graph and hypergraph versions. We shall return to the Hamiltonicity of hypergraphs in Sect. 9. As to ordinary graphs, some generalizations are the Ore-type problems, some other ones are the Pósa–Seymour-type generalizations, discussed in the previous subsection. In an Ore-type theorem we assume that any *two independent* vertices have large degree sums. The first such result was

Theorem 6.21 (Ore (1960), [635]) *If for any two independent vertices x, y of G_n, $\deg(x) + \deg(y) \geq n$, then G_n is Hamiltonian.*

Bondy had an important "meta-theorem" according to which conditions implying Hamiltonicity imply also "pancyclicity", which means that G_n contains cycles of all length between 3 and n.[109] A beautiful illustration is

Theorem 6.22 (Bondy (1971), [134]) *(a) If G_n is Hamiltonian and $e(G_n) \geq \lfloor \frac{n^2}{4} \rfloor$, then either G_n is pancyclic or $G_n = K(n/2, n/2)$.*
(b) Under the Ore condition, for any $k \in [3, n]$, G_n contains a C_k, or $G_n = K(n/2, n/2)$.

Of course, (b) follows from (a) and Ore theorem. Generalizations of these theorems can be found in Broersma, Jan van Heuvel, and Veldman, [149], and in general, there are very many "pancyclicity" theorems, see e.g., Erdős [278], Keevash, Lee, and Sudakov [506, 570], Stacho [781], Brandt, Faudree, and Goddard [147], and many others. (In some sense the Bondy–Simonovits theorem in [136] is also a "weak pancyclic theorem".) There are also very many results on PANCYCLIC DIGRAPHS, see e.g., Häggkvist and Thomassen [461], Krivelevich, Lee, and Sudakov [557].

FOR SOME RELATED RESULTS see e.g., Bollobás and Thomason [132, 133] Brandt, Faudree, and Goddard [147], Favaron, Flandrin, Hao Li, and F. Tian [337], L. Stacho [781], Barát, Gyárfás, Lehel, and Sárközy [90], Barát and Sárközy [91], Kierstead and Kostochka [509], Kostochka[110] and Yu [553] DeBiasio, Faizullah, and Khan [230], and many others.

Weakly pancyclic graphs. There are cases when we cannot hope for all cycles between 3 and n. If a graph G_n contains all cycles between **girth**(G_n) and its circumference $o(G_n)$,[111] then we call it WEAKLY PANCYCLIC, see e.g., Bollobás and Thomason [132, 133]. We mention a theorem of [132], on the girth, answering some questions of Erdős.

Theorem 6.23 (Bollobás and Thomason) *Let G_n be a graph with (at least) two distinct Hamiltonian cycles. Then $n \geq \lfloor (g(G) + 1)^2/4 \rfloor$, and therefore **girth**$(G) \leq \sqrt{4n+1} - 1$.*

There is a pancyclicity defined for bipartite graphs which considers only even cycles. The original Bondy–Simonovits theorem was also about such pancyclicity.

Theorem 6.24 (Bondy and Simonovits (1974), [136]) *If $e(G_n) > 100kn^{1+(1/k)}$, then G_n contains cycles of all lengths 2ℓ, for $\ell = k, \ldots, \lfloor e(G_n)/(100n) \rfloor$.*

---

[109] In some cases we have only weaker conclusions, e.g., in the Bondy–Simonovits theorem [136], and also it may happen in some cases that we get only even cycles! See also the paper of Brandt, Faudree, and Goddard [147] on weakly pancyclic graphs.

[110] Kierstead, Kostochka and others have several results where Ore-type conditions imply Hajnal–Szemerédi-type theorem [509, 512], or a Brooks-type theorem [511].

[111] The circumference is the length of the longest cycle. Here we exclude the trees.

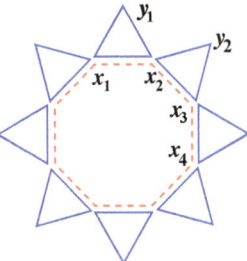

Fig. 8 Triangle-cycle $\mathbb{TC}_{2\ell}$

6.5 Absorbing Method

As we mentioned, when we try to embed a spanning subgraph H_m into a graph G_n, i.e. $m = n$, some difficulties may occur at embedding the "last few vertices". This problem is often solved by using the Blow-up Lemma, or the Absorbing Method, some "Connecting Lemma", or by some Stability Argument. Mostly we combine more than one of them. The stability argument in most of these papers has the form that we distinguish the "nearly extremal" and the "far-from-extremal" cases. Then we handle these two cases separately: mostly we can handle the far-from extremal structure "easily".

We mention among papers combining the Stability and the Absorbing methods the newest results of Hajnal, Herdade, and Szemerédi [446] on the Pósa–Seymour Conjecture, or Balogh, Lo, and Molla, Mycroft, and Sharifzadeh [73–75], on (Ramsey–Turán-) tiling, (see Sect. 6.7). Below we describe the ABSORBING METHOD.

In the Absorbing Method, depending on the problem, we "invent" an ABSORBING STRUCTURE, e.g., in the Erdős-Gyárfás-Pyber theorem [291], the Triangle-Cycle $\mathbb{TC}_{2\ell}$, (defined below) see Fig. 8. Then we start with choosing a special subset of vertices, $\mathcal{A} \subseteq V(G_n)$, defining this special, absorbing substructure $G[\mathcal{A}]$ in our graph/hypergraph, e.g., in [291] an $\mathcal{A}$ spanning a Triangle-Cycle. Next we put aside $\mathcal{A}$ and start building up the whole spanning structure in $G_m = G_n - \mathcal{A}$ as we would do this if m were "noticeably smaller" than n. If the "Absorbing Structure" $G[\mathcal{A}]$ is chosen appropriately, then we will be able to add the last few unembedded vertices of $\mathcal{A}$ at the end: $\mathcal{A}$ will absorb/pick up these remaining, uncovered vertices.

We illustrate this, using a proof-sketch of the Erdős-Gyárfás-Pyber theorem. We shall return to some related newer, sharper results in Sects. 6.7 and 7.4.[112]

Theorem 6.25 (Erdős, Gyárfás, and Pyber [291]) *In any r-colouring of the edges of K_N, we can cover $V(G_n)$ by $p(r) = O(cr^2 \log r)$ vertex-disjoint monochromatic cycles.*

Remark 6.26 The analogous result for bipartite graphs was proved by Haxell [456].

---

[112] A stronger statement is Theorem 7.19.

The basic structure of the proof is as follows. First we define a "Triangle-Cycle" $\mathbb{TC}_{2\ell}$ (Fig. 8). Its vertices are $x_1, \ldots, x_\ell$ and $y_1, \ldots, y_\ell$; and its 3ℓ edges are $x_i x_{i+1}$ (where $x_{\ell+1} := x_1$), $y_i x_i$, and $y_i x_{i+1}$, for $i = 1, \ldots, \ell$.

(a) First, for some $c_1 > 0$, we find a *monochromatic* Triangle-Cycle $\mathbb{TC}_{2\ell}$ in G_n, with $\ell > c_1 n$.
(b) Next we cover $G_m = G_n - \mathbb{TC}_{2\ell}$ with $cr^2 \log r$ vertex-disjoint monochromatic cycles, also allowing to use some vertices from $Y = \{y_1, \ldots, y_\ell\}$.
(c) Finally, we can cover the remaining uncovered vertices with one more monochromatic cycle, since our triangle-cycle $\mathbb{TC}_{2\ell}$ has the nice property that deleting any subset $Y' \subseteq Y$, the remaining $\mathbb{TC}_{2\ell} - Y'$ is still Hamiltonian.

The ABSORBING METHOD was used in the paper of Rödl, Ruciński, and Szemerédi [693], to find a matching in a hypergraph. This seems to be the breakthrough point: soon this method became very popular, both for graphs and hypergraphs. (In Tables 1–3 we list several graph- and hypergraph applications.) Replacing the regularity method by the Absorption Method is discussed, e.g., in Szemerédi [794], Levitt, Sárközy, and Szemerédi [575].

In several cases one uses the ABSORBING METHOD to get sharp results for $n > n_0$ after having already a weaker, asymptotic result. Thus, DeBiasio and Nelsen [232], improving a result of Balogh, Barát, Gerbner, Gyárfás and Sárközy [60] proved a conjecture from [60]:

Theorem 6.27 (DeBiasio and Nelsen [232]) *For any $\gamma > 0$ there exists an $n_0(\gamma)$ such that if $n > n_0(\gamma)$ and $d_{\min}(G_n) > (3/4 + \gamma)n$ and $E(G_n)$ is 2-coloured, then G_n contains two vertex-disjoint monochromatic cycles covering $V(G_n)$.*

Similarly, the 3-uniform hypergraph tiling results of Czygrinow, DeBiasio, and Nagle [221] are the sharp versions of some earlier results of Kühn and Osthus [560] on hypergraph tiling. Let us repeat that the essence of the Absorption Method is to construct certain "advantageous configurations", substructures $G[\mathcal{A}]$, in G_n, called ABSORBING STRUCTURE, covering a large part (say cn vertices) of the host-graph. Next—using the standard methods,—we cover all the vertices of $G_n - G[\mathcal{A}]$ with the given configurations and finally we can expand the embedded configuration into a spanning configuration, using the particular properties of this ABSORBING STRUCTURE. Often the large ABSORBING STRUCTURE consists of many small substructures, and we gain on each of them an uncovered vertex, obtaining at the end a spanning subgraph, as wanted.

We refer the interested reader to the Rödl-Ruciński-Szemerédi paper [696], using the "Absorbing Method", and to the Rödl-Ruciński survey [689], however here we mostly avoid hypergraphs: we return to them in Sect. 9.[113]

— · —

---

[113] In several cases we must distinguish subcases also by some divisibility conditions: not only the proofs but the results also strongly depend on some divisibility conditions.

Table 1 Absorbing method for graphs

Authors	Year	About what? (or title)	Methods	Where
Erdős, Gyárfás, and Pyber	1991	Cycle partition perhaps the first absorbing?	Absorbing	[291] JCTB
Levitt, Sárközy, and Szemerédi	2010	Pósa, How to avoid Regularity Lemma + Stability	Absorbing Connecting Reservoir	DM [575]
Keevash	2014 2018	Existence of designs one of the most celebrated results of these years	Absorption Nibble ...	Arxiv [500, 501]
Ferber, Nenadov, Noever, and Peter Škorić	2014 2014	Robust hamiltonicity of random directed graphs (resilience)	Connecting absorbing	Arxiv [339]
Balogh, Molla, and Sharifzadeh	2016	Triangle factor + small stable sets, weighted graphs	Absorption	RSA [75]
Barber, Kühn, Lo, and Osthus	2016	Edge decomposition of graphs with high mindeg	Iterative absorption	Advances [93]
Balogh-Lo-Molla	2017	Digraph packing in-out-degree $\geq 7n/18$	Stability Absorption	JCTB [73]
DeBiasio, Nelsen	2017	Strengthening Lehel conj.	Absorbing	JCTB [232]
Glock, Kühn, Lo, and Osthus	2018	Existence of designs	Iterative absorption connection	Arxiv [397]
Montgomery	2018	Embedding bounded degree trees into random graphs until $p = \Delta \log^5 n/n$	Iterative absorption?	Arxiv [616]
Hajnal, Herdade, and Szemerédi	2018	Pósa–Seymour without regularity lemma	Absorption connection	Arxiv [446]

In Table 1 we list just a few successful graph-applications of the ABSORBING METHOD. In some other cases, later, we shall just "point out" that the ABSORBING METHOD was successful, when we discuss the corresponding results, e.g., in Sects. 6.3, 6.7, 9,... We collected some papers using the ABSORBING METHOD for hypergraphs in Table 3 (primarily on hypergraph matching) and in Table 2, in Sect. 9 (primarily on Hamiltonian hypergraphs).

These three tables are self-explanatory, however, they contain just a short list of the applications. We could include several further results, like the results of Lo and Markström on multipartite Hajnal–Szemerédi results [576, 579], on graphs and hypergraphs, or [578]…

6.6 Connecting Lemma, Stability, Reservoir

In Table 1 we see several papers using the Absorbing and the Stability Methods. We illustrate this on the example of the new proof of the Pósa–Seymour conjecture [446], by Péter Hajnal, Herdade, and Szemerédi, for sufficiently large n. It has two subcases. A small $\alpha > 0$ is fixed and Case 1 (the "non-extremal" one) is when each $X \subset G_n$ of $\lfloor \frac{n}{k+1} \rfloor$ vertices has $e(X) \geq \alpha n^2$ edges. The remaining situation is Case 2, where we use stability:

Theorem 6.28 (Hajnal–Herdade–Szemerédi: Pósa–Seymour, stability [446]) *Given an integer $k \geq 2$ and an $\alpha > 0$, there exists an $\eta = \eta(\alpha, k) > 0$ such that in Case 1 (called α-non-extremal), if $d_{\min}(G_n) \geq (1 - \frac{1}{k+1} - \eta)k$, then G_n contains the kth power of a Hamiltonian cycle.*

Whenever we use the Absorbing method, mostly we use some other tools as well, tailored specifically to the problem in consideration. The Connecting Lemma and the Reservoir method are combined with the Absorbing Method, e.g., in the earlier paper of Levitt, Sárközy, and Szemerédi [575] on how to eliminate the use of the Regularity Lemma and the Blow-up Lemma in the proof of the Pósa conjecture. The Regularity Lemma, and the Blow-Up Lemma are eliminated in the tree-embedding paper of Csaba, Levitt, Nagy-György, and Szemerédi [215], using stability and some "elementary embedding methods".

The very recent new proof of the Seymour conjecture, by Hajnal, Herdade, and Szemerédi [446] is much more involved and much longer than the original proof. Here the authors use a "Connecting Lemma", asserting that certain parts of G_n can be connected in many advantageous ways. This means that we have a G_n (with large minimum degree) and wish to find a $C_n^{k+1} \subseteq G_n$. We cover most of the vertices by Turán graphs $T_{m,k+1}$ and $T_{m,k+2}$, where $m \to \infty$. Inside these "blocks" we can easily connect some vertices by $(k+1)$th power of a path covering this block, and we must connect these vertices from the various blocks so that altogether we get a $(k+1)$th power of a Hamiltonian cycle. The Connecting Lemma does this.

6.7 Ramsey–Turán Matching

This subsection is about the fifth line of Table 1, about [75]. The question is:

Does there a new, interesting phenomenon appear when we wish to ensure an almost perfect tiling, or a Hamiltonian cycle, or some other (almost) spanning configuration in a graph G_n and assume that

$$\alpha(G_n) = o(n). \tag{26}$$

We have to decide if we wish to use that $e(G_n)$ is large or that $d_{\min}(G_n)$ is large. We know that for ordinary non-degenerate extremal graph problems the edge-extremal and the degree-extremal problems do not differ too much: if **dex**$(n, \mathcal{L})$ is the maximum integer Δ for which, if G_n is $\mathcal{L}$-free, then $d_{\min}(G_n) \leq \Delta$, then

$$\mathbf{dex}(n, \mathcal{L}) \approx \frac{2}{n}\mathbf{ex}(n, \mathcal{L}), \tag{27}$$

and asymptotically $T_{n,p}$ is the degree-extremal graph (where p is defined by (4)). We also saw that if we assume (26) then in some cases (27) can be noticeably improved, see e.g., Theorem 5.20. In other cases (26) changes the maximum only in a negligible way. However, one can also ask what happens if we assume (26) in cases when we wish to ensure a spanning (or an almost-spanning) subgraph, e.g., a 1-factor, or a Hamilton cycle. Anyway, to ensure an (almost) 1-factor, or a Hamilton cycle it is better to have lower bounds on $d_{\min}(G_n)$ than on $e(G_n)$. Without too much explanation, we formulate two related results.

Balogh, McDowell, Molla, and Mycroft studied the minimum degree necessary to guarantee the existence of perfect and almost-perfect triangle-tilings in G_n with $\alpha(G_n) = o(n)$. Among others, they proved

Theorem 6.29 (Balogh, McDowell, Molla, and Mycroft (2018), [74]) *Fix an $\varepsilon > 0$. If (G_n) is a graph sequence with $\alpha(G_n) = o(n)$ and $d_{\min}(G_n) \geq n/3 + \varepsilon n$, then G_n has a triangle-tiling covering all but at most four vertices, if $n > n_0(\varepsilon)$.*

Of course, without the extra condition $\alpha(G_n) = o(n)$ we get back to the Corradi–Hajnal theorem, where $d_{\min}(G_n) \geq \frac{2}{3}n$ is needed. The case when we do not "tolerate" the four exceptional vertices is described by [75]:

Theorem 6.30 (Balogh, Molla, and Sharifzadeh [75]) *For every $\varepsilon > 0$, there exists a $\gamma > 0$ such that if $3|n$ and*

$$d_{\min}(G_n) \geq \left(\frac{1}{2} + \varepsilon\right)n, \quad \text{and} \quad \alpha(G_n) < \gamma n,$$

then G_n has a K_3-factor.[114]

The proofs in [75] also use the Stability method and the Absorbing technique of Rödl, Ruciński, and Szemerédi [694], discussed in Sect. 6.5.

---

[114]The paper has an Appendix written by Reiher and Schacht, about a version of this problem, also using the Absorption technique. In this version they replace the condition that any linear sized vertex-set contains an edge by a condition that any linear sized set contains "many edges".

7 Colouring, Covering and Packing, Classification

The setup. In this section we consider an r-EDGE-COLOURING of a K_N, or of a random graph $R_{n,p}$, or of any G_n satisfying some conditions. We have r families of potential subgraphs, $\mathcal{L}_i$, $(i = 1, \ldots, r)$, and try to cover K_N with as few monochromatic subgraphs $L_i \in \mathcal{L}_i$ in the ith colour as possible. However, there are several types of problems to be considered:

(A) sometimes we wish to cover all or almost all the *edges* of the coloured graph,

(B) in other cases we wish to cover all the *vertices* or almost all the vertices with the vertices of our monochromatic subgraphs.

These are quite different problems and in the next subsection we shall list several versions of these problems, and in some sense, classify them.

There are several problems/results in Extremal Graph Theory simple to formulate, and when we combine some of them, we get very interesting new problems. However, sometimes it is difficult to "classify" these problems. The reader could ask: "Why to classify them?" The answer is that without some classification one may end up with a chaotic picture about the whole field. Mostly,

we have a "host graph" G_n satisfying some conditions, e.g., it may be a complete graph K_N, or a random graph $R_{n,p}$, or a pseudo-random graph with many edges, or with large minimum degree. There is also a family $\mathcal{L}$ of "sample" graphs.[115] *Now, $E(G_n)$ is r-coloured, and*

(a) either we wish to PARTITION THE VERTICES of G_n into a few classes, $U_1, \ldots, U_t$ so that each $G[U_i]$ spans a monochromatic $L_i \in \mathcal{L}$, or

(b) we wish to COVER THE EDGES, $E(G_n)$, by a few copies of monochromatic sample graphs, $L_i \in \mathcal{L}$, or

(c) we wish to approximate the situation (a) or (b), allowing a few uncovered vertices.

Historical remarks. The case of one colour and vertex-disjoint packing goes back to several early papers in Extremal Graph Theory: several proofs, e.g., Erdős [265], or later Simonovits [751], used that large part of the considered G_n can be covered by vertex-disjoint copies of some L.[116] Also this was used in the new proof of the Pósa–Seymour conjecture, in [446].

Mostly we define K_1 and K_2 also as monochromatic graphs from $\mathcal{L}$: this is needed to ensure the existence of suitable colourings. The graphs in $\mathcal{L}$ may have more restricted or less restricted structure, e.g., they may be (A) all the connected graphs, or (B) all the trees, or (C) the sets of independent edges, or (D) all the cycles, or (E) the kth powers of the cycles, or (F) the Hamiltonian graphs, (G) on the other hand, they may be copies of the same fixed graph L.

We start with Covering Problems where we have only one colour. We could continue with Ramsey problems, when we wish to find just one monochromatic subgraph

---

[115] Here we took $\mathcal{L}_i = \mathcal{L}$.

[116] In some sense, this is used also in the original proof of Erdős–Stone theorem [321].

of a particular type, however, we have already written about Ramsey problems, and we shall not consider them here.

7.1 The One-Colour Covering Problem

In this section we consider edge-coverings. Here is an early problem of Gallai, corresponding to the simplest case, to the monochromatic G_n i.e. $r = 1$.

Question 7.1 Given a sample graph L and a graph G_n, how many copies of L and edges are enough to cover $E(G_n)$ (in the worst case)? In principle, we may require that the selected copies of L be (i) edge-disjoint, or (ii) vertex-disjoint, or (iii) we may allow them to overlap.

Of course, the L-free graphs need $e(L)$ edges, so the L-extremal graphs need **ex**(n, L) copies of L or edges. One feels that if $e(G_n)$ is much larger than **ex**(n, L), then one may use many copies of L and only a few edges. This motivates many results in this area. Here one of the first important results is

Theorem 7.2 (Erdős, Goodman, and Pósa (1966), [290]) *Any graph G_n can be edge-covered by $\lfloor \frac{n^2}{4} \rfloor$ complete subgraphs of G_n.*

Remark 7.3 (a) As it is stated in [290], to cover the edges, we may restrict ourselves to edges and triangles in Theorem 7.2.
(b) Theorem 7.2 is sharp, as shown by $T_{n,2}$.
(c) Theorem 7.2 was also proved by Lovász, as remarked in [290].

The extremal number for cycles is $n - 1$. Erdős and Gallai conjectured (see [290]) and Pyber proved

Theorem 7.4 (Pyber (1985), [667]) *Every graph on n vertices can be edge-covered by $n - 1$ cycles and edges.*

Note that for trees this is sharp. Pyber in [667] proved also some stronger results, and mentioned that the crucial tool in his proof was the following

Theorem 7.5 (Lovász (1967), [581]) *A graph on n vertices can be edge-covered by $\lfloor \frac{n}{2} \rfloor$ edge-disjoint path and cycles.*

Lovász proved also several related theorems, e.g.,

Theorem 7.6 (Lovász [581]) *Any graph G_n can be edge-covered by $\frac{2}{3}n$ double-stars (trees of diameter ≤ 3).*

Another result of this (early) Lovász paper is a generalization of the Erdős–Goodman–Pósa theorem.[117]

---

[117] For some related result for random or quasi-random graphs see [395, 398, 547, 558], and many others.

7.2 Embedding Monochromatic Trees and Cycles

In the following parts we consider edge-coloured graphs G_n and try to partition $V(G_n)$ into a few subsets V_i spanning some monochromatic Hamiltonian cycles,[118] or monochromatic powers of Hamiltonian cycles, or spanning trees.

These problems and results are related to Lehel's Conjecture 6.5 about partitioning the vertices of edge-coloured graphs into two monochromatic cycles, and the Gyárfás Conjecture 7.7, (see [429]), about partitioning the vertices of edge-coloured graphs into given monochromatic trees. Here the colourings are always edge-colourings.[119]

Gyárfás Tree-Conjecture

The topic of graph packing has at least two larger parts: the vertex-packing and the edge-packing. Here we are interested in packing some given graphs H_i into a G_n in an edge-disjoint way. This problem is interesting on its own and also was originally motivated by Theoretical Computer Science, more precisely, by COMPUTATIONAL COMPLEXITY, see e.g., the paper of Bollobás and Eldridge [123]. Given some graphs $H_1, \ldots, H_\ell$, their PACKING means finding the automorphisms $\pi_1, \ldots, \pi_\ell$ which map them into K_N in an edge-disjoint way.

For a family $H_1, \ldots, H_l$ of graphs, we say that they PACK into G, if they have edge-disjoint embeddings into G.

Conjecture 7.7 (Gyárfás (1978), [429]) *Let for $i = 1, 2, \ldots, n$, T_i be an i-vertex tree. Then K_N can be decomposed into these trees: $\{T_i\}$ pack into K_N.*

An asymptotic weakening of the conjecture was proved by Böttcher, Hladký, Piguet, and Taraz [142], for bounded degree trees.[120] This was improved to a sharp version:

Theorem 7.8 (Joos, J. Kim, Kühn, and Osthus (2016), [481]) *For any $\Delta > 0$ there exists an n_Δ such that for $n > n_\Delta$, if $T_1, \ldots, T_n$ are trees with $d_{\max}(T_i) < \Delta$ and $v(T_i) = i$ $(i = 1, \ldots, n)$, then $E(K_N)$ has a decomposition[121] into $T_1, \ldots, T_n$.*

Hence the tree packing conjecture of Gyárfás holds for all bounded degree trees, and sufficiently large n. Beside several further results, [481] also contains the following fairly general result.

Theorem 7.9 *Let $\delta > 0$ and Δ be fixed. Let $\mathcal{F}$ be a family of trees T^m of the following properties*[122]:

---

[118] Here "$G[V_i]$ is Hamiltonian" means that it has a spanning cycle.
[119] We used vertex-colouring in connection with colouring properties of excluded subgraphs, or equipartitions in Hajnal–Szemerédi theorem, ...
[120] In fact, one can allow the first $o(n)$ trees to have arbitrary degrees.
[121] I.e. $T_1, \ldots, T_n$ pack into K_N.
[122] We used superscript since here $v(T^m)$ is not necessarily m.

(i) For each $T^m \in \mathcal{F}$, $v(T^m) \leq n$, and $d_{\max}(T^m) \leq \Delta$.
(ii) For at least $(\frac{1}{2} + \delta)n$ values of m $v(T^m) \in [\delta n, (1-\delta)n]$.
(iii) $\sum_m e(T^m) = \binom{n}{2}$.
Then K_N can be decomposed into the trees T^m.

Remark 7.10 There is a vast literature on this topic and we recommend the reader to read the introduction of [481], and Gyárfás [426]. For results where we consider fewer trees, see e.g., Balogh and Palmer [83]. (The main tool of the proof in [83] is the Komlós–Sárközy–Szemerédi theorem [544] on embedding spanning trees.) Several related but slightly different problems are discussed, e.g., in the survey of Kano and Li [486], e.g., it describes several Anti-Ramsey decomposition problems as well.

— · —

One could ask what happens if we wish to embed some trees into a non-complete graph G_n. A nice result on this is

Theorem 7.11 (Gyárfás (1989), [425]) *If a sequence $T_1, T_2, \ldots, T_{n-1}$ of trees can be packed into K_N then they can be packed also into any n-chromatic graph.*

7.3 Bollobás–Eldridge Conjecture

Packing problems (strongly connected to Theoretical Computer Science) were discussed roughly the same time by Bollobás and Eldridge [123], Catlin [175], and Sauer and Spencer [730]. One of the most important conjectures in the field of graph-packing is

Conjecture 7.12 (Bollobás–Eldridge (1978), [123]) *Let H_1 and H_2 be two n-vertex graphs. If*
$$(d_{\max}(H_1) + 1)(d_{\max}(H_2) + 1) \leq n + 1,$$
then there is a packing of H_1 and H_2, i.e., there are two edge-disjoint subgraphs of a complete graph K_N isomorphic to H_1 and H_2, respectively.

The complementary form of this problem is

Conjecture 7.13 (Bollobás–Eldridge) *Let $d_{\max}(H_n) \leq k$. If G_n is a simple graph, with*
$$d_{\min}(G_n) \geq \frac{kn - 1}{k + 1},$$
then it contains H_n.

Aigner and Brandt [3], and Alon and Fischer [32] proved the conjecture for $d_{\max}(H_n) = 2$, i.e. when H_n is the union of cycles. Csaba, Shokoufandeh, and Szemerédi proved this for $d_{\max}(H_n) = 3$.

Theorem 7.14 (Csaba, Shokoufandeh, and Szemerédi [217]) *If G_n is a simple graph, with $d_{\min}(G_n) \geq \frac{1}{4}(3n-1)$, then it contains any H_n for which $d_{\max}(H_n) \leq 3$, if n is sufficiently large.*

Csaba [211] also proved this conjecture for $d_{\max}(H_n) = 4$, and in [212] for bipartite H_n, (where G_n is not necessarily bipartite) and $d_{\max}(H_n) \leq \Delta$.

Improving some results of Sauer and Spencer [730] and of Catlin, P. Hajnal, and Szegedy [447] considered some bipartite packing problems where they proved an asymmetric version: for the first graph they considered the maximum degree but for the second one only the average degree, thus improving the previous results for bipartite graphs.

The reader interested in more details will find a lot of information, e.g., in the survey of Kierstead, Kostochka, and Yu [514].

7.4 Vertex-Partitioning into Monochromatic Subgraphs of Given Type

This area has two parts: one about ordinary graphs and the second one about hypergraphs. The hypergraph results can be found in Sect. 9.8. Here we consider edge-colourings. Gerencsér and Gyárfás [390] proved that the vertices of any 2-coloured K_N can be partitioned into the vertex sets of two monochromatic paths of distinct colours. Here we consider problems where the *vertex set* of an r-coloured graph has to be partitioned into the vertex sets U_i of some *given types of monochromatic subgraphs*. We mention just a few such theorems, to give the flavour of these results.

Problem 7.15 Given a graph G_n and a family $\mathcal{L}$ of graphs. What is the minimum integer t for which every r-colouring of $E(G_n)$ has a vertex partition $V(G_n) = \dot{\bigcup} U_i$ into t vertex sets so that each $G[U_i]$ contains a monochromatic spanning $H_i \in \mathcal{L}$.[123]

Gyárfás, Sárközy, and Selkow [436] discussed a natural but much more general family of problems:

Problem 7.16 Given a family $\mathcal{L}$ of graphs, (trees, connected subgraphs, matchings, cycles,...) and two integers, $r \geq s \geq 1$. At least how many vertices of an r-edge-coloured K_N can be covered by s monochromatic subgraphs $H_i \in \mathcal{L}$ in this K_N?[124]

---

[123] More generally, we may fix for each colour i a family $\mathcal{L}_i$ and may try to cover $V(G_n)$ by vertex disjoint subgraphs $H_i \in \mathcal{L}_i$.

[124] Formally we have here two problems, one when we r-colour $E(K_N)$, the other when we r-colour $E(G_n)$, however, the difference "disappears" if r is large. Further, we may also ask for the largest subgraph $H \subseteq K_N$ that is coloured by at most t colours, which is different from asking for the largest number of edges covered by t monochromatic $H_i \in \mathcal{L}$.

The simplest case is $r = 1$, where we use only one colour. Another simple subcase is when we RED-BLUE-colour the edges of a K_N and study how many monochromatic cycles are needed (in the worst case) to cover the vertices. We cannot always partition the edges into monochromatic cycles, unless we agree that the vertices and the edges are also regarded as cycles.[125] In this case we can always partition $E(G_n)$ into $e(G_n)$ monochromatic "cycles": real cycles, edges and vertices. The next part contains some repetition from Sect. 6, primarily from Sect. 6.5. We start with

Conjecture 7.17 (Gyárfás (1989), [424], proved) *There exists an integer $f(r)$ independent of n such that if $E(K_N)$ is r-coloured, then $V(K_N)$ can always be covered by $f(r)$ vertex-disjoint monochromatic paths.*

Actually, Gyárfás formulated three conjectures in [424]. In the first one he wanted to cover $V(K_N)$ by $f(r) = r$ vertex-disjoint monochromatic paths, in the second, weaker one, the vertex-disjointness was not assumed, and the third, weakest one, was Conjecture 7.17. ($f(r) = r$ would yield the first conjecture.) Gyárfás proved the following weakening of his conjectures:

Theorem 7.18 (Gyárfás [424]) *There exists an integer $f(r) > 0$ such that in any r-colouring of $E(K_N)$ one can cover the vertices by $f(r)$ monochromatic paths.*

This result does not assert that the covering paths are vertex-disjoint. The proof of Gyárfás gave an explicit $f(r) \approx r^4$. This was improved by Erdős, Gyárfás, and Pyber in Theorem 6.25: In any r-edge-colouring of K_N, one can cover $V(G_n)$ by $p(r) = O(r^2 \log r)$ vertex-disjoint monochromatic cycles. This was further improved by

Theorem 7.19 (Gyárfás–Ruszinkó–Sárközy–Szemerédi [431]) *If $E(K_N)$ is r-coloured, then $V(K_N)$ can be partitioned into $O(r \log r)$ vertex-sets of monochromatic cycles.*

We have to point out two things.

(a) While the problems with cycles and paths are basically of the same difficulty in the Erdős–Gallai theorems, here the problems on covering *with cycles* are much more difficult.

(b) Covering with *vertex-disjoint* paths/cycles is significantly more difficult than the case when vertex-disjointness is not assumed.

Here the most important conjecture *was*

Conjecture 7.20 (Erdős, Gyárfás, and Pyber [291]) *In Theorem 7.19 $p(r) = r$ vertex-disjoint monochromatic cycles are enough.*

This was proved for $r = 2$ by Bessy and Thomassé (see Subsection 7.4) but "slightly disproved" for $r \geq 3$ by Pokrovskiy [654], see Remark 6.8. A whole theory emerged around this conjecture, and below we provide a very short description of it, basically following Sárközy's paper [726], which starts with a good "minisurvey" about this area. He extends the problem to covering by powers of cycles, (as Pósa and Seymour extended Dirac's Hamiltonicity theorem). Sárközy proved

---

[125] If for a fixed x, all edges xy are Red, and the other edges are Blue, then we need this.

Theorem 7.21 (Sárközy (2017), [726]) *For every integer $k \geq 1$ there exists a constant $c(k)$ such that in any 2-colouring of $E(K_N)$ at least $n - c(k)$ vertices can be covered by $200k^2 \log k$ vertex-disjoint monochromatic kth powers of cycles.*

The interested reader is referred to [726]. We shall return to the corresponding *hypergraph problems* in Sect. 9.8. We close this part with a result of Grinshpun and Sárközy [422] on a conjecture of Gyárfás where the cycles are replaced by an arbitrary fixed family of bounded degree graphs. (This covers the case of the kth powers of cycles.)

Fix a degree bound Δ and let $\mathcal{F} = \{F_1, F_2, \ldots\}$ be any given family of graphs, where F_r is an r-vertex graph of maximum degree at most Δ.

Theorem 7.22 (Grinshpun and Sárközy [422]) *There exists an absolute constant C such that for any 2-colouring of $E(K_N)$, there is a vertex partition of K_N into (vertex sets of) monochromatic copies of members of $\mathcal{F}$ with at most $2^{C\Delta \log \Delta}$ parts.*

If $\mathcal{F}$ consists of *bipartite graphs* then $2^{C\Delta \log \Delta}$ can be replaced by $2^{c\Delta}$, which is best possible, apart from the value of the constant c:

Theorem 7.23 (Grinshpun and Sárközy [422]) *Let $\mathcal{F}$ be a family of bipartite graphs with maximum degree Δ. There is an absolute constant c such that for any 2-edge colouring of K_N, there is a vertex partition of K_N into (vertex sets of) monochromatic copies of $F_r \in \mathcal{F}$ with at most $2^{c\Delta}$ parts.*

These results are strongly connected to some results of Conlon, Fox and Sudakov [196]. We close this section with an open problem. Related results can be found in the very recent paper of Bustamante, Corsten, D. Frankl, Pokrovskiy, and Skokan [169].

Problem 7.24 Do the Grinshpun–Sárközy theorems extend to three colours?

8 Hypergraph Extremal Problems, Small Excluded Graphs

Until now we primarily concentrated on two Universes: graphs and integers. Here we include a short section on the Universe of hypergraphs. We used hypergraphs, e.g., in estimating independence numbers in Sect. 4.3 in "uncrowded graphs and hypergraphs", to improve a Ramsey number estimate, $R(3, k)$, to disprove the Heilbronn conjecture, see Sect. 4.7, and in case of the infinite Sidon sequences, and several further results, in Sects. 4.3–4.11. The parts in Sect. 5.5 connected to the Removal Lemma were also in some sense hypergraph results. Here we "start again", but go into other directions. We refer the readers interested in more details to the surveys of Sidorenko, [749], Füredi [369, 370], G.O.H. Katona [489], Keevash [498], of Kühn and Osthus [563], and of Rödl and Ruciński [689].

Below, to emphasize that we consider hypergraphs, occasionally we shall use a different typesetting, e.g., for r-uniform hypergraphs, for the excluded hypergraphs we may use $\mathbb{L}^{(r)}$, instead of L, and $\mathcal{L}^{(r)}$ for the family of excluded hypergraphs (Fig. 9).

We start with the case of "Small excluded subhypergraphs". Here "small" means that the excluded hypergraphs have bounded number of vertices. In Sect. 9 we shall consider the case of 1-factors and the problem of ensuring Hamiltonian cycles and other "spanning" or "almost spanning" configurations. For the sake of simplicity, we mostly (but not always) restrict ourselves to 3-uniform hypergraphs. The extremal number for r-uniform hypergraphs will be denoted also by $\mathbf{ex}(n, \mathcal{L})$, or, to emphasise that we consider r-uniform hypergraphs, we may write $\mathbf{ex}_r(n, \mathcal{L})$, or $\mathbf{ex}_r(n, \mathcal{L}^{(r)})$.[126] There are many interesting results on hypergraph extremal problems. We mention the easy theorem of Katona, Nemetz and Simonovits [493], according to which $\mathbf{ex}_r(n, \mathcal{L})/\binom{n}{r}$ is monotone decreasing, non-negative, and therefore convergent.

As for simple graphs, for hypergraphs we can also distinguish DEGENERATE and NON-DEGENERATE extremal graph problems: an r-uniform problem is degenerate if $\mathbf{ex}_r(n, \mathcal{L}) = o(n^r)$. For graphs the Kővári–Sós–Turán theorem was the key in this. A fairly simple result of Erdős [266] generalizes the Kővári–Sós–Turán theorem. Let $\mathbb{K}_r^{(r)}(t_1, t_2, \ldots, t_r)$ denote the r-uniform hypergraph in which the vertex set V is partitioned into $V_1, \ldots, V_r$, $|V_i| = t_i$, and the hyperedges are the "transversals": r-tuples intersecting each V_i in one vertex (Fig. 10).

Theorem 8.1 (Erdős (1964), [266]) *If $t = t_1 \leq t_2 \leq \ldots \leq t_r$, then*

$$\mathbf{ex}_r(n, \mathbb{K}_r^{(r)}(t, t_2, \ldots, t_r)) = O(n^{r-1/t^{r-1}}).$$

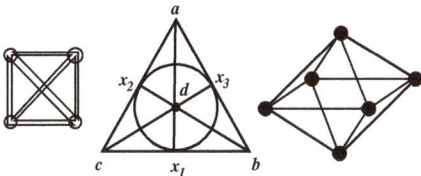

Fig. 9 Small excluded hypergraphs: complete hypergraph, Fano hypergraph and the Octahedron hypergraph

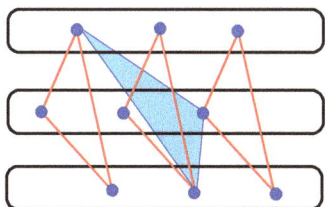

Fig. 10 Three-partite 3-uniform hypergraph

---

[126]Here we use L for an excluded graph, $\mathbb{L}$ for a hypergraph, and $\mathcal{L}$ for a family of graphs or hypergraphs.

This implies, exactly as for $r = 2$, that

Corollary 8.2 $\mathbf{ex}_r(n, \mathcal{L}^{(r)}) = o(n^r)$ *if and only if there is an* $\mathbb{L} \in \mathcal{L}^{(r)}$ *of strong chromatic number* r.[127]

Remark 8.3 As in Corollary 2.12, if the problem is non-degenerate, then the "density constant jumps up":
$$\mathbf{ex}_r(n, \mathcal{L}^{(r)}) > \left(\frac{1}{r^r} + o(1)\right) n^r. \tag{28}$$

A famous problem of Erdős was whether for 3-uniform hypergraphs $1/27$ is a "jumping constant": does there exist a constant $c > 0$ such that if for some $\eta > 0$ $\mathbf{ex}_r(n, \mathcal{L}^{(r)}) > \left(\frac{1}{27} + \eta + o(1)\right) n^r$ then $\mathbf{ex}_r(n, \mathcal{L}^{(r)}) > \left(\frac{1}{27} + c + o(1)\right) n^r$ also holds. There are many related results, here we mention only the breakthrough paper of Frankl and Rödl [356], which however, does not decide if $1/27$ is a "jumping constant" or not. We mention that according to Pikhurko [648] there are continuum many limit densities and there are among them irrational ones even for finite families of excluded r-graphs. We also recommend to read Baber and Talbot [58] and several papers of Y. Peng, e.g., [644], in this area.

Turán's Conjecture

Consider now 3-uniform hypergraphs: $\mathbb{H}^{(3)} = (V, \mathcal{E})$. To formulate the famous hypergraph conjectures of Paul Turán in the two simplest cases, we need two constructions. We shall call an r-uniform hypergraph h-PARTITE if its vertices can be partitioned into h classes, none of which contains hyperedges.

(a) For the excluded 3-uniform complete 4-graph $\mathbb{K}_4^{(3)}$, consider the 3-uniform hypergraph $\mathbb{H}_n^{(3)}$ obtained by partitioning n vertices into 3 classes U_1, U_2 and U_3 as equally as possible and taking the triples of form (x, y, z) where x, y, and z belong to different classes, and the triplets (x, y, z) where x and y belong to U_i and z to U_{i+1}, for $i = 1, 2, 3$, where $U_4 := U_1$, see Fig. 11. Turán conjectured that this is

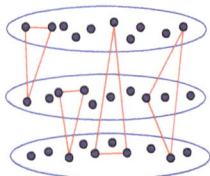

Fig. 11 $\mathbb{K}_4^{(3)}$-extremal?

---

[127] The strong chromatic number $\chi_S(\mathbb{F}^{(r)})$ of $\mathbb{F}^{(r)}$ is the minimum ℓ for which the vertices of $\mathbb{F}^{(r)}$ can be ℓ-coloured so that each hyperedge gets r distinct colours. Our condition is equivalent with that some $\mathbb{F}^{(r)} \in \mathcal{L}^{(r)}$ is a subgraph of $\mathbb{K}_r^{(r)}(a, \ldots, a)$ for some large a.

the extremal hypergraph for $\mathbb{K}_4^{(3)}$. This is unknown, we do not know even if this is asymptotically sharp.

Actually, first Katona, Nemetz, and Simonovits [493] gave examples (for $n = 3m + 1$) showing that if Turán's conjecture holds, then the uniqueness of extremal graphs does not always hold.

A more general construction of Brown [151], generalized by Kostochka [548], shows that if the conjecture holds, then there are many-many extremal graph structures for the extremal hypergraph problem of $\mathbb{K}_4^{(3)}$ and $n = 3t$. For a slightly more detailed description of this see e.g., Fon-der-Flaass [346], Razborov, [672], Simonovits [764].

(b) For the excluded complete 5-graph $\mathbb{K}_5^{(3)}$ Turán had a construction—for the potential extremal hypergraph—with four classes and another one with two classes. The one with two classes is simple, see Fig. 12. It is a COMPLETE BIPARTITE hypergraph: we partition the vertices into $\mathcal{A}$ and $\mathcal{B}$ and consider all the triples intersecting both classes. V.T. Sós observed that the construction with two classes can be obtained from the construction with four classes by moving around some triples in some simple way. Probably J. Surányi found a construction showing that Turán's conjecture for $\mathbb{K}_5^{(3)}$ is false for $n = 9$. Kostochka (perhaps) generalized this, founding counterexamples for every $n = 4k + 1$.[128] Still Turán's conjecture may be sharp, or, at least, asymptotically sharp.[129]

Among the new achievements we mention Razborov's Flag Algebra method and his results on hypergraphs [669, 671, 672].

A Simple Hypergraph Extremal Problem

As we have often emphasized, to solve a hypergraph extremal problem is mostly hopeless, despite that recently many nice results were proved on hypergraphs. Keevash described this situation by writing (in MathSciNet, on [379], in 2008):

"An important task in extremal combinatorics is to develop a theory of Turán problems for hypergraphs. At present there are very few known results, so it is interesting to see a new example that can be solved."

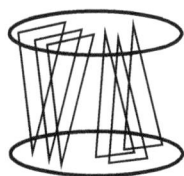

Fig. 12 $\mathbb{K}_5^{(3)}$-extremal?

---

[128]These constructions seem to be forgotten, "lost" and are not that important.

[129]Since for hypergraphs we have at least two popular chromatic numbers, therefore the expression r-uniform ℓ-partite may have at least two meanings in the related literature.

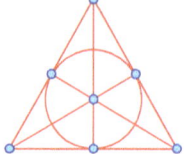

Fig. 13 Fano

Consider 3-uniform hypergraphs. The difficulties are reflected, among others, by that we do not know the extremal graph for $\mathcal{L}_{4,3}^{(3)}$, i.e. when we exclude the 4-vertex hypergraph with 3 triples.[130] The next question is among the easier ones. G.O.H. Katona asked and Bollobás solved the following extremal problem.

Theorem 8.4 (Bollobás (1974), [116]) *If a 3-graph has $3n$ vertices and $n^3 + 1$ triples, then there are two triples whose symmetric difference is contained in a third one.*

$\mathbb{K}_3^{(3)}(n, n, n)$—which generalizes $T_{2n,2}$, to 3-uniform hypergraphs,—shows the sharpness of Theorem 8.4. So Theorem 8.4 is a natural generalization of Turán's theorem: an ordinary triangle-free graph G_n is just a graph where no edge of G_n contains the symmetric difference of two other edges. So the excluded hypergraph can be viewed as a hypergraph-triangle. Bollobás generalized Katona's conjecture to r-uniform hypergraphs. The generalized conjecture was proved for $r = 4$ by Sidorenko [744], for $r = 5, 6$ by Norin and Yepremyan [634]. For related results see e.g., Sidorenko [744], Mubayi and Pikhurko [625], Pikhurko [646].

The Fano Hypergraph Extremal Problem

Here the excluded graph is the 3-uniform Fano hypergraph $\mathbb{F}_7^{(3)}$ on 7 vertices, with seven hyperedges any two of which intersect in exactly one vertex, see Fig. 13: $\mathbb{F}_7^{(3)}$ is the simplest finite geometry. The nice thing about the extremal problem of $\mathbb{F}_7^{(3)}$ is that it is natural, non-trivial, but has a nice solution.

Conjecture 8.5 (V. T. Sós) *Partition n vertices into two classes $\mathcal{A}$ and $\mathcal{B}$ with $||\mathcal{A}| - |\mathcal{B}|| \leq 1$ and take all the triples intersecting both $\mathcal{A}$ and $\mathcal{B}$. The obtained 3-uniform complete bipartite hypergraph $\mathbb{H}[\mathcal{A}, \mathcal{B}]$ is extremal for $\mathbb{F}_7^{(3)}$ (if n is sufficiently large).*

Using some multigraph extremal results of Kündgen and Füredi [378], first de Caen and Füredi proved

---

[130] Generally, $\mathcal{L}_{k,\ell}^{(r)}$ is the family of r-uniform hypergraphs of k vertices and ℓ hyperedges. As we have mentioned, the problem of $\mathbf{ex}(n, \mathcal{L}_{k,\ell}^{(r)})$ was considered in two papers of Brown, Erdős, and Sós [154, 155] and turned out to be very important in this field. Originally Erdős conjectured a relatively simple asymptotic extremal structure, for $\mathcal{L}_{4,3}^{(3)}$ but his conjecture was devastated by a better construction of Frankl and Füredi [355]. This construction made this problem rather hopeless.

Theorem 8.6 (de Caen and Füredi (2000), [172])

$$\mathbf{ex}(n, \mathbb{F}_7^{(3)}) = \frac{3}{4}\binom{n}{3} + O(n^2).$$

Next, applying the stability method, the sharp result was obtained, independently, by Füredi and Simonovits and by Keevash and Sudakov. Since $\chi(\mathbb{F}_7^{(3)}) = 3$, a 3-uniform bipartite hypergraph cannot contain $\mathbb{F}_7^{(3)}$.

Theorem 8.7 (Füredi–Simonovits (2005), [385]/Keevash-Sudakov (2005), [504]) *If $\mathbb{H}_n^{(3)}$ is a triple system on $n > n_1$ vertices not containing $\mathbb{F}_7^{(3)}$ and of maximum number of hyperedges under this condition, then $\mathbb{H}_n^{(3)}$ is bipartite: $\chi(\mathbb{H}_n^{(3)}) = 2$.*

Theorem 8.7 implies that

$$\mathbf{ex}_3(n, \mathbb{F}_7^{(3)}) = \binom{n}{3} - \binom{\lfloor n/2 \rfloor}{3} - \binom{\lceil n/2 \rceil}{3}.$$

There are two important ingredients of the proof. The first one is a multigraph extremal theorem:

Theorem 8.8 (Füredi-Kündgen [378]) *If M_n is an arbitrary multigraph (without restriction on the edge multiplicities, except that they are non-negative) and each 4-vertex subgraph of M_n has at most 20 edges (with multiplicity), then*

$$e(M_n) \leq 3\binom{n}{2} + O(n).$$

The other ingredient of the proof of Theorem 8.7 was that it is enough to prove the theorem for those hypergraphs where the low-degree vertices are deleted, and it is enough to prove a corresponding stability theorem.

Theorem 8.9 *There exist a $\gamma_2 > 0$ and an n_2 such that if $\mathbb{F}_7^{(3)} \not\subseteq \mathbb{H}_n^{(3)}$ and*

$$\deg(x) > \left(\frac{3}{4} - \gamma_2\right)\binom{n}{2} \quad \text{for each} \quad x \in V(\mathbb{H}_n^{(3)}),$$

then $\mathbb{H}_n^{(3)}$ is bipartite.

Recently Bellmann and Reiher [98] proved that Theorem 8.7 holds for any $n \geq 7$. One could ask, what is the essence of their proof. Often when we use stability arguments for extremal problems, one thinks that perhaps induction would also work. Unfortunately, for hypergraphs this mostly breaks down. As an exception, Bellmann and Reiher have found the good way to use here induction. (A similar situation was when the Lehel Conjecture was proved by Bessy and Thomassé [108], yet for hypergraphs we do not know of such cases.)

Füredi, Pikhurko, and Simonovits used the stability method also in [381], to prove a conjecture of Mubayi and Rödl. For further related results see, e.g., Keevash [494], Füredi, Pikhurko, and Simonovits [381, 382], ...and many similar cases.

— · —

We should mention here a beautiful result of Person and Schacht:

Theorem 8.10 (Person and Schacht (2009), [645]) *Almost all Fano-free 3-uniform hypergraphs are bipartite.*

Remark 8.11 (a) One could ask what is the connection between Theorems 8.7 and 8.10. Here one should be careful, e.g., Prömel and Steger [664] proved that almost all Berge graphs are perfect, which means that for most graphs the Berge Strong Perfect Graph Conjecture is true. This was a beautiful result, however, the actual proof of the Perfect Graph Conjecture [179] (which came much later) was much more difficult.

(b) Several similar results are known, where the typical structure of $\mathcal{F}$-free graphs are nicely described. For graphs this was discussed in Sect. 2.13. Several related results were also proved for hypergraphs, e.g., Lefmann, [572].

(c) Some further related results can be found in the papers of Cioabă [191], Keevash [494], Balogh–Morris–Samotij–Warnke [79], or Balogh and Mubayi [81], and in many further cases.

8.1 Codegree Conditions in the Fano Case

As we have already discussed, as soon as we move to hypergraphs, many notions, e.g., the notion of path, cycles and of degrees also can be defined in several different ways. Restricting ourselves to the simplest case of 3-uniform hypergraphs, let $\delta_2(x, y)$ be the number of vertices z for which (x, y, z) is a hyperedge, and $\delta_2(\mathbb{H}_n^{(3)}) = \min \delta_2(x, y)$. Usually $\delta_2(x, y)$ is called the CODEGREE of x and y. Often it is more natural to have conditions on the minimum codegree than on the min-degree.

Theorem 8.12 (Mubayi [622]) *For any $\alpha > 0$, there exists an $n_0(\alpha)$ for which, if $n > n_0$ and $\delta_2(\mathbb{H}_n^{(3)}) \geq (\frac{1}{2} + \alpha)n$, then $\mathbb{H}_n^{(3)}$ contains the Fano plane $\mathbb{F}_7^{(3)}$.*

The sharpness comes from the complete bipartite 3-uniform hypergraph. Mubayi conjectured that, for large n, $(\frac{1}{2} + \alpha)$ can be replaced by $\frac{1}{2}$. This was proved first by Keevash, and then, in a simpler way, by DeBiasio and Jiang:

Theorem 8.13 (Keevash [496]/DeBiasio and Jiang [231]) *There is an n_0 such that if $n > n_0$ and $\delta_2(\mathbb{H}_n^{(3)}) > \lfloor \frac{1}{2}n \rfloor$ then $\mathbb{H}_n^{(3)}$ contains $\mathbb{F}_7^{(3)}$.*

Actually, there are more and more extremal results where we derive the existence of some substructure by knowing that the minimum codegree is high. Later we shall see several further such examples, e.g., in Sect. 9: to ensure a TIGHT Hamiltonian cycle,[131] Theorem 9.12 will use the minimum vertex-degree, while Theorem 9.13 uses that the minimum codegree is large.

Remark 8.14 (a) DeBiasio and Jiang gave a fairly elementary proof, not using the Regularity Method. They had to assume that n is sufficiently large only to use some supersaturation argument.

(b) In their paper DeBiasio and Jiang [231] start with a nice introduction and historical description of the situation.

(c) Both the Keevash paper and the DeBiasio-Jiang paper go beyond just proving the above formulated extremal result.

9 Large Excluded Hypergraphs

While for ordinary graphs the definition of cycles is very natural, for hypergraphs there are several distinct ways to define them, leading to completely different problems and results. Below mostly we shall restrict ourselves to 3-uniform hypergraphs, within that to LOOSE (also called LINEAR) and TIGHT cycles, and we mostly skip Berge cycles.[132] The reader interested in more details about Berge cycles is referred to Gyárfás, Sárközy, and Szemerédi [433, 439] and also to the excellent surveys of Rödl and Ruciński [689], Kühn and Osthus [565].

We start this section with a short subsection on Hypergraph cycle Ramsey problems which could be regarded as medium range extremal problems, since we get in a hypergraph on N vertices a (monochromatic) subgraph of $(1 - c)N$ vertices. Then we shall consider hypergraph extremal problems where the excluded configuration is a spanning subhypergraph, or at least an *almost* spanning one. To make the life of the reader (interested in more details) easier, we follow the definitions and notation of [689], as much as we could.

9.1 Hypergraph Cycles and Ramsey Theorems

We shall need the definition of the so called (k, ℓ)-cycles, covering the matchings, the LOOSE cycles and the TIGHT cycles.

Definition 9.1 (*Tight/loose cycles*) Consider 3-uniform hypergraphs. Let $E_1, \ldots, E_\ell$ be a cyclically arranged family of triples. If the consecutive ones intersect in 2 vertices, and there are no other intersections among them, then this configuration $\widetilde{C}_\ell^3$

---

[131] Defined after Definition 9.2.

[132] Hamiltonicity for Berge cycles were discussed by Bermond, Germa, Heydemann, and Sotteau [104]. Perhaps the first Tight Hamiltonicity was discussed in [492], however, the tight Hamilton cycles were called there Hamilton chains.

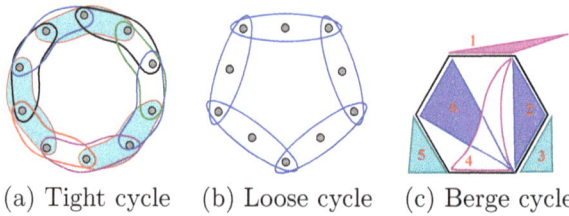

(a) Tight cycle (b) Loose cycle (c) Berge cycle

Fig. 14 Various hypergraph cycles

will be called a TIGHT ℓ-cycle. If the consecutive ones intersect in 1 vertex, and there are no other intersections among them, then this configuration $\mathcal{C}_\ell^3$ will be called a LOOSE ℓ-cycle.

We could have started with the more general

Definition 9.2 ((k, ℓ)-cycle) Fix some $0 \leq \ell \leq k-1$. In a k-uniform hypergraph $\mathbb{H}_n^{(k)}$ the cyclically ordered vertices $a_1, \ldots, a_t$ and the hyperedges $E_1, \ldots, E_t$ form a (k, ℓ)-cycle, if the vertices of E_i are cyclically consecutive (form a segment), and $|E_i \cap E_{i+1}| = \ell$, for $i = 1, \ldots, t$. (Here $E_{t+1} := E_1$).

If $\ell = 1$, we call it a LOOSE cycle, (or LINEAR cycle) and if $\ell = k-1$ it is a TIGHT cycle.

For $\ell = 0$, this reduces to a matching: a family of t/k independent edges.

Divisibility. Mostly we have some hidden divisibility conditions, e.g., speaking of a (k, ℓ)-cycle $\mathcal{C}_m^{k,\ell}$ on m vertices, we assume that m is divisible by $k - \ell$.

— · —

It was known from the beginning that the (ordinary) Ramsey number for cycles strongly depends of the parity: $R(C_n, C_n) = \frac{3}{2}n - 1$ if n is even and $R(C_n, C_n) = 2n - 1$ if n is odd (Bondy [135], Faudree and Schelp [332], Vera Rosta [677]). So it is not too surprising that for hypergraphs the parity also strongly influences the results. (Analogous results are known for ordinary graphs and 3 or more colours, e.g., as we have mentioned, $R(n, n, n) = 4n - 3$ for odd $n > n_0$, see Łuczak, [600], Kohayakawa, Simonovits, and Skokan [530], but it is $2n + o(n)$ if n is even, see Figaj and Łuczak [341, 342].)[133] For hypergraphs the situation is even more involved, since the Ramsey numbers depend on the types of cycles we consider (loose, tight, Berge). Haxell, Łuczak, Peng, Rödl, Ruciński, Simonovits, and Skokan proved the following

Theorem 9.3 (Haxell et al. (2006), [458]) *Consider 3-uniform hypergraphs. If $\mathcal{C}_n^3$ denotes the* LOOSE *n-vertex hypergraph cycle, then* $R(\mathcal{C}_n^3, \mathcal{C}_n^3, \mathcal{C}_n^3) = 5n/4 + o(n)$.

Here the upper bound is the difficult part: an easy construction shows that $R(\mathcal{C}_n^3, \mathcal{C}_n^3, \mathcal{C}_n^3) > 5n/4 - c$, for an appropriate constant $c > 0$. This was generalized

---

[133] The breakthrough result of Łuczak [600] was improved by Kohayakawa, Simonovits, and Skokan [530] and generalized by Jenssen and Skokan [479], see also Gyárfás, Ruszinkó, Sárközy, and Szemerédi [432] and Benevides and Skokan [99] and others.

from 3-uniform to k-uniform graphs and loose cycles by Gyárfás, Sárközy, and Szemerédi [437]. As to the tight cycles, Haxell, Łuczak, Peng, Rödl, Ruciński, and Skokan proved the following

Theorem 9.4 (Haxell et al. (2006), [459]) *Consider 3-uniform hypergraphs. If $\widetilde{C}_n^3$ denotes the* TIGHT *n-vertex hypergraph cycle, then $R(\widetilde{C}_n^3, \widetilde{C}_n^3, \widetilde{C}_n^3) = 4n/3 + o(n)$ when n is divisible by 3, and $\approx 2n$ otherwise.*

On Ramsey numbers for Berge cycles see the papers of Gyárfás, Sárközy, and Szemerédi [428], and Gyárfás and Sárközy [433], or Gyárfás, Lehel, G. Sárközy, and Schelp [430].

9.2 Hamilton Cycles

This research area has two roots:
(A) For simple graphs the extremal graph problem of k independent edges goes back to Erdős and Gallai [289]. It may be surprising, but to ensure k independent edges, or a path P_{2k} requires basically the same degree condition. On the other hand, maybe it is not so surprising. Rödl and Ruciński [689] write: "...for $\ell = 0$ a Hamiltonian ℓ-cycle in a k-graph H becomes a perfect matching in H. Moreover, any Hamiltonian $(k - \ell)$-cycle contains a matching of size $\lfloor n/k \rfloor$. Hence, not surprisingly, the results for Hamiltonian cycles and perfect (or almost perfect) matchings are related." The Erdős–Gallai results were extended to hypergraphs, by Erdős [268]. Bollobás, Daykin, and Erdős [122] ensured t independent hyperedges, Daykin and Häggkvist [229] guaranteed a perfect matching.

(B) G.Y. Katona and Kierstead [492] defined the tight Hamilton cycle and tried to generalize Dirac's theorem to hypergraphs.

The area described in the next few subsections in a fairly concise way is described in much more details, e.g., in the excellent surveys of Kühn and Osthus [563] and of Rödl and Ruciński [689], in the Introduction of the paper of Alon, Frankl, Huang, Rödl, Ruciński, and Sudakov [35], and in the survey of Yi Zhao [825].

The Beginnings: Hamiltonicity

Since hypergraph problems seemed mostly too technical even for combinatorists working outside of hypergraph extremal problems, the research on hypergraph Hamiltonicity started relatively late, with a paper of Katona and Kierstead [492].[134] For k-uniform hypergraphs, for a subset $\mathcal{S} \subseteq V(\mathbb{H}_n^{(k)})$ with $\ell := |\mathcal{S}|$ we define the ℓ-degree $\delta_\ell(\mathcal{S})$ as the number of k-edges of $\mathbb{H}_n^{(k)}$ containing $\mathcal{S}$ and $\delta_\ell(\mathbb{H}_n^{(k)})$ is the min-

---

[134] For Berge cycle Hamiltonicity there were earlier results, e.g., by Bermond, Germa, Heydemann, and Sotteau [104].

imum of $\delta_\ell(\mathcal{S})$ for the ℓ-tuples $\mathcal{S} \subseteq V(\mathbb{H}_n^{(k)})$. G.Y. Katona and Kierstead considered the tight cycles[135] and paths and proved

Theorem 9.5 (Katona, Kierstead (1999), [492]) *If $\mathbb{H}_n^{(k)} = (V, \mathcal{E})$ is a k-uniform hypergraph and*

$$\delta_{k-1}(\mathbb{H}_n^{(k)}) \geq \left(1 - \frac{1}{2k}\right)n + 4 - k - \frac{5}{2k},$$

then it contains a TIGHT *Hamiltonian cycle* $\widetilde{\mathcal{C}}_n^k$.

This is far from being sharp:

Conjecture 9.6 (Katona, Kierstead [492]) *If $\delta_{k-1}(\mathbb{H}_n^{(k)}) > \lfloor \frac{n-k+3}{2} \rfloor$, then $\mathbb{H}_n^{(k)}$ has a* TIGHT *Hamilton cycle.*

Katona and Kierstead also provided the construction supporting this:

Theorem 9.7 (Katona and Kierstead [492]) *For any integers $k \geq 2$ and $n > k^2$ there exists a k-uniform hypergraph $\mathbb{H}_n^{(k)}$ without* TIGHT *Hamilton cycles for which $\delta_{k-1}(\mathbb{H}_n^{(k)}) = \lfloor \frac{n-k+3}{2} \rfloor$.*

Rödl and Ruciński write in [689]:

"In 1952 Dirac [242] proved a celebrated theorem… In 1999, Katona and Kierstead initiated a new stream of research to studying similar questions for hypergraphs, and subsequently, for perfect matchings…".

9.3 Problems: Hypergraph Hamiltonicity

We restrict our attention primarily to 3-uniform hypergraphs. As Rödl, Ruciński, Schacht, and Szemerédi [692] describe, here there are (at least) six different questions:

(i) how large VERTEX- DEGREE ensures a TIGHT Hamiltonian cycle,
(ii) how large VERTEX- DEGREE ensures a LOOSE Hamiltonian cycle,
(iii) how large CO- DEGREE ensures a TIGHT Hamiltonian cycle,
(iv) how large CO- DEGREE ensures a LOOSE Hamiltonian cycle,
(v-vi) and how large degrees/co-degrees ensure a PERFECT MATCHING or an almost perfect matching?

In all these problems some divisibility questions should also be handled.

The goal. As a first step, we would say that most of the research in this area is related to describing two functions defined for k-uniform n-vertex graphs, the thresholds for the Hamiltonicity, $h_d^\ell(k, n)$, and for the Matching, $m_d^r(k, n)$:

What is the minimum integer t for which, if the d-degree $\delta_d(\mathbb{H}_n^{(k)}) \geq t$ in a k-uniform hypergraph $\mathbb{H}_n^{(k)}$, then $\mathbb{H}_n^{(k)}$ contains

---

[135] They called it chain.

(i) a Hamilton ℓ-cycle $\mathcal{C}_n^{k,\ell}$

(ii) an almost-matching $\mathcal{M}_n^{k,r}$, leaving out at most r vertices, respectively.

Before going into details, we formulate two typical theorems, for 3-uniform hypergraphs, for TIGHT cycles.

Theorem 9.8 (Reiher, Rödl, Ruciński, Schacht, and Szemerédi (2016), [674]) *For any $\eta > 0$, there exists an $n_0(\eta)$ such that if $n > n_0$ and in an n-vertex 3-uniform hypergraph $\mathbb{H}_n^{(3)}$ the minimum degree*

$$d_{\min}(\mathbb{H}_n^{(3)}) \geq \left(\frac{5}{9} + \eta\right)\binom{n}{2},$$

then it contains a TIGHT *Hamiltonian cycle. The estimate $(\frac{5}{9} + o(1))\binom{n}{2}$ is sharp.*

The codegree problem is answered by

THEOREM B (RÖDL, RUCIŃSKI, AND SZEMERÉDI (2011), [697]). *There exists an n_0 such that if $n > n_0$ and in a 3-uniform hypergraph $\mathbb{H}_n^{(3)}$, for any $x \neq y$,*

$$\delta_2(x, y) \geq \left\lfloor \frac{n}{2} \right\rfloor,$$

then $\mathbb{H}_n^{(3)}$ contains a TIGHT *Hamiltonian cycle. This is sharp: for any $n > 4$, there exists a 3-uniform hypergraph with $\delta_2(\mathbb{H}_n^{(3)}) = \lfloor \frac{n}{2} \rfloor - 1$, without containing a* TIGHT *Hamiltonian cycle.*

Table 2 contains some of the results, discussed below, and some others.

9.4 Lower Bounds, Constructions

In all cases considered here we have some relatively simple constructions providing a hypergraph not containing the required configuration and having large minimum degree (of a given type). The proofs of that *"the constructed hypergraphs do not contain the excluded configurations"* mostly follow from a Pigeon Hole principle argument or from some parity arguments. A new feature is (compared to ordinary non-degenerate extremal graph problems) that here the *results* and *constructions* often depend on parities, divisibilities, and they may be more complicated.

We start with some constructions, conjectured extremal structures (see e.g., Fig. 15). In the corresponding papers/proofs it turns out that these are indeed, (almost) extremal constructions. They can be found, e.g., in Hàn, Person, Schacht, [449], or earlier, in Daykin and Häggkvist [229], Kühn and Osthus [561], Rödl, Ruciński, and Szemerédi [693], and Pikhurko [647].

Construction 9.9 Assume that n is divisible by k and $0 \leq t < k$. Let $\mathbb{H}_n^{(k)}$ have two vertex classes $\mathcal{A}$ and $\mathcal{B}$, $|\mathcal{A}| = \frac{n}{k} - 1$, $|\mathcal{B}| = \frac{k-1}{k}n + 1$, and the hyperedges be all the k-tuples intersecting $\mathcal{A}$. This hypergraph has no perfect matchings. Further,

Table 2 Using the absorption method (explained in Sect. 6.5), now for hypergraph Hamiltonicity

Authors	Year	About what?	Methods	Journal
Hàn, Schacht	2010	Dirac, loose Hamilton cycles	Absorb.	[450]
Rödl, Ruciński	2010	Dirac-type questions, Survey		[689]
Rödl, Ruciński, Szemerédi	2011	Dirac 3-hyper, approx		Advances [697]
Glebov-Person-Weps	2012	Hypergraph Hamiltonicity	????	[394]
Czygrinow-Molla	2014	Loose Hamilton 3-unif codegree, perfect matching	Absorb. Stabil.	SIDMA [224]
Czygrinow, DeBiasio, Nagle	2014	Tiling with $K_4^{(3)} - 2e$	Absorb.	JGT, [221]
Jie Han, Yi Zhao	2015	Minimum codeg threshold	Absorb.	[451]
Lo, Allan, Markström	2015	F-factors in hypergraphs via Absorb.	Absorb.	GC [579]
Reiher-Rödl-Ruciński-Schacht-Szem	2016	Tight hamiltonian 3-hyper	Absorb. Absorb.	SIDMA [674]
Ferber, Nenadov, Peter		Universality of random graphs	Absorb. Absorb.	RSA [340]
Rödl, Ruciński, Schacht, Szemerédi	2017	Hamiltonicity of triple systems		Annales comb [692]

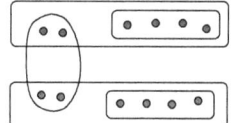
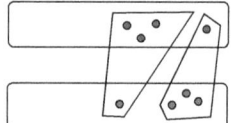

Fig. 15 Extremal structures for 4-uniform graphs and perfect matching in [222]: (a) 4-tuples intersecting both classes in 2 vertices, (b) 4-tuples intersecting one of the classes in 1 vertex

$$\delta_t(\mathbb{H}_n^{(k)}) = \left(1 - \left(\frac{k-1}{k}\right)^{k-t}\right)\binom{n}{k-t} + o(n^{k-t}). \quad (29)$$

Construction 9.10 Assume that n is divisible by k and $0 \le t < k$. Let $\widetilde{\mathbb{H}}_n^{(k)}$ have two vertex classes $\mathcal{A}$ and $\mathcal{B}$, $|\mathcal{A}|$ be the maximum odd integer $\le n/2$, $|\mathcal{B}| = n - |\mathcal{A}|$, and the hyperedges be all the k-tuples intersecting $\mathcal{A}$ in a positive even number of vertices. This hypergraph has no perfect matchings. Further,

$$\delta_t(\widetilde{\mathbb{H}}_n^{(k)}) = \frac{1}{2}\binom{n}{k-t} + O(n^{k-t-1}). \tag{30}$$

These constructions provide the lower bounds in many results in this area. Thus, e.g., for $k = 3$ the obtained coefficients of $\binom{n}{k-t}$ are $\frac{5}{9}$ for mindegree and $\frac{1}{2}$ for codegree, respectively, which will turn out to be sharp in Theorems 9.11, 9.12 and 9.13 below.[136]

9.5 Upper Bounds, Asymptotic and Sharp

Most of the above questions first were solved only in asymptotic forms, then in sharp forms. But, as it is remarked in [692], one of these problems, namely Theorem 9.12 below, the first one, (i), in the 6-item list of Sect. 9.3, is more difficult than the others. There, even the asymptotic result (i.e. to prove $\approx \frac{5}{9}\binom{n}{2} + o(n^2)$ "needed" four steps: first Glebov, Person and Weps improved the trivial $\binom{n-1}{2}$ to $(1-\varepsilon)\binom{n-1}{2}$, then Rödl and Ruciński [690] improved $(1-\varepsilon)$ to $\frac{1}{3}(5-\sqrt{5}) \approx 0.91$, next Rödl, Ruciński, and Szemerédi [695] to 0.8, and only then, they with Reiher and Schacht, obtained the sharp $5/9 = 0.555$, in [674].

— · —

Below we write about this theory, staying mostly with the simplest cases. For $k = 3$ the above codegree threshold is $\lfloor \frac{n}{2} \rfloor$ in (30). One could ask, what is the appropriate degree condition. Cooley and Mycroft [204], using an appropriate Regularity Lemma of Allen, Böttcher, Cooley, and Mycroft [15], proved

Theorem 9.11 (Cooley and Mycroft (2017), [204]) *For any $\eta > 0$ there exists an $n_0(\eta)$ such that if $n > n_0(\eta)$, then any 3-uniform hypergraph $\mathbb{H}_n^{(3)}$ with $d_{\min}(\mathbb{H}_3^{(n)}) \geq (\frac{5}{9} + \eta)\binom{n}{2}$ contains a TIGHT cycle $\widetilde{\mathcal{C}}_m^3$ with $n - m = o(n)$.*

In other words, $\mathbb{H}_n^{(3)}$ contains an almost-Hamiltonian cycle: only $o(n)$ vertices are left out. The constant $\frac{5}{9}$ is asymptotically best possible. Reiher, Rödl, Ruciński, Schacht, and Szemerédi improved Theorem 9.11:

Theorem 9.12 (Reiher, Rödl, Ruciński, Schacht, and Szemerédi (2016), [674]) *For any $\eta > 0$, there exists an $n_0(\eta)$ such that if $n > n_0$ and in an n-vertex 3-uniform hypergraph $\mathbb{H}_n^{(3)}$ the minimum degree*

$$d_{\min}(\mathbb{H}_n^{(3)}) \geq \left(\frac{5}{9} + \eta\right)\binom{n}{2},$$

then it contains a TIGHT Hamiltonian cycle.

---

[136]Identical with Theorems A,B, above.

The sharpness follows from the sharpness of Theorem 9.11. Actually, [674] describes three constructions proving the sharpness of Theorem 9.12. The difference between Theorems 9.11 and 9.12 is that Theorem 9.12 has no left-out vertices. The proof of Theorem 9.12 uses the hypergraph regularity and then the "absorption" method, to pick up the last few vertices.[137]

As to the codegree, we have

Theorem 9.13 (Rödl, Ruciński, and Szemerédi [697]) *There exists an n_0 such that if $n > n_0$ and in a 3-uniform hypergraph $\mathbb{H}_n^{(3)}$, for any $x \neq y$,*

$$\delta_2(x, y) \geq \left\lfloor \frac{n}{2} \right\rfloor,$$

then $\mathbb{H}_n^{(3)}$ contains a TIGHT *Hamiltonian cycle. This is sharp: for any $n > 4$, there exists a 3-uniform hypergraph with $\delta_2(\mathbb{H}_n^{(3)}) = \lfloor \frac{n}{2} \rfloor - 1$, without containing a* TIGHT *Hamiltonian cycle.*

The sharpness follows from the constructions of Sect. 9.4. A similar result holds for Hamiltonian paths.

Matchings

We define a perfect matching in a k-uniform hypergraph H on n vertices as a set of $\lfloor n/k \rfloor$ disjoint hyperedges. To ensure a 1-factor in hypergraphs is a fascinating and important topic and would deserve a much longer survey. Here we again restrict ourselves to just a few related results and references. First we formulate two results on the degree-extremal problem of a 1-factor, for 3-uniform and 4-uniform hypergraphs. Both results are sharp. The following theorem was obtained independently, by Lo and Markström [579], Kühn, Osthus, and Treglown and by Imdadullah Khan:

Theorem 9.14 (Kühn, Osthus, and Treglown [566]/I. Khan [507]) *There is an n_0 such that if $n > n_0$ is divisible by 3 and in a 3-uniform hypergraph $\mathbb{H}_n^{(3)}$*

$$d_{\min}(\mathbb{H}_n^{(3)}) > \binom{n-1}{2} - \binom{2n/3}{2} + 1 \approx \frac{5}{9}\binom{n-1}{2}, \qquad (31)$$

then $\mathbb{H}_n^{(3)}$ contains a 1-factor.

Observe that the LHS of (31) is $\approx \frac{5}{9}\binom{n}{2}$, i.e. the same what we need for a TIGHT Hamilton cycle. Khan also proved the analog theorem for 4-uniform hypergraphs:

---

[137]This, with a sketch of the proof, is nicely explained by Rödl and Ruciński, in [689]. The introduction of [674] and a survey of Yi Zhao [825] are also good descriptions of the otherwise fairly complicated situation.

Theorem 9.15 (Imdadullah Khan [508]) *There exists a threshold n_0 such that if $n > n_0$ is divisible by 4 and in a 4-uniform hypergraph $\mathbb{H}_n^{(4)}$*

$$d_{\min}(\mathbb{H}_n^{(4)}) > \binom{n-1}{3} - \binom{3n/4}{3} + 1,$$

then $\mathbb{H}_n^{(4)}$ contains a 1-factor.

For related results see also Pikhurko [647], Hàn, Person, and Schacht [449], Czygrinow and Kamat [222] providing sharp results and describing the earlier asymptotic results, and Alon, Frankl, Huang, Rödl, Ruciński, and Sudakov [35].

Upper Bounds and the Absorbing Method

In the proofs of the results discussed in this section one often uses the ABSORBING METHOD,— described in Sect. 6.5,—to ensure SPANNING or ALMOST-SPANNING substructures, where ALMOST-SPANNING means a subhypergraph covering the whole hypergraph with the exception of at most $O(1)$ (or, occasionally, $o(n)$) vertices. As we wrote, in most cases considered in these subsections we have a simple, nice conjectured extremal structure, enabling us to use stability arguments. Yet, the actual "upper bounds" (non-construction parts) are fairly complicated.

In Sect. 9.3 we wrote about the thresholds for the Hamiltonicity, $h_d^\ell(k, n)$, and for the Matching, $m_d^r(k, n)$. We have mentioned that sometimes we get the same results for Hamiltonicity and Perfect Matching.

Conjecture 9.16 (Informal/Formal [689]) *The d-degree threshold $\delta_d(\mathbb{H}_n^{(k)})$ for the Hamiltonian problem and the Matching problem are roughy the same. More formally, $h_d(k,n) \approx m_d(k,n)$, if d, k are fixed, $n \to \infty$.*

There are many related results but we skip them, mentioning only that, as it is emphasized in [689], not only these two quantities are often near to each other, but in several cases they are proved to be equal, see e.g., Theorem 3.1 of [689], quoted from [696].

Theorem 9.17 (Hàn, Person, and Schacht (2009), [449]) *For any $\gamma > 0$, if $\delta_1(\mathbb{H}_n^{(3)}) \geq (\frac{5}{9} + \gamma)\binom{n}{2}$ and $n > n_0(\gamma)$, then $\mathbb{H}_n^{(3)}$ contains a 1-factor.*

There is a surprising, sharp difference between the cases when n is divisible by k and when it is not. Consider the simplest hypergraph case, $k = 3$.

Theorem 9.18 *For any $\gamma > 0$, if $\delta_2(\mathbb{H}_n^{(k)}) \geq (\frac{1}{2} + \gamma)\binom{n}{2}$ and $n > n_0(\gamma)$, then $\mathbb{H}_n^{(k)}$ contains a TIGHT Hamiltonian cycle.[138] If n is not divisible by 3 and $\mathbb{H}_n^{(3)}$ does not contain a 1-factor, then $\delta_2(\mathbb{H}_n^{(3)}) \leq \frac{n}{k} + o(n)$.*

---

[138] This implies Theorem 9.17.

So in case of non-divisibility, much smaller degrees ensure a perfect matching.

— · —

One of the first results where the Absorbing Lemma was used is the theorem of Rödl, Ruciński, and Szemerédi [696], on ensuring a perfect matching in a k-uniform hypergraph. As we have mentioned, an appropriate construction of Bollobás, Daykin, and Erdős [122] and further similar constructions show the sharpness of these results.

There is a sequence of papers of Rödl, Ruciński, and Szemerédi on this topic: [693] assumes large minimum degree, [696] assumes large codegree. To ensure Hamiltonian cycles in hypergraphs, see [694, 695, 697], and see also the above authors with Schacht [691], and Reiher [674].

The problem of 1-factors, or almost 1-factors is, of course, connected to (almost perfect) tilings, and there are several results in that direction, as well, see, e.g., Markström and Ruciński [612], Pikhurko [647], Lo and Markström [577–579].[139]

Several results mentioned above or discussed below use the Absorption Method, as is shown in Tables 2, 3. (See also the earlier Rödl-Ruciński survey [689] on these hypergraph problems and results.)

The first line of Table 3 differs from the other ones by that it is connected to Graph Packing, and within that to the Sauer-Spencer theorem. The subsequent lines try to show in time order some important papers using the absorption method, on loose and tight hypergraph Hamilton cycles (Table 2) and matching (Table 3).

These tables also contain some results on random graphs, and also some lines on Universal graphs, but we skip explaining them. For details see the papers of Alon and Capalbo, e.g., [27], or Alon, Capalbo, Kohayakawa, Rödl, Rucinski, and Szemerédi [28]. Further, Table 1 has an item on monochromatic cycle partitions as well.

9.6 Tiling Hypergraphs

Of course, a matching that (almost) covers a hypergraph $V(\mathbb{H}_k^{(n)})$ is an (almost) tiling. One can also ask if for k-uniform hypergraphs do we get interesting results when we wish to tile them for a fixed $\mathbb{F}$ by vertex-disjoint copies of $\mathbb{F}$. The outstanding results of Keevash [500][140] on designs and the related results are also tiling theorems, however, here we mention only that Pikhurko, beside investigating codegree matching problems, in [647] also investigated the problem of ensuring $K_4^{(3)}$-tilings in a hypergraph with large codegree. These problems were connected to each other.

---

[139]The earlier excellent surveys of Füredi [369, 372] are primarily on small excluded subgraphs.

[140]See also Keevash, [502], Barber, Kühn, Lo, and Osthus [93], Barber, Kühn, Lo, Montgomery, and Osthus [92],...

Table 3 Absorbing for hypergraphs, Matching, ...

Authors	Year	About what?	Methods	Where
Rödl, Ruciński, Taraz	1999 1999	Hypergraph packing, embedding	Misleading!	CPC [698]
Kühn and Osthus	2006	Matching, implied by min deg in r-partite hypergraph	Stability ???	[561] JGT
Pikhurko	2008	Matching, Tiling, min codeg	???	GC [647]
Hàn, Person, Schacht	2009	Perfect matching, mindegree	Absorb.	SIDMA [449]
Czygrinow-Kamat	2012	Perfect matching sharp codegree condition, 4-unif	Absorb. Stability	ELECT JC [222]
Alon, Frankl, Huang, Rödl, Ruciński, Sudakov:	2012	Large matchings in uniform hypergraphs and the conjecture of Erdős and Samuels	Fractional matching	JCTA [35]
I. Khan	2013	Perfect matching, min degree 3-uniform	Absorb. Stability	SIDMA [507]
I. Khan	2016	Perfect matching, min degree 4-uniform	Absorb. Stability	JCTB [508]

9.7 Vertex-Partition for Hypergraphs

The vertex-partition results of Sect. 7.4 can also be extended to hypergraphs, see Gyárfás and G.N. Sárközy [434, 435] and of G.N. Sárközy [724]. Thus, e.g., Sárközy proved the following.

Theorem 9.19 (G.N. Sárközy (2014), [724]) *For all integers $k, r \geq 2$, there exists an $n_0 = n_0(k, r)$ such that if $n > n_0(k, r)$ and $\mathbb{K}_n^{(k)}$ is r-edge-coloured, then its vertex set can be partitioned into at most $50rk \log(rk)$ vertex disjoint* LOOSE *monochromatic cycles.*

For the case of two colours and loose/tight cycles Bustamante, Hàn, and M. Stein [168] proved some hypergraph versions of Lehel's conjecture, where, however, a few vertices remain uncovered. Bustamante, Corsten, Nóra Frankl, Pokrovskiy and Skokan [169] proved a tight-cycle version.

Theorem 9.20 *There exists a constant $c(k, r)$ such that if $\mathbb{H}^{(k)}$ is a k-uniform hypergraph whose edges are coloured by r colours, then $V(\mathbb{H}^{(k)})$ can be partitioned into $c(k, r)$ subsets, each defining monochromatic* TIGHT *hypergraph-cycles.*

For some related results see also the surveys of Gyárfás [426], and Fujita, Liu, and Magnant [368].

9.8 Generalizing Gyárfás–Ruszinkó–Sárközy–Szemerédi Theorem to Hypergraphs

We have mentioned that several graph results were generalized to hypergraphs. Theorem 7.19 was generalized to hypergraphs first by Gyárfás and G.N. Sárközy [435]: they covered the coloured $\mathbb{K}_n^{(k)}$ by Berge paths, by Berge cycles, and by LOOSE cycles. Perhaps their LOOSE cycle result was the deepest. Its proof followed the method of Erdős, Gyárfás, and Pyber [291] and used the linearity of Ramsey number for a "crown" that was a generalization of the Triangle-Cycle used in Sect. 6.5. The LOOSE cycle result was improved by G.N. Sárközy, see Theorem 9.19.

— · —

9.9 A New Type of Hypergraph Results, Strong Degree

This subsection is motivated by a paper of Gyárfás, Győri, and Simonovits [427]. Their original motivation was to prove the following conjecture that is still open.

Conjecture 9.21 (Gyárfás–Sárközy, [435]) *One can partition the vertex set of every 3-uniform hypergraph $\mathbb{H}_n^{(3)}$ into $\alpha(\mathbb{H}_n^{(3)})$ linear (i.e. LOOSE) cycles, hyperedges and subsets of hyperedges.*

A theorem of Pósa is

Theorem 9.22 (Pósa (1964), [659]) *For every graph G one can partition $V(G)$ into at most $\alpha(G)$ cycles, where a vertex or an edge is accepted as a cycle.*

Conjecture 9.21 would extend Pósa theorem, see [589] from graphs to 3-uniform hypergraphs. Conjecture 9.21 was proved in [435] for "weak cycles" instead of linear cycles, where "weak cycle" differs from a "loose cycle" by that the consecutive edges must intersect, but not necessarily in 1 vertex. Here one has to consider subsets of hyperedges also as cycles, in Conjecture 9.21, as shown, e.g., by the complete hypergraph $K_5^{(3)}$.

The following weaker version of Conjecture 9.21 was proved recently:

Theorem 9.23 (Ergemlidze, Győri, and Methuku (1964), [325]) *One can cover the vertices of any 3-uniform hypergraph $\mathbb{H}_n^{(3)}$ by $\alpha(\mathbb{H}_n^{(3)})$ **edge-disjoint** linear (i.e. LOOSE) cycles, hyperedges and subsets of hyperedges.*

— · —

Let $\rho(\mathbb{H}_n^{(3)})$ denote the minimum number of linear cycles, hyperedges or subsets of hyperedges needed to partition $V(\mathbb{H}_n^{(3)})$ as described in Conjecture 9.21 and let $\chi(\mathbb{H}_n^{(3)})$ denote the chromatic number of $\mathbb{H}_n^{(3)}$, the minimum number of colors in a vertex coloring of $\mathbb{H}_n^{(3)}$ without monochromatic edges. The following result proves that Conjecture 9.21 is true if there are no linear cycles in $\mathbb{H}_n^{(3)}$.

Theorem 9.24 (Gyárfás–Győri–Simonovits [427]) *If $\mathbb{H}_n^{(3)}$ is a 3-uniform hypergraph without linear cycles, then $\rho(\mathbb{H}_n^{(3)}) \leq \alpha(\mathbb{H}_n^{(3)})$. Moreover, $\chi(\mathbb{H}_n^{(3)}) \leq 3$.*

The family of hypergraphs without linear cycles seems to be intriguing. Gyárfás, Győri, and Simonovits [427] uses a **new degree concept:** the STRONG DEGREE. Let $\mathbb{H}_n^{(3)} = (V, \mathcal{E})$ be a 3-uniform hypergraph, for $v \in V$ the LINK GRAPH of v in $\mathbb{H}_n^{(3)}$ is the graph with vertex set $V - v$ and edge set $\{(x, y) : (v, x, y) \in \mathcal{E}\}$. The STRONG DEGREE $d^+(v)$ for $v \in V$ is the maximum number of independent edges (i.e. the size of a maximum matching) in the link graph of v. The main results of [427] are motivated by the following trivial assertions: a graph of minimum degree 2 contains a cycle; if G_n has no cycles then $\alpha(G_n) \geq n/2$.

Theorem 9.25 (Gyárfás-Győri–Simonovits [427]) *If $\mathbb{H}_n^{(3)}$ is a 3-uniform hypergraph with $d^+(v) \geq 3$ for all $v \in V$, then $\mathbb{H}_n^{(3)}$ contains a linear cycle.*

Theorem 9.25 is "self-improving":

Theorem 9.26 ([427]) *Suppose that $\mathbb{H}_n^{(3)}$ is a 3-uniform hypergraph with $d^+(v) \geq 3$ for all but at most one $v \in V$. Then $\mathbb{H}$ contains a linear cycle.*

The proofs are not easy. One thinks that several interesting embedding results can be proved where one uses $\min_{v \in V(G)} d^+(v)$ instead of $d_{\min}(G)$.

Acknowledgements We thank to several friends and colleagues the useful discussions about the many topics discussed in this survey. We thank above all to József Balogh and András Gyárfás, and also to Zoltán Füredi, János Pach, Jan Hladký, Zoltán L. Nagy, János Pintz, Andrzej Ruciński, Imre Ruzsa, Gábor Sárközy, Andrew Thomason, and Géza Tóth.

Note to References

Below, in the References, (to be more informative) we shall often use the full name in the author's first or second occurrence, and/or in his first occurrence as the first author but this rule is not too strict. Also in some exceptional cases the authors originally were not in alphabetic order and we switched their order to alphabetic.

Mostly we refer to the same Gyula Katona, but occasionally to his son Y.=Younger Gyula Katona, e.g., [492].

References

1. Sarmad Abbasi: How tight is the Bollobás-Komlós conjecture?, Graphs Combin. **16 (2)** (2000), 129–137.
2. Harvey L. Abbott and Andrew C.F. Liu: The existence problem for colour critical linear hypergraphs, Acta Math. Acad Sci Hung Tomus **32 (3–4)** (1978), 273–282.
3. Martin Aigner and Stephan Brandt: Embedding arbitrary graphs of maximum degree two, J. London Math. Soc. (2), **48** (1993), 39–51.

4. Elad Aigner-Horev, David Conlon, Hiep Hàn, and Yury Person: Quasirandomness in hypergraphs, Electronic Notes in Discrete Mathematics **61** (2017) 13–19.
5. Elad Aigner-Horev, Hiep Hàn, and Mathias Schacht: Extremal results for odd cycles in sparse pseudorandom graphs, Combinatorica **34 (4)** (2014), 379–406.
6. Miklós Ajtai, Paul Erdős, János Komlós, and Endre Szemerédi: On Turán's theorem for sparse graphs, Combinatorica **1 (4)** (1981) 313–317.
7. M. Ajtai, J. Komlós, János Pintz, Joel Spencer, and E. Szemerédi: Extremal uncrowded hypergraphs, J. Combin. Theory Ser A **32** (1982), 321–335.
8. M. Ajtai, J. Komlós, Miklós Simonovits, and E. Szemerédi: Erdős-Sós Conjecture. (In preparation, in several manuscripts)
9. M. Ajtai, J. Komlós, and E. Szemerédi: A note on Ramsey numbers. J. Combin. Theory Ser. A **29 (3)** (1980) 354–360.
10. M. Ajtai, J. Komlós, and E. Szemerédi: A dense infinite Sidon sequence, European J. Combin. **2** (1981), 1–11.
11. M. Ajtai, J. Komlós, and E. Szemerédi: Sorting in $c \log n$ parallel steps. Combinatorica **3 (1)** (1983) 1–19.
12. M. Ajtai, J. Komlós, and E. Szemerédi: On a conjecture of Loebl, in Graph Theory, Combinatorics, and Algorithms, Vols. 1, 2 (Kalamazoo, MI, 1992), Wiley-Intersci. Publ., Wiley, New York, (1995), 1135–1146.
13. Vladimir E. Alekseev.: Range of values of entropy of hereditary classes of graphs (in Russian), Diskret. Mat. **4** (1992), 148–157. see also On the entropy values of hereditary classes of graphs. Discrete Math. Appl. **3** (1993) 191–199.
14. Peter Allen: Covering two-edge-coloured complete graphs with two disjoint monochromatic cycles, Combinatorics, Probability and Computing, **17 (4)** (2008) 471–486.
15. P. Allen, Julia Böttcher, Oliver Cooley, and Richard Mycroft: Tight cycles and regular slices in dense hypergraphs. J. Combin. Theory Ser. A **149** (2017), 30–100. arxiv:1411.4597.
16. P. Allen, J. Böttcher, Hiep Hàn, Y. Kohayakawa, and Y. Person: Powers of Hamilton cycles in pseudorandom graphs. LATIN 2014: theoretical informatics, 355–366, Lecture Notes in Comput. Sci., 8392, Springer, Heidelberg, 2014.
17. P. Allen, J. Böttcher, Hiep Hàn, Yoshiharu Kohayakawa, and Yury Person: Powers of Hamilton cycles in pseudorandom graphs. Combinatorica **37 (4)** (2017), 573–616.
18. P. Allen, J. Böttcher, H. Hàn, Y. Kohayakawa, and Y. Person: Blow-up lemmas for sparse graphs, Submitted for publication, arXiv:1612.00622. (2018+)
19. P. Allen, Peter Keevash, Benny Sudakov, and Jacques Verstraëte: Turán numbers of bipartite graphs plus an odd cycle, J. Combin. Theory Ser. B **106** (2014), 134–162.
20. Noga Alon: Eigenvalues and expanders, Combinatorica **6** (1986), 83–96.
21. N. Alon: Independent sets in regular graphs and sum-free subsets of finite groups. Israel J. Math. **73** (1991) 247–256.
22. N. Alon: Tools from Higher Algebra, Chapter 32 of Handbook of Combinatorics (ed. Graham, Lovász, Grötschel), (1995) pp. 1749–1783, North-Holland, Amsterdam.
23. N. Alon: Combinatorial Nullstellensatz, Combinatorics, Probability and Computing **8** (1999), 7–29.
24. N. Alon, József Balogh, Béla Bollobás, and Robert Morris: The structure of almost all graphs in a hereditary property. J. Combin. Theory Ser. B, **101 (2)** (2011) 85–110.
25. N. Alon, J. Balogh, R. Morris, and Wojciech Samotij: A refinement of the Cameron-Erdős conjecture, Proc. London Math. Soc. (3) **108 (1)** (2014), no. 1, 44–72.
26. N. Alon, J. Balogh, R. Morris, and W. Samotij: Counting sum-free sets in Abelian groups, Israel J. Math. **199 (1)** (2014), 309–344. arXiv:1201.6654.
27. N. Alon and M. Capalbo: Sparse universal graphs for bounded-degree graphs, Random Struct Algorithms **31 (2)** (2007), 123–133.
28. N. Alon, M. Capalbo, Y. Kohayakawa, V. Rödl, A. Ruciński, and E. Szemerédi: Near-optimum universal graphs for graphs with bounded degrees (extended abstract), In Approximation, randomization, and combinatorial optimization (Berkeley, CA, 2001), Lecture Notes in Computer Science Vol. 2129, Springer, Berlin, 2001, pp. 170–180.

29. N. Alon and Fan R.K. Chung: Explicit construction of linear sized tolerant networks, Discrete Mathematics **72** (1988), Proceedings of the First Japan Conference on Graph Theory and Applications (Hakone, 1986), 15–19.
30. N. Alon, Richard A. Duke, Hanno Lefmann, Vojtech Rödl, and Raphael Yuster: The algorithmic aspects of the Regularity Lemma. J. Algorithms **16** (1994) 80-109. see also: the extended abstract, 33rd Annu Symp Foundations of Computer Science, Pittsburgh, IEEE Comput Soc. Press, New York, 1992, pp. 473–481.
31. N. Alon and P. Erdős: An application of graph theory to additive number theory, European J. Combin. **6 (3)** (1985) 201–203.
32. N. Alon and Elgar Fischer: 2-factors in dense graphs, Discrete Math., **152** (1996), 13–23.
33. N. Alon, E. Fischer, Michael Krivelevich, and Mário Szegedy: Efficient testing of large graphs, Combinatorica **20** (2000), 451–476. see also its extended abstract, Proc. **40**th Ann. Symp. on Found. of Comp. Sci. IEEE (1999), 656–666.
34. N. Alon, E. Fischer, Ilan Newman, and Asaf Shapira: A combinatorial characterization of the testable graph properties: It's all about regularity, SIAM J. Comput., **39**(1) 143–167, 2009. STOC '06: Proceedings of the 38th Annual ACM Symposium on Theory of Computing, ACM, New York, 2006, pp. 251–260.
35. N. Alon, P. Frankl, H. Huang, V. Rödl, A. Ruciński, and B. Sudakov: Large matchings in uniform hypergraphs and the conjecture of Erdős and Samuels, J. Combin. Theory Ser. A, **119** (2012), pp. 1200–1215.
36. N. Alon and Zoltán Füredi: Spanning subgraphs of random graphs, Graphs and Combin. **8 (1)** (1992) 91–94.
37. N. Alon, Jeong-Han Kim, and J. Spencer: Nearly perfect matchings in regular simple hypergraphs. Israel J. Math. **100** (1997), 171–187.
38. N. Alon, M. Krivelevich, and B. Sudakov: Turán numbers of bipartite graphs and related Ramsey-type questions, Combin. Probab. Comput **12** (2003), 477–494.
39. N. Alon and Vitalij D. Milman: Eigenvalues, expanders and superconcentrators, in Proceedings of the 25th FOCS, IEEE, New York, 1984, pp. 320–322.
40. N. Alon, Ankur Moitra, and B. Sudakov: Nearly complete graphs decomposable into large induced matchings and their applications. J. Eur. Math. Soc. (JEMS) **15 (5)** (2013) 1575–1596. See also in Proc. **44**th ACM STOC,(2012) 1079–1089.
41. N. Alon, János Pach, Ron Pinchasi, Radoš Radoičić, and Micha Sharir: Crossing patterns of semi-algebraic sets, J. Combin. Theory, Ser. A **111** (2005) 310–326.
42. N. Alon, Lajos Rónyai, and Tibor Szabó: Norm-graphs: Variations and application, J. Combin. Theory Ser. B **76** (1999), 280–290.
43. N. Alon and Asaf Shapira: Every monotone graph property is testable, Proc. of the **37**th ACM STOC, Baltimore, ACM Press (2005), 128–137.
44. N. Alon and A. Shapira: A characterization of the (natural) graph properties testable with one-sided error, Proc. **46**th IEEE FOCS (2005), 429–438.
45. N. Alon and A. Shapira: On an extremal hypergraph problem of Brown, Erdős and Sós, Combinatorica **26** (2006), 627–645.
46. N. Alon, A. Shapira, and Uri Stav: Can a graph have distinct regular partitions?, SIAM J. Discrete Math., **23** (2008/09), 278–287.
47. N. Alon and Clara Shikhelman: Many T copies in H-free graphs. J. Combin. Theory Ser. B **121** (2016), 146–172.
48. Noga Alon and Joel Spencer: "The Probabilistic Method", Wiley-Interscience Series in Discrete Mathematics and Optimization, John Wiley & Sons, Inc., Hoboken, NJ, 2016. xiv+375 pp. 4th, updated edition.
49. N. Alon and R. Yuster: H-factors in dense graphs. J. Combin. Theory Ser. B **66** (1996) 269–282.
50. Richard P. Anstee, Balin Fleming, Z. Füredi, and Attila Sali: Color critical hypergraphs and forbidden configurations, in: S. Felsner, (Ed.) Discrete Mathematics and Theoretical Computer Science Proceedings, vol. AE, 2005, pp. 117–122.

51. Tim Austin: Deducing the multidimensional Szemerédi Theorem from an infinitary removal lemma. J. Anal. Math. **111** (2010) 131–150.
52. T. Austin: Deducing the density Hales-Jewett theorem from an infinitary removal lemma, J. Theoret. Probab. 24 (2011), 615–633.
53. T. Austin and Terrence C. Tao: Testability and repair of hereditary hypergraph properties, Random Struct Algor 36 (2010), 373–463.
54. Jacqueline Ayel: Sur l'existence de deux cycles supplémentaires unicolores, disjoints et de couleurs différentes dans un graph complet bicolore, Ph.D. thesis, Univ. Grenoble, 1979.
55. László Babai: An anti-Ramsey theorem, Graphs Combin., **1** (1985), 23–28.
56. L. Babai, M. Simonovits, and J. Spencer: Extremal subgraphs of random graphs, J. Graph Theory **14** (1990), 599–622.
57. L. Babai and Vera T. Sós: Sidon sets in groups and induced subgraphs of Cayley graphs, Europ. J. Combin. **6** (1985) 101–114.
58. Rahil Baber and John Talbot: Hypergraphs do jump. Combin. Probab. Comput. **20 (2)** (2011), 161–171.
59. Béla Bajnok: Additive Combinatorics, a menu of research problems, arXiv:1705:07444, to appear as a book.
60. József Balogh, János Barát, Dániel Gerbner, András Gyárfás and Gábor N. Sárközy: Partitioning 2-edge-colored graphs by monochromatic paths and cycles, Combinatorica **34** (2014), 507–526.
61. J. Balogh, B. Bollobás, and Bhargav Narayanan: Transference for the Erdős-Ko-Rado theorem. Forum of Math. Sigma, (2015), e23, 18 pp.
62. J. Balogh, B. Bollobás, and M. Simonovits: The number of graphs without forbidden subgraphs, J. Combin. Theory Ser. B **91 (1)** (2004) 1–24.
63. J. Balogh, B. Bollobás, and M. Simonovits: The typical structure of graphs without given excluded subgraphs, Random Structures and Algorithms **34** (2009), 305–318.
64. J. Balogh, B. Bollobás, and M. Simonovits: The fine structure of octahedron-free graphs, J. Combin. Theory Ser. B **101** (2011) 67–84.
65. J. Balogh, Neal Bushaw, Maurício Collares, Hong Liu, R. Morris, and Maryam Sharifzadeh: The typical structure of graphs with no large cliques. Combinatorica **37 (4)** (2017) 617–632. arXiv:1406.6961,
66. J. Balogh, Jane Butterfield, Ping Hu, and John Lenz: Mantel's theorem for random hypergraphs. Random Structures and Algorithms **48 (4)** (2016) 641–654.
67. J. Balogh, Béla Csaba, Martin Pei, and W. Samotij: Large bounded degree trees in expanding graphs. Electron. J. Combin. **17 (1)** (2010), Research Paper 6, 9 pp.
68. J. Balogh, B. Csaba, and W. Samotij: Local resilience of almost spanning trees in random graphs. Random Struct. Algor. **38** (2011) 121–139.
69. J. Balogh, P. Hu, and M. Simonovits: Phase transitions in Ramsey-Turán theory, J. Combin. Theory Ser. B **114** (2015) 148–169.
70. J. Balogh and John Lenz: Some exact Ramsey-Turán numbers, Bull. London Math. Soc. **44** (2012), 1251–1258.
71. J. Balogh and J. Lenz: On the Ramsey-Turán numbers of graphs and hypergraphs, Israel J. Math. **194** (2013), 45–68.
72. J. Balogh, Hong Liu, M. Sharifzadeh, and Andrew Treglown: The number of maximal sum-free subsets of integers, Proc. Amer. Math. Soc. **143 (11)** (2015), 4713–4721.
73. J. Balogh, Allan Lo, and Theodore Molla: Transitive triangle tilings in oriented graphs. J. Combin. Theory Ser. B **124** (2017), 64–87.
74. J. Balogh, Andrew McDowell, T. Molla, and Richard Mycroft: Triangle-tilings in graphs without large independent sets, Combinatorics, Probability and Computing, **27 (4)** (2018) 449–474. arXiv:1607.07789
75. J. Balogh, T. Molla, and M. Sharifzadeh: Triangle factors of graphs without large independent sets and of weighted graphs. Random Structures and Algorithms **49 (4)** (2016), 669–693. // Lingli Sun)

76. J. Balogh, R. Morris, and W. Samotij: Random sum-free subsets of Abelian groups, Israel J. Math. **199 (2)** (2014), 651–685.
77. J. Balogh, R. Morris, and W. Samotij: Independent sets in hypergraphs. J. Amer. Math. Soc. **28 (3)** (2015), 669–709. arXiv:1204.6530,
78. J. Balogh, R. Morris, and W. Samotij: The method of Hypergraph containers, ICM 2018, also arXiv:1801.04584v1 (2018)
79. J. Balogh, R. Morris, W. Samotij, and Lutz Warnke: The typical structure of sparse K_{r+1}-free graphs. Trans. Amer. Math. Soc. **368 (9)** (2016), 6439–6485.
80. J. Balogh, Frank Mousset, and Jozef Skokan: Stability for vertex cycle covers. Electron. J. Combin. **24 (3)** (2017), Paper 3.56, 25 pp.
81. J. Balogh and Dhruv Mubayi: Almost all triple systems with independent neighborhoods are semi-bipartite, J. Combin. Theory Ser. A **118 (4)** (2011), 1494–1518.
82. J. Balogh and D. Mubayi: Almost all triangle-free triple systems are tripartite, Combinatorica **32 (2)** (2012), 143–169,
83. J. Balogh and Cory Palmer: On the tree packing conjecture, SIAM Journal on Discrete Mathematics **27** (2013), 1995–2006.
84. J. Balogh and Šárka Petříčková: The number of the maximal triangle-free graphs. Bull. London Math. Soc. **46 (5)** (2014), 1003–1006.
85. J. Balogh and W. Samotij: Almost all C_4-free graphs have fewer than $(1 - \epsilon)\text{ex}(n, C_4)$ edges, SIAM J. Discrete Math. **24** (2010) 1011–1018.
86. J. Balogh and W. Samotij: The number of $K_{m,m}$-free graphs, Combinatorica **31 (2)** (2011) 131–150.
87. J. Balogh and W. Samotij: The number of $K_{s,t}$-free graphs, J. London Math. Soc. (2) **83 (2)** (2011) 368–388.
88. J. Bang-Jensen and G. Gutin: Digraphs. Theory, algorithms and applications, Springer Monographs in Mathematics, Springer-Verlag, London, 2009.
89. Boaz Barak, Anup Rao, Ronen Shaltiel, and Avi Wigderson: 2-source dispersers for $n^{o(1)}$ entropy, and Ramsey graphs beating the Frankl-Wilson construction. Ann. of Math. (2) **176 (3)** (2012), 1483–1543.
90. J. Barát, A. Gyárfás, Jenő Lehel, and G.N. Sárközy: Ramsey number of paths and connected matchings in Ore-type host graphs. Discrete Math. **339 (6)** (2016), 1690–1698.
91. J. Barát and G.N. Sárközy: Partitioning 2-edge-colored Ore-type graphs by monochromatic cycles. J. Graph Theory **81 (4)** (2016), 317–328.
92. Ben Barber, Daniela Kühn, Allan Lo, Richard Montgomery, and Deryk Osthus: Fractional clique decompositions of dense graphs and hypergraphs. J. Combin. Theory Ser. B **127** (2017), 148–186. arXiv:1507.04985.
93. Ben Barber, D. Kühn, Allan Lo, and D. Osthus: Edge-decompositions of graphs with high minimum degree. Advances Math. **288** (2016), 337–385.
94. József Beck: On size Ramsey number of paths, trees, and circuits. I, J. Graph Theory **7 (1)** (1983) 115–129.
95. J. Beck: An algorithmic approach to the Lovász local lemma. Rand. Struct. Algor. **2 (4)** (1991) 343–365.
96. Felix A. Behrend: "On sets of integers which contain no three terms in arithmetical progression," Proc. Nat. Acad. Sci. USA, **32 (12)** (1946), 331–332, 1946.
97. Mathias Beiglböck: A variant of the Hales-Jewett theorem. Bull. London Math. Soc. **40 (2)** (2008), 210–216. (2009c:05247)
98. Louis Bellmann and Christian Reiher: Turán's theorem for the Fano plane, arXiv:1804.07673
99. Fabricio S. Benevides and J. Skokan: The 3-colored Ramsey number of even cycles. J. Combin. Theory Ser. B **99 (4)** (2009) 690–708.
100. Patrick Bennett and Andrzej Dudek: On the Ramsey-Turán number with small s-independence number. J. Combin. Theory Ser. B **122** (2017), 690–718.
101. Clark T. Benson: Minimal regular graphs of girth 8 and 12, Canad. J. Math. **18** (1966) 1091–1094.

102. Vitaly Bergelson and Alexander Leibman: Polynomial extensions of van der Waerden's and Szemerédi's theorems, J. Amer. Math. Soc. **9 (3)** (1996) 725–753.
103. V. Bergelson, A. Leibman: Set-polynomials and polynomial extension of the Hales-Jewett theorem, Ann. of Math. (2) **150 (1)** (1999) 33-75.
104. J.C. Bermond, A. Germa, M.C. Heydemann, D. Sotteau: Hypergraphes hamiltoniens, in: Problèmes combinatoires et théorie des graphes, Colloq. Internat. CNRS, Univ. Orsay, Orsay, 1976, in: Colloq. Internat., vol. 260, CNRS, Paris, 1973, pp. 39–43.
105. J.-C. Bermond and Carsten Thomassen: Cycles in digraphs–a survey, J. Graph Theory **5** (1981), 1–43.
106. Claudia Bertram-Kretzberg, Thomas Hofmeister, H. Lefmann: An algorithm for Heilbronn's problem, *SIAM J. Comput.*, **30 (2)** (2000) 383–390.
107. C. Bertram-Kretzberg and H. Lefmann: The Algorithmic Aspects of Uncrowded Hypergraphs, SIAM Journal on Computing **29** (1999) 201–230.
108. Stéphane Bessy and Stéphan Thomassé: Partitioning a graph into a cycle and an anticycle, a proof of Lehel's conjecture, Journal of Combinatorial Theory, Series B, **100 (2)** (2010) 176–180.
109. Yuri Bilu: Sum-free sets and related sets, Combinatorica **18** (1998), 449–459.
110. Yitzhak Birk, Nati Linial, and Roy Meshulam: On the uniform-traffic capacity of single-hop interconnections employing shared directional multichannels. IEEE Trans. Information Theory **39** (1993) 186–191.
111. G.R. Blakley and P. Roy: A Hölder type inequality for symmetric matrices with nonnegative entries, Proc. Amer. Math. Soc.**16** (1965), 1244–1245.
112. Thomas F. Bloom: A quantitative improvement for Roth's theorem on arithmetic progressions. J. London Math. Soc. (2) **93 (3)** (2016), 643–663. arXiv:1405.5800.
113. Tom Bohman, Alan Frieze, and Ryan Martin: How many random edges make a dense graph Hamiltonian? Random Structures and Algorithms **22 (1)** (2003), 33–42.
114. T. Bohman, A. Frieze, Miklós Ruszinkó, and Lubos Thoma: Vertex covers by edge disjoint cliques. Paul Erdős and his mathematics (Budapest, 1999). Combinatorica **21 (2)** (2001), 171–197.
115. T. Bohman and P. Keevash: Dynamic concentration of the triangle-free process. arXiv:1302.5963 (2013), in The Seventh European Conference on Combinatorics, Graph Theory and Applications, CRM Series 16, Ed. Norm., Pisa, 2013, pp. 489–495.
116. B. Bollobás: Three-graphs without two triples whose symmetric difference is contained in a third, Discrete Mathematics **8** (1974), 21–24.
117. B. Bollobás: Relations between sets of complete subgraphs, in Proceedings of the Fifth British Combinatorial Conference (Univ. Aberdeen, Aberdeen, 1975), Congressus Numerantium, Vol. **15**, Utilitas Mathematica, Winnipeg, MB, (1976), pp. 79–84.
118. B. Bollobás: Complete subgraphs are elusive. J. Combinatorial Theory Ser. B **21 (1)** (1976), no. 1, 1–7.
119. B. Bollobás: *Extremal Graph Theory*, Academic Press, London (1978). / Lond Math. Soc. Monogr **11**, Academic Press, London-New York, 1978, pp. 488.
120. B. Bollobás: *Extremal Graph Theory*, Handbook of Combinatorics, Vol. 2, Elsevier, Amsterdam, (1995) pp. 1231–1292.
121. B. Bollobás: *Random Graphs*, second ed., Cambridge Stud. Advances Math., vol. **73**, Cambridge Univ. Press, Cambridge, (2001).
122. B. Bollobás, David E. Daykin, and P. Erdős: Sets of independent edges of a hypergraph, Quart. J. Math. Oxford **21** (1976) 25–32.
123. B. Bollobás and Stephen E. Eldridge: Packing of Graphs and Applications to Computational Complexity, Journal of Combinatorial Theory, Series B **25** (1978), 105–124.
124. B. Bollobás and P. Erdős: On the structure of edge graphs, Bull. London Math. Soc. **5** (1973) 317–321.
125. B. Bollobás and P. Erdős: Cliques in random graphs. Math. Proc. Cambridge Philos. Soc. **80 (3)** (1976), 419–427.

126. B. Bollobás and P. Erdős: On a Ramsey-Turán type problem, J. Combin. Theory Ser. B **21** (1976), 166–168.
127. B. Bollobás, P. Erdős, and M. Simonovits: On the structure of edge graphs II, *J. London Math. Soc.*, **12 (2)** (1976) 219–224.
128. B. Bollobás, P. Erdős, M. Simonovits, and E. Szemerédi: Extremal graphs without large forbidden subgraphs. Advances in graph theory (Cambridge Combinatorial Conf., Trinity Coll., Cambridge, 1977). Ann. Discrete Math. **3** (1978), 29–41.
129. B. Bollobás, B. Narayanan and Andrei Raigorodskii: On the stability of the Erdős-Ko-Rado theorem, J. Combin. Theory Ser. A **137** (2016), 64–78.
130. B. Bollobás and Andrew G. Thomason: Projections of bodies and hereditary properties of hypergraphs, Bull. London Math. Soc. **27** (1995) 417–424.
131. B. Bollobás and A. Thomason: Hereditary and monotone properties of graphs, in: R.L. Graham and J. Nešetřil (Eds.): The Mathematics of Paul Erdős II, vol **14** of Algorithms Combin. Springer, Berlin Heidelberg, (1997) 70–78.
132. B. Bollobás and A. Thomason: On the girth of Hamiltonian weakly pancyclic graphs. J. Graph Theory **26 (3)** (1997), 165–173.
133. B. Bollobás and A. Thomason: Weakly pancyclic graphs. J. Combin. Theory Ser. B **77 (1)** (1999), 121–137.
134. J. Adrian Bondy: Pancyclic graphs. I. J. Combin. Theory, Ser. B **11** (1971), 80–84.
135. J.A. Bondy and P. Erdős: Ramsey numbers for cycles in graphs. J. Combinatorial Theory Ser. B **14** (1973), 46–54.
136. J.A. Bondy and M. Simonovits: Cycles of even length in graphs, J. Combin. Theory Ser. B **16** (1974) 97–105.
137. Christian Borgs, Jennifer T. Chayes, László Lovász: Moments of two-variable functions and the uniqueness of graph limits, Geom. Func. Anal., **19** (2010), 1597-1619. http://wuw.cs.elte.hu/~lovasz/limitunique.pdf
138. C. Borgs, J. Chayes and L. Lovász, V.T. Sós, Balázs Szegedy, and Katalin Vesztergombi: Graph limits and parameter testing, STOC '06: Proceedings of the 38th Annual ACM Symposium on Theory of Computing, ACM, New York, 2006, pp. 261–270.
139. C. Borgs, J. Chayes, L. Lovász, V.T. Sós, and K. Vesztergombi: Counting graph homomorphisms. Topics in discrete mathematics, 315–371, Algorithms Combin. **26**, Springer, Berlin, 2006.
140. C. Borgs, J. Chayes, L. Lovász, V.T. Sós, and K. Vesztergombi: Convergent sequences of dense graphs I: Subgraph frequencies, metric properties and testing, Advances Math. **219 (6)** (2008), 1801–1851
141. C. Borgs, J.T. Chayes, L. Lovász, V.T. Sós, and K. Vesztergombi: Convergent sequences of dense graphs II. Multiway cuts and statistical physics. Ann. of Math. (2) **176 (1)** (2012), 151–219.
142. J. Böttcher, Jan Hladký, Diana Piguet, Anush Taraz: An approximate version of the tree packing conjecture, Israel J. Math. **211 (1)** (2016) 391–446.
143. J. Böttcher, Y. Kohayakawa, A. Taraz, and Andreas Würfl: An extension of the blow-up lemma to arrangeable graphs, SIAM J. Discrete Math. **29 (2)** (2015) 962-1001. arXiv:1305.2059.
144. J. Böttcher, M. Schacht, and A. Taraz: Proof of the bandwidth conjecture of Bollobás and Komlós. Math. Ann. **343 (1)** (2009), 175–205.
145. Jean Bourgain: On triples in arithmetic progression, Geom. Funct. Anal. **9** (1999), 968–984.
146. J. Bourgain, Alexander Gamburd, and Peter Sarnak: Affine linear sieve, expanders, and sum-product. Invent. Math. **179**, (2010) 559–644.
147. S. Brandt, Ralph Faudree, and W. Goddard: Weakly pancyclic graphs, J. Graph Theory **27** (1998) 141–176.
148. Graham Brightwell, Konstantinos Panagiotou, and Angelika Steger: Extremal subgraphs of random graphs, Random Structures and Algorithms **41** (2012), 147–178. (Extended abstract: pp. 477–485 in SODA '07 (Proc. 18th ACM-SIAM Symp. Discrete Algorithms).
149. Hajo Broersma, Jan van den Heuvel, and Henk J. Veldman: A generalization of Ore's Theorem involving neighborhood unions, Discrete Math. **122** (1993), 37–49.

150. William G. Brown: On graphs that do not contain a Thomsen graph, Can. Math. Bull. **9** (1966), 281–285.
151. W.G. Brown: On an open problem of Paul Turán concerning 3-graphs. In Studies in Pure Mathematics, (1983) Birkhäuser, Basel, pp. 91–93.
152. W.G. Brown, P. Erdős, and M. Simonovits: Extremal problems for directed graphs, J. Combin. Theory Ser B **15** (1973), 77–93.
153. W.G. Brown, P. Erdős, and M. Simonovits: Algorithmic solution of extremal digraph problems, Trans Amer Math. Soc. **292** (1985), 421–449.
154. W.G. Brown, P. Erdős, and V.T. Sós: On the existence of triangulated spheres in 3-graphs and related problems, Period. Math. Hungar. **3** (1973) 221–228.
155. W.G. Brown, P. Erdős, and V.T. Sós: Some extremal problems on r-graphs, in "New Directions in the Theory of Graphs," Proc. 3rd Ann Arbor Conf. on Graph Theory, University of Michigan, 1971, pp. 53–63, Academic Press, New York, (1973).
156. W.G. Brown and Frank Harary: Extremal digraphs, Combinatorial theory and its applications, Colloq. Math. Soc. J. Bolyai, **4** (1970) I. 135–198
157. W.G. Brown and John W. Moon: Sur les ensembles de sommets indépendants dans les graphes chromatiques minimaux. (French) Canad. J. Math. **21** (1969) 274–278.
158. W.G. Brown and M. Simonovits: Digraph extremal problems, hypergraph extremal problems, and the densities of graph structures, Discrete Math., **48** (1984) 147–162.
159. W.G. Brown and M. Simonovits: Extremal multigraph and digraph problems, Paul Erdős and his mathematics, II (Budapest, 1999), pp. 157–203, Bolyai Soc. Math. Stud., **11**, János Bolyai Math. Soc., Budapest, (2002).
160. K. O'Bryant: A complete annotated bibliography of work related to Sidon sequences, Electron. J. Combin. (2004), Dynamic Surveys 11, 39 pp. (electronic).
161. Boris Bukh and David Conlon: Random algebraic construction of extremal graphs. Bull. London Math. Soc. **47** (2015), 939–945.
162. B. Bukh and D. Conlon: Rational exponents in extremal graph theory. J. Eur. Math. Soc. (JEMS) **20 (7)** (2018), 1747–1757.
163. B. Bukh and Zilin Jiang: A bound on the number of edges in graphs without an even cycle. Combin. Probab. Comput. **26 (1)** (2017) 1–15. arXiv:1403.1601, see also the erratum.
164. S.A. Burr and P. Erdős: On the magnitude of generalized Ramsey numbers for graphs, in: Infinite and Finite Sets, vol. 1, Keszthely, 1973, in: Colloq. Math. Soc. János Bolyai, vol. **10**, North-Holland, Amsterdam, 1975, pp. 214–240.
165. S.A. Burr, P. Erdős, Peter Frankl, Ron L. Graham, and V.T. Sós: Further results on maximal antiramsey graphs, In Graph Theory, Combinatorics and Applications, Vol. I, Y. Alavi, A. Schwenk (Editors), John Wiley and Sons, New York, 1988, pp. 193–206.
166. S.A. Burr, P. Erdős, R.L. Graham, and V.T. Sós: Maximal antiramsey graphs and the strong chromatic number, J. Graph Theory **13** (1989), 263–282.
167. S.A. Burr, P. Erdős, and L. Lovász: On graphs of Ramsey type. Ars Combinatoria **1 (1)** (1976) 167–190.
168. Sebastián Bustamante, H. Hàn, and Maya Stein: Almost partitioning 2-coloured complete 3-uniform hypergraphs into two monochromatic tight or loose cycles. arXiv:1701.07806v1.
169. S. Bustamante, Jan Corsten, Nóra Frankl, Alexey Pokrovskiy, and J. Skokan: Partitioning edge-coloured hypergraphs into few monochromatic tight cycles, arXiv:1903.04771v1 (2019)
170. Louis Caccetta and Roland Häggkvist: On diameter critical graphs, Discrete Math. **28** (1979) 223–229.
171. Dominic de Caen: The current status of Turán's problem on hypergraphs, Extremal problems for finite sets (Visegrád, 1991), 187–197, Bolyai Soc. Math. Stud. **3** (1994), János Bolyai Math. Soc., Budapest.
172. Dominic de Caen and Zoltán Füredi: The maximum size of 3-uniform hypergraphs not containing a Fano plane, Journal of Combinatorial Theory. Series B **78** (2000), 274–276.
173. Peter J. Cameron: 'Portrait of a typical sum-free set', Surveys in combinatorics, London Mathematical Society Lecture Note 123 (ed. C. Whitehead; Cambridge University Press, Cambridge, 1987) 13–42.

174. Peter J. Cameron and P. Erdős: 'Notes on sum-free and related sets', Recent trends in combinatorics (Mátraháza, 1995), Combinatorics, Probability and Computing **8** (Cambridge University Press, Cambridge, 1999) 95–107.
175. Paul A. Catlin: Subgraphs of graphs I., Discrete Math. **10** (1974), 225–233.
176. Chan, Tsz Ho, Ervin Győri, and András Sárközy: On a problem of Erdős on integers, none of which divides the product of k others. European J. Combin. **31 (1)** (2010) 260–269.
177. Phong Châu, Louis DeBiasio, and Henry A. Kierstead: Pósa's conjecture for graphs of order at least 8×10^9, Random Struct Algorithms **39 (4)** (2011), 507–525.
178. Sarvadaman Chowla: Solution of a problem of Erdős and Turán in additive-number theory. Proc. Nat. Acad. Sci. India. Sect. A **14 (1–2)** (1944).
179. Maria Chudnovsky, Neil Robertson, Paul Seymour, and Robin Thomas: The strong perfect graph theorem. Ann. of Math. **164** (2006) 51–229.
180. Fan R.K. Chung: Quasi-random classes of hypergraphs. Random Struct. Algorithms **1** (1980) 363–382 MR1138430 Corrigendum: Quasi-random classes of hypergraphs, Random Struct Algorithms **1** (1990), 363–382.
181. F.R.K. Chung: Regularity lemmas for hypergraphs and quasi-randomness. Random Structures and Algorithms **2 (2)** (1991) 241–252.
182. F.R.K. Chung: Quasi-random hypergraphs revisited. Random Structures and Algorithms **40 (1)** (2012), 39–48.
183. F.R.K. Chung and R.L. Graham: Quasi-random tournaments. J. Graph Theory **15 (2)** (1991), 173–198.
184. F.R.K. Chung and R.L. Graham: Quasi-random subsets of $\mathbb{Z}_n$, J. Combin. Theory Ser. A **61 (1)** (1992), 64–86.
185. F.R.K. Chung, R.L. Graham, and Richard M. Wilson: Quasi-random graphs, Combinatorica **9** (1989), 345–362.
186. Vaclav Chvátal: Tree-complete graph Ramsey numbers. J. Graph Theory **1 (1)** (1977), 93.
187. V. Chvátal, V. Rödl, E. Szemerédi, and W. Tom Trotter Jr.: The Ramsey number of a graph with bounded maximum degree, J. Combin. Theory Ser. B **34 (3)** (1983) 239–243.
188. V. Chvátal and E. Szemerédi: On the Erdős-Stone Theorem, J. London Math. Soc. (2) **23** (1981/2), 207–214.
189. Javier Cilleruelo: Infinite Sidon sequences. Advances Math. **255** (2014), 474–486.
190. J. Cilleruelo and Craig Timmons: k-fold Sidon sets. Electron. J. Combin. **21 (4)** (2014) Paper 4.12, 9 pp.
191. Sebastian M. Cioabă: Bounds on the Turán density of $PG_3(2)$, Electronic J. Combin. **11** (2004) 1–7.
192. David Conlon: Hypergraph packing and sparse bipartite Ramsey numbers, Combin. Probab. Comput. **18** (2009) 913-923.
193. D. Conlon: Combinatorial theorems relative to a random set, in: Proceedings of the International Congress of Mathematicians, Vol. **IV** (2014), 303–327. arXiv:1404.3324, 2014.
194. D. Conlon and Jacob Fox: Bounds for graph regularity and removal lemmas, Geom. Funct. Anal. **22** (2012) 1191–1256.
195. D. Conlon and J. Fox: Graph removal lemmas, Surveys in combinatorics 2013, Vol. **409**, London Mathematical Society Lecture Note Series, Cambridge University Press, Cambridge, (2013), 1–49.
196. D. Conlon, J. Fox, and B. Sudakov: On two problems in graph Ramsey theory, Combinatorica **32** (2012), 513–535.
197. D. Conlon, J. Fox, and B. Sudakov: Hereditary quasirandomness without regularity. Math. Proc. Cambridge Philos. Soc. **164 (3)** (2018), 385–399.
198. D. Conlon, J. Fox, and Yufei Zhao: Extremal results in sparse pseudorandom graphs. Advances Math., **256** (2014), 206–290. arXiv:1204.6645v1
199. D. Conlon and W. Timothy Gowers: Combinatorial theorems in sparse random sets, Ann. of Math. (2) **184 (2)** (2016) 367–454. arXiv:1011.4310v1.
200. D. Conlon, W.T. Gowers, W. Samotij, and M. Schacht: On the KŁR conjecture in random graphs. Israel J. Math., **203** (2014), 535–580.

201. D. Conlon and M. Stein: Monochromatic cycle partitions in local edge colorings. J. Graph Theory **81 (2)** (2016), 134–145. arXiv:1403.5975v3.
202. O. Cooley: Proof of the Loebl-Komlós-Sós conjecture for large, dense graphs, Discrete Math., **309** (2009), 6190–6228.
203. O. Cooley, D. Kühn, and D. Osthus: Perfect packings with complete graphs minus an edge, European J. Combin., **28** (2007), 2143–2155.
204. O. Cooley and R. Mycroft: The minimum vertex degree for an almost-spanning tight cycle in a 3-uniform hypergraph. Discrete Math. **340 (6)** (2017), 1172–1179. arXiv:1606.05616.
205. Joshua N. Cooper: Quasirandom permutations. J. Combin. Theory Ser. A **106 (1)** (2004), 123–143.
206. Jeff Cooper, Kunal Dutta, and Dhruv Mubayi: Counting independent sets in hypergraphs. Combin. Probab. Comput. **23 (4)** (2014), 539–550.
207. Jeff Cooper and Dhruv Mubayi: Counting independent sets in triangle-free graphs, Proc. Amer. Math. Soc. **142 (10)** (2014) 3325–3334,
208. J. Cooper and D. Mubayi: Sparse hypergraphs with low independence numbers, Combinatorica, **37 (1)** (2017), 31–40.
209. Keresztély Corrádi and András Hajnal: On the maximal number of independent circuits in a graph, Acta Math. Acad. Sci. Hungar **14** (1963) 423–439.
210. Ernie Croot, Vsevolod F. Lev, and Péter Pál Pach: Progression-free sets in $\mathbb{Z}_4^n$ are exponentially small. Ann. of Math. (2) **185 (1)** (2017) 331–337.
211. Béla Csaba: The Bollobás-Eldridge conjecture for graphs of maximum degree four. Manuscript (2003).
212. B. Csaba: On the Bollobás-Eldridge conjecture for bipartite graphs. Combin. Probab. Comput **16** (2007), 661–691.
213. B. Csaba: Regular spanning subgraphs of bipartite graphs of high minimum degree. Electron. J. Combin. **14 (1)** (2007), Note 21, 7 pp.
214. B. Csaba: On embedding well-separable graphs. Discrete Math. **308 (19)** (2008), 4322–4331.
215. B. Csaba, Ian Levitt, Judit Nagy-György, and E. Szemerédi: Tight bounds for embedding bounded degree trees, in Fete of Combinatorics and Computer Science, Bolyai Soc. Math. Stud. **20**, János Bolyai Math. Soc., Budapest, 2010, pp. 95–137.
216. B. Csaba and Marcello Mydlarz: Approximate multipartite version of the Hajnal-Szemerédi theorem. J. Combin. Theory Ser. B **102 (2)** (2012), 395–410.arXiv:0807.4463v1.
217. B. Csaba, Ali Shokoufandeh, and E. Szemerédi: Proof of a conjecture of Bollobás and Eldridge for graphs of maximum degree three. Paul Erdős and his mathematics (Budapest, 1999). Combinatorica **23 (1)** (2003), 35–72.
218. Andrzej Czygrinow and Louis DeBiasio: A note on bipartite graph tiling. SIAM J. Discrete Math. **25 (4)** (2011), 1477–1489.
219. A. Czygrinow, L. DeBiasio, H.A. Kierstead, and T. Molla: An extension of the Hajnal-Szemerédi theorem to directed graphs. Combin. Probab. Comput. **24 (5)** (2015), 754–773.
220. A. Czygrinow, DeBiasio, Louis; Molla, Theodore; Treglown, Andrew: Tiling directed graphs with tournaments. Forum Math. Sigma **6** (2018), e2, 53 pp.
221. A. Czygrinow, L. DeBiasio, and Brendan Nagle: Tiling 3-uniform hypergraphs with $K_4^3 - 2e$. J. Graph Theory **75 (2)** (2014), 124–136. arXiv:1108.4140.
222. A. Czygrinow, Vikram Kamat: Tight co-degree condition for perfect matchings in 4-graphs. Electron. J. Combin. **19 (2)** (2012) Paper 20, 16 pp.
223. A. Czygrinow, H.A. Kierstead, and T. Molla: On directed versions of the Corrádi-Hajnal corollary. European J. Combin. **42** (2014) 1–14.
224. A. Czygrinow, T. Molla: Tight codegree condition for the existence of loose Hamilton cycles in 3-graphs, SIAM J. Discrete Math. **28** (2014) 67–76.
225. A. Czygrinow, S. Poljak, and V. Rödl: Constructive quasi-Ramsey numbers and tournament ranking, SIAM J. Discrete Math **12** (1999), 48–63.
226. S. Das and T. Tran: Removal and stability for Erdős-Ko-Rado. SIAM J. Discrete Math. **30 (2)** (2016), 1102–1114. arXiv:1412.7885.

227. Giuliana Davidoff, Peter Sarnak, and Alain Valette: Elementary number theory, group theory, and Ramanujan graphs, London Math. Soc. Stud. Texts **55**, Cambridge University Press, Cambridge 2003.
228. Evan Davies, Matthew Jenssen, Will Perkins, and Barnaby Roberts: On the average size of independent sets in triangle-free graphs. Proc. Amer. Math. Soc. **146 (1)** (2018), 111–124. arXiv:1508.04675.
229. David Daykin and R. Häggkvist: Degrees giving independent edges in a hypergraph, Bull. Austral. Math. Soc., **23** (1981), 103–109.
230. L. DeBiasio, Safi Faizullah, and Imdadullah Khan: Ore-degree threshold for the square of a Hamiltonian cycle. Discrete Math. Theor. Comput. Sci. **17 (1)** (2015), 13–32.
231. Louis DeBiasio and Tao Jiang: On the co-degree threshold for the Fano plane. European J. Combin. **36** (2014), 151–158.
232. L. DeBiasio and Luke L. Nelsen: Monochromatic cycle partitions of graphs with large minimum degree, Journal of Combinatorial Theory, Ser. B **122** (2017), 634–667. arXiv:1409.1874v1.
233. Domingos Dellamonica, Kalyanasundaram Subrahmanyam, Daniel Martin, V. Rödl and Asaf Shapira: An optimal algorithm for finding Frieze-Kannan regular partitions. Combin. Probab. Comput. **24 (2)** (2015) 407–437.
234. Bobby DeMarco: Triangles in Random Graphs, thesis defense, Rutgers University, May 3, (2012).
235. B. DeMarco and Jeff Kahn: Mantel's theorem for random graphs. Random Structures and Algorithms **47 (1)** (2015), 59–72.
236. B. DeMarco and J. Kahn: Turán's theorem for random graphs, Submitted for publication, arXiv:1501.01340 (2015).
237. Tristan Denley: The independence number of graphs with large odd girth. Electron. J. Combin. **1** (1994), Research Paper 9, approx. 12 pp.
238. Walther A. Deuber, Alexander V. Kostochka, and Horst Sachs: A shorter proof of Dirac's theorem on the number of edges in chromatically critical graphs, Diskretnyi Analiz Issledovanie Operacii **3 (4)** (1996) 28–34 (in Russian).
239. Pat Devlin and Jeff Kahn: On "stability" in the Erdős-Ko-Rado theorem. SIAM J. Discrete Math. **30 (2)** (2016), 1283–1289.
240. Palahenedi H. Diananda and Hian Poh Yap: Maximal sum-free sets of elements of finite groups, Proceedings of the Japan Academy **45** (1969), 1–5.
241. Rainer Dietmann, Christian Elsholtz, Katalin Gyarmati, and M. Simonovits: Shifted products that are coprime pure powers. J. Combin. Theory Ser. A **111 (1)** (2005), 24–36.
242. Gabor A. Dirac: Some theorems on abstract graphs, *Proc. London Math. Soc.*, **2** (1952), 68–81.
243. G.A. Dirac: A property of 4-chromatic graphs and some remarks on critical graphs, J. London Math. Soc. **27** (1952) 85–92.
244. G.A. Dirac: Extension of Turán's theorem on graphs. Acta Math. Acad. Sci. Hungar **14** (1963) 417–422.
245. G.A. Dirac: The number of edges in critical graphs, J. Reine Angew. Math. **268/269** (1974) 150–164.
246. P. Drineas, P., A. Frieze, R. Kannan, S. Vempala, and V. Vinay: Clustering large graphs via the singular value decomposition. Mach. Learn. **56** (2004) 9–33.
247. Andrzej Dudek, C. Reiher, Andrzej Rucinski, and M. Schacht: Powers of Hamiltonian cycles in randomly augmented graphs, arXiv:1805.10676v1, 2018 May 17
248. Richard A. Duke, H. Lefmann, and V. Rödl: On uncrowded hypergraphs, In: Proceedings of the Sixth International Seminar on Random Graphs and Probabilistic Methods in Combinatorics and Computer Science, "Random Graphs '93" (Poznań, 1993), Random Struct Algorithms **6** (1995), 209–212.
249. R.A. Duke and V. Rödl: On graphs with small subgraphs of large chromatic number, Graphs Combin. **1 (1)** (1985) 91–96.
250. Nancy Eaton: Ramsey numbers for sparse graphs, Discrete Math. **185** (1998) 63–75.

251. Gábor Elek and B. Szegedy: Limits of hypergraphs, removal and regularity lemmas. A nonstandard approach, preprint. arXiv:0705.2179.
252. G. Elek and B. Szegedy: A measure-theoretic approach to the theory of dense hypergraphs. Advances Math. **231 (3-4)** (2012), 1731–1772. arXiv:0810.4062.
253. György Elekes: On the number of sums and products. Acta Arithmetica **81 (4)** (1997): 365–367.
254. Michael Elkin: An improved construction of progression-free sets. Israel J. Math. **184** (2011), 93–128. (See also: ACM-SIAM Symposium on Discrete Algorithms, SODA'10, Austin, TX, USA, (2010) pp. 886–905. arXiv:0801.4310.
255. Jordan Ellenberg and Dion Gijswijt: On large subsets of $\mathbb{F}_q^n$ with no three-term arithmetic progression. Ann. of Math. (2) **185 (1)** (2017) 339–343.
256. David Ellis: Stability for t-intersecting families of permutations, J. Combin. Theory Ser. A **118** (2011) 208–227.
257. D. Ellis, Nathan Keller, and Noam Lifshitz: Stability versions of Erdős-Ko-Rado type theorems, via isoperimetry, J. Eur. Math. Soc. (2016). arXiv:1604.02160.
258. Paul Erdős: On sequences of integers no one of which divides the product of two others and on some related problems, Mitt. Forsch.-Inst. Math. Mech. Univ. Tomsk **2** (1938) 74–82.
259. P. Erdős: On a problem of Sidon in additive number theory and on some related problems. Addendum, J. London Math. Soc. **19** (1944), 208.
260. P. Erdős: On sets of distances of n points, Amer. Math. Monthly **53** (1946), 248–250.
261. P. Erdős, 'Some remarks on the theory of graphs', Bull. Amer. Math. Soc. **53** (1947) 292–294.
262. P. Erdős: On a problem of Sidon in additive number theory, Acta Sci Math. Szeged **15** (1954), 255–259.
263. P. Erdős: Graph theory and probability. Canad. J. Math. **11** (1959) 34–38.
264. P. Erdős: Graph theory and probability. II. Canad. J. Math. **13** (1961) 346–352.
265. P. Erdős: On a theorem of Rademacher-Turán, Illinois J. Math. **6** (1962), 122–127.
266. P. Erdős: On extremal problems of graphs and generalized graphs, Israel J. Math. **2 (3)** (1964) 183–190.
267. P. Erdős: Extremal problems in graph theory, in: M. Fiedler (Ed.), Theory of Graphs and Its Applications, 2nd ed., Proc. Symp., Smolenice, 1963, Academic Press, New York, 1965, pp. 29–36.
268. P. Erdős: A problem on independent r-tuples, Ann. Univ. Sci. Budapest. **8** (1965) 93–95.
269. P. Erdős: Some recent results on extremal problems in graph theory. Results, Theory of Graphs (Internat. Sympos., Rome, 1966), Gordon and Breach, New York, Dunod, Paris, (1967), pp. 117–123 (English); pp. 124–130 (French).
270. P. Erdős: On some applications of graph theory to geometry, Canad. J. Math. **19** (1967), 968–971.
271. P. Erdős: On some new inequalities concerning extremal properties of graphs. In Theory of Graphs (Proc. Colloq., Tihany, 1966), Academic Press, New York, (1968), 77–81.
272. P. Erdős: On some applications of graph theory to number theoretic problems, Publ. Ramanujan Inst. No. (1968/1969), 131–136.
273. P. Erdős: Some applications of graph theory to number theory, The Many Facets of Graph Theory (Proc. Conf., Western Mich. Univ., Kalamazoo, Mich., 1968), pp. 77–82, Springer, Berlin, (1969).
274. P. Erdős: On the number of complete subgraphs and circuits contained in graphs. Časopis Pěst. Mat. **94** (1969) 290–296.
275. P. Erdős: Some applications of graph theory to number theory, Proc. Second Chapel Hill Conf. on Combinatorial Mathematics and its Applications (Univ. North Carolina, Chapel Hill, N.C., 1970), pp. 136–145.
276. P. Erdős: On the application of combinatorial analysis to number theory, geometry and analysis, Actes du Congrès International des Mathématiciens (Nice, 1970), Tome 3, pp. 201–210, Gauthier-Villars, Paris, 1971.
277. P. Erdős: The Art of Counting. Selected writings of P. Erdős, (J. Spencer ed.) Cambridge, The MIT Press (1973).

278. P. Erdős, Some problems in graph theory, in: Hypergraph Seminar, Ohio State Univ., Columbus, Ohio, 1972, in: Lecture Notes in Math., vol. 411, Springer, Berlin, (1974), pp. 187–190.
279. P. Erdős: Problems and results in graph theory and combinatorial analysis, in: Proceedings of the Fifth British Combinatorial Conference, Univ. Aberdeen, Aberdeen, 1975, Congr. Numer. **15** (1976) 169–192.
280. P. Erdős: Paul Turán, 1910–1976: his work in graph theory, J. Graph Theory **1 (2)** (1977) 97–101,
281. P. Erdős: Some personal reminiscences of the mathematical work of Paul Turán. Acta Arith. **37** (1980), 4–8.
282. P. Erdős: Some applications of Ramsey's theorem to additive number theory, European J. Combin. **1 (1)** (1980) 43–46.
283. P. Erdős: Some notes on Turán's mathematical work. P. Turán memorial volume. J. Approx. Theory **29 (1)** (1980), 2–5.
284. P. Erdős: Some applications of graph theory and combinatorial methods to number theory and geometry, Algebraic methods in graph theory, Vol. I, II (Szeged, 1978), Colloq. Math. Soc. János Bolyai, 25, pp. 137–148, North-Holland, Amsterdam-New York, 1981.
285. P. Erdős: Some old and new problems in combinatorial geometry, Applications of discrete mathematics (Clemson, SC, 1986), pp. 32–37, SIAM, Philadelphia, PA, 1988
286. P. Erdős: On some aspects of my work with Gabriel Dirac, in: L.D. Andersen, I.T. Jakobsen, C. Thomassen, B. Toft, P.D. Vestergaard (Eds.), Graph Theory in Memory of G.A. Dirac, Annals of Discrete Mathematics, Vol. 41, North-Holland, Amsterdam, 1989, pp. 111–116.
287. P. Erdős: On some of my favourite theorems. Combinatorics, Paul Erdős is eighty, Vol. 2 (Keszthely, 1993), 97–132, Bolyai Soc. Math. Stud., 2, János Bolyai Math. Soc., Budapest, 1996.
288. P. Erdős, P. Frankl, and V. Rödl: The asymptotic number of graphs not containing a fixed subgraph and a problem for hypergraphs having no exponent, Graphs Combin., **2** (1986), 113–121.
289. P. Erdős and Tibor Gallai: On maximal paths and circuits of graphs, Acta Math. Sci. Hungar. **10** (1959) 337–356.
290. P. Erdős, A.W. Goodman, and Lajos Pósa: The representation of a graph by set intersections, Canad. J. Math, **18** (1966), 106–112.
291. P. Erdős, A. Gyárfás, and László Pyber: Vertex coverings by monochromatic cycles and trees, J. Combin. Theory Ser. B **51** (1991) 90–95.
292. P. Erdős, A. Hajnal, M. Simonovits, V.T. Sós, and E. Szemerédi: Turán-Ramsey theorems and simple asymptotically extremal structures, Combinatorica **13** (1993), 31–56.
293. P. Erdős, A. Hajnal, M. Simonovits, V.T. Sós, and E. Szemerédi: Turán-Ramsey theorems and K_p-independence number, in: Combinatorics, geometry and probability (Cambridge, 1993), Cambridge Univ. Press, Cambridge, 1997, 253–281.
294. P. Erdős, A. Hajnal, M. Simonovits, V.T. Sós, and E. Szemerédi: Turán-Ramsey theorems and K_p-independence numbers, Combin. Probab. Comput. **3** (1994) 297–325.
295. P. Erdős, A. Hajnal, V.T. Sós, and E. Szemerédi: More results on Ramsey-Turán type problem, Combinatorica **3** (1983), 69–81.
296. P. Erdős and Haim Hanani: On a limit theorem in combinatorial analysis, Publ. Math. Debrecen **10** (1963) 10–13.
297. P. Erdős, Daniel J. Kleitman, and Bruce L. Rothschild: Asymptotic enumeration of K_n-free graphs, in Colloquio Internazionale sulle Teorie Combinatorie, Tomo II, Atti dei Convegni Lincei, no. 17 (Accad. Naz. Lincei, Rome) (1976), 19–27.
298. P. Erdős, Chao Ko, and Richard Rado: Intersection theorems for systems of finite sets, Quart. J. Math. Oxford Ser. (2), **12** (1961), 313–320.
299. P. Erdős and L. Lovász: Problems and results on 3-chromatic hypergraphs and some related questions, A. Hajnal (Ed.), Infinite and Finite Sets, North Holland, 1975, pp. 609–628.
300. P. Erdős, Amram Meir, V.T. Sós, and P. Turán: On some applications of graph theory II, Studies in Pure Mathematics (presented to R. Rado), Academic Press, London, (1971) pp. 89-99.

301. P. Erdős, A. Meir, V.T. Sós, and P. Turán: On some applications of graph theory I, Discrete Math. **2 (3)** (1972) 207–228.
302. P. Erdős, A. Meir, V.T. Sós, and P. Turán: On some applications of graph theory III, Canadian Math. Bull. **15** (1972) 27–32.
303. P. Erdős, A. Meir, V.T. Sós, and P. Turán, Corrigendum: 'On some applications of graph theory, I'. [Discrete Math. **2**(3) (1972) 207–228]; Discrete Math. **4** (1973) 90.
304. P. Erdős, S.B. Rao, M. Simonovits, and V.T. Sós: On totally supercompact graphs. Combinatorial mathematics and applications (Calcutta, 1988). Sankhya Ser. A **54** (1992), Special Issue, 155–167.
305. P. Erdős and Alfréd Rényi: Additive properties of random sequences of positive integers, Acta Arith. **6** (1960) 83–110.
306. P. Erdős and A. Rényi: On the evolution of random graphs. Magyar Tud. Akad. Mat. Kut. Int. Közl. **5** (1960) 17–61. (Also see [277], 574–617)
307. P. Erdős, A. Rényi: On the strength of connectedness of a random graph, Acta Math. Acad. Sci. Hungar. **12** (1961) 261–267.
308. P. Erdős, A. Rényi, and V.T. Sós: On a problem of graph theory. Stud. Sci. Math. Hung. **1** (1966) 215–235 (Also see [277], 201–221)
309. P. Erdős and C. Ambrose Rogers: 'The construction of certain graphs', Canad. J. Math. **14** (1962) 702–707.
310. P. Erdős and H. Sachs: Reguläre Graphen gegebener Taillenweite mit minimaler Knotenzahl. (German) Wiss. Z. Martin-Luther-Univ. Halle-Wittenberg Math.-Natur. Reihe **12** (1963) 251–257.
311. P. Erdős, András Sárközy, and V.T. Sós: On product representations of powers, I, European J. Combin. **16** (1995) 567–588.
312. P. Erdős and M. Simonovits: A limit theorem in graph theory, Studia Sci. Math. Hungar. **1** (1966) 51–57. (Reprinted in [277], 1973, pp. 194–200.).
313. P. Erdős and M. Simonovits: Some extremal problems in graph theory, Combinatorial theory and its applications, Vol. I, Proceedings Colloqium, Balatonfüred, (1969), North-Holland, Amsterdam, 1970, pp. 377–390.(Reprinted in [277]).
314. P. Erdős and M. Simonovits: *An extremal graph problem*, Acta Math. Acad. Sci. Hungar. **22** (1971/72), 275–282.
315. P. Erdős and M. Simonovits: On the chromatic number of Geometric graphs, Ars Combin. **9** (1980) 229–246.
316. P. Erdős and M. Simonovits: Supersaturated graphs and hypergraphs, Combinatorica **3** (1983) 181–192.
317. P. Erdős and M. Simonovits: Cube supersaturated graphs and related problems, in: J.A. Bondy, U.S.R. Murty (Eds.), Progress in Graph Theory, Acad. Press, 203–218 (1984)
318. P. Erdős and M. Simonovits: How many colours are needed to colour every pentagon of a graph in five colours? (manuscript, to be published)
319. P. Erdős and V.T. Sós: Some remarks on Ramsey's and Turán's theorem, in: Combinatorial theory and its applications, II (Proc. Colloq., Balatonfüred, 1969), North-Holland, Amsterdam, (1970), 395–404.
320. P. Erdős and J. Spencer: Probabilistic methods in combinatorics, Probability and Mathematical Statistics, Vol. 17, Academic, New York, London, (1974),
321. P. Erdős and Arthur H. Stone: On the structure of linear graphs, Bull. Amer. Math. Soc. **52** (1946) 1087–1091.
322. P. Erdős and E. Szemerédi: On sums and products of integers. In: Studies in Pure Mathematics, Birkhäuser, Basel, 213–218 (1983).
323. P. Erdős and P. Turán: On some sequences of integers, J. London Math. Soc. **11** (1936), 261–264.
324. P. Erdős and P. Turán: On a problem of Sidon in additive number theory, and on some related problems, J. London Math. Soc. **16** (1941), 212–215. Addendum (by P. Erdős) J. London Math. Soc. **19** (1944), 208.

325. Beka Ergemlidze, Ervin Győri, and Abhishek Methuku: A note on the linear cycle cover conjecture of Gyárfás and Sárközy. Electron. J. Combin. **25 (2)** (2018), Paper 2.29, 4 pp.
326. Vance Faber: The Erdős-Faber-Lovász conjecture-the uniform regular case. J. Combin. **1 (2)** (2010), 113–120.
327. V. Faber and L. Lovász: Unsolved problem #18, page 284 in Hypergraph Seminar, Ohio State University, 1972, edited by C. Berge and D. Ray-Chaudhuri.
328. Genghua Fan: On diameter 2-critical graphs, Discrete Math., **67 (3)** (1987) 235–240
329. G. Fan and R. Häggkvist: The square of a Hamiltonian cycle, SIAM J. Discrete Math. (1994) 203–212.
330. G. Fan and H.A. Kierstead: The square of paths and cycles, *Journal of Combinatorial Theory, Ser. B*, **63** (1995), 55–64.
331. R.J. Faudree, Ronald J. Gould, Mike S. Jacobson, and Richard H. Schelp: On a problem of Paul Seymour, Recent Advances in Graph Theory (V. R. Kulli ed.), Vishwa International Publication (1991), 197–215.
332. R.J. Faudree, R.H. Schelp: All Ramsey numbers for cycles in graphs. Discrete Math. **8** (1974) 313–329.
333. R.J. Faudree and R.H. Schelp: Ramsey type results, in: A. Hajnal, et al. (Eds.), Infinite and Finite Sets, in: Colloq. Math. J. Bolyai, vol. **10**, North-Holland, Amsterdam, (1975) pp. 657–665.
334. R. J. Faudree and R.H. Schelp: Path Ramsey numbers in multicolorings, Journal of Combinatorial Theory, Ser. B **19** (1975), 150–160.
335. R.J. Faudree and R.H. Schelp: Path-path Ramsey-type numbers for the complete bipartite graph, J. Combin. Theory Ser. B **19**, (1975), 161–173.
336. R.J. Faudree and M. Simonovits: Ramsey problems and their connection to Turán-type extremal problems. J. Graph Theory **16** (1992) 25–50.
337. Odile Favaron, Evelyne Flandrin, Hao Li, and Feng Tian: An Ore-type condition for pancyclability, Discrete Math. **206** (1999), 139–144.
338. Asaf Ferber, Gal Kronenberg, and Eoin Long: Packing, Counting and Covering Hamilton cycles in random directed graphs. Electron. Notes Discrete Math. **49** (2015) 813–819.
339. A. Ferber, Rajko Nenadov, Andreas Noever, Ueli Peter, and N. škorić: Robust hamiltonicity of random directed graphs, Proceedings of the 26th Annual ACM-SIAM Symposium on Discrete Algorithms (SODA 15) (2015), 1752–1758.
340. A. Ferber, Rajko Nenadov, and U. Peter: Universality of random graphs and rainbow embedding, Random Structures and Algorithms **48** (2016), 546–564.
341. Agnieszka Figaj and Tomasz Łuczak: The Ramsey number for a triple of long even cycles. J. Combin. Theory Ser. B **97 (4)** (2007) 584–596.
342. A. Figaj and T. Łuczak: The Ramsey Numbers for A Triple of Long Cycles. Combinatorica **38 (4)** (2018), 827–845.
343. David C. Fisher: Lower bounds on the number of triangles in a graph. J. Graph Theory **13** (1989) 505–512.
344. Mathew Fitch: Rational exponents for hypergraph Turán problems. arXiv:1607.05788.
345. G. Fiz Pontiveros, Simon Griffiths, R. Morris: The triangle-free process and $R(3, k)$, arXiv:1302.6279.
346. Dmitrii G. Fon-Der-Flaass: "Method for Construction of (3,4)-Graphs", Mat. Zametki **44 (4)** (1998), 546–550; [Math. Notes 44, 781–783 (1988)].
347. Jacob Fox: A new proof of the graph removal lemma, Ann. of Math. **174** (2011) 561–579.
348. J. Fox, Mikhail Gromov, Vincent Lafforgue. Assaf Naor, and J. Pach: Overlap properties of geometric expanders. J. Reine Angew. Math. **671** (2012), 49–83.
349. J. Fox, Po-Shen Loh, and Yufei Zhao: The critical window for the classical Ramsey-Turán problem. Combinatorica **35 (4)** (2015) 435–476.
350. J. Fox, László Miklós Lovász, and Yufei Zhao: On regularity lemmas and their algorithmic applications. Combin. Probab. Comput. **26 (4)** (2017), 481–505.
351. J. Fox, J. Pach, and Andrew Suk: Density and regularity theorems for semi-algebraic hypergraphs. Proceedings of the Twenty-Sixth Annual ACM-SIAM Symposium on Discrete Algorithms, 1517–1530, SIAM, Philadelphia, PA, (2015).

352. J. Fox and B. Sudakov: Density theorems for bipartite graphs and related Ramsey-type results, Combinatorica **29** (2009) 153–196.
353. J. Fox and B. Sudakov: Dependent random choice, Random Structures and Algorithms **38** (2011) 68–99.
354. Peter Frankl: All rationals occur as exponents, J. Combin. Theory Ser. A **42** (1985) 200–206.
355. P. Frankl and Z. Füredi: An exact result for 3-graphs, Discrete Math. **50** (1984) 323–328.
356. P. Frankl and V. Rödl: Hypergraphs do not jump, Combinatorica **4** (1984), 149–159.
357. P. Frankl and V. Rödl: Near perfect coverings in graphs and hypergraphs, European Journal of Combinatorics **6** (1985), 317–326.
358. P. Frankl and V. Rödl: The uniformity lemma for hypergraphs. Graphs Combin. **8 (4)** (1992) 309–312.
359. P. Frankl and R.M. Wilson: Intersection theorems with geometric consequences, Combinatorica **1** (1981) 357–368.
360. Mike Freedman, L. Lovász, and Alex Schrijver: Reflection positivity, rank connectivity, and homomorphism of graphs. J. AMS **20** (2007), 37–51.
361. G.A. Freiman: Foundations of a structural theory of set addition, Kazan Gos. Ped. Inst., Kazan, 1966 (Russian). English translation in Translations of Mathematical Monographs 37, American Mathematical Society, Providence, 1973.
362. Ehud Friedgut: On the measure of intersecting families, uniqueness and stability; Combinatorica **28 (5)** (2008), 503–528.
363. Joel Friedman and Nick Pippenger: Expanding graphs contain all small trees, Combinatorica **7** (1987), 71–76.
364. Alan Frieze and Ravi Kannan: The Regularity Lemma and approximation schemes for dense problems, Proc IEEE Symp Found Comput Sci, 1996, pp. 12–20.
365. A. Frieze and R. Kannan: Quick approximation to matrices and applications, Combinatorica **19** (1999), 175–220.
366. A. Frieze and R. Kannan: A simple algorithm for constructing Szemerédi's regularity partition. Electron. J. Combin., **6**, (1999). 74
367. A. Frieze, R. Kannan, and Santosh Vempala: Fast Monte-Carlo algorithms for finding low-rank approximations. J. ACM **51 (6)** (2004), 1025–1041.
368. Shinya Fujita, Henry Liu, and Colton Magnant: Monochromatic structures in edge-coloured graphs and hypergraphs-a survey, International Journal of Graph Theory and its Applications, **1 (1)** (2015) pp3–56.
369. Zoltán Füredi: Matchings and covers in hypergraphs, Graphs Combin. **4** (1988) 115–206.
370. Z. Füredi: Turán type problems, in Surveys in Combinatorics (1991), Proc. of the 13th British Combinatorial Conference, (A. D. Keedwell ed.) Cambridge Univ. Press. London Math. Soc. Lecture Note Series **166** (1991), 253–300.
371. Z. Füredi: The maximum number of edges in a minimal graph of diameter 2, J. Graph Theory **16** (1992) 81–98.
372. Z. Füredi: Extremal hypergraphs and combinatorial geometry, in: Proc. Internat. Congress of Mathematicians, vols. 1, 2, Zürich, 1994, Birkhäuser, Basel, (1995), pp. 1343–1352.
373. Z. Füredi: An upper bound on Zarankiewicz' problem. Combin. Probab. Comput. **5** (1996) 29–33
374. Z. Füredi: A proof of the stability of extremal graphs, Simonovits' stability from Szemerédi's regularity. J. Combin. Theory Ser. B **115** (2015), 66–71.
375. Z. Füredi, Alexander Kostochka, and Ruth Luo: A stability version for a theorem of Erdős on nonhamiltonian graphs. Discrete Math. **340 (11)** (2017), 2688–2690.
376. Z. Füredi, A. Kostochka, Ruth Luo, and J. Verstraëte: Stability in the Erdős-Gallai Theorem on cycles and paths, II. Discrete Math. **341 (5)** (2018), 1253–1263.
377. Z. Füredi, A. Kostochka, and J. Verstraëte: Stability in the Erdős-Gallai theorems on cycles and paths. J. Combin. Theory Ser. B **121** (2016), 197–228.
378. Z. Füredi and André Kündgen: Turán problems for weighted graphs, J. Graph Theory **40** (2002), 195–225.

379. Z. Füredi, D. Mubayi, and O. Pikhurko: Quadruple systems with independent neighborhoods. J. Combin. Theory Ser. A **115 (8)** (2008), 1552–1560.
380. Z. Füredi, O. Pikhurko, and M. Simonovits: The Turán density of the hypergraph $\{abc, ade, bde, cde\}$. Electronic J. Combin. **10** # R18 (2003)
381. Z. Füredi, O. Pikhurko, and M. Simonovits: On triple systems with independent neighbourhoods, Combin. Probab. Comput. **14 (5-6)** (2005), 795–813,
382. Z. Füredi, O. Pikhurko, and M. Simonovits: 4-books of three pages. J. Combin. Theory Ser. A **113 (5)** (2006), 882–891.
383. Z. Füredi and M. Ruszinkó: Uniform hypergraphs containing no grids. Advances Math. **240** (2013), 302–324. arXiv:1103.1691
384. Z. Füredi and A. Sali: Some new bounds on partition critical hypergraphs. European J. Combin. **33 (5)** (2012), 844–852.
385. Z. Füredi and M. Simonovits: Triple systems not containing a Fano configuration, Combin. Probab. Comput. **14 (4)** (2005), 467–484.
386. Z. Füredi and M. Simonovits: The history of degenerate (bipartite) extremal graph problems, in Erdős Centennial, Bolyai Soc. Math. Stud., **25**, János Bolyai Math. Soc., Budapest, (2013) 169–264. arXiv:1306.5167.
387. Hillel Fürstenberg: Ergodic behavior of diagonal measures and a theorem of Szemerédi on arithmetic progressions, J. Analyse Math. **31** (1977), 204–256.
388. H. Fürstenberg and Yitzhak Katznelson: An ergodic Szemerédi theorem for commuting transformations, J. Analyse Math. **34** (1978) 275–291.
389. H. Fürstenberg and Y. Katznelson: A density version of the Hales-Jewett theorem, J. Analyse Math. **57** (1991), 64–119.
390. László Gerencsér and András Gyárfás: On Ramsey-type problems, Ann. Univ. Sci. Budapest. Eötvös Sect. Math., **10** (1967), 167–170.
391. Stefanie Gerke, Y. Kohayakawa, V. Rödl, and A. Steger: Small subsets inherit sparse ϵ-regularity, J. Combin. Theory Ser. B **97 (1)** (2007) 34–56.
392. S. Gerke and A. Steger: The sparse regularity lemma and its applications, in: B.S. Webb (Ed.), Surveys Combinatorics 2005, University of Durham, 2005, in: London Math. Soc. Lecture Note Ser., vol. 327, Cambridge Univ. Press, Cambridge, 2005, pp. 227–258.
393. Alfred Geroldinger and Imre Z. Ruzsa: Combinatorial number theory and additive group theory. Courses and seminars from the DocCourse in Combinatorics and Geometry held in Barcelona, 2008. Advanced Courses in Mathematics. CRM Barcelona. Birkhäuser Verlag, Basel, 2009. xii+330 pp. ISBN: 978-3-7643-8961-1
394. Roman Glebov, Y. Person, and Wilma Weps: On extremal hypergraphs for Hamiltonian cycles. European J. Combin. **33 (4)** (2012) 544–555.
395. Stephan Glock, D. Kühn, Allan Lo, R. Montgomery, and D. Osthus: On the decomposition threshold of a given graph, arXiv:1603.04724, 2016.
396. S. Glock, D. Kühn, Allan Lo, and D. Osthus: The existence of designs via iterative absorption, arXiv:1611.06827, (2016).
397. S. Glock, D. Kühn, Allan Lo, and D. Osthus: Hypergraph F-designs for arbitrary F, arXiv:1706.01800, (2017).
398. S. Glock, D. Kühn, and D. Osthus: Optimal path and cycle decompositions of dense quasirandom graphs. J. Combin. Theory Ser. B **118** (2016), 88–108.
399. O. Goldreich, S. Goldwasser and D. Ron: Property testing and its connection to learning and approximation, Journal of the ACM **45** (1998), 653–750.
400. Adolph W. Goodman: On sets of acquaintances and strangers at any party. Amer. Math. Monthly **66** (1959) 778–783.
401. W. Timothy Gowers: Lower bounds of tower type for Szemerédi's uniformity lemma, GAFA, Geom. Func. Anal. **7** (1997), 322–337.
402. W.T. Gowers: A new proof of Szemerédi's theorem for arithmetic progressions of length four, Geom. Funct. Anal. **8** (1998), 529–551. MR 1631259 132, 138 MR1631259 // A new proof of Szemerédi's theorem, Geom. Funct. Anal. **11** (2001), 465–588; Erratum, Geom Funct. Anal. **11** (2001), 869

403. W.T. Gowers: Quasirandomness, counting and regularity for 3-uniform hypergraphs, Combinatorics, Probability and Computing **15** (2006), 143–184.
404. W.T. Gowers: Hypergraph regularity and the multidimensional Szemerédi theorem, Ann. of Math. (2) **166 (3)** (2007) 897–946. arXiv:0710.3032.
405. W.T. Gowers: Quasirandom groups. Combin. Probab. Comput. **17**, (2008) 363–387
406. W.T. Gowers: Polymath and the density Hales-Jewett theorem. An irregular mind, 659–687, Bolyai Soc. Math. Stud., 21, János Bolyai Math. Soc., Budapest, (2010).
407. W.T. Gowers: A Geometric Ramsey Problem, *http://mathoverflow.net/ questions/50928/a-geometric-ramsey-problem*, accessed May 2016.
408. W.T. Gowers and Omid Hatami: Inverse and stability theorems for approximate representations of finite groups, arXiv:1510.04085 [math.GR].
409. R.L. Graham and B.L. Rothschild: Ramsey's theorem for n-parameter sets, Trans. Amer. Math. Soc. **159** (1971), 257–292.
410. R.L. Graham, B.L. Rothschild, and J.H. Spencer: Ramsey Theory, Wiley, New York, 1990. 2nd edition
411. R.L. Graham, V. Rödl, and A. Ruciński: On graphs with linear Ramsey numbers, J. Graph Theory **35** (2000) 176–192.
412. Andrew Granville and József Solymosi: Sum-product formulae. Recent trends in combinatorics, 419–451, IMA Vol. Math. Appl., 159, Springer, [Cham], 2016.
413. J. Ben Green: The Cameron-Erdős conjecture, Bull. London Math. Soc. **36** (2004) 769–778.
414. J. B. Green: A Szemerédi-type regularity lemma in abelian groups, with applications, Geom. Funct. Anal. **15 (2)** (2005) 340–376.
415. B. Green and I.Z. Ruzsa: Sum-free sets in abelian groups, Israel J. Math. **147** (2005), 157–188.
416. B. Green and T.C. Tao: The primes contain arbitrarily long arithmetic progressions, Annals of Mathematics 167 (2008), 481–547.
417. B. Green and T.C. Tao: New bounds for Szemerédi's theorem. I. Progressions of length 4 in finite field geometries. Proc. London Math. Soc. (3) **98 (2)** (2009), 365–392.
418. B. Green and T.C. Tao: An arithmetic regularity lemma, an associated counting lemma, and applications. In: I. Bárány, J. Solymosi (eds.) An Irregular Mind. Bolyai Soc. Math. Stud., Vol. 21, pp. 261–334. János Bolyai Mathematical Society, Budapest (2010)
419. B. Green and T.C. Tao: New bounds for Szemerédi's theorem, III: a polylogarithmic bound for $r_4(N)$. Mathematika **63 (3)** (2017) 944–1040.
420. B. Green and Julia Wolf: A note on Elkin's improvement of Behrend's construction. Additive number theory, 141–144, Springer, New York, 2010. arXiv:0810.0732.
421. Jerrold R. Griggs, M. Simonovits, and G. Rubin Thomas: Extremal graphs with bounded densities of small graphs, J. Graph Theory **29** (1998), 185–207.
422. Andrey Grinshpun and G.N. Sárközy: Monochromatic bounded degree subgraph partitions, Discrete Math. **339** (2016) 46–53.
423. Andrzei Grzesik: On the maximum number of five-cycles in a triangle-free graph, J. Combin. Theory Ser. B **102 (5)** (2012) 1061–1066. arXiv:1102.0962.
424. A. Gyárfás: Covering complete graphs by monochromatic paths, in: Irregularities of Partitions, in: Algorithms Combin., **8** (1989), Springer-Verlag, pp. 89–91.
425. A. Gyárfás: Packing trees into n-chromatic graphs. Discuss. Math. Graph Theory **34 (1)** (2014), 199–201.
426. A. Gyárfás: Vertex covers by monochromatic pieces-a survey of results and problems. Discrete Math. **339 (7)** (2016) 1970–1977. Arxiv 2015
427. A. Gyárfás, E. Győri, and M. Simonovits: On 3-uniform hypergraphs without linear cycles. J. Combin. **7 (1)** (2016), 205–216.
428. A. Gyárfás, G.N. Sárközy, and E. Szemerédi: Long monochromatic Berge cycles in colored 4-uniform hypergraphs. Graphs Combin. **26 (1)** (2010) 71–76.
429. A. Gyárfás and J. Lehel: Packing trees of different order into K_n, in: Combinatorics, Proc. Fifth Hungarian Colloq., Keszthely, 1976, in: Colloq. Math. Soc. János Bolyai, vol. **18**, North-Holland, Amsterdam-New York, (1978), pp. 463–469,

430. A. Gyárfás, J. Lehel, G. Sárközy, and R. Schelp: Monochromatic Hamiltonian Berge-cycles in colored complete uniform hypergraphs, J. Combin. Theory Ser. B **98** (2008) 342–358.
431. A. Gyárfás, M. Ruszinkó, G.N. Sárközy, and E. Szemerédi: An improved bound for the monochromatic cycle partition number. J. Combin. Theory Ser. B **96**(6) (2006) 855–873.
432. A. Gyárfás, M. Ruszinkó, G.N. Sárközy, and E. Szemerédi: Three-color Ramsey number for paths, Combinatorica **28**(4) (2008), 499–502; Corrigendum: "Three-color Ramsey numbers for paths" [Combinatorica **27** (**1**) (2007) 35–69].
433. A. Gyárfás and G.N. Sárközy: The 3-colour Ramsey number of a 3-uniform Berge cycle. Combin. Probab. Comput. **20** (**1**) (2011) 53–71.
434. A. Gyárfás and G.N. Sárközy: Monochromatic path and cycle partitions in hypergraphs, Electronic Journal of Combinatorics, **20** (**1**) (2013) #P18.
435. A. Gyárfás and G.N. Sárközy: Monochromatic loose-cycle partitions in hypergraphs, *Electronic J. of Combinatorics* **21** (**2**) (2014) P2.36.
436. A. Gyárfás, G.N. Sárközy, and Stanley M. Selkow: Coverings by few monochromatic pieces: a transition between two Ramsey problems. Graphs Combin. **31** (**1**) (2015), 131–140.
437. A. Gyárfás, G.N. Sárközy, and E. Szemerédi: The Ramsey number of diamond-matchings and loose cycles in hypergraphs. Electron. J. Combin. **15** (**1**) (2008), Research Paper 126, 14 pp.
438. A. Gyárfás, G.N. Sárközy, and E. Szemerédi: Stability of the path-path Ramsey number. Discrete Math. **309** (**13**) (2009), 4590–4595.
439. A. Gyárfás, G.N. Sárközy, and E. Szemerédi: Monochromatic Hamiltonian 3-tight Berge cycles in 2-colored 4-uniform hypergraphs. J. Graph Theory **63** (**4**) (2010), 288–299.
440. Ervin Győri: On the number of C_5 in a triangle-free graph, Combinatorica **9** (1989) 101–102.
441. E. Győri: C_6-free bipartite graphs and product representation of squares, Discrete Math. **165/166** (1997) 371–375.
442. E. Győri and Hao Li: On the number of triangles in C_{2k+1}-free graphs, Comb. Prob. Comput. **21** (**1–2**) (2013) 187–191.
443. E. Győri, Dániel Korándi, Abhishek Methuku, István Tomon, Casey Tompkins, and Máté Vizer: On the Turán number of some ordered even cycles. European J. Combin. **73** (2018), 81–88.
444. András Hajnal and E. Szemerédi: Proof of a conjecture of Erdős, Combinatorial Theory and its Applications vol. II (P. Erdős, A. Rényi and V.T. Sós eds.), Colloq. Math. Soc. J. Bolyai **4**, North-Holland, Amsterdam (1970), pp. 601–623.
445. Péter Hajnal: An $\Omega(n^{4/3})$ lower bound on the randomized complexity of graph properties, Combinatorica **11**(2) (1991) 131–143.)
446. P. Hajnal, Simao Herdade, and E. Szemerédi: Proof of the Pósa-Seymour conjecture (2018), Arxiv
447. P. Hajnal and M. Szegedy: (1992) On packing bipartite graphs. Combinatorica **12** 295–301.
448. P. Hajnal and E. Szemerédi: Two Geometrical Applications of the Semi-random Method, Chapter 8 in G. Ambrus et al. (eds.), New Trends in Intuitive Geometry, Bolyai Society Mathematical Studies **27** (2018).
449. Hiep Hàn, Y. Person, and M. Schacht: On perfect matchings in uniform hypergraphs with large minimum vertex degree. SIAM J. Discrete Math. **23** (**2**) (2009), 732–748
450. H. Hàn and M. Schacht: Dirac-type results for loose Hamilton cycles in uniform hypergraphs, J. Combin. Theory Ser. B **100** (**3**) (2010) 332–346.
451. Jie Han and Yi Zhao: Minimum vertex degree threshold for C_4^3-tiling. J. Graph Theory **79** (2015) 300–317.
452. Christoph Hundack, Hans Jürgen Prömel, and Angelika Steger: Extremal graph problems for graphs with a color-critical vertex. Combin. Probab. Comput. **2** (**4**) (1993) 465–477.
453. Hamed Hatami, Jan Hladký, Daniel Král', Serguei Norine, and Alexander A. Razborov: On the number of pentagons in triangle-free graphs. J. Combin. Theory Ser. A **120** (2013) 722–732.
454. Julie Haviland and A.G. Thomason: Pseudo-random hypergraphs, in "Graph Theory and Combinatorics (Cambridge, 1988)" Discrete Math. **75** (**1-3**) (1989), 255–278.

455. J. Haviland and A.G. Thomason: On testing the "pseudo-randomness" of a hypergraph, Discrete Math. **103 (3)** (1992), 321–327.
456. Penny E. Haxell: Partitioning complete bipartite graphs by monochromatic cycles, Journal of Combinatorial Theory, Ser. B **69** (1997) pp. 210–218.
457. P.E. Haxell: Tree embeddings, J. Graph Theory, **36** (2001), 121–130.
458. P.E. Haxell, T. Łuczak, Yuejian Peng, V. Rödl, A. Ruciński, M. Simonovits, and J. Skokan: The Ramsey number for hypergraph cycles I, J. Combin. Theory Ser. A **113** (2006) 67–83.
459. P.E. Haxell, T. Łuczak, Y. Peng, V. Rödl, A. Ruciński, and J. Skokan: The Ramsey number for 3-uniform tight hypergraph cycles. Combin. Probab. Comput. **18 (1-2)** (2009) 165–203.
460. Teresa Haynes, Michael Henning, Lucas C. van der Merwe, and Anders Yeo: Progress on the Murty-Simon Conjecture on diameter-2 critical graphs: a survey, J. Comb. Optim. **30 (3)** (2015) 579–595.
461. R. Häggkvist and C. Thomassen: On pancyclic digraphs, J. Combin. Theory Ser. B **20** (1976) 20–40.
462. D.R. Heath-Brown: Integer sets containing no arithmetic progressions, J. London Math. Soc. **35** (1987), 385–394.
463. Anthony J.W. Hilton and Eric C. Milner: Some intersection theorems for systems of finite sets, Quart. J. Math. Oxf. 2 **18** (1967) 369–384.
464. Neil Hindman: On a conjecture of Erdős, Faber, and Lovász about n-colorings. Canad. J. Math. **33 (3)** (1981), 563–570.
465. Jan Hladký, J. Komlós, D. Piguet, M. Simonovits, Maya Stein, and E. Szemerédi: An approximate version of the Loebl-Komlós-Sós Conjecture for sparse graphs, arXiv:1211.3050.v1 (2012) Nov 13.
466. J. Hladký, J. Komlós, D. Piguet, M. Simonovits, M. Stein, and E. Szemerédi: The approximate Loebl-Komlós-Sós Conjecture I: The sparse decomposition, SIAM J. Discrete Math., **31** (2017), 945–982, arXiv:1408.3858.
467. J. Hladký, J. Komlós, D. Piguet, M. Simonovits, M. Stein, and E. Szemerédi: The approximate Loebl-Komlós-Sós Conjecture II: The rough structure of LKS graphs, SIAM J. Discrete Math., **31** (2017), 983–1016,
468. J. Hladký, J. Komlós, D. Piguet, M. Simonovits, M. Stein, and E. Szemerédi: The approximate Loebl-Komlós-Sós Conjecture III: The finer structure of LKS graphs, SIAM J. Discrete Math., **31** (2017), 1017–1071,
469. J. Hladký, J. Komlós, D. Piguet, M. Simonovits, M. Stein, and E. Szemerédi: The approximate Loebl-Komlós-Sós conjecture IV: Embedding techniques and the proof of the main result. SIAM J. Discrete Math. **31 (2)** (2017) 1072–1148.
470. J. Hladký, D. Piguet, M. Simonovits, M. Stein, and E. Szemerédi: The approximate Loebl-Komlós-Sós conjecture and embedding trees in sparse graphs, Electron. Res. Announc. Math. Sci., **22**, (2015), 1–11.
471. J. Hladký and M. Schacht: Note on bipartite graph tilings. SIAM J. Discrete Math. **24 (2)** (2010), 357–362.
472. Alan J. Hoffman and Robert R. Singleton: Moore graphs with diameter 2 and 3, IBM Journal of Res. Develop. **4** (1960) 497–504.
473. W. Imrich: Explicit construction of graphs without small cycles, Combinatorica **4** (1984), 53–59.
474. B. Jackson and O. Ordaz: Chvátal-Erdős condition for path and cycles in graphs and digraphs. A survey, Discrete Math. **84** (1990), 241–254.
475. S. Janson: Quasi-random graphs and graph limits, European J. Combin. **32** (2011), 1054–1083. arXiv:0905.3241
476. Svante Janson, Tomasz Łuczak, and A. Ruciński: Random graphs. Wiley-Interscience Series in Discrete Mathematics and Optimization. Wiley-Interscience, New York, (2000). xii+333 pp.
477. Janson, Svante and Sós, Vera T.: More on quasi-random graphs, subgraph counts and graph limits. European J. Combin. **46** (2015), 134–160.

478. S. Janson and A. Ruciński: The infamous upper tail. Probabilistic methods in combinatorial optimization. Random Structures and Algorithms **20 (3)** (2002) 317–342.
479. Matthew Jenssen and J. Skokan: Exact Ramsey numbers of Odd cycles via nonlinear optimization, arXiv:1608.05705V1
480. Anders Johansson, J. Kahn, and Van H. Vu: Factors in random graphs, Random Structures and Algorithms, **33** (2008), 1–28.
481. F. Joos, Jaehoon Kim, D. Kühn, and D. Osthus: Optimal packings of bounded degree trees. arXiv:1606.03953 (Submitted on 13 Jun 2016)
482. Jeff Kahn: Coloring nearly-disjoint hypergraphs with $n + o(n)$ colors, J. Combin. Theory Ser. A **59** (1991) 31–39.
483. Jeff Kahn and Paul Seymour: A fractional version of the Erdős-Faber-Lovász conjecture. Combinatorica **12 (2)** (1992), 155–160.
484. Gil Kalai: Designs exist! [after Peter Keevash]. Astérisque No. 380, Séminaire Bourbaki. Vol. 2014/2015 (2016), Exp. No. 1100, 399–422.
485. Ravi Kannan: Markov chains and polynomial time algorithms. 35th Annual Symposium on Foundations of Computer Science (Santa Fe, NM, 1994), 656–671, IEEE Comput. Soc. Press, Los Alamitos, CA, 1994.
486. Mikio Kano and Xueliang Li: Monochromatic and heterochromatic subgraphs in edge colored graphs – A Survey, Graphs and Combinatorics **24** (2008) 237–263.
487. Roman Karasev, Jan Kynčl, Pavel Paták, Zuzana Patáková, and Martin Tancer: Bounds for Pach's selection theorem and for the minimum solid angle in a simplex. Discrete Comput. Geom. **54 (3)** (2015), 610–636.
488. Gyula Katona: Graphs, vectors and inequalities in probability theory. (Hungarian) Mat. Lapok **20** (1969) 123–127.
489. G.O.H. Katona: Extremal problems for hypergraphs, in: M. Hall, J.H. van Lint (Eds.), Combinatorics Part II, in: Math. Centre Tracts, vol. **56** (1974), Mathematisch Centre Amsterdam, pp. 13–42.
490. G. Katona: Probabilistic inequalities from extremal graph results (a survey), Random Graphs '83, Poznan, 1983, Ann. Discrete Math. **28** (1985) 159–170.
491. G.O.H. Katona: Turán's graph theorem, measures and probability theory. Number theory, analysis, and combinatorics, 167–176, De Gruyter Proc. Math., De Gruyter, Berlin, (2014).
492. G.Y. Katona and H.A. Kierstead: Hamiltonian chains in hypergraphs. J. Graph Theory **30 (3)** (1999) 205–212
493. Gy. Katona, Tibor Nemetz, and M. Simonovits: On a graph problem of Turán, Mat. Fiz. Lapok **15** (1964) 228–238 (in Hungarian).
494. Peter Keevash: The Turán problem for projective geometries, J. Combin. Theory Ser. A, **111** (2005), 289–309.
495. P. Keevash: Shadows and intersections: stability and new proofs, Advances Math. **218 (5)** (2008) 1685–1703.
496. P. Keevash: A hypergraph regularity method for generalised Turán problems, Random Struct Algorithm **34** (2009), 123–164.
497. P. Keevash: A hypergraph blow-up lemma, Random Struct. Algorithms **39** (2011) 275–376.
498. P. Keevash: Hypergraph Turán problems, in: R. Chapman (Ed.), Surveys in Combinatorics, Cambridge Univ. Press, 2011, pp. 83–140.
499. P. Keevash and R. Mycroft: A multipartite Hajnal-Szemerédi theorem. The Seventh European Conference on Combinatorics, Graph Theory and Applications, 141–146, CRM Series, 16, Ed. Norm., Pisa, 2013. arXiv:1201.1882.
500. P. Keevash: The existence of designs, arXiv:1401.3665. (see also P. Keevash – "The existence of designs", videotaped lectures https://www.youtube.com/watch?v=tN6oGXqS2Bs, 2015.)
501. P. Keevash: The existence of designs, II. arXiv:1802.05900.
502. P. Keevash: Counting designs. J. Eur. Math. Soc. (JEMS) **20 (4)** (2018), 903–927. arXiv:1504.02909.
503. P. Keevash and Dhruv Mubayi: Stability theorems for cancellative hypergraphs. J. Combin. Theory Ser. B **92 (1)** (2004), 163–175.

504. P. Keevash and B. Sudakov: The Turán number of the Fano plane. Combinatorica **25 (5)** (2005) 561–574.
505. P. Keevash and B. Sudakov: Triangle packings and 1-factors in oriented graphs. J. Combin. Theory Ser. B **99 (4)** (2009) 709–727. arXiv:0806.2027
506. P. Keevash and B. Sudakov: Pancyclicity of Hamiltonian and highly connected graphs. J. Combin. Theory Ser. B **100 (5)** (2010), 456–467.
507. Imdadullah Khan: Perfect matchings in 3-uniform hypergraphs with large vertex degree. SIAM J. Discrete Math. **27 (2)** (2013) 1021–1039 arXiv:1101.5830.
508. I. Khan: Perfect matchings in 4-uniform hypergraphs. J. Combin. Theory Ser. B **116** (2016), 333–366. arXiv:1101.5675
509. Henry A. Kierstead and A.V. Kostochka: An Ore-type theorem on equitable coloring, J. Combinatorial Theory Series B **98** (2008), 226–234.
510. H.A. Kierstead and A.V. Kostochka: A short proof of the Hajnal-Szemerédi Theorem on equitable coloring, Combinatorics, Probability and Computing **17** (2008), 265–270.
511. H.A. Kierstead and A.V. Kostochka: Ore-type versions of Brooks' theorem. J. Combin. Theory Ser. B **99 (2)** (2009) 298–305.
512. H.A. Kierstead, A.V. Kostochka, T. Molla, and Elyse C. Yeager: Sharpening an Ore-type version of the Corrádi-Hajnal theorem. Abh. Math. Semin. Univ. Hambg. **87 (2)** (2017) 299–335. (Reviewer: Colton Magnant)
513. H.A. Kierstead, A.V. Kostochka, M. Mydlarz, and E. Szemerédi: A fast algorithm for equitable coloring. Combinatorica **30 (2)** (2010) 217–224.
514. H.A. Kierstead, A.V. Kostochka and Gexin Yu: Extremal graph packing problems: Ore-type versus Dirac-type. In Surveys in Combinatorics 2009 Vol. 365 of London Mathematical Society Lecture Note Series, Cambridge Univ. Press, 113–136.
515. Jeong Han Kim: The Ramsey number $R(3, t)$ has order of magnitude $t^2 / \log t$. Random Struct. Algorithms **7 (3)** (1995) 173–207.
516. Jaehoon Kim, Younjin Kim, and Hong Liu: Two conjectures in Ramsey-Turán theory, SIAM J. Discrete Math. **33 (1)** (2019) 564–586. arXiv:1803.04721.
517. Valerie King: "Lower bounds on the complexity of graph properties," In Proceedings of the 20th Annual ACM Symposium on the Theory of Computing, Chicago, IL, 1988, pp. 468–476.
518. V. King: A Lower Bound for the Recognition of Digraph Properties. Combinatorica, 1990, 10: 53–59
519. D.J. Kleitman and B.L. Rothschild: The number of finite topologies, Proc Amer Math. Soc. **25 (2)** (1970), 276–282.
520. Daniel J. Kleitman and B.L. Rothschild: Asymptotic enumeration of partial orders on a finite set, Trans Amer Math. Soc. **205** (1975), 205–220.
521. D.J. Kleitman, B.L. Rothschild, and J.H. Spencer: The number of semigroups of order n. Proc. Amer. Math. Soc. **55 (1)** (1976) 227–232.
522. D.J. Kleitman and David B. Wilson: On the number of graphs which lack small cycles, preprint, 1997.
523. D.J. Kleitman and Kenneth J. Winston: On the number of graphs without 4-cycles, Discrete Math. **41 (2)** (1982) 167–172.
524. Yoshiharu Kohayakawa: Szemerédi's regularity lemma for sparse graphs, in Foundations of Computational Mathematics (Rio de Janeiro, 1997), Springer, Berlin, 1997, pp. 216–230.
525. Y. Kohayakawa, T. Łuczak, and V. Rödl: On K_4-free subgraphs of random graphs. Combinatorica **17 (2)** (1997), 173–213.
526. Y. Kohayakawa; Guilherme Oliveira Mota, and M. Schacht; A. Taraz: Counting results for sparse pseudorandom hypergraphs II. European J. Combin. **65** (2017), 288–301.
527. Y. Kohayakawa, B. Nagle, V. Rödl, and M. Schacht: Weak hypergraph regularity and linear hypergraphs, Journal of Combinatorial Theory. Series B **100** (2010), 151–160.
528. Y. Kohayakawa and V. Rödl: Szemerédi's regularity lemma and quasi-randomness, in: Recent Advances in Algorithms and Combinatorics, in: CMS Books Math./Ouvrages Math. SMC, vol. **11**, Springer, New York, 2003, pp. 289–351.

529. Y. Kohayakawa, V. Rödl, M. Schacht, and J. Skokan: On the triangle removal lemma for subgraphs of sparse pseudorandom graphs, in: An Irregular Mind (Szemerédi is 70), in: Bolyai Soc. Math. Stud., vol. **21** (2010) Springer, Berlin, pp. 359–404.
530. Y. Kohayakawa, M. Simonovits, and J. Skokan: *The 3-colored Ramsey number of odd cycles*, Proceedings of GRACO2005, pp. 397–402 (electronic), Electron. Notes Discrete Math., 19, Elsevier, Amsterdam, (2005).
531. Phokion G. Kolaitis, Hans Jürgen Prömel, and B.L. Rothschild: $K_{\ell+1}$-free graphs: asymptotic structure and a 0-1 law, Trans. Amer. Math. Soc. **303 (2)** (1987) 637–671.
532. János Kollár, L. Rónyai, and Tibor Szabó: Norm-graphs and bipartite Turán numbers. Combinatorica **16** (1996), 399–406.
533. János Komlós: The Blow-up lemma, Combin. Probab. Comput. **8** (1999) 161–176.
534. J. Komlós: Tiling Turán theorems, Combinatorica **20** (2000), 203–218.
535. J. Komlós, János Pintz, and E. Szemerédi: On Heilbronn's triangle problem. J. London Math. Soc. (2) **24 (3)** (1981), 385–396.
536. J. Komlós, J. Pintz, and E. Szemerédi: A lower bound for Heilbronn's problem. J. London Math. Soc. (2) **25 (1)** (1982) 13–24.
537. J. Komlós, G.N. Sárközy, and E. Szemerédi: Proof of a packing conjecture of Bollobás, Combinatorics, Probability and Computing **4** (1995), 241–255.
538. J. Komlós, G.N. Sárközy, and E. Szemerédi: On the square of a Hamiltonian cycle in dense graphs, Random Structures and Algorithms, **9**, (1996), 193–211.
539. J. Komlós, G.N. Sárközy, and E. Szemerédi: Blow-up Lemma, Combinatorica, **17 (1)** (1997), pp. 109–123.
540. J. Komlós, G.N. Sárközy, and E. Szemerédi: On the Pósa-Seymour conjecture, Journal of Graph Theory **29** (1998) 167–176.
541. J. Komlós, G.N. Sárközy, and E. Szemerédi: An algorithmic version of the blow-up lemma, Random Structures and Algorithms **12 (3)** (1998) 297–312.
542. J. Komlós, G.N. Sárközy and E. Szemerédi: Proof of the Seymour conjecture for large graphs, Annals of Combinatorics, **2**, (1998), 43–60.
543. J. Komlós, G.N. Sárközy, and E. Szemerédi: Proof of the Alon-Yuster conjecture. Combinatorics (Prague, 1998). Discrete Math. **235 (1-3)** (2001) 255–269.
544. J. Komlós; G.N. Sárközy; E. Szemerédi: Spanning trees in dense graphs. Combin. Probab. Comput. **10 (5)** (2001) 397–416.
545. J. Komlós, Ali Shokoufandeh, M. Simonovits, and E. Szemerédi: The regularity lemma and its applications in graph theory, in Theoretical Aspects of Computer Science (Tehran, 2000), Lecture Notes in Comput. Sci., 2292, Springer, Berlin, (2002), 84–112.
546. J. Komlós and M. Simonovits: Szemerédi's regularity lemma and its applications in graph theory, in Combinatorics, Paul Erdős Is Eighty, Vol. **2** (Keszthely, 1993), Bolyai Soc. Math. Stud. 2, János Bolyai Math. Soc., Budapest, 1996, pp. 295–352.
547. Dániel Korándi, M. Krivelevich, and B. Sudakov: Decomposing random graphs into few cycles and edges. Combin. Probab. Comput. **24 (6)** (2015) 857–872.
548. Alexander V. Kostochka: "A Class of Constructions for Turán's (3,4)-Problem," Combinatorica **2 (2)** (1982) 187–192.
549. A.V. Kostochka, D. Mubayi, and J. Verstraëte: Hypergraph Ramsey numbers: triangles versus cliques. J. Combin. Theory Ser. A **120 (7)** (2013) 1491–1507.
550. A.V. Kostochka, K. Nakprasit, and S. Pemmaraju: On equitable coloring of d-degenerate graphs, SIAM J. Discrete Math. **19** (2005) 83–95.
551. A.V. Kostochka, M.J. Pelsmajer, and Doug B. West: A list analogue of equitable coloring. J. Graph Theory **44 (3)** (2003) 166–177.
552. A.V. Kostochka and Michael Stiebitz: A list version of Dirac's theorem on the number of edges in colour-critical graphs. J. Graph Theory **39 (3)** (2002), 165–177.
553. A.V. Kostochka and G. Yu: Ore-type graph packing problems, Combin. Probab. Comput., **16** (2007), pp. 167–169.
554. Tamás Kővári, V.T. Sós, and P. Turán: On a problem of Zarankiewicz, Colloq Math. **3** (1954), 50–57.

555. Daniel Král' and Oleg Pikhurko: Quasirandom permutations are characterized by 4-point densities. Geom. Funct. Anal. **23 (2)** (2013) 570–579.
556. Michael Krivelevich: On the minimal number of edges in color-critical graphs, Combinatorica **17** (1997) 401–426.
557. M. Krivelevich, Choongbum Lee, and B. Sudakov: Resilient pancyclicity of random and pseudorandom graphs. SIAM J. Discrete Math. **24 (1)** (2010) 1–16.
558. M. Krivelevich and W. Samotij: Optimal packings of Hamilton cycles in sparse random graphs, SIAM J. Discrete Math. **26 (3)** (2012) 964–982.,
559. M. Krivelevich and B. Sudakov: Pseudo-random graphs, in: More Sets, Graphs and Numbers, in: Bolyai Soc. Math. Stud., vol. **15** (2006) Springer, pp. 199–262.
560. D. Kühn and D. Osthus: Loose Hamilton cycles in 3-uniform hypergraphs of high minimum degree. J. Combin. Theory Ser. B **96 (6)** (2006) 767–821.
561. D. Kühn, D. Osthus: Matchings in hypergraphs of large minimum degree, J. Graph Theory **51** (2006) 269–280.
562. D. Kühn and D. Osthus: The minimum degree threshold for perfect graph packings, Combinatorica **29** (2009), 65–107.
563. D. Kühn and D. Osthus: Embedding large subgraphs into dense graphs. Surveys in combinatorics 2009, 137–167, London Math. Soc. Lecture Note Ser., **365** (2009), Cambridge Univ. Press, Cambridge.
564. D. Kühn and D. Osthus: Hamilton decompositions of regular expanders: A proof of Kelly's conjecture for large tournaments, Advances Math., **237** (2013), pp. 62–146.
565. D. Kühn and D. Osthus: Hamilton cycles in graphs and hypergraphs: an extremal perspective. Proceedings of the International Congress of Mathematicians 2014. Vol. IV, 381–406, Kyung Moon Sa, Seoul.
566. D. Kühn, D. Osthus, and A. Treglown: Matchings in 3-uniform hypergraphs. J. Combin. Theory Ser. B **103 (2)** (2013) 291–305. arXiv:1009.1298.
567. Felix Lazebnik, Vasyl A. Ustimenko, and Andrew J. Woldar: Properties of certain families of $2k$-cycle-free graphs. J. Combin. Theory Ser. B **60 (2)** (1994), 293–298.
568. F. Lazebnik, V.A. Ustimenko, and A.J. Woldar: A new series of dense graphs of high girth. Bull. Amer. Math. Soc. (N.S.) **32 (1)** (1995) 73–79.
569. F. Lazebnik and A.J. Woldar: General properties of some families of graphs defined by systems of equations. J. Graph Theory **38 (2)** (2001) 65–86.
570. Choongbum Lee and B. Sudakov: Hamiltonicity, independence number, and pancyclicity. European J. Combin. **33 (4)** (2012) 449–457.
571. Hanno Lefmann: On Heilbronn's problem in higher dimension. Combinatorica **23 (4)** (2003), 669–680.
572. H. Lefmann, Y. Person, V. Rödl, and M. Schacht: On colourings of hypergraphs without monochromatic Fano planes, Combin. Probab. Comput., **18** (2009), pp. 803–818.
573. H. Lefmann and Niels Schmitt: A deterministic polynomial-time algorithm for Heilbronn's problem in three dimensions. SIAM J. Comput. **31 (6)** (2002) 1926–1947.
574. Vsevolod F. Lev, T. Łuczak, and Tomasz Schoen: Sum-free sets in abelian groups, Israel Journal of Mathematics **125** (2001), 347–367.
575. Ian Levitt, G.N. Sárközy, and E. Szemerédi: How to avoid using the regularity lemma: Pósa's conjecture revisited, Discrete Math., **310** (2010), 630–641.
576. Allan Lo and Klas Markström: A multipartite version of the Hajnal-Szemerédi theorem for graphs and hypergraphs. Combin. Probab. Comput. **22 (1)** (2013) 97–111. arXiv:1108.4184.
577. Allan Lo and K. Markström: Minimum codegree threshold for $(K_4^3 - e)$-factors. J. Combin. Theory Ser. A **120** (2013) 708–721. arXiv:1111.5334v1
578. Allan Lo and K. Markström: Perfect matchings in 3-partite 3-uniform hypergraphs, J. Combin. Theory Ser. A **127** (2014) 22–57. arXiv:1103.5654.
579. Allan Lo and K. Markström: F-factors in hypergraphs via absorption. Graphs Combin. **31 (3)** (2015), 679–712. arXiv:1105.3411v1
580. Po-Shen Loh and Jie Ma: Diameter critical graphs. J. Combin. Theory Ser. B **117** (2016) 34–58.

581. László Lovász: On covering of graphs. Theory of Graphs (Proc. Colloq., Tihany, 1966) pp. 231–236 (1967) Academic Press, New York
582. L. Lovász: On chromatic number of finite set-systems, Acta Math. Acad. Sci. Hungar. **19** (1968) 59–67.
583. L. Lovász: The factorization of graphs. Combinatorial Structures and their Applications (Proc. Calgary Internat. Conf., Calgary, Alta., 1969) pp. 243–246 Gordon and Breach, New York (1970).
584. L. Lovász: Subgraphs with prescribed valencies. J. Combinatorial Theory **8** (1970) 391–416.
585. L. Lovász: On the sieve formula. (Hungarian) Mat. Lapok **23** (1972), 53–69 (1973).
586. L. Lovász: Independent sets in critical chromatic graphs, Stud. Sci. Math. Hung. **8** (1973) 165–168.
587. L. Lovász: Chromatic number of hypergraphs and linear algebra. Studia Sci. Math. Hungar. **11 (1-2)** (1976) 113–114 (1978).
588. L. Lovász: On the Shannon capacity of a graph, IEEE Transactions on Information Theory **25** (1979) 1–7.
589. L. Lovász: Combinatorial Problems and Exercises, Akadémiai Kiadó, Budapest. North-Holland Publishing Co., Amsterdam-New York, (1979). 551 pp. 2nd Edition, American Mathematical Society, Providence, Rhode Island, 2007.
590. L. Lovász: Very large graphs. Current developments in mathematics, (2008) 67–128, Int. Press, Somerville, MA, 2009.
591. L. Lovász: Large networks and graph limits. American Mathematical Society Colloquium Publications, **60** . American Mathematical Society, Providence, RI, (2012) xiv+475 pp.
592. L. Lovász and M. Simonovits: On the number of complete subgraphs of a graph, in Proc. 5th British Combinatorial Conference, Aberdeen 1975. Congress. Numer. **XV**(1976) 431–441.
593. L. Lovász and M. Simonovits: On the number of complete subgraphs of a graph II, Studies in Pure Math. (dedicated to P. Turán) (1983) 459–495 Akadémiai Kiadó+Birkhäuser Verlag.
594. L. Lovász and V.T. Sós: Generalized quasirandom graphs, J. Combin. Theory Ser B **98** (2008), 146–163.
595. L. Lovász and B. Szegedy: Szemerédi's lemma for the analyst. Geom. Funct. Anal. **17 (1)** (2007) 252–270.
596. L. Lovász and B. Szegedy: Testing properties of graphs and functions. Israel J. Math. **178** (2010), 113–156. arXiv:0803.1248.
597. Alex Lubotzky: Discrete Groups, Expanding Graphs and Invariant Measures. Progr. Math. **125** (1994) Birkhäuser, Basel (1994)
598. A. Lubotzky, Ralph Phillips, and P. Sarnak: Ramanujan Conjecture and explicit construction of expanders (Extended Abstract), Proceedings of the STOC (1986), pp. 240–246.
599. A. Lubotzky, R. Phillips, and P. Sarnak: Ramanujan graphs, Combinatorica **8** (1988), 261–277.
600. Tomasz J. Łuczak: $R(C_n, C_n, C_n) \leq (4 + o(1))n$, Journal of Combinatorial Theory, Ser. B **75 (2)** (1999) 174–187.
601. T. Łuczak: On triangle-free random graphs, Random Struct. Algorithms **16 (3)** (2000), 260–276.
602. T. Łuczak, V. Rödl, and E. Szemerédi: Partitioning two-colored complete graphs into two monochromatic cycles, Combinatorics, Probability and Computing, **7** (1998) pp. 423–436.
603. T. Łuczak and T. Schoen: On the maximal density of sum-free sets. Acta Arith. **95 (3)** (2000) 225–229.
604. T. Łuczak and T. Schoen: On the number of maximal sum-free sets, Proc. Amer. Math. Soc. **129** (2001) 2205–2207.
605. Clara Lüders and C. Reiher: The Ramsey–Turán problem for cliques, Isr. J. Math. **232 (2)** (2019), 613–652.
606. W. Mantel: Problem 28 (solution by H. Gouwentak, W. Mantel, J. Texeira de Mattes, F. Schuh and W. A. Wythoff), Wiskundige Opgaven **10** (1907), 60–61.
607. Gregory A. Margulis: Explicit constructions of expanders. (Russian) Problemy Peredači Informacii **9 (4)** (1973), 71–80. English transl.: Problems Inform. Transmission **9** (1975) 325–332.

608. G.A. Margulis: Explicit constructions of graphs without short cycles and low density codes, Combinatorica **2** (1982), 71–78.
609. G.A. Margulis: Arithmetic groups and graphs without short cycles, 6th Internat. Symp. on Information Theory, Tashkent 1984, Abstracts, Vol. **1**, pp. 123–125 (in Russian).
610. G.A. Margulis: Some new constructions of low-density paritycheck codes. 3rd Internat. Seminar on Information Theory, convolution codes and multi–user communication, Sochi 1987, pp. 275–279 (in Russian).
611. G.A. Margulis: Explicit group theoretic constructions of combinatorial schemes and their applications for the construction of expanders and concentrators, Journal of Problems of Information Transmission, **24** (1988) 39–46, (in Russian).
612. K. Markström and A. Ruciński: Perfect matchings (and Hamilton cycles) in hypergraphs with large degrees. European J. Combin. **32 (5)** (2011), 677–687
613. Ryan Martin and E. Szemerédi: Quadripartite version of the Hajnal-Szemerédi theorem. Discrete Math. **308 (19)** (2008), (special edition in honor of Miklós Simonovits), 4337–4360.
614. Michael Molloy: The Probabilistic Method. Book chapter in "Probabilistic Methods for Algorithmic Discrete Mathematics", M. Habib, C. McDiarmid, J. Ramirez-Alfonsin and B. Reed, editors. pp. 1–35. Springer, 1998.
615. M. Molloy and Bruce Reed: Graph colouring and the probabilistic method, Algorithms and Combinatorics, vol. **23** (2002), Springer-Verlag, Berlin.
616. Richard Montgomery: Embedding Bounded Degree Spanning Trees in Random Graphs, preprint, https://arxiv.org/abs/1405.6559, 2014.
617. John W. Moon and Leo Moser: On a problem of Turán. Magyar Tud. Akad. Mat. Kutató Int. Közl. **7** (1962) 283–286.
618. Robert Morris and David Saxton: The number of $C_{2\ell}$-free graphs, Advances in Mathematics **298** (2016), 534–580. arXiv:1309.2927.
619. Robin A. Moser: Derandomizing the Lovász local lemma more effectively. (2008) Eprint arXiv:0807.2120v2.
620. R.A. Moser and Gábor Tardos: A constructive proof of the general Lovász local lemma. J. ACM **57 (2)** (2010), Art. 11, 15 pp.
621. Guy Moshkovitz and A. Shapira: A short proof of Gowers' lower bound for the regularity lemma. Combinatorica **36** (2016) 187–194.
622. Dhruv Mubayi: The co-degree density of the Fano plane, J. Combin. Theory Ser. B **95 (2)** (2005) 333–337.
623. D. Mubayi: Structure and stability of triangle-free set systems, Trans. Amer. Math. Soc., **359** (2007), 275–291.
624. D. Mubayi: Counting substructures I: color critical graphs, Advances Math. **225 (5)** (2010) 2731–2740.
625. D. Mubayi and O. Pikhurko: A new generalization of Mantel's theorem to k-graphs. J. Combin. Theory Ser. B **97 (4)** (2007), 669–678.
626. D. Mubayi and V.T. Sós: Explicit constructions of triple systems for Ramsey–Turán problems. J. Graph Theory **52 (3)** (2006), 211–216.
627. Marcello Mydlarz and E. Szemerédi: Algorithmic Brooks' theorem, 2007, manuscript.
628. Brendan Nagle and V. Rödl: The asymptotic number of triple systems not containing a fixed one, in: Combinatorics, Prague, 1998, Discrete Math. **235** (2001) 271–290.
629. B. Nagle, V. Rödl, and M. Schacht: An algorithmic hypergraph regularity lemma. Random Structures Algorithms **52 (2)** (2018), 301–353.
630. Jarik Nešetřil and V. Rödl: A short proof of the existence of highly chromatic hypergraphs without short cycles. J. Combin. Theory Ser. B **27 (2)** (1979), 225–227.
631. Rajko Nenadov, B. Sudakov, and Mykhaylo Tyomkyn: Proof of the Brown-Erdős-Sós conjecture in groups, arXiv:1902.07614
632. Vladimir Nikiforov: The number of cliques in graphs of given order and size. Trans. Amer. Math. Soc. **363** (2011) 1599–1618. arXiv:0710.2305v2
633. V. Nikiforov and R.H. Schelp: Cycles and stability, J. Combin. Theory Ser. B **98** (2008) 69–84.

634. S. Norin and L. Yepremyan: Turán number of generalized triangles. J. Combin. Theory Ser. A **146** (2017), 312–343.
635. Oystein Ore: Note on Hamilton circuits, Amer. Math. Monthly **67** (1960) 55.
636. Deryk Osthus, H.J. Prömel, and A. Taraz: 'For which densities are random triangle-free graphs almost surely bipartite?', Combinatorica **23** (2003), 105–150.
637. János Pach: A Tverberg-type result on multicolored simplices. Comput. Geom. **10 (2)** (1998) 71–76.
638. J. Pach: Geometric intersection patterns and the theory of topological graphs. Proceedings of the International Congress of Mathematicians–Seoul 2014. Vol. **IV**, 455–474, Kyung Moon Sa, Seoul, (2014).
639. J. Pach and Pankaj K. Agarwal: Combinatorial geometry. Wiley-Interscience Series in Discrete Mathematics and Optimization. A Wiley-Interscience Publication. John Wiley & Sons, Inc., New York, (1995). xiv+354 pp.
640. Pach, János:; Sharir, Micha: On the number of incidences between points and curves. Combin. Probab. Comput. **7 (1)** (1998) 121–127.
641. J. Pach and József Solymosi: Crossing patterns of segments, Journal of Combinatorial Theory, Ser. A **96** (2001), 316–325.
642. Balázs Patkós: Supersaturation and stability for forbidden subposet problems, J. Combin. Theory Ser. A **136** (2015) 220–237.
643. Michael S. Payne and David R. Wood: On the general position subset selection problem, *SIAM J. Discrete Math.*, **27 (4)** (2013), 1727–1733. arXiv:1208.5289.v2
644. Y. Peng and C. Zhao: Generating non-jumping numbers recursively, Discrete Applied Mathematics **156** (2008), 1856–1864.
645. Yury Person and M. Schacht: Almost all hypergraphs without Fano planes are bipartite, Proceedings of the Twentieth Annual ACM–SIAM Symposium on Discrete Algorithms, SIAM, Philadelphia, PA, 2009, pp. 217–226.
646. Oleg Pikhurko: An exact Turán result for the generalized triangle, Combinatorica **28** (2008), 187–208.
647. O. Pikhurko: Perfect matchings and K_4^3-tilings in hypergraphs of large codegree, Graphs Combin. **24 (4)** (2008) 391–404.
648. O. Pikhurko: On possible Turán densities. Israel J. Math. **201 (1)** (2014), 415–454. arXiv:1204.4423
649. O. Pikhurko and Zelealem B. Yilma: Supersaturation problem for color-critical graphs. J. Combin. Theory Ser. B **123** (2017), 148–185.
650. Oleg Pikhurko and Alexandr Razborov: Asymptotic structure of graphs with the minimum number of triangles. Combin. Probab. Comput. **26 (1)** (2017) 138–160.
651. Mark S. Pinsker: On the complexity of a concentrator, in: Proc. of the 7th International Teletraffic Conference, 1973, pp. 318/1–318/4.
652. Nick Pippenger and J. Spencer: Asymptotic behavior of the chromatic index for hypergraphs, J. Combin. Theory Ser A **51** (1989), 24–42.
653. Ján Plesník: Critical graphs of given diameter. (Slovak, Russian summary) Acta Fac. Rerum Natur. Univ. Comenian. Math. **30** (1975), 71–93.
654. Alexey Pokrovskiy: Partitioning edge-coloured complete graphs into monochromatic cycles and paths, J. Combin. Theory Ser. B **106** (2014) 70–97. arXiv:1205.5492v1.
655. D. H.J. Polymath, Density Hales-Jewett and Moser numbers, in An Irregular Mind: Szemerédi is 70, Springer–Verlag, New York, 2010, pp. 689–753.
656. D.H.J. Polymath: A new proof of the density Hales-Jewett theorem. Ann. of Math. (2) **175 (3)** (2012), 1283–1327.
657. Carl Pomerance and A. Sárközy: Combinatorial number theory, Handbook of combinatorics, Vol. 1, 2, pp. 967–1018, Elsevier, Amsterdam, 1995.
658. Lajos Pósa: A theorem concerning Hamilton lines, Publ. Math. Inst. Hung. Acad. Sci. **7** (1962), 225–226.
659. L. Pósa: On the circuits of finite graphs. Magyar Tud. Akad. Mat. Kutató Int. Közl. **8** (1963) 355–361 (1964)

660. L. Pósa: Hamiltonian circuits in random graphs, Discrete Math., **14** (1976), pp. 359–364.
661. H.J. Prömel and V. Rödl: Non-Ramsey graphs are $c \log n$-universal, J. Comb. Th. (A) **88** (1999) 378–384.
662. Hans Jürgen Prömel and Angelika Steger: Excluding induced subgraphs: quadrilaterals, Random Structures and Algorithms **2** (1991) 55–71.
663. H.J. Prömel and A. Steger: A. Excluding induced subgraphs. III. A general asymptotic. Random Structures and Algorithms **3 (1)** (1992) 19–31.
664. H.J. Prömel and A. Steger: Almost all Berge graphs are perfect, Combin. Probab. Comput. **1** (1992) 53–79.
665. H.J. Prömel and A. Steger: 'The asymptotic number of graphs not containing a fixed colorcritical subgraph', Combinatorica **12** (1992), 463–473.
666. H.J. Prömel and A. Steger: On the asymptotic structure of sparse triangle free graphs, J. Graph Theory **21 (2)** (1996), 137–151,
667. László Pyber: An Erdős-Gallai conjecture. Combinatorica **5 (1)** (1985), 67–79.
668. L. Pyber: Covering the edges of a connected graph by paths, J. Comb. Theory B. (**66**) 1996, 52–159.
669. Alexander A. Razborov: Flag algebras. J. Symbolic Logic **72** (2007) 1239–1282.
670. A.A. Razborov: On the minimal density of triangles in graphs. Combinatorics, Probability and Computing, **17 (4)** (2008) 603–618.
671. A.A. Razborov: On 3-hypergraphs with forbidden 4-vertex configurations, SIAM J. Discrete Math. **24 (3)** (2010) 946–963.
672. A.A. Razborov: On the Fon-der-Flaass interpretation of extremal examples for Turán's (3,4)-problem. Proceedings of the Steklov Institute of Mathematics **274** (2011), 247–266 Algoritmicheskie Voprosy Algebry i Logiki, 269–290; translation in Proc. Steklov Inst. Math. **274 (1)** (2011), 247–266.
673. Christian Reiher: The clique density theorem, Ann. of Math. **184** (2016) 683–707. arXiv:1212.2454v1.
674. C. Reiher, V. Rödl, A. Ruciński, M. Schacht, and E. Szemerédi: Minimum vertex degree condition for tight Hamiltonian cycles in 3-uniform hypergraphs (2016) arXiv:1611.03118v1
675. István Reiman: Über ein Problem von K. Zarankiewicz, Acta Math. Acad. Sci. Hungar. **9** (1958), 269–278.
676. D. Romero and A. Sánchez-Arroyo: "Advances on the Erdős-Faber-Lovász conjecture", in G. Grimmet and C, McDiarmid, Combinatorics, Complexity, and Chance: A Tribute to Dominic Welsh, Oxford Lecture Series in Mathematics and Its Applications, Oxford University Press, (2007) pp. 285–298.
677. Vera Rosta: On a Ramsey-type problem of J.A. Bondy and P. Erdős. I, J. Combin. Theory Ser. **15** (1973) 94–104; and V. Rosta: On a Ramsey-type problem of J.A. Bondy and P. Erdős. II, J. Combin. Theory Ser. B **15** (1973) 105–120.
678. Klaus F. Roth: On a problem of Heilbronn, *J. London Math. Soc.* **26** (1951), 198–204.
679. K.F. Roth: On certain sets of integers, J. London Math. Soc. **28** (1953), 104–109.
680. K.F. Roth: On a problem of Heilbronn II, *Proc. London Math. Soc.* **25 (3)** (1972), 193–212.
681. K.F. Roth: On a problem of Heilbronn III, *Proc. London Math. Soc.* **25 (3)** (1972), 543–549.
682. K.F. Roth: Estimation of the Area of the Smallest Triangle Obtained by Selecting Three out of n Points in a Disc of Unit Area, AMS, Providence, Proc. of Symposia in Pure Mathematics **24** (1973) 251–262.
683. K.F. Roth: Developments in Heilbronn's triangle problem, *Advances in Math.* **22 (3)** (1976), 364–385.
684. Vojtěch Rödl: On a packing and covering problem, European J. Combin. **6** (1985) 69–78.
685. V. Rödl: Note on a Ramsey-Turán type problem, Graphs and Combin. **1 (3)** (1985), 291–293.
686. V. Rödl: Quasi-randomness and the regularity method in hypergraphs. Proceedings of the International Congress of Mathematicians–Seoul 2014. Vol. 1, 573–601, Kyung Moon Sa, Seoul, (2014).
687. V. Rödl, B. Nagle, J. Skokan, M. Schacht, and Y. Kohayakawa: The hypergraph regularity method and its applications, Proc. Natl. Acad. Sci. USA **102 (23)** (2005) 8109–8113.

688. V. Rödl and A. Ruciński: Perfect matchings in ϵ-regular graphs and the blow-up lemma, Combinatorica **19 (3)** (1999) 437–452.
689. V. Rödl and A. Ruciński: Dirac-type questions for hypergraphs – a survey (or more problems for Endre to solve). In: I. Bárány, J. Solymosi, G. Sági (eds.) An Irregular Mind, pp. 561–590. János Bolyai Math. Soc., Budapest (2010)
690. V. Rödl and A. Ruciński: Families of triples with high minimum degree are Hamiltonian. Discuss. Math. Graph Theory **34 (2)** (2014), 361–381.
691. V. Rödl, A. Ruciński, M. Schacht, and E. Szemerédi: A note on perfect matchings in uniform hypergraphs with large minimum collective degree. Comment. Math. Univ. Carolin. **49 (4)** (2008) 633–636.
692. V. Rödl, A. Ruciński, M. Schacht, and E. Szemerédi: On the Hamiltonicity of triple systems with high minimum degree. Annales Comb. **21 (1)** (2017), 95–117.
693. V. Rödl, A. Ruciński, and E. Szemerédi: Perfect matchings in uniform hypergraphs with large minimum degree, European J. Combin. **27 (8)** (2006) 1333–1349.
694. V. Rödl, A. Ruciński, and E. Szemerédi: A Dirac-type theorem for 3-uniform hypergraphs, Combinatorics, Probability, and Computing **15 (1-2)** (2006) 229–251.
695. V. Rödl, A. Ruciński, and E. Szemerédi: An approximate Dirac-type theorem for k-uniform hypergraphs. Combinatorica **28 (2)** (2008) 229–260.
696. V. Rödl, A. Ruciński, and E. Szemerédi: Perfect matchings in large uniform hypergraphs with large minimum collective degree. J. Combin. Theory Ser. A **116 (3)** (2009) 613–636.
697. V. Rödl, A. Ruciński, and E. Szemerédi: Dirac-type conditions for Hamiltonian paths and cycles in 3-uniform hypergraphs. Advances Math. **227 (3)** (2011), 1225–1299. arXiv:1611.03118.
698. V. Rödl, A. Ruciński, and A. Taraz: Hypergraph packing and graph embedding, Combinatorics, Probab. Comput. **8** (1999) 363–376.
699. V. Rödl, A. Ruciński, and Michelle Wagner: An algorithmic embedding of graphs via perfect matchings. Randomization and approximation techniques in computer science (Barcelona, 1998), 25–34, Lecture Notes in Comput. Sci., 1518, Springer, Berlin, 1998.
700. V. Rödl, M. Schacht, E. Tengan and N. Tokushige: Density theorems and extremal hypergraph problems, Israel J. Math. **152** (2006), 371–380.
701. V. Rödl and M. Schacht: Regular partitions of hypergraphs: Regularity lemmas, Combin. Probab. Comput. **16** (2007), 833–885.
702. V. Rödl and M. Schacht: Generalizations of the removal lemma, Combinatorica **29 (4)** (2009) 467–501.
703. V. Rödl and M. Schacht: Extremal results in random graphs, Erdős centennial, Bolyai Soc. Math. Stud., vol. **25**, János Bolyai Math. Soc., Budapest, (2013), pp. 535–583, arXiv:1302.2248.
704. V. Rödl and Mark H. Siggers: Color critical hypergraphs with many edges, J. Graph Theory **53** (2006) 56–74.
705. V. Rödl and J. Skokan: Regularity lemma for k-uniform hypergraphs, Random Structures and Algorithms **25 (1)** (2004) 1–42.
706. V. Rödl and J. Skokan: Applications of the regularity lemma for uniform hypergraphs, Random Structures and Algorithms **28 (2)** (2006) 180–194.
707. V. Rödl and Zsolt Tuza: On color critical graphs. J. Combin. Theory Ser. B **38 (3)** (1985), 204–213.
708. Imre Z. Ruzsa: Solving a linear equation in a set of integers. I, Acta Arith. **65 (3)** (1993) 259–282.
709. I.Z. Ruzsa: Solving a linear equation in a set of integers II, Acta Arith. **72** (1995) 385–397.
710. I.Z. Ruzsa: An infinite Sidon sequence, J. Number Theory **68** (1998) 63–71.
711. I.Z. Ruzsa and E. Szemerédi: Triple systems with no six points carrying three triangles. Combinatorics (Proc. Fifth Hungarian Colloq., Keszthely, 1976), Vol. II, pp. 939–945, Colloq. Math. Soc. János Bolyai, 18, North-Holland, Amsterdam–New York, 1978.
712. Horst Sachs and Michael Stiebitz: Construction of colour-critical graphs with given major-vertex subgraph. Combinatorial mathematics (Marseille–Luminy, 1981), 581–598, North-Holland Math. Stud., 75, Ann. Discrete Math., **17**(1983) North-Holland, Amsterdam.

713. H. Sachs and M. Stiebitz: Colour-critical graphs with vertices of low valency. Graph theory in memory of G.A. Dirac (Sandbjerg, 1985), 371–396, Ann. Discrete Math., **41**, North-Holland, Amsterdam, 1989.
714. H. Sachs and M. Stiebitz: On constructive methods in the theory of colour-critical graphs. Graph colouring and variations. Discrete Math. **74 (1-2)** (1989) 201–226.
715. Wojciech Samotij: Stability results for discrete random structures, Random Structures & Algorithms **44** (2014), 269–289.
716. Tom Sanders: On a non-abelian Balog-Szemerédi-type lemma, J. Aust. Math. Soc. **89** (2010), 127–132.
717. T. Sanders: On Roth's theorem on progressions. Ann. of Math. (2) **174 (1)** (2011), 619–636.
718. T. Sanders: Roth's theorem: an application of approximate groups. Proceedings of the International Congress of Mathematicians–Seoul 2014. Vol. III, 401–423, Kyung Moon Sa, Seoul, (2014).
719. Alexander A. Sapozhenko: On the number of sum-free sets in Abelian groups, Vestnik Moskov. Univ. Ser. Mat. Mekh. **4** (2002) 14–18.
720. A. A. Sapozhenko: Asymptotics of the number of sum-free sets in abelian groups of even order, Rossiiskaya Akademiya Nauk. Dokladi Akademii Nauk **383** (2002), 454–457.
721. A.A. Sapozhenko: 'Solution of the Cameron-Erdős problem for groups of prime order', Zh. Vychisl. Mat. Mat. Fiz. **49** (2009) 1435–1441.
722. Gábor N. Sárközy: Cycles in bipartite graphs and an application in number theory. J. Graph Theory **19 (3)** (1995), 323–331.
723. G.N. Sárközy: The regularity method, and the Blow-up method and their application, Thesis for Doctor of Sciences of Hungarian Acad Sci (with an Introduction in Hungarian).
724. G.N. Sárközy: Improved monochromatic loose cycle partitions in hypergraphs. Discrete Math. **334** (2014), 52–62.
725. G.N. Sárközy: A quantitative version of the Blow-up Lemma, arXiv:1405.7302, submitted for publication. (2014)
726. G.N. Sárközy: Monochromatic cycle power partitions. Discrete Math. **340 (2)** (2017) 72–80.
727. G.N. Sárközy and Stanley M. Selkow: An extension of the Ruzsa-Szemerédi theorem, Combinatorica **25 (1)** (2004) 77–84.
728. G.N. Sárközy and S.M Selkow: On a Turán-type hypergraph problem of Brown, Erdős and T. Sós. Discrete Math. **297 (1-3)** (2005), 190–195.
729. G.N. Sárközy and S.M. Selkow: On an anti-Ramsey problem of Burr, Erdős, Graham, and T. Sós. J. Graph Theory **52 (2)** (2006), 147–156.
730. Norbert Sauer and J. Spencer: Edge disjoint placement of graphs. J. Combin. Theory Ser. B **25** (1978) 295–302.
731. David Saxton and A. Thomason: Hypergraph containers, Invent. Math. **201** (2015), 925–992. arXiv:1204.6595v2
732. Mathias Schacht: Extremal results for random discrete structures, Ann. Math. **184** (2016) 333–365. arXiv:1603.00894.
733. Richard H. Schelp: Some Ramsey-Turán type problems and related questions, Discrete Math. **312** (2012) 2158–2161.
734. Wolfgang M. Schmidt: On a Problem of Heilbronn, Journal of the London Mathematical Society (2) **4** (1972) 545–550.
735. Alex Scott: Szemerédi's Regularity Lemma for matrices and sparse graphs, Combin. Probab. Comput. **20** (2011) 455–466.
736. Paul Seymour: Problem section, *Combinatorics: Proceedings of the British Combinatorial Conference 1973* (T. P.McDonough and V.C. Mavron eds.), Lecture Note Ser., No. **13** (1974), pp. 201–202. London Mathematical Society, Cambridge University Press, London.
737. Asaf Shapira and R. Yuster: The effect of induced subgraphs on quasi-randomness. Random Structures and Algorithms **36 (1)** (2010), 90–109.
738. James B. Shearer: A note on the independence number of triangle-free graphs. Discrete Math. **46 (1)** (1983) 83–87.

739. J.B. Shearer: A note on the independence number of triangle-free graphs. II. J. Combin. Theory Ser. B **53 (2)** (1991) 300–307.
740. J.B. Shearer: The independence number of dense graphs with large odd girth, Electron. J. Combin. **2** (1995), Note 2 (electronic).
741. J.B. Shearer: On the independence number of sparse graphs. Random Struct. Alg. **7** (1995) 269–271.
742. Ilja D. Shkredov: Szemerédi's theorem and problems of arithmetic progressions. Uspekhi Mat. Nauk **61 (6)** (2006) 6(372), 111–178; translation in Russian Math. Surveys **61 (6)** (2006), 1101–1166
743. Alexander F. Sidorenko: Extremal estimates of probability measures and their combinatorial nature. (Russian) Izv. Akad. Nauk SSSR Ser. Mat. **46 (3)** (1982), 535–568,
744. A.F. Sidorenko: The maximal number of edges in a homogeneous hypergraph containing no prohibited subgraphs, Mathematical Notes **41** (1987), 247–259. Translated from Matematicheskie Zametki.
745. A.F. Sidorenko: An unimprovable inequality for the sum of two symmetrically distributed random vectors. (Russian) Teor. Veroyatnost. i Primenen. **35 (3)** (1990), 595–599; translation in Theory Probab. Appl. **35 (3)** (1990), 613–617 (1991)
746. A.F. Sidorenko: Inequalities for functional generated by bipartite graphs, Discrete Math. Appl. **3 (3)** (1991) 50–65 (in Russian); English transl., Discrete Math. Appl. **2 (5)** (1992) 489–504.
747. A.F. Sidorenko: A correlation inequality for bipartite graphs, Graphs Combin. **9** (1993) 201–204.
748. A.F. Sidorenko: Boundedness of optimal matrices in extremal multigraph and digraph problems, Combinatorica **13** (1993), 109–120.
749. A.F. Sidorenko: What we know and what we do not know about Turán numbers, Graphs and Combinatorics **11** (1995), 179–199.
750. A.F. Sidorenko: An Erdős-Gallai-type theorem for keyrings, Graphs Comb. **34 (4)** (2018) 633–638. arXiv:1705.10254.
751. M. Simonovits: A method for solving extremal problems in graph theory, Theory of Graphs, Proc. Colloq. Tihany, (1966), (Ed. P. Erdős and G. Katona) Acad. Press, N.Y., (1968) pp. 279–319.
752. M. Simonovits: On the structure of extremal graphs, PhD thesis (1969), (Actually, "Candidate", in Hungarian).
753. M. Simonovits: On colour-critical graphs. Studia Sci. Math. Hungar. **7** (1972), 67–81.
754. M. Simonovits: Note on a hypergraph extremal problem, in: C. Berge, D. Ray-Chaudury (Eds.), Hypergraph Seminar, Columbus, Ohio, USA, 1972, Lecture Notes in Mathematics, Vol. 411, Springer, Berlin, (1974), pp. 147–151.
755. M. Simonovits: The extremal graph problem of the icosahedron. J. Combinatorial Theory Ser. B **17** (1974) 69–79.
756. M. Simonovits: Extremal graph problems with symmetrical extremal graphs, additional chromatic conditions, Discrete Mathematics **7** (1974), 349–376.
757. M. Simonovits: On Paul Turán's influence on graph theory, J. Graph Theory **1 (2)** (1977), 102–116.
758. M. Simonovits: Extremal graph theory, in: L.W. Beineke, R.J. Wilson (Eds.), Selected Topics in Graph Theory II., Academic Press, London, (1983), pp. 161–200.
759. M. Simonovits: Extremal graph problems and graph products, in Studies in Pure Mathematics, Birkhäuser, Basel, 1983, pp. 669–680.
760. M. Simonovits: Extremal graph problems, Degenerate extremal problems and Supersaturated graphs, Progress in Graph Theory (Acad. Press, ed. Bondy and Murty) (1984) 419–437.
761. M. Simonovits: Paul Erdős' influence on extremal graph theory, The mathematics of Paul Erdős, II, 148–192, Algorithms Combin., **14** (1997), Springer, Berlin. (See also with the same title an updated version, in "The Mathematics of Paul Erdős" (eds Graham, Butler, Nešetřil, 2013)
762. M. Simonovits: How to solve a Turán type extremal graph problem? (linear decomposition), Contemporary trends in discrete mathematics (Stirin Castle, 1997), pp. 283–305, DIMACS Ser. Discrete Math. Theoret. Comput. Sci., **49** (1999), Amer. Math. Soc., Providence, RI.

763. M. Simonovits: Paul Erdős' Influence on Extremal Graph Theory (Updated/Extended version of [760]), Springer, (2013) (eds Baker, Graham and Nešetřil).
764. M. Simonovits: Paul Turán's influence in Combinatorics, in Turán Memorial: Number Theory, Analysis, and Combinatorics 309–392, De Gruyter, (2013).
765. M. Simonovits and V.T. Sós: Szemerédi's partition and quasirandomness, Random Struct. Algorithms **2** (1991), 1–10.
766. M. Simonovits and V.T. Sós: Hereditarily extended properties, quasi-random graphs and not necessarily induced subgraphs, Combinatorica **17** (1997), 577–596.
767. M. Simonovits and V.T. Sós: Ramsey-Turán theory. Combinatorics, graph theory, algorithms and applications. Discrete Math. **229 (1-3)** (2001) 293–340.
768. M. Simonovits and V.T. Sós: Hereditarily extended properties, quasi-random graphs and induced subgraphs, Combin. Probab. Comput. **12** (2003), 319–344.
769. Jozef Skokan and Luboš Thoma: Bipartite subgraphs and quasi-randomness. Graphs Combin. **20 (2)** (2004), 255–262.
770. József Solymosi: Regularity, uniformity, and quasirandomness. Proc. Natl. Acad. Sci. USA **102 (23)** (2005), 8075–8076.
771. J. Solymosi: On the number of sums and products. Bulletin of the London Mathematical Society **37** (2005) 491–494.
772. J. Solymosi: The (7, 4)-conjecture in finite groups. Combin. Probab. Comput. **24 (4)** (2015) 680–686.
773. Vera T. Sós: On extremal problems in graph theory, Combinatorial Structures and their Applications (Proc. Calgary Internat. Conf., Calgary, Alta., 1969) pp. 407–410 Gordon and Breach, New York (1970)
774. V.T. Sós: Remarks on the connection of graph theory, finite geometry and block designs; in: Teorie Combinatorie, Tomo II, Accad. Naz. Lincei, Rome, (1976), 223–233.
775. V.T. Sós: An additive problem in different structures, in: Proc. of the Second Int. Conf. in Graph Theory, Combinatorics, Algorithms, and Applications, San Fra. Univ., California, July 1989, SIAM, Philadelphia, (1991), pp. 486–510.
776. V.T. Sós: Turbulent years: Erdős in his correspondence with Turán from 1934 to 1940. Paul Erdős and his mathematics, I (Budapest, 1999), 85–146, Bolyai Soc. Math. Stud., **11** (2002), János Bolyai Math. Soc., Budapest.
777. Joel Spencer: Ramsey's theorem – a new lower bound. J. Combinatorial Theory Ser. A **18** (1975), 108–115.
778. J. Spencer: "Ten Lectures on the Probabilistic Method," SIAM, Philadelphia, (1987).
779. J. Spencer: Eighty years of Ramsey $R(3, k)$...and counting! Ramsey theory, 27–39, Progress in Math., **285** (2011) Birkhäuser/Springer, New York.
780. J. Spencer: Erdős magic. The mathematics of Paul Erdős. I, 43–46, Springer, New York, (2013).
781. Ladislas Stacho: Locally pancyclic graphs, J. Combin. Theory Ser. B **76** (1999), 22–40.
782. Angelika Steger: The determinism of randomness and its use in combinatorics. Proceedings of the International Congress of Mathematicians–Seoul 2014. Vol. IV, 475–488, Kyung Moon Sa, Seoul, (2014).
783. Michael Stiebitz: Subgraphs of colour-critical graphs. Combinatorica **7 (3)** (1987) 303–312.
784. M. Stiebitz, Zs. Tuza, and Margit Voigt: On list critical graphs. Discrete Math. **309 (15)** (2009) 4931–4941.
785. Anne P. Street: Sum-free sets. In: Lecture Notes in Math. **292** (1972), 123–272. Springer.
786. Benny Sudakov: A few remarks on Ramsey-Turán-type problems, J. Combin. Theory Ser. B **88 (1)** (2003) 99–106.
787. B. Sudakov and V.H. Vu: Local resilience of graphs. Random Structures and Algorithms **33 (4)** (2008), 409–433.
788. Balázs Szegedy: Gowers norms, regularization and limits of functions on abelian groups, preprint. arXiv:1010.6211.
789. E. Szemerédi: On sets of integers containing no four elements in arithmetic progression. Acta Math. Acad. Sci. Hungar. **20** (1969) 89–104.

790. E. Szemerédi: On graphs containing no complete subgraph with 4 vertices (Hungarian), Mat. Lapok **23** (1972), 113–116.
791. E. Szemerédi: On sets of integers containing no k elements in arithmetic progression, Acta Arith. **27** (1975) 199–245, Collection of articles in memory of Jurii Vladimirovich Linnik.
792. E. Szemerédi: Regular partitions of graphs. In Problèmes combinatoires et théorie des graphes (Colloq. Internat. CNRS, Univ. Orsay, Orsay, 1976), volume **260** of Colloq. Internat. CNRS, pages 399–401. CNRS, Paris, 1978.
793. E. Szemerédi: Integer sets containing no arithmetic progressions, Acta Math. Hungar, **56** (1990), 155–158.
794. E. Szemerédi: Is laziness paying off? ("Absorbing" method). Colloquium De Giorgi 2010–2012, 17–34, Colloquia, 4, Ed. Norm., Pisa, 2013.
795. E. Szemerédi: Arithmetic progressions, different regularity lemmas and removal lemmas. Commun. Math. Stat. **3** (**3**) (2015), 315–328.
796. E. Szemerédi: Erdős's unit distance problem. Open problems in mathematics, 459–477, Springer, [Cham], 2016.
797. E. Szemerédi and W.T. Trotter Jr.: Extremal problems in discrete geometry. Combinatorica **3** (**3-4**) (1983), 381–392.
798. Terrence C. Tao: Szemerédi's regularity lemma revisited. Contrib. Discrete Math. **1** (**1**) (2006), 8–28.
799. T.C. Tao: A variant of the hypergraph removal lemma, Journal of Combinatorial Theory, Ser. A **113** (2006), 1257–1280.
800. T.C. Tao: The Gaussian primes contain arbitrarily shaped constellations, J. Anal. Math. **99** (2006), 109–176.
801. T.C. Tao and V.H. Vu: Additive Combinatorics, volume **105** of Cambridge Studies in Advanced Mathematics, Cambridge University Press, 2006.
802. T.C. Tao and V.H. Vu: Sum-free sets in groups: a survey. J. Comb. **8** (**3**) (2017) 541–552.
803. Andrew Thomason: Pseudorandom graphs, Random graphs '85 (Poznań, 1985), North-Holland Math. Stud., **144** (1987) North-Holland, Amsterdam, 307–331.
804. A. Thomason: Random graphs, strongly regular graphs and Pseudorandom graphs, in: Surveys in combinatorics 1987 (New Cross, 1987), 173–195.
805. C. Thomassen: Long cycles in digraphs with constraints on the degrees, in: B. Bollobás (Ed.), Surveys in Combinatorics, in: London Math. Soc. Lecture Notes, vol. **38** (1979) pp. 211–228. Cambridge University Press.
806. Craig Timmons and Jacques Verstraëte: A counterexample to sparse removal. European J. Combin. **44** (2015), part A, 77–86. arXiv:1312.2994.
807. Bjarne Toft: Two theorems on critical 4-chromatic graphs. Studia Sci. Math. Hungar. **7** (1972), 83–89.
808. Paul Turán: On an extremal problem in graph theory (in Hungarian). Mat. Fiz. Lapok **48** (1941) 436–452. (For its English version see [810].)
809. P. Turán: On the theory of graphs, Colloq. Math. **3** (1954) 19–30.
810. P. Turán: Applications of graph theory to geometry and potential theory. Combinatorial Structures and their Applications (Proc. Calgary Internat. Conf., Calgary, Alta., 1969) pp. 423–434 Gordon and Breach, New York (1970)
811. P. Turán: Collected papers. Akadémiai Kiadó, Budapest, 1989. Vol. **1-3**, (with comments of Simonovits on Turán's papers in Combinatorics and by others on other topics).
812. Michail Tyomkyn and Andrew J. Uzzell: Strong Turán stability. Electron. J. Combin. **22** (**3**) (2015), Paper 3.9, 24 pp.
813. Bartel L. van der Waerden: Beweis einer Baudetschen Vermutung, Nieuw Arch. Wisk. **15** (1927), 212–216.
814. Richard Wenger: Extremal graphs with no C^4's, C^6's or C^{10}'s J. Combin. Theory Ser. B, **52** (1991), 113–116.
815. Richard M. Wilson: An existence theory for pairwise balanced designs. I. Composition theorems and morphisms. J. Combinatorial Theory Ser. A **13** (1972), 220–245.

816. R.M. Wilson: An existence theory for pairwise balanced designs. II. The structure of PBD-closed sets and the existence conjectures. J. Combinatorial Theory Ser. A **13** (1972), 246–273.
817. R.M. Wilson: An existence theory for pairwise balanced designs. III. Proof of the existence conjectures. J. Combinatorial Theory Ser. A **18** (1975) 71–79.
818. R.M. Wilson: Decompositions of complete graphs into subgraphs isomorphic to a given graph, in: Proceedings of the Fifth British Combinatorial Conference, Univ. Aberdeen, Aberdeen, 1975, Congressus Numerantium, No. XV, Utilitas Math., Winnipeg, Man., 1976, pp. 647–659.
819. Hian Poh Yap: Maximal sum-free sets of group elements, Journal of the London Mathematical Society **44** (1969), 131–136.
820. H.P. Yap: Maximal sum-free sets in finite abelian groups. V., Bull. Austral. Math. Soc. **13 (3)** (1975) 337–342.
821. Raphael Yuster: Tiling transitive tournaments and their blow-ups. Order **20 (2)** (2003), 121–133.
822. R. Yuster: Quasi-randomness is determined by the distribution of copies of a fixed graph in equicardinal large sets, In Proceedings of the 12th International Workshop on Randomization and Computation (RANDOM), Springer Verlag, Boston, MA, 2008, pp. 596–601.
823. Yi Zhao: Bipartite graph tiling. SIAM J. Discrete Math. **23 (2)** (2009), 888–900.
824. Yi Zhao: Proof of the $(n/2 - n/2 - n/2)$ Conjecture for large n, Electron. J. Combin., **18** (2011), 27. 61 pp.
825. Yi Zhao: Recent advances on Dirac-type problems for hypergraphs, in: Recent Trends in Combinatorics, the IMA Volumes in Mathematics and its Applications, vol. 159, Springer, New York, 2016.
826. Alexander A. Zykov: On some properties of linear complexes. Mat. Sb. **24 (66)** (1949), 163–188. (in Russian); English translation in Amer. Math. Soc. Transl., 1952 (1952), 79.
827. Paul Erdős' homepage is: www.renyi.hu/~p_erdos
828. Miklós Simonovits' homepage is: www.renyi.hu/~miki

Curriculum Vitae of László Lovász

László Lovász was born March 9, 1948 in Budapest, Hungary. He received his diploma in Mathematics and degree of Dr. Rher. Nat. at the Eötvös Loránd University, Budapest, Hungary, 1971. Interestingly and in an extremely unusual way he obtained the degree of Candidate of Mathematical Sciences one year before finishing his university studies in 1970. He was awarded the degree of Doctor of Mathematics Sciences by the Hungarian Academy of Sciences in 1977. He became a corresponding member of the Hungarian Academy of Sciences in 1979, and a regular member in 1985.

Besides of the Hungarian Academy of Sciences, he is a member of several academies, including the European Academy of Sciences, Arts and Humanities; Academia Europaea; Nordrhein-Westfälische Akademie der Wissenschaften und der Künste (corresponding member); Deutsche Akademie der Naturforscher Leopoldina; Russian Academy of Sciences; Royal Dutch Academy of Science; Royal Swedish Academy of Sciences; National Academy of Sciences of the U.S..

He started his professional carrier as Research Associate at Eötvös Loránd University, Budapest, 1971–75. Then he moved to József Attila University, Szeged, where he worked as associate professor between 1975–78 and then Professor and Chair of Geometry, 1978–82. After these years, he returned to the Eötvös Loránd University, Budapest as a Professor and the Chair of Computer Science, for the period 1983–93. He continued as Professor, Department of Computer Science, Yale University, 1993–2000. He served as Senior Researcher at Microsoft Research, 1999–2006. He returned to Hungary and became the Director of Mathematical Institute, Eötvös Loránd University, Budapest, 2006–2011, where he is a Professor currently.

László Lovász is not only a brilliant mathematician, but also plays an important role in the organization of the scientific community. He was a member of the Executive Committee of the International Mathematical Union, 1987–1994, the Presidium of the Hungarian Academy of Sciences 1990–1993, and the Abel Prize Committee 2004–2006. He served as President of the International Mathematical Union, 2006–2010. Since 2014, he has been the President of the Hungarian Academy of Sciences.

Another aspect of his community service is that he is a member of the editorial boards of several journals, including Combinatorica (Editor-in-Chief), Advances in Mathematics, J. Combinatorial Theory (B), Discrete Math., Discrete Applied Math., J. Graph Theory, Europ.J. Combinatorics, Discrete and Computational Geometry, Random Structures and Algorithms, Acta Mathematica Hungarica, Acta Cybernetica, Electronic Journal of Combinatorics, and Geometric and Functional Analysis.

His work was acknowledged by several international awards, including the George Pólya Prize, Soc. Ind. Appl. Math., 1979; Ray D.Fulkerson Prize, Amer. Math. Soc. 1982 and 2012; Brouwer Medal, Royal Netherl. Acad. Sci., 1993; Bolzano Medal, Czech Mathematical Society, 1998; Wolf Prize, Israel, 1999; Knuth Prize, ACM, 1999; Goedel Prize, ACM-EATCS, 2001; John von Neumann Medal, IEEE, 2005; John von Neumann Theory Prize, INFORMS, 2006; Kyoto Prize, Inamori Foundation, 2011.

He also received prestigious Hungarian awards as State Prize, 1985; National Order of Merit of Hungary, 1998; Corvin Chain, 2001; Bolyai Prize, 2007 and Széchenyi Grand Prize, 2008.

He holds the degree of Doctor Honoris Causa from University of Waterloo, Ontario, Canada, 1992; University of Szeged, Hungary, 1999; Budapest University of Technology, 2002; and the University of Calgary, 2006.

László Lovász is happily married and has four children.

Publications of László Lovász

Books:

1. Lovász L., Pelikán J., Vesztergombi K.: *Kombinatorika*, Tankönyvkiadó, Budapest, 1977 (German translation: Teubner, 1977; Japanese translation: 1985)
2. Gács P., Lovász L.: *Algoritmusok*, Műszaki Könyvkiadó, Budapest, 1978; Tankönyvkiadó, Budapest, 1987.
3. L. Lovász: *Combinatorial Problems and Exercises,* Akadémiai Kiadó–North-Holland, Budapest, 1979 (Japanese translation: Tokai Univ.Press, 1988; Hungarian translation: Typotech, 1999; Second edition: North-Holland Publishing Co., Amsterdam, 1993. Reprinted by AMS Chelsea Publishing, 2007. Chinese translation: CIP 2017).
4. L. Lovász, M.D. Plummer: *Matching Theory*, Akadémiai Kiadó - North Holland, Budapest, 1986 (Russian translation: Mir, 1998; reprinted by AMS Chelsea Publishing, 2009).
5. L. Lovász: *An Algorithmic Theory of Numbers, Graphs, and Convexity*, CBMS-NSF Regional Conference Series in Applied Mathematics **50**, SIAM, Philadelphia, Pennsylvania 1986.
6. M. Grötschel, L. Lovász, A. Schrijver: *Geometric Algorithms and Combinatorial Optimization*, Springer, 1988; Chinese edition: World Publishing Corp., Beijing, 1990.
7. B. Korte, L. Lovász, R. Schrader: *Greedoids*, Springer, 1991.
8. R. Graham, M. Grötschel, L. Lovász (eds.): *Handbook of Combinatorics* Elsevier Science B.V. (1995) 1740–1748.
9. L. Lovász, J. Pelikán, K. Vesztergombi: *Discrete Mathematics: Elementary and Beyond*, Springer, New York (2003); Portuguese translation: Sociedade Braziliera de Matemática, Rio de Janiero (2005); German Translation: Springer, Heidelberg (2005); Hungarian Translation: Typotex, Budapest (2006).
10. L. Lovász: *Large networks and graph limits*, Amer. Math. Soc. Colloquium Publ. **60**, Providence, R.I. (2012).

Research Papers:

1. Lovász L.: Független köröket nem tartalmazó gráfokról (On graphs containing no independent circuits), *Mat. Lapok* **16** (1965), 289–299.
2. L. Lovász: On decomposition of graphs, *Studia Math. Hung.* **1** (1966), 237–238.
3. L. Lovász: On connected sets of points, *Annales Univ. R. Eötvös* **10** (1967), 203–204.
4. L. Lovász: Über die starke Multiplikation von geordneten Graphen, *Acta Math. Hung.* **18** (1967), 235–241.
5. L. Lovász: Operations with structures, *Acta Math. Hung.* **18** (1967), 321–328.
6. L. Lovász: Graphs and set-systems, in: *Beiträge zur Graphentheorie*, Teubner, Leipzig (1968), 99–106.
7. L. Lovász: On chromatic number of graphs and set-systems, *Acta Math. Hung.* **19** (1968), 59–67.
8. L. Lovász: On covering of graphs, in: *Theory of Graphs* (ed. P. Erdös, G. Katona), Akad. Kiadó, Budapest (1968), 231–236.
9. Lovász L.: Kapcsolatok polinomoknak és helyettesítési értékeiknek számelméleti tulajdonságai között, *Mat. Lapok* **20** (1969), 129–132.
10. L. Lovász: Generalized factors of graphs, in: *Combinatorial Theory and its Applications*, Coll. Math. Soc. J. Bolyai **4** (1970), 773–781.
11. L. Lovász: Subgraphs with prescribed valencies, *J. Comb. Theory* **8** (1970), 391–416.
12. L. Lovász: A generalization of König's theorem, *Acta Math. Hung.* **21** (1970), 443–446.
13. L. Lovász: A remark on Menger's theorem, *Acta Math. Hung.* **21** (1970), 365–368.
14. L. Lovász: The factorization of graphs, in: *Combinatorial Struc. Appl.*, Gordon and Breach (1970), 243–246.
15. L. Lovász: Representation of integers by norm-forms II, (K. Győry), *Publ. Math. Debrecen* **17** (1970), 173–181.
16. L. Lovász: On the cancellation law among finite relational structures, *Periodica Math. Hung.* **1** (1971), 145–156.
17. L. Lovász: On finite Dirichlet series, *Acta Math. Hung.* **22** (1971), 227–231.
18. L. Lovász: On the number of halving lines, *Annales Univ. Eötvös* **14** (1971), 107–108.
19. L. Lovász: Normal hypergraphs and the perfect graph conjecture, *Discrete Math.* **2** (1972), 253–267; reprinted *Annals of Discrete Math.* **21** (1984) 29–42.
20. L. Lovász: On the structure of factorizable graphs, *Acta Math. Hung.* **23** (1972), 179–195.
21. L. Lovász: The factorization of graphs II, *Acta Math. Hung.* **23** (1972), 223–246.
22. L. Lovász: On the structure of factorizable graphs II, *Acta Math. Hung.* **23** (1972), 465–478.

23. L. Lovász: Direct product in locally finite categories, *Acta Sci. Math. Szeged* **23** (1972), 319–322.
24. L. Lovász: A characterization of perfect graphs, *J. Comb. Theory* **13** (1972), 95-98; reprinted in: *Classic Papers in Combinatorics* (ed. I. Gessel and G.C. Rota), Birkhäuser, 1987, 447–450.
25. L. Lovász: A note on the line reconstruction problem, *J. Comb. Theory* **13** (1972), 309–310; reprinted in: *Classic Papers in Combinatorics* (ed. I. Gessel and G.C. Rota), Birkhäuser, 1987, 451–452.
26. L. Lovász: A note on factor-critical graphs, *Studia Sci. Math.* **7** (1972), 279–280.
27. L. Lovász, J. Pelikán: On the eigenvalues of trees, *Periodica Math. Hung.* **3** (1973), 175–182.
28. L. Lovász: Antifactors of graphs, *Periodica Math. Hung.* **4** (1973), 121–123.
29. P. Erdős, L. Lovász, G.J. Simmons, E.G. Strauss: Dissection graphs of planar point sets, in: *A Survey of Comb. Theory* (ed. S. Srivastava), Springer (1973), 139–149.
30. L. Lovász, P. Major: A note to a paper of Dudley, *Studia Sci. Math.* **8** (1973), 151–152.
31. L. Babai, L. Lovász: Permutation groups and almost regular graphs, *Studia Sci. Math.* **8** (1973), 141–150.
32. L. Babai, W. Imrich, L. Lovász: Finite homeomorphism groups of the 2-sphere, in: *Topics in Topology*, Coll. Math. Soc. J. Bolyai **9** (1973), 61–75.
33. L. Lovász: Connectivity in digraphs, *J. Comb. Theory* **15** (1973), 174–177.
34. L. Lovász: Independent sets in critical chromatic graphs, *Studia Sci. Math.* **8** (1973), 165–168.
35. L. Lovász, A. Recski: On the sum of matroids, *Acta Math. Hung.* **24** (1973), 329–333.
36. L. Lovász: Coverings and colorings of hypergraphs, in: *Proc. 4th Southeastern Conf. on Comb.*, Utilitas Math. (1973), 3–12.
37. L. Lovász: Factors of graphs, in: *Proc. 4th Southeastern Conf. on Comb.*, Utilitas Math. (1973), 13–22.
38. L. Lovász: Valencies of graphs with 1-factors, *Periodica Math. Hung.* **5** (1974), 149–151.
39. L. Lovász: Minimax theorems for hypergraphs, in: *Hypergraph Seminar* (ed. C. Berge and D.K.Ray-Chaudhuri), Lecture Notes in Math. **411** (1974) Springer, 111–126.
40. V. Chvátal, L. Lovász: Every directed graph has a semi-kernel, in: *Hypergraph Seminar* (ed. C. Berge and D.K. Ray-Chaudhuri), Lecture Notes in Math. **411** (1974), Springer, 175.
41. D. Greenwell, L. Lovász: Applications of product coloring, *Acta Math. Hung.* **25** (1974), 335–340.
42. L. Lovász, M.D. Plummer: A family of planar bicritical graphs (M.D.Plummer), in: *Combinatorics*, London Math. Soc. Lecture Notes **13** (1974), 103-108; journal version: *Proc. London Math. Soc.* **30** (1975), 160–176.

43. P. Erdős, L. Lovász: Problems and results on 3-chromatic hypergraphs and some related questions, in: *Infinite and Finite Sets*, Coll. Math. Soc. J. Bolyai **11** (1975), 609-627.
44. L. Lovász, M.D. Plummer: On bicritical graphs, in: *Infinite and Finite Sets*, Coll. Math. Soc. J. Bolyai **11** (1975), 1051–1079.
45. R. Appleson, L. Lovász: A characterization of cancellable k-ary structures, *Periodica Math. Hung.* **6** (1975), 17–19.
46. L. Lovász: Three short proofs in graph theory, *J. Comb. Theory* **19** (1975), 269–271.
47. L. Lovász: Spectra of graphs with transitive groups, *Periodica Math. Hung.* **6** (1975), 191–195.
48. L. Lovász: 2-matchings and 2-covers of hypergraphs, *Acta Math. Hung.* **26** (1975), 433–444.
49. L. Lovász: On the ratio of optimal fractional and integral covers, *Discrete Math.* **13** (1975), 383–390.
50. L. Lovász: A kombinatorika minimax tételeiről (On the minimax theorems of combinatorics), *Mat. Lapok* **26** (1975), 209–264.
51. D.E. Daykin, L. Lovász: The number of values of Boolean functions, *J. London Math. Soc.* **30** (1976), 160–176.
52. L. Lovász: On two minimax theorems in graph theory, *J. Comb. Theory B* **21** (1976), 93-103.
53. S.A. Burr, P. Erdős, L. Lovász: On graphs of Ramsey type, *Ars Combinatoria* **1** (1976), 167–190.
54. L. Lovász: On some connectivity properties of Eulerian graphs, *Acta Math. Hung.* **28** (1976), 129–138.
55. L. Lovász: Covers, packings and some heuristic algorithms, in: *Combinatorics*, Proc. 5th British Comb. Conf. (ed. C.St.J.A.Nash-Williams, J.Sheehan), Utilitas Math. (1976), 417–429.
56. L. Lovász, M. Simonovits: On the number of complete subgraphs of a graph, in: *Combinatorics*, Proc. 5th British Comb. Conf. (ed. C.St.J.A.Nash-Williams, J.Sheehan), Utilitas Math. (1976), 439–441.
57. L. Lovász, M. Marx: A forbidden substructure characterization of Gauss codes, *Bull. Amer. Math. Soc.* **82** (1976), 121–122; full version *Acta. Sci. Math. Szeged* **38** (1976), 115–119.
58. L. Lovász: Chromatic number of hypergraphs and linear algebra, *Studia Sci. Math.* **11** (1976), 113–114.
59. P. Gács, L. Lovász: Some remarks on generalized spectra, *Zeitschr. f. math. Logik u. Grundlagen d. Math.* **23** (1977), 547–554.
60. L. Lovász: Certain duality principles in integer programming, *Annals of Discrete Math.* **1** (1977), 363–374.
61. L. Lovász: A homology theory for spanning trees of a graph, *Acta Math. Hung.* **30** (1977), 241–251.
62. L. Lovász, M.D. Plummer: On minimal elementary bipartite graphs, *J. Comb. Theory B* **23** (1977), 127–138.

63. L. Lovász: Flats in matroids and geometric graphs, in: *Combinatorial Surveys*, Proc. 6th British Comb. Conf., Academic Press (1977), 45–86.
64. R.L. Graham, L. Lovász: Polynomes de la matrice des distences d'un arbre, in: *Problemes Combinatoires et Theorie de Graphes*, CNRS (1977), 189–190.
65. R.L. Graham, L. Lovász: Distance matrices of trees, in: *Theory and Appl. of Graphs*, Lecture Notes in Math. **642** (1978), Springer, 186-190; journal version: Distance matrix polynomials of trees, *Adv. in Math.* **29** (1978), 60–88.
66. L. Lovász, V. Neumann-Lara, M.D. Plummer: Mengerian theorems for paths with bounded length, *Periodica Math. Hung.* **9** (1978), 269–276.
67. L. Lovász: Some finite basis theorems in graph theory, in: *Combinatorics*, Coll. Math. Soc. J. Bolyai **18** (1978), 717–729.
68. L. Lovász, K. Vesztergombi: Restricted permutations and the distribution of Stirling numbers, in: *Combinatorics*, Coll. Math. Soc. J. Bolyai **18** (1978), 731–738.
69. L. Lovász: Kneser's conjecture, chromatic number, and homotopy, *J. Comb. Theory A* **25** (1978), 319–324.
70. L. Lovász: Topological and algebraic methods in graph theory, in: *Graph Theory and Related Topics*, Academic Press (1979), 1–14.
71. P. Erdős, L. Lovász, J. Spencer: Strong independence of graphcopy functions, in: *Graph Theory and Related Topics*, Academic Press (1979), 165–172.
72. L. Lovász: Graph theory and integer programming, *Annals of Discrete Math.* **4** (1979), 141–158.
73. L. Lovász: On the Shannon capacity of graphs, *IEEE Trans. Inform. Theory* **25** (1979), 1–7.
74. A. Hajnal, L. Lovász: An algorithm to prevent the propagation of certain diseases at minimum cost, in: *Interfaces between Computer Science and Operations Research*, Amsterdam Math. Centr. Tract **99** (1979), 105–108.
75. L. Lovász: Gráfelmélet és diszkrét programozás (Graph theory and discrete programming), *Mat. Lapok* **27** (1979), 69–86.
76. L. Lovász: Determinants, matchings, and random algorithms, in: *Fundamentals of Computation Theory, FCT'79* (ed. L. Budach), Akademie-Verlag Berlin (1979), 565–574.
77. R. Aleliunas, R.M. Karp, R.J. Lipton, L. Lovász, C.W. Rackoff: Random walks, universal travelling sequences, and the complexity of maze problems, *Proc. 20th IEEE Ann. Symp. on Found. of Comp. Sci.* (1979), 218–223.
78. L. Lovász, J. Nesetril, A. Pultr: On a product dimension of graphs, *J. Comb. Theory B* **29** (1980), 47–67.
79. L. Lovász: Selecting independent lines from a family of lines in a space, *Acta Sci. Math. Szeged* **42** (1980), 121–131.
80. L. Lovász: Matroid matching and some applications, *J. Comb. Theory B* **28** (1980), 208–236.
81. L. Lovász: The matroid matching problem, in: *Algebraic Methods in Graph Theory*, Coll. Math. Soc. J. Bolyai **25** (1980), 495-517.
82. L. Lovász: Matroids and Sperner's Lemma, *Europ. J. Combin.* **1** (1980), 65–66.

83. L. Lovász: Efficient algorithms: an approach by formal logic, in: *Studies on Math. Programming* (ed. A. Prékopa), Akadémiai Kiadó (1980), 119–126.
84. B. Korte, L. Lovász: Mathematical structures underlying greedy algorithms, in: *Fundamentals of Computation Theory* (F. Gécseg, ed.) Lecture Notes in Comp. Sci. **117** (1981), Springer, 205–209.
85. L. Lovász, A. Sárközi, M. Simonovits: On additive arithmetic functions satisfying a linear recursion, *Annales Univ. R. Eötvös* **24** (1981), 205–215.
86. M. Grötschel, L. Lovász, A. Schrijver: The ellipsoid method and its consequences in combinatorial optimization, *Combinatorica* **1** (1981), 169–197; Corrigendum *Combinatorica* **4** (1984), 291–295.
87. L. Lovász, A. Schrijver: Remarks on a theorem of Rédei, *Studia Sci. Math. Hung.* **16** (1981), 449–454.
88. A.J. Bondy, L. Lovász: Cycles through given vertices of a graph, *Combinatorica* **1** (1981), 117–140.
89. P. Gács, L. Lovász: Khachiyan's algorithm for linear programming, *Math. Prog. Study* **14** (1981), 61–68.
90. A.K. Lenstra, H.W. Lenstra, L. Lovász: Factoring polynomials with rational coefficients, *Math. Annalen* **261** (1982), 515–534.
91. J. Edmonds, L. Lovász, W.R. Pulleyblank: Brick decompositions and the matching rank of graphs, *Combinatorica* **2** (1982), 247–274.
92. L. Lovász, Y. Yemini: On generic rigidity in the plane, *SIAM J. Alg. Discr. Methods* **1** (1982), 91–98.
93. L. Lovász: Some combinatorial applications of the new linear programming algorithms, in: *Combinatorics and Graph Theory* (ed. S.B.Rao), Lecture Notes in Math. **885** (1982), Springer, 33–41.
94. L. Lovász: Bounding the independence number of a graph, *Ann. of Discr. Math.* **16** (1982), 213–223.
95. L. Lovász, A. Recski: Selected topics of matroid theory and its applications, *Suppl. Rendiconti del Circ. Mat. Palermo* **2** (1982), 171–185.
96. I. Bárány, L. Lovász: Borsuk's Theorem and the number of facets of centrally symmetric polytopes, *Acta Math. Hung.* **40** (1982), 323–329.
97. L. Lovász: Perfect graphs, in: *More Selected Topics in Graph Theory* (ed. L. W. Beineke, R. L. Wilson), Academic Press (1983), 55–67.
98. L. Lovász: Ear-decompositions of matching-covered graphs, *Combinatorica* **3** (1983) 105–117.
99. B. Korte, L. Lovász: Structural properties of greedoids, *Combinatorica* **3** (1983) 359–374.
100. L. Lovász: Submodular functions and convexity, in: *Mathematical Programming: the State of the Art* (ed. A.Bachem, M.Grötschel, B.Korte), Springer (1983), 235–257.
101. L. Lovász: Self-dual polytopes and the chromatic number of distance graphs on the sphere, *Acta Sci. Math. Szeged* **45** (1983), 317–323.
102. L. Lovász, M. Simonovits: On the number of complete subgraphs of a graph II, in: *Studies in Pure Math.*, To the memory of P. Turán (ed. P. Erdös), Akadémiai Kiadó (1983), 459–495.

103. L. Lovász: Algorithmic aspects of combinatorics, geometry and number theory, in: *Proc. Int. Congress Warsaw 1982*, Polish Sci. Publishers – North-Holland (1984) 1591–1595.
104. M.Grötschel, L. Lovász, A. Schrijver: Geometric methods in combinatorial optimization, in: *Progress in Combinatorial Optimization* (ed.W.R.Pulleyblank), Academic Press (1984), 167–183.
105. B. Korte, L. Lovász: Greedoids - a structural framework for the greedy algorithm, in: *Progress in Combinatorial Optimization* (ed.W.R.Pulleyblank) Academic Press (1984), 221–243.
106. M.Grötschel, L. Lovász, A. Schrijver: Polynomial algorithms for perfect graphs, *Annals of Discrete Math.* **21** (1984), 325–256.
107. W. Cook, L. Lovász, A. Schrijver: A polynomial-time test for total dual integrality in fixed dimension, *Math. Programming Study* **22** (1984), 64–69.
108. B. Korte, L. Lovász: Greedoids and linear objective functions, *SIAM J. on Algebraic and Discrete Methods* **5** (1984), 229–238.
109. R. Kannan, A.K. Lenstra, L. Lovász: Polynomial factorization and the nonrandomness of bits of algebraic and some transcendental numbers, in: *Proc. 16th ACM Symp. on Theory of Computing* (1984), 191–200.
110. B. Korte, L. Lovász: Shelling structures, convexity, and a happy end, in: *Graph Theory and Combinatorics* (ed. B. Bollobas), Acad. Press (1984), 219–232.
111. B. Korte, L. Lovász: A note on selectors and greedoids, *Eur. J. Combinatorics* **6** (1985), 59–67.
112. B. Korte, L. Lovász: Posets, matroids, and greedoids, in: *Matroid Theory,* Coll. Math. Soc. J. Bolyai **40** (ed. L. Lovász, A. Recski), North-Holland (1985), 239–265.
113. B. Korte, L. Lovász: Polymatroid greedoids, *J. Comb. Theory B* **38** (1985), 41–72.
114. B. Korte, L. Lovász: Basis graphs of greedoids and 2-connectivity, *Math. Prog. Study* **24** (1985), 158–165.
115. L. Lovász: Computing ears and branchings in parallel, *26th IEEE Annual Symp. on Found. of Comp. Sci.* (1985), 464–467.
116. L. Lovász: Some algorithmic problems on lattices, in: *Theory of Algorithms*, (cds. L. Lovász and E. Szemerédi), Coll. Math. Soc. J. Bolyai 44, North-Holland (1985), 323–337.
117. L. Lovász: Vertex packing algorithms, *Proc. Int. Coll. Automata, Languages, and Programming*, Springer (1985), 1–14.
118. A. Björner, B.Korte, L. Lovász: Homotopy properties of greedoids, *Advances in Appl. Math.* **6** (1985), 447–494.
119. B. Korte, L. Lovász: Relations between subclasses of greedoids, *Zeitschr. f. Oper. Res. A: Theorie* **29** (1985), 249–267.
120. L. Lovász: Algorithmic aspects of some notions in classical mathematics, in: *Mathematics and Computer Science* (ed. J.W. de Bakker, M. Hazewinkel, J.K. Lenstra), CWI Monographs **1**, North-Holland, Amsterdam (1986), 51–63.

121. M. Grötschel, L. Lovász, A.Schrijver: Relaxations of vertex packing, *J. Combin. Theory B* **40** (1986), 330–343.
122. L. Lovász, J. Spencer, K. Vesztergombi: Discrepancy of set-systems and matrices, *Europ. J. Combin.* **7** (1986), 151–160.
123. B. Korte, L. Lovász: Non-interval greedoids and the transposition property, *Discrete Math.* **59** (1986), 297–314.
124. K. Cameron, J. Edmonds, L. Lovász: A note on perfect graphs, *Periodica Math. Hung.* **17** (1986), 173–175.
125. N. Linial, L. Lovász, A. Wigderson: A physical interpretation of graph connectivity, *Proc. 27th Annual IEEE Symp. on Found. of Comp. Sci.*, (1986), 39–48.
126. R. Kannan, L. Lovász: Covering minima and lattice point free convex bodies, *Proc. Conf. on Foundations of Software Technology and Theoretical Comp. Sci.*, Lecture Notes in Comp. Science **241**, Springer (1986) 193–201.
127. L. Lovász: Connectivity algorithms using rubber bands, *Proc. Conf. on Foundations of Software Technology and Theoretical Comp. Sci.*, Lecture Notes in Comp. Science **241**, Springer (1986) 394–411.
128. U. Faigle, L. Lovász, R. Schrader, G. Turán: Searching in trees, series-parallel and interval orders, *SIAM J. Computing* **15** (1986) 1075–1084.
129. B. Korte, L. Lovász: Homomorphisms and Ramsey properties of antimatroids, *Discrete Appl. Math.* **15** (1986), 283–290.
130. N. Alon, P. Frankl, L. Lovász: The chromatic number of Kneser hypergraphs, *Trans. Amer. Math. Soc.* **298** (1987), 359–370.
131. A. Dress, L. Lovász: On some combinatorial properties of algebraic matroids, *Combinatorica* **7** (1987), 39–48.
132. L. Lovász: Matching structure and the matching lattice, *J. Comb. Theory B* **43** (1987), 187–222.
133. A. Björner, L. Lovász: Pseudomodular lattices and continuous matroids, *Acta Sci. Math. Szeged* **51** (1987), 295–308.
134. N. Linial, L. Lovász, A. Wigderson: Rubber bands, convex embeddings, and graph connectivity, Combinatorica 8 (1988), 91–102.
135. L. Lovász: Geometry of numbers: an algorithmic view, in: *ICIAM '87: Proc. 1st Internatl. Conf. on Industr. Appl. Math.* (ed. J. McKenna, R. Teman), SIAM, Philadelphia (1988), 144–152.
136. L. Lovász, M. Saks: Lattices, Möbius functions and communication complexity, *29th IEEE Annual Symp. on Found. of Comp. Sci.* (1988), 81–90.
137. P. Erdős, L. Lovász, K. Vesztergombi: The chromatic number of the graph of large distances, in: *Combinatorics*, Proc. Coll. Eger 1987, Coll. Math. Soc. J. Bolyai **52**, North-Holland (1988), 547–551.
138. C.A.J. Hurkens, L. Lovász, A. Schrijver, É. Tardos: How to tidy up your set-system? in: *Combinatorics*, Proc. Coll. Eger 1987, Coll. Math. Soc. J. Bolyai **52**, North-Holland (1988), 309–314.
139. B. Korte, L. Lovász: The intersection of matroids and antimatroids, *Discrete Math.* **73** (1988), 143–157.

140. R. Kannan, L. Lovász: Covering minima and lattice point free convex bodies, *Annals of Math.* **128** (1988), 577–602.
141. L. Lovász, K. Vesztergombi: Extremal problems for discrepancy, in: *Irregularities of Partitions*, (ed. G. Halász, V.T.Sós) Algorithms and Combinatorics **8** (1989), Springer, 107–113.
142. L. Lovász, M. Saks and A. Schrijver: : Orthogonal representations and connectivity of graphs, *Linear Alg. Appl.* **114/115** (1989), 439-454. A correction: *Linear Algebra Appl.* **313** (2000), 101–105.
143. B. Korte, O. Goecke, L. Lovász: Examples and algorithmic properties of greedoids, in: Combinatorial Optimization (ed. B. Simeone), *Lecture Notes in Math.* **1403**, Springer (1989), 113–161.
144. R. Anderson, L. Lovász, P. Shor, J. Spencer, É. Tardos, S. Winograd: Disks, balls and walls: The analysis of a combinatorial game, *Amer. Math. Monthly* **96** (1989), 481–493.
145. L. Lovász, M. Saks, W. T. Trotter: An on-line graph coloring algorithm with sublinear performance ratio, *Discrete Math.* **75** (1989), 319-325.
146. L. Lovász: Faster algorithms for hard problems, *Information Processing '89* (ed. G. X. Ritter), Elsevier (1989), 135–141.
147. L. Lovász, B. Korte: Polyhedral results for antimatroids, in: *Combinatorial Mathematics*, Proc. 3rd Intern. Conf., (ed. G.S. Bloom, R. L. Graham, J. Malkevitch), Annals of the NY Academy of Sciences **555** (1989), 283–295.
148. P. Erdös, L. Lovász, K. Vesztergombi: On the graph of large distances, *Discr. Comput. Geometry* **4** (1989) 541–549.
149. L. Lovász: Geometry of numbers and integer programming, in: *Mathematical Programming, Recent Developments and Applications*, Kluwer Academic Publishers (1989), 177–201.
150. L. Lovász: Singular spaces of matrices and their application in combinatorics, *Bol. Soc. Braz. Mat.* **20** (1989), 87–99.
151. L. Lovász, M.D. Plummer: Some recent results on graph matching, in: *Graph Theory and its Applications: East and West*, (ed. M. F. Capobianco, M. Guan, D. F. Hsu, F. Tian), Ann. NY Acad. Sci. **576** (1989), 389–398.
152. I. Bárány, Z. Füredi, L. Lovász: On the number of halving planes, *Proc. 5th Symp. Comp. Geom.*, (1989), 140–144; journal version: *Combinatorica* **10** (1990), 175–183.
153. L. Lovász, A. Schrijver: Matrix cones, projection representations, and stable set polyhedra, in: *Polyhedral Combinatorics*, DIMACS Series in Discrete Mathematics and Theoretical Computer Science I (1990), 1–17.
154. R. Kannan, L. Lovász, H. E. Scarf: The shapes of polyhedra, *Math. of Oper. Res.* **15** (1990), 364–380.
155. L. Lovász, M. Simonovits: The mixing rate of Markov chains, an isoperimetric inequality, and computing the volume, *Proc. 31st IEEE Annual Symp. on Found. of Comp. Sci.* (1990), 346–354.
156. L. Lovász: Communication complexity: a survey, in: *Paths, flows, and VLSI-Layout*, (ed. B. Korte, L. Lovász, H. J. Prömel, A. Schrijver), Springer (1990), 235–265.

157. I. Csiszár, J. Körner, L. Lovász, K. Marton, G. Simonyi: Entropy splitting for antiblocking pairs and perfect graphs, *Combinatorica* **10** (1990), 27–40.
158. L. Babai, A.J. Goodman, L. Lovász: Graphs with given automorphism group and few edge orbits, *Europ. J. Combin.* **12** (1991), 185–203.
159. A. Björner, L. Lovász, P. Shor: Chip-firing games on graphs, *Europ. J. Comb.* **12** (1991), 283–291.
160. L. Lovász, A. Schrijver: Cones of matrices and set-functions, and 0-1 optimization, *SIAM J. Optim.* **1** (1991), 166–190.
161. L. Lovász: Geometric algorithms and algorithmic geometry, *Proc. of Int. Congress of Math, Kyoto, 1990*, Springer-Verlag (1991), 139–154.
162. L. Lovász, M. Naor, I. Newman, A. Wigderson: Search problems in the decision tree model, *Proc. 32nd IEEE Annual Symp. on Found. of Comp. Sci.* (1991), 576–585; journal version: *SIAM J. Disc. Math.* **8** (1995), 119–132.
163. J. Csima, L. Lovász: A matching algorithm for regular bipartite graphs, *Discrete Appl. Math.* **35** (1992), 197–203.
164. L. Lovász: How to compute the volume? *Jber. d. Dt. Math.-Verein, Jubiläumstagung 1990*, B. G. Teubner, Stuttgart (1992), 138–151.
165. A. Björner, L. Lovász, A. Yao: Linear decision trees, hyperplane arrangements, and Möbius functions, in: *Proc. 24th ACM Symp. on Theory of Computing* (1992), 170–177.
166. U. Feige, L. Lovász: Two-prover one-round proof systems: their power and their problems, in: *Proc. 24th ACM Symp. on Theory of Computing* (1992), 733–744.
167. I. Bárány, R. Howe, L. Lovász: On integer points in polyhedra: a lower bound, *Combinatorica* **12** (1992), 135–142.
168. L. Lovász, M. Simonovits: On the randomized complexity of volume and diameter, *Proc. 33rd IEEE Annual Symp. on Found. of Comp. Sci.* (1992), 482–491.
169. L. Lovász, H. Scarf: The generalized basis reduction algorithm, *Math. of OR* **17** (1992), 751–764.
170. A. Björner, L. Lovász: Chip-firing games on directed graphs, *J. Algebraic Combinatorics* **1** (1992), 305–328.
171. J. Csima, L. Lovász: Dating to marriage, *Discrete Appl. Math.* **41** (1993), 269–270.
172. L. Lovász, Á. Seress: The cocycle lattice of binary matroids, *Europ. J. Comb.* **14** (1993), 241–250.
173. L. Lovász, M. Simonovits: Random walks in a convex body and an improved volume algorithm, *Random Structures and Alg.* **4** (1993), 359–412.
174. L. Lovász, M. Saks: Communication complexity and combinatorial lattice theory, *J. Comp. Sys. Sci.* **47** (1993), 322–349.
175. L. Lovász, P. Winkler: Note: on the last new vertex visited by a random walk, *Journal of Graph Theory* **17** (1993), 593–596.
176. N. Karmarkar, R.M. Karp, R. Lipton, L. Lovász, M. Luby: A Monte-Carlo algorithm for estimating the permanent, *SIAM J. Comp.* **22** (1993), 284–293.
177. L. Lovász: Stable sets and polynomials, *Discrete Math.* **124** (1994), 137–153.

178. A. Björner, L. Lovász: Linear decision trees, subspace arrangements, and Möbius functions, *Journal of the Amer. Math. Soc.* **7** (1994), 677–706.
179. A. Björner, L. Lovász, S. Vrecica, R. Zivaljevic: Chessboard complexes and matching complexes, *Journal of the London Math. Soc.* **49** (1994), 25–39.
180. R. Kannan, L. Lovász, M. Simonovits: Isoperimetric problems for convex bodies and a localization lemma, *Disc. Comput. Geometry*, **13** (1995), 541–559.
181. L. Lovász: Randomized algorithms in combinatorial optimization, in: Combinatorial Optimization, Papers from the DIMACS Special Year, (ed. W. Cook, L. Lovasz, P. Seymour), DIMACS Series in Discrete Mathematics and Combinatorial Optimization **20**, Amer. Math. Soc., Providence (1995), 153–179.
182. L. Lovász, P. Winkler: Exact mixing in an unknown Markov chain, *Electronic Journal of Combinatorics* **2** (1995), paper R15, 1–14.
183. L. Lovász, P. Winkler: Mixing of random walks and other diffusions on a graph, in: *Surveys in Combinatorics*, 1995 (ed. P. Rowlinson), London Math. Soc. Lecture Notes Series **218**, Cambridge Univ. Press (1995), 119–154.
184. L. Lovász, P. Winkler: Efficient stopping Rules for Markov Chains, *Proc. 1995 ACM Symp. Theory of Computing* (1995), 76–82.
185. L. Lovász, Á. Seress: The cocycle lattice of binary matroids II, *Linear Algebra and its Applications*, **226–228** (1995), 553–566.
186. H. v.d.Holst, L. Lovász, A. Schrijver: On the invariance of Colin de Verdière's graph parameter under clique sums, *Linear Algebra and its Applications*, **226–228** (1995), 509–518.
187. A. Kotlov, L. Lovász: The rank and size of graphs, *J. Graph Theory* **23** (1996), 185–189.
188. L. Lovász: Random walks on graphs: a survey, in: *Combinatorics, Paul Erdős is Eighty*, Vol. 2 (ed. D. Miklos, V. T. Sos, T. Szonyi), János Bolyai Mathematical Society, Budapest, (1996), 353–398.
189. U. Feige, S. Goldwasser, L. Lovász, S. Safra and M. Szegedy: Approximating clique is almost NP-complete, *Proc. 32nd IEEE Annual Symp. on Found. of Comp. Sci.* (1991), 2–12; journal version: *Journal of the ACM* **43** (1996), 268–292.
190. L. Lovász: The membership problem in jump systems, *J. Comb. Theory* (B) **70** (1997), 45–66.
191. L. Lovász, J. Pach, M. Szegedy: On Conway's thrackle conjecture, *Discrete and Computational Geometry* **18** (1997), 369–376.
192. R. Kannan, L. Lovász, M. Simonovits: Random walks and an $O^*(n^5)$ volume algorithm for convex bodies, *Random Structures and Algorithms* **11** (1997), 1–50.
193. A. Kotlov, L. Lovász, S. Vempala: The Colin de Verdière number and sphere representations of a graph, *Combinatorica* **17** (1997), 483–521.
194. D. Aldous, L. Lovász, P. Winkler: Mixing times for uniformly ergodic Markov chains, *Stochastic Processes and their Applications* **71** (1997), 165–185.
195. L. Lovász, A. Schrijver: A Borsuk theorem for antipodal links and a spectral characterization of linklessly embeddable graphs, *Proceedings of the Amer. Math. Soc.* **126** (1998), 1275–1285.

196. L. Lovász, P. Winkler: Reversal of Markov chains and the forget time, *Combinatorics, Probability and Computing* **7** (1998), 189–204.
197. L. Lovász, P. Winkler: Mixing times, in: *Microsurveys in Discrete Probability* (ed. D. Aldous and J. Propp), DIMACS Series in Discrete Math. and Theor. Comp. Sci., Amer. Math. Soc. (1998), 85–133.
198. A. Brieden, P. Gritzmann, R. Kannan, V. Klee, L. Lovász, M. Simonovits: Approximation of diameters: randomization doesn't help, *Proceedings 39th Ann. Symp. on Found. of Comp. Sci.* (1998) 244–251.
199. A. Beveridge, L. Lovász: Random walks and the regeneration time, J. Graph Theory **29** (1998), 57–62.
200. L. Lovász, A. Schrijver: The Colin de Verdière graph parameter, Proc. Conf. Combinatorics, in: *Graph Theory and Combinatorial Biology*, Bolyai Soc. Math. Stud. **7**, János Bolyai Math. Soc., Budapest (1999), 29–85.
201. F. Chen, L. Lovász, I. Pak: Lifting Markov chains to speed up mixing, in: *Proc. 31st Annual ACM Symp. on Theory of Computing* (1999), 275–281.
202. R. Kannan, L. Lovász: Faster mixing via average conductance, in: *Proc. 31st Annual ACM Symp. on Theory of Computing* (1999), 282–287.
203. L. Lovász, A. Schrijver: On the null space of a Colin de Verdière matrix, *Annales de l'Institute Fourier* **49** (1999), 1017–1026.
204. L. Lovász: Hit-and-run mixes fast, *Math. Programming, series A* **86** (1999), 443–461.
205. L. Lovász: Integer sequences and semidefinite programming *Publ. Math. Debrecen* **56** (2000), 475–479.
206. L. Lipták, L. Lovász: Facets With Fixed Defect of the Stable Set Polytope, *Math. Programming*, Series A **88** (2000), 33–44.
207. J. Kahn, J.H. Kim, L. Lovász, V.H. Vu: The cover time, the blanket time, and the Matthews bound, *Proc. 41st IEEE Ann. Symp. on Found. of Comp. Sci.* (2000).
208. L. Lipták, L. Lovász: Critical Facets of the Stable Set Polytope, *Combinatorica* **21** (2001), 61–88.
209. L. Lovász: Steinitz representations of polyhedra and the Colin de Verdière number, *J. Comb. Theory B* **82** (2001), 223–236.
210. L. Lovász: Energy of convex sets, shortest paths and resistance *J. Comb. Theory A* **94** (2001), 363–382.
211. N. Alon, L. Lovász: Unextendible product bases, *J. Combin. Theory A* **95** (2001), 169–179.
212. A. Brieden, P. Gritzmann, R. Kannan, V. Klee, L. Lovász, M. Simonovits: Deterministic and randomized approximation of radii, *Mathematika* **48** (2001), 63–105.
213. I. Benjamini, L. Lovász: Global Information from Local Observation, *Proc. 43rd Ann. Symp. on Found. of Comp. Sci.* (2002), 701–710.
214. S. Arora, B. Bollobás, L. Lovász: Proving integrality gaps without knowing the linear program *Proc. 43rd Ann. Symp. on Found. of Comp. Sci.* (2002), 313–322; journal version: S. Arora, B. Bollobás, L. Lovász, I. Tourlakis, *Theory of Computing*, **2** (2006), 19–51.

215. L. Lovász, K. Vesztergombi: Geometric representations of graphs, in: *Paul Erdős and his Mathematics,* (ed. G. Halász, L. Lovász, M. Simonovits, V.T. Sós), Bolyai Soc. Math. Stud. **11**, J. Bolyai Math. Soc., Budapest (2002), 471–498.
216. L. Lovász: Harmonic and analytic functions on graphs, *J. of Geometry* **76** (2003), 3–15.
217. L. Lovász, S. Vempala: Logconcave Functions: Geometry and Efficient Sampling Algorithms *Proc. 43rd Ann. Symp. on Found. of Comp. Sci.* (2003), 640–649.
218. L. Lovász, S. Vempala: Simulated Annealing in Convex Bodies and an $O*(n^4)$ Volume Algorithm, *Proc. 43rd Ann. Symp. on Found. of Comp. Sci.* (2003), 650–659; journal version *J. Comput. System Sci.* **72** (2006), 392–417.
219. N. Harvey, E. Ladner, L. Lovász, T. Tamir: Semi-Matchings for Bipartite Graphs and Load Balancing, *Algorithms and Data Structures*, Lecture Notes in Comput. Sci. *2748*, Springer, Berlin (2003), 294–306; journal version *Journal of Algorithms* **59** (2006), 53–78.
220. L. Lovász: Semidefinite programs and combinatorial optimization, in: *Recent Advances in Algorithms and Combinatorics,* CMS Books Math./Ouvrages Math. SMC *11*, Springer, New York (2003), 137–194.
221. L. Lovász, K. Vesztergombi, U. Wagner, E. Welzl: Convex quadrilaterals and k-sets, in: *Towards a Theory of Geometric Graphs*, (J. Pach, Ed.), AMS Contemporary Mathematics **342** (2004), 139–148.
222. J. Chen, R.D. Kleinberg, L. Lovász, R. Rajaraman, R. Sundaram, A. Vetta: (Almost) tight bounds and existence theorems for confluent flows, *Proc. 36th Annual ACM Symposium on Theory of Computing*, ACM, New York (2004), 529–538.
223. L. Lovász, S. Vempala: Hit-and-run from a corner, *Proc. 36th Annual ACM Symposium on Theory of Computing*, 310–314 (electronic), ACM, New York (2004).
224. U. Feige, L. Lovász, P. Tetali: Approximating min sum set cover, *Proc. Conf. APPROX* (2002), 94–107; journal version: *Algorithmica* **40** (2004), 219–234.
225. L. Lovász: Discrete Analytic Functions: An Exposition, in: *Surveys in Differential Geometry IX, Eigenvalues of Laplacians and other geometric operators* (Ed. A. Grigoryan, S.T. Yau), Int. Press, Somerville, MA (2004), 241–273.
226. M. Bordewich, M. Freedman, L. Lovász, D. Welsh: Approximate counting and quantum computation, *Combinatorics, Probability and Computing* **14** (2005), 737–754.
227. C. Borgs, J.T. Chayes, L. Lovász, V.T. Sós, B. Szegedy and K. Vesztergombi: Graph Limits and Parameter Testing, Proc. 38th Annual ACM Symp. on Theory of Computing 2006, 261–270.
228. K. Jain, L. Lovász, P.A. Chou: Building scalable and robust peer-to-peer overlay networks for broadcasting using network coding, *Ann. ACM Symp. Principles of Dist. Comp.* (2005) 51–59.

229. R. Kannan, L. Lovász, R. Montenegro: Blocking conductance and mixing in random walks, *Combinatorics, Probability and Computing* **15** (2006), 541–570.
230. C. Borgs, J. Chayes, L. Lovász, V.T. Sós, K. Vesztergombi: Counting graph homomorphisms, in: *Topics in Discrete Mathematics* (ed. M. Klazar, J. Kratochvil, M. Loebl, J. Matoušek, R. Thomas, P. Valtr), Springer (2006), 315–371.
231. L. Lovász, M. Saks: A localization inequality for set functions, *J. Comb. Theory A* **113** (2006), 726–735.
232. L. Lovász: The rank of connection matrices and the dimension of graph algebras, *Eur. J. Comb.* **27** (2006), 962–970.
233. L. Lovász: Graph minor theory, *Bull. Amer. Math. Soc.* **43** (2006), 75–86.
234. I. Benjamini, G. Kozma, L. Lovász, D. Romik, G. Tardos: Waiting for a bat to fly by (in polynomial time), *Combinatorics, Probability and Computing* **15** (2006), 673–683.
235. L. Lovász, B. Szegedy: Limits of dense graph sequences, *J. Comb. Theory B* **96** (2006), 933–957.
236. S. Arora, L. Lovász, I. Newman, Y. Rabani, Y. Rabinovich, S. Vempala: Local versus Global Properties of Metric Spaces, *Proc. 17th Ann. ACM-SIAM Symp. Disc. Alg.* (2006), 41–50. Journal version: *SIAM J. Comput.* **41** (2012), 250–271.
237. L. Lovász: Connection matrices, in: *Combinatorics, Complexity and Chance, A Tribute to Dominic Welsh* Oxford Univ. Press (2007), 179–190.
238. M. Freedman, L. Lovász, A. Schrijver: Reflection positivity, rank connectivity, and homomorphisms of graphs *J. Amer. Math. Soc.* **20** (2007), 37–51.
239. L. Lovász, S. Vempala: The Geometry of Logconcave Functions and Sampling Algorithms, *Random Struct. Alg.* **30** (2007), 307–358.
240. L. Lovász, B. Szegedy: Szemerédi's Lemma for the analyst, *Geom. Func. Anal.* **17** (2007), 252–270.
241. A. Blokhuis, L. Lovász, L. Storme, T. Szőnyi: On multiple blocking sets in Galois planes, *Adv. Geom.* **7** (2007), 39–53.
242. L. Lovász, V.T. Sós: Generalized quasirandom graphs, *J. Comb. Th. B* **98** (2008), 146–163.
243. L. Lovász, A. Schrijver: Graph parameters and semigroup functions, *Europ. J. Comb.* **29** (2008), 987–1002.
244. C. Borgs, J.T. Chayes, L. Lovász, V.T. Sós, and K. Vesztergombi: Convergent Graph Sequences I: Subgraph frequencies, metric properties, and testing, *Advances in Math.* **219** (2008), 1801–1851.
245. A. Gács, L. Lovász and T. Szőnyi: Directions in $AG(2, p^2)$, *Innovations in Incidence Geometry* **6-7** (2007-2008), 189–201.
246. L. Lovász, B. Szegedy: Contractors and connectors in graph algebras, *J. Graph Theory* **60** (2009), 11–31.
247. L. Lovász, A. Schrijver: Semidefinite functions on categories, *Electron. J. Combin.* **16** (2009), no. 2, Special volume in honor of Anders Björner, Research Paper 14, 16 pp.

248. L. Lovász: Very large graphs, in: *Current Developments in Mathematics 2008* (eds. D. Jerison, B. Mazur, T. Mrowka, W. Schmid, R. Stanley, and S. T. Yau), International Press, Somerville, MA 2009, 67–128.
249. C. Borgs, J. Chayes, L. Lovász: Moments of Two-Variable Functions and the Uniqueness of Graph Limits, *Geom. Func. Anal.* **19** (2009), 1597–1619.
250. A. Beveridge, L. Lovász: Exit Frequency Matrices for Finite Markov Chains, *Combinatorics, Probability and Computing* **19** (2010), 541–560.
251. G. Palla, L. Lovász and T. Vicsek: Multifractal network generator, *Proc. NAS* **107** (2010), 7640–7645.
252. L. Lovász, A. Schrijver: Dual graph homomorphism functions, *J. Combin. Theory Series A*, **117** (2010), 216–222.
253. L. Lovász, B. Szegedy: Testing properties of graphs and functions, *Israel J. Math.* **178** (2010), 113–156.
254. L. Lovász, B. Szegedy: Regularity partitions and the topology of graphons, in: *An Irregular Mind, Szemerédi is 70*, J. Bolyai Math. Soc and Springer-Verlag (2010) 415–446.
255. C. Borgs, J.T. Chayes, L. Lovász, V.T. Sós, and K. Vesztergombi: Limits of randomly grown graph sequences, *Eur. J. Combin.* **32** (2011), 985–999.
256. L. Lovász, B. Szegedy: Finitely forcible graphons *J. Comb. Theory B* **101** (2011), 269–301.
257. R. Kang, L. Lovász, T. Müller and E. Scheinerman: Dot product representations of planar graphs, *Electr. J. Combin.* **18** (2011), P216.
258. L. Lovász: Subgraph densities in signed graphons and the local Sidorenko conjecture, *Electronic J. of Combinatorics*, **18** (2011), P127.
259. C. Borgs, J.T. Chayes, L. Lovász, V.T. Sós, and K. Vesztergombi: Convergent Graph Sequences II: Multiway Cuts and Statistical Physics, *Annals of Math.* **176** (1912), 151–219.
260. J. Draisma, D.C. Gijswijt, L. Lovász, G. Regts, A. Schrijver: Characterizing partition functions of the vertex model, *Journal of Algebra* **350** (2012), 197–206.
261. L. Lovász, B. Szegedy: Random Graphons and a Weak Positivstellensatz for Graphs, *J. Graph Theory* **70** (2012), 214–225.
262. I. Deák, L. Lovász: Computational results of an $O^*(n^4)$ volume algorithm, *Eur. J. Oper. Res.* **216** (2012), 152–161.
263. C. Borgs, J.T. Chayes, J. Kahn, L. Lovász: Left and right convergence of graphs with bounded degree, *Random Struc. Alg.* **42** (2013), 1–28.
264. L. Lovász and K. Vesztergombi: Nondeterministic property testing, *Combinatorics Probability and Computing* **22** (2013), 749–762.
265. H. Hatami, L. Lovász and B. Szegedy: Limits of locally-globally convergent graph sequences, *Geom. Func. Anal.* **24** (2014), 269–296.
266. L. Lovász, B. Szegedy: The automorphism group of a graphon, *Journal of Algebra* **421** (2015), 136–166.
267. O.A. Camarena, E. Csóka, T. Hubai, G. Lippner and L. Lovász: Positive graphs, *European Journal of Combinatorics* **52**, Part B (2016), 290–301.

268. L. Lovász, A. Schrijver: Nullspace embeddings for outerplanar graphs, in: *Journey Through Discrete Mathematics. A Tribute to Jiri Matousek*, Springer (2017), 571–591.
269. D. Kunszenti-Kovács, L. Lovász, B. Szegedy: Measures on the square as sparse graph limits, *J. Comb. Theory B* (2019), online: https://www.sciencedirect.com/science/article/pii/S009589561930005X

Expository Papers:

1. L. Lovász: A matroidelmélet rövid áttekintése (A short survey of matroid theory), *Mat. Lapok 22* (1971), 249–267.
2. L. Lovász: A szitaformuláról (On the sieve formula), *Mat. Lapok* **23** (1972), 53–69.
3. L. Lovász: Kombinatorikus optimalizáció (Combinatorial optimization), *Magyar Tudomány* **25** (1980), 736–742.
4. L. Lovász: A new linear programming algorithm: better or worse than Simplex Method? *Math. Intelligencer* **2** (1980), 141–146.
5. L. Lovász: Mit ad a matematikának és mit kap a matematikától a számítógéptudomány? (What does computer science get from mathematics and what does it give to it?) *Magyar Tudomány* **35** (1990), 1041–1047.
6. L. Lovász: The mathematical notion of complexity, *Proc. IFAC Symposium*, Budapest (1984).
7. L. Lovász: Algorithmic mathematics: an old aspect with a new emphasis, in: *Proc. 6th ICME, Budapest, J. Bolyai Math. Soc.* (1988), 67–78.
8. L. Lovász: The work of A. A. Razborov, *Proc. of Int. Congress of Math, Kyoto*, Springer-Verlag (1989), 37–40.
9. L. Lovász: Features of computer language: communication of computers and its complexity, *Acta Neurochirurgica* **56** [Suppl.] (1994) 91–95.
10. L. Lovász: Random walks, eigenvalues, and resistance, Appendix to Chapter 31: Tools from linear algebra, in: Handbook of Combinatorics (ed. R. Graham, M. Grötschel, L. Lovász), Elsevier Science B.V. (1995), 1740–1748.
11. M. Grötschel, L. Lovász: Combinatorial optimization, Chapter 28 in: *Handbook of Combinatorics* (ed. R. Graham, M. Grötschel, L. Lovász), Elsevier Science B.V. (1995), 1541–1597.
12. L. Lovász, D.B. Shmoys, É. Tardos: Combinatorics in computer science, Chapter 40 in: *Handbook of Combinatorics* (ed. R. Graham, M. Grötschel, L. Lovász), Elsevier Science B.V. (1995), 2003–2038.
13. L. Lovász, L. Pyber, D.J.A. Welsh, G.M. Ziegler: Combinatorics in pure mathematics, Chapter 41 in: *Handbook of Combinatorics* (ed. R. Graham, M. Grötschel, L. Lovász), Elsevier Science B.V. (1995), 2039–2082.
14. L. Lovász: Information and complexity (how to measure them?) in: *The Emergence of Complexity in Mathematics, Physics, Chemistry and Biology* (ed. B. Pullman), Pontifical Academy of Sciences, Vatican City, Princeton University Press (1996), 65–80.
15. L. Lovász: One mathematics, The Berlin Intelligencer, Mitteilungen der Deutschen Math.-Verein, Berlin (1998), 10–15.

16. L. Lovász: Egységes tudomány-e a matematika? (Is mathematics a single science?) *Természet Világa, Special issue on Mathematics*, (1998).
17. L. Lovász: Discrete and Continuous: Two sides of the same? *GAFA, Geom. Funct. Anal., Special volume – GAFA2000*, Birkheuser, Basel (2000), 359–382.
18. L. Lovász: Véletlen és álvéletlen (Randomness and pseudo-randomness) *Természet Világa*, Special issue in Informatics, (2000), 5–7.
19. L. Lovász: Nagyon nagy gráfok (Very large graphs), *Természet Világa* **138** (2007), 98–103.
20. L. Lovász: The "Little Geometer" and the difficulty of computing the volume, *Annales Univ. R. Eötvös* **52** (2009), 31–36.
21. Lovász L.: Prímek, számítógépek és Abel-díj (Primes, computers and Abel Prize), *Természet Világa* **143** (2012), 242–244.
22. L. Lovász: 45 Jahre Graphentheorie, in: *Eine Einladung in die Mathematik* (D. Schleicher, M. Lackmann, eds.), 87–98, Springer, Berlin–Heidelberg (2013).
23. L. Lovász: Trends in mathematics: how they could change education, *ICCM Notices* **1**(2) (2013), 79–84.

The List of the Former Volumes

(1) Higher Dimensional Varieties and Rational Points (2003)
    Károly Böröczky Jr, Alfréd Rényi, Tamás Szamuely
(2) Surgery on Contact 3-Manifolds and Stein Surfaces (2004)
    Burak Ozbagci, András I. Stipsicz
(3) A Panorama of Hungarian Mathematics in the Twentieth Century I (2006)
    János Horváth
(4) More Sets, Graphs and Numbers (2006)
    Ervin Győri, Gyula O. H. Katona, László Lovász
(5) Entropy, Search, Complexity (2007)
    Imre Csiszár, Gyula O. H. Katona, Gábor Tardos, Gábor Wiener
(6) Building Bridges (2008)
    Martin Grötschel, Gyula O. H. Katona, Gábor Sági
(7) Handbook of Large-Scale Random Networks (2008)
    Béla Bollobás, Robert Kozma, Dezső Miklós
(8) Horizons of Combinatorics (2008)
    Ervin Győri, Gyula O. H. Katona, László Lovász, Gábor Sági
(9) Fete of Combinatorics and Computer Science (2010)
    Gyula O. H. Katona, Alexander Schrijver, Tamás Szőnyi, Gábor Sági
(10) An Irregular Mind (2010)
    Imre Bárány, József Solymosi, Gábor Sági
(11) Erdos Centennial (2013)
    László Lovász, Imre Z. Ruzsa, Vera T. Sós
(12) Cylindric-like Algebras and Algebraic Logic (2013)
    Hajnal Andréka, Miklós Ferenczi, István Németi
(13) Geometry - Intuitive, Discrete, and Convex (2013)
    Imre Bárány, Károly J. Böröczky, Gábor Fejes Tóth, János Pach
(14) Deformations of Surface Singularities (2013)
    András Némethi, Ágnes Szilárd

(15) Contact and Symplectic Topology (2014)
    Frédéric Bourgeois, Vincent Colin, András Stipsicz
(16) New Trends in Intuitive Geometry (2018)
    Gergely Ambrus, Imre Bárány, Károly J. Böröczky, Gábor Fejes Tóth, János Pach